Methods in Cell Biology

VOLUME 63
Cytometry
Third Edition, Part A

Series Editors

Leslie Wilson
Department of Biological Sciences
University of California, Santa Barbara
Santa Barbara, California

Paul Matsudaira
Whitehead Institute for Biomedical Research and
Department of Biology
Massachusetts Institute of Technology
Cambridge, Massachusetts

Methods in Cell Biology

Prepared under the Auspices of the American Society for Cell Biology

VOLUME 63
Cytometry
Third Edition, Part A

Edited by

Zbigniew Darzynkiewicz
Brander Cancer Research Institute
New York Medical College
Hawthorne, New York

Harry A. Crissman
Cell and Molecular Biology Group
Los Alamos National Laboratory
Los Alamos, New Mexico

J. Paul Robinson
Purdue Cytometry Laboratories
Purdue University
West Lafayette, Indiana

 ACADEMIC PRESS
A Harcourt Science and Technology Company

San Diego   San Francisco   New York   Boston   London   Sydney   Tokyo

This book is printed on acid-free paper.

Copyright © 2001, 1994, 1990 by ACADEMIC PRESS

All Rights Reserved.
No part of this publication may be reproduced or transmitted in any form or by any means, electronic or mechanical, including photocopy, recording, or any information storage and retrieval system, without permission in writing from the Publisher.

The appearance of the code at the bottom of the first page of a chapter in this book indicates the Publisher's consent that copies of the chapter may be made for personal or internal use of specific clients. This consent is given on the condition, however, that the copier pay the stated per copy fee through the Copyright Clearance Center, Inc. (222 Rosewood Drive, Danvers, Massachusetts 01923), for copying beyond that permitted by Sections 107 or 108 of the U.S. Copyright Law. This consent does not extend to other kinds of copying, such as copying for general distribution, for advertising or promotional purposes, for creating new collective works, or for resale. Copy fees for pre-2001 chapters are as shown on the title pages. If no fee code appears on the title page, the copy fee is the same as for current chapters.
0091-679X/01 $35.00

Explicit permission from Academic Press is not required to reproduce a maximum of two figures or tables from an Academic Press chapter in another scientific or research publication provided that the material has not been credited to another source and that full credit to the Academic Press chapter is given.

Academic Press
A Harcourt Science and Technology Company
525 B Street, Suite 1900, San Diego, California 92101-4495, USA
http://www.academicpress.com

Academic Press
Harcourt Place, 32 Jamestown Road, London NW1 7BY, UK
http://www.academicpress.com

International Standard Book Number: 0-12-544166-5 (case)
International Standard Book Number: 0-12-203053-2 (combbound)

PRINTED IN THE UNITED STATES OF AMERICA
00 01 02 03 04 05 EB 9 8 7 6 5 4 3 2 1

CONTENTS

Contents of Volume 64 — xii
Contributors — xvii
Preface to the Third Edition — xxi
Preface to the Second Edition — xxv
Preface to the First Edition — xxix

PART I Principles of Cytometry and General Methods

1. A Brief History of Flow Cytometry and Sorting
Myron R. Melamed

    I. Introduction — 3
    II. Instrumentation — 4
    III. Applications — 9
    References — 13

2. Principles of Flow Cytometry: An Overview
Alice L. Givan

    I. Introduction — 19
    II. The Illumination of a Particle — 20
    III. Fluidics: Centering Particles in the Illuminating Beam — 26
    IV. Collection of Light Signals from Particles — 33
    V. From Light Signals to a Data File — 37
    VI. From Data to Information — 41
    VII. Sorting — 44
    VIII. Conclusions — 48
    References — 48

3. Laser Scanning Cytometry
Louis A. Kamentsky

    I. Introduction — 51
    II. Background — 52
    III. Description of the Instrument — 55
    IV. The Utility and Operational Characteristics of Some Laser Scanning Cytometry List Mode Features — 62
    V. Utility of Solid Phase Cytometry for Cell Preparation — 80

	VI. Future Directions	82
	References	84

4. Principles of Confocal Microscopy
J. Paul Robinson

	I. Brief History of Microscope Development	89
	II. Development of Confocal Microscopy	90
	III. Image Formation in Confocal Microscopy	92
	IV. Useful Fluorescent Probes for Confocal Microscopy	96
	V. Applications of Confocal Microscopy	100
	VI. Conclusions	105
	References	105

5. Optical Measurements in Cytometry: Light Scattering, Extinction, Absorption, and Fluorescence
Howard M. Shapiro

	I. Introduction	108
	II. Signal Processing Tasks in Flow Cytometry: An Overview	108
	III. The Optical Signal: Interaction of Light with Cells	110
	IV. Detection: Converting Optical Signals to Current	118
	V. Electronics: Converting Current to Voltage	123
	VI. Fluorescence Compensation and Logarithmic Amplification	124
	VII. Peak Detection, Integration, and Pulse Width Measurement; Triggering	127
	VIII. Measurement Sensitivity: Changing Concepts and the Bottom Line	127
	References	129

6. Flow Cytometric Fluorescence Lifetime Measurements
Harry A. Crissman and John A. Steinkamp

	I. Introduction	131
	II. Applications of the Technology	134
	III. Cell Preparation and Staining	137
	IV. Fluorescence Lifetime Flow Cytometry Instrumentation	138
	V. Results	141
	VI. Critical Aspects of the Technology	144
	VII. Future Directions	146
	References	147

7. Principles of Data Acquisition and Display
Howard M. Shapiro

	I. Introduction	149
	II. Pulse Characterization Using Analog and Hybrid Circuits	150
	III. Analog Data Display	154

IV.	Analog-to-Digital Conversion	155
V.	Pulse Characterization by Digital Signal Processing	159
VI.	Data Storage and Display with Digital Computers	161
	References	167

8. Time as a Flow Cytometric Parameter
Larry Seamer and Larry A. Sklar

I.	Introduction	169
II.	Historical Overview	171
III.	Sample Mixing and Delivery	171
IV.	Data Analysis	175
V.	Applications	179
VI.	Conclusions	181
	References	181

9. Protein Labeling with Fluorescent Probes
Kevin L. Holmes and Larry M. Lantz

I.	Introduction	185
II.	Labeling of Proteins with Organic Fluorescent Dyes	186
III.	Labeling of Proteins with Phycobiliproteins	194
IV.	Conclusion	202
	References	202

PART II Cell Preparation

10. Preparation of Cells from Blood
J. Philip McCoy, Jr.

I.	Introduction	207
II.	Collection, Transport, and Storage of Blood	208
III.	Fixation and Preservation	210
IV.	Separation of Erythrocytes from Leukocytes	210
V.	Assessment of Cell Viability	211
VI.	Staining	212
VII.	Summary	212
	Appendix 1	213
	Appendix 2	214
	Appendix 3	214
	References	214

11. Cell Preparation for the Identification of Leukocytes
Carleton C. Stewart and Sigrid J. Stewart

I.	Introduction	218
II.	Antibodies	219
III.	Tandem Fluorochromes	234

IV.	Cell Preparation and Staining Procedures	237
V.	Titering Antibodies	244
VI.	Solutions and Reagents	249
	References	250

12. Strategies for Cell Permeabilization and Fixation in Detecting Surface and Intracellular Antigens
Steven K. Koester and Wade E. Bolton

I.	Introduction	253
II.	Application	254
III.	Materials and Methods	262
IV.	Concluding Remarks	266
	References	266

PART III Standardization, Quality Assurance

13. Stoichiometry of Immunocytochemical Staining Reactions
James W. Jacobberger

I.	Introduction	271
II.	Structure of Immunoglobulin G	273
III.	Cell Structure	274
IV.	Permeabilized Cell Structure	274
V.	Antibody–Antigen Reactions	281
VI.	Multiparametric Analyses	292
VII.	Summary	294
	References	295

14. Standardization and Quantitation in Flow Cytometry
Robert A. Hoffman

I.	Introduction	300
II.	General Issues	300
III.	Performance Characteristics—Dynamic Range, Linearity, Resolution, and Sensitivity	302
IV.	Standardization and Calibration of Common Cytometry Measurements	312
V.	Examples of Applications Using Calibrated Measurements	330
VI.	Issues in Quantitation of Fluorochromes and Other Molecules	332
	References	334

PART IV Cell Proliferation

15. Methods to Identify Mitotic Cells by Flow Cytometry
Gloria Juan, Frank Traganos, and Zbigniew Darzynkiewicz

I.	Introduction	343
II.	Materials	345

III.	Cell Preparation and Staining	345
IV.	Instruments	346
V.	Critical Aspects of the Procedure	346
VI.	Results and Discussion	347
VII.	Comparison of Anti-H3-P Monoclonal Antibody with Other Markers of Mitotic Cells	350
	References	353

16. Cell Cycle Kinetics Estimated by Analysis of Bromodeoxyuridine Incorporation

Nicholas H. A. Terry and R. Allen White

I.	Introduction	355
II.	Applications	357
III.	Materials	358
IV.	Methods	360
V.	Critical Aspects of the Procedure	369
	References	372

17. Flow Cytometric Analysis of Cell Division History Using Dilution of Carboxyfluorescein Diacetate Succinimidyl Ester, a Stably Integrated Fluorescent Probe

A. Bruce Lyons, Jhagvaral Hasbold, and Philip D. Hodgkin

I.	Introduction and Background	376
II.	Reagents and Solutions	378
III.	Preparation and Labeling of Cells	378
IV.	Gathering of Information Concurrent with Division	381
V.	Analysis of Data	386
VI.	Application of Carboxyfluorescein Diacetate Succinimidyl Ester to *in Vitro* Culture of Lymphocytes	390
VII.	Monitoring Lymphocyte Responses *in Vivo*	392
VIII.	Antigen Receptor Transgenic Models	395
	References	397

18. Antibodies against the Ki-67 Protein: Assessment of the Growth Fraction and Tools for Cell Cycle Analysis

Elmar Endl, Christiane Hollmann, and Johannes Gerdes

I.	Introduction	399
II.	Application	401
III.	Materials and Methods	402
IV.	Critical Aspects	408
V.	Controls and Standards	409
VI.	Examples of Results	410
	References	414

19. Detection of Proliferating Cell Nuclear Antigen

Jørgen K. Larsen, Göran Landberg, and Göran Roos

I. Introduction	419
II. Molecular Biology of Proliferating Cell Nuclear Antigen	420
III. Methods for Immunochemical Detection and Quantification of Proliferating Cell Nuclear Antigen	421
IV. Results of Cytometric Analysis of Proliferating Cell Nuclear Antigen Expression	423
V. Applications in Toxicology, Pathology, and Oncology	426
References	428

20. Lymphocyte Activation Associated Antigens

Andrea Fattorossi, Alessandra Battaglia, and Cristiano Ferlini

I. Introduction	433
II. Methodological Aspects	437
III. To Flow or Not to Flow for Assessing Lymphocyte Activation/Proliferation? And, If Yes, How Reliable Is Immunophenotyping?	447
IV. Additional Approaches	453
V. Concluding Remarks	455
References	457

PART V Cell Death/Apoptosis

21. Analysis of Mitochondria during Cell Death

Andrea Cossarizza and Stefano Salvioli

I. Introduction	467
II. Scientific Background	468
III. Apoptosis and Mitochondria	470
IV. Method	472
V. Results	473
VI. Pitfalls and Misinterpretation of the Data	476
VII. Comparison with Other Methods	477
VIII. Reviews of the Applications	478
IX. Biological and Biomedical Information	479
X. Future Directions	480
References	481

22. Cytometry of Caspases

Steven K. Koester and Wade E. Bolton

I. Introduction	487
II. Materials and Methods: Caspase Peptide Inhibitors and Methods to Monitor Responses	490
III. Results and Discussion	497
References	502

Contents

23. Analysis of Apoptosis in Plant Cells
Iona E. Weir

I. Introduction	505
II. Apoptosis in Plants	506
III. Problems Associated with Analyzing Plant Cells Using Flow Cytometry	509
IV. Morphological Changes of Plant Cells	512
V. Physiological Changes during Apoptosis	520
VI. Conclusion	524
References	524

24. Difficulties and Pitfalls in Analysis of Apoptosis
Zbigniew Darzynkiewicz, Elżbieta Bedner, and Frank Traganos

I. Introduction	527
II. Apoptotic Index May Not Be Correlated with Incidence of Cell Death	529
III. Difficulties in Estimating Frequency of Apoptosis by Analysis of DNA Fragmentation	531
IV. The Lack of Evidence Is Not Evidence for the Lack of Apoptosis	532
V. Misclassification of Apoptotic Bodies or Nuclear Fragments as Single Apoptotic Cells	533
VI. Apoptosis versus Necrosis versus "Necrotic Stage" of Apoptosis	535
VII. Selective Loss of Apoptotic Cells during Sample Preparation	537
VIII. Live Cells Engulfing Apoptotic Bodies Masquerade as Apoptotic Cells	538
IX. The Problems with Commercial Kits and Reagents	538
X. Cell Morphology Is Still the Gold Standard for Identification of Apoptotic Cells	539
XI. Laser Scanning Cytometry: Have Your Cake and Eat It Too	541
References	544

PART VI Cell–Cell, Cell–Environment Interactions

25. Analysis of Cell Migration
Nicole Dodge Zantek and Michael S. Kinch

I. Introduction and Application	549
II. General Strategies to Measure Cell Migration	550
References	558

26. Three-Dimensional Extracellular Matrix Substrates for Cell Culture
Sherry L. Voytik-Harbin

I. Introduction	561
II. Application	563
III. Methods	565
IV. Application of Intestinal Submucosa as a Three-Dimensional Extracellular Matrix Substrate	572
V. Summary	576
References	578

27. Three-Dimensional Imaging of Extracellular Matrix and Extracellular Matrix–Cell Interactions

Sherry L. Voytik-Harbin, Bartlomiej Rajwa, and J. Paul Robinson

I.	Introduction	583
II.	Three-Dimensional Imaging of Extracellular Matrix and Extracellular Matrix–Cell Interactions: Current Techniques and Their Limitations	584
III.	Three-Dimensional Microscopy of Living Systems: Extracellular Matrix and Extracellular Matrix–Cell Interactions	587
IV.	Summary	593
	References	596

28. Cytometric Analysis of Cell Contact and Adhesion

Michael S. Kinch

I.	Introduction and Application	599
II.	General Strategies to Measure Cell–Cell Adhesions	600
III.	General Strategies to Measure Cell–Ligand Adhesions	606
IV.	Specificity of Cell Adhesion	608
V.	Optimization of Experimental Conditions	609
	References	611

29. Invadopodia: Unique Methods for Measurement of Extracellular Matrix Degradation *in Vitro*

Emma T. Bowden, Peter J. Coopman, and Susette C. Mueller

I.	Introduction	613
II.	Invadopodia Activity, a Measurement for Localized Membrane Degradation	615
III.	Fluorescent Activated Cell Sorting—Phagocytosis, a Measurement for Internalization of Proteolyzed Extracellular Matrix	619
IV.	Protocols	623
	References	626

Index	629
Volumes in Series	645

Contents of Volume 64
Cytometry, Third Edition, Part B

PART VII Cytogenetics and Molecular Genetics

30. Sorting of Plant Chromosomes

Jaroslav Doležel, Martin A. Lysák, Marie Kubaláková, Hana Šimková, Jiří Macas, and Sergio Lucretti

Contents

31. Quantitative DNA Fiber Mapping
    Heinz-Ulli G. Weier

32. Primed *in Situ* Labeling
    Johnny Hindkjaer, Lars Bolund, and Steen Kølvraa

33. Measurements of Telomere Length on Individual Chromosomes by Image Cytometry
    Steven S. S. Poon and Peter M. Lansdorp

34. Detection of Chromosome Translocation Products in Single Interphase Cell Nuclei
    Jingly Fung, Santiago Munné, and Heinz-Ulli G. Weier

PART VIII Cell Function and Differentiation

35. Analysis of Mitochondria by Flow Cytometry
    Martin Poot and Robert H. Pierce

36. Analysis of RNA Synthesis by Cytometry
    Peter Østrup Jensen, Jacob Larsen, and Jørgen K. Larsen

37. Flow Cytometry of Erythropoiesis in Culture: Bivariate Profiles of Fetal and Adult Hemoglobin
    Ralph M. Böhmer

38. Flow Cytometric Analysis of Human Hemopoietic Progenitor Differentiation by Assessing Cell Division Rate and Phenotypic Profile
    Luca Pierelli, Giovanni Scambia, and Andrea Fattorossi

PART IX Experimental Oncology

39. Cytometry of Antitumor Drug-Intracellular Target Interactions
    Paul J. Smith and Marie Wiltshire

40. Monitoring of Cellular Resistance to Cancer Chemotherapy: Drug Retention and Efflux
    Awtar Krishan

41. Resistance of Tumor Cells to Chemo- and Radiotherapy Modulated by the Three-Dimensional Architecture of Solid Tumors and Spheroids
   Ralph E. Durand and Peggy L. Olive

42. Analysis of DNA Damage in Individual Cells
   Peggy L. Olive, Ralph E. Durand, Judit P. Banáth, and Peter J. Johnston

43. Cytometric Methods to Analyze Ionizing-Radiation Effects
   William D. Wright, Isabelle Lagroye, Peng Zhang, Robert S. Malyapa, and Joseph L. Roti Roti

44. Cytometric Methods to Analyze Thermal Effects
   Robert P. VanderWaal, Ryuji Higashikubo, Mai Xu, Douglas R. Spitz, William D. Wright, and Joseph L. Roti Roti

PART X Clinical Oncology

45. Multiparameter Data Acquisition and Analysis of Leukocytes by Flow Cytometry
   Carleton C. Stewart and Sigrid J. Stewart

46. Immunophenotyping of Hematological Malignancies by Laser Scanning Cytometry
   Richard J. Clatch

47. Immunophenotyping of Acute Leukemia: Utility of CD45 for Blast Cell Identification
   J-P. Vial and F. Lacombe

48. Cell Proliferation Markers in Human Solid Tumors: Assessing Their Impact in Clinical Oncology
   Maria Grazia Daidone, Aurora Costa, and Rosella Silvestrini

49. Detection of Minimal Residual Disease
   Andrzej Deptala and Sharon P. Mayer

50. Analysis of Human Tumors by Laser Scanning Cytometry
   Wojciech Gorczyca, Andrzej Deptala, Elżbieta Bedner, Xun Li, Myron R. Melamed, and Zbigniew Darzynkiewicz

51. Laser Cytometry of Human Tissues and Tumors: Proliferation and Therapeutic Applications
    David A. Rew

52. Prediction and Precise Diagnosis of Diseases by Data Pattern Analysis in Multiparameter Flow Cytometry: Melanoma, Juvenile Asthma, and Human Immunodeficiency Virus Infection
    Günter Valet, Hanna Kahle, Friedrich Otto, Edeltraut Bräutigam, and Luc Kestens

PART XI Microorganisms and Infectious Diseases

53. Flow Cytometric Analysis of Microorganisms
    S. A. Sincock and J. Paul Robinson

54. Staining and Measurement of DNA in Bacteria
    Harald B. Steen

55. Flow Cytometric Monitoring of Bacterial Susceptibility to Antibiotics
    Mette Walberg and Harald B. Steen

56. Flow Cytometry for Evaluation and Investigation of Human Immunodeficiency Virus Infection
    Thomas W. Mc Closkey

CONTRIBUTORS

Numbers in parentheses indicate the pages on which the authors' contributions begin.

Alessandra Battaglia (433), Institute of Obstetrics and Gynecology, Università Cattolica del Sacro Cuore, 00136 Rome, Italy

Elżbieta Bedner (527), Department of Pathology, Pomeranian School of Medicine, Szczecin, Poland

Wade E. Bolton (253, 487), Advanced Technology, Beckman Coulter, Inc., Miami, Florida 33196

Emma T. Bowden (613), Lombardi Cancer Center, Georgetown University Medical Center, Washington, DC 20007

Peter J. Coopman (613), Lombardi Cancer Center, Georgetown University Medical Center, Washington, DC 20007

Andrea Cossarizza (467), Chair of Immunology, Department of Biomedical Sciences, University of Modena and Reggio, Emilia School of Medicine, 41100 Modena, Italy

Harry A. Crissman (131), Biosciences Division, Los Alamos National Laboratory, Los Alamos, New Mexico 87545

Zbigniew Darzynkiewicz (343, 527), Brander Cancer Research Institute, New York Medical College, Hawthorne, New York 10532

Nicole Dodge Zantek (549), Department of Basic Medical Sciences, Purdue University, West Lafayette, Indiana 47907

Elmar Endl (399), Division of Molecular Immunology, Research Center Borstel, D-23845 Borstel, Germany

Andrea Fattorossi (433), Institute of Obstetrics and Gynecology, Università Cattolica del Sacro Cuore, 00136 Rome, Italy

Cristiano Ferlini (433), Institute of Obstetrics and Gynecology, Università Cattolica del Sacro Cuore, 00136 Rome, Italy

Johannes Gerdes (399), Division of Molecular Immunology, Research Center Borstel, D-23845 Borstel, Germany

Alice L. Givan (19), Englert Cell Analysis Laboratory of the Norris Cotton Cancer Center, and Department of Physiology, Dartmouth Medical School, Lebanon, New Hampshire 03756

Jhagvaral Hasbold (375), The Centenary Institute of Cancer Medicine and Cell Biology, Sydney, Australia

Philip D. Hodgkin (375), Discipline of Pathology, Faculty of Health Science, The University of Tasmania, Hobart, Australia; The Centenary Institute of Cancer Medicine and Cell Biology, Sydney, Australia; and Medical Foundation, University of Sydney, Sydney, Australia

Robert A. Hoffman (299), BD Biosciences, San Jose, California 95131

Christiane Hollmann (399), Division of Molecular Immunology, Research Center Borstel, D-23845 Borstel, Germany

Kevin L. Holmes (185), Flow Cytometry Section, National Institute of Allergy and Infectious Diseases, National Institutes of Health, Bethesda, Maryland 20892

James W. Jacobberger (271), Cancer Research Center and Department of Genetics, Case Western Reserve University, Cleveland, Ohio 44106

Gloria Juan (343), Brander Cancer Research Institute, New York Medical College, Hawthorne, New York 10532

Louis A. Kamentsky (51), CompuCyte Corporation, Cambridge, Massachusetts 02139

Michael S. Kinch (549, 599), Department of Basic Medical Sciences, Purdue University, West Lafayette, Indiana 47907

Steven K. Koester (253, 487), Advanced Technology, Beckman Coulter, Inc., Miami, Florida 33196

Göran Landberg (419), Department of Pathology, University of Umeå, S-90187 Umeå, Sweden

Larry M. Lantz (185), Flow Cytometry Section, National Institute of Allergy and Infectious Diseases, National Institutes of Health, Bethesda, Maryland 20892

Jørgen K. Larsen (419), Finsen Laboratory, Finsen Center, Rigshospitalet, Copenhagen University Hospital, DK-2100 Copenhagen, Denmark

A. Bruce Lyons (375),* Discipline of Pathology, Faculty of Health Science, The University of Tasmania, Hobart, Australia

J. Philip McCoy, Jr. (207), Cooper Hospital/UMC, Robert Wood Johnson Medical School at Camden, Camden, New Jersey 08103

Myron R. Melamed (3), New York Medical College, Valhalla, New York 10595

Susette C. Mueller (613), Lombardi Cancer Center, Georgetown University Medical Center, Washington, DC 20007

Bartlomiej Rajwa (583), Department of Biophysics, Institute of Molecular Biology, Jagiellonian University, 31-120 Krakow, Poland

J. Paul Robinson (89, 583), Department of Basic Medical Sciences, School of Veterinary Medicine, and Department of Biomedical Engineering, Purdue University, West Lafayette, Indiana 47907

Göran Roos (419), Department of Pathology, University of Umeå, S-90187 Umeå, Sweden

Stefano Salvioli (467), Department of Experimental Pathology, University of Bologna, 40126 Bologna, Italy

Larry Seamer (169), Bio-Rad Laboratories, Hercules, California 94547

Howard M. Shapiro (107, 149), Howard M. Shapiro, M.D., P.C., West Newton, Massachusetts 02465

Larry A. Sklar (169), Department of Pathology and Cancer Research and Treatment Center, University of New Mexico, Albuquerque, New Mexico 87131; and National Flow Cytometry Resource, Los Alamos National Laboratory, Los Alamos, New Mexico 87545

*Current address: Leukemia Research Laboratory, The Hanson Centre for Cancer Research, Institute of Medical and Veterinary Science, Adelaide SA 5000, Australia.

John A. Steinkamp (131), Biosciences Division, Los Alamos National Laboratory, Los Alamos, New Mexico 87545

Carleton C. Stewart (217), Laboratory of Flow Cytometry, Roswell Park Cancer Institute, Buffalo, New York 14263

Sigrid J. Stewart (217), Laboratory of Flow Cytometry, Roswell Park Cancer Institute, Buffalo, New York 14263

Nicholas H. A. Terry (355), Department of Experimental Radiation Oncology, The University of Texas M. D. Anderson Cancer Center, Houston, Texas 77030

Frank Traganos (343, 527), Brander Cancer Research Institute, New York Medical College, Hawthorne, New York 10532

Sherry L. Voytik-Harbin (561, 583), Department of Basic Medical Sciences, School of Veterinary Medicine, and Department of Biomedical Engineering, Purdue University, West Lafayette, Indiana 47907

Iona E. Weir (505), Horticulture and Food Research Institute of New Zealand, Ltd., Auckland, New Zealand

R. Allen White (355), Department of Biomathematics, The University of Texas M. D. Anderson Cancer Center, Houston, Texas 77030

PREFACE TO THE THIRD EDITION

This is the third edition of cytometry volumes in the *Methods in Cell Biology* series. The first, single-volume edition (*Flow Cytometry, Methods in Cell Biology,* Volume 33, 1990) appeared a decade ago. The continuing rapid growth of this methodology prompted us to prepare the second, two-volume edition (*Flow Cytometry, Methods in Cell Biology,* Volumes 41 and 42, 1994), which introduced a variety of new methods developed since the publication of the first edition. The growth and applications of this methodology have continued at an accelerating pace. This progress and the demand for the first two editions, which have become the "bible" for researchers who utilize the presented methods in a variety of fields of biology and medicine, prompted us to prepare the third edition.

This two-volume set differs from the earlier editions in several respects. The title is changed to *Cytometry* to indicate its wider scope. Several chapters describe methods and instrumentation that are not particular to cell analysis in "flow." Also changed are the scope and specifics of many chapters. Specifically, with the appearance of similar series of books on methods by other publishers (e.g., *Current Protocols in Cytometry* by Wiley-Liss or the "Practical Approach" books by Oxford Press), there was no point in duplicating them by focusing on presentation of individual methods in a cookbook form only. The authors, therefore, were requested to prepare their chapters in a form that presented not only technical protocols but also different aspects of the methodology that cannot be included in the protocols format. Thus, theoretical foundations of the described methods, their applicability in experimental laboratory and clinical settings, traps and pitfalls common to particular methods, problems with data interpretation, comparison with alternative assays, etc., are all presented in greater detail in many chapters. Furthermore, some chapters review applications of cytometry and complementary methodologies to particular biological problems or clinical tasks.

With few exceptions, nearly all 56 chapters in the present edition are novel, describing methods that were not included in the earlier editions. The present edition thus complements rather than merely updates the earlier edition. Because most of the methods described in Volumes 41 and 42 have not changed much since publication and are still in wide use, the combination of the earlier two volumes and these two new volumes becomes the most comprehensive collection of all methods in cytometry ever published.

The chapters presented in *Cytometry* cover a wide range of topics. The first several chapters are introductory. They describe principles of flow cytometry, laser scanning cytometry (LSC), confocal microscopy, and general approaches in cell measurement and data acquisition. Newcomers to the field of cytometry may find these chapters particularly useful, as they provide the foundation needed to understand specific methods and more complex data analysis. The next chapters address the issue of cell preparation for analysis by cytometry, quality assurance, and standardization. Of special interest may be the chapters focused on strategies for cell permeabilization and fixation to detect intracellular components, quantitation of the immunocytochemical staining reactions, and standardization in cytometry in general. Unfortunately, these important issues are neglected in many studies utilizing cytometric methods.

Analysis of cell proliferation is the subject of several other chapters. All the methods presented in these chapters are used extensively in experimental and clinical research. The methods include measurements of mitotic activity by flow cytometry, assays of cell kinetics by analysis of BrdU incorporation, analysis of the history of cell proliferation of the progeny cells from geometric dilution of the probe integrated into parent cells, applications of Ki-67 and PCNA antibodies as proliferation markers, and analysis of the lymphocyte activation antigens. Further chapters are devoted to methods of analysis of cell death, primarily by apoptosis. They include probing of mitochondria, activation of caspases, and analysis of apoptosis in the plant kingdom. A review of common problems, difficulties, and pitfalls encountered in analysis of apoptosis, with the key information of how to avoid them, also is provided.

Another group of methods is focused on analysis of cell-to-cell interactions and interactions of cells with the extracellular matrix. This is an exciting cross-disciplinary area that is revealing new directions for cytometry. This section includes cell migration assays, cytometric analysis of cell contact and adhesion, and measurement of extracellular matrix degradation. Also included in this group is a review of extracellular matrix substrates for culturing cells and analysis of cell interactions in three dimensions.

The field of cytogenetics and molecular genetics is represented by several chapters that present methods for sorting plant chromosomes, quantitative DNA fiber mapping, primed *in situ* (PRINS) methodology, individual chromosome telomere length analysis, and approaches to detecting products of chromosomal translocation in individual interphase cells. Functional cell assays, such as probing mitochondria with new markers of the electrochemical transmembrane potential, and measurement of RNA synthesis by immunocytochemical detection of the incorporated BrU, as well as new approaches to monitoring erythropoiesis or proliferation and differentiation of human progenitor cells, are all the subjects of additional chapters.

A large group of chapters is devoted to applications of cytometry in experimental oncology. Presented here are the methods to study interactions between antitumor drugs and intracellular targets, monitoring cellular resistance to chemotherapy in solid tumors and in spheroids related to their three-dimensional architecture, and analyzing DNA damage in individual cells caused by ionizing radiation. The methods specifically designed for studying effects of hyperthermia on tumor cells are also presented.

The most numerous chapters are those on applications of cytometry in the clinic. Indisputably, immunophenotyping is the most common application of cytometry in the clinical setting, and three chapters are devoted to this subject. A very exhaustive chapter on multiparametric analysis of human leukocytes describes the approaches to identifying the cells in different hematological malignancies. Adaptation of laser scanning cytometry to achieve similar tasks is the topic of another chapter on this subject. The third chapter addresses the specific issue of utility of a CD45 gate for identification of malignant cells in acute leukemias. Two insightful and exhaustive reviews, one that critically assesses clinical impact of analysis of different proliferation markers in human solid tumors and another that covers applications of flow cytometry and complementary methodologies in detection of minimal residual disease in leukemias, will be of great value for oncologists. There are also two reviews on applications of laser scanning cytometry to analysis of human tumors, one presented from the perspective of the pathologist and another from the surgeon's perspective. Applications of flow cytometry to monitoring HIV-infected patients is also a subject of thorough review. The last chapter in this group (Chapter 53) may be of particular interest to all researchers, regardless of the discipline. This chapter presents a unique approach to multiparameter data analysis in the clinic that often reveals unexpected correlations with high impact on disease prognosis.

The last group of chapters describes the methods and applications of cytometry in studies of microorganisms. They include flow cytometric analysis of microorganisms and monitoring of bacterial susceptibility to antibiotics. Applications of these assays are expected to rapidly expand and become routine tools in microbiology with wide application in the field of infectious diseases as well as in monitoring environmental contaminations.

As in the earlier editions, the chapters were prepared by colleagues who developed the described methods, contributed to their modification, or found new applications and have extensive experience in their use. The list of authors, as before, represents a "Who's Who" directory in the field of cytometry. On behalf of the readers, we express our gratitude to all contributing authors for the time they devoted to sharing their knowledge and experience.

Zbigniew Darzynkiewicz
Harry A. Crissman
J. Paul Robinson

PREFACE TO THE SECOND EDITION

The first edition of this book appeared four years ago (*Methods in Cell Biology,* Vol. 33, *Flow Cytometry,* Z. Darzynkiewicz and H. A. Crissman, Eds., Academic Press, 1990). This was the first attempt to compile a wide variety of flow cytometric methods in the form of a manual designed to describe both the practical aspects and the theoretical foundations of the most widely used methods, as well as to introduce the reader to their basic applications. The book was an instant publishing success. It received laudatory reviews and has become widely used by researchers from various disciplines of biology and medicine. Judging by this success, there was a strong need for this type of publication. Indeed, flow cytometry has now become an indispensable tool for researchers working in the fields of virology, bacteriology, pharmacology, plant biology, biotechnology, toxicology, and environmental sciences. Most applications, however, are in the medical sciences, in particular immunology and oncology. It is now difficult to find a single issue of any biomedical journal without an article in which flow cytometry has been used as a principal methodology. This book on methods in flow cytometry is therefore addressed to a wide, multidisciplinary audience.

Flow cytometry continues to rapidly expand. Extensive progress in the development of new probes and methods, as well as new applications, has occurred during the past few years. Many of the old techniques have been modified, improved, and often adapted to new applications. Numerous new methods have been introduced and applied in a variety of fields. This dramatic progress in the methodology, which occurred recently, and the positive reception of the first edition, which became outdated so rapidly, were the stimuli that led us to undertake the task of preparing a second edition.

The second edition is double the size of the first one, consisting of two volumes. It has a combined total of 71 chapters, well over half of them new, describing techniques that had not been presented previously. Several different methods and strategies for analysis of the same cell component or function are often presented and compared in a single chapter. Also included in these volumes are selected chapters from the first edition. Their choice was based on the continuing popularity of the methods; chapters describing less frequently used techniques were removed. All these chapters are updated, many are extensively modified, and new applications are presented.

From the wide spectrum of chapters presented in these volumes it is difficult to choose those methods that should be highlighted because of their novelty, possible high demand, or wide applicabilities. Certainly those methods that offer new tools for molecular biology belong in this category; they are presented in chapters on fluorescence *in situ* hybridization (FISH), primed *in situ* labeling

(PRINS), mRNA species detection, and molecular phenotyping. Detection of intracellular viruses and viral proteins and analysis of bacteria, yeasts, and plant cells are broadly described in greater detail than before in separate chapters. The chapter on cell viability presents and compares ten different methods for identifying dead cells and discriminating between apoptosis and necrosis, including a new method of DNA gel electrophoresis designed for the detection of degraded DNA in apoptotic cells. The chapter describing analysis of enzyme kinetics by flow cytometry is very complete. The subject of magnetic cell sorting is also described in great detail.

Numerous chapters that focus on the analysis of cell proliferation also should be underscored. The subjects of these chapters include univariate DNA content analysis (using a variety of techniques and fluorochromes applicable to cell cultures, fresh clinical samples, or paraffin blocks), the deconvolution of DNA content frequency histograms, multivariate (DNA vs protein or DNA vs RNA content) analysis, simple and complex assays of cell cycle kinetics utilizing BrdUrd and IdUdr incorporation, and studies of the cell cycle based on the expression of several proliferation-associated antigens, including the G_1- and G_2-cyclin proteins. Approaches to discriminating between cells having the same DNA content but at different positions in the cell cycle (e.g., noncycling G_0 vs cycling G_1, G_2 vs M, and G_2 of lower DNA ploidy vs G_1 of higher ploidy) are also presented.

Many of the methods described in these volumes will be used extensively in the fields of toxicology and pharmacology. Among these are the techniques designed for analysis of somatic mutants, formation of micronuclei, DNA repair replication, and cumulative DNA damage in sperm cells (DNA *in situ* denaturability). The latter is applicable as a biological dosimetry assay. A plethora of methods for analysis of different cell functions (functional assays) will also find application in toxicology and pharmacology.

The largest number of chapters is devoted to methods having clinical applications, either in medical research or in routine practice. Chapters dealing with lymphocyte phenotyping, reticulocyte and platelet analysis, analysis and sorting of hemopoietic stem cells, various aspects of drug resistance, DNA ploidy, and cell cycle measurements in tumors are very exhaustive. Diagnosis and disease progression assays in HIV-infected patients, as well as sorting of biohazardous specimens, new topics of current importance in the clinic, are also represented in this book.

Individual chapters are written by the researchers who developed the described methods, contributed to their modification, or found new applications and have extensive experience in their use. Thus, the authors represent a "Who's Who" directory in the field of flow cytometry. This ensures that the essential details of each methodology are included and that readers may easily learn these techniques by following the authors' protocols. We express our gratitude to all contributing authors for sharing their knowledge and experience.

The chapters are designed to be of practical value for anyone who intends to use them as a methods handbook. Yet, the theoretical bases of most of the techniques are presented in detail sufficient for teaching the principle underlying the described methodology. This may be of help to those researchers who want to modify the techniques, or to extend their applicability to other cell systems. Understanding the principles of the method is also essential for data evaluation and for recognition of artifacts. A separate section of most chapters is devoted to the applicability of the described method to different biological systems. Another section of most chapters covers the critical points of the procedure, possible pitfalls, and experience of the author(s) with different instruments. Appropriate controls, standards, instrument adjustments, and calibrations are the subjects of still another section of each chapter. Typical results, frequently illustrating different cell types, are presented and discussed in yet another section. The Materials and Methods section of each chapter is exhaustive, providing a detailed, step-by-step description of the procedure in a protocol or cookbooklike format. Such exhaustive treatment of the methodology is unique; there is no other publication on the subject of similar scope.

We hope that the second edition of *Flow Cytometry* will be even more successful than the first. The explosive growth of this methodology guarantees that soon there will be the need to compile new procedures for a third edition.

Zbigniew Darzynkiewicz
Harry A. Crissman
J. Paul Robinson

PREFACE TO THE FIRST EDITION

Progress in cell biology has been closely associated with the development of quantitative analytical methods applicable to individual cells or cell organelles. Three distinctive phases characterize this development. The first started with the introduction of microspectrophotometry, microfluorometry, and microinterferometry. These methods provided a means to quantitate various cell constituents such as DNA, RNA, or protein. Their application initiated the modern era in cell biology, based on quantitative—rather than qualitative, visual—cell analysis. The second phase began with the birth of autoradiography. Applications of autoradiography were widespread and this technology greatly contributed to better understanding of many functions of the cell. Especially rewarding were studies on cell reproduction; data obtained with the use of autoradiography were essential in establishing the concept of the cell cycle and generated a plethora of information about the proliferation of both normal and tumor cells.

The introduction of flow cytometry initiated the third phase of progress in methods development. The history of flow cytometry is short, with most advances occurring over the past 15 years. Flow cytometry (and associated with it electronic cell sorting) offers several advantages over the two earlier methodologies. The first is the rapidity of the measurements. Several hundred, or even thousands, of cells can be measured per second, with high accuracy and reproducibility. Thus, large numbers of cells from a given population can be analyzed and rare cells or subpopulations detected. A multitude of probes have been developed that make it possible to measure a variety of cell constituents. Because different constituents can be measured simultaneously and the data are recorded by the computer in list mode fashion, subsequent bi- or multivariate analysis can provide information about quantitative relationships among constituents either in particular cells or between cell subpopulations. Still another advantage of flow cytometry stems from the capability for selective physical sorting of individual cells, cell nuclei, or chromosomes, based on differences in the variables measured. Because some of the staining methods preserve cell viability and/or cell membrane integrity, the reproductive and immunogenic capacity of the sorted cells can be investigated. Sorting of individual chromosomes has already provided the basis for development of chromosomal DNA libraries, which are now indispensable in molecular biology and cytogenetics.

Flow cytometry is a new methodology and is still under intense development, improvement, and continuing change. Most flow cytometers are quite complex and not yet user friendly. Some instruments fit particular applications better than others, and many proposed analytical applications have not been extensively tested on different cell types. Several methods are not yet routine and a certain degree of artistry and creativity is often required in adapting them to new biological material, to new applications, or even to different instrument designs. The methods published earlier often undergo modifications or improvements. New probes are frequently introduced.

This volume represents the first attempt to compile and present selected flow cytometric methods in the form of a manual designed to be of help to anyone interested in their practical applications. Methods having a wide immediate or potential application were selected, and the chapters are written by the authors who pioneered their development or who modified earlier techniques and have extensive experience in their application. This ensures that the essential details are included and that readers may easily master these techniques in their laboratories by following the described procedures.

The selection of chapters also reflects the peculiarity of the early phase of method development referred to previously. The most popular applications of flow cytometry are in the fields of immunology and DNA content–cell cycle analysis. While the immunological applications are now quite routine, many laboratories still face problems with the DNA measurements, as is evident from the poor quality of the raw data (DNA frequency histograms) presented in many publications. We hope that the descriptions of several DNA methods in this volume, some of them individually tailored to specific dyes, flow cytometers, and material (e.g., fixed or unfixed cells or isolated cell nuclei from solid tumors), may help readers to select those methods that would be optimal for their laboratory setting and material. Of great importance is the standardization of the data, which is stressed in all chapters and is a subject of a separate chapter.

Some applications of flow cytometry included in this volume are not yet widely recognized but are of potential importance and are expected to become widespread in the near future. Among these are methods that deal with fluorescent labeling of plasma membrane for cell tracking, flow microsphere immunoassay, the cell cycle of bacteria, the analysis and sorting of plant cells, and flow cytometric exploration of organisms living in oceans, rivers, and lakes.

Individual chapters are designed to provide the maximum practical information needed to reproduce the methods described. The theoretical bases of the methods are briefly presented in the introduction of most chapters. A separate section of each chapter is devoted to applicability of the described method to different biological systems, and when possible, references are provided to articles that review the applications. Also discussed under separate subheads are the critical points of procedure, including the experience of the authors with different instruments, and the appropriate controls and standards. Typical results, often illustrating different cell types, are presented and discussed in the "Results" section. The "Materials and Methods" section of each chapter is the most extensive, giving a detailed description of the method in a cookbook format.

Flow cytometry and electronic sorting have already made a significant impact on research in various fields of cell and molecular biology and medicine. We hope that this volume will be of help to the many researchers who need flow

Preface to the First Edition

cytometry in their studies, stimulate applications of this methodology to new areas, and promote progress in many disciplines of science.

Zbigniew Darzynkiewicz
Harry A. Crissman

PART I

Principles of Cytometry and General Methods

CHAPTER 1

A Brief History of Flow Cytometry and Sorting

Myron R. Melamed
New York Medical College
Valhalla, New York 10595

I. Introduction
II. Instrumentation
III. Applications
   References

I. Introduction

For those of us who have watched the origins and maturation of flow cytometry from laboratory curiosity to an essential instrument of cell biology and clinical patient care, the events since the 1960s have been nothing less than spectacular. They could not have happened without an extraordinary confluence of events that included not only the astonishing developments in computer science and laser light sources, simultaneous with the discovery of monoclonal antibody technology and specific DNA fluorochromes, but also the coincident political will to fight (and fund) a "war on cancer." In the 1960s and early 1970s, the value of the Pap smear had been recognized, but the technique was not yet widely implemented because of a lack of trained cytotechnologists. Earlier work primarily by Caspersson (1936, 1950) and Mellors (Mellors et al., 1950, 1952; Mellors and Silver, 1951) held out hope that the labor intensive slide screening process could be accomplished by an automated image analyzing microscope, alleviating the shortage of skilled cytotechnologists. A major effort to develop such an instrument was undertaken in this country by the National Cancer Institute in the 1970s, by some technology oriented private companies, and by government and industry in Japan and Europe. Although the goal to develop a machine that could screen Pap smears and replace the cytotechnologist was never realized, the flow cytometry instruments that were developed have had a much broader, much greater impact on biology and medicine than was ever anticipated.

These new instruments and a growing number of monoclonal antibody probes of cell structure, cell constituents, and cell function have forever changed the classification of hemopoietic and other cell types, and recast the study of cell biology. It is difficult now to find a single clinical or biomedical laboratory that does not make use of this technology. During the 1980s and 1990s there have been dramatic advances resulting from (i) improved instrumentation, increasing the precision of measurements and the number of simultaneously measurable parameters including, most recently, combining flow with cell scanning image cytometry; (ii) new antibodies and other probes of cellular constituents, with new techniques to measure a variety of cell attributes and cell functions; and (iii) new clinical and research applications. Advances in each of these areas depends on and influences the others. This brief review can only highlight some of the key events in the development of flow cytometry; for a more detailed description the reader is referred to chapters on the history of flow cytometry in texts by Melamed *et al.* (1990) and Shapiro (1988).

II. Instrumentation

Efforts to quantify constituents of individual cells began, understandably, with measurements of cells on slides. The cells to be measured could be selected visually, and the measurements obtained were readily correlated with cell type. Pioneering this work, as noted earlier, was Tjorborn Caspersson at the Karolinska Institute in Stockholm and Robert Mellors at the Sloan-Kettering Institute in New York. Caspersson's studies followed the discovery of nucleic acids by Miescher (1897), the invention of the ultraviolet (UV) microscope by Kohler (1904), and the demonstration by Dhere (1906) that purine and pyrimidine bases of nucleic acids absorbed UV light at about 260 nm. By measuring UV absorption of intact, individual cells photometrically, Caspersson introduced the concept and began to develop the techniques of quantitative cytochemistry. He demonstrated that metabolically active cells, including cancer cells, had increased nucleic acid content (Caspersson, 1936); he also showed banding in chromosomes by UV absorption. His monograph on cell growth and cell function summarized this seminal work (Caspersson, 1950). Mellors was the first to use fluorescent nucleic acid dyes to stain and measure intracellular nucleic acids (Mellors *et al.*, 1950). He invented a microfluorometric scanner to scan and measure the cells in smears of exfoliated cells (Mellors and Silver, 1951), and he was the first to identify cancer cells in a Pap smear by quantifying cellular nucleic acid content (Mellors *et al.*, 1952). It was the work by Mellors in collaboration with Dr. Papanicolaou (Mellors *et al.*, 1952) that excited interest in automated cytology and led to the funding of this effort, first by the American Cancer Society and then by the National Cancer Institute.

But the early instruments and even the later, better engineered cell scanning, image analyzing microscopes were very slow, and they suffered from a number of then insurmountable physical limitations. Laser light sources had yet to be developed, computer data analysis systems were too slow for the huge amounts of data that had to be analyzed from each cell image, and conventionally prepared slides presented a confusing picture of overlapping cells and artifact. Flow cytometry was to overcome at least some of these difficulties by global measurements of a limited number of key features on each cell, and by measurement rates in the hundreds of cells per second.

The first attempt to count or measure cells flowing in suspension is generally credited to Moldavan (1934). He described an apparatus consisting simply of a microscope that was focused on a capillary glass tube through which he forced a suspension of cells, recording the passage of each cell by a photoelectric device at the eyepiece. Improved versions of this instrument were described later by Cornwall and Davison (1950), and by Bierne and Hutcheon (1957). The capillary tubes were easily clogged by cell aggregates and debris in these instruments, which Kamentsky (1965) was able to minimize by using a bow-tie-shaped channel. However, it was Crosland-Taylor (1953) who showed that clogging could be prevented by using a large diameter channel and centering the cell stream in a fluid sheath using the principle of laminar flow. All flow channels are now designed this way. The basic hydrodynamic principles that control particles in flow is described in detail by Kachel *et al.* (1990).

The first practical, clinically successful instrument for counting (and later for analyzing) blood cells in suspension was designed and filed for patent by Walter Coulter in 1949 (Coulter, 1953). For the next three decades, and longer, the Coulter Counter was used routinely in clinical laboratories worldwide to do white and red blood cell counts. The cells suspended in an ionic solution are made to pass singly through a constricted orifice between two electrodes maintained at a constant potential difference. The electrical conductivity of the suspending medium is intentionally different from that of the cell so that as a cell passes between the electrodes it alters conductivity and generates an electrical signal. Red blood cell counts were performed on diluted whole blood; white blood cell counts were obtained by first lysing the red cells. The Coulter Counter underwent a number of improvements over the years, most notably in the electronics and design of the counting chamber so that cells could be differentiated by size as they were counted (Coulter, 1956). In principle it was also possible to use frequency modulated electric currents to distinguish different types of cells based on differences in their cellular constituents (Coulter and Hogg, 1970), but this was never incorporated into a commercially successful instrument.

At about this same time, Parker and Horst (1959) described a photometry based blood cell counter that could be used to distinguish and separately count red and white blood cells. They used a blue dye to stain the nucleated white blood cells in a suspension of whole blood, while the hemoglobin containing erythrocytes were red. They then passed the cells in single file through a small diameter glass channel that was illuminated by narrowly focused red and blue beams of light. It appears that blood cells were less likely to clog the channel than other cell suspensions, and the transmitted red and blue wavelength light was measured by separate photomultipliers. This was the first automatic differential cell counter based on differences in optical properties of the cells.

These early optically based cell counters remained laboratory curiosities until the 1960s when Louis Kamentsky, then working at the IBM Watson Laboratory on the campus of Columbia University, designed and built the first multiparameter flow cytometer (Kamentsky, 1965; Kamentsky et al., 1965). That instrument, originally intended to identify cancer cells in uterine cervical cytology specimens, simultaneously measured DNA and total protein content per cell, in unstained cells, at their respective wavelengths of maximum absorption, and did so at rates up to a thousand cells per second. A field trial evaluating the instrument yielded surprisingly good sensitivity, but the specificity was too low at that time to be clinically acceptable (Koenig et al., 1968). A second instrument was built and sent to Leonard Herzenberg at Stanford University for research in cellular immunology.

These first flow cytometers had a bow-tie-shaped quartz flow channel illuminated at the waist by a narrowly focused mercury arc or halogen light beam with excitation wavelengths in the UV as well as the visible spectrum. The flow channel was located on the stage of a conventional research microscope. Light absorption at the desired wavelengths and light scatter were measured by appropriately filtered photomultipliers and scatter sensors. This instrument was the first to measure and record in list mode several simultaneous measurements on each cell, opening the possibility of directly correlating measurements of different constituents or different cell attributes for individual cells on a cell by cell basis in large populations of cells. It was the first flow cytometer to be produced commercially (Cytograf and Cytofluorograf, manufactured by Biophysics Systems, Inc., Mahopac, NY), and led to the family of advanced flow cytometry instruments later produced and sold by the Ortho Instrument division of Johnson & Johnson.

The excitation beam of light in Kamentsky's flow cytometer was orthogonal to the flow chamber. A different approach was taken by Dittrich and Gohde (1969, 1972), who filed for a patent on a flow cytometer in which the axis of cell flow was aligned with that of the excitation beam of light. The cells were made to flow though the focus of the excitation beam instead of orthogonal to it; this eliminated slight variances in measurement due to differences in excitation when cells deviated slightly from the center of the stream. They used a mercury arc light source and high numerical aperture microscope objective to assure uniform distribution of excitation light intensity across the whole cell at the focal plane of the objective. The mercury arc light source made it possible to select from many different excitation wavelengths by appropriate choice of filters. The coefficient of variation of DNA measurements with this instrument (and later versions of it) was the best ever achieved. Users were able to measure DNA content of sperm and some somatic cells with accuracy sufficient to distinguish XY from XX cells. The Phywe Impulscytophotometer, produced commercially in Europe, was based on the design by Dittrich and Gohde (1972). It was manufactured in the United States later by the Ortho Instruments division of Johnson & Johnson and sold as the ICP 22. A more advanced instrument with improved optics and electronics, and options that include laser excitation and enclosed channel cell sorting, is now available in Europe by PARTEC.

In the late 1960s laser light sources, fluorescent dyes, and immunofluorescent reagents were introduced, greatly increasing the sensitivity, precision, and range of possible measurements (Kamentsky and Melamed, 1969; Dittrich and Gohde, 1969; Van Dilla et al., 1969; Hulett et al., 1969). With few exceptions almost all of the flow cytometry instruments now use laser light sources for excitation. Lasers provide high intensity single wavelength beams of light that can be precisely focused on the flow channel. The blue emitting (488 nm) argon ion laser has been ideal for excitation of fluorescein, propidium iodide, and acridine orange, three of the most commonly used fluorochromes, and it is now used in all the commercial orthogonal flow cytometers. Ultraviolet lasers were and still are very expensive and much less reliable, and they have been used only in a few research instruments. More recently, red excited fluorescent dyes have been developed for flow cytometry (Oi et al., 1982; Stryer and Glazer, 1985; Stryer et al., 1985; Waggoner, 1990), and instruments are being designed to use inexpensive red emitting helium neon lasers, usually as a second laser with the argon ion laser in a dual excitation instrument.

Soon after the first flow cytometers were designed it was recognized that fluorescence measurements had many advantages over measurements of absorbing dyes. Fluorescence offered much greater sensitivity, and at the dye concentrations used for studies of biological cells there was no distributional error to contend with. Fluorescence was essential if one wanted to quantify antigen expression of cellular constituents, a common requirement in experimental and clinical laboratories, and two or more different constituents could be measured simultaneously—something that was difficult or impossible to do with absorbing dyes. Thus, with rare exceptions, light sources and optics of all flow cytometers are now optimized for fluorescence measurements.

Research flow cytometers were designed and developed at Los Alamos under the leadership of Marvin Van Dilla. Like the Kamentsky instruments, excitation was orthogonal to the flow stream, as were fluorescence detectors and scatter sensors, but high intensity laser light sources were emphasized. Other modifications were made, including design changes in the flow channel, and a Coulter sensor incorporated with the optical sensors. A multiangle scatter sensor was used to study light scatter as a means of cell classification. Salzman et al. (1975) demonstrated that the major types of white blood cells in peripheral blood could be distinguished by a combination of forward and right-angle light scatter, and these measurement features are now the basis for automated differential white blood cell counters in the clinical laboratory. Light scatter is also used to identify lymphocytes in whole blood or marrow for subclassification according to surface antigen expression.

A flow cytometer with dual beam excitation was described by Stohr (1976), and multilaser excitation by Curbelo *et al.* at Block Engineering (1976). The Cytomat flow cytometer developed and manufactured at Block Engineering simultaneously measured at least four fluorescence parameters, UV absorption, and light scatter. It was intended to do red and white blood cell counts and with very sophisticated cytochemistry to do white cell differential counts (Shapiro *et al.*, 1977). It was a complex engineering marvel but not a commercial success. Dual laser excitation was subsequently incorporated in the Los Alamos flow cytometers (Steinkamp *et al.*, 1979), but it was not available in commercial instruments until many years later.

Cell sorting was added to the early flow cytometers for two purposes: (i) diagnostic, that is, for visual identification of cells selected by flow cytometry measurement; and (ii) preparative, that is, to collect a pure population of cells based on some selected flow cytometry features. Kamentsky was intent on demonstrating that he had correctly identified cancer cells in cervical cytology specimens, and he devised a fluidic sorter that briefly diverted the flow into a side channel when a presumed cancer cell was detected (Kamentsky and Melamed, 1967). Sorting was slow, and the desired cell was almost always mixed among many others as contaminants. A sorter described by Mack Fulwyler (1965) could separate single cells from the cell stream by electrostatic deflection as they emerged from a Coulter cell sizing nozzle. It was based on an earlier invention of a printing method by Sweet (1965) in which ink jet droplets were deflected electrostatically. In 1969, Hulett *et al.* successfully adapted electrostatic cell sorting to a flow cytometer and separated cells according to fluorescence measurements. The laser excitation beam was focused on the cell stream as it emerged from the nozzle of the flow channel, before breaking into droplets. These measurements of the cell stream in air were surprisingly good. The electrostatic sorter was further refined by Hulett, Sweet, and their associates (Hulett *et al.*, 1973), and it was incorporated into what became known as the fluorescence activated cell sorter (FACS), manufactured by Becton Dickinson. High speed electrostatic cell sorters with dual laser excitation were subsequently designed and built at the Los Alamos (Steinkamp *et al.*, 1979) and Lawrence Livermore National Laboratories (Gray *et al.*, 1975, 1987; Carrano *et al.*, 1979), and used to sort single chromosomes for preparation of gene libraries in plasmids.

One commercial instrument, Hemalog, was designed by Ansley and Ornstein (1970) and refined later by them (Ornstein and Ansley, 1974) and by Mansberg *et al.* (1974) and their colleagues to identify and classify stained cells by light absorption and scatter rather than fluorescence. It was manufactured by Technicon Corp. in Tarrytown, New York, and made use of parallel sample processing through different channels in each of which a specific cell type was stained and counted. This instrument was used in a number of clinical laboratories.

Still another flow cytometer design was proposed by Steen (Steen, 1980; Steen et al., 1982). In his instrument the fluid stream of cells is directed at a specific angle onto the surface of a glass coverslip at the focus of the microscope objective. A mercury or xenon lamp is used for excitation of the fluorescence stained cells. The Steen instrument is attractive because of its simplicity, low cost, and high sensitivity.

One of the advantages of image over flow cytometry, perhaps the most significant advantage, is ready correlation with cell classification by conventional light microscopy. A major goal in cytometry has been to combine this advantage of image cytometry with the speed and precision of measurements by flow cytometry. It appears now that this has been accomplished with a new microscope based instrument described by Kamentsky and Kamentsky (1991). This instrument, a laser scanning cytometer (LSC) manufactured by CompuCyte Inc., Cambridge, Massachusetts, makes flow cytometry type measurements of cells on slides in imprints, smears, cell suspensions, or tissue sections. Speed and precision of measurement are comparable to that of conventional flow cytometry. The cells are retained on the slide and can be remeasured at desired time intervals to derive kinetics of an enzymatic or other reaction for each of any number of cells, and since cell position also is recorded the cells can be restained and examined for some other feature, or for classification by conventional light microscopy. The instrument is described in detail in Chapter 3 of this volume.

III. Applications

The earliest and most consistently measured cell feature has been nucleic acid content, first as total nucleic acid measured by UV absorption, and then as DNA content measured by the fluorescence of cells stained with DNA specific fluorescent dyes. The staining techniques for DNA in intact cells were developed early and paralleled the development of flow cytometry instruments; performance standards for the instruments were built around the precision of measurements of DNA in standard cells in which DNA content was considered to be invariable. From the work of Caspersson, Mellors, and others (Atkin and Kay, 1979; Barlogie et al., 1980; Buchner et al., 1985; Dressler et al., 1988; Raber and Barlogie, 1990) it was evident that most (though not all) cancer cells had abnormal values for DNA content.

Fluorescence staining and measurement techniques were introduced at almost the same time in several different laboratories working independently, which led to an immediate great improvement in the precision of measurements and range of applications. Kamentsky reported dual DNA and protein measurements using a fluorescent Feulgen stain for DNA and Naphthol Yellow S for protein. He also was the first to use a computer to record and analyze two-dimensional data displayed in a dot plot (Kamentsky and Melamed, 1969). Van Dilla and colleagues (1969) at Los Alamos also used a fluorescent Feulgen reaction to measure DNA distribution which they displayed in univariate frequency histograms and identified G_1, S, and G_2/M phases of the cell cycle. Gohde and Dittrich (1970) were the first to use ethidium bromide or ethidium bromide and mithramycin as specific DNA stains in dual parameter measurements with protein fluorochromes, and they pioneered the application of flow cytometry to studies of antitumor drugs on the cell cycle (Gohde and Dittrich, 1971). Hulett, Bonner, and their associates at Stanford University used fluorescence measurements of cells to select those they wished to sort on the fluorescence activated cell sorter (Hulett *et al.*, 1969, 1973).

Of the many different DNA dyes that were subsequently introduced, the most precise measurements were obtained with 4,6-diamidino-2-phenylindole (DAPI) (Otto and Oldiges, 1978), but because it is excited in the UV, and there are no reliable, inexpensive UV lasers, it is used almost exclusively with flow cytometers that have mercury arc, halogen, or xenon light sources. Propidium iodide, a DNA (and RNA) intercalating dye excited by the blue light of the argon ion laser, when used with RNase became a favored DNA dye. In evaluating the various DNA fluorochromes, and measurements of cellular DNA content obtained with them, it must be remembered that stainability of DNA *in situ* depends on its accessibility to the dye, which in turn is influenced by chromatin structure and nuclear proteins.

Interest in DNA measurements of human tumor cells was sparked by a report of Hedley *et al.* (1983) who described a method for extracting cell nuclei from archived paraffin embedded tissues and measuring nuclear DNA content. The DNA histograms they obtained correlated well with histograms of corresponding fresh specimens, and opened the way for numerous studies of almost all types of human tumors over the succeeding years. In general, some proportion of virtually all human tumors were aneuploid, and aneuploidy correlated with tumor grade (anaplasia).

We found the metachromatic fluorescent dye acridine orange (AO) to be one of the most interesting. As a vital stain it is a lysosomal probe that can be used to distinguish polymorphonuclear leukocytes in peripheral blood from monocytes, and both from lymphocytes (Melamed *et al.*, 1972). In patients with infections the polymorphonuclear leukocytes exhibit increased variance of lysosomal staining (Melamed *et al.*, 1974). In fixed cells, or cells permeabilized with a mild detergent, AO can be used to differentially stain double versus single stranded DNA (Darzynkiewicz *et al.*, 1974), or DNA versus RNA (Darzynkiewicz *et al.*, 1976; Traganos *et al.*, 1977). A change in chromatin structure that affects resistance to denaturation can be demonstrated by a change in the ratio of native to denatured DNA under appropriate denaturing conditions, and such studies have been effective in identifying abnormal sperm in subfertile individuals (Evenson *et al.*, 1980). Simultaneous DNA and RNA measurements in cells stained with AO have allowed noncycling G_0 cells to be distinguished from G_1 (Darzynkiewicz *et al.*, 1976), and cells in mitosis from cells in G_2 (Darzynkiewicz *et al.*, 1977).

RNA content has also been used to subclassify human acute leukemias (Andreeff *et al.*, 1980).

Cell sizing by the Coulter principle was replaced by forward light scatter measurements, which roughly reflect the cell diameter. More precise estimates of cell size were derived from the fluorescence pulse shape by Sharpless and Melamed (1976).

Tumor growth rates have long been valued prognostic indicators, whether measured grossly or inferred from mitotic index counts or thymidine labeling. With increasingly sophisticated flow cytometry techniques of cell cycle analysis, new methods have come available for directly assaying tumor cell proliferation. The simplest of these is the S-phase (proliferative) fraction of cells as determined by deconvolution of the DNA histogram (Gray *et al.*, 1990), and the S-phase fraction not surprisingly appears to be a better indicator of prognosis than DNA ploidy (Meyer and Lee, 1980).

A major advance in flow cytometry assays of cell proliferation was made by Gratzner *et al.* (1975), who developed polyclonal and then monoclonal (Gratzner, 1982) antibodies to the thymidine analog bromodeoxyuridine (BrdUrd), providing for rapid, direct detection of newly synthesized DNA that has incorporated BrdUrd. Thin sections or cell suspensions of tumor may be incubated with BrdUrd *in vitro*, or it may be given intravenously to label DNA synthesizing cells *in vivo*. BrdUrd was used as a radiation enhancer for many years in patients undergoing treatment for cancer, and it is nontoxic in the doses used. The method is near ideal for study of cell cycle kinetics, particularly when combined with analysis of the cell cycle distribution by simultaneous measurement of DNA content (Dolbeare *et al.*, 1983).

There is growing awareness that cell death by apoptosis may play as important a role as cell proliferation in the control of tumor growth, and a number of new flow cytometry techniques have been developed more recently to quantify this potentially important descriptor (for review see Darzynkiewicz *et al.*, 1992). The most direct and probably the most specific method is by labeling the 3'-OH termini of DNA breaks in the nucleus, permitting the apoptotic cells to be identified and counted by flow cytometry (Gorczyca *et al.*, 1993).

The most widely used clinical applications of flow cytometry are in the classification of hemopoietic cells by cell surface immunofluorescence. Lymphocytes once thought to be of a single class are now known to represent many diverse classes of cells, and more than 150 monoclonal antibodies to cell surface antigens subcategorize these and other hemopoietic cells functionally and by lineage and differentiation. Although many have contributed conceptually, and by generating useful new antibodies, major credit must go first to Reinherz (1979a,b), Kung, Goldstein, and co-workers (Kung *et al.*, 1979) for their seminal work in subclassifying T lymphocytes according to cell surface antigen expression. The classification of leukemias (Andreeff, 1990a) and lymphomas (Andreeff, 1990b) is, in fact, now determined to a large extent by cell surface antigen expression, as is hemopoietic cell maturation (Ault, 1990; Loken *et al.*, 1977). Typically, combinations of two or three (Loken *et al.*, 1987; Hardy *et al.*, 1983) different antigens, or more, may be measured simultaneously or in sequence to arrive at a final classification. With newly available phycobiliproteins (Waggoner, 1990) and other fluorochromes that have emission spectra separable from that of the fluorescent dyes now in use, and with dual laser flow cytometers that can accommodate a second laser light source to match their excitation peak, we can look forward to more routine use of complex multiparameter analyses that may include functional assays simultaneously with cell classification.

More recently, techniques have been developed to quantify intracellular antigens by immunofluorescence. These require that the cellular antigens be made accessible to antibodies by cell fixation or permeabilization without disrupting the conformation of their epitopes (Clevenger *et al.*, 1987; Jacobberger *et al.*, 1986). Of initial and persisting interest are the cell proliferation antigens, in particular Ki-67 (Gerdes *et al.*, 1983), proliferating cell nuclear antigen (Celis *et al.*, 1986; Tan *et al.*, 1987), nucleolar p120 antigen (Busch *et al.*, 1979), and most recently the nucleolar organizing region NOR (W. Gorczyca, personal communication, 1999). Cyclins and cyclin dependent kinases are currently of great interest and are discussed in detail elsewhere in this volume. Oncogene protein products studied by flow cytometry include p53 (Levine, 1997) and total and phosphorylated pRB (Juan *et al.*, 1998). Other intracellular components that have been studied include the estrogen and progesterone receptors in tumor cells of the breast (Gorczyca *et al.*, 1998) and Erb-B2 expression.

Flow cytometry karyotyping was reported by Gray *et al.* (1975), who first measured total DNA content of single chromosomes stained in suspension with ethidium bromide and could distinguish some but not all chromosomes by DNA content alone. To separate chromosomes of similar DNA content, Gray and colleagues at the Livermore laboratory used dual staining with Hoechst 33258, which is specific for AT base pairs, and chromomycin A3, which is specific for GC base pairs. With this dye combination, which is now the preferred staining methodology, they were able to distinguish most of the human chromosomes. Flow karyotyping has been used to sort chromosomes to establish gene libraries in plasmids, as noted previously; it is also used to obtain pure populations of particular chromosomes for sequencing as part of the human genome project. Chromosomes or segments of chromosomes can be stained in the interphase nucleus by *in situ* hybridization with fluoresceinated oligonucleotide probes and identified by flow cytometry (Trask *et al.*, 1985).

Finally, there are many probes that have been developed to measure functional properties of the cell, including the formation of oxidative products in the stimulated neutrophil (Bass *et al.*, 1983). Plasma membrane potential and mitochondrial membrane potential have been measured with cationic cyanine dyes (Shapiro *et al.*, 1979) or rhodamine 123 (Darzynkiewicz *et al.*, 1982), giving evidence of cell activation. Other cell function determinations include calcium ion concentration (Rabinovitch *et al.*, 1986), intracellular pH (Visser *et al.*, 1979), cell surface charge (Valet *et al.*, 1979), phagocytosis (Bassoe *et al.*, 1983), drug uptake and retention (Krishan *et al.,* 1985), and a variety of enzymatic reactions (Watson, 1980).

This brief survey of the development and applications of flow cytometry is focused primarily on human cell biology and clinical medicine. It does not include important applications in microbiology, studies of plant chloroplasts or algae, or flow cytometry assays of soluble antigens using antibody coated microspheres. Nor does it include descriptions or references to more recently developed techniques and applications that are described in detail elsewhere in this monograph.

References

Andreeff, M. (1990a). Flow cytometry of leukemias. *In* "Flow Cytometry and Sorting" (M. R. Melamed, T. Lindmo, and M. L. Mendelsohn, eds.), 2nd Ed., Chap. 35. Wiley-Liss, New York.

Andreeff, M. (1990b). Flow cytometry of lymphomas. *In* "Flow Cytometry and Sorting" (M. R. Melamed, T. Lindmo, and M. L. Mendelsohn, eds.), 2nd Ed., Chap. 36. Wiley-Liss, New York.

Andreeff, M., Darzynkiewicz, Z., Sharpless, T. K., Clarkson, B. D., and Melamed, M. R. (1980). Discrimination of human leukemia subtypes by flow cytometric analysis of cellular DNA and RNA. *Blood* **55,** 282–293.

Ansley, H., and Ornstein, L. (1970). Enzyme histochemistry and differentiated white counts on the technicon hemalog D. *In* "Advances in Automated Analysis" Technicon International Congress, Vol. 1, pp. 437–446. Thuman Associates, Miami, Florida.

Atkin, N. B., and Kay, R. (1979). Prognostic significance of modal DNA value and other factors in malignant tumors based on 1465 cases. *Br. J. Cancer* **40,** 210–221.

Ault, K. A. (1990). Applications in immunology and lymphocyte analysis. *In* "Flow Cytometry and Sorting" (M. R. Melamed, T. Lindmo, M. L. Mendelsohn, eds.), 2nd Ed., Chap. 34. Wiley-Liss, New York.

Barlogie, B., Drewinko, B., Schumann, J., Gohde, W., Dosik, G., Latreille, J., Johnston, D. A., and Freireich, E. J. (1980). Cellular DNA content as a marker of neoplasia in man. *Am. J. Med.* **69,** 195–203.

Bass, D. A., Parce, J. W., DeChatelet, L. R., Szejda, P., Seeds, M. C., and Thomas, M. (1983). Flow cytometric studies of oxidative product formation by neutrophils: A graded response to membrane stimulation. *J. Immunol.* **130,** 1910–1917.

Bassoe, C. F., Laerum, O. D., Solberg, C. O., and Honeberg, B. (1983). Phagocytosis of bacteria by human leukocytes measured by flow cytometry. *Proc. Soc. Exp. Biol. Med.* **174,** 182–186.

Bierne, T., and Hutcheon, J. M. (1957). A photoelectric particle counter for use in the sieve range. *J. Sci. Instrum.* **34,** 196–200.

Buchner, T., Hiddemann, W., Wormann, B., Kleinmeir, B., Schumann, J., Gohde, W., Ritter, J., Muller, K. M., Von Bassewitz, D. B., and Roessner, A. (1985). Differential pattern of DNA aneuploidy in human malignancies. *Path. Res. Proctol.* **179,** 310–317.

Busch, H., Gyorkey, F., Busch, R. K., Davis, F. M., Gyorkey, P., and Smetana, K. (1979). A nucleolar antigen found in a broad range of human malignant tumor specimens. *Cancer Res.* **39,** 3024–3030.

Carrano, A. V., Van Dilla, M. A., and Gray, J. W. (1979). Flow cytogenetics: A new approach to chromosome analysis. *In* "Flow Cytometry and Sorting," (M. R. Melamed, P. F. Mullaney, and M. L. Mendelson, eds), 1st Ed., Chap 23. Wiley, New York.

Caspersson, T. (1936). Uber den Chemischen Aufbau der Strutkusen des Zellkernes. *Skand. Arch. Physiol.* **73**(Suppl. 8), 1–151.

Caspersson, T. (1950). "Cell Growth and Cell Function." Norton, New York.

Celis, J. E., Madsen, P., Nielsen, S., and Celis, A. (1986). Nuclear patterns of cyclin (PCNA) antigen distribution subdivide S phase in cultured cells—some applications of PCNA antibodies. *Leuk. Res.* **10,** 237–249.

Clevenger, C. V., Epstein, A. L., and Bauer, K. D. (1987). Modulation of the nuclear antigen p105 in lymphocytes as a function of cell cycle progression. *J. Cell. Physiol.* **130,** 130–136.

Cornwall, J. B., and Davison, R. M. (1950). Rapid counter for small particles in suspension. *J. Sci. Instrum.* **37,** 414–417.

Coulter, W. H. (1953). U.S. Patent 2,656,508, Means for counting particles suspended in a fluid. Filed Aug. 27, 1949. Issued October 20, 1953.

Coulter, W. H. (1956). High speed automatic blood cell counter and cell size analyzer. *Proc. Natl. Electron Conf.* **12,** 1034–1042.

Coulter, W. H., and Hogg, W. R. (1970). U.S. Patent 3,502,974, Signal modulated apparatus for generating and detecting sensitive and reactive changes in a modulated current path for particle classification and analysis. Filed May 23, 1966. Issued March 24, 1970.

Crosland-Taylor, P. J. (1953). A device for counting small particles suspended in a fluid through a tube. *Nature (London)* **171,** 37–38.

Curbelo, R., Schildkraut, E. R., Hirschfeld, T., Webb, R. H., Block, M. J., and Shapiro, H. M. (1976). A generalized machine for automated flow cytology system design. *J. Histochem. Cytochem.* **24,** 388–395.

Darzynkiewicz, Z., Traganos, F., Sharpless, T., and Melamed, M. R. (1974). Thermally-induced changes in chromatin of isolated nuclei and of intact cells as revealed by acridine orange staining. *Biochem. Biophys. Res. Commun.* **59,** 392–399.

Darzynkiewicz, Z., Traganos, F., Sharpless, T., and Melamed, M. R. (1976). Lymphocyte stimulation: A rapid multiparameter analysis. *Proc. Natl. Acad. Sci. U.S.A.* **73,** 2881–2884.

Darzynkiewicz, Z., Traganos, F., Sharpless, T., and Melamed, M. R. (1977). Recognition of cells in mitosis by flow cytofluorometry. *J. Histochem. Cytochem.* **25,** 875–880.

Darzynkiewicz, Z., Traganos, F., Staiano-Coico, L., Kapuscinski, J., and Melamed, M. R. (1982). Actions of Rhodamine 123 with living cells studied by flow cytometry. *Cancer Res.* **42,** 799–806.

Darzynkiewicz, Z., Bruno, S., Del Bino, G., Gorczyca, W., Hotz, M. A., Lassota, P., and Traganos, F. (1992). Features of apoptotic cells measured by flow cytometry. *Cytometry* **13,** 795–808.

Dhere, C. (1906). Sur l'absorption des rayons ultraviolets par l'acide muscleique extrait de la levure de biere. *Compt. Rend. Soc. Biol.* **1,** 34.

Dittrich, W., and Gohde, W. (1969). Impulsfluorometrie bei einzelzellen in suspensionen. *Z. Naturforsch.* **24b,** 360–361.

Dittrich, W., and Gohde, W. (1972). British Patent 1,300,585, Automatic measuring and counting device for particles in a dispersion. Filed Dec. 18, 1968 in Germany. Issued Dec. 20, 1972.

Dolbeare, F., Gratzner, H., Pallavicini, M. G., and Gray, J. W. (1983). Flow cytometric measurement of total DNA content and incorporated bromodeoxyuridine. *Proc. Natl. Acad. Sci. U.S.A.* **80,** 5573–5577.

Dressler, L. G., Seamer, L. C., Owens, M. A., Clark, G. M., and McGuire, W. L. (1988). DNA flow cytometry and prognostic factors in 1331 frozen breast cancer specimens. Cancer **61,** 420–427.

Evenson, D. P., Darzynkiewicz, Z., and Melamed, M. R. (1980). Relation of mammalian sperm heterogeneity to fertility. Science **210,** 1131–1133.

Fulwyler, M. J. (1965). Electronic separation of biological cells by volume. *Science* **150,** 910–911.

Gerdes, J., Schwab, U., Lemke, H., and Stein, H. (1983). Production of a mouse monoclonal antibody reactive with a human nuclear antigen associated with cell proliferation. *Int. J. Cancer* **31,** 13–20.

Gohde, W., and Dittrich, W. (1970). Simultane Impulsfluorimetrie des DANS-und Proteingehaltes von Tumorzellen. *Z. Anal. Chem.* **252,** 328–330.

Gohde, W., and Dittrich, W. (1971). Die cytostatische wirkung von daunomyein in Impuls cytophotometrictest. *Arzneim.-Forsch. (Drug Res.)* **21,** 1656–1658.

Gorczyca, W., Gong, J., and Darzynkiewicz, Z. (1993). Detection of DNA strand breaks in individual apoptotic cells by the in situ terminal deoxynucleotidyl transferase and nick translation assays. *Cancer Res.* **53,** 1945–1951.

Gorczyca, W., Davidian, M., Gherson, J., Ashikari, R., and Darzynkiewicz, Z. (1998). Laser scanning cytometry quantification of estrogen receptors in breast cancer. *Anal. Quant. Cytol. Histol.* **20,** 470–476.

Gratzner, H. G. (1982). Monclonal antibody to 5 bromo and 5 iododeoxyuridine: A new reagent for detection of DNA replication. *Science* **218**, 474–475.

Gratzner, H. G., Leif, R. C., Ingram, D. J., and Castro, A. (1975). The use of antibody specific for bromodeoxyuridine for the immunofluorescent determination of DNA replication in single cells and chromosomes. *Exp. Cell Res.* **95**, 84–88.

Gray, J. W., Carrano, A. V., Steinmetz, L. L., VanDilla, M. A., Moore, D. H., Mayall, B. H., and Mendelsohn, M. L. (1975). Chromosome measurement and sorting by flow systems. *Proc. Natl. Acad. Sci. U.S.A.* **72**, 1231–1234.

Gray, J. W., Dean, P. N., Fuscoe, J. C., Peters, D. C., Trask, B. J., Van den Engh, G. J., and Van Dilla, M. A. (1987). High-speed chromosome sorting. *Science* **238**, 323–329.

Gray, J. W., Dolbeare, F., and Pallavicini, M. G. (1990). Quantitative cell-cycle analysis. *In* "Flow Cytometry and Sorting" (M. R. Melamed, T. Lindmo, and M. L. Mendelsohn eds.), 2nd Ed., Chap. 23. Wiley-Liss, New York.

Hardy, R. R., Hayakawa, K., Parks, D. R., and Herzenberg, L. A. (1983). Demonstration of B-cell maturation in x-linked immunodeficient mice by simultaneous three-colour immunofluorescence. *Nature* **306**, 270–272.

Hedley, D. W., Freidlander, M. L., Taylor, I. W., Rugg, C. A., and Musgrove, E. A. (1983). Method for analysis of cellular DNA content of paraffin-embedded pathological material using flow cytometry. *J. Histochem. Cytochem.* **31**, 1333–1335.

Hulett, H. R., Bonner, W. A., Barret, J., and Herzenberg, L. A. (1969). Cell sorting: Automated separation of mammalian cells as a function of intracellular fluorescence. *Science* **166**, 747–749.

Hulett, H. R., Bonner, W. A., Sweet, R. G., and Herzenberg, L. A. (1973). Development and application of a rapid cell sorter. *Clin. Chem.* **19**, 813–816.

Jacobberger, J. A. W., Fogelman, D., and Lehman, J. M. (1986). Analysis of intracellular antigens by flow cytometry. *Cytometry* **7**, 356–364.

Juan, G., Gruenwald, S., and Darzynkiewicz, Z. (1998). Phosphorylation of retinoblastoma susceptibility gene protein assayed in individual lymphocytes during their mitogenic stimulation. *Exp. Cell. Res.* **239**, 104–110.

Kachel, V., Fellner-Feldegg, H., and Menke, E. (1990). Hydrodynamic properties of flow cytometry instruments. *In* "Flow Cytometry and Sorting" (M. R. Melamed, T. Lindmo, and M. L. Mendelsohn, eds.), 2nd Ed., Chap. 4. Wiley-Liss, New York.

Kamentsky, L. A. (1965). Rapid biological cell identification by spectroscopic analysis. *Proc. 18th Ann. Conf. Eng. Biol. Med.* **7**, 178.

Kamentsky, L. A., and Kamentsky, L. D. (1991). Microscope-based multiparameter laser scanning cytometer yielding data comparable to flow cytometry data. *Cytometry* **12**, 381–387.

Kamentsky, L. A., and Melamed, M. R. (1967). Spectrophotometer cell sorter. *Science* **156**, 1364–1365.

Kamentsky, L. A., and Melamed, M. R. (1969). Rapid multiple mass constituent analysis of biological cells. *Ann. N.Y. Acad. Sci.* **157**, 310–323 (Presented June 6, 1967 at New York Academy of Science Conference on Data Extraction and Processing of Optical Images in the Medical and Biological Sciences. New York).

Kamentsky, L. A., Melamed, M. R., and Derman, H. (1965). Spectrophotometer: New instrument for ultrarapid cell analysis. *Science* **150**, 630–631.

Koenig, S. H., Brown, R. D., Kamentsky, L., Sedlis, A., and Melamed, M. R. (1968). A report of the efficacy of a rapid cell spectrophotometer in screening for cervical cancer. *Cancer* **21**, 1019–1026.

Kohler, A. (1904). Mikrophotographische Untersuchungen Mit Ultraviolettenlicht. *Z. Wiss. Mikroskopie* **21**, 129–165.

Krishan, A., Sauerteig, A., and Wellham, L. (1985). Flow cytometric studies on modulation of cellular adriamycin retention by phenothiazines. *Cancer Res.* **45**, 1046–1051.

Kung, P. C., Goldstein, G., Reinherz, E. L., and Schlossman, S. F. (1979). Monoclonal antibodies defining distinctive human T cell surface antigens. *Science* **206**, 347–349.

Levine, A. (1997). p53, the cellular gatekeeper for growth and division. *Cell* **88**, 323–331.

Loken, M. R., Shah, V. O., Dattilio, K. L., and Civin, C. I. (1987). Flow cytometric analysis of human bone marrow. II Normal B lymphocyte development. *Blood* **70**, 1316–1324.

Loken, M. R., Parks, D. R., and Herzenberg, L. A. (1977). Two-color immunofluorescence using a fluorescence activated cell sorter. *J. Histochem. Cytochem.* **25**, 899–907.

Mansberg, H. P., Saunders, A. M., and Groner, W. (1974). The Hemalog D white cell differential system. *J. Histochem. Cytochem.* **22**, 711–724.

Melamed, M. R., Adams, L. R., Zimring, A., Murnick, J. G., and Mayer, K. (1972). Preliminary evaluation of acridine orange as a vital stain for automated differential leukocyte counts. *Am. J. Clin. Pathol.* **57**, 95–102.

Melamed, M. R., Adams, L. R., Traganos, F., and Kamentsky, L. A. (1974). Blood granulocyte staining with acridine orange: Changes with infection. *J. Histochem. Cytochem.* **22**, 525–530.

Melamed, M. R., Lindmo, T., and Mendelsohn, M., eds. (1990). "Flow Cytometry and Sorting." 2nd Ed. Wiley-Liss, New York.

Mellors, R. C., Keane, J. F., Jr., and Papanicolaou, G. N. (1952). Nucleic acid content of the squamous cancer cell. *Science* **116**, 265–269.

Mellors, R. C., and Silver, R. (1951). A microfluorometric scanner for the differential detection of cells: Application to exfoliative cytology. *Science* **114**, 356–360.

Mellors, R. C., Berger, R. E., and Streim, H. G. (1950). Ultraviolet microscopy and microspectroscopy of resting and dividing cells: Studies with reflecting microscope. *Science* **111**, 627–632.

Meyer, J. S., and Lee, J. W. (1980). Relationship of S-phase fraction of breast carcinoma in relapse to duration of remission, estrogen receptors content, therapeutic responsiveness and duration of survival. *Cancer Res.* **40**, 1890–1896.

Miescher, F. (1897). "Die Histochemischen Und Physiologischen Arbeiten." Vogel, Leipzig.

Moldavan, A. (1934). Photo-electric technique for the counting of microscopical cells. *Science* **80**, 188–189.

Oi, V. T., Glazer, A. N., and Stryer, L. (1982). Fluorescent phycobiliprotein conjugates for analysis of cells and molecules. *J. Cell Biol.* **93**, 981–986.

Ornstein, L., and Ansley, H. R. (1974). Spectral matching of classical cytochemistry to automated cytology. *J. Histochem. Cytochem.* **22**, 453–460.

Otto, F., and Oldiges, H. (1978). Requirements and procedures for chromosomal DNA measurements for rapid karyotype analysis in mammalian cells. *In* "Pulse-Cytophotometry" (D. Lutz, ed.), pp. 393–400. European Press, Ghent, Belgium.

Parker, J. C., and Horst, W. R. (1959). U.S. Patent 2,875,666, Method of simultaneously counting red and white blood cells. Filed July 13, 1953. Issued March 3, 1959.

Raber, M. N., and Barlogie, B. (1990). DNA flow cytometry of human solid tumors. *In* "Flow Cytometry and Sorting" (M. R. Melamed, T. Lindmo, and M. L. Mendelsohn, eds.), 2nd Ed., Chap. 37. Wiley-Liss, New York.

Rabinovitch, P. S., June, C. H., Grossmann, A., and Ledbetter, J. (1986). Heterogeneity among T cells in intracellular free calcium response after mitogen stimulation with PHA or anti-CD3. Simultaneous use of indo-1 and immunofluorescence with flow cytometry. *J. Immunol.* **137**, 952–961.

Reinherz, E. L., Kung, P. C., Goldstein, G., and Schlossman, S. F. (1979a). A monoclonal antibody with selective reactivity with functionally mature human thymocytes and all peripheral human T cells. *J. Immunol.* **123**, 1312–1317.

Reinherz, E. L., Kung, P. C., Goldstein, G., and Schlossman, S. F. (1979b). Separation of functional subsets of human T cells by a monoclonal antibody. *Proc. Natl. Acad. Sci. U.S.A.* **76**, 4061–4065.

Salzman, G. C., Crowell, J. M., Martin, J. C., Trujillo, T. T., Romero, A., Mullaney, P. F., and LaBauve, P. M. (1975). Cell classification by laser light scattering: Identification and separation of unstained leukocytes. *Acta Cytol.* **19**, 374–377.

Shapiro, H. (1988). "Practical Flow Cytometry," 2nd Ed. Alan R. Liss, New York.

Shapiro, H. M., Schildkraut, E. R., Curbelo, R., Turner, R. B., Webb, R. H., Brown, D. C., and Block, M. J. (1977). Cytomat-R: A computer-controlled multiple laser source multiparameter flow cytophotometer system. *J. Histochem. Cytochem.* **25**, 836–844.

Shapiro, H. M., Natale, P. J., and Kamentsky, L. A. (1979). Estimation of membrane potentials of individual lymphocytes by flow cytometry. *Proc. Natl. Acad. Sci. U.S.A.* **76**, 5728–5730.

Sharpless, T. K., and Melamed, M. R. (1976). Estimation of cell size from pulse shape in flow cytofluorometry. *J. Histochem. Cytochem.* **24,** 257–264.

Steen, H. B. (1980). Further developments of a microscope-based flow cytometer: Light-scatter detection and excitation intensity compensation. *Cytometry* **1,** 26–31.

Steen, H. B., Boye, E., Skarstad, K., Bloom, B., Godal, T., and Mustafa, S. (1982). Applications of flow cytometry on bacteria: Cell-cycle kinetics, drug effects, and quantitation of antibody binding. *Cytometry* **2,** 249–257.

Steinkamp, J. A., Orlicky, D. A., and Crissman, H. A. (1979). Dual-laser flow cytometry of single mammalian cells. *J. Histochem. Cytochem.* **27,** 273–276.

Stohr, M. (1976). Double beam application in flow techniques and recent results. *In* "Pulse Cytophotometry" (W. Gohde, J. Schumann, and T. H. Buchner, eds.), Second International Symposium, pp. 36–45. European Press, Ghent, Belgium.

Stryer, L., and Glazer, A. N. (1985). Phycobiliprotein fluorescent conjugates. U.S. Patent 4,542,104.

Stryer, L., Glazer, A. N., and Oi, V. T. (1985). Fluorescent conjugates for analysis of molecules and cells. U.S. Patent 4,520,110.

Sweet, R. G. (1965). High frequency recording with electrostatically deflected ink jets. *Rev. Sci. Instrum.* **36,** 131–136.

Tan, E. M., Ogata, K., and Takasaki, Y. (1987). PCNA/cyclin: A lupus antigen connected with DNA replication. *J. Rheumatol.* **14**(Suppl. 13), 89–96.

Traganos, F., Darzynkiewicz, Z., Sharpless, T., and Melamed, M. R. (1977). Simultaneous staining of ribonucleic and deoxyribonucleic acids in unfixed cells using acridine orange in a flow cytometric system. *J. Histochem. Cytochem.* **25,** 46–56.

Trask, B., Van den Engh, G., Landegent, J., in de Wal, N. J., and van der Ploeg M. (1985). Detection of DNA sequences in nuclei in suspension by *in situ* hybridization and dual beam flow cytometry. *Science* **230,** 1401–1403.

Valet, G., Bamberger, S., Hofmann, H., Schindler, R., and Rukenstroth-Bauer, G. (1979). Flow cytometry as a new method for the measurement of electrophoretic mobility of erythrocytes using membrane charge staining by fluoresceinated polycations. *J. Histochem. Cytochem.* **27,** 342–349.

Van Dilla, M. A., Trujillo, T. T., Mullaney, P. F., and Coulter, J. R. (1969). Cell microfluorometry: A new method for rapid fluorescence measurement. *Science* **163,** 1213–1214.

Visser, J. W. M., Jongeling, A. A. M., and Tanke, H. J. (1979). Intracellular pH determination by fluorescence measurements. *J. Histochem. Cytochem.* **27,** 32–35.

Waggoner, A. S. (1990). Fluorescent probes for cytometry. *In* "Flow Cytometry and Sorting" (M. R. Melamed, T. Lindimo, and M. L. Mendelsohn, eds.), 2nd Ed., Chap. 12. Wiley-Liss, New York.

Watson, J. U. (1980). Enzyme kinetic studies in cell populations using fluorogenic substrates and flow cytometric techniques. *Cytometry* **1,** 143–151.

CHAPTER 2

Principles of Flow Cytometry: An Overview

Alice L. Givan

Englert Cell Analysis Laboratory of the Norris Cotton Cancer Center and
Department of Physiology
Dartmouth Medical School
Lebanon, New Hampshire 03756

I. Introduction
II. The Illumination of a Particle
III. Fluidics: Centering Particles in the Illuminating Beam
IV. Collection of Light Signals from Particles
V. From Light Signals to a Data File
VI. From Data to Information
VII. Sorting
VIII. Conclusions
References

I. Introduction

In 1934, Andrew Moldavan in Montreal took a step from classic microscopy toward a flowing analysis system. He suggested the development of an apparatus to count red blood cells and neutral-red-stained yeast cells as they were forced through a capillary tube on a microscope stage. A photodetector attached to the microscope eyepiece would register each passing cell. In a trail of innovation, leading from the work by Moldavan, through work on automated microscope-based cervical cytometry (Kamentsky and Melamed, 1965, 1967) and DNA fluorescence (Dittrich and Göhde, 1969), then picking up threads from hydrodynamic focusing (Crosland-Taylor, 1953), from drop charging for ink jet printing (Sweet, 1965), and from sizing of blood cells (Coulter, 1956), flow cytometry gradually took form in the 1960s with work coming primarily from the Lawrence Livermore and Los Alamos National Laboratories (e.g., Fulwyler, 1965; Van Dilla *et al.*,

1969) and from the Herzenberg group at Stanford University (e.g., Hulett *et al.*, 1969). By the early 1970s, flow cytometric instrumentation had reached a stage of development that, essentially, defines it still.

Flow cytometry is a system for detecting and then analyzing the light signals generated by particles as they flow in a liquid stream past an illuminating beam. By this definition, it is apparent that a chapter providing a description of flow cytometry will involve a confluence of the disciplines of electronics (detecting); computational data analysis (analyzing); optics (light signals, illuminating beam); and fluidics (flow in a liquid stream). Excellent books, articles, and chapters, elsewhere (e.g., Melamed *et al.*, 1990; Robinson, 1997; Shapiro, 1995; Van Dilla *et al.*, 1985; Watson, 1991, 1992) and in this series, explain the disciplines behind flow cytometry in detail. This chapter attempts, simply, to present a cohesive overview of the theory of flow cytometric instrumentation; I cite a selection of representative references from the literature and describe a generalized flow cytometer in an attempt to provide an introduction into the field—with the background necessary for a critical understanding of the strengths, pitfalls, and potential of this technique and its applications.

II. The Illumination of a Particle

Although some flow cytometers still use arc lamps (and these provide the decided advantage of flexibility of illumination wavelength), most current instruments require a laser as a source of illumination. The reason for the use of lasers involves their ability to provide intense illumination that can be focused to a small point (the analysis point or interrogation point). Particles in a stream of fluid can move through this analysis point rapidly, but receive enough illumination during their short time in the laser beam to produce fluorescence or scattered light of detectable intensity.

Basic ion laser design (Fig. 1) involves a plasma tube containing a noble gas; a high voltage supply which provides electromagnetic energy to the gas, in order to ionize the atoms and then raise ground state electrons into excited orbitals; and carefully aligned mirrors at both ends of the plasma tube to reflect, back and forth, the light emitted by the atoms as the electrons fall back to orbitals with lower energy. Because the electrons in the atoms are maintained in an excited state by the current (a population inversion refers to this condition where more electrons are in the higher than lower orbitals), each time a photon of light interacts with an excited atom, more light is emitted and joins the beam. In this way, the reflection or oscillation of the light between the front and rear mirrors amplifies the intensity of the light with each pass of the beam through the plasma tube. In order to make use of this continually amplified light, the mirror at the back end of the plasma tube is highly reflective, but the output mirror at the front end of the tube transmits a small fraction (commonly 2–10%) of this oscillating light out of the tube for use in the flow cytometer. Because of the

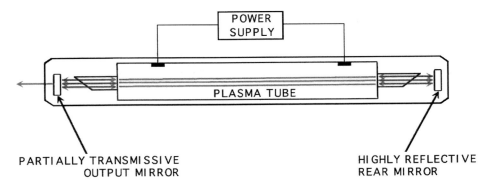

Fig. 1 Diagram of an ion laser, indicating the plasma tube with angled Brewster windows which transmit light from the tube to the reflective mirrors at each end. A power supply provides the energy for exciting the electrons in the gas atoms. The partially transmissive output mirror at the front end of the laser provides the light beam used in the flow cytometer.

way laser light is generated (with each photon producing, when it interacts with an atom with an electron in an excited orbital, another photon exactly like itself), lasers provide a uniquely coherent light source: that is, laser light is closely restricted with respect to wavelength, polarization, and direction. For flow cytometry, the coherent direction of laser light is of most importance because it provides high intensity over a small area. Unfortunately, the coherent (and restricted) wavelength from lasers is a decided disadvantage.

The wavelength of the light from a laser is determined by the gas in the plasma tube (according to the energy differences between its ground state and excited electron orbitals). For this reason, the wavelengths of light provided by different lasers are defined and inflexible. The single most common laser found on the optical bench of a flow cytometer is the argon ion laser; it was chosen for early flow cytometers because it provides turquoise light at 488 nm, a wavelength that is absorbed efficiently by fluorescein, a fluorochrome that had long been used for fluorescence microscopy. Argon ion lasers also can produce light at several other wavelengths, most usefully green light at 514 nm, as well as ultraviolet (UV) light (at 351 and 364 nm) if the power of the laser is high enough. Typically, from an argon ion laser, the power of light produced at 488 nm is about 75% of that at 514 nm; light at the UV wavelengths is about 10% of the 514 nm power. In some research flow cytometers, these other argon ion wavelengths can be selected from the laser output by using wavelength-selective mirrors or by adjusting a prism at the rear end of the plasma tube.

Whereas the earliest flow cytometers used simply an argon ion laser to excite fluorescein fluorescence [and, later, to include phycoerythrin (PE), propidium iodide (PI), peridinin chlorophyll protein (PerCP), and phycoerythrin–tandem (energy transfer) dyes—all excited sufficiently by 488 nm light], the increasing demand for a wider array of fluorochromes has led to an increase in the number of lasers on the flow cytometer optical bench (see Shapiro, 1997). Table I indicates some of the lasers commonly found in flow cytometers, along with their useful wavelengths.

Research flow cytometers now may include, for example, the following multiple laser options:

- Two argon ion lasers, one tuned to 488 nm and the other to the UV;
- An argon ion laser at 488 nm and a red diode or He-Ne laser (633/635 nm);
- Two argon ion lasers (488 nm and UV) along with a third He-Ne or red diode laser (633/635 nm);
- An argon ion laser producing both 488 nm and UV light as well as a krypton laser emitting in the far red (647 nm);
- One or two argon ion lasers (488 nm and UV) plus a rhodamine 6G dye laser (570–620 nm).

Flow cytometers with more than one laser focus the beams from each laser at different spots along the flow stream, so that cells will pass through each laser beam in turn. In this way, the signals elicited from the cells by the different lasers will arrive at the photodetectors in a spatially or temporally defined sequence and can thus be associated with a particular excitation wavelength.

Lasers are inefficient users of energy and give off a great deal of heat as well as light. Bench-top cytometers (without sorting ability) and even some sorters can function with relatively low illumination light intensity because of their efficient light collection optics and/or low stream velocity. Lasers providing light for these cytometers are, therefore, often of low overall power (15 mW) and can be cooled by circulating air from a fan. Most sorters, because of the inefficiency with which they illuminate cells and collect light signals, require higher laser intensity; the lasers on the optical benches of these research/sorting cytometers are usually of high overall power (2–5 W) and require circulating water to remove the resulting heat.

Table I
Lasers Commonly Found in Flow Cytometers

Laser type	Usual laser wavelengths (nm)	Examples of compatible fluorochromes
Argon ion	488	Fluorescein, phycoerythrin, PerCP, PE–tandems, propidium iodide
	514	Rhodamine
	UV(351/364)	Hoechst dyes, DAPI, Indo-1, AMCA, Cascade Blue
Helium–neon (He-Ne)	Usually 633	Allophycocyanin, Cy5
Krypton ion	568	Cy3, Texas Red
	647	Allophycocyanin, Cy5
Red diode	635	Allophycocyanin, Cy5
Rhodamine 6G dye	570–620	Texas Red, allophycocyanin

Everything that a flow cytometrist learns about a cell comes from the interrogation or analysis point where the laser beam intersects the flow stream. Therefore, in addition to the color of the beam, its shape is also important (see Shapiro, 1997). Lasers can operate in different modes—sometimes producing beams with cross-sectional profiles looking like doughnuts (or worse). Although unavoidable in some types of lasers, these higher order modes are not ideal for the uniform spot required for cytometry. Closing down the iris at the front end of an ion laser usually allows selection of the fundamental (TEM_{00}) mode (TEM stands for transverse electrical and magnetic), which has a circular, radially symmetrical cross-sectional profile as the beam leaves the plasma tube, with a diameter of approximately 1–2 mm. Focusing lenses in the cytometer itself are used to shape the laser beam and focus it to a smaller diameter as it meets the cell stream. Simple spherical lenses can provide a round spot about 60 μm in diameter, with its most intense region in the center, and the intensity decreasing rapidly out toward the edge. The decrease in intensity is Gaussian (Fig. 2) so that cells passing through the middle of the beam will be much more intensely illuminated than those passing near the periphery. The nominal diameter of a beam of light refers, by convention, to the distance, across the center of the beam, between the $1/e^2$ points (e = 2.718)—where the intensity drops to 13.5% of its maximal, central value. At a position 30 μm from the center of a nominal 60 μm beam, a cell will receive only 13.5% of the intensity it would receive at the center of that beam. At 10 μm from the center, the intensity is at about 78% and at 3 μm

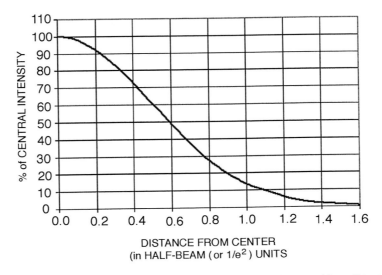

Fig. 2 A Gaussian curve, illustrating the decrease in light intensity at positions distant from the center of a light beam. Distance is expressed in half-beam units; a nominal diameter of a beam is expressed as twice the distance from the center to the point of 13.5% ($1/e^2$) of the central intensity.

from the center, the intensity is 98% of the central intensity. Because intensity falls off abruptly at even small distances from the center of the beam, cells need to be confined to a well-defined path if they are going, one by one, to receive closely similar illumination as they flow through a round beam of light. To alleviate this stringency, it is possible to have the laser beam focused to a much larger round spot. This strategy would have two major disadvantages: First, a large beam decreases the speed with which cells can be analyzed because it increases the probability that two or more cells following each other closely in a stream will be found in the beam at the same time and their signals combined (coincidence). Second, a large beam spreads the light energy over a large area, decreasing the intensity illuminating the cell at any one point. For these reasons, the beam-focusing lenses most frequently used in today's cytometers are a compromise; cylindrical lens combinations provide an elliptical spot, for example, 20 μm by 60 μm in size, with the short dimension parallel to the direction of stream flow and the longer dimension perpendicular to the stream (Fig. 3). By retaining a wide beam diameter across the stream, an elliptical illumination spot can provide considerable side-to-side tolerance, thus illuminating cells more-or-less identically even if they stray from the exact center of the beam, but at the same time provide temporal resolution between cells as they pass one by one into and out of the beam in its narrow dimension (Fig. 4). As far as the ideal dimension of the beam in the direction of flow, the narrower the beam is, the more quickly will a cell pass through it—giving opportunity for the signal from that cell to drop off before the start of the signal from the next cell in line and

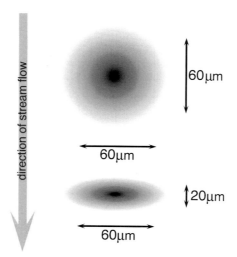

Fig. 3 Comparison of a circular (60 μm by 60μm) and an ellipsoidal (20 μm by 60 μm) beam cross-sectional profile, with the short dimension of the ellipsoidal beam parallel to the direction of stream flow.

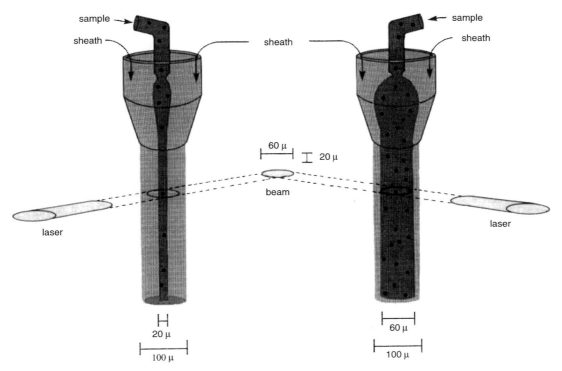

Fig. 4 The flow of particles within the core of sheath fluid through an ellipsoidal laser beam. When the sample is injected slowly into the sheath stream (left), the core remains narrow and cells are confined to the center of the beam and illuminated one at a time. At higher sample injection pressures (right), the central core widens: cells may then be illuminated erratically and multiple cells may coincide in the beam.

avoiding the coincidence of two cells in the beam simultaneously. In other words, if a beam is only 20 μm in height, then cells need to be only a bit more than 20 μm apart (the beam dimension plus one cell diameter) to be identified as single cells rather than one large particle. Because flow cytometry is a technique that depends on analysis of individual cells, a beam that is narrow in the direction of flow has strong implications for the rate at which cells can be analyzed. There are, in addition, further implications of the size of the beam in the direction of flow. A beam that is wide in the direction of flow will allow cells to be illuminated for a longer period of time and, if the light intensity is saturating and photobleaching is not an issue, the resulting signal may then be greater. This can improve the sensitivity of the system for detection of weak fluorescence. Conversely, as illustrated in Fig. 5, a beam dimension that is close to or shorter than the diameter of the cell will allow the duration of the signal to give an indication of the diameter of the cell (see Wheeless, 1990).

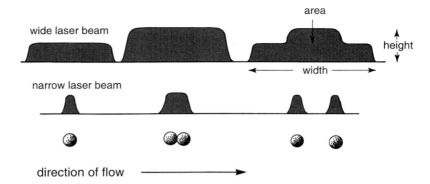

Fig. 5 The signal pulses that result from single and doublet cells passing through a laser beam that is either wide (upper pulses) or narrow (lower pulses) in the direction of flow. For the purpose of this illustration, it is assumed that the laser beam photon density is the same for both the wide and narrow examples. In fact, for any given laser, focusing the beam to a wider profile results in lower photon density.

The implications of beam focusing can be summarized as follows. With regard to the width of the laser beam perpendicular to the direction of flow, it is beneficial to have the beam wide so as to provide closely similar illumination to all cells, with side-to-side positional tolerance; however a very wide beam will spread the laser power over a larger area and therefore can result in dimmer signals. With regard to the height of the laser beam in the direction of stream flow, if sensitivity is an issue, it may be beneficial, under saturating conditions, to have the beam large so that the cell will spend more time in the beam and therefore produce a greater signal; however a beam that is narrow in the direction of flow will permit cells to follow each other closely without two cells coinciding in the beam at the same time. The compromise that is found in most cytometers is brought about by the use of an elliptical beam profile. The dimension of the beam at right angles to the direction of flow is large enough to include most of the width of the stream. The dimension along the direction of flow can be large if sensitivity is an issue and it becomes important to illuminate the cell or particle for a longer period of time; alternatively, it can be shorter if fast rates of sorting or analysis are important.

III. Fluidics: Centering Particles in the Illuminating Beam

The distinguishing aspect of flow cytometry, as opposed to microscopy, is that particles flow through the cytometer; that is, particles to be analyzed by flow cytometry need to be suspended in fluid and each particle is then analyzed only once, over a brief and defined period of time. The downside of the requirement for a flowing suspension is that the particles need to be single; blood cells, bacteria, and small plankton are naturally single, but tissues need to be disaggregated before being subjected to flow cytometry and, in the process, information about tissue architecture is lost. The upside of the requirement for flow is that many cells can be analyzed in a short period of time, and statistical information about large populations of cells can be obtained. Bearing these ups and downs in mind, a flow cytometrist requires a fluidics system (Fig. 6) for getting particles, once suspended in liquid in a test tube, into the center of the (usually) elliptical laser beam where the cells will be illuminated. Flow cytometers may deliver the cell into the laser beam in different ways, for example, by applying air pressure to the test tube containing the cell suspension and forcing the cells out of the tube or, alternatively, by drawing the cell suspension into a syringe and then applying pressure to the syringe plunger. In either case, the flow characteristic for optimal optical sensitivity is that the cells need to be positioned, when they get to the laser beam analysis point, as beads spaced out along a string—in the

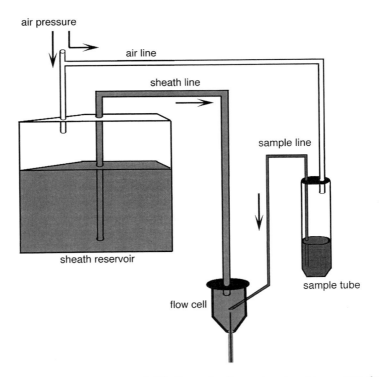

Fig. 6 A schematic diagram of the fluidic lines of a flow cytometer. Air pressure forces both particle suspension from the sample vessel and sheath fluid from the sheath reservoir into the flow cell. In the flow cell, the particle suspension forms a central core within the sheath. The diameters of both the sheath and the core narrow as the velocity of the fluid increases when it is forced out of the narrow orifice of the flow cell.

center of the beam so as to be brightly and uniformly illuminated one at a time. In addition, cells should flow at a slow enough rate so that, given the height of the beam, they will each dwell in the light for a long enough period of time to produce a detectably intense signal (but should also pass in and out of the laser beam quickly enough to permit rapid cell analysis).

With regard to the issue of confining cells to the center of the laser beam, although an elliptical beam profile relieves some of the stringency of this requirement, unequal illumination will result in unequal signals if the illumination is not saturating and will therefore widen the variation (usually expressed as coefficient of variation (CV), the standard deviation/mean) of the output response. This is particularly true for the intensity of scatter signals which do not saturate and are, therefore, always directly related to the illumination intensity. One potential strategy for confining cells to the center of a laser beam would be to inject them through the beam in an optically clear chamber with a very narrow diameter or, alternatively, to squirt them through the beam from a nozzle with a narrow orifice. The problem with pushing cells from a narrow orifice or through a narrow chamber is that the cells, if large or clumped, tend to clog the pathway. The way around this dilemma was first suggested by Crosland-Taylor in 1953. He noted that "attempts to count small particles suspended in fluid flowing through a tube have not hitherto been very successful. With particles such as red blood cells the experimenter must choose between a wide tube which allows particles to pass two or more abreast across a particular section, or a narrow tube which makes microscopical observation of the contents of the tube difficult due to the different refractive indices of the tube and the suspending fluid. In addition, narrow tubes tend to block easily." Crosland-Taylor's strategy for confining the cells in a narrow flow stream but preventing blockage through a narrow chamber or orifice involved injecting the cell suspension into the center of a wide, rapidly flowing stream (the sheath stream), where, according to hydrodynamic principles, the cells will remain confined to a narrow core at the center of the wider stream as it flows through a wide orifice. The characteristics of this hydrodynamic focusing result in coaxial flow (a narrow core stream of cells flowing within a wider sheath stream) and were first applied to cytometry by Crosland-Taylor who realized that this was a way to confine cells to a precise position without requiring a narrow overall stream that was susceptible to obstruction.

The place where the cell suspension is injected into the center of the sheath stream is the flow cell (or flow chamber or flow nozzle, loosely interchangeable terminology). The design of this chamber is critical so as to minimize turbulence and resulting background scatter of light, to maximize optical efficiency for the collection of signals from the cells, to create an increase in the velocity of the stream as it progresses to the orifice, and to provide conditions for maintaining a narrow stream of cells within the wider sheath. There are many different approaches to flow cell design (see Pinkel and Stovel, 1985). Some cytometers illuminate the stream within an optically clear region of the flow cell (as in a cuvette). In other systems (jet-in-air), the light beam intersects the cell stream after it emerges from the flow chamber. The outlet orifice of a jet-in-air flow cell is usually about 70 μm, but can vary between 50 and 400 μm for specialized applications. Cuvette-style flow chambers are generally of 150–250 μm diameter. A small orifice will become obstructed easily. A large orifice puts limits on the rate of drop formation and the speed of sorting.

When, within the flow chamber, a stream of cell suspension is injected into the center of a wide sheath stream, the velocity of the cell suspension (in m per sec) increases or decreases so that it becomes equal to the velocity of the wider sheath stream (see Kachel *et al.*, 1990; Pinkel and Stovel, 1985; Stovel, 1997). This means that the cross-sectional diameter of the core stream containing the cells will either increase or decrease to bring about this change in the velocity of flow while maintaining the same sample volume flow rate (in ml per sec). The injection rate of the cell suspension will, therefore, directly affect the width of the core (sample) stream and the stringency by which cells are confined, when they get to the laser analysis point, to the center of the illumination beam and are illuminated one at a time. If the cell suspension is injected at a slow velocity, the core diameter will narrow as its velocity increases to match the sheath velocity; if it is injected at a velocity faster than the sheath stream, the core diameter will widen as its velocity decreases (Figs. 4 and 7). In order to have cells uniformly

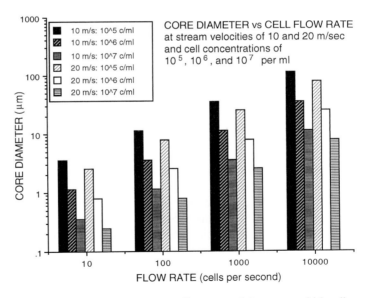

Fig. 7 The relationship between sample core diameter and the rate at which cells pass the laser beam. As cells are forced at faster speeds, the core diameter widens. Increasing the stream velocity from 10 to 20 m/sec or increasing the cell concentration (from 10^5 to 10^6 to 10^7 cells/ml) both permit faster cell flow rate with narrow core diameters.

illuminated one at a time, it is, therefore, necessary to inject them into the flow chamber at a slow rate. This is particularly true for DNA analysis where the width of the core stream may be a major factor in affecting the deviation from the "true" value of signal intensity, which is proportional to the closely defined DNA content of a cell. The rate at which the cell suspension is injected into the flow stream is more or less under user control.

The velocity of the outer sheath stream may or may not be under user control. In many flow cytometers the sheath stream flows at approximately 10 m/sec as it passes the laser beam. At a sheath flow velocity of 10 m/sec, approximately 140 ml of fluid will exit a 70 μm orifice in an hour (for each second, calculate the volume of a cylinder with a 35 μm radius and a height of 10 m). Of this 140 ml, approximately 2.8 ml will be of the cell suspension if the core diameter within the sheath is 10 μm. The rest will be from the sheath reservoir. High speed sorters provide for increase in stream velocity by allowing for variable pressure on the sheath stream, with resulting velocities of 15–50 m/sec (see Leary, 1997; Peters, *et al.,* 1985). One advantage of higher sheath stream velocity is that cells can be analyzed faster while a narrow core diameter is maintained; there is, despite the fast flow rate, sufficient distance between cells in the stream to avoid coincidence in the laser beam.

The flow rate of the cells past the laser beam (in cells/sec) will be affected by the velocity of the sheath stream (in m/sec), by the diameter of the sample core, and also by the concentration of cells in the original suspension (in cells per milliliter). If cells are close to each other in the core stream (either because the cell concentration is high or because the core diameter is wide), two or more may dwell in the laser beam simultaneously; they will be indistinguishable from a single particle and their signals will be recorded together as a single signal (with approximately double the intensity of a single cell). For any given cell concentration, this will happen more frequently if the core stream has a wide diameter and also if the laser beam is wide in the direction of flow. With a given laser beam size and core stream width, it can be avoided by decreasing the concentration of cells in the original suspension.

In relation to the sheath velocity and the cell concentration, it is important to distinguish between the *distance* separating particles and the *time* separating particles. The distance is measured in μm and needs to be long enough so that only rarely do two cells find themselves in the laser beam at the same time. This means that, if the laser beam is 20 μm wide in the direction of flow, that cells need to be 20 μm plus one cell diameter apart in the flow stream if they are not going to be considered a single particle. Poisson statistics need to be considered on this issue. Just because cells are, on average, a certain distance apart does not mean that all cells will be exactly that distance apart. Probability predicts that many cells will be closer than the average. Figure 8 [based on calculations of coincidence probability in an excellent discussion on the subject from James Watson's book (Watson, 1991)] indicates the probability of a flow cytometric "event" actually resulting from multiple cells in the laser beam at the same time.

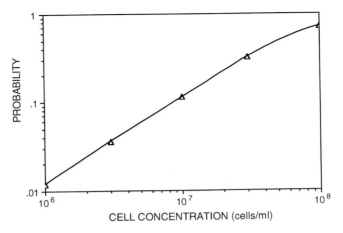

Fig. 8 The probability of the signals from an "event" in a flow cytometer having been generated by multiple cells in the laser beam simultaneously. The data come from an adaptation of the analysis by James Watson in "Introduction to Flow Cytometry" (Cambridge University Press, 1991). For this example, the laser beam was considered to be 30 μm in height (in the direction of flow) and the core diameter 10 μm—these dimensions defining the volume in which the probability of coincidence has been calculated.

The time separating particles, on the other hand, is expressed as the number of seconds between particles as they pass the laser beam (or more conventionally by its inverse, the number of particles per second or the rate of cell flow). The time between particles needs to be long enough so that the electronics of the cytometer have time to register and clear all the signals from the first cell before the signals from the second cell begin. The distance separating cells and the time separating cells (the flow rate) are related. The shorter the distance between cells in the stream, the faster will be their flow rate past the laser beam. Increasing the concentration of cells in their original suspension will increase their flow rate by decreasing the distance separating them in the narrow sample core. Increasing the pressure pushing the cell suspension into the sheath stream will increase the cell flow rate because the increased pressure will decrease the distance between cells by widening the sample core. But it is also possible to increase the flow rate without decreasing the distance between cells. This can be done by increasing total sheath stream velocity and is the strategy employed in high speed sorting, where cells remain well separated from each other but move in and out of the illumination beams quickly so that coincidence is not a problem. In applications requiring sorting, high stream velocities also permit rapid drop generation.

The signal emitted by a cell as it passes through a laser beam is a light pulse with a height, area, and width (Fig. 5). Pulse-processing electronics can provide information not only about the height of the signal from the cell (the maximum intensity of the signal pulse as the cell passes in and then out of the beam), but also about the integrated area of that signal (the total amount of light that comes from the cell over the time that it spends in the laser beam), and about the width of the signal (the time that elapses from the entry of the cell into the laser beam until its exit). When the beam is narrow in the direction of flow relative to the diameter of the particles, the width of the signal provides information about particle size (Wheeless, 1990). Aggregates of cells will take longer to pass through the laser beam than will single cells; their signal width will be long and the peak height of the signal from an aggregate will be less than the peak height from a single cell with the same total integrated fluorescence. This is particularly useful for distinguishing two clumped G_0/G_1 cells from single cells in the G_2 or mitotic stage of the cell cycle.

When a cell flows at a rate of about 10 m/sec past a beam of light that has a width of 20 μm in the direction of flow, the cell will spend 2 μsec in the laser beam. The time between absorption of light by a fluorochrome and emission of this absorbed energy as fluorescent light is generally on the order of nanoseconds. Therefore, the fluorochromes on or in a cell go through, by rough estimate, approximately 100–1000 cycles of excitation and fluorescence as the cell passes through the laser beam of a cytometer. The slower the flow rate, the more times will each fluorochrome be excited and the more light will be collected from each cell. This is one way to increase the sensitivity of fluorescent light collection. However, repeated excitation of electrons will result in significant photobleaching (permanent destruction) of many fluorochromes. So brighter excitation and a longer dwell time in the laser beam are not always a formula for more sensitive fluorescent light collection (see van den Engh and Farmer, 1992).

A series of calculations at the cytometer itself may help to reinforce some of the relationships discussed in this section:

Knowing the diameter of the flow cell orifice, the velocity of the flow stream can be calculated by measuring the volume of fluid coming out of the flow cell in a given period of time:

$$v = (10^6)(m/\pi r^2)$$

where v is the stream velocity in m/sec, m is the stream flow rate in ml/sec, and r is the radius of the orifice in μm.

The stream velocity can also be calculated, in a sorter, by measuring the distance between drops:

$$v = f\lambda$$

where v is the stream velocity in m/sec, f is the drop drive frequency in drops (or cycles)/sec (hertz), and λ is the distance between drop centers in meters.

To calculate the diameter of the central cell stream core, based on knowledge of the sheath velocity, sample concentration, and cell flow rate (Pinkel and Stovel, 1985):

$$d = (1.13 \times 10^3)[(u/nv)]^{1/2}$$

where d is the core diameter in μm, u is the particle flow rate in cells/sec, n is the concentration of particles in the original cell suspension in cells/ml, and v is the sheath velocity in m/sec.

Alternatively, the core diameter can be calculated by determining the total volume flowing from the flow cell in a given period of time and also the volume of cell suspension that has been used in that period of time. Knowing the flow cell orifice (sheath) diameter, the diameter of the core can be derived from the proportion between cell suspension volume and total volume:

$$d_c = [(V_c/V_t)d_t^2]^{1/2}$$

where d_c is the diameter of the core in μm, V_c is the volume of cell suspension consumed in ml, V_t is the total volume through the flow cell in ml, and d_t is the diameter of the orifice in μm.

To calculate the time a cell spends in the laser beam:

$$t = w/v$$

where t is the time in the laser beam in μsec, w is the height of the laser beam parallel to flow in μm, and v is the velocity of the sheath stream in m/sec.

These estimates will all be somewhat inaccurate if the stream diameter as it emerges from the nozzle differs from the nominal orifice—because of viscosity or surface tension (see Pinkel and Stovel, 1985)—but they provide approximations that will allow an appreciation of the geometric and temporal considerations that affect signal production and collection.

IV. Collection of Light Signals from Particles

Laser-based flow cytometers are engineered so that the illuminating beam direction, the flow stream direction, and the direction from which fluorescence signals are collected are all orthogonal to each other; that is, they are at right angles to each other as if on three intersecting edges of a cube (Fig. 9). When alignment is maintained, signal collection from cells that have been illuminated in the laser beam is consistent and reproducible. It is this orthogonal orientation (Van Dilla *et al.,* 1969) that has facilitated the development of flow cytometers into multiparameter instruments with the ability to collect multiple signals from each cell and to sort cells based on those signals.

Collection of the light signals produced by illumination of cells begins with the positioning of two lenses (Figs. 9 and 10). One is in the so-called forward direction; that is, it is positioned head on along the direction of the laser beam. The other is to the side, at right angles to that direction. The lens in the forward direction focuses light onto a photodiode. Across the front of this lens is a bar (the "obscuration bar") approximately 1 mm wide, positioned so as to block the laser beam itself as it passes through the stream. Only light from the laser that

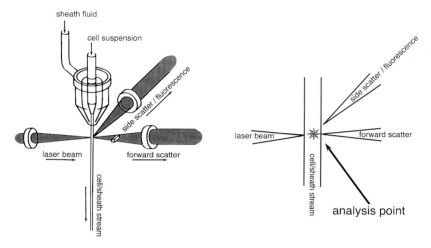

Fig. 9 Orthogonal alignment of the laser beam, the sheath stream, and the right angle light collection direction, as they converge on/diverge from the flow cytometric analysis point. An obscuration bar blocks the laser beam so that only light diverging around the bar strikes the forward scatter lens. Modified from Givan, A. L., "Flow Cytometry: First Principles" (1992). Reprinted by permission of Wiley-Liss, a subsidiary of John Wiley & Sons, Inc.

has been refracted or scattered as it goes through a particle in the stream will be bent enough from its original direction to avoid the obscuration bar and strike the forward-positioned lens. Light hitting the lens is therefore light that has been bent to small angles by the cell: the three-dimensional range of angles collected by this lens falls between those obscured by the bar and those lost at the limits of the outer diameter of the lens. Light hitting this forward lens is focused onto the photosensitive surface of a photodiode where it is converted to an electrical current. The burst of electrical current from the photodiode is proportional to the intensity of the light. The light striking this photodiode is called forward scatter light (fsc) or forward angle light scatter (fals). Although precisely defined in terms of the optics of light collection for any given cytometer, forward scatter light is not at all precisely defined in terms of the biology or chemistry of the cell by means of which it has been generated. A cell with a large cross-sectional area will refract a large amount of light onto the photodiode. But a large cell with a refractive index quite close to that of the medium (i.e., a dead cell with a permeable outer membrane) will refract light less than a similarly large cell with a refractive index quite different from that of the medium; less light will reach the photodiode from around the obscuration bar. Because of the rough relationship between the amount of light refracted and the size of the particle, the forward scatter signal is sometimes (misleadingly) referred to as a "volume" signal. This term belies its complexity (see Salzman *et al.*, 1990).

The lens at right angles to the direction of the laser beam collects light that has been scattered to wide angles from the original direction of that beam. The

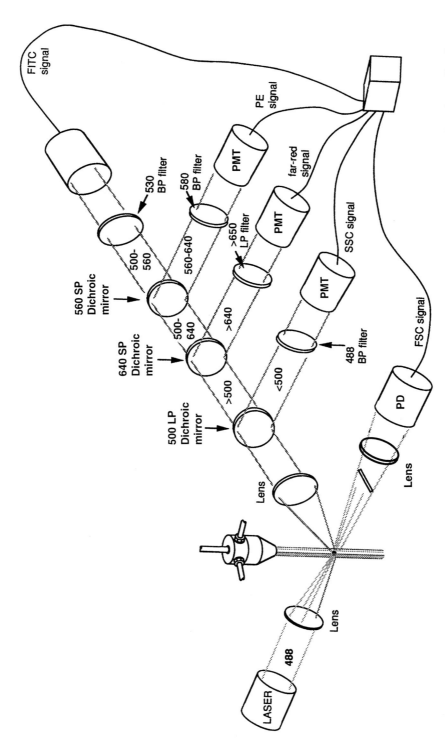

Fig. 10 Lenses, filters, and dichroic mirrors that collect and then partition the light signals that are generated by the cell in the laser beam. Signals are partitioned according to their wavelength and directed toward photodetectors that are, therefore, specific for light of different colors. Modified from Givan, A. L. "Flow Cytometry: First Principles" (1992). Reprinted by permission of Wiley-Liss, a subsidiary of John Wiley & Sons, Inc.

angles collected are defined by the diameter of this collecting lens and its distance from the analysis point. Light is scattered to these angles primarily by irregularities on or within the cell. Irregular nuclei and fibroblastic processes, for example, scatter large amounts of light to these wide angles. Laser light scattered to the right-angle collecting lens is termed side scatter light (ssc) or sometimes a granularity signal. It is focused on a photodetector, which converts the light to an electrical current that is proportional to the intensity of the light.

Forward and side scatter light are light signals of the same color as the laser beam striking the cell. This is usually 488 nm light in a one-laser system. It is also usually 488 nm light in a multilaser system, because the scattered light is conventionally used as a trigger signal and must therefore be collected from the first (primary) laser in the flow path. This scattered light provides information about the physical characteristics of a cell. The additional photodetectors in a flow cytometer are used to provide information about the biochemical characteristics of a cell; they collect light of different colors, that is, light that has been given off by fluorescent molecules within or on the cell when the molecules absorb laser light and emit light of a longer wavelength. The fluorescent molecules can be, for example, naturally occurring endogenous chemicals (like chlorophyll or pyridine or flavin nucleotides); fluorescently tagged monoclonal antibodies; calcium ion or pH indicators; DNA probes; or substrates that become fluorescent on enzymatic cleavage. The intensity of their fluorescence is, to an arguable extent, related to their abundance in the cell. The fluorescent light from the cell is collected by the same right-angle lens that collects side scatter light. The light hitting this lens is subsequently partitioned according to its color, so that scattered (488 nm) light is directed to one photodetector, green light to another, orange to another, and so forth depending on the number of "parameters" that the cytometer is configured to detect.

The partitioning of this multicolor light is accomplished by a series of optical interference filters and mirrors (see Waggoner, 1997) that have been vacuum-coated with layers of metallic compounds so that they transmit and reflect only well-defined wavelengths (Fig. 10). Dichroic mirrors with specific short pass cutoffs are used at 45° angles to transmit (pass) short wavelengths to photodetectors straight ahead and to reflect wavelengths longer than the cutoff to phototubes at right angles. Long-pass dichroic mirrors transmit wavelengths above the cutoff and reflect shorter wavelengths. Band pass filters in front of each detector further restrict the wavelengths of light that will be detected by that tube. A basic, bench-top flow cytometer may have, in addition to a photodetector for forward scatter light, one photodetector for side scatter detection and three for fluorescence detection. For example (follow the 90° light path from the analysis point in Fig. 10), an initial 45° angled long-pass dichroic mirror might reflect 488 nm laser light toward the side scatter detector and transmit light with wavelengths longer than 500 nm straight ahead. The second dichroic mirror in the chain might have a short-pass cutoff and reflect long wavelength (greater than 640 nm) light toward a detector with a 650 nm long-pass filter in front of it; light of less than

640 nm will continue through this dichroic mirror until it strikes a third angled mirror with a 560 nm cutoff; light longer than 560 nm will at this position be reflected toward a photodetector with a 580 nm band pass filter; light less than 560 nm will pass straight through the third dichroic mirror and strike a photodetector with a 530 nm band pass filter in front of it. In this example, the four right angle photodetectors will have been made specific for turquoise side scatter light at 488 nm, red (650 nm) light (e.g., from PerCP), orange (580 nm) light (e.g., phycoerythrin), and green (530 nm) light (e.g., fluorescein). On research cytometers, additional phototubes can be present, and dichroic mirrors and filters can be changed to suit different combinations of fluorochromes (see Roederer et al., 1997). With knowledge of the optical filter arrangement in a cytometer, the current from a given photodetector can be precisely related to the wavelengths of light coming from a cell. A flow cytometer may have one detector for fluorescence or may have six or more detectors for fluorescence. The number of detectors is not limited by engineering considerations so much as by the ability of the experimenter to find a range of fluorochromes with absorption wavelengths appropriate to the available lasers and with different emission wavelengths distinguishable by mirrors and band pass filters. In addition, the optimal number of photodetectors may also be limited by our intellectual ability to correlate and comprehend the data after the experiment is finished.

V. From Light Signals to a Data File

While a cell is in the laser beam, it scatters light and/or fluoresces; these light signals generated from the cell strike one or another of the photodetectors, dependent on the direction of the scattered light and the wavelength of the fluorescence. Photodiode detectors are generally used for detection of forward scatter light, because it is bright and sensitivity is not an issue. Photomultiplier tubes are generally used for detection of side scatter light and of fluorescence because the high voltage applied to them increases their gain. In either case, photons strike the photocathode of the detector and produce electrons that result in a current. The current is fed into a preamplifier, which produces a voltage pulse that is proportional in size to the number of photons that originally reached the detector. By electronic pulse-processing, the output voltage pulse can be proportional to the width of the light signal, the height of the signal, or to the integrated area of the signal. Peak and hold circuits in some cytometers keep their voltage for a period of time so that signals from the cell in the primary laser beam can be maintained until signals are collected from secondary and tertiary laser beams. While the peak and hold circuit is operating, no other signals can be processed. In other multilaser systems, "bucket brigade" electronics permits signals from a second cell to be collected from the primary laser beam before the preceding cell has completed its pass through the final laser. In any case, this length of time sets a requirement for the electronic "dead time" of the cytometer (the time during which the generation of signals from a second cell will cause both the first and second cell to be ignored rather than risking the addition of signals from two different cells). If cells are too close to each other in the core stream, then this eliminating (aborting) of cells from analysis will happen relatively frequently and will strongly affect the yield of cells in sorting applications. Two cells that actually overlap in a single laser beam will, by contrast, produce a signal that will not be aborted but will be misinterpreted as one large cell.

Once the light from cells has been converted to voltage pulses, the signal processing electronics in the cytometer may apply compensation subtraction or ratio calculations. Compensation circuitry is used to subtract spectral overlap when fluorochromes are close to each other in emission wavelength and band pass filters cannot entirely separate the signals from each according to color. Ratio circuits can be used, for example, to normalize intensity changes at one wavelength relative to stable or reciprocal intensity changes at another wavelength (calcium flux measurements with Indo-1 or pH measurements with Snarf are examples). If an instrument does not have electronic circuitry for calculating ratios or for applying cross-spectral compensation, these calculations can be made from the raw data in a stored data file.

In the penultimate signal processing step, amplification of the peak height or area intensity signals (compensated or not), of the width of the signal, or of the ratio of two signals can be applied either linearly or logarithmically. Linear amplification displays the signal in such a way that the output voltage is directly proportional to the input voltage from the preamplifier (and also to the original light signal). A logarithmic amplifier, by contrast, has an output voltage that is proportional to the logarithm of the input signal. Questions about the quality of these amplifiers, about consistency over their entire range, and about ways to correct for inconsistencies have been discussed elsewhere in detail (see Bagwell *et al.,* 1989; Muirhead *et al.,* 1983; Wood, 1997). They are of particular importance in procedures requiring calibration or standardization of fluorescence scales. Without considering those questions here, it is important to understand the reasons for choice of linear or logarithmic amplification. Linear amplification displays a limited range of intensities at any one setting and will convert an absolute change in intensity to the same voltage difference, at any point along the intensity scale. A logarithmic amplifier, by contrast, will display a much larger range of intensities at one time and will convert a certain "fold increase" in intensity to the same voltage difference, at any point along the scale. For example, at a given linear amplifier setting, two signals with intensities of 100 and 200 will be the same voltage distance apart as two signals with intensities of 800 and 900. By contrast, with a log amplifier at a given setting, two signals with intensities of 100 and 200 (twofold increase in intensity) will be much farther apart than two signals with intensities of 800 and 900 (1.125-fold increase in intensity). Logarithmic amplifiers are described as being three, four, or more decades full scale.

On a four-decade log amplifier, a signal displayed at the top of the scale will be 10^4 times as bright as a signal at the bottom of the scale. A four-decade full scale logarithmic amplifier is common on flow cytometers because this range is useful for most proteins on cell surfaces: stained cells are commonly 10 to 1000 times as bright as unstained cells. The other important advantage of logarithmic amplification is that a cell population with a given variation of intensities around a mean will appear, when displayed on the output scale, identical to any other (brighter or dimmer) population of cells with the same variation around its different mean. This makes it easy to compare characteristics of stained cells without regard to their absolute intensities. It also makes it easy to evaluate, visually, the area under a curve; with log amplification, for distributions with identical CVs, the area is represented by the height of the distribution (compare the upper and lower histograms in Fig. 11). The use of logarithmic amplification for chromosome analysis makes it easy to see if any one chromosome in a flow karyotype is present at less than or more than the expected number.

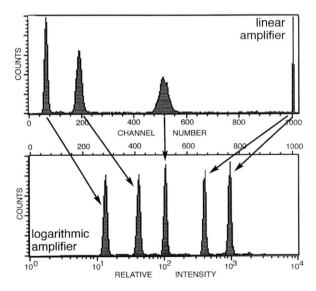

Fig. 11 A histogram plot of the fluorescence data from latex beads of five different intensities. The histogram acquired with linear amplification (top) displays effectively the three sets of beads with lowest intensity, but the two sets of brightest beads are "off scale" above 1000. Logarithmic amplification (bottom) displays all five sets of beads effectively, despite their 100-fold range of intensities. Using log amplification, beads with similar CVs and similar concentrations in the sample suspension give similar height and width profiles; this is not true for linear amplification. The intensity scale is expressed as "channel number" (0–1023), or, alternatively, with logarithmic amplification, the intensity scale can also be expressed as "relative intensity." The relative intensity scale can be calculated from the channel numbers because it is known that the log amplifier displays four decades for the full scale.

The last step in the electronic processing of light signals from the cell is analog-to-digital conversion. The original light signal is a so-called analog signal. It is continuously variable (right down to the photon level, at which point, all light signals are in fact digital) and the amplifier produces a voltage that can have any value on a scale of 0 to 10 volts. Because this can be any value, the number of different values is infinite and would require infinitely large resolution in the data file. In order to keep the size of data files under control, the analog signal is converted to a digital signal, that is, the voltage signals, after amplification, are binned so that ranges of intensity values are lumped together to give the same output. Although there are proposals to use 16-bit analog-to-digital converters (ADCs) (Shapiro *et al.*, 1998) because these may give sufficient resolution over a large enough range to reduce the requirement for log amplification, most flow cytometers today use 10-bit ADCs. On a 10-bit ADC, there are 2^{10} or 1024 separate bins or channels, and signals can have a range of values from 0 to 1023. With linear amplification, a signal appearing in channel 768 of a 10-bit ADC will have 10 times the intensity of a signal appearing in channel 77 (691 channels apart). If the detector voltage is changed so that the bright signal is in channel 512, then the dimmer signal will appear in channel 51 (461 channels apart). If, on the other hand, this were a four-decade full scale logarithmic amplifier, then two signals, one 10 times the intensity of the other, would always be 256 channels apart. If the brighter signal were assigned to channel 768, the dimmer one would appear in channel 512. If the voltage on the detector were decreased so that the brighter signal was in channel 512, then the dimmer one would be in channel 256.

After processing through a 10-bit ADC, each light signal will have been converted to a number on a scale of 0–1023. If the flow cytometer is, for example, a five-parameter instrument with five photodetectors, then each cell will be described flow cytometrically by five numbers, each on a scale of 0–1023, with each number representing the intensity of light striking each of the five photodetectors. A flow cytometric data file from this cytometer is, then, a long string of numbers, with five numbers describing the first cell, the next five numbers describing the second cell, and so on. The header for the data file gives information about the number of parameters recorded for each cell, so that the data file can be interpreted correctly with the relevant numbers applying to the appropriate cell in the temporal sequence. Flow cytometric data file structure conforms more-or-less to a published flow cytometry standard (FCS) format (Seamer *et al.*, 1997). Although the data file is actually a continuous string of numbers, it can be visualized as a list of cells with five different fields applying to each cell in the list (Fig. 12). Because each number has a 10-bit range, it requires two bytes of eight bits each to specify it. If the information from five photodetectors has been stored about each cell (a five-parameter data file), then each cell is described by 10 bytes of information. A 10,000 cell flow cytometric data file will therefore take up 100,000 bytes of space (plus a bit extra for a text header). More parameters for each cell and/or more cells described in the data file will take up proportionally more data storage space.

2. Principles of Flow Cytometry

CELL #	FSC	SSC	FL1	FL2	FL3
1	264	65	49	749	0
2	192	86	152	769	0
3	214	252	25	803	0
4	266	71	555	552	0
5	306	87	94	681	272
6	293	59	431	585	0
7	268	63	187	740	156
8	57	195	0	713	0
9	241	52	64	713	147
10	304	53	87	668	223
11	1002	1023	291	861	155
12	211	59	108	673	239
13	93	8	123	691	0
14	327	103	106	617	197
15	262	120	6	721	0
16	232	108	47	758	146
17	282	108	158	776	0
18	267	67	131	748	255
19	201	161	99	746	0
20	289	97	141	642	307
21	255	97	134	824	0
22	260	34	82	427	0
23	302	112	0	606	301
24	274	91	4	698	293
25	275	84	521	708	0
26	246	33	69	717	0

etc

Fig. 12 Visualization of a five-parameter flow cytometric data file as a list of cells, in the temporal order of which they passed through the laser beam. Each cell has five numbers associated with it, representing its intensity (on a 0–1023 scale) as registered on the forward scatter, side scatter, and three fluorescence detectors.

VI. From Data to Information

All the rest of flow cytometry is computing. There are many versions of software available to do this computing. They vary in price, in computer platform, in what kind of graphics they will display, and in the ease with which they manipulate the data. But they all do essentially the same thing. With usually three–eight numbers (each on a scale of 0–1023) describing each cell, histograms can be displayed for the distribution of the values of parameter number 1 (e.g., forward scatter intensity) for all the cells in the data file. Similar histograms, describing intensity distributions, can be drawn for all of the parameters in turn

(Fig. 13). Such a histogram distribution can then be analyzed for the mean, mode, or median intensity (still on a scale of 0–1023) of all the cells in that distribution. Or it can be analyzed for the percentage of the total number of cells above a certain value. Or the cells with intensities above a certain value can be analyzed for their mean or mode or median intensity. In some software applications, when a logarithmic amplifier has been used, the intensity scale on a histogram will be converted from channel numbers (0–1023) to "relative intensity units." (Refer back to Fig. 11 for the use of these alternative scales.) The accuracy of this software conversion to relative intensity units depends on the accuracy of the logarithmic amplifier in its nominal calibration at three, or four, or five

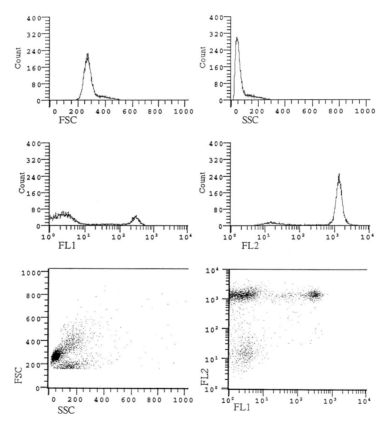

Fig. 13 Four histograms and two dot plots, representing the information for the 10,000 cells stored in a four-parameter flow cytometric data file. The four histograms display the intensity distributions for fsc, ssc, green fluorescence (FL1), and orange fluorescence (FL2). The two dot plots display fsc data correlated with ssc data and FL1 data correlated with FL2 data. Data for this figure were graphed using a prerelease copy of FCSPress software from Ray Hicks, Cambridge University.

decades full scale. The "hard" units are always the original channel numbers. However, relative intensity units allow some appreciation of the range of intensities encompassed by the cells in an experiment.

In addition to one-parameter histograms describing the distribution of the cells in a data file with respect to the intensity of the signals from one particular photodetector, two different parameters can be analyzed with respect to each other. For display of this correlation, two parameters can be plotted on a two-dimensional dot plot, for all the cells in the data file (Fig. 13, lower plots). Information can then be derived about the intensity of cells with respect to these two characteristics and about the correlation of these two parameters with each other. Conventionally, the cells can be grouped (by quadrant analysis) into those that are single positive (brighter than the control) for parameter 1, single positive for parameter 2, double positive for both parameters, or double negative for both parameter. Although cluster analysis (see Dean, 1991; Murphy, 1985) can provide this kind of correlation for three or more parameters simultaneously, the usual analysis of flow data for more than two parameters involves "gating" on several parameters before analysis of the final two parameters.

Gating is the flow cytometric term for the designation of cells of interest for further analysis. Although relatively mundane in terms of computer programming, it provides a conceptual leap that accounts for much of the power of flow cytometry (see Loken, 1997). In terms of the computer analysis of a list-mode data file, gating simply means going through the list of cells, one by one; selecting those cells that fulfill certain criteria with respect to their intensities on any of several, initial (gating) parameters; and then analyzing the cells that fulfill those criteria for their characteristics with respect to one or two other analysis parameters. Traditionally, cells have been gated according to their forward- and side-scatter characteristics so that cells with different physical characteristics can be analyzed separately from within a mixed cell suspension (Fig. 14). For example, lymphocytes scatter light differently from monocytes and neutrophils so that the fluorescence of stained lymphocytes can be analyzed in a mixed population of peripheral blood leukocytes without actually separating the leukocyte classes from each other. Gating on scatter characteristics can also allow exclusion of debris or aggregated or dead cells from analysis; this can be useful because a flow cytometer, unlike a microscopist, sees all particles as equally important unless told otherwise. Gating on scattered light does have its limits—because light scatter characteristics of cells are poorly understood and especially because cells of very different types may overlap in their scatter properties. For this reason, much current cytometric analysis avoids decisions based on scatter signals and involves, instead, staining cells simultaneously with fluorescent probes for multiple biochemical properties (surface membrane proteins, nuclear DNA content, soluble cytoplasmic constituents, and/or calcium concentration, e.g.) and then gating on some of these characteristics before analysis of others (e.g., Roederer et al., 1997).

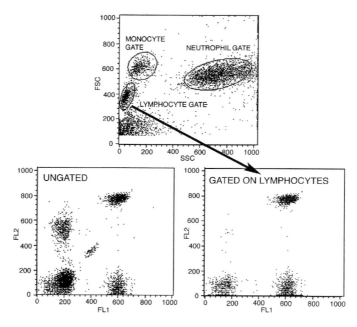

Fig. 14 Dot plots showing correlated FL1 versus FL2 data for leukocytes from the peripheral blood. The dot plot on the lower right was gated so as to display the fluorescence intensities of only those cells that fulfill the fsc versus ssc gating criteria set for lymphocytes in the upper plot.

VII. Sorting

Although early flow cytometers were designed as instruments that could separate or "sort" cells from mixed populations based on the intensity of their light signals (see Fulwyler, 1965; Hulett *et al.*, 1969; Kamentsky and Melamed, 1967), most of the cytometers in laboratories today are used only for analysis and do not possess a sorting capability. In addition, many techniques more rapid than flow cytometry are currently available for sorting cells in conjunction with monoclonal antibody staining (e.g., magnetic bead separation or complement depletion). Nevertheless, flow sorting is an impressive technology (see Lindmo *et al.*, 1990) and may be the best separation technique available when cell phenotypes are based on dim marker expression, on multiparameter criteria, or on light scatter characteristics. Cells sorted by flow cytometry are routinely used for functional assays, for PCR replication of cell type-specific DNA sequences, and for cloning of transfected cells; in addition, sorted chromosomes are used for the generation of DNA libraries. The strategy for electronic flow sorting involves the enclosure of cells in individual drops of fluid and, then, the applying of an electrostatic charge to the drops containing cells with desirable characteristics so that these designated drops can be deflected out of the main stream of flow and into separate collection vessels. The challenges of flow sorting involve the formation of drops enclosing no more than one cell each and the charging of only those drops containing the appropriate cells.

When a stream of fluid is vibrated along its axis, it will break up into drops. These drops will form according to the following equation:

$$v = f\lambda$$

defining the fixed relationship between the velocity of the stream (v), the frequency of the drop generation (f), and the distance between the drops (λ). Stable droplet formation occurs when the distance between the drops is equal to 4.5 times the diameter of the stream. For example, with a 70 μm stream, droplets will form most easily at an interval of about 315 μm. The implication of this relationship is that, if the stream velocity is 10 m per sec (as in a usual flow cytometer), drops need to be generated by stream vibration at about 31 kHz to provide stable drop formation. If cells in a sample core are moving into the laser beam at a rate of 3000 cells per sec, there will then be, on average, one cell in every tenth drop in the stream flowing at 10 m per sec. If the stream velocity were 20 m per sec (as in a high speed sorter), then drop drive vibration would need to be at 62 kHz, and, on average, there would be one cell in every 20 drops. The actual distribution of cells among the drops will obey Poisson statistics.

Because drops break off from a vibrating stream at a distance from the fixed point of vibration, a stream of cells can be illuminated by a laser beam and the signals collected from those cells with minimal perturbation as long as the analysis point is relatively close to the point of vibration and far enough away from the point of drop formation (Fig. 15). In a sorting cytometer, cells are illuminated close to or within the flow chamber and their signals collected, amplified, and digitized in ways similar to those in nonsorting cytometers. It is only when the cell has moved down the stream that it will be enclosed in a drop as the drop breaks away from the stream. At points below the position of droplet formation, the stream will consist of a series of drops, all separate from each other, with some drops containing cells. The number of drops containing cells, the number of empty drops, and the distribution of cells among the sequential drops will depend on the distance between cells in the sample core as it emerges from the flow cell.

In order to sort the cells, an electrostatic charge is applied to the drops containing cells of interest. The sorting operator sets parameter gates, indicating the flow cytometric characteristics that describe the cells to be selected. The operator then determines the time that elapses between the illumination of a cell in the laser beam and its enclosure in a drop, downstream. This drop delay time can be determined empirically (by trying different times on a test sort with beads) or can be measured directly using drop separation units (knowing the drop generation frequency in kilohertz, the reciprocal is equal to sec per drop; therefore the distance between drops has an equivalence in time units). Given the flow characteristics of the desired cells and the time that it takes between analysis

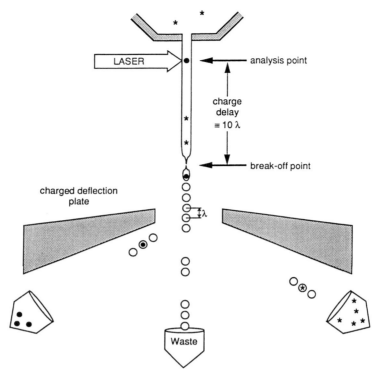

Fig. 15 Droplet formation for sorting. The elapsed time between the analysis of a cell in the laser beam and its passage to the position where it will be enclosed in a drop is measured in drop (λ) units. Charging of the stream is delayed for that period of time. Cells that are enclosed in charged drops are deflected toward the right or left deflection plates. From Givan, A. L., "Flow Cytometry: First Principles" (1992). Reprinted by permission of Wiley-Liss, a subsidiary of John Wiley & Sons, Inc.

of a cell at the laser beam intercept and the enclosure of that cell from the main stream into its own self-contained drop, the flow cytometer can be programmed to apply an electrostatic charge to the stream for a short interval, starting at the time just before the cell of interest is about to detach from the main stream into a drop. If the charge is applied for a short interval, only the drop containing the cell of interest will be charged. By applying the charge for a longer time, a drop on either side of the selected drop can be charged for security, in case there is some fluctuation (due to changes in temperature and sheath pressure) in the time between analysis and drop formation. The charge on the stream can be positive or negative—and therefore the drops containing two mutually exclusive classes of desired cells can be deflected by high voltage plates to the left or to the right away from the center stream of drops and into separate collecting vessels.

In this way, cell sorting puts one additional set of constraints on the fluidic characteristics of a flow cytometer. If cells flow any faster than the rate of drop formation, then multiple cells will be enclosed in a single drop. In practice, sort operators generally keep cells flowing at speeds of about one-tenth the rate of drop formation, so that, on average, every tenth drop will contain a cell (but remember Poisson). Therefore, the rate of drop formation sets a limit on the rate of sorting of cells. Because the frequency of drop formation is directly related to the stream velocity, higher stream velocities permit cells to flow at faster rates yet still be enclosed in separate drops so that they can be sorted separately. The use of higher stream velocities (20–50 m per sec) is the strategy for high speed sorting (see Leary, 1997; Peters *et al.,* 1985).

Cell sorting is usually validated by the purity of the sorted cells, by the efficiency of sorting of the selected cells from the original mixed suspension, and also by the time taken to obtain a given number of sorted cells. In many cytometers, purity is protected by the aborting of drop charging when cells are so close together at the analysis point that they may be enclosed in a single drop downstream (or in three adjacent drops, if three drops are being deflected). In this way, high cell flow rates will lead to lost cells and will compromise the sorting efficiency but may not compromise purity until cells are so close together that they are coincident in the laser beam. As cell flow rates are increased by using more concentrated suspensions, more sort decisions may be aborted, but the speed of sorting of the desired cells will increase until the cell flow rate gets so high and aborted sorts become so frequent that the actual speed of sorting of the desired cells starts to drop off (Fig. 16). In most cytometers, with strict abort

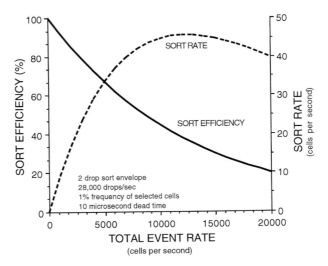

Fig. 16 The effect of cell analysis rate on sorting rate and sorting efficiency (yield). For the purpose of this modeled data (for the model, see Hoffman and Houck 1998), it was assumed that two drops were charged for each sort decision, the drop drive frequency was 28,000 drops per sec, the selected cells were 1% of the original cell population, and the instrument had a 10 μsec dead time.

conditions, it is only when the cell flow rate is so high that multiple cells coincide in the laser beam that purity of the sorted cells becomes compromised. Every sorting experiment is, in this way, a procedure that needs to be optimized for its own requirements. Speed, purity, and sorting efficiency interact and are not, all three, optimizable under the same conditions. For example, a sort for a minor subpopulation of cells from a large, easily obtainable cell suspension may require optimization of speed, without concern about low sorting efficiency. On the other hand, a requirement for as many specific cells as possible from among a scarce mixed suspension will need to maximize recovery without much concern with speed.

VIII. Conclusions

Webster's New World Dictionary (1958) says that a "minimizer" is "one who tries to make religious or philosophical problems appear easily explained." It is therefore not without some humility that we need to admit that flow cytometry is, essentially, a minimizing technique. It simplifies the uniqueness and elegance and complexity of a cell down to a set of five or so numbers. The power of flow cytometry comes from rapid, objective, and quantitative computing, allowing calculations, correlations, and statistical conclusions from those few numbers derived from each of many cells. Our intelligence and training should, however, make us aware of all those important things that flow cytometry does *not* tell us about a cell. It is only with this perspective that flow cytometric analysis really comes into its own: we use flow cytometric data to give us remarkably useful information about cells but are not lulled into thinking that this is all that needs to be known.

Acknowledgments

I thank Joseph Trotter for a lucid discussion on "bucket brigade" electronics; Howard Shapiro, Nancy Perlmutter, Lillian Burke, and Gary Ward for helpful reviews of the manuscript; Gary Ward, Jan Fisher, and Mary Waugh for the production of some exemplary data files; Ray Hicks for the prerelease use of his FCSPress software; and Joan Thomson (of Dartmouth Medical School) and Ray Joyce (of Newcastle University Medical School) for elegant, as well as accurate, interpretations of some fairly messy sketches.

References

Bagwell, C. B., Baker, D., Whetstone, S., Munson, M., Hitchox, S., Ault, K. A., and Lovett, E. J. (1989). A simple and rapid method of determining the linearity of a flow cytometer amplification system. *Cytometry* **10,** 689–694.
Coulter, W. H. (1956). High speed automatic blood cell counter and size analyzer. *Proc. Natl. Electronics Conf.* **12,** 1034–1040.
Crosland-Taylor, P. J. (1953). A device for counting small particles suspended in a fluid through a tube. *Nature* **171,** 37–38.

Dean, P. N. (1991). Data processing. *In* "Flow Cytometry and Sorting" (M. R. Melamed, T. Lindmo and M. L. Mendelsohn, eds.), pp. 415–444. Wiley-Liss, New York.

Dittrich, W., and Göhde, W. (1969). Impulsfluorometrie dei einzelzellen in suspensionen. *Z. Naturforsch.* **24b,** 360–361.

Fulwyler, M. J. (1965). Electronic separation of biological cells by volume. *Science* **150,** 910–911.

Hoffman, R. A., and Houck, D. W. (1998). High speed sorting efficiency and recovery: Theory and experiment. *Cytometry* 9 (suppl.), 142.

Hulett, H. R., Bonner, W. A., Barrett, J., and Herzenberg, L. A. (1969). Cell sorting: Automated separation of mammalian cells as a function of intracellular fluorescence. *Science* **166,** 747–749.

Kachel, V., Fellner-Feldegg, H., and Menke, E. (1990). Hydrodynamic properties of flow cytometry instruments. *In* "Flow Cytometry and Sorting" (M. R. Melamed, T. Lindmo, and M. L. Mendelsohn, eds.), pp. 27–44. Wiley-Liss, New York.

Kamentsky, L. A., and Melamed, M. R. (1965). Spectrophotometer: New instrument for ultrarapid cell analysis. *Science* **150,** 630–631.

Kamentsky, L. A., and Melamed, M. R. (1967). Spectrophotometric cell sorter. *Science* **156,** 1364–1365.

Leary, J. F. (1997). High-Speed Cell Sorting. "Current Protocols in Cytometry" (J. P. Robinson, Ed.), pp. 1.7.1–1.7.6 Wiley, New York.

Lindmo, T., Peters, D. C., and Sweet, R. G. (1990). Flow sorters for biological cells. "Flow Cytometry and Sorting" (M. R. Melamed, T. Lindmo, and M. L. Mendelsohn, eds.), pp. 145–169. Wiley-Liss, New York.

Loken, M. R. (1997). Multidimensional data analysis in immunophenotyping. *In* "Current Protocols in Cytometry" (J. P. Robinson, ed.), pp. 10.4.1–10.4.7. Wiley, New York.

Melamed, M. R., Lindmo, T., and Mendelsohn, M. L., eds. (1990). "Flow Cytometry and Sorting" 2nd Ed. Wiley-Liss, New York.

Moldavan, A. (1934). Photo-electric technique for the counting of microscopical cells. *Science* **80,** 188–189.

Muirhead, K. A., Schmitt, T. C., and Muirhead, A. R. (1983). Determination of linear fluorescence intensities from flow cytometric data accumulated with logarithmic amplifiers. *Cytometry* **3,** 251–256.

Murphy, R. F. (1985). Automated identification of sub-populations in flow cytometric list mode data using cluster analysis. *Cytometry* **6,** 302–309.

Peters, D., Branscomb, E., Dean, P., Merrill, T., Pinkel, D., Van Dilla, M., and Gray, J. W. (1985). The LLNL high-speed sorter: Design features, operational characteristics, and biological utility. *Cytometry* **6,** 290–301.

Pinkel, D., and Stovel, R. (1985). Flow chambers and sample handling. *In* "Flow Cytometry: Instrumentation and Data Analysis" (M. A. Van Dilla, P. N. Dean, O. D. Laerum, and M. R. Melamed, eds.), pp. 3–128. Academic Press, London.

Robinson, J. P., ed. (1997). "Current Protocols in Cytometry." Wiley, New York.

Roederer, M., De Rosa, S., Gerstein, R., Anderson, M., Bigos, M., Stovel, R., Nozaki, T., Parks, D., and Herzenberg, L. (1997). 8 color, 10-parameter flow cytometry to elucidate complex leukocyte heterogeneity. *Cytometry* **29,** 328–39.

Salzman, G. C., Singham, S. B., Johnston, R. G., and Bohren, C. F. (1990). Light scattering and cytometry. *In* "Flow Cytometry and Sorting" (M. R. Melamed, T. Lindmo, and M. L. Mendelsohn, eds.), pp. 81–107. Wiley-Liss, New York.

Seamer, L. C., Bagwell, C. B., Barden, L., Redelman, D., Salzman, G. C., Wood, J. C., and Murphy, R. F. (1997). Proposed new data file standard for flow cytometry, version FCS 3.0. *Cytometry* **28,** 118–122.

Shapiro, H. M. (1995). "Practical Flow Cytometry." Wiley-Liss, New York.

Shapiro, H. M. (1997). Laser beam shaping and spot size. *In* "Current Protocols in Cytometry" (J. P. Robinson, ed.), pp. 1.6.1–1.6.5. Wiley, New York.

Shapiro, H. M. (1997). Lasers for flow cytometry. *In* "Current Protocols in Cytometry" (J. P. Robinson, ed.), pp. 1.9.1–1.9.13. Wiley, New York.

Shapiro, H. M., Perlmutter, N. G., and Stein, P. G. (1998). Recent advances in cytometry: Building what can't be bought. *Cytometry* **9** (suppl.), 122.

Stovel, R. (1997). Fluidics. *In* "Current Protocols in Cytometry" (J. P. Robinson, ed.), pp. 1.2.1–1.2.7. Wiley, New York.

Sweet, R. G. (1965). High frequency recording with electrostatically deflected ink jets. *Rev. Sci. Instrum.* **36,** 131.

van den Engh, G., and Farmer, C. (1992). Photo-bleaching and photon saturation in flow cytometry. *Cytometry* **13,** 669–677.

Van Dilla, M. A., Trujillo, T. T., Mullaney, P. F., and Coulter, J. R. (1969). Cell microfluorimetry: A method for rapid fluorescence measurement. *Science* **163,** 1213–1214.

Van Dilla, M. A., Dean, P. N., Laerum, O. D., and Melamed, M. R., eds. (1985). "Flow Cytometry: Instrumentation and Data Analysis." Academic Press, London.

Waggoner, A. (1997). Optical filter sets for multiparameter flow cytometry. *In* "Current Protocols in Cytometry" (J. P. Robinson, ed.), pp. 1.5.1–1.5.8. Wiley, New York.

Watson, J. V. (1991). "Introduction to Flow Cytometry." Cambridge Univ. Press, Cambridge.

Watson, J. V. (1992). "Flow Cytometry Data Analysis: Basic Concepts and Statistics." Cambridge Univ. Press, Cambridge.

Wheeless, L. L., Jr. (1990). Slit-scanning. *In* "Flow Cytometry and Sorting" (M. R. Melamed, T. Lindmo, and M. L. Mendelsohn, eds.), pp. 109–125.

Wood, J. C. S. (1997). Establishing and maintaining system linearity. *In* "Current Protocols in Cytometry" (J. P. Robinson, ed.), pp. 1.4.1–1.4.12. Wiley, New York.

CHAPTER 3

Laser Scanning Cytometry

Louis A. Kamentsky

CompuCyte Corporation
Cambridge, Massachusetts 02139

    I. Introduction
   II. Background
  III. Description of the Instrument
  IV. The Utility and Operational Characteristics of Some Laser Scanning Cytometry List Mode Features
        A. Total Value Feature
        B. Maximum Pixel Value Feature
        C. Extranuclear Total Value Feature
        D. Area Feature
        E. Cell Position Feature
        F. Time Feature
        G. Probe Spot Counting Feature
        H. Multiple Cell Exclusion Feature
   V. Utility of Solid Phase Cytometry for Cell Preparation
  VI. Future Directions
      References

I. Introduction

Laser scanning cytometry (LSCM) automatically measures laser excited fluorescence at multiple wavelengths and light scatter from cells on slides that have been treated with one or more fluorescent dyes in order to rapidly determine multiple cellular constituents and other features of the cells. A large body of literature has evolved since the 1970s specifying techniques using fluorescence to characterize cells by flow cytometry (FCM). I will describe a specific laser scanning cytometer, the LSC, which can use these techniques perfected for FCM to provide data comparable to FCM. Because it is microscope based and measures cells on the surface of a slide, records position of each cell on the slide, and has higher resolution, it can provide a number of benefits, which I will describe, that may make it a more suitable cytometer for certain applications.

FCM has used fluorescent dyes to quantify cell constituents because if dye concentration is low, and the dye is specifically bound to a constituent, fluorescence emission from the dye is directly proportional to the mass of the constituent. Because it is unnecessary to account for the spatial distribution of fluorescence, it is necessary only to illuminate the cell uniformly with light absorbed by the dye and to measure the total fluorescence emission of a cell resulting from that dye. It is therefore not necessary to dissect elements of the cell by imaging in order to account for constituent distributions in the cell. Because of this, FCM has used zero resolution imaging, and LSCM too can use low resolution scanning to obtain accurate constituent determinations at rapid cell processing rates. Additionally, because light scatter emits at the excitation wavelength, is usually much stronger, and can be distinguished from fluorescence, cell light scatter can be used to further characterize cells.

LSCM is not comparable to confocal microscopy. Because LSCM must uniformly illuminate cells throughout their volume to obtain accurate whole cell constituent measurements, its optical components are designed to be nonconfocal. LSCM uses large field depths, and confocal microscopy emphasizes short field depth to provide detailed images at a narrow depth focal plane through each cell that is imaged. Additionally, LSCM is designed to automatically measure large heterogeneous populations of cells, unlike the detailed single cell analysis for which confocal microscopy is most useful.

II. Background

Three technologies, illustrated in Fig. 1, can quantify cell constituents using fluorescence. Fluorescence image analysis (FIA), FCM, and LSCM can each automatically measure fluorescence at multiple wavelengths of cells that have been treated with one or more fluorescent dyes to rapidly assay multiple cellular constituents. Two of these FCM and LSCM can measure light scatter as well as fluorescence. In FCM and LSCM, fluorescence and scatter result from interaction of the cells with a laser beam comparable in spot size to the cell. The laser optics is designed to produce a nearly collimated object excitation to achieve accurate constituent measurements that are not too sensitive to cell position in the FCM flow channel or to LSCM slide focal position. In FIA the cells are uniformly illuminated, preferably by a mercury or xenon arc epi-illuminator. Fluorescence is imaged at high resolution and low depth of field to a sensitive CCD camera. A wavelength band pass filter is used to isolate the fluorescence, and this filter can be mechanically changed to measure cell fluorescence at multiple wavelengths. The lasers of FCM and LSCM provide an intense concentration of excitation energy at the optical plane of the cell as well as monochromatic energy to allow better separation of fluorescence emission from excitation. Therefore,

3. Laser Scanning Cytometry

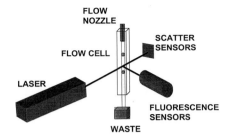

FLOW CYTOMETRY (FCM)

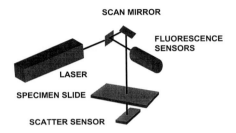

LASER SCANNING CYTOMETRY (LSCM)

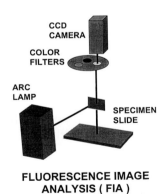

FLUORESCENCE IMAGE ANALYSIS (FIA)

Fig. 1 Simplified representations of the three technologies for cellular constituent measurements using multiple fluorescence wavelength detection, flow cytometry (FCM), laser scanning cytometry (LSCM), and fluorescence image analysis (FIA).

cell constituents can generally be detected with higher measurement sensitivity than FIA. For each cell found, commercial FIA, FCM, and LSCM instruments generate and store a set of feature values characterizing attributes of each cell in standardized format computer list mode files.

LSCM is different than FCM in the following three ways. (1) In LSCM, the cells are measured and retained on a solid support such as a slide. In FCM cells in suspension flow past the laser beam in a flow cell ending in a waste container. (2) The LSCM slide and laser beam are moved under computer control to excite the cells. Because the computer controls the positions of the slide and laser beam, cell position on the slide can be a measurement feature. This is impossible with a flow system, because cell position is not maintained. (3) In LSCM, the interactions of each cell and the laser are measured and recorded many times in a two-dimensional pattern and features computed from these multiple interactions are derived, whereas in FCM properties of a single analog pulse are recorded as each cell flows past the laser focus.

Because cells are prepared and measured on a slide, it is not necessary to provide single cell or nuclei, only cell suspensions. Touch or needle biopsy specimens made as imprints or smears or tissue can be measured directly. Cytoplasmic as well as nuclear constituents are thus available for characterization. Solid phase specimen preparation techniques may be employed, measuring cells directly on the solid phase after preparation, eliminating centrifugation steps to separate phases. Specific assays such as *in situ* hybridizations are best done on a slide because they are difficult to perform in suspension due to loss of cell integrity. Preparations requiring amplifications or specific fixatives can be employed without agglutination or cell clumping. The complete area encompassing a specimen can be scanned to allow all cells in a small specimen to be measured. There is no hazardous waste and no carryover of specimen or dyes and no dilution of specimen staining during runs. Finally, multiple specimens can be placed in different areas of a slide, they can be analyzed as a single assay, and the data from all specimens can be analyzed and reported without user intervention.

Because the coordinates on the slide of each measured cell are recorded in the list of features of that cell, the position feature can be used to relocate cells with a defined range of properties for visual observation or CCD camera image capture. Images may be included in reports or used for high-resolution analysis of selected cells. Conversely, cells may be visually located and features of observed cells displayed. Additionally, cells may be observed and categorized and these categories used for subsequent data gating. Slides can be rerun, and feature values of each cell from each run can be combined using the position of each cell on the slide as the merge key to create multirun feature sets. The slide may be restained, or the laser excitation wavelength may be changed between runs. Specimens may be assayed multiple times to measure and record the kinetic properties of each cell of heterogeneous populations. The position feature can be displayed as any other feature and used for quality control of staining by displaying a fluorescence feature versus X or Y position. It can also be used to trace out the locations on specimens, such as tissue sections or touch preparations, of individual cells with specific features, by scattergram gating on the features to another scattergram of X position versus Y position. Conversely, on sections, X position versus Y position scattergrams may define gross morphology gating regions on the section to determine characteristics of cells within delimited tissue areas.

The LSC laser scanning cytometer makes measurements on each cell at 0.5-μ spatial intervals. This two-dimensional array of values replaces the pulse shape analysis of FCM. There is sufficient resolution so that new features can be computed from these arrays that have utility in characterizing specimens. Features can be computed such as area, perimeter, the maximum value found in the array, and texture all of which may give additional information useful in characterizing cells with fewer dyes and sensors. Constituents that are localized to regions of the cell such as to the cytoplasmic compartment or fluorescence *in situ* hybridization (FISH) probe spots can be independently characterized yielding further features not obtainable with FCM. As will be described, the total fluorescence, area, and maximum pixel fluorescence of the individual probe spots are used as LSC features, allowing the LSC to more accurately count probe spots in cells of FISH specimens.

The first description of a large spot fluorescence scanning cytometer was made by Mellors and Silver (1951), and applied to the automation of cervical cancer screening. Mansberg and Ohringer (1969) described design principles of a fluorescence cytometer using a galvanometer scanner. Laser excited fluorescence scanning cytometers have been developed by Burger and Gershman (1988) in which a Bragg cell was used to move a laser beam, by de Grooth *et al.* (1985) in which specimens were placed in grooves on a rotating disk, by Deutsch and Weinreb (1994) in which specimen cells were placed in fixed locations in a grid, and by Dietz *et al.* (1996) in which specimens were measured in a cuvette. I will describe the design and operational characteristics of a specific laser scanning cytometer the LSC, which we have designed to assay specimens on standard microscope slides (Kamentsky and Kamentsky, 1991; Kamentsky *et al.*, 1997a; Gorczyca *et al.*, 1997; Darzynkiewicz *et al.*, 1999).

III. Description of the Instrument

Figure 2 shows a photograph of the CompuCyte LSC laser scanning cytometer (CompuCyte Inc., Cambridge, Massachusetts), which we designed using a standard research microscope and a personal computer (PC). In the block diagram of its optical system, shown in Fig. 3, the beams from an argon ion and a HeNe laser are combined at a first dichroic mirror and steered to a second dichroic mirror that reflects the laser wavelengths and transmits other wavelengths. Each laser beam is shuttered under computer control. The combined beam is steered to a computer controlled scanning mirror moving in a sawtooth motion at a

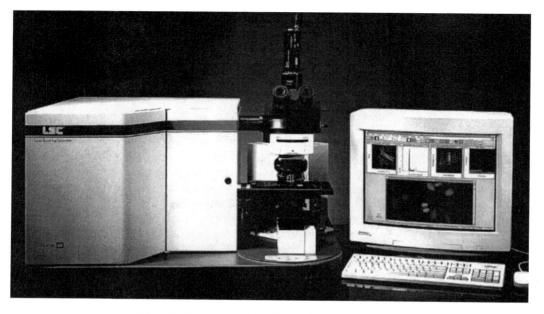

Fig. 2 Photograph of the CompuCyte Corporation LSC laser scanning cytometer.

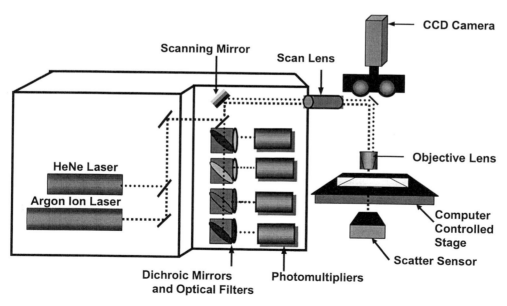

Fig. 3 Block diagram of the LSC laser scanning cytometer optical system.

nominal rate of 350 Hz, to create a line scan at the microscope slide. After passing through a scan lens, the beam enters the epi-illumination port of an Olympus BX 50 microscope. After it is reflected off a partially silvered mirror placed in the optical train to allow CCD imaging of relocated cells, the laser beam is imaged by a tube lens and the objective lens of the microscope to the focal plane at the specimen. The line scan at the focal plane has a 10-μ diameter $1/e^2$ width and 685-μ extent using a 10X objective, a 5-μ diameter width and 342-μ extent using a 20X objective, or a 2.5-μ diameter width and 171-μ extent using a 40X objective. The specimen slide is mounted in a holder on the stage of a computer controlled stepper motor stage equipped with home position sensors. By initially homing the stage, the computer can maintain a record of the position of the slide throughout the scan. Because the scan mirror is computer controlled, the stage and scan mirror positions can be combined by software to record the coordinates of the object with respect to the laser beam. The stage is moved in steps of 0.5 μ per scan line perpendicular to the line scan during assays until a strip of 1000 steps is completed at which time the data from this strip is analyzed. Contiguous strips are processed covering a specified specimen width on the slide. The stage is then stepped in the direction of the scan to process contiguous sets of strips in the scan direction until a complete specified specimen area is covered or a specified number of events are detected.

With currently available PCs, for many experiments, data acquisition and processing rates are high enough so that the limiting factor in the cell processing rate is mechanical. It results from the limits inherent in moving the microscope stage and the laser scan mirror. The nominal area covered each minute by movement of the scan beam at the slide and the stage is 4.25 mm^2 using a 20X objective. Optimum cell rates are achieved with cell densities in the range 100 to 1000 cells per mm^2 on the slide, providing cell-processing rates in excess of 1000 cells per min. Because the scan length is inversely proportional to objective lens power, cell acquisition rate depends on the objective used. It can be increased with a lower power objective and the rate will decrease as objective power is increased.

As the laser beam intersects a cell, scattered light is imaged by the condenser lens of the microscope to an assembly containing a beam blocking bar and solid state photo sensor. When not scanning, this assembly is moved so that the bright field source of the microscope can be used for viewing objects through the eyepiece or the CCD camera. Fluorescent energy emitted by each cell is collected by the objective lens, reflected by the partially silvered CCD camera mirror, and steered through the scan lens to the scanning mirror. The then collimated fluorescence emission passes through a series of dichroic mirrors and optical interference filters to up to four photomultipliers, each detecting a specific fluorescence wavelength range.

As shown diagrammatically in Fig. 4, sensor signals are simultaneously digitized at 625,000 Hz. By adjusting both the scan beam velocity and the scan length, sampling is made at 0.5-μ spatial intervals along the scan. Because of the 0.5-μ

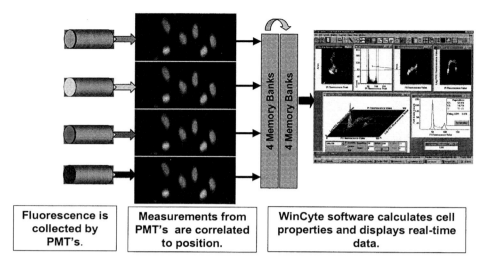

Fig. 4 Block diagram of the LSC laser scanning cytometer data acquisition and processing system.

stage movement, the fluorescence and scatter sensor signals are sampled each 0.5 μ in both X and Y directions. The sensors are digitized by an analog to digital converter to 14-bit digital pixel values. These pixel values are stored directly in six of twelve banks of memory in the PC. Acquired data is placed into six data banks while data is simultaneously analyzed in the other six data banks. Data bank sets are interchanged after each scan strip to allow the LSC to simultaneously acquire and analyze data. Twelve memory banks allow simultaneous digitization and storage of five sensor signals and a signal resulting from the sum of two or more sensor signals. If required, the user can set the LSC software to sum sensor pixel values from two or more successive scans into each memory location representing the data along each scan line, while slowing the step rate, effectively increasing digitization precision and instrument dynamic range and sensitivity but slowing data acquisition.

A software program, WinCyte, which is described next, has been written to process the scan data so as to derive sets of features characterizing cells in the specimen. Shown in Fig. 5 is a scan data display of part of the digitized signal of one fluorescence sensor after scanning a strip. This scan data display provides a representation of the scan strip data in computer memory for direct user observation of data and parameter setting. The darkness at each pixel location is proportional to the data value of the pixel. To process events to obtain cell features, the scan data is first segmented in order to associate specific pixels with each event. The computer program draws a contour around each event at a value set by the user on data from one sensor or data from the sum of two or more sensors. These threshold contours can be visualized on the scan data display

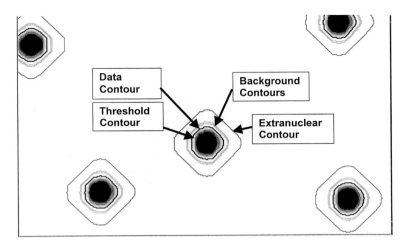

Fig. 5 Laser scan image of cells in which the darkness of each pixel is proportional to its fluorescence or scatter value. The software finds each event having fluorescence or scatter above a threshold value. It draws a contour around the event at the threshold value. Other contours are drawn outside the threshold contour, namely, the data contour used to calculate total fluorescence or scatter, two background contours used to determine a background value for each cell, and the extranuclear contour defining an area outside of the data contour to calculate total fluorescence or scatter.

as geometric shapes surrounding the data corresponding to each event. In Fig. 5, the first inner contour around the cell image is this threshold contour. The same contour locations are used to segment data from each event for all other sensor data, as well as the sensor used for the contours. The number of pixels within each contour is used as the *area* feature of the event. The values of each sensors maximum pixel within the contour are used as *maximum pixel* features. The number of pixels in the contour line is used as the *perimeter* feature of the event.

Because the Gaussian shaped laser excitation beam is larger in diameter than the sampling distance the threshold contour will not coincide with the cell boundary. Cell data images are generated in which all of the pixels representing each cell are not completely contained within its threshold contour. For this reason, a second contour, a fixed number of pixels outside the threshold contour, is established so as to include all cell pixels in the total fluorescence computation. The pixels for each sensor within this data contour are summed to compute the *total scatter or fluorescence* feature of the cell, the estimate of the amount of cell constituent resulting from scatter or the fluorescent dye measured by the sensor. By increasing the spacing between threshold and data contours, the user can set a higher threshold to better isolate cell events when cells are touching.

The instrument was designed to provide accurate determinations of cell constituent values. To this end the software provides two necessary computations on the summed pixel values. The first is to correct for differences in excitation intensity and emission detection efficiency as a function of scan mirror position. The laser beam is off the optical axis at all but central pixel values of the scan motion. This and movement of the scan beam over optical surfaces creates irregularities in light excitation and collection. For this reason the efficiency of excitation and emission detection is established by scanning uniform calibration particles. The total fluorescence of the particles is measured as a function of scan position and used to generate a permanently stored table to automatically correct the total fluorescence of each cell based on its position along the scan line.

To accurately determine cell constituent and maximum pixel values in the presence of variable background fluorescence or scatter, it is necessary to establish a background value for each sensor for each event. At the option of the user, background levels for each sensor are either set on a specimen basis with one background level per sensor or automatically with a background level determined independently for each event. If the user chooses automatic background determination, as shown in Fig. 5, third and fourth contours, set some number of pixels outside the data contour, are constructed. A function of all of the pixel values lying between these contours is used to determine the background of each event for each sensor. *Total fluorescence or scatter* is computed as the sum of pixel values within the data contour minus the background values for each sensor. The user can select background algorithms that subtract either averaged pixel values within ranked minimum or median ranges. Subtracting minimum values assures that background determinations are free of data from abutting events but may bias the summed value if background values are close to data values. Algorithms are also included to reject the background determination of the event and either reject the event, or use the previous background of the event, if the background data does not meet specific regularity criteria.

The pixel values within the data contour are further examined by the software to determine the presence of data saddles indicating possible multiple cell events within the data contour. For most cells, single cell events should create data that decreases in pixel value monotonically from the maximum value to the data contour. If a pixel value saddle is found, that event is tagged, and the event can be excluded using the *multiple cell exclusion* feature. A measure of the deviation of averages of pixel values of adjacent areas within the data contour is also determined and used as the *texture* feature.

As will be described later, the software can merge data from multiple assays using event maximum pixel location as the merge key to associate additional measurements with previous measurements of each event. In applications in which the only change between assays is a change in laser excitation, there is provision for using contours generated by excitation from one laser to derive features determined from excitation by a second laser. In this mode of operation, a strip is scanned with one of the lasers of the LSC, contouring events and determining and storing all features of all active sensors as well as contours for the event set. The same strip is then rescanned with the second laser, and a second set of features is determined for active sensors using the contours of the previous scan. The feature sets from both scans are combined for each event found and stored in list mode.

The software can generate a second set of contours within each data contour using a sensor signal other than that used to establish the data contour. This was provided to analyze specimens, primarily those with probe spots from FISH preparations, which contain localized constituents that can be independently stained and their fluorescence differentiated from the fluorescence of a counterstain. Prior to generating probe spots, the values within each data contour are processed by convoluting each with a matrix designed to produce a spatial derivative of the original data. Spatial derivatives of the scan data, rather than the values, are used to obtain the contours of the probe spots, because of the considerable differences found in background fluorescence resulting from fluorescence leakage from the counterstain. Once the probe spot contours are determined, the data within these contours is summed to determine the *probe total fluorescence* feature. The number of pixels within each contour is used as its *probe area* feature. The maximum pixel value within the probe spot contour is used as *probe maximum pixel* feature.

A second alternative to compartmentalizing the data to estimate cytoplasmic constituents is also available. An algorithm is included in the software to sum pixel values between two contours of user determined spacing set in position outside the data contour. The sum of these pixel values determines the *extranuclear* feature and can be used to estimate constituents outside the nuclear membrane of each cell if contouring was made using nuclear fluorescence.

A list of feature values is computed for each cell found by the software and stored as a flow cytometry standard (FCS) format list mode file on the PC disk. Any of these features or the ratio of any two features can be used as the axes of histogram, scattergram, or isometric displays. This list contains the following feature values:

For the sensor used for contouring:

1. *Total value:* The corrected sum of the pixel values in the data contour—equivalent to the FCM constituent value,
2. *Extranuclear total value:* The sum of pixel values in an area outside of and surrounding the data contour,
3. *Max pixel:* The maximum pixel value within the data contour,
4. *Area:* The area of the thresholding contour,
5. *Perimeter:* The perimeter of the thresholding contour,
6. *Texture:* A measure of the variation of local pixel values,
7. *X Position, Y Position, Scan position:* The slide position and scan position of the maximum pixel value of the event,
8. *Time:* The computer clock time when the event was measured,
9. *Spot Count:* The number of probe spots within the data contour of the cell,

10. *Annotation:* A binary valued annotation feature that the user determines as cells are relocated and visually observed used as a gating criterion,
11. *Multiple cell exclusion:* The structure of the data within the contour is analyzed to determine if the event represents a single or a multiple cell.

For every other sensor:

1. *Total value:* The corrected sum of the pixel values in the data contour,
2. *Extranuclear total value:* The sum of pixel values in an area outside of and surrounding the data contour,
3. *Max pixel:* The peak value within the data contour,
4. *Texture:* Measure of the variation of local pixel values.

For every probe spot:

1. *Total value:* The sum of all pixel values in the probe contour,
2. *Max pixel:* The maximum pixel value within the probe contour,
3. *Area:* The area of the probe contour.

As in FCM, any number of windows containing scattergrams of any two features, or histograms of one feature can be displayed. Feature ratios can also be used as well. Because of the large number of features available to LSCM, any number of gating regions can be drawn on scattergrams or histograms, and other scattergrams or histograms can be designated to inherit data from a region. Both gating regions and relationships are established by using the computer mouse. In this way complex data selections involving any sets of features can be developed and used to display subsequent scattergrams, isometric displays, or histograms or to select events for statistical summaries, cell cycle analysis, or cell relocation and visualization.

The WinCyte LSC software, written for Microsoft Windows 3.1 and NT4, includes the capability to generate, display, and print reports with displays, statistics, and galleries of cell images inserted dynamically into user designed report forms as data is generated from a specimen. All instrument operating parameters and display setup parameters are stored in user accessible protocol files. All list mode data files include headers describing how the instrument was set up. Data files follow the FCS format for compatibility with other analysis software.

IV. The Utility and Operational Characteristics of Some Laser Scanning Cytometry List Mode Features

A. Total Value Feature

As in FCM, total value, the corrected sum of all pixel values within the data contours, are used to report mass constituent values. As is done for FCM, sensitivity of laser scanning cytometry for constituent measurements is determined by running sets of particles of known equivalent molecules of each of fluorescence marker molecules such as fluorescein isothiocyanate (FITC), phycoerythrin (PE), or CY5. To evaluate LSC sensitivity, Spherotech Inc. (Libertyville, IL) RCP-30-5, 3.0-μ particles in suspension were placed on a standard microscope slide, cover slipped, and sealed with Cytoseal. The slide was run on the LSC, measuring forward angle scatter and fluorescence within three wavelength ranges corresponding to the dyes, FITC, PE, and CY5. Particles were found and contoured based on forward angle scatter. Single particles were isolated based on gating a region using the features particle peak scatter and area. Total fluorescence values from each of the three sensors were determined during each run. For each histogram, the mean value of the fluorescence of each cluster was determined using WinCyte statistics software. These values were plotted versus the Spherotech published molecules of equivalent dye molecule. The values are linearly related on a logarithmic plot. FCM instrument sensitivity has been characterized by extrapolating the plotted lines to the mean values of the blank particles and setting the molecules of equivalent dye of the blank values as the minimum detectable values. The LSC minimum detectable values using one scan per pixel digitization in each of the three wavelength ranges were approximately 500 equivalent fluorescein molecules, 200 equivalent phycoerytherin molecules, and 300 equivalent CY5 molecules. These equivalent molecule values can be reduced using the scan line addition feature with a lower event rate.

The sensitivity of the instrument can be improved assaying these calibration particles by using apertures in front of each of the photomultipliers, to establish confocal imaging for each fluorescence emission. However, confocal imaging reduces depth of field, increasing variation of constituent determinations somewhat. Computer calculated base 10 logarithmic plots of per particle fluorescence using apertures are shown in Fig. 6 in which the published mean equivalent fluorescence molecule values for each particle set are also shown.

A variety of cell culture and clinical specimens have been run to establish that LSCM determinations of per cell DNA are equivalent to FCM determinations and to evaluate tumors (Martin-Reay *et al.*, 1994; Sasaki *et al.*, 1996; Clatch and Walloch, 1996; Clatch *et al.*, 1997; Kamada *et al.*, 1997; Rew *et al.*, 1998; Kamiya *et al.*, 1999; Kawamura *et al.*, 2000). Results from a typical DNA determination are shown in Fig. 7 in which a mixture of diploid and triploid fibroblasts have been dropped on a slide, air dried, ethanol fixed, and stained with 5 μg/ml of PI and mounted in PBS buffered glycerol containing 5 μg/ml PI. The coefficients of variations (CV) and means of the distributions of single cells and cell clusters are shown in the statistics window. CV is reported by the full width half maximum (FWHM) method for cells. CV values between 2 and 3% can be achieved with cell culture specimens or particles such as FLOW-CHECK (Coulter Corporation, Miami, FL). It should be noted that each of the triploid and multiple cell mean DNA values fall on close multiples of the mean diploid DNA value. Satoh *et al.* (1999) have also used LSCM to study DNA content in individual chromosomes.

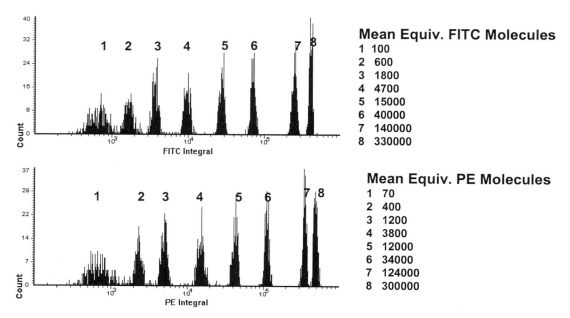

Fig. 6 Plots of per particle fluorescence of Spherotech Inc. RCP-30-5, 3.0-μm particles in suspension placed on a standard microscope slide and run on the LSC equipped with a field limiting aperture, measuring forward angle scatter and fluorescence within two wavelength ranges corresponding to the dyes, FITC and PE. Particles were found and contoured based on forward angle scatter. Single particles were gated based on the features particle peak scatter and area. The published mean equivalent fluorescence molecule values for each particle set are also shown. The computer using base 10 calculates fluorescence logarithmic values.

B. Maximum Pixel Value Feature

The maximum fluorescence value computed by selecting the brightest pixel within each cell contour is similar in concept to the pulse height of FCM; however, the increased resolution of laser scanning has been shown to provide unique utility primarily, a means of characterizing the phase of the cell cycle. This is illustrated in Fig. 8 which shows scattergrams of integrated fluorescence versus maximum pixel value of PI stained HeLa cells that had been grown on the slide. The slide was restained with hematoxylin and eosin after assaying the cells and replaced on the stage of the LSC for relocation of cells based on gating regions. Two gating regions were established at high values of maximum pixel, the first at diploid and the second at tetraploid DNA values. Cells with DNA and maximum pixel values within these regions were relocated to create the corresponding cell images of tetraploid mitotic cells and diploid newly divided cells. This result can be contrasted to Fig. 9, which shows scattergrams of integrated fluorescence versus cell texture and relocated cell images of the same cells gated on the feature

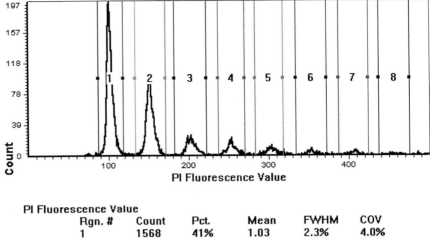

Fig. 7 DNA distribution of a mixture of diploid and triploid fibroblasts stained with 5 μg/ml of PI. The coefficients of variations and means of the distributions of single cells and cell clusters are shown in the statistics window. CV is reported by the full width half maximum (FWHM) method for cells.

high texture value. The maximum pixel value has been shown to provide for characterization of the phase of the cell cycle using a single dye, for a variety of cell types including tissue sections (Luther and Kamentsky, 1996; Kawasaki *et al.*, 1997; Gorczyca *et al.*, 1996; Juan and Darzynkiewicz, 1998). Bedner *et al.* (1997) to classify leukocytes, Musco *et al.* (1998) to differentiate cells with intracellular adenoviral infectivity and p53 protein expression, and Kawamura *et al.* (1999) to study PCNA localization during the cell cycle, have also used the maximum pixel feature.

C. Extranuclear Total Value Feature

Extranuclear integrated value is computed by adding pixel values within two concentric contours drawn outside of the thresholding contour. Its use is illustrated in the following example.

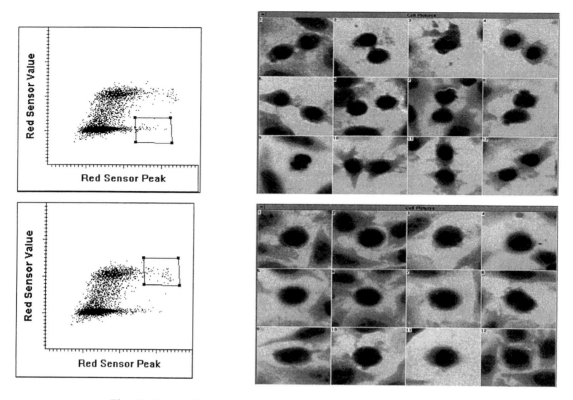

Fig. 8 HeLa cells were grown on a chamber slide, formalin/methanol fixed *in situ,* stained with PI, and assayed. Total PI fluorescence was plotted versus maximum pixel value, and two gating regions were drawn in areas corresponding to high maximum pixel values and either G_0/G_1 or G_2/M DNA values. After staining the slide with hematoxylin and eosin and replacing it on the LSC stage, cells were relocated and imaged using each gating region, resulting in the two CCD camera galleries.

As described by Luther and Kamentsky (1997), HeLa cells were grown on a chamber slide, formalin/methanol fixed *in situ,* and then stained with FITC conjugated antibodies to cyclin B1 and counterstained with propidium iodide (PI). This slide was assayed measuring PI red fluorescence and FITC green fluorescence. Cells were contoured based on PI fluorescence, and extranuclear integrated value was computed for each cell. A scattergram resulting from this determination is shown in Fig. 10. Cells were relocated based on the gating region shown on the scattergram resulting in the epi-illuminated fluorescence image gallery shown in the figure. Deptala *et al.* (1998) have used this feature to show activation of nuclear factor Kappa B (NF-κB) by measuring translocation of NF-κB from cytoplasm to nucleus. Deptala *et al.* (1999) have used this feature

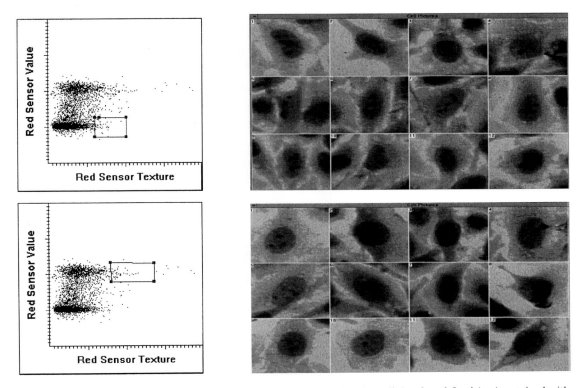

Fig. 9 HeLa cells were grown on a chamber slide, formalin/methanol fixed *in situ*, stained with PI, and assayed. Total PI fluorescence was plotted versus texture value, and two gating regions were drawn in areas corresponding to high maximum pixel values and either G_0/G_1 or G_2/M DNA values. After staining the slide with hematoxylin and eosin and replacing it on the LSC stage, cells were relocated and imaged using each gating region, resulting in the two CCD camera galleries.

to study the location of p53, p21^{WAF1}, and Bax in MCF-7 cells after induction of apoptosis.

D. Area Feature

The area feature equal to the number of pixels within the threshold contour of each event is similar to the pulse width feature of FCM. Its primary utility is to distinguish single cells from cell clusters as well as single probe spots from overlapping spots in FISH assays. Area versus maximum pixel scattergrams of many cell culture and clinical specimens show a characteristic pattern in which single cells tend to form clusters distinguishable from multiple cell events. The relative number of multiple events of any assay will depend greatly on cell density and the threshold level that is used to establish the contours for cell isolation.

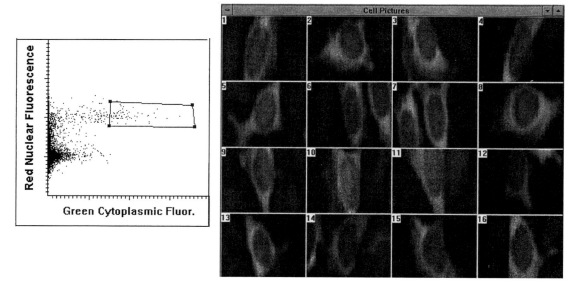

Fig. 10 HeLa cells were grown on a chamber slide, formalin/methanol fixed *in situ*, stained with FITC conjugated antibody to cyclin B1, counterstained with PI, and assayed. Total PI fluorescence was plotted versus extranuclear FITC fluorescence, and a gating region was drawn in an area corresponding to G_2/M DNA values and high extranuclear FITC total fluorescence values. Cells were relocated and imaged using the gating region, resulting in the CCD camera gallery of the microscope epi-illuminator excited images.

Figure 11 shows a scattergram of area versus maximum pixel fluorescence with a gating region believed to include single events, from an assay of HeLa cells stained with PI. Comparison of histograms corresponding to all events and events inherited from the gating region show the reduction in events equal to or greater than G_2/M DNA when area gating is used.

The area feature can also be used with the maximum pixel feature to reduce artifactual probe spot counts due to touching or overlapping spots in FISH assays.

Fig. 11 HeLa cells were grown on a chamber slide, formalin/methanol fixed *in situ*, stained with PI, and assayed. Area was plotted versus PI fluorescence maximum pixel. The "All Cells" scattergram and histogram of per cell total fluorescence show data without gating. The "Region Gated Cells" scattergram and histogram of per cell total fluorescence show the scattergram gating region used and the resulting per cell total fluorescence distribution. The "Multiple Cell Exclusion Gated Cells" scattergram and histogram of per cell total fluorescence show the scattergram and histogram resulting when the multiple cell exclusion feature is used for gating.

3. Laser Scanning Cytometry

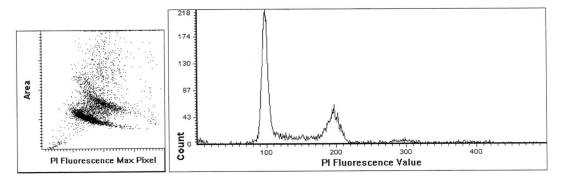

All Cells

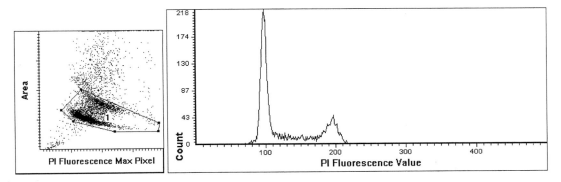

Region Gated Cells

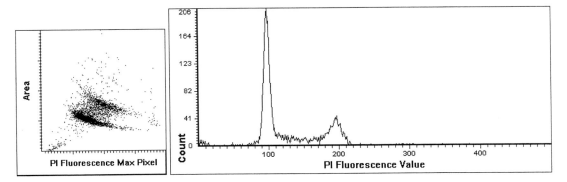

Multiple Cell Exclusion Gated Cells

Data from a study (Kamentsky et al., 1997b) was used for Fig. 12 and illustrates the result of a FISH assay of a mixture of human male and female lymphocytes counterstained with PI and hybridized with a centromeric probe to the X chromosome conjugated to FITC. Single cells were isolated by gating on area versus

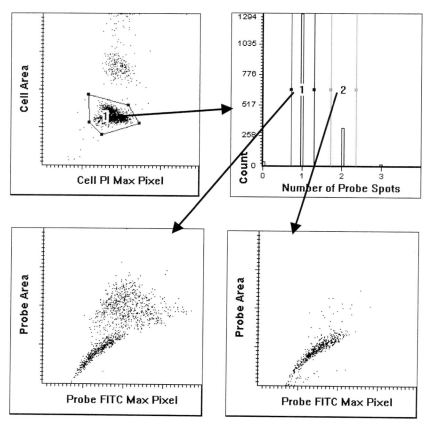

Fig. 12 A mixture of female and male human lymphocytes, using a FISH probe to the X chromosome conjugated to FITC and counterstained with PI, was assayed. Single cells are distinguished using the gating region in the first scattergram of per cell PI area versus per cell PI fluorescence maximum pixel. Single cells inherited from the gating region are displayed as a histogram of the feature per cell probe count. The one-count region of this histogram is gated to a scattergram displaying probe area versus per probe FITC fluorescence maximum pixel, whereas the second two-count histogram region is gated to a second identical scattergram display. The two-count region, which I believe resulted from female cells with two well-segmented spots, produced a probe scattergram with almost all probe values in a well-defined cluster. The one-count region produced a probe scattergram with most cells in the same well-defined cluster, which I believe resulted from the male cells, as well as values diffusely scattered outside this cluster, which I believe result from female cells with overlapping spots, and artifacts.

maximum pixel of PI fluorescence, as described previously. The inherited histogram of Fig. 12 shows the distribution of per cell spot count of these cells. Two gating regions were drawn on the histogram. The first region encompassing the one-count bar should include male cells with a single probe spot and female cells in which the two probe spots cannot be separated from each other. The second region about the two-count bar should include female cells in which the probe spots are separable. Two scattergrams of probe area versus probe maximum pixel, inheriting data from each of the histogram regions, are shown in the figure. The probe scattergram data from the one-count region shows both a compact dot cluster from the male cells and a diffuse area from the touching and overlapping spots of female cells. The probe scattergram inheriting data from the two-count region shows dots primarily within the compact cluster. The software provides a means of utilizing this finding to reduce artifactual counts resulting from spot overlap by gating cells to an inheriting scattergram only when *all* of its probe spots fall within a probe scattergram gating region.

E. Cell Position Feature

Relocation functionality has been used in four different ways. (1) A specimen can be assayed two or more times, changing laser excitation or stains between assays or for determining features in kinetic measurements, and data for the events of each assay combined by matching events based on their location in each assay. (2) Cells relocated based on distinguishing features may be viewed or their images captured in files immediately after an assay, or after staining the slide with chromatic dyes and replacing the slide on the stage. (3) A specimen containing cells with a known and measurable constituent difference may be assayed and scattergrams displayed in which Y position is plotted versus X position, to indicate the positions on the slide of these cells. A region can be drawn on this scattergram and used as a gating area for reanalysis of the assay data. (4) Special slides containing multiple wells may be used in a single assay with each well containing either a different specimen or a specimen with different reagents. The multiple wells can be serially measured, their data analyzed using position gating regions to partition the data from each well, and the results from the multiple wells combined in a report.

To demonstrate list file merging capability, a suspension mixture of chicken erythrocytes and calf thymocytes were stained with PI, dropped on to a slide, and cover slipped. Two assays were run measuring red PI fluorescence and green autofluorescence of the formalin fixed erythrocytes. The Fig. 13 scattergram of red versus green fluorescence of the first run shows a heterogeneous distribution for the two cell types and complexes formed of their various combinations. The two data files were merged into a single list mode file using the position of each event on the slide as the merge key to combine both data sets for every event. Figure 13 shows the run 1 versus run 2 red fluorescence scattergram and the run 1 versus run 2 green fluorescence scattergram of the combined data. Most events

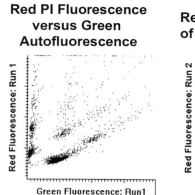
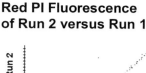
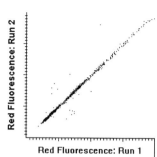
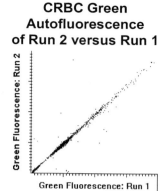

Fig. 13 A mixture of glutaraldehyde fixed calf thymocytes and chicken red blood cells (RBC) were dropped on a slide, stained with PI, and assayed twice. The scattergram from the first run of total PI fluorescence versus green fluorescence resulting from chicken RBC autofluorescence indicates the heterogeneity of the specimen. The data files from each run were merged into one list mode file using cell position as the merge key. The total PI fluorescence of run 2 versus run 1 and the total autofluorescence of run 2 versus run 1 show the high correlation of the data between the runs.

are highly correlated between runs 1 and 2. Utilizing merging of laser scanning data, Clatch and Foreman (1998) have demonstrated five-color immunophenotyping plus DNA content measurements in lymphocytes. Li and Daryzynkiewicz (1999) have used the merge capability to quantify, on the same cells, both functional properties of live cells as well as features requiring cell fixation.

Luther et al. (1996), have studied the application of an LSC to Cytyc thinprep prepared slides of various pleural effusions and fine needle aspirates of cancers of various organs. The objective of this study was to assist in characterizing tumors by providing pathologists with images of cells with abnormal ploidy. Slides were stained with PI for DNA as well as AE1/AE3 conjugated to FITC to measure cytokeratins. Galleries of cell images were obtained by selecting cells with higher values of both DNA and cytokeratin. The images shown in Fig. 14,

Fig. 14 A Cytyc thinprep prepared slide of a FNA of a kidney from a study by Luther et al. (1996) of a patient with renal cell carcinoma. The slide was stained with PI for DNA as well as the AE1/AE3 antibody conjugated to FITC to measure cytokeratins. Shown are images generated by relocating cytokeratin positive cells with high DNA shown on each image. The first gallery of photonegative fluorescent images show selected cells imaged using the LSC's epi-illumination excitation of the PI stain. The next set of photo-positive images was obtained by staining the slide with hematoxylin and eosin, replacing the slide on the stage of the LSC, and relocating the same cells. All but one cell remained on the slide.

3. Laser Scanning Cytometry

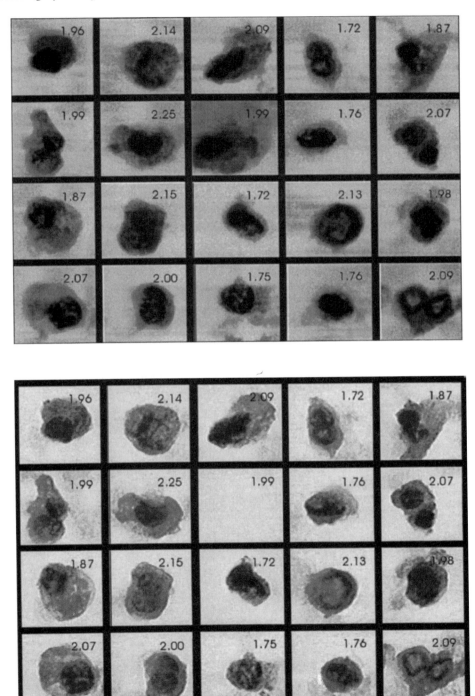

from this study, are from a FNA specimen of kidney renal cell carcinoma. The upper gallery of photo-negative CCD camera fluorescence images was obtained during the assay using epi-illumination of the PI stain. The lower set of photo-positive bright field images was obtained by removing and staining the slide with hematoxylin and eosin, replacing the slide on the instrument stage, and relocating the same cells as noted earlier. All but one cell remained on the slide, and the bright field images of the restained set of relocated cells correspond with their fluorescent images. The relocation feature has been used in studies to resolve ambiguities in data by identification of cancer cells (Clatch et al., 1997; Reeve and Rew, 1997), for identification of specific cells in sputum (Woltmann et al., 1999), for verification of apoptotic cells (Furuya et al., 1997; Bedner et al., 1999a; Darzynkiewicz et al., 1998; Darzynkiewicz and Bedner, 1999; Gorczyca et al., 1998), to study the localization of cyclin B1 during the cell cycle (Kakino et al., 1996), and to show binding of fluorescein isothiocyanate to eosinophils (Bedner et al., 1999b).

The cell position feature can also be used in scattergram displays. Fluorescence features can be displayed versus X or Y position to control the quality of the staining procedure. Improper staining of constituents such as DNA, for which many cells have a constant value, can result in uneven total fluorescence curves when plotted against position. The position feature can also be used to display the location on the slide of events having specific feature sets by gating from regions of scattergrams of fluorescence or scatter features to a display of Y position versus X position. An example of application of this technique is illustrated using data derived from a study by Gorczyca et al. (1998) of estrogen receptors in histology tissue sections from breast cancer patients. Five-micrometer sections were cut from paraffin embedded, formalin fixed tissue stained with FITC conjugated antibody to estrogen receptor or to an irrelevant antibody, and counterstained with PI. PI fluorescence was used to find and contour cellular events. The distributions of the ratio of per cell FITC to PI fluorescence of an ER stained section and a control antibody stained section are shown in Fig. 15. The positions of all events of the ER stained section are shown in the scattergram, and the positions of events with high FITC/PI fluorescence inherited from the higher value histogram gating region are also shown as darker dots, indicating those regions of the scanned section with high ER. This technique may be further extended to partition areas of tissue or touch specimens for subsequent analysis of other data features. This technique has been used by Grace et al. (1999) to determine the depth of penetration of apoptosis resulting from adenovirus p53 gene therapy in human xenograft biopsies.

Clatch et al. (1996, 1997, 1998) have developed techniques using specially prepared slides containing multiple wells for immunophenotyping profiles of human lymphocytes. As shown in Fig. 16, well slides are prepared by sandwiching a 200-μ thick mylar sheet, containing bonding material on each of its sides and cut into the pattern as shown in the figure, between a microscope slide and two coverslips. The 12 chambers formed by the mylar sheet are each 2 mm wide by

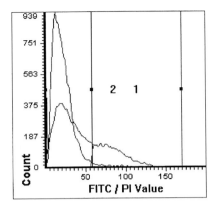 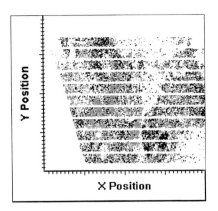

Fig. 15 Data from a study by Gorczyca *et al.* (1998) of estrogen receptors (ER) in histology tissue sections from breast cancer patients. Five-micrometer sections were cut from paraffin embedded, formalin fixed tissue stained with FITC conjugated antibody to estrogen receptor and counterstained with PI. The ratio of per cell FITC to PI fluorescence is shown for an isotype control and estrogen receptor antibody. The position of each event of the ER stained section is shown in the scattergram, and the position of events with high per cell FITC/PI fluorescence using the histogram gating region is also shown as darker dots, indicating areas of the scanned section with high ER.

2.5 cm in length. They are each filled with the same ficoll-hypaque density centrifugation, or ammonium chloride red cell lysed prepared samples. Clatch has shown that the lymphocytes of the specimen will adhere to the microscope slide surface after a 15-min incubation period. After incubation, 15 μl of a mixture of different combinations of two antibodies conjugated to either FITC, or PE, and a third antibody CD45 conjugated to PE/CY5, is added to each well displacing the liquid. This is accomplished by wicking one end of the chamber as the antibody mixture is added to each chamber. After a 30-min incubation the chambers are simultaneously washed with 50 μl of PBS. The slide is then placed on the stage of the LSC cytometer programmed to scan each of the 12 chamber areas in turn, measuring light scatter and three fluorescence wavelengths and detecting and contouring events based on light scatter. After the assay, the data is analyzed by first isolating lymphocytes by gating using scatter area versus total scatter, then gating using scatter area versus CD45 total fluorescence. The data inherited from this gate is displayed as a scattergram of Y versus X position. Twelve gating regions have been drawn on this display to correspond to positions of each of the 12 chamber areas. The resulting data falling in each of the 12 regions is inherited by each of 12 different spectral overlap compensated logarithmic scale scattergrams displaying FITC total fluorescence versus PE total fluorescence. The resulting profiles of a typical analysis, shown in Fig. 16, were obtained from R. Clatch. We have also used this technique with rectangular chamber slides with each chamber containing either different specimens stained alike or the same specimen treated differently.

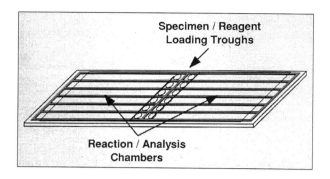

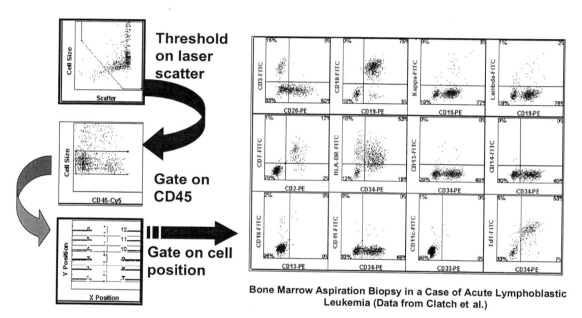

Fig. 16 Well slides are prepared for immunophenotyping by sandwiching a patterned 200-μm-thick mylar sheet between a microscope slide and two coverslips, creating 12 chambers, 2 mm wide by 2.5 cm in length. Specimens are prepared in the wells and assayed as described by Clatch *et al.* (1998). The data (Clatch *et al.,* 1998) are analyzed by first isolating lymphocytes by gating on area and total scatter, then scatter area versus CD45 total fluorescence, followed by gating on position to delineate the 12 well areas. The resulting panel of 12 scattergrams of logarithmically displayed compensated FITC and PE total fluorescence is shown.

F. Time Feature

Flow cytometry has been used for kinetic studies of populations in which some characteristic of the population is measured as a function of time. However, in FCM it is not possible to follow the time course of individual cells in heterogeneous populations. Because each cell can be identified by its location, if the cells are stationary, the kinetics of each cell of heterogeneous populations of cells can be quantified, and specific cells with specific constituent kinetic behaviors can be identified. These cells can be relocated and observed or other properties can be displayed. This technique has been used by Bedner *et al.* (1998) to measure the kinetic activity of L-aminopeptidase and esterases, and data obtained from their laboratory is used in the following example to demonstrate the utility of the time feature.

In this example, 50 μl of a suspension of HL-60 cells at a concentration of 1×10^6 cells per ml were placed in a microscope slide well prepared from parafilm. Maintained at room temperature, the cells were allowed to adhere to the slide surface, the slide was washed with HBSS, and 25 μl of the substrate di-(leucyl)-rhodamine 110 was added, resulting in green fluorescence from the activity of L-aminopeptidase. The slide was immediately placed on the stage of the LSC programmed to measure light scatter and green fluorescence of approximately 400 cells, then to repeat the measurements of the same cells, immediately afterwards or at fixed time intervals. Data was contoured using light scatter, and values of green total fluorescence were displayed versus time in scattergrams such as shown in Fig. 17. Two scattergrams are shown, the first of

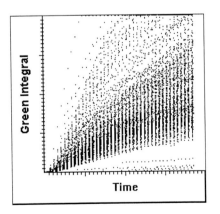 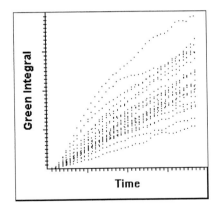

Fig. 17 L-Aminopeptidase activity of HL-60 cells was measured by Bedner *et al.* (1999a) using the substrate di-(leucyl)-rhodamine 110. The slide was placed on the stage of the LSC programmed to measure light scatter and green fluorescence of approximately 400 cells, then to repeat the measurements of the same cells. The scattergrams from their data file show total fluorescence versus time of all cells assayed and of a smaller number of cells selected using an X versus Y position scattergram, to better show the heterogeneity of the individual cell trajectories.

all cells assayed and, to better show the heterogeneity of the individual cell trajectories, a smaller number of cells were selected using an X versus Y position scattergram. It would be possible to gate events with specific kinetic properties to observe them or to display additional characteristics of these selected cells.

G. Probe Spot Counting Feature

The LSC has been used for cytogenetic studies with metaphase chromosomes and interphase FISH specimens. Numa *et al.* (1996), Satoh *et al.* (1996), and Sakamoto *et al.* (1996) have reported using the total fluorescence feature of PI stained dispersed chromosome spreads to demonstrate resolution of the DNA differences of individual chromosomes. They have applied this technique to study chemically induced mutations. Kamentsky *et al.* (1997b) have described studies using the LSC to count interphase FISH probe spots.

As previously described, the LSC analyzes FISH preparations or other specimens containing localized constituents that can be independently stained and sensed, by constructing a second set of contours within the cell contours based on this second sensor signal. Figure 18 shows sensor data from a run of mixed diploid and triploid fibroblasts stained with PI and hybridized with a FISH probe to chromosome 8 conjugated to Spectrum Green (Vysis, Downers Grove, IL). Laser scan images of the green sensor data, measuring probe Spectrum Green fluorescence, are shown without and with contouring of the PI stained cell nuclei and the FITC conjugated probe spots.

The probe spot count feature relies on use of a counterstain such as PI to segment scan data by contouring pixels of each cell nucleus. The array of probe fluorescence data within each nuclear contour is then convoluted with a specific matrix described by Kamentsky *et al.* (1997b) to enhance the image contrast between each probe and its local background. The transformed data is then contoured with a fixed threshold value. The original data within this contour is used to generate the *total fluorescence area,* and *maximum pixel* features of each probe spot. As previously described, in order to reduce artifactual spot counts do to overlapping images of probes, the features *probe area* versus *probe maximum pixel* are displayed on a scattergram. As previously described, spot counts are recorded only on cells with all probe spots within a specific gating region of the scattergram. Figure 18 also shows a two-dimensional histogram of probe spot count versus PI total fluorescence resulting from an assay of this mixture of diploid and triploid fibroblasts prepared with a probe to the centromeric region of chromosome 8 conjugated to Spectrum Green and counterstained with PI.

The capability to measure features of events resulting from multiple concentrations of a constituent stained with one fluorochrome occurring within a larger area event resulting from a second constituent stained with a different fluorochome, can be used for other types of assays. An example of this has been described by Bedner *et al.* 2000, in which number of cells, DNA, p53 expression, and estrogen receptor expression of individual cells within clusters of cells growing on a slide were reported.

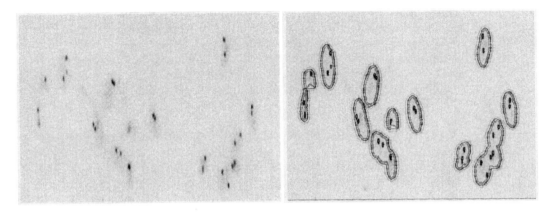

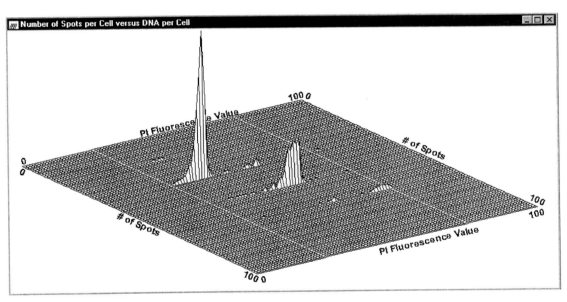

Fig. 18 Images produced from a mixture of diploid and triploid fibroblasts prepared using a FISH probe to chromosome 8 conjugated to Spectrum Green and counterstained with PI. The brightness at each pixel position of the top left image shows the green fluorescence from Spectrum Green staining probes and background fluorescence of the PI stained nuclei. The right image shows the contours used to segment the cell nuclei and to segment the probes. A two-parameter histogram of spot count versus DNA per cell from an assay of this specimen is shown below.

H. Multiple Cell Exclusion Feature

For most cells, single cell events should create data that decreases in pixel value monotonically from the maximum value to the data contour. To determine the *multiple cell exclusion* feature, pixel data within the data contour of each cell is analyzed for saddles as a function of pixel position. If saddles are found in the data, the event is considered to result from multiple cells, and the event is tagged as a multiple cell event. Shown in Fig. 11 are per cell DNA histograms of an assay of PI stained HeLa cells, comparing histograms of nonselected events with events selected using a gating region on an area versus maximum pixel scattergram, and events gated using the multiple cell exclusion feature.

V. Utility of Solid Phase Cytometry for Cell Preparation

I will next describe additional differences between LSCM and FCM that are related to the utilization of solid phase rather than liquid phase analysis with LSCM. It should be noted that LSCM can analyze samples prepared in liquid phase that have been pipetted on to a slide and coverslipped. However, LSCM, because it measures cells on a solid substrate, can be used to simplify specimen preparation.

In many assays, including all of those that use heterogeneous specimens such as blood, the cells of interest for the assay may be a small fraction of the total cells of the specimen. The presence of irrelevant cells will increase FCM processing time to analyze a given number of cells of interest if irrelevant cells must be measured and analyzed. In liquid phase FCM preparations, irrelevant cells are removed by lysis or separated by centrifugation. For LSCM larger scan areas would be required to place all the cells on a surface without overlap of the cells, requiring lengthy scan times. Using a small scan area, the specimen itself will have multiple layers of irrelevant cells present overlaying the cells of interest and interfering with the measurement process itself. It is advantageous if only cells of interest remain in the area to be scanned and irrelevant cells are removed prior to measurement.

I have described how Clatch *et al.* (1996, 1997, 1998) have used cell adhesion to a slide surface to separate lymphocytes from other blood components. I have used techniques based on application of magnetic particles or pores in a membrane to place relevant cells extracted from a heterogeneous population on to a small area of a surface for efficient laser scanning. Next, use of these techniques is illustrated in a whole blood assay determining the activation of granulocytes by the agonist f-Met-Leu-Phe (fMLP).

Two samples of the same whole blood specimen were obtained. To one sample at 37°C, 10 μM fMLP was added and that sample was incubated for 10 min to cause activation of the granulocytes of the sample. Both samples were mixed 1:1 with PBS containing two mouse antihuman antibodies. CD15 was conjugated to FITC and binds to granulocytes, and CD11b which binds to an activation antigen of granulocytes was conjugated to the dye PE. After an incubation period, paraformaldehyde was added followed by addition of magnetic particles coated with an antimouse antibody (Polysciences, Inc., Warrington, PA). Each mixture was incubated and then added to one end of a flow chamber followed by a PBS wash. The flow chamber was constructed by cementing two 200-μ thick mylar spacers between a standard microscope slide and coverslip to define a 5 mm wide channel. A bar magnet was cemented below the slide to create a magnetic field perpendicular to the fluid flow. As the specimen or wash was pipetted into one end of the chamber, excess fluid was wicked from the other end. After separation, each chamber slide was placed on the LSC stage and assayed, detecting and contouring events based on green fluorescence from the FITC bound to the granulocytes. The orange PE fluorescence resulting from the binding of CD11b to the granulocyte activation antigen was measured. A scattergram of the positions of each event (Fig. 19) shows that most events are located at the edges of the magnet where the magnetic field gradient is largest. The superimposed histogram display of CD11b PE fluorescence per cell from assays of these two specimens is also shown.

Another two samples of a whole blood specimen were prepared as earlier, except both samples were mixed 1:1 with PBS containing CD15 conjugated to PE and CD11b conjugated to PECy5. After an incubation period, 100 μl of the specimen was pipetted into the center of an Osmotics Catalog 10572, 5 μm pore size 13-mm filter placed over absorbent material. The pipetting of 200 μl of PBS on to the membrane center surface immediately followed this. The filter was removed with absorber and placed on the stage of an LSC equipped with a 543-mμ emitting HeNe laser and assayed. Granulocytes were detected and contoured based on orange fluorescence from the PE bound to the granulocytes.

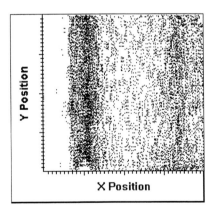
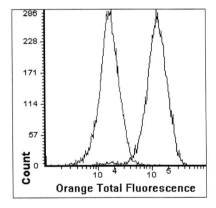

Fig. 19 Two samples of the same blood specimen with and without the activator 10 μM fMLP were prepared as described with CD15 antibody conjugated to FITC that binds to granulocytes and CD11b antibody conjugated to PE that binds to activated granulocytes. Antimouse antibody conjugated magnetic particles were then added and each mixture was pipetted into one end of a 5 mm × 0.2 mm flow chamber, followed by a PBS wash. A bar magnet was cemented below the chamber to create a magnetic field. The specimens were assayed by detecting and contouring on granulocyte green fluorescence. A scattergram of the positions of each event shows that most events are located at the edges of the magnet where the magnetic field gradient is largest. The superimposed per cell PE orange total fluorescence histograms of the activated and control specimen are also shown.

The red fluorescence resulting from the binding of CD11b to the granulocyte activation antigen was measured for every contoured event. The position of each granulocyte on a filter is displayed in Fig. 20 with resulting superimposed LSC histogram displays of CD11b-PeCy5 fluorescence per cell, for the resting and activated specimen.

VI. Future Directions

I have described LSCM and a specific laser scanning cytometer, the LSC, conceived to provide multiparameter cellular fluorescence and scatter data, that is comparable to data from many flow cytometry measurements and can thus utilize much of the methodology developed for FCM since the 1970s. Because LSCM automatically measures cells on slides not in a flow stream, can determine cell position on a slide, and is not zero resolution, I was able to show how it may provide additional measurement features, additional ways of preparing specimens, and allow cell relocation for visual observation. Current LSCM utility will be expanded as it has for FCM by providing additional sensors for light scatter, providing other laser excitation wavelengths particularly ultraviolet excitation, resolving fluorescence lifetimes, and increasing cell analysis rates. Because fluorescence from each cell is excited over an array of many points per cell, excitation can be more uniform than that of the single Gaussian excitation profile

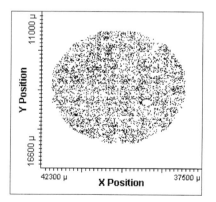
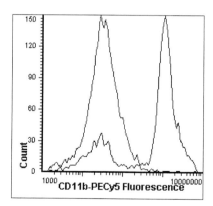

Fig. 20 Two samples of the same blood specimen with and without the activator 10 μM fMLP were prepared as described so as to react with CD15 antibody conjugated to PE that binds to granulocytes and CD11b antibody conjugated to PeCy5 that binds to activated granulocytes. Each mixture was pipetted on to a 5 μ pore size membrane filter with absorbent below, followed by a PBS wash. The filters on the absorber were assayed by detecting and contouring on granulocyte orange fluorescence. A scattergram of the positions of each event shows their distribution on the filter surface. The superimposed per cell PECy5 total fluorescence histograms of the activated and control specimen are also shown.

of FCM, and LSCM should in principal provide better measurement accuracy. The present limits on LSCM accuracy will be studied and likely reduced as they have been for FCM.

Because excitation energies appear to be at or near saturation for many dyes and the total amount of time spent illuminating each cell is greater than FCM, sensitivity should be better than FCM. This is not the case presently because of the additional optical elements currently required for LSCM. We are studying very different optical designs in order to eliminate those optical elements known to increase background fluorescence in order to substantially decrease minimum detectable fluorescence.

The unique utility provided by solid rather than liquid phase assays will be further exploited. The ability to prepare and assay specimens on the same solid phase can form the basis for simple to use clinical point of care cytometric instruments because, as in clinical immunodiagnostic instrumentation, required separation steps can be eliminated and easy to use disposable devices for specimen containment can be employed. Additionally, control specimens can be added to the device and these assayed with the same preparation and measurement conditions as the specimen.

LSCM is able to relocate and provide images of cells meeting specific criteria among its feature set values. High quality CCD cameras and image analysis software will be used to generate features of higher resolution image data, providing a more comprehensive feature set for selected cells. This could provide additional data for decision algorithms such as neural nets and could be used for more precise cell classification.

My original impetus for developing a laser scanning cytometer was to solve the problem of making kinetic measurements of subsets of cells in heterogeneous populations. This resulted from our unsuccessful earlier attempt to measure the transmembrane potential of subsets of activated lymphocytes (Shapiro *et al.*, 1979). It did not work because we needed to assay fluorescence in the same individual cells at different times to account for intrinsic variation of initial fluorescence among the cells. Bedner *et al.* (1998) have shown it is possible to use the LSC to measure and display the kinetic behavior of individual cells. Methods to stabilize and add reagents to preparations, track cell movement if necessary, and display kinetic data should be developed to exploit this capability.

One major benefit of FCM is its capability for high discrimination preparative sorting of specific cells in heterogeneous populations. Techniques have been described more recently for isolating and retrieving specific cells from tissue using a laser that can be visually controlled to precisely position the laser beam. In the first technique, a tissue section is covered with a film and cells of interest are made to adhere to the film by means of a focused laser pulse to cause fusion (Emmert-Buck *et al.*, 1996). Other techniques use a laser to obliterate material surrounding a cell of interest. Adapting techniques such as these to LSCM so that relocated cells can be viewed then isolated for subsequent genetic analysis would enhance the utility of laser scanning cytometry.

Finally, there are some potential applications of the capability of the laser scanning cytometry to analyze cells on microscope slides that have not to my knowledge been adequately explored or software has not been completed to make their application feasible. Among these are use of *in situ* amplification techniques such as PCR for DNA sequence multiplication and a variety of reporter amplification techniques. Using appropriate dyes it should be possible to extend reported single probe analyses to two or more probes, to develop techniques to assay for translocations by adding features related to the distance between probe spots, and to combine genotypic and phenotypic markers.

Acknowledgments

The LSC cytometer evolved through the joint efforts of many individuals whom I would like to acknowledge. Lee Kamentsky was responsible for writing the WinCyte software and conceived many of its analysis techniques. Richard Clatch, Edmund Cibas, Zbiginiew Darzynkiewicz, Jonathan Fletcher, Myron Melamed, and Kohsuke Sasaki and their staff provided many suggestions, provided specimens, and conducted studies that aided the development. The LSC instrument was originally designed jointly with Olympus Optical Co., Ltd., Japan engineers led by Hisao Kitagawa. The CompuCyte LSC instrument was designed jointly with Douglas Burger, Russell Gershman, Richard Parker, and Arthur Daniels. Ed Luther provided much of the biology support during its evolution and generated many of the applications described in this review. Finally, I acknowledge Ann Byrne's assistance with specimen preparation and interpretation of data.

References

Bedner, E., Gorczyca, W., Melamed, M., and Darzynkiewicz, Z. (1997). Laser scanning cytometry distinguishes lymphocytes, monocytes and granulocytes by differences in their chromatin structure. *Cytometry* **29,** 191–196.

Bedner, E., Melamed, M. R., and Darzynkiewicz, Z. (1998). Enzyme kinetic reactions and fluorochrome uptake rates measured in individual cells by laser scanning cytometry. *Cytometry* **33,** 1–9.

Bedner, E., Li, X., Gorczyca, W., Melamed, M. R., and Darzynkiewicz, Z. (1999a). Analysis of apoptosis by laser scanning cytometry. *Cytometry* **35,** 181–195.

Bedner, E., Halicka, H., Cheng, W., Salomon, T., Deptala, A., Gorczyca, W., Melamed, M., and Darzynkiewicz, Z. (1999b). High affinity binding of fluorescein isothiocyanate to eosinophils detected by laser scanning cytometry: A potential source of error in analysis of blood samples utilizing fluorescein-conjugated reagents in flow cytometry. *Cytometry* **36,** 77–82.

Bedner, E., Ruan, Q., Chen, S., Kamentsky, L. A., and Darzynkiewicz, Z. (2000). Multiparameter analysis of progeny of individual cells in clonogenicity assays by laser scanning cytometry (LSC). *Cytometry,* in press.

Burger, D., and Gershman, R. (1988). Acousto-optic laser scanning cytometer. *Cytometry* **9,** 101–110.

Clatch, R. J., and Foreman, J. R. (1998). Five-color immunophenotyping plus DNA content, analysis by laser scanning cytometry. *Cytometry* **34,** 36–38.

Clatch, R. J., and Walloch, J. L. (1996). Sensitivity of laser scanning cytometric DNA content analysis in detecting minor subpopulations of aneuploid tumor cells. *Anal. Quant. Cytol. Histol.* **18,** 64.

Clatch, R. J., and Walloch, J. L. (1997). Multiparameter immunophenotypic analysis of fine needle aspiration biopsies and other hematologic specimens by laser scanning cytometry. *Acta Cytol.* **41,** 109–122.

Clatch, R. J., Walloch, J. L., Zutter, M. M., and Kamentsky, L. A. (1996). Immunophenotypic analysis of hematologic malignancy by laser scanning cytometry. *Am. J. Clin. Pathol.* **105,** 744–755.

Clatch, R. J., Walloch, J. L., Foreman, J. R., and Kamentsky, L. A. (1997). Multiparameter analysis of DNA content and cytokeratin expression in breast carcinoma by laser scanning cytometry. *Arch. Pathol. Lab. Med.* **121,** 585–592.

Clatch, R. J., Foreman, J. R., and Walloch, J. L. (1998). Simplified immunophenotypic analysis by laser scanning cytometry. *Cytometry* **34,** 3–16.

Darzynkiewicz, Z., and Bedner, E. (1999). Analysis of apoptotic cells by flow and laser scanning cytometry. In "Methods in Enzymology: Methods in Apoptosis Research." (J. C. Reed, ed.), in press. Academic Press, San Diego.

Darzynkiewicz, Z., Bedner, E., Traganos, F., and Murakami, T. (1998). Critical aspects in the analysis of apoptosis and necrosis. *Hum. Cell* **11,** 3–12.

Darzynkiewicz, Z., Bedner, E., Li, X., Gorczyca, W., and Melamed, M. R. (1999). Laser scanning cytometry. A new instrument with many applications. *Exp. Cell Res.* **249,** 1–12.

de Grooth, B. G., Geerken, T. H., and Greve, J. (1985). The Cytodisk: A cytometer based upon a new principal of cell alignment. *Cytometry* **6,** 226–233.

Deptala, A., Bedner, E., Gorczyca, W., and Darzynkiewicz, Z. (1998). Activation of nuclear Kappa B (NF-κB) assayed by laser scanning cytometry. *Cytometry* **33,** 376–382.

Deptala, A., Li, X., Bedner, E., Cheng, W., Traganos, F., and Darzynkiewicz, Z. (1999). Differences in induction of p53, p21 WAF1, and apoptosis in relation to cell cycle phase of MCF-7 cells treated with camptothecin. *Int. J. Oncol.* **15,** 861–871.

Deutsch, M., and Weinreb, A. (1994). Apparatus for high-precision repetitive sequential optical measurement of living cells. *Cytometry* **16,** 14–26.

Dietz, L. J., Dubrow, R. S., Manian, B. S., and Sizto, N. L. (1996). Volumetric capillary cytometry: A new method for absolute cell enumeration. *Cytometry* **23,** 177–186.

Emmert-Buck, M. R., Bonner, R. F., Smith, P. D., Chuaqui, R. F., Zhuang, Z., Goldstein, S. R., Weiss, R. A., and Liotta, L. A. (1996). Laser capture microdissection. *Science* **274,** 998–1001.

Furuya, T., Kamada, T., Murakami, T., Kurose, A., and Sasaki, K. (1997). Laser scanning cytometry allows detection of cell death with morphological features of apoptosis in cells stained with PI. *Cytometry* **29,** 173–177.

Gorczyca, W., Melamed, M. R., and Darzynkiewicz, Z. (1996). Laser scanning cytometer (LSC) analysis of labeled mitoses (FLM). *Cell Prolif.* **29,** 539–547.

Gorczyca, W., Darzynkiewicz, Z., and Melamed, M. R. (1997). Laser scanning cytometry in the pathology of solid tumors. *Acta Cytol.* **41,** 98–108.

Gorczyca, W., Davidian, M., Gherson, J., Ashikari, R., Darzynkiewicz, Z., and Melamed, M. R. (1998). Laser scanning cytometry quantification of estrogen receptors in breast cancer. *Anal. Quant. Cytol. Histol.* **20,** 470–476.

Gorczyca, W., Bedner, E., Burfeind, P., Darzynkiewicz, Z., and Melamed, M. R. (1999). Analysis of apoptosis in solid tumors by laser scanning cytometry. *Mod. Pathol.* **11,** 1052–1058.

Grace, M. J., Xie, L., Musco, M. L., Cui, S., Gurnani, M., DiGiacomo, R., Chang, A., Indelicato, S., Syed, J., Johnson, R., and Nielsen, L. L. (1999). The use of laser scanning cytometry to assess depth of penetration of adenovirus p53 gene therapy in human xenograft biopsies. *Am. J. Pathol.* **155,** 1869–1878.

Juan, G., and Darzynkiewicz, Z. (1998). Cell cycle analysis by flow and laser scanning cytometry. In "Handbook of Cell Biology" (J. Celis, ed.), 2nd Edition, pp. 261–273. Academic Press, San Diego.

Kakino, S., Sasaki, K., Kurose, A., and Ito, H. (1996). Intracellular localization of cyclin B1 during the cell cycle of glioma cells. *Cytometry* **24,** 49–54.

Kamada, T., Sasaki, K., Tsuji, T., Todoroki, T., Takahashi, M., and Kurose, A. (1997). Sample preparation from paraffin-embedded tissue specimens for laser scanning cytometric DNA analysis. *Cytometry* **27,** 290–294.

Kamentsky, L. A., and Kamentsky, L. D. (1991). Microscope-based multiparameter laser scanning cytometer yielding data comparable to flow cytometry data. *Cytometry* **12,** 381–387.

Kamentsky, L. A., Burger, D. E., Gershman, R. J., Kamentsky, L. D., and Luther, E. (1997a). Slide-based laser scanning cytometry. *Acta Cytol.* **41,** 123–143.

Kamentsky, L. A., Kamentsky, L. D., Fletcher, J. A., Kurose, A., and Sasaki, K. (1997b). Methods for automatic multiparameter analysis of fluorescence in situ hybridized specimens with a laser scanning cytometer. *Cytometry* **27,** 117–125.

Kamiya, N., Yokose, T., Kiyomatsu, Y., Fahey, M. T., Kodama, T., and Mukai, K. (1999). Assessment of DNA content in formalin-fixed, paraffin-embedded tissue of lung cancer by laser scanning cytometer. *Pathol. Int.* **49,** 695–701.

Kawamura, K., Kobayashi, Y., Tanka, T., Ikeda, R., Fujikawa-Yamamoto, K., and Suzuki, K. (1999). Intranuclear localization of proliferating cell nuclear antigen during the cell cycle in renal cell carcinoma. *Anal. Quant. Cytol. Histol.* **22,** 107–113.

Kawamura, K., Tanaka, T., Ikeda, R., Fujikawa-Yamamoto, K., and Suzuki, K. (2000). DNA ploidy analysis of urinary tract epithelial tumors by laser scanning cytometry. *Anal. Quant. Cytol. Histol.* **22,** 26–30.

Kawasaki, M., Sasaki, K., Satoh, T., Kurose, A., Kamada, T., Furuya, T., Murakami, T., and Todoroki, T. (1997). Laser scanning cytometry (LSC) allows detailed analysis of the cell cycle in PI stained human fibroblasts (TIG-7). *Cell Prolif.* **3–4,** 139–147.

Li, X., and Darzynkiewicz, Z. (1999). The Schrödinger's cat quandry in cell biology: Integration of live cell functional assays with measurements of fixed cells in analysis of apoptosis. *Exp. Cell. Res.* **249,** 404–412.

Luther, E., and Kamentsky, L. A. (1996). Resolution of mitotic cells using laser scanning cytometry. *Cytometry* **23,** 272–278.

Luther, E., and Kamentsky, L. A. (1997). Laser scanning microscopy applied to studies of the cell cycle. *Proc. Microsc. Microanal.* **1997,** 235–236.

Luther, E., Kamentsky, L. A., and Cibas, E. S. (1996). Laser scanning cytometry in the cytology laboratory. *Anal. Quant. Cytol. Histol.* **18,** 76.

Mansberg, H. P., and Ohringer, P. (1969). Design considerations for electronic and electromechanical flying spot scanners. *Ann. N. Y. Acad. Sci.* **157,** 5–37.

Martin-Reay, D. G., Kamentsky, L. A., Weinberg, D. S., Hollister, K., and Cibas, E. (1994). Evaluation of a new slide-based laser scanning cytometer for DNA analysis of tumors: Comparison with flow cytometry and image analysis. *Am. J. Clin. Pathol.* **102,** 432–438.

Mellors, R. C., and Silver, R. (1951). A microfluorometric scanner for the differential detection of cells: Application to exfoliative cytology. *Science* **114,** 356–360.

Musco, M. L., Cui, S., Small, D., Nodelman, M., Sugarman, B., and Grace, M. (1998). Comparison of flow cytometry and laser scanning cytometry for intracellular evaluation of adenoviral infectivity and p53 protein expression in gene therapy. *Cytometry* **33,** 290–296.

Numa, Y., Matsudaira, K., Tsukazaki, H., Kawamoto, K., Sato, T., and Kiyomatsu, Y. (1996). Analysis of cell nuclei and the quantity of chromosomal DNA by laser scanning cytometer (LSC). *Hum. Cell* **9,** 237–243.

Reeve, L., and Rew, D. A. (1997). New technology in the analytical cell sciences: The laser scanning cytometer. *Eur. J. Surg. Oncol.* **23,** 445–450.

Rew, D. A., Reeve, L. J., and Wilson, G. D. (1998). Comparison of flow and laser scanning cytometry for the assay of cell proliferation in human solid tumors. *Cytometry* **33,** 355–361.

Sakamoto, M., Miura, F., Sakamoto, H., Kikuchi, Y., Suehiro, Y., Akiya, T., Iwabuchi, H., Sakunaga, H., Muroya, T., Ishidate, M., Sugishita, T., and Tenjin, Y. (1996). Study on the elucidation of the method for cisplatin-resistance in ovarian cancer using comparative genomic hybridization and laser scanning cytometry. *Cytometry Res.* **6,** 9–20.

Sasaki, K., Kurose, A., Miura, Y., Sato, T., and Ikeda, E. (1996). DNA ploidy analysis by laser scanning cytometry (LSC) in colorectal cancers and comparison with flow cytometry. *Cytometry* **23,** 106–109.

Satoh, T., Yamamoto, K., Sasaki, K., and Ishidate, M. Jr. (1996). Laser-scan karyotyping: Analysis of chromosomal alteration by laser-scan karyotyping. *Cytometry Res.* **6,** 5–8.

Satoh, T., Yamamoto, K., Miura, K., Sasaki, K., and Ishidate, M. (1999). Application of laser scanning cytometry to the analysis of chromosomal aberrations induced by benzo[a]pyrene in CHO-WBLT cells. *Cytometry* **35,** 363–368.

Shapiro, H. M., Natale, P. J., and Kamentsky, L. A. (1979). Estimation of membrane potentials of individual lymphocytes by flow cytometry. *Proc. Natl. Acad. Sci. U.S.A.* **76,** 5728–5730.

Woltman, G., Ward, R. J., Symon, F. A., Rew, D. A., Pavord, I. D., and Wardlaw, A. J. (1999). Objective quantitative analysis of eosinophils and bronchial epithelial cells in induced sputum by laser scanning cytometry. *Thorax* **54,** 124–130.

CHAPTER 4

Principles of Confocal Microscopy

J. Paul Robinson

Department of Basic Medical Sciences
School of Veterinary Medicine and
Department of Biomedical Engineering
Purdue University
West Lafayette, Indiana 47907

I. Brief History of Microscope Development
II. Development of Confocal Microscopy
III. Image Formation in Confocal Microscopy
    A. Benefits of Confocal Microscopy
    B. Excitation Sources
    C. Line Scanners
IV. Useful Fluorescent Probes for Confocal Microscopy
    A. Fluorochrome Photobleaching
    B. Antifade Reagents
V. Applications of Confocal Microscopy
    A. Cell Biology
    B. Microscopy of Living Cells
    C. Calcium Imaging
    D. Cell Adhesion Studies
    E. Colocalization Studies
    F. Fluorescence Recovery after Photobleaching
VI. Conclusions
    References

I. Brief History of Microscope Development

Microscopy techniques have taken over 200 years to mature into technologies capable of the measurements now possible using confocal microscopy. Prior to 1800, production microscopes using simple lens systems were of higher resolution than compound microscopes despite the chromatic and spherical aberrations present in the double convex lens design. Microscopy did not really prosper until W. H. Wollaston made a significant improvement to the simple lens in 1812.

Soon after, Brewster improved on this design in 1820, and in 1827 Giovanni Battista Amici introduced the first matched achromatic microscope. Key features of the Amici design were the recognition of the importance of coverglass thickness and the development of the concept of "water immersion." Then came Carl Zeiss and Ernst Abbé, who introduced oil-immersion systems by developing oils that matched the refractive index of glass. By 1886, Dr. Otto Schott formulated glass lenses that color-corrected objectives and produced the first "apochromatic" objectives. Just after the turn of the twentieth century, Köhler illumination revolutionized brightfield microscopy. This discovery has been considered one of the most significant developments in microscopy prior to the electronic age.

Later developments such as the use of phase-contrast illumination, Nomarski illumination, and epi-illumination have each had significant impact on cell biology. In more recent years, the advent of confocal microscopy has changed the way cell biologists prepare and examine material, because we now have more options. Using this technology, the biologist can pinpoint the location of labeled molecules (e.g., a growth factor) in relatively thick specimens. This allows him to identify the organelle or location for the synthesis of the molecule. It is also possible to reconstruct the three-dimensional structure of many cells, organs, and even small organisms or animals quite accurately. Such information has given us a great deal of insight into the structure and function of many biological systems. This chapter discusses the basic principles of confocal microscopy. Detailed texts on the subject include Pawley (1995), Matsumoto (1993), Paddock (1999), and Sheppard and Shotton (1997).

II. Development of Confocal Microscopy

Marvin Minsky, then at Harvard University, filed the first patent for a confocal microscopy in 1957 (Minsky, 1988). Many scientists contributed to the enhancement and practical application of the technology, including Brackenhoff (Brackenhoff *et al.*, 1979, 1985), Wijnaendts van Resandt (Wijnaendts van Resandt *et al.*, 1985), Carlsson (Carlsson *et al.*, 1985), Amos (Amos *et al.*, 1987; Amos, 1988), and White (White *et al.*, 1987) among others. A confocal microscope achieves crisp images of structures even within thick tissue specimens by a process known as optical sectioning. The image source is primarily the photon emission from fluorescent molecules within—or attached to structures within—the object being sectioned. An alternative to fluorescence emission, reflectance, is discussed later. A point source of laser light illuminates the back focal plane of the microscope objective and is subsequently focused to a diffraction-limited spot within the specimen. Within this spot fluorescent molecules are excited and emit light in all directions. However, the emitted light is refocused in the objective image plane and any out-of-focus light is essentially removed from the image by passing the light through a pinhole aperture, so only a thin optical section of the specimen is formed. The effective removal of out-of-focus light from the emission creates an essentially background-free image—as opposed to the traditional fluorescent

4. Principles of Confocal Microscopy

microscope that includes all of this out-of-focus light. The comparison between traditional fluorescence microscopy and confocal microscopy is illustrated in Fig. 1. To create "depth" of the optical section, the diameter of the pinhole is reduced, thereby decreasing the light collection from the specimen but reducing the "thickness" of the optical sections. Although this effectively decreases the light collection for the images obtained, it allows a greater definition of the structure of the sample because more sections are collected over the sample thickness.

The resolution of a point light source is defined by the circular Airy diffraction pattern formed on the image plane. This pattern consists of a central bright region enclosed by an outer dark ring. The radius r_{Airy} of this central bright region is defined as $r_{Airy} = 0.61\lambda/NA$, where λ is the wavelength of the excitation source and NA is the numerical aperture of the objective lens. To increase the signal-to-noise (background) ratio and decrease the background light, it is necessary to decrease the pinhole to a size slightly less than r_{Airy}; a correct adjustment can decrease the background light by a factor of 10^3 over conventional fluorescence microscopy. Thus, achieving the correct pinhole diameter is crucial for achieving maximum resolution in a thick specimen. This becomes a trade-off, however, between optimizing axial resolution (optimum = $0.7\ r_{Airy}$) and lateral resolution (optimum = $0.3\ r_{Airy}$).

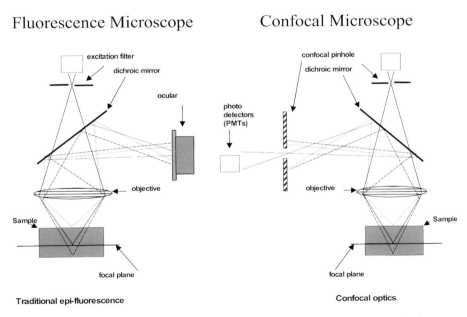

Fig. 1 The light pathway in conventional fluorescence microscopy versus in confocal microscopy. It should be noted that the principle of confocality is that the emission signal recorded will be exclusive for each focal plane within the specimen that is imaged. This feature is essentially the role of the confocal iris as described in the text.

Although the image collection optics removes the background light and creates a nice clean section, it is important to realize that the entire image is still being bathed in excitation light. By the time a thick section has been imaged a number of times, the impact of substantial photobleaching must be considered.

III. Image Formation in Confocal Microscopy

There are several methods for achieving a confocal image. The most common method scans the point source of light (a laser beam) over the image using a pair of galvanometer mirrors. One galvanometer scans in the x direction and the other in the y direction. The emitted fluorescence traverses the reverse pathway, is separated from the excitation source by a beam-splitting dichroic mirror, and is reflected to a photomultiplier tube amplifying the signal. After passing through an analog-to-digital converter (ADC), the signal is displayed as a sequential raster scan of the image. Depending on the desired measurements within the imaging requirement, it is possible to collect very small scan ranges from 50×50 points (or even smaller) up to rather large scanning areas with as many as 4096×4096 points. Most current systems utilize 16-bit ADCs, allowing an effective image of 1024×1024 points or more with at least 256 gray levels. Some confocal microscopes can collect high-speed images at video rates (30 frames/sec), whereas others achieve faster scanning by slit scanning. Just because an instrument can collect more points (and thus often claims higher resolution) does not necessarily mean it is useful. For example, a 512×512 image might require 0.3 sec to collect, and a 1024×1024 image perhaps 1.5 sec. A single scan of a 4096×4096 image might take 15 sec. To collect a relatively small number of sections (50) with signal averaging of three scans per image would take nearly 40 min, an impractical time constraint with most biological specimens. Regardless, the perfect image could take several runs to acquire, and so the "high-resolution" mode is less practical for three-dimensional imaging than the commercial literature might suggest.

Frequently, practical operation of the confocal microscope will be 512×512 image collection using the fastest possible point scanning available on the instrument. Once the imaging area is selected, the top and bottom (in the z-axis) of the image sections are identified; if desired, the image collection parameters can be changed at this point to obtain higher resolution. Electronic magnification is one of the most useful components of the confocal collection system and is universally available on all microscopes. The principle of electronic magnification is that the imaging area is reduced, but the number of pixels in the collection area remains constant. This effectively magnifies the image. However, it is generally not feasible to magnify the image beyond the point exceeding the Nyquist criterion (2.3 f), since beyond this is considered empty magnification—although there are cases where "super-resolution" is possible (Plasek and Reischig, 1998). An important point to consider is that the power delivered to the specimen increases with the square of the magnification. Therefore, a zoom factor of 2 places four times the laser power onto the object. This could cause serious bleaching or physically heat the specimen beyond a reasonable level. An example of electronic magnification is shown in Fig. 2. The primary advantage is that one can view a larger sample field and zoom in to areas of particular interest using the zoom feature.

Investigators have demonstrated two-photon excitation, in which a fluorophore simultaneously absorbs two photons each having half the energy—and twice the wavelength—normally required to raise the molecule to its excited state. A significant advantage of this system is that only the fluorophore molecules in the focal plane are excited, as this is the only area with sufficient light intensity. The higher wavelengths mean that considerably less background noise is collected and the efficiency of imaging thick specimens is significantly increased. Those probes requiring ultraviolet (UV) excitation can be excited by means of three-photon excitation (Sako *et al.*, 1997), which may have the advantage of causing less tissue damage (particularly when imaging live cells); this is still subject to verification, since little to no supportable data have been published on this matter. What is clear, is that two-photon microscopy has decided advantages in imaging to a greater depth. Although resolution in most thin (<70 μm) tissues

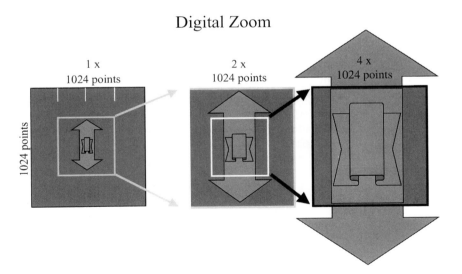

Fig. 2 The principle of electronic zoom is based on the notion that by reducing the area of the scan, but not reducing the number of points within that scan, the image will be electronically magnified. This is shown in the cartoon above. The first box represents a 1024 × 1024 matrix. If a box half this size is now imaged using 1024 × 1024 points, the resultant image will be magnified by a factor of 2. If this is repeated the resultant image will be magnified by a factor of 4. The impact is that the area of the object to be image is reduced as demonstrated by the arrowlike object in the cartoon.

is actually worse than in conventional confocal microscopy, it is actually improved in thick tissues, where conventional confocal microscopy is unable to image well at all.

A. Benefits of Confocal Microscopy

A now-familiar tool in the research laboratory, confocal microscopy has a number of significant advantages over conventional fluorescence microscopy. Among these are:

Increased effective resolution: A point source of light imaged through a pinhole provides increased resolution.

Reduced blurring of the image from light scattering: Because out-of-focus light is excluded from the image plane, collected images are sharper than those obtained with regular fluorescent microscopes. Photons emitted from points outside the image plane are rejected.

Improved signal-to-noise ratio: Compared to fluorescence microscopy, the background light is decreased, allowing for significantly improved signal-to-noise.

z-axis scanning: A series of optical sections can be obtained at regular distances by progressively moving the objective through the specimen from bottom to top.

Depth perception in z-sectioned images: Reconstruction techniques are used to reconstruct an image of the fluorescence emission of the specimen through the entire depth of the specimen.

Electronic magnification adjustment: By reducing the scanned area of the excitation source, but retaining the effective resolution, it is possible to magnify the image electronically. This has a number advantages over conventional microscopy.

B. Excitation Sources

The most common light sources for confocal microscopes are lasers. The acronym LASER stands for light amplification by stimulated emission of radiation. Most lasers on conventional confocal microscopes are continuous wave lasers (CW) and are either gas, dye, or solid-state lasers. Argon ion (Ar) lasers are the most popular gas lasers, followed by either krypton ion (Kr) or a mixture of argon and krypton (Kr-Ar). Helium–neon (He-Ne) and helium–cadmium (He-Cd) are also used in confocal microscopy. The He-Cd laser can provide UV lines at 325 or 441 nm, although use of the 325-nm line is very difficult in most microscopes because of loss of signal transmission at wavelengths below 350 nm. The most common source of UV excitation for the confocal microscope is the argon ion laser, which can emit 350- to 363-nm UV light. Frequently the UV excitation source for UV–vis confocal systems is the Coherent "Enterprise." The power necessary to excite fluorescent molecules at a specific wavelength can be calculated. For instance, consider a standard system with 1 mW of power at 488 nm focused via a 1.25 NA objective to a Gaussian spot whose radius at $1/e^2$ intensity is 0.25 μm. The peak intensity at the center will be 10^{-3} W/$[\pi(0.25 \times 10^{-4}$ cm$)^2] = 5.1 \times 10^5$ W/cm^2 or 1.25×10^{24} photons/(cm^2 sec^1) (Sheppard and Shotton, 1997). If fluorescein isothiocyanate (FITC) were the fluorochrome used in such a system, 63% of its molecules would be in an excited state and 37% in the ground state at any one point in time. This photon flux would be sufficient to obtain efficient excitation of this probe. For optimal confocal microscopy, the power delivered to the fluorescent probe must be sufficient to saturate the fluorescent molecules in the specimen.

Most confocal microscopes are designed around conventional microscopes, with the modification of the light source, which can be one of several lasers. An example of the layout of a typical confocal microscope is shown in Fig. 3. Here the instrument is divided into three essential components: (A) the light sources, (B) the optical components for manipulating the signals, and (C) the microscope itself. For most cell biology studies, arc lamps are not adequate sources of illumination for confocal microscopy. When using multiple laser beams, it is vital to expand the laser beams using a beam-expander telescope so that the back focal aperture of the objective is always completely filled. The beam widths from several different lasers must also be matched if simultaneous excitation is required. The most important feature in selecting the laser line is the absorption maximum of the fluorescent probe.

C. Line Scanners

One of the limiting factors that must be addressed in confocal microscopy is photobleaching of the fluorophore because of the intense illumination. An alternative to a moving-spot microscope is one in which the laser spot is "extruded" to form a line of light which is then used to scan the sample. Such scanning is naturally considerably faster than a moving-spot scan, so these systems can produce very rapid kinetic image sets. The principal reason for using a line scanner is to obtain rapid successive images of a fluorescence emission. Lower-intensity, even illumination can be applied at high rates by scanning a "line" of laser light instead of a point of light across the specimen. In general, these instruments are referred to as "slit-scanners" because they utilize a slit aperture, which can either scan or remain stationary. Quite different from most point scanners, line scanners commonly use a sensitive video camera—either silicon intensified target (SIT), intensified silicon intensified target (ISIT), or cooled charge coupled device (CCD)—to capture the fluorescence signal. An example of the light path of one such slit-scanner is shown in Fig. 4. Figure 5 is provided to demonstrate how this system might actually image cells in culture. In this figure, the cells are scanned by the light source using an inverted microscope.

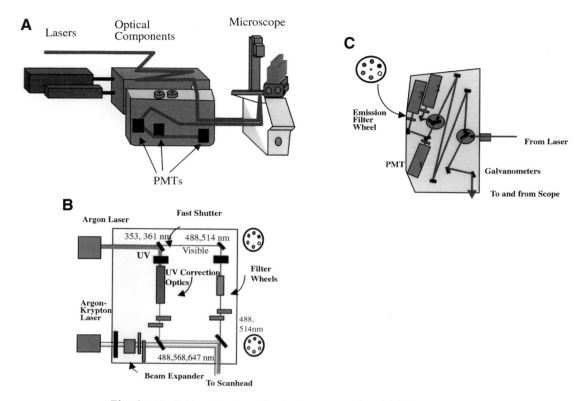

Fig. 3 The light paths of a confocal microscope with multiple lasers and using an inverted microscope. This shows three components: (A) the light sources, (B) the optical components that manipulate the signal, and (C) the microscope system. Shown are several laser lines that can be used together or independently to excite an object. Resultant signals are collected in the detector region where several PMTs reside. A computer system controls the system and creates the images. (See color plates).

The figure also demonstrates how the point source is converted to a line source for such scanning.

IV. Useful Fluorescent Probes for Confocal Microscopy

The essential requirement for a fluorescent molecule is an appropriate excitation source. Because most lasers can be successfully used in confocal systems, the number of fluorescent probes available for use in confocal microscopy is very broad. A series of tables is provided detailing the properties of fluorescent

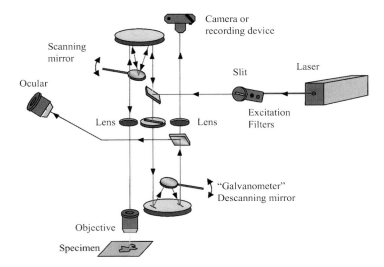

Fig. 4 The line scanning confocal microscope showing the laser light source, scanning and descanning mirrors, and the light path in the system. This system was originally designed for the BioRad DVC 250 line scanning confocal microscope (Hercules, CA). In the line scanning microscope, the confocal image is visible to the viewer at the ocular. Normally, a CCD camera is attached to the ocular to record the image.

probes for proteins (Table I), for intracellular organelles (Table II), for nucleic acids (Table III), for ions (Table IV), and for measuring intracellular changes in oxidation state (Table V). The excitation properties of each probe depend on its chemical composition. Ideal fluorescent probes will have high quantum yield, large Stokes shift, and nonreactivity with the molecules to which they are bound. It is vital to match the absorption maximum of each probe to the appropriate laser excitation line. For fluorochrome combinations, it is desirable to have fluorochromes with similar absorption peaks but significantly different emission peaks, enabling use of a single excitation source. It is common in confocal microscopy to use two, three, or even four distinct fluorescent molecules simultaneously, although it is usually necessary to image each probe independently and combine the images postcollection.

A. Fluorochrome Photobleaching

Photobleaching is defined as the irreversible destruction of an excited fluorophore by light. Uneven bleaching throughout the thickness of a specimen will bias the detection of fluorescence, causing a significant problem in confocal microscopy. Methods for countering photobleaching include shorter scan times, high magnification, high-NA objectives, and wide emission filters, as well as reduced excitation intensity. A number of "antifade" reagents are available (see

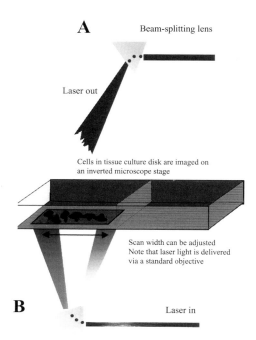

Fig. 5 The light source for a line scanning confocal microscope is a laser; however, instead of a point source of light, a line must be delivered to the specimen. This can be achieved by stretching the point into a line by means of a prism as shown in the diagram. As shown in (A), the line source is scanned (B) across the specimen. This allows very fast scanning of the specimen since the scanning requires a change in the y direction (not both x and y as in spot scanning).

Table I
Probes for Proteins

Probe	Excitation (nm)	Emission (nm)
FITC	488	525
PE	488	575
APC	630	650
PerCP	488	680
Cascade Blue	360	450
Coumarin-phalloidin	350	450
Texas Red	610	630
Tetramethylrhodamine-amines	550	575
CY3 (indotrimethinecyanines)	540	575
CY5 (indopentamethinecyanines)	640	670

Table II
Organelle-Specific Stains

Probe	Specificity	Excitation (nm)	Emission (nm)
BODIPY	Golgi	505	511
NBD	Golgi	488	525
DPH	Lipid	350	420
TMA-DPH	Lipid	350	420
Rhodamine 123	Mitochondria	488	525
DiO	Lipid	488	500
DiI-Cn-(5)	Lipid	550	565
DiO-Cn-(3)	Lipid	488	500

later); unfortunately, many are not compatible with viable cells. In the absence of an antifade reagent, FITC in particular is very susceptible to photobleaching.

B. Antifade Reagents

Many quenchers act by reducing oxygen concentration to prevent formation of excited species of oxygen. Antioxidants such as propyl gallate, hydroquinone, and p-phenylenediamine can be used for fixed specimens but are not useful for live-cell studies. Quenching fluorescence in live cells is possible using either systems with reduced O_2 concentration or singlet-oxygen quenchers such as carotenoids (50 mM crocetin or etretinate in cell cultures), ascorbate, imidazole, histidine, cysteamine, reduced glutathione, uric acid, or trolox (vitamin E analog). Photobleaching can be calculated for a particular fluorochrome to determine the maximum scan time possible for that molecule. For example, the most commonly used fluorescent probe, FITC, bleaches with a quantum efficiency Q_b of 3×10^{-5}. A standard laser intensity would pump 4.4×10^{23} photons cm^{-2} sec^{-1} and FITC would be bleached with a rate constant of 4.2×10^3 sec^{-1}. After 240 μsec of irradiation, only 37% of the molecules would remain. In a single plane, 16 scans would cause 6–50% bleaching (Sheppard and Shotton, 1997).

Table III
DNA Dyes

Probe	Excitation (nm)	Emission (nm)
Hoechst 33342	350	460
DAPI	350	470
PI	530	620
Acridine orange	500	520
TOTO-1	514	530
TOTO-3	640	660
Thiazole orange (vis)	510	525

Table IV
Probes for Ionic Fluxes

Molecule	Probe	Excitation (nm)	Emission (nm)
Calcium	Indo-1	351	405, >460
Magnesium	Mag-Indo-1	351	405, >460
Calcium	Fluo-3	488	525
Calcium	Fura-2	363	>500
Calcium	Calcium green	488	515
P1 A	Acyl pyrene	351	405, >460

V. Applications of Confocal Microscopy

A. Cell Biology

The applications in cell biology are expanding on a daily basis, in no small part owing to a new generation of simple-to-use confocal microscopes that have been designed to remove the technical difficulties previously associated with operating these instruments. Currently one of the more frequent applications is cell tracking using green fluorescent protein (GFP), a naturally occurring protein from the jellyfish *Aequorea victoria* which fluoresces when excited by UV or blue light (Moerner et al., 1999; Jordan et al., 1999; Sullivan and Shelby, 1999). A fluorescent protein such as GFP can be transfected into cells so that subsequent replication of the organism carries with it the fluorescent reporter molecule, providing a valuable tool for tracking the presence of that protein in developing tissue or differentiated cells. This is particularly useful for identifying regulatory genes in developmental biology, and for identifying the biological impact of alterations to normal growth and development processes. In almost any application, multiple fluorescent wavelengths can be detected simultaneously. If a UV–vis confocal microscope is available, Hoechst 33342 (460 nm), FITC (525 nm),

Table V
Probes for Oxidative States

Probe	Oxidant	Excitation (nm)	Emission (nm)
DCFH-DA[a]	(H_2O_2)	488	525
HE[b]	(O_2^-)	488	590
DHR 123[c]	(H_2O_2)	488	525

[a] DCFH-DA, dichlorofluorescin diacetate.
[b] HE, hydroethidine.
[c] DHR-123, dihydrorhodamine 123.

and Texas Red (630 nm) can be simultaneously collected to create a three-color image, providing excellent information regarding the location of the labeled molecules and the structures they identify, and the relationships between them.

B. Microscopy of Living Cells

Evaluation of live cells using confocal microscopy presents some difficult challenges. One is the need to maintain a stable position while imaging a live cell. For example, a viable respiring cell may be constantly changing shape, preventing a finely resolved three-dimensional image reconstruction. Fluorescent probes must be found which are not toxic to the cell. For example, it is possible to evaluate cells attached to an extracellular matrix. In such a situation, the cells can be accurately identified and enumerated, and their relative locations within the matrix determined as well. This requires not just excellent instrumentation, but also an array of analytical tools for image analysis. Confocal microscopy is an effective tool for qualitative and quantitative assessment and for creating three-dimensional image reconstructions of live cells. In most cases it is possible to image several fluorescent markers within a single cell system, making powerful use of the correlative tools of confocal microscopy. As shown in Fig. 6, another example of three-dimensional imaging of live cells demonstrates how the three

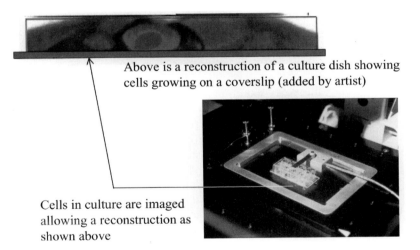

Above is a reconstruction of a culture dish showing cells growing on a coverslip (added by artist)

Cells in culture are imaged allowing a reconstruction as shown above

Fig. 6 Endothelial cells growing on a coverslip–tissue culture chamber (shown in the photograph) were imaged using a confocal microscope mounted on an inverted scope. Approximately 60 sections were collected at 0.1-μm z-axis steps. The images were then analyzed using VoxelView (a three-dimensional software package, Vital Images, Inc., Plymouth, MN), and the cell image was reconstructed from an electronic slice through the cells. The cells appear to be sitting on the coverslip (which was added to the figure to demonstrate its position). (See color plates.)

dimensionality of the image can provide powerful information. In this case endothelial cells were grown on glass in a tissue-culture dish. Thirty image sections were taken 0.2 μm apart; the image plane presented shows an x–z plane with the cells attached to the coverglass. The coverglass is represented by a cartoon insertion to show where the cells would be actually attached.

C. Calcium Imaging

Confocal microscopy can be used for evaluation of physiological processes within cells. Examples are changes in cellular pH, changes in free Ca^{2+} ions, and changes in membrane potential and oxidative processes within cells. One of the most successful methods for evaluating these phenomena is emission ratioing in real time. Usually the molecules under study are excited at one wavelength but emit at two wavelengths depending on the change in properties of the molecule. Changes in cellular pH can be identified using SemiNaphtho-Rhodafluor (SNARF-1) (Edwards *et al.*, 1998) or 2',7'-*bis*-(2-carboxyethyl)-5-(and-6)-carboxyfluorescein (BCECF) (Yip and Kurtz, 1995; Stephano and Gould, 1997), which is excited at 488 nm and emits at 525 and 590 nm. The ratio of 590/525 signals reflects the intracellular pH. Calcium changes can be detected using Indo-1, which can be excited at 350 nm (Niggli *et al.*, 1994; Sako *et al.*, 1997). Indo-1 can bind Ca^{2+}, and the fluorescence of the bound molecule is preferentially at the lower emission wavelength; the ratio of emission signals at 400/525 nm reflects the concentration of Ca^{2+} in the cell. Rapid changes in Ca^{2+} can be detected by kinetic imaging—taking a series of images at both emission wavelengths in quick succession. An example is shown in Fig. 7.

D. Cell Adhesion Studies

One early example of the power of confocal microscopy was the study of chondrocytes essentially *in vivo* (Errington *et al.*, 1997). In addition, studies of osteoblastic cell adhesion have been performed using confocal microscopy. Investigators in those studies were interested in the cell attachment and release mechanisms of human osteoblasts to orthopedic devices used for bone or joint replacement (Shah *et al.*, 1999).

E. Colocalization Studies

One of the routine uses for confocal microscopy is the colocalization of or distribution of molecules produced within living organisms. For example, studies of the distribution of HMG-1 protein, a high-mobility group protein which interacts *in vitro* with the minor groove of AT-rich B-DNA, have demonstrated that it is found exclusively in the nucleus (Amirand *et al.*, 1998). Other examples of colocalization have been shown in studies of the TR6 protein produced by equine herpes virus. Confocal microscopy was able to determine that the IR6 protein

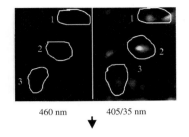

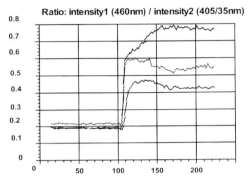

Fig. 7 Changes in the fluorescence of cells loaded with a calcium-sensitive dye were measured using a confocal microscope system with calcium ratioing software. The same region in each wavelength was measured, and the relative change was recorded and exported to a spreadsheet for analysis. The ratio of fluorescence signals can be plotted on a cell-by-cell basis as shown in the graph.

of wild-type RacL11 virus colocalizes with nuclear lamins very late in infection, whereas the mutant IR6 protein encoded by the RacM24 strain did not colocalize with the lamin proteins (Osterrieder *et al.*, 1998).

Similarly, colocalization studies using confocal microscopy have determined that gene 1 products associated with murine hepatitis virus (MHV) are directly associated with the viral RNA synthesis. Confocal microscopy revealed that all the viral proteins detected by these antisera colocalized with newly synthesized viral RNA in the cytoplasm, particularly in the perinuclear region of infected cells. Several cysteine and serine protease inhibitors—E64d, leupeptin, and zinc chloride—inhibited viral RNA synthesis without affecting the localization of viral proteins, suggesting that the processing of the MHV gene 1 polyprotein is tightly associated with viral RNA synthesis. Dual labeling with antibodies specific for cytoplasmic membrane structures showed that RNA and MHV gene 1 products colocalized with the Golgi apparatus in HeLa cells. However, in murine 17CL-1 cells, the viral proteins and viral RNA did not colocalize with the Golgi apparatus but, instead, partially colocalized with the endoplasmic reticulum (Shi *et al.*, 1999). It is fair to say that despite the many alternative technologies available, only confocal microscopy, via its unique ability to create accurate

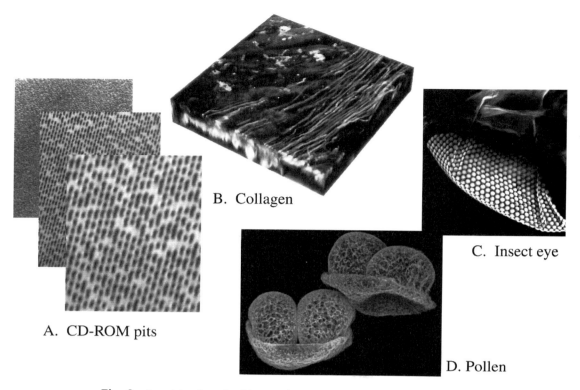

Fig. 8 A variety of confocal images demonstrating (A) reflected light imaging—the pits from a CD-ROM; (B) collagen autofluorescence—an image reconstructed from approximately 150 sections; (C) a reconstructed image of the eye of an insect; and (D) a three-dimensional representation of pollen grains from a pine tree.

three-dimensional structural representations of cells and their organelles, was able to demonstrate the location of the viral RNA.

F. Fluorescence Recovery after Photobleaching

Fluorescence recovery after photobleaching (FRAP) is a measure of the dynamics of the chemical changes in a fluorescent molecule within an object such as a cell. A small area of the cell is bleached by exposure to an intense laser beam, and the recovery of fluorescent species in the bleached area is measured. The recovery time (t) can be calculated from the equation $t = W^2/4D$, where W is the diameter of the bleached spot and D is the diffusion coefficient of the fluorescent molecule under study. An alternative technique can measure interactions between cells; one of two attached cells is bleached and the recovery of the pair as a whole is monitored. To perform FRAP experiments satisfactorily, it is necessary to be able to park the laser beam over a particular cell, or part of a cell, and effectively bleach the fluorescence of this component of the cells. A number of studies designed to explore multistep signal-transducing events can be performed using FRAP apparatus, e.g., tracking G-protein segments to localized regions of the cell (Kwon *et al.*, 1994). FRAP has also been used to determine macromolecular diffusion of biological polymers (Gribbon and Hardingham, 1998).

VI. Conclusions

Confocal microscopy has reached the point that instruments are now effective and inexpensive compared to the early 1990s when commercial technologies were introduced. More complex systems such as UV microscopes are still relatively rare; however, these may eventually be superseded by the multiphoton microscopes now becoming almost turnkey in operation. For routine one-, two-, or three-color fluorescence microscopy where three dimensionality is important, conventional confocal microscopes surpass multiphoton resolution and will be around for many more years. A new breed of low-cost instruments is now available, making this technology a simple tool requiring no significant technical or engineering knowledge. New fluorescent dyes are also available, increasing the applications for confocal microscopy. Finally, new applications are being developed that utilize reflectance (Fig. 8) (Brightman *et al.*, 2000) or autofluorescence (see Chapter 27 of this volume), which while previously considered a problem can be used a tool for extracting three-dimensional information from within thick tissue.

References

Amirand, C., Viari, A., Ballini, J. P., Rezaei, H., Beaujean, N., Jullien, D., Kas, E., and Debey, P. (1998). Three distinct sub-nuclear populations of HMG-I protein of different properties revealed by co-localization image analysis. *J. Cell Sci.* **111**, 3551–3561.

Amos, W. B. (1988). Results obtained with a sensitive confocal scanning system designed for epifluorescence. *Cell Motil. Cytoskeleton* **10**, 54–61.

Amos, W. B., White, J. G., and Fordham, M. (1987). Use of confocal imaging in the study of biological structures. *Appl. Opt.* **26**, 3239–3243.

Brackenhoff, G. J., Blom, P., and Barends, P. (1979). Confocal scanning light microscopy with high aperture immersion lenses. *J. Microsc.* **117**, 219–232.

Brackenhoff, G. J., van der Voort, H. T. M., van Spronsen, E. A., Linnemans, W. A. M., and Nanninga, N. (1985). Three dimensional chromatin distribution in neuroblastoma nuclei shown by confocal scanning laser microscopy. *Nature* **317**, 748–749.

Brightman, A. O., Rajwa, B. P., Sturgis, J. E., McCallister, M. E., Robinson, J. P., and Voytik-Harbin, S. L. (2000). Time-lapse confocal reflection microscopy of collagen fibrillogenesis and ECM assembly in vitro. *Biopolymers* in press.

Carlsson, K., Danielsson, P., Lenz, R., Liljeborg, A., Majlof, L., and Aslund, N. (1985). Three-dimensional microscopy using a confocal laser scanning microscope. *Opt. Lett.* **10**, 53–55.

Edwards, L. J., Williams, D. A., and Gardner, D. K. (1998). Intracellular pH of the preimplantation mouse embryo: Effects of extracellular pH and weak acids. *Mol. Reprod. Dev.* **50,** 434–442.

Errington, R. J., Fricker, M. D., Wood, J. L., Hall, A. C., and White, N. S. (1997). Four-dimensional imaging of living chondrocytes in cartilage using confocal microscopy: A pragmatic approach. *Am. J. Physiol.* **272**(3 Pt. 1), C1040–C1051.

Gribbon, P., and Hardingham, T. E. (1998). Macromolecular diffusion of biological polymers measured by confocal fluorescence recovery after photobleaching. *Biophys. J.* **75,** 1032–1039.

Jordan, K., Solan, J. L., Dominguez, M., Sia, M., Hand, A., Lampe, P. D., and Laird, D. W. (1999). Trafficking, assembly, and function of a connexin43-green fluorescent protein chimera in live mammalian cells. *Mol. Biol. Cell* **10,** 2033–2050.

Kwon, G., Axelrod, D., and Neubig, R. R. (1994). Lateral mobility of tetramethylrhodamine (TMR) labelled G protein α and $\beta\gamma$ subunits in NG 108-15 cells. *Cell Signal.* **6,** 663–679.

Matsumato, B. (1993). "Cell Biological Applications of Confocal Microscopy." Methods in Cell Biology, Vol. 38. Academic Press, San Diego.

Minsky, M. (1988). Memoir on inventing the confocal scanning microscope. *Scanning* **10,** 128–138.

Moerner, W. E., Peterman, E. J., Brasselet, S., Kummer, S., and Dickson, R. M. (1999). Optical methods for exploring dynamics of single copies of green fluorescent protein. *Cytometry* **36,** 232–238.

Niggli, E., Piston, D. W., Kirby, M. S., Cheng, H., Sandison, D. R., Webb, W. W., and Lederer, W. J. (1994). A confocal laser scanning microscope designed for indicators with ultraviolet excitation wavelengths. *Am. J. Physiol.* **266,** C303–C310.

Osterrieder, N., Neubauer, A., Brandmuller, C., Kaaden, O. R., and O'Callaghan, D. J. (1998). The equine herpesvirus 1 IR6 protein that colocalizes with nuclear lamins is involved in nucleocapsid egress and migrates from cell to cell independently of virus infection. *J. Virol.* **72,** 9806–9817.

Paddock, S. W. (1999). "Confocal Microscopy: Methods and Protocols." Humana Press, Totowa, New Jersey.

Pawley, J. B. (1995). "Handbook of Biological Confocal Microscopy," 2nd Ed. Plenum, New York.

Plasek, J., and Reischig, J. (1998). Transmitted-light microscopy for biology: A physicist's point of view. *Proc. RMS* **33,** 196–205.

Sako, Y., Sekihata, A., Yanagisawa, Y., Yamamoto, M., Shimada, Y., Ozaki, K., and Kusumi, A. (1997). Comparison of two-photon excitation laser scanning microscopy with UV-confocal laser scanning microscopy in three-dimensional calcium imaging using the fluorescence indicator Indo-1. *J. Microsc.* **185,** 9–20.

Shah, A. K., Sinha, R. K., Hickok, N. J., and Tuan, R. S. (1999). High-resolution morphometric analysis of human osteoblastic cell adhesion on clinically relevant orthopedic alloys. *Bone* **24,** 499–506.

Sheppard, C. J. R., and Shotton, D. M. (1997). "Confocal Laser Scanning Microscopy." Springer, New York.

Shi, S. T., Schiller, J. J., Kanjanahaluethai, A., Baker, S. C., Oh, J. W., and Lai, M. M. (1999). Colocalization and membrane association of murine hepatitis virus gene 1 products and de novo-synthesized viral RNA in infected cells. *J. Virol.* **73,** 5957–5969.

Stephano, J. L., and Gould, M. C. (1997). The intracellular calcium increase at fertilization in *Urechis caupo* oocytes: Activation without waves. *Dev. Biol.* **191,** 53–68.

Sullivan, K. F., and Shelby, R. D. (1999). Using time-lapse confocal microscopy for analysis of centromere dynamics in human cells. *Methods Cell Biol.* **58,** 183–202.

White, J. G., Amos, W. B., and Fordham, M. (1987). An evaluation of confocal versus conventional imaging of biological structures by fluorescence light microscopy. *J. Cell Biol.* **105,** 41–48.

Wijnaendts van Resandt, R. W., Marsman, H. J. B., Kaplan, R., Davoust, J., Stelzer, E. H. K., and Strickler, R. (1985). Optical fluorescence microscopy in three dimensions: Microtomoscopy. *J. Microsc.* **138,** 29–34.

Yip, K. P., and Kurtz, I. (1995). NH3 permeability of principal cells and intercalated cells measured by confocal fluorescence imaging. *Am. J. Physiol.* **269,** F545–F550.

CHAPTER 5

Optical Measurements in Cytometry: Light Scattering, Extinction, Absorption, and Fluorescence

Howard M. Shapiro

Howard M. Shapiro, M.D., P.C.
West Newton, Massachusetts 02465

    I. Introduction
   II. Signal Processing Tasks in Flow Cytometry: An Overview
  III. The Optical Signal: Interaction of Light with Cells
      A. Flow Cytometry Optics
      B. Illumination Power: How Many Photons Reach the Cell?
      C. Extinction and Absorption Signals: How Many Photons Are Removed from the Illuminating Beam?
      D. Forward Scatter Signals
      E. Large Angle Scatter, Absorption, and Fluorescence Signals
      F. Signal, Background, and Sensitivity
      G. Light Collection Optics
      H. Wavelength Selection: Filters and Dichroics and Alternatives
  IV. Detection: Converting Optical Signals to Current
      A. Photodiodes
      B. Photomultiplier Tubes
      C. Avalanche Photodiodes
      D. Single Photon Counting
      E. Charge-Coupled Devices
   V. Electronics: Converting Current to Voltage
  VI. Fluorescence Compensation and Logarithmic Amplification
 VII. Peak Detection, Integration, and Pulse Width Measurement; Triggering
VIII. Measurement Sensitivity: Changing Concepts and the Bottom Line
      References

I. Introduction

Flow, image, and static cytometry provide users with information about various characteristics of individual cells, including physical attributes such as cell size and internal granularity, chemical properties such as DNA and RNA content and the amount of antibody bound to sites on the cell surface or inside the cell, and physiological characteristics, such as calcium ion concentration, membrane permeability, and enzyme activity. Most of this information is derived from measurements of scattering, absorption, or extinction of incident electromagnetic radiation and of fluorescence emitted by cellular components and/or by fluorescent dyes bound to substances in or on cells.

Under most circumstances, users do not concern themselves with fluxes of photons incident on and emanating from cells. However, as the capabilities of cytometry are extended to making measurements at the single molecule level, and as interest increases in quantification of chemical measurements in cells in such terms as numbers of bound antibody molecules, some basic understanding of the measurement processes involved can facilitate the design and interpretation of experiments. This chapter will follow the path of flow cytometric signals from the cell through collection and wavelength selection optics, detectors, and analog and hybrid electronics, with some mention made of significant differences between flow and static cytometry. Chapter 7 will deal with data acquisition, that is, conversion to digital form, and display.

II. Signal Processing Tasks in Flow Cytometry: An Overview

Flow cytometry (Shapiro, 1995) characterizes a cell in terms of the numerical values of one or more measured optical characteristics, or parameters. These numerical values are obtained by creating an optical signal, transforming it into an electrical signal, and processing it using a combination of analog, hybrid, and digital circuitry.

Cytometry is often described as involving the measurement of light, or photometry. However, light, according to the precise definition of physics, is electromagnetic radiation perceptible to the human eye, which is unresponsive below about 400 nm and above about 750 nm and most sensitive at around 550 nm. Cytometers actually measure radiant energy, or electromagnetic radiation; the process is defined as radiometry, and the results are expressed in familiar units, joules and watts (joules per second). Since old habits die hard, "light" will be used as a synonym for "electromagnetic radiation" in the following discussion.

In flow cytometry, cells, after exposure to appropriate reagents, or probes, are injected into a fluid stream, typically contained within a capillary or cuvette, and pass through one or more focused high-intensity illuminating beams, scattering and absorbing light at the illumination wavelength and emitting fluorescence at longer wavelengths during their transit of each beam. The optical signals, that is, scattered, transmitted, and emitted light, are collected by lenses and directed, through such optical filters as may be needed to select spectral regions of interest, to photodiode or photomultiplier tube (PMT) detectors. Light reaching a detector produces a pulse of electrical current at its output. The information about the cell represented by the width, peak amplitude (or pulse height), integral (or area), and shape of this electrical signal pulse must then be extracted.

In the first step of this process, current-to-voltage conversion, the output current of each detector goes to a preamplifier, which produces a voltage output made up of the cell-associated pulse superimposed on a baseline signal with a slowly varying (or DC) component and a more rapidly varying (or AC) component due to fluctuations in the output of the light source and to electronic noise. Next, baseline restoration removes the DC component of the baseline. Fluorescence is usually measured in several spectral regions, to determine the amounts of two or more fluorescent materials in or on cells. The signal in each spectral band is usually dominated by emission from one fluorochrome; for example, when 488 nm excitation is used, and fluorescein, phycoerythrin (PE) and a phycoerythrin tandem conjugate with Cy5 (PE-Cy5) are employed as labels, emissions near 525, 585, and 670 nm, respectively, come predominantly from fluorescein, PE, and PE-Cy5. However, because there is some overlap of emission spectra between different fluorochromes, each at least potentially contributes some emission in all spectral bands. Thus, in order to determine the actual amount of each fluorochrome present, it is necessary to perform fluorescence compensation, by subtracting an appropriate fraction of the signal in each spectral region from the signals in the other spectral regions.

Many flow cytometric measurements require a large (three to four decades) dynamic range, making it useful to express data values on a logarithmic scale. In most flow cytometers, transformation from a linear to a logarithmic scale is done using logarithmic amplifiers, or log amps, analog circuits that, over some range of input levels, deliver an output which is, at least ideally, proportional to the logarithm of the input. However, fluorescence compensation requires linear data and thus must be done before the signals are transformed.

Information about the measured particle may be extracted from the peak amplitude (height), the integral (area), the duration (width), and the shape of signal pulses. It has been customary to capture pulse height, integral, and width in hybrid circuits, which combine analog and digital electronics to store the appropriate analog values for long enough to permit analog-to-digital conversion of the data. These circuits must be reset as each particle passes through the illuminating beam, so new analog signal levels can be acquired, and the outputs must then be held constant, so data can be digitized. Digital logic reset and hold signals are generated by additional hybrid "front end" electronics, which compare one or more trigger signal levels with preset threshold values to determine when a cell is present. Once held signals have been digitized, further analysis is accomplished with a digital computer, typically either an Apple Macintosh or an Intel/Microsoft-based personal computer; the necessary software is available from both flow cytometer manufacturers and third parties, in some cases at no cost.

III. The Optical Signal: Interaction of Light with Cells

The illuminating beam in a cytometer directs a stream of photons at particles in the sample. The total number of photons scattered and absorbed by a particle obviously cannot exceed the number of photons incident on it, and the number of photons emitted as fluorescence must be less than or equal to the number absorbed. The number of photons incident on a particle during its transit of the illuminating beam can be calculated relatively precisely; the numbers of photons involved in scattering, absorption, and fluorescence can be estimated, with somewhat less precision, from the amplitudes of the measured signals.

A. Flow Cytometer Optics

A schematic (definitely not to scale!) of the light collection optics and detectors of a typical flow cytometer appears in Fig. 1. This apparatus has an orthogonal geometry, meaning that the axis of the illuminating beam, the optical axis of the orthogonal collection lens, and the direction in which cells flow are mutually perpendicular. The incoming laser beam (a), shown in light gray, is focused by a pair of cylindrical lenses, not shown, to form an elliptical spot in the center of the square cuvette (b) in which a cell (c) is observed; the sample stream flows in a direction perpendicular to the plane of the paper.

Light from the illuminating beam is scattered in all directions; the figure shows a cone of light scattered at relatively small angles to the beam, representing the forward scatter signal, and a blocker or obscuration bar (d), which intercepts the transmitted beam. An iris diaphragm (e) defines the upper limit of the angle at which forward scattered light is collected by the lens (f), which converges this light on the forward scatter detector (g). A photodiode is typically used as a forward scatter detector.

Light scattered at large angles (between about 45° and about 135°; also called orthogonal scatter) (scattered light is shown in light gray) and fluorescence (shown in dark gray) are collected by the orthogonal collection lens (h). A dichroic mirror (i), which, in this instance, reflects short wavelengths and transmits longer wavelengths, separates scatter and fluorescence signals, directing light in each spectral band through a band pass filter (j) to a detector (k), usually a photomultiplier tube. In most instruments, additional dichroics, filters, and detectors are used to allow measurement of fluorescence in three or four spectral regions.

B. Illumination Power: How Many Photons Reach the Cell?

The energy [E, in joules (J)] of a single photon is

$$E = hc/\lambda,$$

5. Optical Measurements in Cytometry

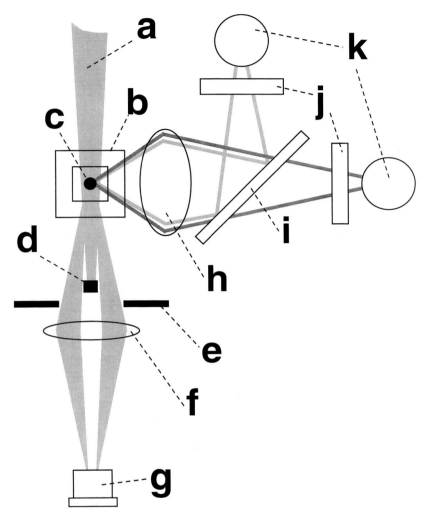

Fig. 1 Light collection optics and detectors of a typical flow cytometer. Illuminating light and scattered light are shown in light gray, fluorescence emission is shown in dark gray. The incoming laser beam (a) converges to a focus in the center of the square cuvette (b) in which a cell (c) is observed; the sample stream flows in a direction perpendicular to the plane of the paper. A cone of light scattered at relatively small angles to the illuminating beam represents the forward scatter signal; the blocker, or obscuration bar, at (d) intercepts the transmitted beam. An iris diaphragm (e) limits the angle at which forward scattered light is collected by the lens (f), which converges this light on the forward scatter detector (g). Light scattered at large angles, and fluorescence, are collected by the lens at (h). A dichroic mirror (i), reflecting short wavelengths and transmitting longer wavelengths, separates scatter and fluorescence signals, directing light in each spectral band through band-pass filters (j) to detectors (k).

where h is Planck's constant (6.63×10^{-34} J-sec), c is the speed of light, (3×10^8 m/sec), and λ is the wavelength of the photon (in meters). Most flow cytometers and many confocal microscopes and scanning cytometers use an argon ion laser emitting at a wavelength of 488 nm as a light source. A 488 nm photon has an energy of approximately 4.08×10^{-19} J; the photon flux from a laser emitting 25 mW at 488 nm is 6.13×10^{16} photons per sec. If the beam from such a laser is focused to an elliptical spot 100 μm wide $\times$ 20 μm high, it can be determined from the Gaussian energy distribution in the beam that a particle 10 μm in diameter passing through the central region of a beam will be exposed to about 5% of the power in the beam, or 1.25 mW. If it takes 5 μsec to traverse the beam, some 1.53×10^{10} photons will be incident on the particle during this observation period.

C. Extinction and Absorption Signals: How Many Photons Are Removed from the Illuminating Beam?

The extinction, or light loss, signal represents the diminution in intensity of the illuminating beam during the passage of a particle, and thus is a measure of the total number of photons removed by absorption and scattering. Although most flow cytometers now in use do not make extinction measurements, it is easy to visualize, using Fig. 1, how this can be done. If the obscuration bar (d) is removed, and the iris diaphragm (e) closed down to pass only the transmitted beam, the detector (g) captures an extinction signal rather than a forward scatter signal. When such measurements are made on cultured mammalian cells or plastic particles, the intensity of the transmitted beam is typically seen to diminish by no more than about 5% during the passage of a cell. Therefore, the large and small angle scatter and fluorescence signals together could not include more than 7.7×10^8 photons; the equivalent power in all of these signals would be no more than 63 μW.

Absorption signals are not readily measured with the collection optics used for forward scatter or extinction measurements in most flow cytometers, because these optics will not collect light scattered at large angles to the illuminating beam, making it impossible to discriminate light loss by this mechanism from light loss by absorption. Flow cytometers specially designed for absorption measurement use lenses that will collect light over a larger angle; the chemical composition of sheath and core fluid is also adjusted to match the refractive index of the cells, minimizing light loss by scattering. Incandescent rather than laser sources are typically used in such apparatus (Mansberg et al., 1974).

D. Forward Scatter Signals

According to Mie's theory, most of the 488 nm light scattered by a 10-μm particle will be at small angles (5° or less) to the illuminating beam. Forward scattered light at angles nearest the beam axis may be blocked by the obscuration bar, but the rest of the scattered light passing through the iris diaphragm will reach the detector. Knowing the detector quantum efficiency and gain, and the output signal amplitudes, it can be calculated that at least 10^8 photons are contained in the forward scatter signal, corresponding to a power of about 8 μW.

E. Large Angle Scatter, Absorption, and Fluorescence Signals

At most, 6.7×10^8 photons can be scattered at large angles to the beam or absorbed. The orthogonal scatter signal probably accounts for about half of these; thus, no more than 3.4×10^8 photons can be absorbed. For any particular absorber, the number of photons emitted as fluorescence is the product of the number of photons absorbed and the quantum yield of fluorescence; assuming an average quantum yield for all the absorbers of 0.5, combined fluorescence signals in all spectral regions could not include more than 1.7×10^8 photons.

F. Signal, Background, and Sensitivity

Successful cytometry does not depend solely on collection of as much light as possible from the cells of interest; it is equally important to minimize collection of background light. The background signal for a fluorescence measurement may include several components, some of which may be reduced or removed more readily than others. Stray light from the outside world is generally excluded by placing the flow cell and collection optics in an appropriate enclosure and/or by operating the instrument in a darkened room. Optical filters with inappropriate band pass characteristics may allow stray illuminating light to reach the detector; illuminating light may also induce fluorescence in filters containing absorptive elements. Proper filter selection largely eliminates both problems. Raman scattering from water in the sample and sheath streams is more difficult to deal with. The Raman emission occurs at a frequency that is the difference of the illumination frequency and the frequency of stretching of O—H bonds. When 488 nm illumination is used, Raman emission appears at 592 nm, a wavelength within the pass band of many filters used for phycoerythrin fluorescence measurements, and may interfere with such measurements. Autofluorescence, that is, fluorescence of intrinsic cellular constituents in the same spectral region as the fluorescence of a dye or label, is another source of background fluorescence that is hard to remove. Although a change of illumination wavelength, for example, from 488 nm to 532 nm, can shift Raman fluorescence out of the phycoerythrin emission region and diminish autofluorescence from most mammalian cells, this option is not available in most flow cytometers.

Background fluorescence from unbound dye, labeled antibody, etc. in the sample and/or sheath solution is possibly the most difficult to deal with. This is minimized by selecting collection optics that collect as much light as possible from a cell and as little as possible from regions even as near as a few micrometers from the cell; this explains why microscope objectives and similar lenses are used for orthogonal collection in most flow cytometers. Although the intensities of forward and orthogonal scatter signals are directly proportional to the power in the illuminating beam, the relationship between fluorescence signal intensities and illumination power is more complex. For fluorescence to occur, a molecule must first absorb a photon at the excitation wavelength, which results in promotion of an orbital electron from a lower to a higher energy state. After some of the absorbed energy is lost by various nonradiative processes that occur over a short time period, the fluorescence lifetime (usually nanoseconds), the electron returns to the lower energy level, and the remainder of the absorbed energy is emitted as a photon of lower energy (i.e., longer wavelength) than the photon initially absorbed. Absorbed energy can be lost by means other than fluorescence, including nonradiative intra- and intermolecular energy transfer, photolysis or bleaching (breakage of a chemical bond, destroying the molecule), and phosphorescence (which also results in emission at wavelengths longer than the excitation wavelength, typically after a period much longer than the fluorescence lifetime). The fluorescence quantum yield is the probability that absorption of a photon will result in fluorescence rather than the alternatives just mentioned; for many of the fluorescent materials commonly measured in flow cytometry, this value is between 0.25 and 0.5.

If the illumination power, that is, the photon flux through the specimen, is increased, photon saturation is reached; dye molecules in the specimen are in the excited state most of the time. A further increase in power will not result in increased fluorescence emission, because the specimen is incapable of absorbing more of the excitation light. Also, as the molecules in a specimen spend proportionately more time in the excited state, or undergo more excitations per unit time, the probability that they will be bleached increases. As van den Engh and Farmer (1992) pointed out, when illumination power is increased by a factor n, both photon saturation and bleaching result in fluorescence signals being increased by a factor less than n; they and others have noted, based on dual-beam measurements, that a substantial fraction of dye molecules may be bleached during the transit of even a relatively low-power illuminating beam.

Thus, while increasing illumination power may result in increased signal levels, it will not always increase measurement sensitivity. Background for both scatter and fluorescence signals typically increases by the same factor as the signal, and one faces a law of diminishing returns in terms of increasing fluorescence signal amplitude. It is more efficient, and, usually, less expensive to collect more of the photons coming from the cell than to throw more photons at it.

G. Light Collection Optics

Whereas the forward scatter collection optics of a typical flow cytometer may capture more than half of a forward scatter or extinction signal, orthogonal collection is less efficient. As was previously mentioned, cells scatter and emit light in all directions; the best conventional lens, with a field of view equivalent to that of a "fisheye" or 180° photographic lens, could collect only half of this at best.

The light gathering power of a lens is a function of numerical aperture (NA). Lenses, as seen in Fig. 1, capture a "cone" of light. If the projection of this cone in the plane of the paper subtends an angle of $2u$, the NA of the lens is defined as $n \sin u$, where n is the refractive index of the medium in which the lens is operating. A lens with a 180° field of view in air has an NA of 1.0; typical "high-dry" microscope objectives have NA between 0.65 and 0.85. Light gathering power is proportional to $(NA)^2$. The collection of forward scatter or extinction signals is best accomplished using a lens with a low NA, because the light of interest is dispersed over a small angle; for collection of large angle scatter and fluorescence signals, a higher NA is desirable.

Because the sine of an angle cannot exceed 1.0, NA above this value can be achieved only by coupling the lens to the specimen with a medium of refractive index higher than 1.0. This is what is done when oil immersion or gel-coupled objectives are used; in this case, NA of 1.3–1.4 is readily obtainable. Flow cytometers equipped with such lenses can collect approximately half of the light scattered at large angles by, and the fluorescence emitted from, cells; lenses with NA 0.65–0.68 collect only about one-eighth of the large angle scatter and fluorescence signals. The lenses typically used with large stream-in-air cell sorters, with NA 0.5–0.55, could collect no more than one-tenth of these signals; light collection is typically further reduced because an obscuration bar is placed in front of the orthogonal collection lens to block light scattered from the air–stream interface. Collection may also be reduced by transmission losses in lenses; these are usually more severe in highly corrected, multielement lenses designed for high-resolution imaging than in simpler, less expensive optics.

In principle, more of the light emitted from or scattered by cells could be collected by a parabolic or ellipsoidal mirror with the observation point at its focus than by an immersion lens. However, this arrangement discriminates poorly against background light, and, in practice, decreases measurement sensitivity. In some systems, a spherical mirror is placed on the side of the flow cell opposite the collection lens to increase light collection; this may or may not increase background.

In most flow cytometers, a magnified image of the observation point is formed at some point in the orthogonal collection system, and an aperture, or field stop, is placed in the plane of this image, effectively preventing light collected from outside the immediate vicinity of the observation point from reaching the large angle scatter and fluorescence detectors; in the author's instruments, a separate field stop is used for each detector.

H. Wavelength Selection: Filters and Dichroics and Alternatives

In a spectrophotometer or spectrofluorometer, the measured wavelength region is typically defined using a monochromator, an optical system composed of an entrance slit, a grating or prism for spectral dispersion, and an exit slit. The grating or prism is rotated to divert the selected spectral component of the entering light to the exit slit. The spectral bandwidth is varied by adjusting the width of the slits; most instruments allow bandwidths between 1 and 10 nm.

Flow cytometers employing gratings or prisms for spectral dispersion have been built in several laboratories; however, this arrangement has some practical disadvantages. When detectors in multiple spectral bands are used, a relatively long light path is required in order to allow light from different spectral regions to be separated enough in space to reach different detectors of substantial size, for example, photomultiplier tubes. Precise placement of the detectors is also necessary to insure that each is responding to light in the desired spectral region. Polychromatic detectors using prisms or gratings and photodiode arrays or charge-coupled devices (CCDs) as detectors are somewhat easier to build because of the smaller size of these detectors, but, because the former are not as sensitive as PMTs, and the latter do not respond nearly as rapidly, most flow cytometers instead rely on combinations of optical filters for wavelength selection.

An interference filter, also called a dielectric, reflective, or dichroic, filter, is composed of a transparent glass or quartz substrate on which are deposited multiple thin layers of dielectric material, usually separated by spacer layers. Constructive and destructive optical interference occur between reflections from the various layers. The wavelength range(s) transmitted by an interference filter is determined primarily by the thickness of the layers; the selectivity is determined by the number of layers. The unwanted wavelengths are reflected. The spectral response of an interference filter varies with the angle of incidence of light on the filter; because the distance light must travel between layers varies with this angle, the wavelengths at which constructive and destructive interference occur also change.

In an absorptive, or color glass filter, unwanted light is removed by absorption, which can be substantially more effective than interference. Light transmission outside the passband, that is, the region in which the filter is designed to transmit light, is commonly a few percent for an interference filter and a few hundredths of a percent or less for a color glass filter. Transmission in the passband is usually higher for color glass filters than for interference filters. As a general rule, interference filters allow sharper transitions between rejected regions and the passband(s), and permit greater selectivity; filters with a FWHM [full width at half maximum (transmission)] bandwidth of 3 nm are readily available.

Unlike interference filters, absorptive filters may fluoresce, and, as was mentioned previously, such filter fluorescence can be a significant source of background in low-level fluorescence measurements. Although almost all flow cytometers now use interference filters for wavelength selection, most of these incorporate absorptive elements as well as dielectric layers in order to obtain highly effective blocking of light outside the passband. As a result, the filters have a "right side" and a "wrong side." The side with the absorptive layer is pigmented, whereas the side without it is shinier and less pigmented. When the filter is mounted in the instrument, the absorptive side should face away from the collection lens and toward the detector.

Band-pass, short-pass, and long-pass interference filters are available. The pertinent characteristics of a band-pass filter are its center wavelength, usually specified in nanometers (nm), its maximum transmission, and its bandwidth, usually given by the FWHM in nanometers. The band-pass filters used in flow cytometry typically have a FWHM of 20–30 nm and a maximum transmission of 70–80%. Long-pass filters are characterized by their maximum transmission and by their "cut on" wavelength, at which light transmission is 50% of maximum. Absorptive long-pass filters are also available; these have transmission curves similar to those of long-pass interference filters, but typically have a less sharp cut on and a much lower transmission outside the passband. Short-pass filters are characterized by their maximum transmission and cutoff wavelength, where transmission is 50% of maximum. They have practical limitations; one is that transmission is typically poor below about 420 nm due to absorption of some of the filter materials. A more significant problem is that transmission generally increases substantially at 150–200 nm above the cutoff wavelength and beyond.

Returning to Fig. 1, it can be seen that a dichroic, placed at a 45° angle to the axis of the collection optics, is used to separate collected light into spectral bands; in the figure, the dichroic reflects the short wavelength light in the orthogonal scatter signal and transmits the long wavelength light in the fluorescence signal. In the collection optics of most flow cytometers, additional dichroics are used to divide fluorescence into spectral bands; dichroics are also used in epi-illuminated fluorescence microscopes, and in arc source flow cytometers, to reflect band-pass filtered excitation light from an arc lamp through a lens toward the specimen, while transmitting light collected by the same lens to a fluorescence detector or detectors.

A dichroic is simply an interference filter, made without absorptive layers, and characterized for its transmission and reflectance characteristics at a 45° angle of incidence. Both long-pass (short-reflect) and short-pass (long-reflect) types are available. In more recent years, several manufacturers have developed filter sets for fluorescence excitation in microscopy incorporating excitation filters and dichroics that respectively, transmit and reflect in several discrete wavelength regions. This allows simultaneous viewing of (e.g.) blue, green, and red fluorescence excited by ultraviolet (UV), blue-green, and yellow light; such dichroics have not been widely applied in flow cytometry but have been helpful in high-resolution image cytometry, where the mechanical process of changing dichroics is likely to result in enough movement of the specimen and/or optics to cause problems in image registration.

In some commercial flow cytometers, the mechanical layout of the instrument places limits on the choice of dichroics, making it difficult to optimize light delivery to all fluorescence detectors. It is common practice to place a short wavelength reflecting dichroic first in the chain, as is shown in Fig. 1, because this will reflect scattered light and stray illuminating light out of the path of fluorescence detectors, minimizing the likelihood that it will pass through and/or induce fluorescence in the band-pass or long-pass filters associated with the detectors. However, it could be argued that at least some of the subsequent dichroics used to separate fluorescence into spectral bands should be of the short-pass or long-reflect type.

In a typical instrument with 488 nm excitation, in which measurements are to be made of cells binding roughly equal amounts of antibodies labeled with fluorescein (green fluorescent), phycoerythrin (yellow fluorescent), and phycoerythrin-Cy5 (red fluorescent), the emission spectral characteristics of the labels indicate that the yellow fluorescence signal should be strongest and the red weakest. The thin film deposition process involved in making interference filters dictates that reflectance can almost always be made higher than transmission (reflectance near 99% is readily obtainable; transmission higher than 90% is not). It thus makes sense for the weakest signals to be reflected from, rather than transmitted through, dichroics; once the first dichroic has separated 488 nm scattered light from fluorescence at longer wavelengths, it would probably be optimal to place a red (long) reflecting dichroic second and a green (short) reflecting dichroic third. For maximum efficiency in light transmission, the dichroics and detectors can be placed in a branched configuration, so no signal has to pass through more than two dichroics; an alternative approach, used in the author's instruments, is to place a second set of orthogonal collection optics on the side of the flow cell opposite the first set, making it unnecessary to use more than two dichroics in any signal path.

Color glass filters may be degraded by exposure to strong light in regions in which they absorb; putting such a filter in the path of an undiverged laser beam can produce discernable bleached areas in the filter. Interference filters typically absorb moisture over a period of years, resulting in changes in the spacing between dielectric layers and consequential changes in spectral response. This delamination is detectable by the mottled appearance which develops in the filter, and interference filters should be checked periodically and replaced if evidence of delamination is noted.

IV. Detection: Converting Optical Signals to Current

The numbers of photons contained in the various signals collected from cells vary over a wide range. In an absorption or extinction measurement, a substantial fraction of the illumination energy may impinge on the detector, and the intensity of forward scatter signals is typically at least a few percent of the intensity of the illuminating beam. It would be counterproductive to use a detector capable of responding to a small number of photons to measure these signals; in fact, it is often necessary to place a neutral density (N.D.) filter (preferably reflective, rather than absorptive) in the optical path to prevent high light levels from damaging the detector.

Orthogonal scatter signals and strong fluorescence signals typically do not include more than one- or two-hundredths as many photons as are incident on a cell during its transit, and, as has already been noted, only a small percentage of these may reach the detector due to the limited efficiency of light collection and transmission losses in the dichroics and filters. When fluorescence measurements are made of cells bearing only a few hundred to a few thousand molecules of cell-associated dye, no more than a few tens of photons may be available to the detector, and it is essential that the detector be able to respond to them. It has thus become common practice to use photodiodes as detectors for absorption, extinction, and forward scatter signals and PMTs as detectors for orthogonal scatter and fluorescence signals.

A. Photodiodes

Photodiodes, which are generally made of silicon, produce current when photons impinge on their photocathodes. A solar cell is a silicon photodiode, or an assembly of photodiodes, with a surface area sufficiently large enough to provide enough current to do useful work. Photodiodes do not require an external power source in order to operate. The peak sensitivity of silicon photodiodes is at about 900 nm, where they produce about 500 milliamperes (mA) of current per watt of incident radiant power. The quantum efficiency of photodiodes, that is, the number of photoelectrons produced at the cathode per hundred photons incident on it, is somewhat higher at shorter wavelengths; recall that a watt of radiant energy contains more photons as wavelength increases and the energy per photon decreases. A "UV-enhanced" silicon diode, with dopants added to the silicon to increase its response at shorter wavelengths, has a quantum efficiency of 61% at 450 nm, which rises to a maximum of 77% at 600 nm, and drops only to 72% at 800 nm. What are referred to in the industry as "blue-enhanced" and UV-enhanced devices actually exhibit improved performance over most of the visible spectrum compared to "standard" photodiodes.

A photodiode can be operated in the photovoltaic mode, in which no external voltage is applied to the diode, or in the photoconductive mode, with a bias voltage applied; bias does not increase responsivity or quantum efficiency, but shortens the response time of the device. Whether in the photovoltaic or photoconductive mode, a photodiode has unity gain, meaning it lacks an internal amplification mechanism. The current it produces in response to incident light must be amplified by external electronic components.

Photodiodes are typically small; the active area of devices used for cytometry is anywhere from a fraction of a square millimeter to several square millimeters. The devices are often packaged in integrated circuits that include components necessary to provide bias and amplification, and they are available for as little as $10.

B. Photomultiplier Tubes

Photomultiplier tubes, like photodiodes, produce current at their anodes when photons impinge on their light-sensitive photocathodes. However, unlike photodiodes, PMTs incorporate internal gain, and can produce output currents equivalent to millions of electrons out for each photon in. A PMT is a vacuum tube, which contains a photosensitive cathode, an anode, and a series of intermediate series of electrodes called dynodes, to each of which is applied an electrical potential (voltage) higher than that on the neighboring dynode nearer the photocathode. The anode is kept at a higher potential than any of the other electrodes; in practice, the anode is usually maintained at ground potential, and the dynodes and cathode are connected to progressively negative voltages, with the total voltage between cathode and anode anywhere from a few hundred to one to two thousand volts.

The electrodes in a PMT are contained in a transparent quartz or glass envelope, allowing photons passing through the envelope to reach the cathode, causing emission of electrons by the photoelectric effect. These photoelectrons are accelerated toward the first dynode by the electric field established by the potential difference between these two electrodes. The electrons, which have had their kinetic energy raised in proportion to the applied voltage difference, strike the dynode, each impact causing secondary emission of several relatively low energy electrons from the dynode; this process is repeated between each succeeding pair of dynodes and between the last dynode and the anode. Depending on the voltage applied, anywhere between 10^3 and 10^8 electrons may reach the anode for every electron that left the cathode.

The gain mechanism of a PMT is termed noise-free because, at least to a first approximation, the dynodes only emit electrons when struck by electrons. There is some low level "dark current" noise in PMTs, most of which results from thermionic (i.e., due solely to temperature) emission of electrons from the photocathode in the absence of incident light. The dark current and photoelectron and secondary electron statistics determine the signal-to-noise ratio obtainable from the PMT. When extremely low light levels must be detected, as in single photon counting, it is common practice to refrigerate the detector, reducing dark current by decreasing thermionic emission; in flow cytometry, this is usually unnecessary, because extraneous light of one kind or another, reaching the detector, rather than dark current, is typically the principal noise source.

The PMT spectral response is determined by the composition of the photocathode, which usually includes various alkali metals. Tubes with bialkali photocathodes have peak (photocathode) responsivity of about 40 mA/W at about 400 nm with corresponding quantum efficiencies in the range of 20–30%; their responsivity falls off sharply above 550 nm and is low enough at 600–650 nm to make these tubes unusable in this region. Multialkali photocathodes can extend the usable wavelength range to beyond 750 nm. Although the longest wavelength response is obtained using gallium arsenide (GaAs) cathodes, which provide a relatively flat responsivity curve between 300 and 850 nm, making them useful in spectrophotometers and spectrofluorometers, tubes with GaAs photocathodes have relatively low gain. The newest high-sensitivity multialkali tubes have quantum efficiencies exceeding those of many GaAs tubes even at wavelengths as high as 800 nm (6 versus 5%), and much higher gain. PMTs are, in general, larger than photodiodes; although some new subminiature types occupy a volume of only about a cubic centimeter, the devices used for cytometry are usually about 30 mm in diameter by 80 mm long. A PMT must be operated in a light-tight housing, preferably with electrostatic and magnetic shielding, and requires an external power source to provide the voltages for the dynodes and photocathode or anode. At present, only one manufacturer, Hamamatsu Corporation (Hamamatsu City, Japan; Bridgewater, NJ), makes PMTs with high red sensitivity and gain characteristics well suited to flow cytometry; these cost several hundred dollars each.

Hamamatsu also makes sockets that incorporate high voltage power supplies and detector modules that contain a PMT and a high voltage supply in a shielded housing. Newer PMT power supply designs use a voltage multiplier instead of a resistive voltage divider; this allows higher anode current to be drawn from the PMT while keeping response linear.

C. Avalanche Photodiodes

The avalanche photodiode (APD) is a detector that combines some desirable properties of photodiodes and PMTs; like the former, it has high cathode quantum efficiency, like the latter, it has gain. When a bias voltage ranging between a few hundred and a few thousand volts is applied across an APD, some electron acceleration occurs within the device, leading to secondary electron emission. The resulting gain, typically on the order of 100–1000, is considerably less than one can get with photomultipliers; however, because the cathode quantum efficiency of the diode is roughly an order of magnitude higher, an APD at a gain of 1000 is, in principle, as good as a PMT at a gain of 10,000.

Although the bias current drawn from a high voltage supply by an APD is substantially lower than the current needed by a PMT, the APD's requirement for a bias voltage remains something of a disadvantage. APDs also have relatively high dark currents, and their gain fluctuates considerably with slight variations in temperature; both of these problems can be dealt with by controlled cooling of the devices. Calculations and experimental results indicate that, under most circumstances, an APD assembly is less sensitive than a PMT as a detector for flow cytometry.

The "avalanche" in avalanche photodiode describes a phenomenon that occurs at a relatively high applied bias voltage. At lower voltages, the anode current remains proportional to the number of photons reaching the cathode. Above the "breakdown voltage," liberation of a single electron at the cathode results in a relatively substantial discharge of current; this is followed by what a neurophysiologist would call a refractory period, during which the device is unresponsive to light. The response of the APD in this "Geiger" mode is thus markedly nonlinear; the Geiger metaphor refers to the Geiger counter, in which ionization of gas by radiation produces a similar electrical breakdown phenomenon. APDs operating above breakdown voltage are incorporated in highly sensitive modules for single photon counting.

D. Single Photon Counting

The ultimate light detection task is single photon counting, in which each photoelectron released from the detector cathode generates a current pulse, and the individual pulses are detected and counted. Sensitivity is limited by dark current, and is improved by refrigerating the detector, which decreases thermionic emission. Single photon counting is used for bioluminescence measurements; it has also been employed (Nguyen et al., 1987; Goodwin et al., 1993) in flow cytometers using extremely low flow rates to detect single phycoerythrin molecules and small fragments of DNA labeled with fluorescent dyes. As long as photons reach the detector at relatively low rates, an APD-based detection module offers advantages over a PMT for single photon counting, because the APD's higher cathode quantum efficiency results in conversion of more incident photons to detectable current pulses.

In flow cytometry, or any other situation in which all cells spend the same amount of time in the illuminating beam, different numbers of fluorescence photons are collected from cells bearing different amounts of fluorescent dyes. The process follows Poisson statistics; if n photons are counted, the expected standard deviation of the measurement is $n^{1/2}$, that is, the square root of n, and the coefficient of variation, which is the standard deviation divided by the mean, is $n^{1/2}/n$. Since the precision of any individual measurement is a function of the number of photons counted, measurements of lower fluorescence intensities made under these conditions are less precise than measurements of higher fluorescence intensities.

In a static cytometer, it is feasible to measure fluorescence with uniform precision, independent of intensity, by illuminating each cell until a preset number of photons have been counted; for example, if the preset count is 10,000 photons, all measurements will have a coefficient of variation of (100/10,000), or 1%. Fluorescence intensity can be determined from the time required to accumulate the preset photon count. This technique has been used to improve the precision of fluorescence polarization measurements, which require computation of ratios of relatively low fluorescence intensities (Deutsch and Weinreb, 1994).

E. Charge-Coupled Devices

The charged-coupled device, or CCD, is the detector most commonly used in camcorders and digital cameras; cooled CCDs are also found in imaging cytometers designed for low light level measurements. A CCD typically contains anywhere from hundreds to millions of individual light sensing elements, which are usually arranged in a linear or rectangular array. Exposure of any element to light causes it to accumulate electric charge; attached electronic circuitry senses the amount of stored charge in each element at regular intervals. CCDs do not have gain, but, because they integrate the detected signal over relatively long time intervals (usually at least a few milliseconds), they are useful for measurement of low light intensities, especially when cooled. They would not be sensitive detectors for conventional flow cytometers, in which measurement times are on the order of microseconds.

V. Electronics: Converting Current to Voltage

While a flow cytometer is operating, its detectors run continuously, and, because they respond to stray light and generate some dark current, always generate some output in the form of electrical current. This is converted to voltage by preamplifiers. In principle, this could be accomplished by running the current through a grounded resistor, and using an electronic circuit to amplify the voltage across the resistor. In practice, current-to-voltage conversion, and most other analog electronic processing in flow cytometers and other instruments, is done using circuits composed of operational amplifiers (op amps), which use feedback to improve measurement accuracy. Preamplifiers in flow cytometers typically include not only a current-to-voltage converter, but also a feedback circuit, incorporating two or more additional op amps, to provide baseline restoration.

The baseline signal from the detector has a constant, or DC, and a variable, or AC, component. In a scatter channel, the baseline signal is due primarily to stray illumination light; in a fluorescence channel, the baseline signal may include contributions from autofluorescence, free dyes in the sample stream, and Raman scattering of the illuminating beam by water. At gain settings normally used for flow cytometry, PMT dark current does not contribute a substantial amount to the baseline signal in either scatter or fluorescence channels, and the baseline signal is typically proportional to the output of the light source.

Baseline restoration circuitry removes the DC, but not the AC, component of the baseline signal. The preamplifier output thus contains pulses representative of the signals from cells, with a superimposed AC component, including contributions due to fluctuations in the output of the light source and to electronic noise, with the former generally dominant. There is also a small DC offset voltage at a preamplifier output, due to the real-world performance limitations of operational amplifiers.

Because photodiodes have no gain, and APDs have relatively low gain, it is usually necessary to use one or more stages of electronic amplification in addition to a preamplifier to convert the outputs of these detectors to voltages that span the full input range of the analog-to-digital conversion circuitry used for data acquisition (in older instruments, this range was typically 0 to 10 V; in newer apparatus, the upper end of the range is more likely to be between 2.5 and 5 V). At least one stage of amplification typically has adjustable gain, allowing signals of interest to be placed in an appropriate range on the measurement scale.

In some flow cytometers, the output signals of PMT preamplifiers are subjected to one or more stages of electronic amplification with variable gain; this is likely to be counterproductive, because an electronic amplifier always adds some noise to the amplified signal, whereas increasing the gain of the PMT, at least in principle, does not.

VI. Fluorescence Compensation and Logarithmic Amplification

Cytometry often deals with signals that encompass a wide dynamic range. When immunofluorescence is measured, it is not uncommon to find some cells binding no more than a few thousand molecules of an antibody, while other cells in the same sample may bind hundreds of thousands of molecules. It is therefore common practice to display flow cytometric data on a logarithmic scale, with conversion from a linear scale usually being accomplished by routing the output signal from the preamplifier through a logarithmic amplifier or log amp. When fluorescence is measured in several spectral bands, accurate measurement of the signals from individual labels requires fluorescence compensation to correct for the effects of overlaps in emission spectra of different labels. This is usually accomplished using analog electronics; the signal in each spectral band has a fraction of the signals from the other spectral bands subtracted from it. The amount of a given signal that is subtracted is typically set by the operator interactively and intuitively, using either a knob or a computer pointing device. The compensation circuit itself is similar to those used in tone control and equalization circuits in audio equipment. As the number of spectral bands increases, the number of adjustments needed for compensation and the resulting complexity of the circuit increase. Two-color compensation requires two adjustments; complete three-color compensation would need six, although four are more commonly used, and four-color compensation requires twelve. Each additional adjustment inevitably increases the level of electronic noise at the output of the compensation circuit, which is the log amp input; compensation circuitry for four-color fluorescence typically restricts the available dynamic range to three decades or so, simply because of the higher noise level. Even if there were no noise, it would be difficult to adjust compensation empirically.

There is, however, an algorithmic method for determining the correct amount of fluorescence subtraction to be applied for each of any number of channels. Fluorescence compensation for n colors requires the solution of n equations in n unknowns, which is a relatively trivial problem for a personal computer. The coefficients of the equations are obtained from measurements, on a linear scale, of samples stained with each of the labels involved. Bagwell and Adams (1993) have described the procedure, and it has been implemented in several software packages. Fluorescence compensation, whether accomplished by hardware or software, requires data on a linear scale. Thus, in a system in which both fluorescence compensation and logarithmic transformation of data are accomplished by electronic circuits, the compensation circuit must precede the logarithmic amplifier in the signal path.

The compensation process is based on the assumptions that the fluorescence signals are dominated by fluorescence from the probes involved, and that the contributions from a fixed amount of any probe to any given fluorescence signal remain constant. The first assumption tends to break down for cells bearing relatively small amounts of label or exhibiting relatively high levels of autofluorescence; the circuitry or algorithm ends up inappropriately subtracting one portion of an autofluorescence signal from another, often resulting in negative values of the compensated signal. These must be adjusted to zero, or positive values near zero, either electronically or computationally, because one cannot apply the logarithmic transformation to negative data values. The electronic circuits used to adjust compensated signals to values above zero are called clamp circuits; they may exhibit deviations from ideal behavior for small input signal values, and thus introduce inaccuracies into measurement results in the lowest decade of the logarithmic scale.

The assumption that the fraction of fluorescence contributed by any given probe to any given signal remains constant can break down whenever probe molecules are close enough to one another for energy transfer to occur; in this circumstance, emission from the probe with the shorter wavelength emission peak is decreased, and emission from the probe with the longer wavelength emission peak is increased. Since, in typical situations, there is little spectral overlap between red- and green-emitting fluorescent antibody labels, the compensation circuitry in some flow cytometers makes no provision for subtracting red from green or green from red fluorescence. If, for example, a green-emitting fluorescein-labeled antibody to a nuclear antigen is used in conjunction with a red-emitting nuclear DNA stain such as propidium or 7-aminoactinomycin D, there may be substantial energy transfer from the fluorescein to the DNA stain; this is difficult to compensate for according to the classic model, and impossible to compensate for in an instrument which has not got the adjustments built in.

Logarithmic amplifiers are of two basic types. The first is an operational amplifier circuit with a transistor in its feedback loop, and various other components added to minimize the effects of temperature change and other extraneous variables on the output. The second is a more complex circuit incorporating a series of amplifiers connected in such a way that the sum of their outputs approximates the logarithm of the input signal. Both types of circuits are incorporated into modules by electronic component manufacturers; builders of instruments usually buy log amp modules rather than attempt the unenviable task of putting log amps together from scratch. At a minimum, it is necessary to add gain and offset adjustments to a circuit incorporating a log amp module, in order to produce outputs in the appropriate measurement range (0 to 2.5–10 V) from the output of the module, which usually spans only a few hundred millivolts, and may be negative or bipolar.

The response of even the best log amp modules is typically specified as within 0.5 decibel of the true logarithmic value over the usable range. Although this sounds like impressive fidelity to casual purchasers of audio equipment, a decibel represents a change in intensity of $10^{1/20}$, or about 12%, and the 0.5 decibel error amounts to 6%. In the real world, log amps may be far less accurate, exhibiting systematic deviations from ideal response, sharp peaks or dips, etc. It is fairly simple to determine the response characteristics of log amps using a procedure described by Parks et al. (1988).

Measurements are made of bright fluorescent beads, the ratio of fluorescence intensities of which is first determined on a linear scale. A ratio between 1.5 and 2 is desirable. Next, a series of fluorescence measurements is made using the log amp, using scatter parameters to set an acquisition gate including only single beads. PMT gain is initially adjusted to place the median channel of the peak representing the brighter beads near the top channel of the histogram, and this channel position and the number of channels between the medians of the two peaks are recorded. The PMT gain is then lowered to move the median channel of the brighter peak down by a few channels (5–10 on a 256-channel scale; 20–50 on a 1024-channel scale), and its new location and the number of channels between the peaks are again recorded. The procedure is continued, moving the peaks to lower channels; if PMT voltage must be set below 300 V to place the bead peaks at the lower end of the range, it is advisable to lower laser power or to place a neutral density filter in front of the PMT, because PMT response may itself be nonlinear at low voltages. The response curve is a plot of median channel of the upper peak, along the horizontal axis, versus the number of channels' difference between the medians of the bright and dim peaks, along the vertical axis. Since the ratio of bead fluorescence intensities is constant, the difference between the medians of the two peaks, which is proportional to the logarithm of this ratio, should also be constant. The ideal response curve is thus a line parallel to the horizontal axis. Log amps incorporating a single op amp circuit tend to deviate systematically from this curve at high and low signal values; log amps incorporating multistage modules exhibit a "wavy" response curve, with the number of "bumps" usually equally to the number of stages, but with less deviation from ideal at the extremes than is seen with single-circuit modules.

Whereas DC offsets of even a few tens of millivolts have little effect on the accuracy of flow cytometric measurements made on a linear scale, they can profoundly affect the accuracy of measurements made on a logarithmic scale, regardless of whether analog or digital computation is used for logarithmic transformation of data. If the highest voltage on a four-decade logarithmic scale is 10 V, the range of the highest decade is 1 to 10 V, and that of the next lower decade is 0.1 (100 mV) to 1 V. Signals in the next lower decade range from 10 to 100 mV; those in the lowest decade range from 1 to 10 mV. A 1 mV offset on a 1 V signal introduces an error of 1%, which is essentially negligible; a 1 mV offset on a 1 mV signal introduces an error of 100%. It is thus necessary to minimize offsets in preamplifiers, fluorescence compensation circuitry, and other linear electronics in order to maintain accuracy when data are transformed to a logarithmic scale.

VII. Peak Detection, Integration, and Pulse Width Measurement; Triggering

The processing and storage of information about cells represented by the height, area, and width of signal pulses requires the conversion of this information from an analog to a digital form. However, the pulses produced during the passage of a cell through the measurement system typically last for only a few microseconds at most, and, until more recently, the only analog-to-digital converters which could practically be used in flow cytometers required more time than this to digitize signals. As a result, it has been common practice to use hybrid circuits, which combine analog and digital electronics, to store the appropriate analog values for long enough to permit analog-to-digital conversion. These peak detector, integrator, and pulse width measurement circuits must be reset as each particle passes through the illuminating beam, allowing new analog signal levels to be acquired, and their outputs must then be held constant until digitization is complete.

In order for "reset" and "hold" signals to be delivered to the analog storage circuits at the proper times, it is necessary to use additional hybrid "front end" electronics, which compare one or more trigger signal levels with preset threshold values to determine when a cell is present. As a general rule, the signal selected for triggering should be relatively strong and available from all cells in the sample. In immunofluorescence analysis, for example, the forward scatter signal is usually chosen as the trigger signal, because at least some cells can be expected to have low-amplitude fluorescence signals.

Front end electronics cannot generate reset or hold signals until the trigger signal level rises above the set threshold; this usually occurs some hundreds of nanoseconds after what would be described as the leading edge, or "start," of the signal pulse. This does not pose problems for peak detection, since the pulse typically does not reach its peak value until some time after this. However, accurate integration and pulse width measurement require that the circuits used be reset at or shortly before the start of the pulse. This is usually accomplished by routing the input signals to integrators and pulse width measurement circuits through analog delay lines. The mechanics of peak detection, integration, and pulse width measurement will be discussed in more detail in Chapter 7 on data acquisition and display.

VIII. Measurement Sensitivity: Changing Concepts and the Bottom Line

The sensitivity of flow cytometers has commonly been expressed in terms of the smallest number of molecules of a fluorescent dye or labeled antibody detectable above background noise. This is estimated using mixtures of beads directly labeled with different amounts of dye or directly labeled antibodies and beads with defined antibody-binding capacities. Such mixtures usually include a blank bead, which does not contain dye or bind antibody. The fluorescence signal from a blank bead will have some finite value, and a comparison between signal intensities from blank and stained beads allows the fluorescence signal from the blank to be defined in terms of the numbers of molecule-equivalents of dye or antibody which that signal would represent. The background fluorescence level in many modern instruments is equivalent to a few hundred molecules of fluorescein or phycoerythrin.

More recently (Chase and Hoffman, 1998; Wood, 1998; Wood and Hoffman, 1998), it has been appreciated that sensitivity is not solely a function of the minimal detectable signal level. At the lowest signal levels encountered in flow cytometry, those obtained from weakly fluorescent cells or particles, it can be expected that most of the variance in signal intensity will be contributed by photon or photoelectron statistics. The resolution of two weakly fluorescent populations with slightly different mean fluorescent intensities or particles will thus be improved by collecting more photons from each particle measured, and by reducing the number of background photons. Whereas it is not possible to characterize the ability of a cytometer to resolve weakly fluorescent populations in terms of a single number, procedures do exist for determining the efficiency of light collection and for measuring the background light level. Using beads bearing known amounts of dye or labeled antibody, the fluorescence measurement scale can be calibrated in units of molecules of dye or antibody; once light collection efficiency and background level are known, it is also possible to calibrate the scale in units of photoelectrons. This was done for an instrument built in the author's laboratory (Shapiro *et al.*, 1998).

It was found that beads nominally bearing 1400 phycoerythrin molecules generated emission of 82 photoelectrons from the PMT cathode. At the normal PMT operating voltage, the four-decade range of output signals between 1 mV and 10 V spanned a range roughly corresponding to emission of between 2 and 20,000 photoelectrons, equivalent to photocurrents between 650 pA and 650 nA, and presumably resulting from between 15 and 150,000 580 nm photons impinging on the photocathode during a particle's transit of the beam. Because the energy of a 580 nm photon is 3.4×10^{-19} J, the optical power reaching the photocathode during a 5-μsec pulse is between 1.4×10^{-13} W and 1.4×10^{-9} W. Modern bench-top flow cytometers, which typically have background light levels at least as low, and light collection efficiencies at least as high, as the instrument described, would be expected to have approximately equivalent capability to detect and resolve weakly fluorescent populations.

It is appropriate to note, in conclusion, that in three decades' time, and four decades' range, fluorescence flow cytometry has progressed from counting cells with difficulty to counting molecules with relative ease. One can only hope that progress in solving the biological and medical problems to which the technology is now applied will be as rapid.

References

Bagwell, C. B., and Adams, E. G. (1993). Fluorescence spectral overlap compensation for any number of flow cytometry parameters. *Ann. N.Y. Acad. Sci.* **677,** 167–184.

Chase, E. S., and Hoffman, R. A. (1998). Resolution of dimly fluorescent particles, a practical measure of fluorescence sensitivity. *Cytometry* **33,** 267–279.

Deutsch, M., and Weinreb, A. (1994). An apparatus for high-precision repetitive sequential optical measurement of living cells. *Cytometry* **16,** 214–226.

Goodwin, P. M., Johnson, M. E., Martin, J. C., Ambrose, W. P., Marrone, B. L., Jett, J. H., and Keller, R. A. (1993). Rapid sizing of individual fluorescently stained DNA fragments by flow cytometry. *Nucleic Acids Res.* **21,** 803–806.

Mansberg, H. P., Saunders, A. M., and Groner, W. (1974). The Hemalog D white cell differential system. *J. Histochem. Cytochem.* **22,** 711–724.

Nguyen, D. C., Keller, R. A., Jett, J. H., and Martin, J. C. (1987). Detection of single molecules of phycoerythrin in hydrodynamically focused flows by laser-induced fluorescence. *Anal. Chem.* **59,** 2158–2160.

Parks, D. R., Bigos, M., and Moore, W. A. (1988). Logarithmic amplifier transfer function evaluation and procedures for log amp optimization and data correction. *Cytometry* **Suppl. 2,** 27.

Shapiro, H. M. (1995). "Practical Flow Cytometry," 3rd Ed. Wiley-Liss, New York.

Shapiro, H. M., Perlmutter, N. G., and Stein, P. G. (1998). A flow cytometer designed for fluorescence calibration. *Cytometry* **33,** 280–287.

van den Engh, G., and Farmer, C. (1992). Photo-bleaching and photon saturation in flow cytometry. *Cytometry* **13,** 669–677.

Wood, J. C. S. (1998). Fundamental flow cytometer properties governing sensitivity and resolution. *Cytometry* **33,** 260–266.

Wood, J. C. S., and Hoffman, R. A. (1998). Evaluating fluorescence sensitivity on flow cytometers, an overview. *Cytometry* **33,** 256–259.

CHAPTER 6

Flow Cytometric Fluorescence Lifetime Measurements

Harry A. Crissman and John A. Steinkamp

Biosciences Division
Los Alamos National Laboratory
Los Alamos, New Mexico 87545

I. Introduction
    A. Excited-State Lifetime Measurements
    B. Fluorescence Lifetime Image Cytometry
    C. Fluorescence Lifetime Flow Cytometry
II. Applications of the Technology
    A. Fluorescence Lifetime Measurements
    B. Phase-Resolved Fluorescence Measurements
III. Cell Preparation and Staining
IV. Fluorescence Lifetime Flow Cytometry Instrumentation
    A. Conventional Flow Cytometry Measurements
    B. Phase-Resolved Separation of Fluorescence Emission Signals
    C. Fluorescence Lifetime Measurements
    D. Instrument Initialization for Lifetime Measurements
V. Results
VI. Critical Aspects of the Technology
    A. Analysis of Heterogeneous Fluorescence Decays
    B. Homodyne versus Heterodyne Signal Detection
VII. Future Directions
    References

I. Introduction

    Excited-state lifetimes (fluorescence decay times) provide a means to discriminate among fluorescent markers in flow cytometric measurements and can be used to study the interaction of fluorescent markers with their cellular targets, with the each other, and the surrounding microenvironment. Fluorescence lifetime measurements by flow cytometry (FCM) are important because they yield additional information about fluorophore–cell interactions at the molecular level. An advantage of fluorescence lifetime measurements is that, in some instances, lifetimes can be considered as absolute quantities. However, the lifetimes of fluorophores bound to cellular macromolecules can be influenced by physical and chemical factors near the binding site, such as solvent polarity, cations, pH, energy transfer, and excited-state reactions. Such changes are often accompanied by a change in the temporal nature of the fluorescence decay (e.g., single-exponential, multiexponential, or nonexponential). Therefore, it is expected that lifetime measurements can be used to probe cellular complexes and subcompartments.

Fluorescence lifetime is defined as the characteristic time, ranging from a few hundredths to hundreds of nanoseconds (nsec), that a fluorophore molecule remains in an excited state prior to returning to the ground state. During the lifetime of the excited state the fluorophore can undergo conformational changes as well as interacting with its local environment. If a uniform population of fluorescent molecules is excited with a brief pulse of excitation light, the decay of the fluorescence intensity as a function of time can be described by the exponential function:

$$I(t) = I_o\, e^{-t/\tau}, \tag{1}$$

where $I(t)$ is the intensity measured at time t, I_o is the initial intensity immediately after the excitation pulse, and τ is the fluorescence lifetime. The fluorescence lifetime is defined as the time in which the fluorescence intensity decays to the 1/e of the initial intensity. A number of other deactivation or energy depleting processes can compete with fluorescence for return of the excited state electrons to the ground state. These include internal conversion, phosphorescence, and quenching. Other than fluorescence and phosphorescence, the processes for return of the excited state electrons to the ground state represent nonfluorescent mechanisms.

A. Excited-State Lifetime Measurements

There are two general methods for measuring fluorescence lifetimes: time-domain and frequency-domain analysis. In the time-domain method, the sample is excited by a series of short light pulses, and the time evolution of the fluorescence emission is measured directly by time-correlated single photon counting or a high-speed digital storage oscilloscope/sampling device (Demas, 1983; Lakowicz, 1983; O'Connor and Phillips, 1984). In the time-correlated single photon counting method the number of photo detector output pulses within a series of discrete sequential time intervals after the excitation pulse are counted, and a computer-generated histogram approximating the fluorescence decay curve is obtained. The fluorescence decay time(s) is determined from these data through the use of a computer algorithm.

Phase-modulation measurement is a frequency-domain alternative to the time-domain methods. In contrast to pulsed techniques, which record the amplitude of the fluorescence decay directly, the phase-modulation method determines the phase and amplitude of the sample fluorescence emission relative to a periodically modulated excitation source. If the excitation source is sinusoidally modulated [i.e., similar to an alternating current (AC) source], then the phase and amplitude of the fluorescence emission are a function of both the input frequency ω and the fluorescence lifetime. For a single exponential decay lifetime, the phase angle ϕ with respect to a reference signal and the relative modulation index m, defined as the ratio of the peak-to-peak amplitude of the AC component divided by the DC [direct current (average)] level of the fluorescence emission [modulation of fluorescence (m_{em})] divided by the ratio of the peak-to-peak amplitude of the AC component divided by the DC level of the excitation [modulation of excitation (m_{ex})] (Spencer and Weber, 1996), are related to the input frequency by the expressions:

$$\tan \phi = \omega \tau \quad \text{and} \quad m = m_{em}/m_{ex} = \cos \phi = [1 + (\omega \tau)^2]^{-1/2.} \quad (2)$$

By measuring phase and modulation over a wide range of frequencies, fluorescence lifetime(s) can be readily determined (Jameson *et al.*, 1984; Lakowicz *et al.*, 1984); however, phase-modulation methods even at fixed single modulation frequency, have a number of advantages: (1) short lifetimes (subnanosecond) are easily measured, (2) phase and modulation measurements are rapid, requiring only seconds of data acquisition, and (3) the technique of phase-sensitive detection of fluorescence increases the usefulness of phase-modulation methods in the analysis of heterogeneous samples (Lakowicz, 1983). A review by Bright *et al.* (1990) summarizes some of the past advances in frequency-domain spectrofluorometric developments.

B. Fluorescence Lifetime Image Cytometry

Fluorescence lifetime imaging provides a new way to unite functional and structural information for a more complete understanding of cellular processes by adding a new dimension to conventional fluorescence microscopy. Conventional fluorescence measurements are thus enhanced by the addition of lifetime imaging, along with spatial lifetime measurement within the cell. The sensitivity of fluorescence lifetime to the microenvironment within the cell makes lifetime imaging useful in measuring numerous cellular features, and lifetime itself can serve as a contrast enhancing mechanism. The utilization of fluorescence lifetimes enables numerous novel imaging applications that are particularly useful in cell biology. A review by French *et al.* (1998) summarizes the developments of fluorescence lifetime imaging techniques for cytometry.

C. Fluorescence Lifetime Flow Cytometry

Flow cytometers have been described which are capable of measuring lifetimes by phase shift using real time analog phase-sensitive detection electronic (Pinsky *et al.*, 1993; Steinkamp *et al.*, 1993); by amplitude demodulation using a high-speed digital storage oscilloscope followed by computer processing to determine lifetimes (Deka *et al.*, 1994) and real time analog electronics (Steinkamp *et al.*, 1998); by phase shift using high-speed digital oscilloscope to record the data which was then analyzed by the fast Fourier transform (Beisker and Klocke, 1997) and by phase shift and amplitude demodulation of heterodyned fluorescence emission signals that were acquired by a computer-controlled data acquisition system for subsequent lifetime calculation (Durack *et al.*, 1998). The flow cytometric technology for combined fluorescence lifetime measurements by phase shift on fluorophore-labeled particles and the surrounding fluorophore solution also has been described (Steinkamp and Keij, 1999a). Time-domain fluorescence lifetime measurements by FCM also have been accomplished using a high-speed digital storage oscilloscope to record the fluorescence decay signals from fluorochrome-labeled cells excited with a pulsed laser (Deka and Steinkamp, 1996). Lifetimes were then calculated from stored data by interative reconvolution analysis.

II. Applications of the Technology

A. Fluorescence Lifetime Measurements

The autofluorescence lifetime histograms of cultured viable human lung fibroblasts (HLFs) and of ethanol-, methanol-, paraformaldehyde-, and formaldehyde-fixed HLFs have been measured and compared (Steinkamp *et al.*, 1999a). The results illustrated that the broadened lifetime histogram CVs (coefficients of variation: standard deviations divided by the means) of both viable and the fixed HLFs partially overlaps the lifetimes of fluorescent antibody probe labeling, such as fluorescein isothiocyanate (FITC) labeled antibody, thus making phase-resolved fluorescence measurements to eliminate autofluorescence background in immunofluorescently labeled cells, based on lifetime differences, difficult to accomplish. To partially alleviate this problem, an approach in which low-concentration glutaraldehyde is used as a fixative for phase-resolved, cell-surface immunofluorescence measurements was developed. When compared to viable HLFs and HLFs fixed with the other fixatives, the fluorescence lifetime of glutaraldehyde-fixed cells is considerably shorter, the lifetime histogram CV is smaller, and phase-resolved fluorescence measurement of FITC-labeled antibody probe is permitted (Steinkamp *et al.*, 1999a).

Fluorescence quenching was first demonstrated using microspheres labeled on their surface with varying numbers of FITC molecules in which the results showed lifetime values ranging from 2.0 to 3.8 nsec (Steinkamp *et al.*, 1996). Detailed studies by Deka *et al.* (1996) further demonstrated the FITC self-quenching phenomena as a function of antibody labeling dilution and fluorescence to protein (F/P) ratio on murine thymus cells labeled with FITC-αThy 1.2 having F/P ratios of 13.7 (maximum quenching), 8.7, and 4.5 (minimum quenching). The measurement of fluorescence lifetimes for immunofluorescence cell-surface markers on murine thymus cells labeled with the fluorophores FITC, phycoerythrin (PE), and the PE/Texas Red (PE/Tx Red) tandem conjugate conjugated to anti(α)-Thy 1.2 antibody also demonstrated lifetime histograms with less than the expected values when compared to solution measurements and suggested fluorescence quenching (Steinkamp et al., 1999b).

Propidium iodide (PI) and ethidium bromide (EB) were used in studies designed to assess the sensitivity of fluorescence lifetime in detection of chromatin structural changes in apoptotic cells (Sailer et al., 1996). HL-60 cells induced into apoptosis by treatment with camptothecin (CAM) were stained with PI or EB to obtain bivariate DNA content and fluorescent lifetime profiles. Gated analysis showed the lifetime of both of these intercalating dyes was significantly reduced in apoptotic subpopulations compared to nonapoptotic cells. Several other classes of DNA-specific fluorochromes, with different modes of base-pair binding, were also evaluated for their assessment of chromatin damage during apoptosis (Sailer et al., 1997a). Results obtained indicated that the lifetime of some of the intercalating fluorochromes, such as EB, PI, and ethidium homodimer II (EthD II), but not TOTO or YOYO (Molecular Probes, Inc., Eugene, OR), are sensitive to apoptotic-induced chromatin damage. Results show that the excited state lifetime of some fluorochromes is not affected by (a) mode of binding to damaged DNA, (b) the DNA-fluorochrome microenvironment, (c) the solvent environment, and (d) size and structure of the fluorochrome molecule. Mithramycin lifetime was not affected even though EthD II, which also binds to G-C, had a reduced lifetime.

Deuterium oxide (D_2O) is known to increase both the fluorescence lifetime and the fluorescence intensity of the intercalating dyes by reducing the proton transfer rate between dye bound to DNA and the solvent system. In the studies cited earlier (Sailer et al., 1996, 1997a), spectroscopic analysis and FCM were performed to compare the alterations in intensity and lifetime of various DNA-binding fluorochromes bound to DNA in solution and in cells in the presence of D_2O/saline versus phosphate-buffered saline (PBS). Results of these studies indicate that the different modes of fluorochrome–DNA interaction, as well as differences in solvent accessibility to the fluorochromes, lead to a differential enhancement effect of D_2O on intensity and lifetime.

Combined conventional flow cytometric and lifetime analysis were used to monitor changes in residual chromatin in apoptotic HL-60 cell populations following treatment with camptothecin, cycloheximide, genistein, H7, and γ radiation (Sailer et al., 1998a). Results showed that all of these metabolic inhibitors, acting through different signaling cascades, induced apoptotic subpopulations with decreased, but different lifetime values for DNA-bound EB. Lifetime data appear to reflect difference in the extent of degradation, the potential for DNA repair, and the conformation of residual chromatin. Such information can be important in the evaluation and design of new apoptotic-inducing agents.

Other studies further demonstrated the capability to distinguish DNA and double-stranded (ds) RNA by the differences in the fluorescence lifetime of either PI or EB (Sailer et al., 1998b). The differences in lifetime of dyes bound to DNA and RNA could relate to the structural properties of the ds nucleic acids that influence differential intercalation of dyes and/or to changes in the dye–solvent interactions. Future lifetime applications will involve simultaneous measurement of both dsRNA and DNA correlated with immunofluorescent assay of cell cycle regulating proteins. This would allow for direct comparison of DNA and dsRNA metabolism in cycle-perturbed populations.

Ellipticine, like adriamycin and daunomycin, is a fluorescent, antitumor agent that has a high affinity for DNA and to a lesser extent RNA. Because ellipticine can bind to both DNA and RNA the capability to distinguish and resolve these drug-complexes in viable cells would provide information regarding its mode of binding. The fluorescence intensity and lifetime of ellipticine in viable HL-60 cells was monitored as a function of exposure time and/or ellipticine concentration (Sailer et al., 1997b). Data obtained showed a differential in the lifetime values of ellipticine bound to DNA, RNA, and other subcellular components. Phase-sensitive flow cytometry will potentially provide quantitative analysis of the binding of ellipticine to these components. At present, such determinations cannot be made by conventional methods for analyzing viable cells.

B. Phase-Resolved Fluorescence Measurements

Fluorescence lifetimes also can be used to resolve overlapping spectral emissions from fluorescent probes based on differences in their lifetimes expressed as phase shifts using phase-sensitive detection. The flow cytometric resolution of signals from fluorescence emissions by phase-sensitive detection was first demonstrated on cells stained with PI and FITC for total cellular DNA and protein content, respectively (Steinkamp and Crissman, 1993). Although the PI and FITC fluorescence emission signals are readily separable by conventional FCM methods, they were separated electronically using a single photomultiplier tube (PMT) detector, a long-pass (barrier) filter to block scattered laser excitation light, and two phase-sensitive detection channels, one for PI and the other for FITC.

Rat thymus cells labeled with the PE–anti(α)Thy 1.1 antibody and suspended in PBS containing PI (partially overlapping emission spectra) for discriminating "dead cells" also have analyzed by phase-resolved methods to identify cells that were PI-positive only, PE–αThy 1.1 labeled and PI-positive, and PE–αThy 1.1 labeled only; along with murine thymus cells labeled with PE/Tx Red–αThy 1.2 antibody and stained with PI as described later (Steinkamp et al., 1999b). Phase-sensitive detection also has been used to eliminate autofluorescence from glutaraldehyde-fixed lung fibroblasts labeled with a cell-surface FITC-antibody (Steinkamp et al., 1999a), and viable cells labeled with Hoechst 33342 and monobromobimane have been analyzed by phase-sensitive flow cytometry to determine relative DNA and glutathione content, respectively (Keij et al., 1999). Phase-resolved fluorescence measurements have been used to determine propidium uptake in damaged/dead macrophages containing phagocytized microspheres (Steinkamp et al., 2000). In addition, background interferences, such as light scatter, may be reduced or eliminated by phase-sensitive detection (Steinkamp et al., 1997; Steinkamp and Keji, 1999b).

III. Cell Preparation and Staining

Mouse thymocytes were obtained from deeply anesthetized 9- to 10-weeks-old C3H/HEJ mice (Harlan Sprague Dawley, Indianapolis, IN) after intraperitoneal injection of 10–15 mg of pentobarbital sodium (Abbott Laboratories, North Chicago, IL). The thymus from each animal was excised and rinsed in PSB (Gibco BRL, Grand Island, NY), Ca^{2+} and Mg^{2+} free, and the extraneous tissue was removed. Four to five thymuses were pooled, minced into small pieces, suspended in PBS, and passed four to five times through an 18-gauge needle to further disperse the cells. The cell suspensions were then passed through a 100-μm nylon mesh filter to remove large debris and clumps of cells, and the cells were centrifuged at 300 g, 4°C, for 10 min. The cell pellet was resuspended in PBS–1% bovine serum albumin (PBS–BSA) (Sigma, St. Louis, MO), washed, and resuspended in the same solution at 10^6 cells/50 μl for antibody labeling per manufacturer's instructions (Gibco BRL) at 4 μl/10^6 cells.

To demonstrate the ability of phase-sensitive FCM to resolve fluorescence signals from PE/Tx Red and/or PI labeled mouse thymocytes, three cell samples were prepared, the final cell concentration being 2.0–2.5 × 10^6 cells/ml in 3-ml tubes containing PBS–BSA. The samples consisted of (a) unlabeled washed mouse thymocytes to which PI (Sigma) was added (2 μg/ml) (control sample), (b) thymocytes that were incubated on ice for 30 min while labeling with PE/Tx Red–αThy 1.2 (CD90.2) (control sample), and (c) PE/Tx Red–αThy 1.2 labeled thymocytes to which PI was added (2 μg/ml). All sample tubes were kept on ice prior to phase-resolved analysis by FCM using 488 nm wavelength laser excitation, a 10 MHz modulation frequency, and a RG610 long-pass (barrier) filter (Melles Griot, Irvine, CA) in the fluorescence detector.

All protocols involving the use of animals in this study were reviewed and approved by the Los Alamos National Laboratory Animal Care and Use Committee. Furthermore, the conduct and reporting of the study was in accordance with the Guide for the Care and Use of Laboratory Animals prepared by the Committee on Care and Use of Laboratory Animals of the Institute of Laboratory Animal Resources, National Research Council, National Academy Press, Washington, DC.

Human promyelocytic (line HL-60) cells were grown in suspension culture at 37°C in T-75 flasks containing RPMI-1640 (Gibco BRL) supplemented with 20% fetal bovine serum (Hyclone Laboratories Inc., Logan, UT), 100 units/ml, 100 mg/ml streptomycin (Gibco BRL), and 1.25 mM L-glutamine. Cell densities were maintained between 0.2 and 1×10^6 cells/ml. The population doubling time ranged from 14 to 18 hr.

For induction of apoptosis, HL-60 cells in culture were treated with 0.15 mM camptothecin (Sigma) for 3 hr prior to fixation. Cells were harvested from suspension culture by centrifugation at room temperature for 5 min at 200 g and thoroughly resuspended in one volume of cold Puck's Saline A, without Ca^{2+} and Mg^{2+}, containing 1.0 mM EDTA (Gibco BRL). Three volumes of cold 95% ethanol were then added to yield a final cell concentration of 1.0×10^6/ml in 70% ethanol. After at least 24 hr of fixation, the cells were centrifuged at room temperature for 5 min at 200 g and the fixative aspirated. Cell pellets were resuspended in 2 ml of phosphate-citrate buffer (PC buffer; 192 parts 0.2 M Na_2HPO_4, 8 parts 0.1 M citric acid, pH 7.8) and incubated at room temperature for 45 min to extract low molecular weight DNA. The cells were centrifuged at room temperature for 5 min at 200 g, and the PC buffer was aspirated. Stock solutions of PI (Sigma) or EB (Molecular Probes, Eugene, OR) were prepared at 1 mg/ml in phosphate-buffered saline (PBS; pH 7.0). Cells were stained at 1×10^6 cells/ml with 3 μg/ml EB in PBS containing RNase at a concentration of 50 μg/ml.

Preparation of HL-60 cells for lifetime analysis of DNA and double-stranded RNA included fixation of cells in 70% ethanol as described previously, and then treatment in PBS with either RNase (Sigma) for DNA analysis, or DNase I (Sigma) for dsRNA analysis, at a concentration of 50 μg enzyme/10^6 cells at 37°C in the presence of 5 mM $MgCl_2$ for 1 hr. Untreated samples were treated in the same solution but without the enzymes. Following treatment, all samples were stained with 5 μg/ml PI in PBS. Lifetime measurements were performed using an excitation wavelength of 488 nm, a 10 MHz modulation frequency, and a OG515 long-pass (barrier) filter (Melles Griot) in the fluorescence detector.

IV. Fluorescence Lifetime Flow Cytometry Instrumentation

Fluorochrome-labeled cells are analyzed as they flow through a chamber and pass across an optically focused cw (continuous wave) laser beam from an argon laser which has a Gaussian-shaped intensity profile that is intensity-modulated [high frequency (ω) sine wave] as previously described (Steinkamp and Crissman, 1993) (see Fig. 1A). An RF signal synthesizer (low phase noise) is used as the sine wave generator for the modulator drive electronics and as the reference frequency source for homodyne signal detection. The Gaussian-shaped modulated fluorescence emission and light scatter signals, which consist of a low (cw-excited) and a high (sine wave modulated) frequency signal component, are detected by PMTs configured with high-speed, operational amplifiers (transimpedance mode) that serve as preamplifiers.

A. Conventional Flow Cytometry Measurements

The cw-excited (DC) fluorescence signal is extracted from the modulated fluorescence emission signal using low-pass filtering (0 to 180 kHz bandwidth)

6. Flow Cytometric Fluorescence Lifetime Measurements

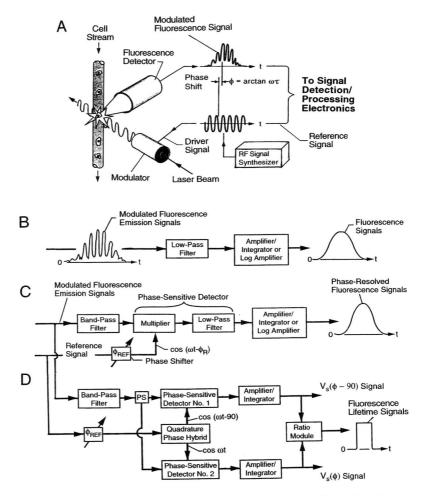

Fig. 1 Conceptual diagram of the fluorescence lifetime flow cytometer illustrating the laser excitation beam, the modulator, the modulated laser beam, the cell-stream laser beam intersection point in the flow chamber, the fluorescence detector, the modulated fluorescence emission and reference signals, and sine wave synthesized signal generator (A). The fluorescence signals and reference signal are input to analog signal processing electronics to give conventional FCM signals (B), phase-resolved fluorescence signals (C) and fluorescence lifetimes (D), prior to acquisition by a computer-based data acquisition system.

to give conventional FCM fluorescence-intensity information as shown in Fig. 1B. The conventional fluorescence signals are amplified/integrated or logarithmically amplified prior to acquisition by a computer-based data acquisition system. Similarly, orthogonal and forward light scatter signals are obtained using low-pass filtering followed by linear or logarithmic amplification.

The high-frequency sine wave modulated fluorescence signal component, which is shifted in phase (ϕ_s) by an amount

$$\phi_s = \arctan \omega\tau \tag{3}$$

relative to the excitation frequency, is processed by phase-sensitive detectors to resolve fluorescence emission signals based on differences in their lifetimes (expressed as phase shifts) and to quantify fluorescence lifetimes directly as a parameter, as illustrated in Fig. 1C,D, respectively.

B. Phase-Resolved Separation of Fluorescence Emission Signals

The principle of phase suppression, as applied to flow cytometry, for separating two fluorescence emission signals having different lifetimes, that is, phase shifts ϕ_1 and ϕ_2, is based on the theory of Veselova *et al.* (1970) for individually recording fluorescence spectra in systems containing two luminescent centers and the phase-sensitive fluorescence spectroscopy work of Lakowicz and Cherek (1981). By use of phase-sensitive detection, two superimposed modulated fluorescence signals each having different phase shifts are multiplied by a sine wave reference signal (phase shift ϕ_R) using a signal mixer, and are low-pass filtered (0 to 180 kHz bandwidth) to remove the high frequency signal components as shown in Fig. 1C. Switchable nanosecond delay lines are used to shift the phase of the reference signal with respect to the modulated fluorescence signals that are input to the phase detectors. The detector output signals are amplified, including logarithmic, or integrated. To resolve either of the two signals, the reference phase is shifted by an amount $\pi/2 + \phi_1$ or $-\pi/2 + \phi_2$ degrees. This results in one signal being passed and the other being nulled. When fluorescence signals are processed by two phase-sensitive detectors operating in parallel, then by setting one detector reference to $\pi/2 + \phi_1$ degrees and the other detector reference to $-\pi/2 + \phi_2$ degrees, the contributions to the total fluorescence signal are resolved (Steinkamp and Crissman, 1993).

C. Fluorescence Lifetime Measurements

Fluorescence lifetimes are measured by the two-phase, phase comparator method using two reference sine wave signals that are 90° out of phase with each other and that are each multiplied by the band-pass filtered fluorescence emission signal using the phase-sensitive detectors (Fig. 1D). The detector output signals are divided on a cell-by-cell basis, which results in the $V_s(\phi - 90)/V_s(\phi)$ ratio signal expression which is proportional to the sin ϕ_s divided by the cos ϕ_s (Meade, 1982). The fluorescence decay time, which equals the tan ϕ_s from Equation (3), is directly proportional to the $V_s(\phi - 90)/V_s(\phi)$ ratio expressed as

$$\tau = \frac{1}{\omega}\tan\phi_s = \frac{1}{\omega}[V_s(\phi - 90)/V_s(\phi)]. \tag{4}$$

The conventional fluorescence and light scatter, phase-resolved fluorescence, and lifetime signals are recorded as list mode data for display as frequency distribution histograms or as bivariate dot/contour diagrams using a computer-based data acquisition system.

Fluorescence lifetime also may be determined by measuring the relative depth of amplitude modulation (m) of the emission signal (m_{em}) with respect to the excitation signal (m_{ex}). The relative modulation, or demodulation factor m is determined from the ratio

$$m = m_{em}/m_{ex} = \frac{\text{modulation of fluorescence}}{\text{modulation of excitation}} = \cos \phi = 1/[1 + (\omega\tau)^2]^{1/2}. \quad (5)$$

In the steady-state system it is only necessary to measure the AC and DC fluorescence emission and excitation signal components and determine the relative modulation by ratio calculations (Spencer and Weber, 1969). In flow cytometry, the amplitude demodulation factor can be determined by measuring the maximum and minimum signal components at the peak height of the Gaussian-shaped fluorescence detector output signal and the AC and DC components of the steady-state laser excitation either by digital (Deka *et al.*, 1994) or analog (Steinkamp *et al.*, 1998) methods. The real-time ratio results in the cos ϕ that is proportional to the fluorescence decay lifetime expressed as

$$\tau = 1/\omega \cdot \tan (\cos^{-1} m). \quad (6)$$

D. Instrument Initialization for Lifetime Measurements

The ability to quantify fluorescence decay times on cells labeled with fluorescent probes by direct phase-shift measurement is illustrated next. The outputs of phase-sensitive detector numbers 1 and 2 of Fig. 1D, that is, $V_s(\phi - 90)$ and $V_s(\phi)$, are first initialized by removing the long-pass barrier filter in the fluorescence detector, adjusting the reference phase shift (ϕ_R) to zero (null), and maximizing the $V_s(\phi - 90)$ and $V_s(\phi)$ output signals, respectively, using nonfluorescent microspheres. The barrier filter is then replaced and fluorescent labeled particles of known lifetime, for example, Flow-Check alignment fluorospheres (Coulter Corp., Miami, FL) lifetime ~7.0 nsec, are analyzed at the same PMT and phase-sensitive detector amplifier gain settings. The ratio module gain is adjusted to center the microspheres histogram typically in channel 70 or 140 prior to analyzing labeled cell samples at fixed gain settings. Neutral density filters are used in the fluorescence detector to compensate for differences in light scatter (nonfluorescent particles) and fluorescence (microspheres and labeled cells) signal intensities when required to maintain the PMT voltage constant during lifetime measurement.

V. Results

An example illustrating phase-resolved measurements on murine thymus cells labeled with PE/Tx Red–αThy 1.2 and suspended in PBS containing PI (for labeling PI-positive damaged/dead cells) using two phase-sensitive detector (PSD) channels operating in parallel is shown in Fig. 2. Since the fluorescence emission spectra of PE/Tx Red and PI completely overlap (see Fig. 2A), the separation of PE/Tx Red and PI signals cannot be achieved using conventional methods employing electronic compensation (Loken et al., 1977), however, the measured fluorescence lifetime histograms of PE/Tx Red and PI individually labeled cells are well separated (see Fig. 2B). Based on the lifetimes of PE/Tx Red (2.4 nsec) and PI (16.5 nsec), murine thymus cells were labeled separately with PE/Tx Red–αThy 1.2 and PI (controls), and the phase shifts of the two phase-sensitive detector channels were first adjusted to: (1) null PE/Tx Red signals in the PSD channel number 2 output and (2) null PI signals in the PSD channel number 1 output. Thymus cells were labeled with both PE/Tx Red–αThy 1.2 and suspended in PBS containing PI and then analyzed at the same reference phase shift and gain settings as earlier, and the phase-resolved histograms (see Fig.

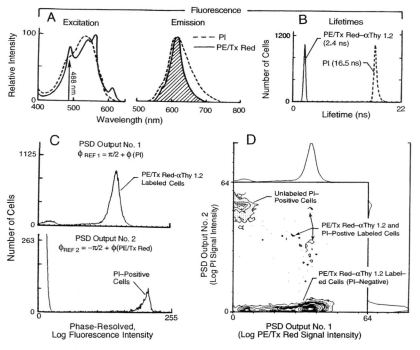

Fig. 2 Fluorescence excitation and emission spectra of PI and PE/Tx Red (A) and fluorescence lifetime histograms recorded on murine thymus cells labeled separately with PE/Tx Red–αThy 1.2 and PI (B). Phase-resolved, log fluorescence frequency distribution histograms recorded from the PSD number 1 and number 2 outputs of two phase-sensitive detector (PSD) channels operating in parallel (Steinkamp et al., 1999b) on unfixed murine thymus cells labeled with Tx Red–αThy 1.2 and suspended in PBS containing PI (C) and corresponding bivariate contour diagram (D).

2C) and corresponding bivariate contour diagram (see Fig. 2D) were recorded. Approximately 89% of the total cell population were labeled only with PE/Tx Red–αy 1.2, 6% were only PI-positive, 2% of the PE/Tx Red labeled cells were PI-positive, and 3% were unlabeled null cells.

The induction of apoptosis results in the activation of cellular endonucleases that cleave DNA at nucleosomal linker regions in chromatin. DNA fragments diffuse from apoptotic cells following ethanol fixation and a subsequent wash with aqueous buffer. The DNA remaining in apoptotic cells represents fragments of DNA loops still attached to the nuclear matrix and long stretches of nucleosomes that cannot be extracted under these conditions. Apoptotic chromatin fragments, depleted in histone H1, show an accumulation of HMG1 and HMG2 proteins and probably derive from chromatin in a transcriptionally active configuration.

Fluorescence lifetime of EB bound to apoptotic subpopulations was noted to be significantly reduced compared to the nonapoptotic portion of the population. Figure 3 shows the fluorescence lifetime distributions and bivariate distributions of fluorescence intensity (DNA content) versus fluorescence lifetime for HL-60 cells treated with camptothecin, fixed, PC buffer-treated, and then stained with stained with EB (3 μg/ml). Gated analysis revealed that the nonapoptotic cells had a lifetime of 22 nsec, while apoptotic subpopulations had lifetime values of 18 nsec. Results indicate that the binding of these probes within the chromatin of the apoptotic cells decreases the fluorescence lifetime. The ability to discriminate this lower lifetime population denotes the ability to potentially monitor changes in chromatin structure by altered lifetime values.

Flow cytometry lifetime studies provided the distinction of DNA and dsRNA from the observed differences in the fluorescence lifetime of either PI or EB when bound to either nucleic acid. Figure 4 shows a lifetime value of 15.6 nsec for DNA-bound PI in ethanol-fixed HL-60 cells treated with RNase prior to staining and analysis, while a value of 17.2 nsec is noted for dsRNA-bound PI in cells treated with DNase. An intermediate lifetime value of 16.4 nsec was obtained for PI bound to untreated samples. Lifetime differences of a similar magnitude were noted when EB was used as the probe in similar studies (data not shown). Results obtained from the untreated sample represent the average lifetime value for both DNA and dsRNA, and since they are present in cells in somewhat equal portions, an intermediate lifetime value is obtained. These differences in lifetime relate to the differences in the structure of the nucleic acid complexes, and reflect the dissimilarities in the intercalation of these fluorochromes into DNA or dsRNA. Simultaneous measurement of both DNA and dsRNA by phase-resolved separation of the fluorescence lifetimes allows direct correlation of DNA and predominately ribosomal RNA metabolism on a cell-by-cell basis. These data also demonstrate the sensitivity of FCM lifetime analysis, which can reproducibly detect significant lifetime changes as small as 0.2 nsec in DNA-binding fluorochromes.

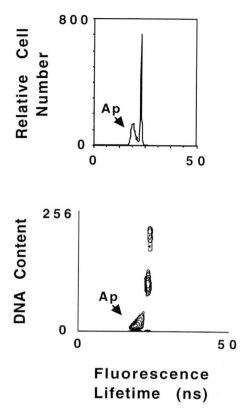

Fig. 3 Fluorescence lifetime histogram of EB bound to HL-60 cells induced to undergo apoptosis (top) and a corresponding bivariate distribution of lifetime versus DNA content (bottom). Cells were treated with 0.15 mM camptothecin (Sigma) for 3 hr prior to fixation in 70% ethanol. Fixed cells were treated with PC buffer (192 parts 0.2 M Na$_2$HPO$_4$, 8 parts 0.1M citric acid) for 45 min at room temperature to remove small molecular weight DNA produced during apoptosis. Samples were subsequently stained at a concentration of 10^6 cells/ml in a solution containing 3 μg/ml EB and 50 mg/ml RNase in PBS.

VI. Critical Aspects of the Technology

A. Analysis of Heterogeneous Fluorescence Decays

The phase and modulation fluorescence lifetime equations noted earlier are derived on the assumption of a single-component exponential decay of fluorescence from a homogeneous emitting fluorophore population. This is often cited as the major shortcoming of the single-frequency method, because the existence of a unique single-component decay is presupposed, but not demonstrated by

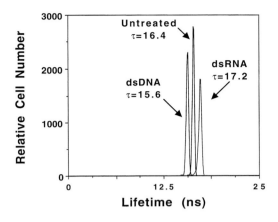

Fig. 4 Histogram showing the fluorescence lifetime of PI bound to DNA or dsRNA, or both (untreated) in HL-60 cells. Cells were fixed in 70% ethanol, and then treated in PBS with either RNase (Sigma) for DNA analysis, or DNase I (Sigma) for dsRNA analysis, at a concentration of 50 μg enzyme/10^6 cells at 37°C in the presence of 5 mM MgCl$_2$ for 1 hr. Untreated samples were stained without any prior enzyme treatment. Following treatment, all samples were stained with 5 μg/ml PI in PBS.

the measurement. This is indeed true if only one of the two quantities, that is, lifetime by phase shift or amplitude demodulation, is measured. However, if both are measured, the existence of an exponential can be demonstrated. In the heterogeneous fluorophore population, the lifetime measured by the degree of amplitude demodulation will almost always be larger than the weighted average of the individual component lifetime values, whereas the lifetime determined by phase shift will always be shorter than the weighted average (Spencer and Weber, 1969). It is only when there is a single exponential decay that both methods give the same result. Also, phase shift measurements at two or more frequencies can be used to detect heterogeneous fluorescence decays (Deka et al., 1995). If the decay is multiexponential, measurement by time-domain (Demas, 1983; Lakowicz, 1983; O'Connor and Phillips, 1984) or by frequency-domain (Jameson et al., 1984; Lakowicz et al., 1984) methods is required to characterize the decay.

The longest lifetime that can be measured in phase-modulation flow cytometry depends on the lowest usable excitation frequency, which is about 0.5 MHz. This corresponds to a 318-nsec lifetime calculated at a 45° phase shift, but in practice will be somewhat higher in value. For modulation frequencies lower than 0.5 MHz, the cw-excited, low-frequency signal component interferes with the 0.5 MHz high-frequency signal. The shortest measurable lifetime depends on the maximum highest modulation frequency usable and the bandwidth of the signal detection/processing electronics and data acquisition system. Lifetime measurement capabilities by FCM of a few tenths of nanoseconds have been reported (Pinksy et al., 1993; Steinkamp et al., 1993).

B. Homodyne versus Heterodyne Signal Detection

Signal homodyning relies on direct measurement of the phase shift by multiplying the modulated fluorescence signal with a sine wave or other suitable reference signal of the same frequency using any number of electronic devices, such as a double-balanced mixer. This is conceptually the simplest form of signal processing and is performed using analog or digital methods. To avoid the need for high-frequency signal processing electronics and to better isolate the signal from noise interference, a frequency-heterodyning technique has been developed for static spectrofluorometric frequency-domain lifetime measurements (Spencer and Weber, 1969) and is adaptable to flow cytometers (Pinsky and Ladasky, 1994; Durack *et al.*, 1998). This technique works by mixing the fluorescence signal, at the detector PMT base or an external mixer, with a second signal of different frequency, namely, frequency heterodying. The resulting difference frequency contains the same information as the original high-frequency modulated signal, but the difference can be set to any suitable lower value to suit the measurement conditions, for example, signal processing (digital) speed. By replacing the analog processing electronics with a computer-controlled digital signal processing and acquisition system and heterodyne signal detection, lifetime data can be processed by software (Feddersen *et al.*, 1989). The digital-acquisition technology provides new capabilities for measuring excited-state lifetimes by both time- and frequency-domain methods.

VII. Future Directions

Time-resolved fluorescence measurements by flow cytometry is so new that many of its potential applications have not been fully explored or developed. This technology will add a new dimension to multiparameter flow cytometric analyses through the development of techniques for measuring fluorescence lifetime of probes bound to macromolecular complexes in cells, will increase the range of fluorescent markers that can be used in multilabeling applications, will reduce background interferences, and, will thus enhance measurement precision to yield more accurate results. In the past, procedures were limited in some cases by the availability of fluorescent markers with common excitation regions (so that a single laser excitation source could be used) and emission spectra that were sufficiently separated using optical color-separating filters. Because the lifetime-based sensing technology can separate fluorescence emissions electronically (and also optically), quantify fluorescence lifetimes directly, and make conventional flow cytometric measurements, it has a wide range of technically possible applications. The technology will significantly expand understanding of the biological processes at the cellular, subcellular, and molecular level. In addition, the technology can be adapted to commercial flow cytometry systems where it can be used for virtually any clinical or research application involving the analysis of cells, cell function, or subcellular components through the use of fluorescent markers directed to specific targets.

Acknowledgments

This work was performed at the Los Alamos National Laboratory, Los Alamos, New Mexico, under the joint sponsorship of the U.S. Department of Energy, the Los Alamos National Flow Cytometry Resource (National Institutes of Health Grant P41-RR013150), and National Institutes of Health (Grants R01-RR07855 and R01-RR06758). We thank Nancy M. Lehnert, Carolyn Bell-Prince, and Joseph G. Valdez for their assistance in cell preparation and staining.

References

Beisker, W., and Klocke, A. (1997). Fluorescence lifetime measurement in flow cytometry. *Proc. SPIE* **2982,** 436–446.

Bright, F. V., Betts, T. A., and Litwiler, K. S. (1990). Advances in multifrequency phase and modulation fluorescence analysis. *Anal. Chem.* **21,** 389–405.

Deka, C., and Steinkamp, J. A. (1996). Time-resolved fluorescence-decay measurement and analysis on single cells by flow cytometry. *Appl. Opt.* **35,** 4481–4489.

Deka, C., Sklar, L. A., and Steinkamp, J. A. (1994). Fluorescence lifetime measurements in a flow cytometer by amplitude demodulation using digital data acquisition technique. *Cytometry* **17,** 94–101.

Deka, C., Cram, L. A., Habbersett, R., Martin, J. C., Sklar, L. A., and Steinkamp, J. A. (1995). Simultaneous dual-frequency phase-sensitive flow cytometric measurements for rapid identification of heterogeneous fluorescence decays in fluorochrome-labeled cells and particles. *Cytometry* **21,** 318–328.

Deka, C., Lehnert, B. E., Lehnert, N. M., Jones, G. M., Sklar, L. A., and Steinkamp, J. A. (1996). Analysis of fluorescence lifetime and quenching of FITC-conjugated antibodies on cells by phase-sensitive flow cytometry. *Cytometry* **25,** 271–279.

Demas, J. N. (1983). "Excited State Lifetime Measurements." Academic Press, New York.

Durack, G., Yu, W., Mantulin, W., and Gratton, E. (1998). Fluorescence lifetime flow cytometry: a phase-sensitive detection system utilizing a cross correlation technique and digital signal processing. *Cytometry* **9**(Suppl.), 39.

Feddersen, B. A., Piston, D. W., and Gratton, E. (1989). Digital parallel acquisition in frequency domain fluorometry. *Rev. Sci. Instrum.* **60,** 2929–2936.

French, T., So, P. T. C., Dong, C. Y., Berland, K., and Gratton, E. (1998). Fluorescence lifetime imaging techniques for microscopy. *Methods Cell Biol.* **56,** 277–304.

Jameson, D. M., Gratton, E., and Hall, R. D. (1984). The measurement and analysis of heterogeneous emissions by multifrequency phase and modulation fluorometry. *Appl. Spectrosc. Rev.* **20,** 55–106.

Keij, J. F., Bell-Prince, C., and Steinkamp, J. A. (1999). Simultaneous analysis of relative DNA and glutathione content in viable cells using phase-sensitive flow cytometry. *Cytometry* **35,** 48–54.

Lakowicz, J. R. (1983). "Principles of Fluorescence Spectroscopy." Plenum, New York.

Lakowicz, J. R., and Cherek, H. (1981). Resolution of heterogeneous fluorescence from proteins and aromatic amino acids by phase-sensitive detection of fluorescence. *J. Biol. Chem.* **256,** 6348–6353.

Lakowicz, J. R., Laczko, G., Cherek, H., Gratton, E., and Limkeman, M. (1984). Analysis of fluorescence decay kinetics from variable-frequency phase shift and modulation data. *Biophys. J.* **46,** 463–477.

Loken, M. R., Parks, D. R., and Herzenberg, L. A. (1977). Two-color immunofluorescence using a fluorescence activated cell-sorter. *J. Histochem. Cytochem.* **25,** 899–907.

Meade, M. L. (1982). Advances in lock-in amplifiers. *J. Phys. E: Sci. Instrum.* **15,** 395–403.

O'Connor, D. V., and Phillips, D. (1984). "Time-Correlated Single Photon Counting." Academic Press, London.

Pinsky, B. G., and Ladasky, J. J. (1994). Heterodyning of modulated pulses for fluorescence lifetime measurements in flow cytometry. *Proc. SPIE* **2137**, 794–799.

Pinsky, B. G., Ladasky, J. J., Lakowicz, J. R., Berndt, K., and Hoffman, R. A. (1993). Phase-resolved fluorescence lifetime measurements for flow cytometry. *Cytometry* **14**, 123–135.

Sailer, B. L., Nastasi, A. J., Valdez, J. G., Steinkamp, J. A., and Crissman, H. A. (1996). Interactions of intercalating fluorochromes with DNA analyzed by conventional and fluorescence lifetime flow cytometry utilizing deuterium oxide. *Cytometry* **25**, 164–172.

Sailer, B. L., Nastasi, A. J., Valdez, J. G., Steinkamp, J. A., and Crissman, H. A. (1997a). Differential effects of deuterium oxide on the fluorescence lifetimes and intensities on dyes with different modes of binding to DNA. *J. Histochem. Cytochem.* **45**, 165–175.

Sailer, B. L., Valdez, J. G., Steinkamp, J. A., Darzynkiewicz, Z., and Crissman, H. A. (1997b). Monitoring uptake of ellipticine and its fluorescence lifetime in relation to the cell cycle by flow cytometry. *Exp. Cell Res.* **236**, 259–267.

Sailer, B. L., Valdez, J. G., Steinkamp, J. A., and Crissman, H. A. (1998a). Apoptosis induced with different cycle perturbing agents produces differential changes in the fluorescence lifetime of DNA-bound ethidium bromide. *Cytometry* **31**, 208–216.

Sailer, B. L., Steinkamp, J. A., and Crissman, H. A. (1998b). Flow cytometric fluorescence lifetime analysis of DNA-binding probes. *Eur. J. Histochem.* **42**, 19–28.

Spencer, R. D., and Weber, G. (1969). Measurements of subnanosecond fluorescence lifetimes with a cross-correlation phase fluorometer. *Ann. N.Y. Acad. Sci.* **158**, 361–376.

Steinkamp, J. A., and Crissman, H. A. (1993). Resolution of fluorescence signals from cells labeled with fluorochromes having different lifetimes by phase-sensitive flow cytometry. *Cytometry* **14**, 210–216.

Steinkamp, J. A., and Keij, J. F. (1999a). Fluorescence intensity and lifetime measurement of free and particle-bound fluorophore in a sample stream by phase-sensitive flow cytometry. *Rev. Sci. Instrum.* **70**, 4682–4688.

Steinkamp, J. A., and Keij, J. F. (1999b). Elimination of light scatter interference in dual-laser flow cytometry by synchronous detection of emitted fluorescence: theory and demonstration us simulated signals. *Proc. SPIE* **3604**, 170–176.

Steinkamp, J. A., Yoshida, T. M., and Martin, J. C. (1993). Flow cytometer for resolving signals from heterogeneous fluorescence emissions and quantifying lifetime in fluorochrome-labeled cells/particles by phase-sensitive detection. *Rev. Sci. Instrum.* **64**, 3440–3450.

Steinkamp, J. A., Deka, C., Lehnert, B. E., and Crissman, H. A. (1996). Fluorescence lifetime as a new parameter in analytical cytology measurements. *Proc. SPIE* **2678**, 221–230.

Steinkamp, J. A., Lehnert, B. E., and Keij, J. F. (1997). Phase-sensitive detection as a means to recover fluorescence signals from interfering backgrounds in analytical cytology measurements. *Proc. SPIE* **2982**, 447–455.

Steinkamp, J. A., Parson, J. D., and Keij, J. F. (1998). Progress towards combined phase and amplitude demodulation fluorescence lifetime measurements by flow cytometry. *Proc. SPIE* **3260**, 236–244.

Steinkamp, J. A., Lehnert, N. M., Keij, J. F., and Lehnert, B. E. (1999a). Enhanced immunofluorescence measurement resolution of surface antigens on highly autofluorescent, glutaraldehyde-fixed cells by phase-sensitive flow cytometry. *Cytometry* **37**, 275–283.

Steinkamp, J. A., Lehnert, B. E., and Lehnert, N. M. (1999b). Discrimination of damaged/dead cells by propidium iodide uptake in immunofluorescently labeled populations analyzed by phase-sensitive flow cytometry. *J. Immunol. Methods* **226**, 59–70.

Steinkamp, J. A., Valdez, Y. E., and Lehnert, B. E. (2000). Flow cytometric, phase-resolved fluorescence measurement of propidium iodide uptake in macrophages containing fluorescent microspheres. *Cytometry* **39**, 45–55.

Veselova, T. V., Cherkasov, A. S., and Shirokov, V. I. (1970). Fluorometric method for individual recording of spectra in systems containing two types of luminescent centers. *Opt. Spectrosc.* **29**, 617–618.

CHAPTER 7

Principles of Data Acquisition and Display

Howard M. Shapiro
Howard M. Shapiro, M.D., P.C.
West Newton, Massachusetts 02465

I. Introduction
II. Pulse Characterization Using Analog and Hybrid Circuits
    A. Triggering
    B. Peak Detector Timing and Performance
    C. Timing and Performance of Integrators and Pulse Width Measurement Circuits
    D. Integration by Pulse Shaping
III. Analog Data Display
IV. Analog-to-Digital Conversion
    A. Single Slope Analog-to-Digital Converters
    B. Successive Approximation Analog-to-Digital Converters
    C. Flash and Hybrid Converters
    D. Analog-to-Digital Converter Resolution: How Much Is Enough?
V. Pulse Characterization by Digital Signal Processing
VI. Data Storage and Display with Digital Computers
    A. The Flow Cytometry Standard Storage Format
    B. Dot Plots: The Simplest Displays
    C. Single-Parameter Histograms
    D. Two-Parameter Histograms: Contour, Density, and Isometric Plots
    E. Three-Dimensional Plots and Histograms
    F. Axes and Scales: Can Quantification and Simplification Coexist?
References

I. Introduction

The analog signal pulses produced during a passage of a cell through the measurement system of a flow cytometer typically last for only a few microseconds at most, and the information about cells represented by the height, area, and width of these pulses would be lost if it were not converted from an analog to a digital form. Until more recently, the analog-to-digital conversion process itself required at least several microseconds, necessitating the use of hybrid electronic peak detectors, integrators, and pulse width measurement circuits, which provided short-term analog storage of the data prior to digitization. An alternative approach is now technically and economically feasible, in which the analog signal is digitized directly at rates of several megahertz, and subsequent processing is accomplished using digital computation. This chapter will discuss analog and digital methods of pulse characterization, and techniques for capture, display, and storage of data in digital form.

In laser source flow cytometers, an illuminating beam is typically focused to an elliptical spot. If the spot height is substantially larger than the particle diameter, both the peak amplitude and the integral of a fluorescence pulse are proportional to the total amount of fluorescent material in or on the particle. However, when the spot height is small compared to the particle diameter, only the integral provides an accurate indication of the total amount of fluorescent material in or on the particle; the peak amplitude or pulse height is a measure of relative brightness. Comparisons of the relative values of pulse height and integral provide additional information about particles; they are commonly used to discriminate single cells from doublets, which may be cells that are actually attached to one another or cells passing through the measurement system in close proximity in space and time.

When spot height is substantially larger than particle diameter, the widths of pulses generated by particles of different sizes are substantially equal; under these circumstances, pulse width measurements are not useful. However, when spot height is near or smaller than particle diameter, particles of different sizes generate pulses of different widths, and pulse width measurements can be used as indicators of particle size.

II. Pulse Characterization Using Analog and Hybrid Circuits

Analog peak detectors, integrators, and pulse width measurement circuits provide short-term (microseconds to tens of microseconds) storage of the appropriate analog values for long enough to permit analog-to-digital conversion of the data. A capacitor typically serves as the storage medium in all three types of circuits, which incorporate operational amplifiers and other active electronic components which prevent the capacitor from discharging unless a digital logic "reset" signal is applied, and which allow the input to the capacitor to be disconnected, holding the voltage on the capacitor approximately constant, when a digital logic "hold" signal is applied.

In a peak detector, the capacitor, after being discharged by a reset signal, is charged by applying the signal to it through a circuit which includes a diode, which allows the voltage on the capacitor to increase, but not to decrease. Thus, the voltage reaches a value equal to the peak value of the signal and remains at that level until a reset signal is applied. The capacitor in an integrator is charged whenever neither a reset nor a hold signal is present, and, after a signal pulse has passed, the voltage on the capacitor is increased by an amount proportional to the integral of the pulse. The capacitor in a pulse width measurement circuit is charged not by the input signal, but by the output of a circuit known as a linear ramp generator. This is turned on at the beginning of the pulse, in response to the reset signal; its output starts at ground and increases linearly with time. At the end of the pulse, the hold signal is applied; at this point the stored voltage on the capacitor is proportional to the duration of the pulse. Peak detectors, integrators, and pulse width measurement circuits operate on baseline-restored signals, which may be on a linear or a logarithmic scale, with or without fluorescence compensation applied.

A. Triggering

Before the height, integral, or width of a pulse from a cell can be determined using analog and hybrid circuits, it is necessary to establish that a cell is indeed present in the measurement system; this is the function of a "front end" or "trigger" circuit, which operates on analog inputs derived from one or more detectors. In general, it is preferable to use a relatively strong signal for triggering. For example, when immunofluorescence is measured, the forward scatter signal is customarily selected as the trigger signal, because all of the cells in the sample will scatter a substantial amount of light. Although the fluorescence signals from "positive" cells bearing tens of thousands of molecules of labeled antibody may also be usable for triggering, those from "negative" cells bearing little or no antibody may not rise sufficiently above the baseline to generate a trigger signal. Because the objective is usually to determine the percentages of positive and negative cells, it is necessary to use a signal that will be strong in either case to insure that the measurement system detects all cells of interest. Electronic signals associated with trigger circuits are shown in Fig. 1.

The critical element in a trigger circuit is a device called a comparator, which has two (positive and negative) analog inputs and a digital logic output, which is a logical "one" (typically 5 V) when the voltage on the positive input is greater than that on the negative input and a logical "zero" (ground or 0 V) when the voltage on the negative input is greater than that on the positive input. The trigger signal, which is typically a preamplifier output signal with the baseline restored to 0, is applied to the positive input. A constant "threshold" voltage, set by the operator, usually to a value slightly above the baseline signal level, is applied to the negative input.

When a cell is not passing through the observation point, the trigger signal is at baseline level, and the comparator output is at logical 0. As a cell passes through the observation point, a pulse appears on the trigger signal, and, once the pulse voltage rises above the threshold level, the comparator output changes

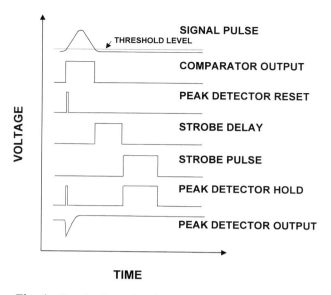

Fig. 1 Signal pulse and peak and front end logic waveforms.

to logical one, remaining there until the trailing edge of the pulse, when the trigger signal again drops below threshold. In some instruments, it is possible to use logical or Boolean combinations ("and," "or," etc.) of the outputs of comparators connected to two or more input signals for triggering. Comparators take a finite time to change output states, that is, to change their output levels from logical 0 to logical one, after the input signal goes above threshold; the fastest ones may respond in a few nanoseconds, but those used in many instruments require 100 nsec or more. This change of state is used to generate the reset signals for peak detectors, integrators, and pulse width measurement circuits.

The electronics in a front end or trigger circuit are also used to generate a "data ready" or "strobe" logic signal, when the outputs of peak detectors, integrators, and/or pulse width measurement circuits have reached stable values after the passage of a cell through the measurement system. The strobe signal is transmitted to the data acquisition system and is used to initiate the analog-to-digital conversion process. A brief time interval, indicated by the strobe delay pulse in Fig. 1, is typically allowed to pass before the strobe pulse is generated; this interval is increased when multiple illuminating beams are used, to provide for the transit time between beams.

In the example given in Fig. 1, the interval between the point at which the preamplifier output pulse signal first rises above threshold and the point at which the strobe pulse returns to logical 0 represents the minimum time required to process a pulse; this is on the order of a few tens of microseconds for most flow cytometers. The arrival of a second cell in the measurement region during this interval represents a "coincidence," and the signal from the second cell can potentially interfere with that from the original cell to produce values of peak height, integral, and/or pulse width not accurately representative of either cell. Different instruments incorporate different circuit arrangements for dealing with coincidences; in simpler systems, a coincidence results in signals from both cells being aborted, whereas, in more sophisticated apparatus, "pipeline" processing of pulses allows resolution of almost all but the most closely separated cells.

B. Peak Detector Timing and Performance

A peak detector is typically reset by a brief logic pulse such as that shown in Fig. 1, generated in response to the change in comparator output from logical 0 to logical one. A hold signal, during which the peak detector input is disconnected or grounded, is applied during the reset pulse, eliminating interference by the input signal with the discharging of the capacitor, and also during the interval in which the held peak signal is digitized, to prevent a change in the stored voltage during the digitization process.

Peak detector performance is affected by the capacitance value chosen for the hold capacitor; lower values allow the circuit to charge to the correct voltage faster, and to discharge to 0 more rapidly in response to a reset pulse. However, there is always some leakage of charge from the hold capacitor, and this and the resulting "droop" in output signal levels occur more rapidly when capacitance is lower.

Reset and hold signals, which reach levels of at least 5 V in a matter of a few nanoseconds, are applied to circuit components in close physical proximity to those which carry the analog signals in peak detectors, and a small amount of cross talk is inevitable. This results in the injection of small amounts of additional charge into the hold capacitor, introducing some inaccuracy into the held signal value. Charge injection errors are relatively inconsequential for signal levels above 100 mV, but seriously compromise the function of peak detectors at lower signal levels.

C. Timing and Performance of Integrators and Pulse Width Measurement Circuits

The timing of reset signals for integrators and pulse width measurement circuits is more critical than the timing for peak detectors. In the sequence just described for peak detectors, and illustrated in Fig. 1, the reset signal is not generated until the comparator changes output state, but this does not occur until the signal level exceeds threshold, which occurs some way up the leading edge of the pulse. An integrator or a pulse width measurement circuit reset at this time would "miss" the first portion of the pulse. This problem is generally solved by introducing an analog delay line in the input to an integrator or pulse width measurement circuit; the output of the delay line reproduces its input value a short time (typically a few microseconds) earlier. The real-time trigger signal is applied to the comparator, and the resultant reset pulse reaches the integrator(s) and/or pulse width measurement circuit(s) just as the delayed pulse "begins."

If hold signals for integrators and pulse width measurement circuits are not applied precisely at the end of a pulse, some error is introduced into the output. A peak detector signal, however, can be held for an arbitrary time, because it reaches its maximum value in the middle of the pulse. Thus, as is also the case for reset signals, the timing of hold signals is more critical for integrators and pulse width measurement circuits than for peak detectors. When delay lines are used, reasonably accurate timing can be achieved by using the rise and fall of a delayed comparator output signal to generate reset and hold signals.

Integrators and pulse width measurement circuits, like peak detectors, store analog signal values in capacitors, and their dynamic characteristics, such as response time and output droop rate, are similarly affected by the capacitance of the hold capacitor. Charge injection introduces inaccuracies into the output signals of integrators and pulse width measurement circuits in the same manner as occurs in peak detectors.

D. Integration by Pulse Shaping

For several reasons, integration in some instruments is accomplished simply by pulse shaping rather than by integrator circuits. Pulse shaping is accomplished by routing the baseline-restored signal through an electronic low pass filter; the resultant output has a peak amplitude proportional to the integral of the input pulse, and a peak detector can then be used to store the integral value. This approach eliminates the need for delay lines and critical timing.

Integration by pulse shaping is also necessary when logarithmic amplifiers (log amps) are used, because the desired output signal is the log of the integral of the pulse. If the output of a log amp is routed to an integrator, the integrator output will be the integral of the log, which is not the same as the log of the integral. However, if the input to the log amp is a low pass filtered signal, with the peak proportional to the integral of the original pulse, a peak detector operating on the log amp output will acquire the right signal, that is, one proportional to the log of the integral.

III. Analog Data Display

In the early days of flow cytometry, when the computer hardware required for the task would have been prohibitively expensive, the most practical way of visualizing two-parameter data from cells was to display the held analog signal values from peak detectors, integrators, or pulse width measurement circuits on a cathode ray oscilloscope. In an oscilloscope, electrons are accelerated from a heated cathode toward a phosphor-coated screen by applying a high voltage between the cathode and the screen. A magnetic field focuses the electrons into a beam; the horizontal and vertical locations at which the beam hits the screen are determined by voltages applied to pairs of deflection plates inside the tube. A modulation voltage may also be applied to control how much of the beam reaches the screen. Electrons which reach the screen produce light emission from the phosphor.

If the amplified signal from, for example, the orthogonal scatter peak detector is connected to the horizontal deflection plate drive electronics, and that from the forward scatter peak detector is connected to the vertical deflection plate drive electronics, and the strobe signal is connected to the modulation voltage input, an illuminated spot on the screen will be produced whenever a cell passes through the measurement system. If a large number of these spots are visualized, either by taking a time-lapse photograph of the screen or by using a special storage cathode ray tube (CRT), in which emission from illuminated spots can be made to persist, what appears is a dot plot, or cytogram, in which the x- and y-positions of spots, respectively, indicate orthogonal and forward scatter signal values of cells.

Both cytograms and raw pulse profiles, also displayed on oscilloscopes, were useful in that they provided a rapid indication of changes in instrument performance, and these types of data displays are still valuable in aligning and troubleshooting flow cytometer optics and fluidics. However, since digital computers are now less expensive than oscilloscopes, cytograms and pulse profiles are likely to be generated by computers following rapid digital conversion of data, and displayed on CRT or flat-panel monitors.

IV. Analog-to-Digital Conversion

The analog peak detectors, integrators, and pulse width measurement circuits used in flow cytometry can store signal values for periods of time on the order of microseconds, and some droop typically occurs even in these short intervals. Longer term storage of data requires conversion to digital form, which is accomplished using analog-to-digital converters (ADCs).

A comparator is essentially a one-bit ADC; it has a digital output, which is 1 when the analog signal at the positive output is greater than the analog signal at the negative input, and 0 otherwise. A practical ADC usually divides its analog input range into more than two parts, or channels, and the corresponding digital output is a binary number with more than one bit. An n-bit ADC has an output range of $0-(2^n - 1)$; the output of an eight-bit ADC ranges between 0 and 255, that of a 10-bit converter, between 0 and 1023, that of a 16-bit ADC, between 0 and 65,535, and so on. The complexity of the circuit required increases with the number of bits. It should not come as a surprise that ADC circuits contain comparators; some may incorporate as few as two, whereas others include large numbers.

ADCs can be unipolar, with the binary output scale corresponding to input voltage values between 0 (ground) and some positive voltage, typically 2.5, 5, or 10 V, or bipolar, with the scale representing a range of -2.5 to 2.5 V, -5 to 5 V, etc. Because the signals of interest in flow cytometry typically span a range between ground and a positive voltage, unipolar converters are used to digitize them.

A. Single Slope Analog-to-Digital Converters

The multichannel pulse height analyzers used in early flow cytometers incorporated what are known as single slope ADCs. A single slope ADC incorporates a linear ramp generator similar to that used in a pulse width measurement circuit. The heart of the converter is a binary counter, a digital circuit that can represent any number between 0 and $(2^n - 1)$, where n is the number of bits' resolution of the converter, in binary form. The output of the counter is reset to 0 by the "start convert" signal, which might be the rising edge of the strobe pulse from the flow cytometer; the start convert signal also resets the ramp generator.

The converter also incorporates a "clock," an oscillator that generates logic pulses at regular intervals, and comparators, which control the input to the counter. When the ramp generator output goes above ground, a comparator signal connects the output of the clock to the counter, which adds 1 to its count every time it receives a clock pulse. When the ramp voltage reaches the input voltage, another comparator disconnects the clock from the counter, which, at this point, holds a count proportional to the input voltage.

Single slope ADCs produce relatively smooth histograms, or distributions of signal values, because the linear character of the voltage ramp insures that the voltage differences between any two adjacent channels are very nearly equal. However, the absolute accuracy of single slope ADCs is not outstanding; they incorporate a lot of parts, which makes them expensive, and they are relatively slow, with speed determined by the clock frequency. Even with a clock oscillator operating at 100 MHz, a 10-bit conversion takes slightly longer than 10 μsec.

B. Successive Approximation Analog-to-Digital Converters

Many of the ADCs now used for flow cytometry are the successive approximation type; they generate comparison voltages using a digital-to-analog converter (DAC), which takes a digital input and generates an analog voltage or current output. The DAC incorporates a voltage divider made up of a string of resistors in series, so that points between any two resistors will be at one-half, one-quarter, one-eighth, and successively smaller fractions of an applied voltage equal to the highest voltage in the input range of the converter.

Conversion is a multistep process. A test voltage, initially equal to the ADC input voltage, is first compared with one-half the maximum DAC output voltage. If the test voltage is greater, the most significant bit (MSB) of the ADC output it set to 1, and the test voltage is replaced by the difference between the previous test voltage and one-half the maximum DAC output. Otherwise, the test voltage remains as it was, and the MSB of the ADC output is set to 0. Next, the test voltage is compared to one-quarter the maximum DAC output voltage; if the test voltage is greater, it is replaced by the difference between the two voltages, and the next most significant bit of the ADC output is set to 1; if not, the test voltage remains as it was and the appropriate bit is set to 0. After as many cycles as there are bits' resolution in the ADC, the ADC output will contain a digital value representing the input voltage.

Successive approximation ADCs suffer from a problem known as differential nonlinearity. This arises from real-world characteristics of the resistors used in the DAC voltage divider. If the smallest resistor in the voltage divider of a successive approximation ADC, that is, the resistor in the "1's" bit, has a precision of 1%, in order to maintain the same voltage accuracy (not precision) in the resistor in the "128's" bit, this resistor has to be precise to better than 0.01%. Although the resistors in a typical successive approximation ADC are "trimmed" to the necessary precision during fabrication of the circuit, most such ADCs are only specified to be linear to within one-half LSB (least significant bit). In an 8-bit (256 channel; output range 0–255) ADC meeting this specification, the distance between adjacent channels, nominally 40 mV, could vary between 20 and 60 mV. By comparison, the differential nonlinearities in an 8-bit single slope ADC, which arise from slight deviations from true linear response in the ramp generator, typically result in distances between channels varying by less than 1 mV. A histogram of values obtained from the successive approximation ADC which should ideally show, say, 60 cells in channel 41 and 60 cells in channel 42, could show 90 cells in 41 and 30 in 42, or 30 in 41 and 90 in 42, or anything in between.

The quickest fix for this problem is to use an ADC with higher resolution, and throw away the less significant bits. The resistors in a 12-bit ADC, which has an output range of 0–4095, are more precisely trimmed than those in an 8-bit ADC. Taking the upper 8 bits (0–255) or 10 bits (0–1023) of the output from the 12-bit ADC will produce smoother histograms.

C. Flash and Hybrid Converters

A successive approximation ADC incorporates an internal clock, and derives its output only after its internal workings have gone through multiple clock cycles. As techniques for producing large scale analog and hybrid circuits have improved, it has become possible to produce converters incorporating large numbers of comparators and voltage references, allowing conversions to be accomplished in a single cycle. Although it remains difficult to produce these so-called "flash" converters with more than 8 bits' resolution, it is possible to produce high-speed, high-resolution ADCs which combine flash and successive approximation elements. One such device produces a 14-bit output in only three clock cycles, each of which is about 100 nsec in duration; however, since a new input value can be acquired every clock cycle, the overall digitization rate can be as high as 10 MHz. The device is priced below $50; a 14-bit converter half as fast cost nearly $1000 only a few years ago.

D. Analog-to-Digital Converter Resolution: How Much Is Enough?

In the early days of flow cytometry, 256-channel (8-bit) ADCs were the norm; multichannel pulse height analyzers used to collect single-parameter histograms might offer 512- (9-bit) or 1024-channel (10-bit) resolution. Converters were expensive enough that data acquisition systems typically used a single ADC to digitize held signals from several channels in sequence. As is indicated by the discussion in the preceding paragraph, ADCs are now inexpensive enough that price is not the primary consideration in deciding on specifications; the rational selection of converters can now be based on the characteristics of the data.

For data traditionally analyzed and displayed on linear scales, for example, forward and orthogonal scatter signals and fluorescence of dyes which stoichiometrically stain DNA, it is usually unnecessary to retain more than 10 bits worth of information. If data with a large dynamic range, such as the results of immunofluorescence measurements, are transformed to a logarithmic scale electronically, that is, by using logarithmic amplifiers, digitizing to 10 bits' resolution is also more than adequate.

To increase the accuracy of measurements encompassing a large dynamic range, it is desirable to digitize to high enough resolution to permit fluorescence compensation and logarithmic transformation to be accomplished by digital computation. If the four-decade logarithmic scale now standard in flow cytometry is retained, this requires a converter capable of responding to signals as low as 1/10,000 of full scale (e.g., 1 mV if the largest signal is 10 V). The lowest resolution ADC that meets this criterion is a 14-bit converter (range 0–16,383); for a 0–10 V input range, such a device will have an output of 1 if the input signal is over 610 μV. However, the bottom decade of the range (1 to 10 mV) will be represented by channels 1–16. The inaccuracy represented by a change of one bit in digital signal value is 100% at the low end of this range; at the high end, it is just over 6%.

It would be preferable to use a 16-bit converter (range 0–65,535), in the output of which a 1 mV signal would appear in channel 7 and a 10 mV signal in channel 65. A 1-bit change in digital signal is about 14% of the value at the low end and about 2% of the value at the high end of the bottom decade; this is acceptable because the uncertainty introduced into measurements in this range by photon statistics is typically at least this great.

Some laboratory-built (Shapiro *et al.*, 1998) and commercial (EPICS XL, Beckman Coulter, Fullerton, CA) flow cytometers digitize the outputs of peak detectors or integrators to 16 bits' precision or more, allowing fluorescence compensation and logarithmic transformation of data to be accomplished by software, and improving the accuracy and precision of high dynamic range measurements. However, this approach requires that the peak detectors or integrators respond accurately to low level signals, and achieving this necessarily increases the complexity of the analog and hybrid electronics. With very fast, high-resolution converters now available, instrument designers have turned their attention to digital techniques for pulse characterization.

V. Pulse Characterization by Digital Signal Processing

Digital signal processing, or DSP, usually accomplished by specialized microprocessors known as DSP chips, has become a ubiquitous technology. Personal computers incorporate auxiliary DSP chips to do a wide variety of things, including fast audio (speech and music processing and synthesis) and video (image processing) operations, and echo cancellation and other tasks required for high-speed data transmission over telephone lines. In DSP, signals are sampled continuously at a relatively high rate. For digital compact disk recording, the sampling frequency is 44.1 kHz, which means that two conversions (one for each stereo channel), each of at least 16 bits' resolution, are done every 22.7 μsec; this generates a minimum of 176,400 bytes of data every second, or 620 megabytes per hour.

In order to accurately capture periodic components of frequency f in a signal, the sampling frequency must be at least $2f$. The 44.1 kHz sampling rate for digital audio preserves frequency components as high as 22 kHz in the digitized signal. The sampling rate required for digital characterization of pulses from flow cytometers depends on the pulse duration, which, in different instruments, ranges from about 500 nsec to over 10 μsec.

Figure 2 shows what samples of a simulated flow cytometer signal, a Gaussian pulse with some superimposed random noise, would look like if taken at a frequency of about 17 MHz. The baseline is not restored; the first step in analysis of a data stream such as this using DSP would be to examine the slowly varying regions of the signal, that is, the intervals between pulses, to determine the baseline level and the threshold above which the signal level would have to rise to indicate the presence of a pulse. The peak values of pulses can then be found simply by searching for maxima and subtracting the baseline value. The pulse integral is determined by taking the sum of all sample values (minus baseline) captured between the time the signal level rises above threshold and the time it falls below threshold, and this time interval represents the pulse duration.

Examination of Fig. 2 reveals that noise introduces some uncertainty into all of the measurements; the maximum value of the signal occurs at a point distinct from that which an observer would identify as the peak of the Gaussian pulse, and baseline noise makes it difficult to determine where the pulse begins and ends. The presence of baseline fluctuations means that an integral computed by taking a constant value of baseline is likely to deviate from the true value.

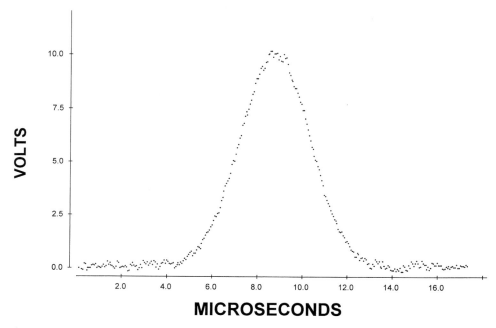

Fig. 2 Representation of a simulated Gaussian pulse with added noise sampled at approximately 17 MHz.

In theory, taking a large number of digital samples of each pulse should provide a very large dynamic range for the value of the integral. An 8-bit converter sampling at 40 MHz (one sample every 25 nsec) will acquire 256 samples of a 6.4-μsec pulse, each with a value between 0 and 255. The integral, that is, the sum of the 256 values, can range between 0 and 65,280. This, at first glance, appears to yield the same dynamic range as a 16-bit ADC. However, an 8-bit converter, which operates over a range of 0–10 V, has channels spaced about 40 mV apart. Unless the amplitude of a pulse exceeds 40 mV at some point, the converter output and the integral will remain 0. The converter will not detect 1-mV or even 10-mV pulses, and its response will thus be restricted to the upper 2.5 decades of what appeared to be a four-decade dynamic range. A 14-bit converter, sampling at 2.5 MHz, will collect 16 samples of a 6.4-μsec pulse; values of the integral could range between 0 and 262,128, and the converter would respond to a 1-mV pulse, since channel 1 corresponds to just over 600 μV.

Although digital pulse processing has been used for some years in laboratory-built flow cytometers, it has only more recently become possible, thanks to improvements in converter and processor design, to achieve several decades' worth of dynamic range without the use of analog logarithmic amplification. The first commercial instrument to incorporate the electronics necessary to do this is the Luminex 100 (Luminex Corporation, Austin, TX) flow cytometer, a device designed to do multiplexed ligand binding assays using color-coded fluorescent beads as substrates (Fulton et al., 1997). In this instrument, 12-bit ADCs provide a three decade range for the color coding channels, whereas a 14-bit converter allows readout of assay results over a four decade range.

DSP techniques for pulse processing can eliminate the need for analog and hybrid baseline restorers, threshold sensing and front end electronics, integrators, peak detectors, and pulse width measurement circuits. Laser noise can be compensated for on a point-by-point basis. In slit-scanning flow cytometers (Wheeless, 1990) and some experimental systems (e.g., Zilmer et al., 1995), digitization at rates up to tens of megahertz allows collection of enough data points to provide detailed information on the shapes of pulses as well as the more conventional measures of pulse height, width, and area. Pulse shape information can aid in identification of cell populations and in discrimination of single cells from doublets.

VI. Data Storage and Display with Digital Computers

A. The Flow Cytometry Standard Storage Format

Whether conventional methods or data storage and display (DSP) techniques are used, the data acquisition systems in flow cytometers ultimately convert the information about cells contained in signal pulse heights, areas, and widths to digital numerical form. Most instruments store raw data in disk files conforming to the Flow Cytometry Standard, which is now in a second revision (FCS 2.0) (Dean et al., 1990; Data File Standards Committee of the Society for Analytical Cytology, 1990), and awaiting a third (Seamer et al., 1997). FCS files contain a header, which may provide information identifying the sample, instrument, operator, time and date of acquisition, parameters measured, instrument settings, etc., and which specifies the numerical format and location of the data that make up the major portion of the file.

FCS files are most commonly used to store data in so-called list mode, with the data section of the file consisting of a sequence of sets of data points, each set representing values of all parameters measured for one cell. However, one- and multidimensional statistical distributions as well as raw data may be stored in FCS files. Data analysis software from flow cytometer manufacturers and third parties can read most or all files conforming to the standard.

B. Dot Plots: The Simplest Displays

Some display formats commonly used in flow cytometry are shown in Figs. 3 and 4. The dot plot, or cytogram, which was discussed in the section on analog

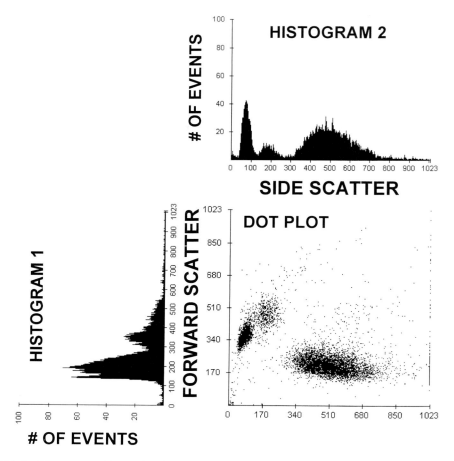

Fig. 3 Histograms of forward and side scatter signals and a dot plot of forward versus side scatter for leukocytes in peripheral blood analyzed after red cell lysis.

data display, is the simplest form of data display, in that it shows raw data, rather than the results of computation. Values of one measured parameter are represented on the horizontal ($x-$) axis; values of a second measured parameter are represented on the vertical ($y-$) axis. If any "event" (a category which includes cells, noise, and anything else that generates a valid trigger signal) among those included in the dot plot has a value of a for the first parameter and b for the second, a dot will be plotted at the coordinates (a,b). If no events have the corresponding parameter values, a dot will not be plotted.

Because they show all events, dot plots provide an excellent illustration of the range of data in both the one-dimensional measurement spaces associated with each parameter and the two-dimensional space in which the plot appears. How-

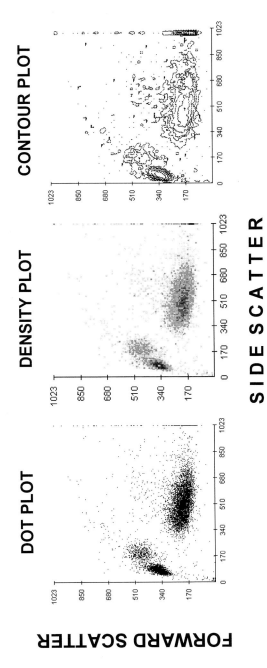

Fig. 4 A dot plot, a density plot, and a contour plot of forward versus side scatter for the leukocyte population shown in Fig. 3.

ever, a dot plot cannot provide an indication of the frequency of occurrence of a given parameter value or set of values; this information can only be obtained from additional computation.

C. Single-Parameter Histograms

A single-parameter histogram, or univariate frequency distribution, is a plot of signal intensity, typically on the horizontal axis, against the number of events with the corresponding measured values of signal intensity, typically on the vertical axis. The procedure, or algorithm, for calculating a histogram is fairly simple. A number of storage locations, usually corresponding to the number of channels to which data are digitized, are reserved in computer memory.

"Storage location" here generally does not mean a single byte of memory, because a single byte of memory can only store a number between 0 and 255. Because even a histogram representing data from only a few thousand cells may have more than 255 events in one channel, it was common, even when computer memory was expensive, to use at least two bytes per storage location, allowing as many as 65,535 events per channel. At present, the computers used for flow cytometric data analysis typically operate on 32-bit numbers, which occupy four bytes; this accommodates as many as 4,294,967,295 events per channel.

As a first step, the contents of all storage locations are set to 0. Values of the appropriate parameter for all events to be included are taken in sequence from list mode data stored in memory or on disk; whenever the value c appears, the contents of the cth memory location reserved for the histogram are increased by one. Calculation is typically stopped when the total number of cells reaches a preset value, or when all cells in a file have been analyzed.

Figure 3 shows histograms of forward (small-angle) and side (large-angle) light scatter signal intensities and a dot plot of forward versus side scatter intensity for a population of human peripheral blood leukocytes. The forward scatter histogram is rotated to put it into correspondence with the appropriate axis on the dot plot. It is apparent that, although the histograms provide information unavailable from the dot plot about the frequency with which different parameter values occur, each is multimodal, that is, contains several peaks. These correspond to signals from discrete subclasses of leukocytes, with lymphocytes, monocytes, and granulocytes, respectively, showing low, intermediate, and high side scatter signals. There is substantial overlap between peaks on the forward scatter histogram, and less on the side scatter histogram, whereas the two-dimensional dot plot shows clearly separated clusters corresponding to the three leukocyte types.

D. Two-Parameter Histograms: Contour, Density, and Isometric Plots

Figure 4 shows the same dot plot seen in Fig. 3, with data from a corresponding two-parameter histogram, or bivariate distribution, displayed in two alternative formats. In principle, one can compute a two-parameter histogram by setting aside n^2 storage locations, where n is the number of channels resulting from analog-to-digital conversion. In practice, this is rarely done, for two reasons. First, the memory requirements are substantial. If each parameter has values ranging from 0 to 1023, it is necessary to use 1,048,576 storage locations for a single histogram; this accounts for 2 megabytes using 16-bit numbers and 4 megabytes using 32-bit numbers. Second, when a two-parameter histogram is computed at high resolution, it is usually necessary to include a very large number of events in order to have more than a few events in each storage location.

For a relatively long time, it was common to compute two-parameter histograms with a resolution of 64 × 64; these require 4096 storage locations per histogram, which was a manageable amount of memory even in the early days of personal computers. Now, resolutions of 128 × 128 (16,384 storage locations) and 256 × 256 (65,536 storage locations) are available. The density plot and contour plot shown in Fig. 4 have 64 × 64 resolution; values on a 1024-channel scale are simply divided by 16 to produce the appropriate value on a 64-channel scale.

On the density plot, the number of cells with any given pair of values is indicated by the shade of gray shown at the corresponding point in the display. This type of density plot is called a gray scale plot; a similar plot, in which different frequencies of occurrence are represented by different colors instead of different shades of gray, is referred to as a chromatic or color plot. Since the computers now used for flow cytometric data analysis have color displays and are often equipped with color printers, chromatic plots are more common than gray scale plots.

In a contour plot, such as that shown in Fig. 4, a direct indication of frequency of occurrence is not given for each point in the x–y plane. Instead, a series of contour lines, or isopleths, are drawn; each of these connects points for which data values occur with equal frequency. A contour plot is essentially the same type of display as appears on a topographic map, on which the contour lines connect points at equal altitudes; increasing frequencies of occurrence on the density plot correspond to increasing altitudes on the map.

It is also possible to display data in a density plot in an analog of a relief map; this is called an isometric plot, or three-dimensional projection, also commonly known as a peak-and-valley plot. In an isometric plot, a simulated "surface" is created; the apparent "height," or z-value, corresponding to any pair of x- and y-coordinates is made proportional to the frequency of occurrence of the corresponding paired data values in the sample. The primary disadvantage of isometric plots lies in the unpredictability of where peaks and valleys will appear; smaller peaks toward the "rear" of the display may be masked by larger peaks toward the "front." Although this problem may be alleviated by changing the angle from which the plot is viewed, this requires extra computations.

However, the isometric plot does have an advantage in that the simulated third dimension clearly indicates the frequency of occurrence of events. A gray scale or chromatic plot ought to be accompanied by a key relating each gray level or color used to the corresponding frequency of occurrence, similar to the key on a map that correlates ranges of altitudes with the corresponding colors. In a gray scale plot, it makes sense to use progressively darker shades of gray to indicate higher frequencies, but no similarly intuitive scheme exists for chromatic plots.

Contour plots require more computation than density plots, because the raw two-parameter histogram data is typically smoothed, by combining each data point with an average of its neighboring points, in order to generate contours which are not jagged. This makes contour plots appear to have higher resolution; this appearance is deceiving. Contour plots also typically omit single occurrences, although some software packages allow these to be superimposed as dots.

Many people prefer to use contour plots in publications, because they provide more detail than dot plots and often reproduce better than density plots, particularly when color printing is not an option. However, a well chosen gray scale plot is no less informative than a contour plot and can be more informative, particularly if a binary logarithmic intensity scale is used. On this intensity scale, one color or gray level indicates single occurrences, the next, 2–3 occurrences, and subsequent colors or gray levels indicate 4–7, 8–15, 16–31, 32–63, 63–127, and more than 127 occurrences. This provides frequency information while still facilitating identification of cells that occur with frequencies of less than one in 10,000.

E. Three-Dimensional Plots and Histograms

Some software packages produce a "three-dimensional cytogram," also described as a cloud plot. Cloud plots have the same disadvantage as isometric plots; when one cloud obscures another, it becomes necessary to recompute and change the viewing angle to locate different cell clusters. Three-parameter histograms are more problematic. Although there are some tricks that can reduce the storage requirements, a $64 \times 64 \times 64$ three-parameter histogram nominally requires 262,144 memory locations, and also poses serious problems in data display. Isometric plots would require four dimensions, and contour and chromatic plots must be shown as "slices," each of which represents a two-parameter histogram of events for which values of the third parameter lie within a relatively narrow range. The resolution along the "sliced" axis is often lower than that of the two-parameter histograms, so that there might be four to sixteen 64×64 histograms rather than 64.

F. Axes and Scales: Can Quantification and Simplification Coexist?

Most flow cytometric data analysis packages display single- and two-parameter histograms with labeled axes, often providing more data than the reader wants or needs. However, the parameter axes almost always represent arbitrary quantities, for example, a linear scale extending from channel 0 to channel 1023, or a four-decade logarithmic scale ranging from 1 to 10,000, are shown in a publication. When multiple single- or two-parameter displays are shown as a panel in a publication, the axis labels and associated tick marks are often absent, although a grid may be superimposed to facilitate comparisons between individual displays in a panel.

Although an extended discussion of the subject is beyond the scope of this chapter, it should be pointed out that it is now possible to relate the signal intensities measured in a flow cytometer to such biologically relevant quantities as the volumes of cells, in femtoliters, cellular DNA content, in picograms, and the number of molecules of a particular antibody bound to the cell surface and/ or the interior. The more sophisticated data analysis packages now available facilitate this process. Although quantification may at first require additional planning and additional labor, it provides the only reliable means for comparing data taken from different samples, in different laboratories, using different instruments, at different times, and thus is likely to be widely adopted in the future, to the benefit of all who use it.

References

Data File Standards Committee of the Society for Analytical Cytology (1990). Data file standard for flow cytometry. *Cytometry* **11,** 323–332.

Dean, P. N., Bagwell, C. B., Lindmo, T., Murphy, R. F., and Salzman, G. C. (1990). Introduction to flow cytometry data file standard. *Cytometry* **11,** 321–322.

Fulton, R. J., McDade, R. L., Smith, P. L., Kienker, L. J., and Kettman, J. R., Jr. (1997). Advanced multiplexed analysis with the FlowMetrix system. *Clin. Chem.* **43,** 1749–1756.

Seamer, L. C., Bagwell, C. B., Barden, L., Redelman, D., Salzman, G. C., Wood, J. C. S., and Murphy, R. F. (1997). Proposed new data file standard for flow cytometry, version FCS 3.0. *Cytometry* **28,** 118–122.

Shapiro, H. M., Perlmutter, N. G., and Stein, P. G. (1998). A flow cytometer designed for fluorescence calibration. *Cytometry* **33,** 280–287.

Wheeless, L. L., Jr. (1990). Slit-Scanning. *In* "Flow Cytometry and Sorting" M. R., Melamed, T. Lindmo, and M. L. Mendelsohn, eds., 2nd Ed., pp. 109–125. Wiley-Liss, New York.

Zilmer, N. A., Godavarti, M., Rodriguez, J. J., Yopp, T. A., Lambert, G. M., and Galbraith, D. W. (1995). Flow cytometric analysis using digital signal processing. *Cytometry* **20,** 102–117.

CHAPTER 8

Time as a Flow Cytometric Parameter

Larry Seamer* and Larry A. Sklar[†]

*Bio-Rad Laboratories
Hercules, California 94547

[†]Department of Pathology and Cancer Research and Treatment Center
University of New Mexico
Albuquerque, New Mexico 87131; and
National Flow Cytometry Resource
Los Alamos National Laboratory
Los Alamos, New Mexico 87545

I. Introduction
II. Historical Overview
III. Sample Mixing and Delivery
IV. Data Analysis
    A. Data Storage
    B. Data Display
    C. Quality Control
    D. Pseudotime
    E. Flow Cytometry Standard Support
V. Applications
VI. Conclusions
    References

I. Introduction

Time is measured by change (Hawking, 1988), meaning without change in the universe time would have no definition. Although the preceding statement may be a source for philosophical debate, in biological systems the argument is true in a very practical sense. Time is a valuable parameter because it provides insight into how biological systems change. Correlating time with other, physiological measurements, allows a process to be viewed in a meaningful sequence, yielding information on the sequence of the changes. Information about biological mechanism can often be ascertained from the sequence of information as well as the rate and the amplitude of biological response. In the early twentieth century, time based physiological measurements were made on whole organs, for example, the classic smoked cylinder measurements to study muscle contraction. More recently, flow cytometry and similar technologies such as laser scanning microscopy have made it possible to analyze individual cellular responses against the backdrop of time.

Flow cytometry is a technology that allows rapid analysis of individual cells, encompassing several correlated measurements. Many of the properties of flow cytometry that make it valuable for time resolved measurements are similar to those that are traditionally associated with static flow cytometry. First, is the ability with this technology to discriminate fluorescence associated with a cell or particle from the fluorescence in the medium, often completely eliminating the requirement of removing unbound fluorescent indicator from the analysis system. Homogeneous detection schemes shorten the time between mixing and the first measurement significantly over other, bulk methods. Second, because each cell is analyzed individually, flow cytometry can analyze single populations of cells in a heterogeneous mixture. This is a considerable advantage over bulk measurements such as fluorometry, a technology that averages the signal from all cells into a single measurement. This feature also allows multiple correlated analysis of individual cells. When these individual events are viewed in rapid succession, kinetic measurements can accurately be made. A third advantage to flow based measurements is that 10,000 or more cells per second can be analyzed. Therefore, the events can be parsed into very short time intervals and still allow the response of many cells to be viewed and, when appropriate averaged, yielding increased statistical accuracy.

Flow cytometry can also impose intrinsic limitations for time-dependent analyses. The time resolution of the measurements in flow cytometry depends on several factors. First is the time that it takes to mix cells with reagents and the time it takes to deliver them to the analysis point. A second issue is cell throughput or acquisition rate that statistically determines minimal sample size and time interval. This problem is further exacerbated when heterogeneous populations are being analyzed, and the cells representing any given population are analyzed less frequently than the overall-sampling rate.

This chapter describes the technical elements that must be present in order achieve successful continuous time-dependent analyses. Several innovative sample handling devices that shorten the mixing and delay time inherent in commercial flow cytometers are compared. Many applications in cell biology that have benefited from time-based analyses such as cell activation, functional studies, and molecular assemblies, are examined. Several aspects of on-line flow cytometry for kinetic analysis have been described by Nooter *et al.* (1994). The analysis of time-dependent ligand–receptor interactions has been reviewed by Nolan *et al.* (1998).

II. Historical Overview

Since Van Dilla *et al.* (1968) first described a method of static analysis based on a flowing stream, flow cytometry has developed into a valuable tool for studying biological processes at the level of the individual cells. It did not take long before this powerful technology was applied to dynamic systems. Watson *et al.* (1977) published one of the first papers describing the use of flow cytometry to make time resolved measurements. This work described a method utilizing serial discrete analyses, that is, first measuring the baseline static conditions of the system, then perturbing the system and collecting individual data files at regular intervals. The kinetics of cellular response was then calculated by analyzing and comparing many sequentially archived files. Using this simple system, time resolution can be no better than the time required to acquire and to save a static data file. Serial acquisition also has the problem of data smearing as dynamic systems continue to change during the collection of each data file. Sklar and co-workers subsequently used this approach to analyze ligand–receptor interactions in a homogeneous assay format, showing its applicability to binding, dissociation, and internalization of a leukocyte peptide receptor system (Sklar *et al.*, 1984; Fay *et al.*, 1991; Hoffman *et al.*, 1996).

Martin and Swartzendruber (1980) first described a flow cytometer hardware modification that allowed a relative time measure to be collected along with other analysis parameters. Time-based measurements are improved by adding a time parameter to the acquisition, in essence creating a time-stamp to each event. Data is then acquired continuously throughout the period of the reaction and stored in a single list-mode file. Their method used a linear voltage ramp generator, the steadily increasing voltage throughout the sample acquisition period was digitized and archived along with other analysis parameters for each cell. Since this publication, voltage ramps have found wide use in commercial instruments.

Although it is preferable to time stamp each event, the system of multiple snapshot analyses remains useful even today if an instrument is incapable of collecting a time parameter or the reaction time of the biological system is sufficiently long that continual sampling would create tremendously large data files. Durack and colleagues (1991) proposed an improvement to the sequential static data acquisition paradigm, time interval gating, to better deal with prolonged data acquisition. This is a technique where the sequential acquisitions are stored in a single data file by concatenating the new data to the old.

III. Sample Mixing and Delivery

The time resolution of the early flow cytometry systems was limited by the need to remove the sample tube from the instrument in order to add appropriate reagent to the sample to start the reaction after acquiring baseline data. Several sample-mixing stations have been described to shorten that time to the first measurement (Table I). The design and implementation of such systems is not trivial because sample-mixing stations must balance diametrically conflicting requirements. First, the sample and reagent must be completely and rapidly mixed. If reagent is not brought into contact with all of the cells almost simultaneously, individual cells will reflect different reaction points when they are analyzed. However, if too much time is used to insure complete mixing, the time to the first measurement will be lengthened and early reaction events will be missed. The second conflicting requirement results from the need to move the mixture of cells and reagent to the analysis point of the instrument quickly while maintaining stable laminar flow. Of course, stable flow is required to achieve accurate cytometric measurements. When sample is injected too forcefully into the sheath stream, chaotic flow occurs and erroneous measurements result. If sample is moved too timidly, once again the time to the first measurement will suffer and early events will be missed. Failure to achieve any of these requirements severely limits the usefulness of the system.

Adequate mixing can be verified with a simple experiment that exploits the ability of heavy atoms to quench fluorescence (Nolan *et al.,* 1995). When fluorescein labeled beads are mixed with iodide, the fluorescence quenching is essentially instantaneous. Therefore, uniform mixing will be detected as a stable fluorescence signal over time from the earliest to the latest times of sample delivery.

As illustrated in Fig. 1, sample-mixing systems can be broadly divided into two groups, conveniently referred to as "time-window" and "time-zero" devices. Time-window systems are those which continually analyze cells at a single time point after reagent mixing. These systems are useful when a large number of measurements are required at a given time point or when sorting cells at a fixed point in time during a response. For example, one can imagine sorting cells that respond to a stimulus at time X after reagent addition from those that were unresponsive at that same time. Using a time-window device, any number of cells can be collected, all at time point X. However, to achieve a kinetic response sequence many individual measurements must be made, adjusting the sample delivery apparatus between measurements.

Dunne (1991) described a type of time-window apparatus that utilizes a mixing "T." In this system, sample is injected from one arm of the T and reagent from

Table I
Rapid Mix Devices

Instrument	Type	Mixing	Time
Mixing T	Time window	T	1–3 sec
Coaxial	Time window	Fluid flow	0.055 sec
Cuvet	Time zero	Magnetic stir bar	1 sec
Syringe	Time zero	T	0.3 sec

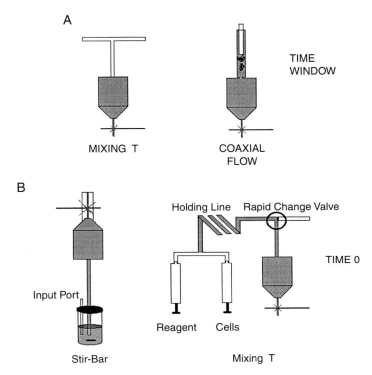

Fig. 1 Rapid delivery systems. (A) Time-window delivery systems; mixing T and coaxial flow systems that deliver particles to the cytometer at a constant time after reagent addition. (B) Time-zero delivery systems; stir bar and mixing T systems that deliver particles to the cytometer over time following a single reagent addition event.

the other. Both sample and reagent are delivered at a constant slow rate, which together provide the mixed sample to the flow cell. Small volumes or low flow rates through the T can compromise adequate mixing. However, the sample volumetric flow rates can only be increased to the extent that they can be accommodated under the constraints imposed by orderly laminar flow.

When chaotic flow conditions exist at the point of analysis, some cells will be outside of the optimal optical detection point and, therefore, will be poorly illuminated, yielding significantly lower level signals than when positioned optimally. This situation will result in data distributions with artifactual left shoulders. One technique for eliminating these events from the analysis was described by Nolan *et al.* (1995). Because turbulent flow will likely cause these events to exhibit artificially low measurements on all parameters they can be eliminated from the analysis by gating on light scatter, allowing only those events with unperturbed scatter into the final analysis. In the mixing T system, the time point being measured is a function of the distance of the T to the sample analysis point. The limitation imposed by mechanical constraints of positioning the T make very early time point measurements difficult. The time to the first measurement reported for these systems typically range from 1 to 3 sec.

The constraints imposed by the mixing T have been avoided with the development a coaxial mixing device. Coaxial mixing is a system in which sample is injected through a capillary tube into the center of a flowing stream of the reagent with which the cells are to be mixed. Fluid flow at the point of injection is responsible for cell and reagent mixing. Scudder *et al.* (1993) modified a coaxial mixing device previously used in chemical analysis to microscope based cell studies. Later that same year the device was adapted to flow cytometry (Lindberg *et al.*, 1993). In a subsequent study this system was shown to provide measurements of intracellular calcium in as little as 100 msec after reagent addition (Blankenstein *et al.*, 1996). Once again, the time point analyzed is a function of the distance from the end of the sample insertion tube to the cell analysis point. By systematically adjusting the sample capillary during reagent addition it is possible to vary the time between reagent addition and sample analysis. Because cell and reagent volumetric flow rates remain constant, this system also has the important advantage of allowing stable laminar flow. It is the very close positioning of the coaxial mixing device to the analysis point that has shortened the time to the first measurement to under 100 msec.

Time-zero devices mix sample and reagent at a single time point and follow the reaction sequence over time. These devices are often used when kinetic data are desired or very early time point measurements are required. A time-zero device was described that used a pressurized cuvette and magnetic stir bar for shortening the time between mixing and the first measurement to ~3 sec (Omann *et al.*, 1985). In this system, reagent and cell suspension are brought together in a mixing cuvette by injecting either reagent or cell suspension into the cuvette containing the other. The mixed sample is then pushed with pressure from the instrument lines to the instrument analysis point. In 1989 a simplification of this basic system was described that achieved 1- to 2-sec time to first measurement (Kelley, 1989). Both time-window and time-zero approaches have become available in commercial components. It is worth noting, however, that the relatively large volumes required in magnetic stir bar mixing make very rapid mixing problematic. In fact, uniform mixing in a stir bar system can require several seconds.

In 1995 a syringe driven mixing and delivery system was described that achieved 300-ms time to first measurement by using a mixing T as in the time-window device (Nolan *et al.*, 1995). The system used three automated syringes to mix and deliver sample. Two syringes were typically used to mix relatively large volumes (up to several hundred microliters) of reagent and cell suspension and to deliver them into a holding or "sample delay" line. One can imagine that moving these large volumes of liquids very rapidly in the sample line of a flow cytometer would cause catastrophic instability in the flow cell and small orifice used in most commercial cytometers. To avoid such problems, the sample line was isolated from the instrument flow cell by a small volume, fast action valve during this push and mix phase. On completion of mixing, the valve changes position and the mixed sample is then moved from the holding line to the instrument analysis point at a rate compatible with stable laminar flow. In fact, the third syringe moved the sample in two phases. The first, or boost phase, rapidly cleared the dead volume between the valve and the analysis points by pushing a few microliters of sample in a time of a few milliseconds. During the second, or delivery phase, the sample was delivered at an acceptable analysis rate. Surprisingly, unstable flow could be observed in the particle scatter distributions which, although generated in ~100 msec, appeared to persist for 1 sec or longer.

The time to the first measurement limitations of such systems were dependent on several factors: the mixing step; the distance between the holding-line valve and the sample analysis point; and the rate at which mixed sample can be delivered without perturbing laminar flow. We have begun to address the question of sample instability. We have used a separate syringe to automate control of the sheath, compensating for the increase in flow rate through the flow cell during the injection of sample with a simultaneous and parallel reduction in sheath flow. This has allowed the mixed sample to be moved quickly to the instrument analysis point. Once there, the sample is slowed and sheath returned to normal flow rates, reestablishing laminar flow more rapidly than otherwise possible. Using relatively inexpensive computer controlled syringes and valves, 600-msec time to first measurement was obtained (Seamer et al., 1999).

In summary, the time-window coaxial devices appear to offer the earliest time resolution, but must be adjusted for each time point desired. The automated syringes in time-zero devices offer continuous analysis and have considerable promise in complex sampling application in multimix, multistep, or automated reactions. The ability to sample over time, both quickly and repetitively, is beginning to be addressed in both commercial and research applications (Sklar et al., 1998; Edwards et al., 1999). Although details of this work are beyond the scope of this chapter, the potential of flow cytometry for high throughput screening is just beginning to become apparent.

IV. Data Analysis

The data derived from flow cytometric, multiparametric time resolved analysis is extremely rich and seldom exploited fully. Although other approaches may yield more rapid analysis or more sensitive measurements, the ability to detect and follow individual subsets of related cells, simultaneously exposed to the same stimulus, within milliseconds of contact, provides a valuable view of reactions that complements many other techniques. With such complex data, analysis techniques must be developed to allow insightful interpretation.

A. Data Storage

Any discussion of data analysis must be preceded by a description of data organization and storage. The type of analysis that can be accomplished is affected by the format of data storage. Commercial flow cytometers can store data in two primary formats. The first is as a processed histogram, which is simply an intensity distribution curve. A more complex version of this format is the two-parameter correlated data file in which the data are stored as an array representing the two-parameter measurement space. Once data are stored in histogram format, no improvement in resolution can be extracted. For example, if a 1-min time course is acquired with hardware resolution of 100 msec and saved as a 64 × 64 channel correlated plot, the analysis resolution has been reduced to the 60-sec acquisition divided among 64 channels, which would yield only slightly better than 1-sec resolution. Any higher resolution inherent in the original data would be lost.

The second format provided by commercial instruments is as list-mode data. As the name implies, list mode is a list of the events acquired, retaining all measurements collected on each event with their association to each event remaining intact. From this list, correlations can later be drawn and the data reduced as necessary for display; however, the significant feature is that the raw data can remain unchanged for further analysis, and any necessary mathematical manipulations of the data can draw on the resolution of the raw data. In the early years of flow cytometry, the large size of list-mode files limited their usefulness. However, with the low cost of data storage in more recent years, this is no longer a problem, and routine list-mode storage has become common.

B. Data Display

Whether the data are stored as histograms or by list mode, it must be displayed in a meaningful way for analysis. Viewing a time parameter, as a single-parameter histogram may seem to provide little useful information, however, such a plot can sometimes prove interesting. As seen in Fig. 2, the vertical axis represents the number of events in each time channel and the horizontal axis of course represents time, left to right reflecting data acquisition from beginning to end. This type of plot can provide quick information on flow rate stability during acquisition.

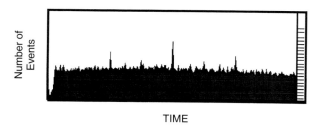

Fig. 2 Time as a parameter. The x-axis represents time and the y-axis represents the number of events. This type of histogram provides a graphical display of the consistency of the sample acquisition rate.

The two-parameter correlated plot, with time on one axis (usually the horizontal) and an analytical parameter on the other axis, is much more information rich. The resolution of the time axis is limited by the resolution of the raw data. However, advances in data collection hardware have created a situation where the data in the list-mode file is almost always much greater than can be accommodated by either the computer monitor or the software creating the plot. Therefore the data is reduced to some manageable number of slices. The three most common resolutions are 64×64 (4096 channel array), 128×128 (16,384 channel array), or 256×256 (65,536 channel array). It should be noted that even though the original data may have been collected at higher resolution, the binning requirements of the display software will inevitably reduce the apparent time resolution of the data. Analysis software can then either use the display resolution to calculate statistics or return to the original resolution in the raw data for those calculations. If higher time resolution analysis is required, one must consider how the software is performing those calculations.

Once the events are plotted, methods to quantify the response are needed. Reactions can be measured either through an intrinsic parameter such as size and morphology changes measured by light scatter or current interruption (as in Coulter sizing devices), or through extrinsic parameters such as fluorescent indicators. The most straightforward technique for data reduction is to simply divide the data into time intervals and calculate the average reaction for all cells in each interval. Commercial software is now available through the instrument manufacturers or third-party providers that provides this function. However, this has not always been the case. Investigators often rely on in-house software to calculate mean fluorescence as a function of time in order to retain the original time resolution of the raw data (Hoffman *et al.*, 1996). Average fluorescence can be a useful metric but also can be an oversimplification of the data. For example, it is not uncommon in cell activation studies to see more than one population of cells responding to the stimulus, to different degrees. As in the simple example in Fig. 3, a large number of cells are responding with sizable increases in intracellular calcium while another, smaller population shows almost no response. Calculating a mean response of all events does not accurately represent the biology represented in the data. Therefore, an enumeration of the cells responding and those not responding such as percent of events above some threshold level at each time point would create a more accurate picture.

As noted earlier, caution must be exercised when drawing conclusions about the number of cells responding during the course of an experiment from these data. If at every time point, 80% of the cells are seen in the reacted population, a claim that 80% of all cells responded to the stimulus over the course of the experiment assumes that all cells responded at the same time. However, from the data alone no such assumption can be made because some of the cells might be responding over different time courses. With the caveats noted, once the time resolved data has been reduced to a line graph (time vs. mean, median, or

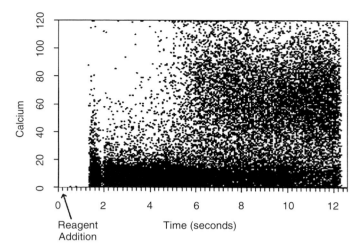

Fig. 3 Heterogeneous response. The *x*-axis represents time and the *y*-axis represents fluorescence intensity. Note that shortly after reagent addition at least two populations of events are seen. This type of response heterogeneity will impact cell response statistics.

percentage), computer modeling of that line can provide insights into the nature of the reaction being studied (Nolan *et al.*, 1998).

C. Quality Control

Incorporating a time parameter can help to identify time dependent artifacts in flow cytometric acquisition. These artifacts are particularly troublesome in DNA cell-cycle analyses because imprecise measurements or drifts in the apparent fluorescent intensity of a population can cause misidentification of ploidy and errors in estimates of a proliferating cell fraction. This problem was first attacked by using a dot plot (time vs. fluorescence) to visually identify unstable fluorescence over time (Watson, 1987; Horan *et al.*, 1990), as seen in Fig. 4. In

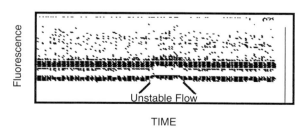

Fig. 4 Instrument stability. The *x*-axis represents time and the *y*-axis represents fluorescence intensity. Note the small increase in fluorescence intensity in the area noted.

1994, Kusuda and Melamed proposed methods of correcting the data, when the artifact is constant over time, by applying a correction factor to individual fluorescence data points based on their position in the acquisition sequence. Alternately, they proposed eliminating events acquired during transient instabilities through time gating, that is, removing the events acquired during the unstable period of acquisition. In 1995, Seamer and Altobelli applied standard statistical techniques, specifically, regression analysis, runs-based analysis, and standard deviation to identify and quantify time-dependent instability. These objective tests remove from such determinations the subjective ambiguities of visual inspection. Determinations of the stability of the acquisition can be performed either on time-stamped data or by creating a pseudotime plot when real-time acquisition is not available.

D. Pseudotime

Some older flow cytometers do not have the ability to store time-stamped data. On other occasions, data are stored without time information, even if it is available. List-mode acquisition will allow an estimation of time to be imposed on the data. List-mode data are typically stored as a sequential list of cells acquired with relative time being preserved. By parsing the file into equally sized fractions, this feature can be exploited to estimate a time parameter. This feature is called pseudotime. If the total file acquisition time is known and the flow rate during acquisition was relatively constant, pseudotime will yield reasonably accurate results, and it is adequate when real-time measurements are not required. There are currently several commercial software packages that include this feature.

E. Flow Cytometry Standard Support

In 1997, the Data File Standards committee of the International Society of Analytical Cytology proposed version 3.0 of the Flow Cytometry data file standard (Seamer et al., 1997). That version, for the first time, recognized the special requirements of time-based analysis and added support for the inclusion of time as an analytical parameter. The $PnN keyword that specifies the name of a parameter "n" must be set to the string "TIME" to identify it as a time measurement. The keyword $TIMESTEP then shows the resolution of the time measurement in seconds or fractions of a second. These variables allow reconstruction of the time sequence with real-time results.

V. Applications

Time-based measurements have found many applications in cell biology (Table II). The study of enzyme kinetics in individual cells was a very early target of

Table II
Time Applications

Application	Examples	Resolution time required
Ion concentration	Ca^{2+}, Mg^{2+}, H^+	Subsecond to seconds
Functional assay	NK cell—tumor killing	Seconds to minutes
Ligand–receptor binding	Formylpeptide	Subsecond
Enzyme kinetics in cells	Esterase	Seconds
Macromolecular assembly	Flap endonuclease	Subsecond to minutes
Quality control	Instrument drift analysis	Minutes
Pump activity	Daunorubicin elimination	Seconds to minutes

flow cytometric evaluation. By using fluorescent intensity as an indicator of substrate or product concentration, standard kinetic calculations such as Michaelis–Menton and Lineweaver–Burk plots can be obtained (Watson et al., 1977). A novel approach to measuring enzyme kinetics has been shown by attaching substrate to a solid phase composed of cell-sized beads. The loss of fluorescence substrate associated with the beads is then measured. Such noncell systems broaden the use of flow cytometry to new areas of biochemistry and molecular analysis (Shen et al., 1996, 1997; Nolan et al., 1996; Nolan and Sklar, 1998).

The study of macromolecular assembly is another area of investigation where flow cytometry has made an impact. The simple technique of labeling one molecular component with a fluorescent dye and measuring its accumulation on or departure from a particle over time is the most general technique. Another technique is that employed in f-actin quantization (Omann et al., 1987). Using fluorescently tagged phalloidin, a molecule that preferentially binds to actin polymers, the kinetics of f-actin assembly can be elucidated. This technique has also been used to study neutrophil activation in response to formylpeptides. One specific type of assembly, the study of the association of ligands with their cellular receptors, has been studied extensively by flow cytometry (Sklar et al., 1984; Fay et al., 1991; Hoffman et al., 1996; Nolan et al., 1998). As noted earlier, a well-recognized strength of flow cytometry is its ability to discriminate fluorescent ligand bound to a cell from that at much lower concentration in the surrounding media. This eliminates the normal requirement for washing the cells free of extraneous ligand. When examined as a function of time, association and dissociation kinetics can be determined.

The study of cell activation has been one of the most fertile areas for flow cytometric investigation. As cells become activated, they begin a sequence of events usually ending in altered functional states. One of the most studied responses is that of rapid and dramatic changes in ion concentration. Intracellular Ca^{2+} concentration is an example of one of the first applications of flow cytometry to the study the dynamic processes of cell activation (Rabinovitch et al., 1986; Montero et al., 1993; Bedner et al., 1998).

If cells are examined in minutes rather than seconds, later activation events and cell function assays can also be fertile ground for kinetic measurements by flow cytometry. For example, there are studies exploring alterations in the activity of natural killer lymphocytes (NK cells) against tumor cells as a function of time post activation (Edwards *et al.*, 1989; Zamai *et al.*, 1998) and the analysis of granule secretion (Fletcher and Seligmann, 1985) in neutrophils, as well as a variety of other cell adhesion responses.

Other disciplines have found value in the ability of flow cytometry to measure individual parameters over time. For example, multidrug resistance has been studied using the loss of daunarubicin over time as an indication of a functional pump (Nooter *et al.*, 1989).

VI. Conclusions

Improvements in hardware and software have extended the time frame of flow cytometric analysis of cellular and biochemical events from minutes to milliseconds. Flow cytometric time stamping of data was once only possible by collecting individual records at the interval of minutes. Improvements were made when a voltage ramp was applied, allowing measurements to be binned into time channels. As analog to digital converters improved, those time-bins were shortened, yielding higher resolution. Innovations in hardware have allowed the time to the first measurement to be reduced from 10 sec to 100 msec. Data storage standards have provided a mechanism to reconstruct real-time information from data collected at remote sites. However, with existing hardware, more improvements are possible. Current data collection software continues to bin cells. However, by using the computer clock to assign time values to each event, the time parameter for each cell can be unique, allowing time resolution to approach the computer clock speed. The challenge would then shift to software as the time parameter would be stored at much higher resolution than other analytical parameters. Now that current data file standards allow storage of each parameter at a different resolution, few current file readers are beginning to accept such files.

References

Bedner, E., Melamed, M., and Darzynkiewicz, Z. (1998). Enzyme kinetic reactions and fluorochrome uptake rates measured in individual cells by laser scanning cytometry. *Cytometry* **33**, 1–9.

Blankenstein, G., Scampavia, L. D., Ruzicka, J., and Christian, G. D. (1996). Coaxial flow mixer for real-time monitoring of cellular responses in flow injection cytometry. *Cytometry* **25**, 200–204.

Dunne, J. (1991). Time window analysis and sorting. *Cytometry* **12**, 597–601.

Durack, G., Lawler, G., Kelley, S., Ragheb, K., Roth, R. A., Ganey, P., and Robinson, J. P. (1991). Time interval gating for analysis of cell function using flow cytometry. *Cytometry* **12**, 701–706.

Edwards, B. S., Nolla, H. A., and Hoffman, R. R. (1989). Relationship between target cell recognition and temporal fluctuations in intracellular Ca^{2+} of human NK cells. *J. Immunol.* **143**, 1058–1065.

Edwards, B. E., Kuckuck, F., and Sklar, L. A. (1999). Plug-flow flow cytometry: an automated coupling device for rapid sequential flow cytometric sample analysis. *Cytometry* **37**, 156–159.

Fay, S. P., Posner, R. G., Swann, W., and Sklar, L. A. (1991). Real-time analysis of the assembly of ligand, receptor and G protein by quantitative fluorescence flow cytometry. *Biochemistry* **30,** 5066–5075.

Fletcher, M. P., and Seligmann, B. E. (1985). Monitoring human neutrophil granule secretion by flow cytometry: Secretion and membrane potential changes assessed by light scatter and a fluorescent probe of membrane potential. *J. Leukocyte Biol.* **37,** 431–447.

Hawking, S. J. (1998). "A Brief History of Time." Bantam Books, New York.

Hoffman, J. F., Linderman, J. J., and Omann, G. M. (1996). Receptor up-regulation, internalization and interconverting receptor states: Critical components of a quantitative description of N-formyl peptide-receptor dynamics in the neutrophil. *J. Biol. Chem.* **271,** 18394–18404.

Horan, P. K., Muirhead, K. A., and Slezak, S. E. (1990). Standards and controls in flow cytometry. *In* "Flow Cytometry and Sorting" (M. R. Melamed, T. Lindmo, and M. L. Mendelsohn, eds.), 2nd Ed., pp. 397–414. Wiley-Liss, New York.

Kelley, K. (1989). Sample station modification providing on-line reagent addition and reducing sample transit time for flow cytometers. *Cytometry* **10,** 796–800.

Kusuda, L., and Melamed, M. R. (1994). Display and correction of flow cytometric time-dependent fluorescence changes. *Cytometry* **17,** 340–342.

Lindberg, W., Ruzicka, J., and Christian, G. D. (1993). Flow injection cytometry: A new approach for sample and solution handling in flow cytometry. *Cytometry* **14,** 230–236.

Martin, J. C., and Swartzendruber, D. E. (1980). Time: A new parameter for kinetic measurement in flow cytometry. *Science* **207,** 199–201.

Montero, M., Garcia-Sanch, J., and Alvarez, J. (1993). Transient inhibition by chemotactic peptides of a store-operated Ca^{2+} entry pathway in human neutrophils. *J. Biol. Chem.* **268,** 13055–13061.

Nolan, J. P., and Sklar, L. A. (1998). Emergence of flow cytometry for measurements of molecular interactions. *Nature Biotechnol.* **16,** 633–638.

Nolan, J. P., Posner, R. G., Martin, J. C., Habbersett, R., and Sklar, L. A. (1995). A rapid mix flow cytometer with subsecond kinetic resolution. *Cytometry* **21,** 223–229.

Nolan, J., Shen, B., Park, M., and Sklar, L. A. (1996). Kinetic analysis of human flap endonuclease by flow cytometry. *Biochemistry* **35,** 11668–11676.

Nolan, J., Chambers, J. D., and Sklar, L. A. (1998). Flow cytometric analysis of ligand–receptor interactions. *In* "Cytometry Approaches to Cellular Analysis" (G. Babcock and P. Robinson, eds.), pp. 19–46. Wiley-Liss, New York.

Nooter, K., Oostrum, R., Jonker, R., van Dekken, H., Stokdijk, W., and van den Engh, G. (1989). Effect of cyclosporin A on daunorubicin accumulation in multidrug-resistant P388 leukemia cells measured by real-time flow cytometry. *Cancer Chemother. Pharmacol.* **23,** 296–300.

Nooter, K., Herweijer, H., Jonker, R. R., and van den Engh, G. J. (1994). On-line flow cytometry: A versatile method for kinetic measurements. *Methods Cell Biol.* **41,** 509–525.

Omann, G. M., Coppersmith, W., Finney, D. A., and Sklar, L. A. (1985). A convenient in-line device for reagent addition, sample mixing and temperature control of cell suspensions in flow cytometry. *Cytometry* **6,** 69–73.

Omann, G. M., Swann, W. N., Oades, Z., Parkos, C. A., Jesaitis, A. J., and Sklar, L. A. (1987). N-Formylpeptide-receptor dynamics, cytoskeletal activation and intracellular calcium response in human neutrophil cytoplasts. *J. Immunol.* **139,** 3447–3455.

Rabinovitch, P. S., June, C. H., and Ledbetter, J. A. (1986). Heterogeneity among T cells in the intracellular free calcium responses after mitogen stimulation with PHA or anti-CD3. Simultaneous use of indo-1 and immunofluorescence with flow cytometry. *J. Immunol.* **137,** 952–962.

Seamer, L. C., and Altobelli, K. K. (1995). Fluorescence drift detection as a novel QC procedure for DNA cell-cycle analysis. *Cytometry* **22,** 60–64.

Seamer, L. C., Bagwell, C. B., Barden, L., Redelman, D., Salzman, G. C., Wood, J. C. S., and Murphy, R. F. (1997). Proposed new data file standard for flow cytometry, version FCS 3.0. *Cytometry* **28,** 118–122.

Seamer, L. C., Kuckuck, F., and Sklar, L. A. (1999). Sheath fluid control to permit stable flow in rapid mix flow cytometry. *Cytometry* **35,** 75–79.

Scudder, K. M., Christian, G. D., and Ruzicka, J. (1993). Flow injection fluorescence microscopy: A novel tool for the study of cells through controlled perfusion. *Exp. Cell Res.* **205,** 197–204.

Shen, B., Nolan, J. P., Sklar, L. A., and Park, M. S. (1996). Essential amino acids for substrate binding & catalysis of human flap endonuclease-1. *J. Biol. Chem.* **271,** 9173–9176.

Shen, B., Nolan, J. P., Sklar, L. A., and Park, M. S. (1997). Functional analysis of point mutations in human flap endonuclease-1 active site. *Nucleic Acids Res.* **25,** 3332–3338.

Sklar, L. A., Finney, D. A., Oades, Z., Jesaitis, A. J., Painter, R. G., and Cochrane, C. G. (1984). The dynamics of ligand–receptor interactions: real-time analysis of association, dissociation and internalization of an N-formyl peptide and its receptors on the human neutrophil. *J. Biol. Chem.* **259,** 5661–5669.

Sklar, L. A., Seamer, L. C., Kuckuck, F., Posner, R. G., Prossnitz, E., Edwards, B., and Nolan, J. P. (1998). Sample handling for kinetics and molecular assembly in flow cytometry. *Adv. Opt. Biophys.* **3256,** 144–153.

Van Dilla, M. A., Trujillo, T. T., Mullaney, P. F., and Coulter, J. R. (1968). Cell microfluorometry: A method for rapid fluorescence measurement. *Science* **163,** 1213–1214.

Watson, J. V. (1987). Time, a quality-control parameter in flow cytometry. *Cytometry* **8,** 646–649.

Watson, J. V., Chambers, S. H., Workman, P., and Horsnell, T. S. (1977). A flow cytofluorimetric method for measuring enzyme reaction kinetics in intact cells. *FEBS Lett.* **81,** 179–182.

Zamai, L., Mariani, A. R., Zauli, G., Rodella, L., Rezzani, R., Manzoli, F. A., and Vitale, M. (1998). Kinetics of in vitro natural killer activity against K562 cells as detected by flow cytometry. *Cytometry* **32,** 280–285.

CHAPTER 9

Protein Labeling with Fluorescent Probes

Kevin L. Holmes and Larry M. Lantz

Flow Cytometry Section
National Institute of Allergy and Infectious Diseases
National Institutes of Health
Bethesda, Maryland 20892

I. Introduction
II. Labeling of Proteins with Organic Fluorescent Dyes
    A. Conjugation Chemistry
    B. Practical Considerations: Optimizing Conditions for Labeling
    C. Specific Organic Dyes
    D. Choice of Reactive Group
III. Labeling of Proteins with Phycobiliproteins
    A. Background
    B. Principle of Coupling and the Use of Heterobifunctional Reagents
IV. Conclusion
    References

I. Introduction

Conjugation of fluorescent molecules to proteins is a subset of the much larger field of bioconjugation chemistry. It has nevertheless developed into an important field itself, since the initial descriptions of the use of fluorescently labeled antibodies in tissue sections (Coons *et al.,* 1942; Coons and Kaplan, 1950). With the advent of monoclonal antibodies and flow cytometry, the value of labeling proteins and, in particular, antibodies with fluorescent molecules has become quite evident in the biomedical and basic research communities. The following is a concise overview of the very large field of fluorochrome bioconjugation, including a discussion of the basis of the procedures, optimization of the conjugation, and fluorochrome specific topics. The conjugation of fluorochromes to antibody molecules is emphasized, but the same principles can be applied to other large proteins. There have been many excellent reviews of bioconjugation chemistry in general (Brinkley,

1992; Hermanson, 1996a; Wong, 1991a) and detailed methods of fluorochrome labeling (Hardy, 1986; Haugland, 1995; Haugland and You, 1995; Holmes *et al.*, 1997; Johnson and Holborow, 1986). The reader is directed to the latter sources for specific protocols in fluorochrome conjugation.

II. Labeling of Proteins with Organic Fluorescent Dyes

A. Conjugation Chemistry

The coupling of fluorescent probes to proteins, in particular antibodies, can be best accomplished by an understanding of the chemistry involved in the conjugation. Proteins are polymers of amino acids that contain various side chains. These side chains are utilized as reactive groups to attach dyes and fluorochromes. The reactivity of the protein/antibody will, therefore, be determined by the amino acid composition and the sequence location of the individual amino acids in the three-dimensional structure of the molecule. Thus the nonpolar hydrophobic amino acids (glycine, alanine, valine, leucine, isoleucine, methionine, proline, phenylalanine, and tryptophan) are usually found on the interior of the protein and unavailable for modification. Amino acids with ionizable side chains (arginine, aspartic acid, glutamic acid, cysteine, histidine, lysine, and tyrosine) and with polar groups (glutamine, serine and threonine) are usually located on the protein surface and are available for modification (Wong, 1991b). Protein modification reactions are nucleophilic substitution reactions (Fig. 1). In this type of reaction, a nucleophile (Nu:) with a lone pair of electrons attacks an electron deficient (electrophilic) center, resulting in the displacement of a leaving group (X:). The nucleophile in protein modification is the amino acid side chain. The electrophilic center is most commonly a carbon atom in which a more electronegative atom, such as oxygen has been attached. The relative reactivity is determined by the nucleophilicity of the amino acid side chain (see later). For protein modification with fluorochromes, there are two general classes of agents that are most commonly used: acylating and alkylating agents (Fig. 2).

In alkylation, an alkyl group is transferred to the nucleophile (amino acid side chain for proteins), and in acylation, an acyl group is bonded. Figure 3 shows a listing of commonly used reactive groups, the functional group on the protein

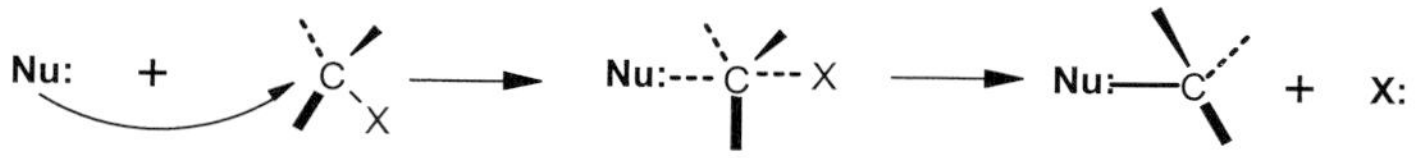

Fig. 1 Schematic representation of a nucleophilic substituion reaction. In this type of reaction, a nucleophile (Nu:) with a lone pair of electrons attacks an electron-deficient (electrophilic) center resulting in a covalent coupling of the nucleophile and the electrophile along with the displacement of a leaving group (X:).

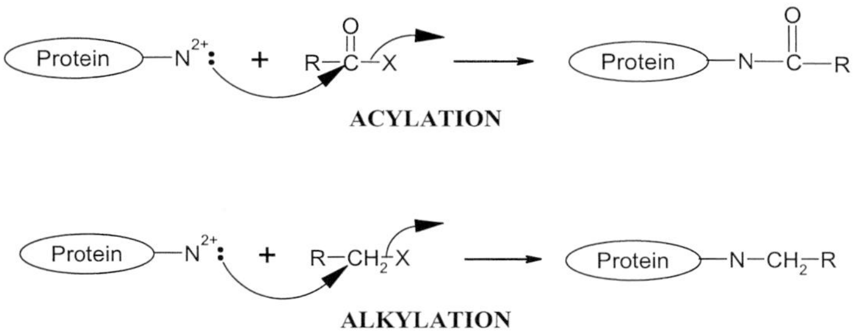

9. Protein Labeling

Fig. 2 Schematic representation of alkylation and acylation reactions. In acylation, an active carbonyl group undergoes addition to the amino acid side chain, and in alkylation, an alkyl group is transferred to the nucleophile. Both reactions result in the displacement of a leaving group (X:) as shown in Fig. 1.

or antibody that is targeted, the linkage formed, and the dyes that are available that use this linkage for conjugation. The acylating agents are amine reactive and form amide, thiourea, or sulfonamides. The alkylating agents most commonly used are the chlorinated *s*-triazine and *N*-maleimide groups. Chlorinated triazines react with amines forming amino triazines with loss of a chloride ion. Maleimides react primarily with sulfhydryl groups at neutral pH to form a thioether bond with proteins. Sulfhydryls already present on the protein, such as cytsteines, can be utilized or sulfhydryls can be added to the protein with thiolating reagents such as *N*-succinimidyl 3-(2-pyridyldithio)propionate (SPDP) or *N*-succinimidyl *S*-acetylthioacetate (SATA) (see later). These may provide an alternative to amine reactive agents, when it has been found that this coupling results in a loss of biological activity (Imam, 1979).

B. Practical Considerations: Optimizing Conditions for Labeling

When performing coupling of fluorochromes to antibodies, the procedure used to couple is generally less dependent on the fluorophore and more dependent on the reactive moiety attached to it. Therefore all dyes having *N*-hydroxysuccinimide (NHS) ester reactive groups, for example, can utilize the same conjugation protocol. The differences in the reaction conditions between the commonly used reactive groups depend on the characteristics of the functional group(s) that is targeted as well as of the reactive group on the fluorochrome.

1. pH and Buffers

The rate of the labeling reaction is governed by several factors, but is dependent on the nucleophilicity of the amino acid side chains, which in turn is dependent on their pK_a (Table I). Ionizable groups in proteins such as carboxylic and

PROTEIN FUNCTIONAL GROUP	REACTIVE GROUP ON DYE	LINKAGE FORMED	DYES AVAILABLE WITH THESE REACTIVE GROUPS
 Protein—NH$_2$ **PRIMARY AMINE**	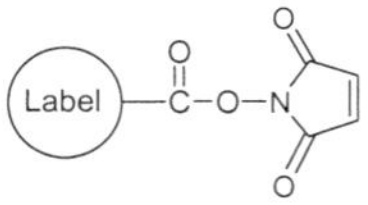 Label—C(=O)—O—N(maleimide) **NHS-ESTER**	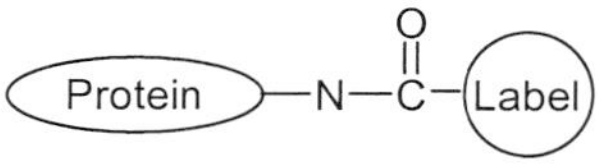 Protein—N—C(=O)—Label **AMIDE BOND**	fluorescein, AMCA, carboxy-fluorescein, biotin, cyanine and Alexa dyes, rhodamine,
 Protein—NH$_2$ **PRIMARY AMINE**	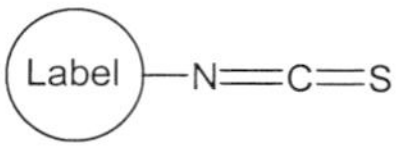 Label—N=C=S **ISOTHIO-CYANATE**	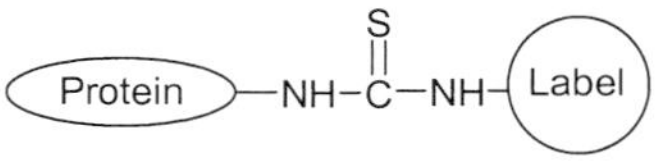 Protein—NH—C(=S)—NH—Label **THIOUREA**	fluorescein, rhodamine, Oregon Green
 Protein—NH$_2$ **PRIMARY AMINE**	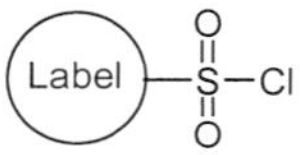 Label—S(=O)$_2$—Cl **SULFONYL HALIDE**	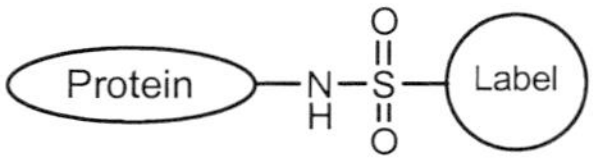 Protein—N(H)—S(=O)$_2$—Label **SULFONAMIDE BOND**	Texas Red, Lissamine rhodamine B, sulfonyl chloride
 Protein—NH$_2$ **PRIMARY AMINE**	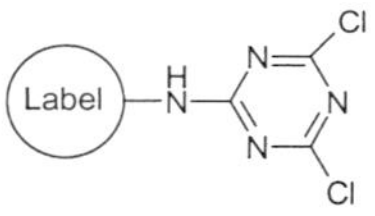 Label—NH—(triazine-Cl$_2$) **CHLORINATED S-TRIAZINES**	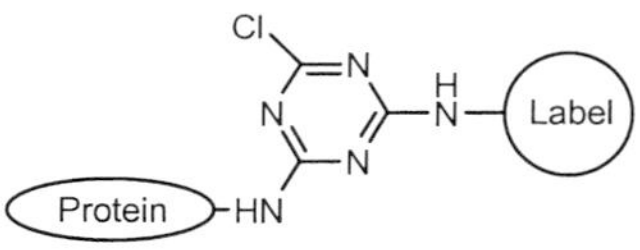 Protein-HN—(triazine-Cl)—NH—Label **AMINO TRIAZINES**	5-DTAF
 Protein—SH **SULFHYDRYL**	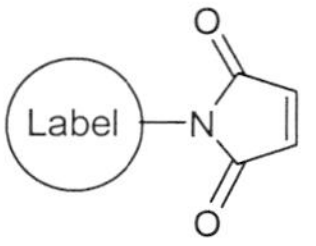 Label—N(maleimide) **MALEIMIDE**	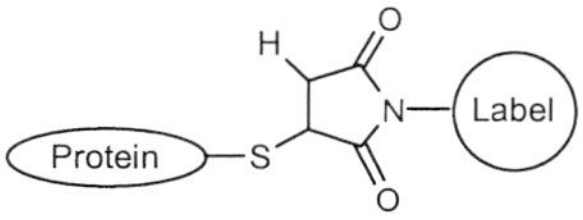 Protein—S—(succinimide)—N—Label **THIOETHER BOND**	biotin, fluorescein, Alexa dyes, Texas Red, rhodamine, Oregon Green

Fig. 3 Reactive groups and reactions of commonly used fluorescent dyes. Listing of commonly used reactive groups for protein labeling, the functional group on the protein or antibody that is targeted, the linkage formed, and the dyes that are available that use this linkage for conjugation.

amine groups exist either in protonated or unprotonated forms. The degree of protonation is dependent on their pK_a and the pH. Carboxylic groups at pH values above their pK_a will be unprotonated and carry a negative charge. At pH

Table I
pK_a of Amino Acid Functional Groups[a]

Functional group	pK_a in free amino acids	pK_a in proteins
ε-Amino	10.5–10.8	9.11–10.7
α-Amino	8.8–10.8	6.72–8.14
Sulfhydryl	8.3–8.4	8.5–8.8
α-Carboxyl	1.8–2.6	3.1–3.7

[a] Data compiled from (Wong, 1991b) and (Botelho and Gurd, 1989).

values below their pK_a, carboxylic groups will be protonated and carry no charge. Amine groups, however, are protonated and positively charged at pH values below their pK_a, and they are unprotonated and neutrally charged at pH values above their pK_a. Because protonation decreases nucleophilicity, pH will affect the rate and specificity of the reaction. This can be used to direct reactivity of agents toward particular groups. For example, at neutral pH N-ethylmaleimide reacts more readily to the sulfhydryl group of cysteine (having a pK_a of 8.5–8.8), than with the ε-amino group of lysine (having a pK_a of 9.11–10.7). It is therefore important to choose a buffer that has a buffering capacity within the pH range that is optimal for the reaction. Isothiocyanate labeling which occurs optimally at a pH of 9.0–9.5 requires a borate or carbonate buffer, whereas N-ethylmaleimide coupling, which occurs at near neutral pH requires a phosphate buffer. In addition, amine-containing buffers such as Tris should not be used because of competition between the amine groups in the buffer and the amino acid side chains of the protein. It should be noted, however, that attempts to selectively target particular amino acid groups *exclusively,* using pH alone, is not possible. This is because the microenvironmental effects of the protein, in which the group exists, modify the pK_a of amino acid groups. This means that there may be significant overlap in the pK_a ranges of the different reactive amino acid side chains, and, therefore, overlap in the nucleophilicity of these groups.

2. Protein Concentration and Hydrolysis

The speed and degree of substitution of the coupling reaction is dependent on the concentration of the protein. This is primarily due to hydrolysis of the acylating (or alkylating) agent, because water competes as a nucleophile with the amino acid side chains. This is especially noticeable with NHS esters, which are more reactive with amines at alkaline pH but also show an increased rate of hydrolysis with increasing pH. Therefore, a pH range of 7.5–8.5 has been recommended (Haugland and You, 1995). Protein concentrations of at least 1 mg/ml are desirable; higher concentrations give higher rates of reaction and higher fluorophore/protein (F/P) ratios. The F/P ratio is the quantitative measure of the level of fluorophore modification of the protein, is determined spectrophotometrically, and is expressed as moles of fluorophore/moles of protein. (For a more detailed discussion of this topic see Brinkley, 1992; Haugland, 1995.) Brinkley recommends concentrations of 7.5 to 15 mg/ml (Brinkley, 1992). Ideally, the protein concentration should be kept constant to ensure reproducible results. In this regard, protein concentration is more important than total amount of protein when labeling small amounts of antibody.

3. Degree of Substitution and Activity/Quenching Effects

The final F/P ratio achieved in fluorochrome labeling is dependent on the factors previously listed, that is, pH and protein concentration, but also time and temperature. In general, most reactions can be performed at room temperature, with some reactions showing more or less sensitivity to temperature changes (Wong, 1991c; and see later). It has been suggested that a convenient procedure is to add the fluorochrome to a stirred protein solution in an ice-bath and allow the bath to warm to room temperature over a period of about 2 hr (Brinkley, 1992). Most protocols are designed to provide an optimal F/P within usually 1–2 hr. Longer incubation time will provide increased F/P ratios, but the final F/P ratio *desired* may vary, dependent on the dye and the antibody or protein that is being labeled. It is best to systematically vary conditions, usually either time of incubation or amount of dye added to the protein solution, to achieve optimal labeling. Optimal labeling will be below the maximum substitution possible for three reasons. First, unless the antibody-combining site is somehow protected (Imam, 1979) it may be labeled, and the degree of substitution is limited by the desire to retain the biological activity of the antibody. Second, the intensity of fluorescence of the conjugate will reach a maximum and then decrease with increasing F/P ratios due to fluorescence quenching effects (Der-Balian *et al.*, 1988; Haugland, 1996). Third, increases in F/P may result in higher nonspecific binding of the antibody. For example, fluorescein isothiocyanate (FITC) shows increased negative charge with increases in F/P resulting in binding to positively charged cellular ions, similar to the binding of eosin dyes (The and Feltkamp, 1970a). Therefore, a balance must be achieved between obtaining the highest degree of substitution that is consistent with the preservation of activity and reduction of nonspecific binding.

C. Specific Organic Dyes

1. Fluorescein Derivatives

Since its introduction as a fluorescent label for immune serum, FITC has been used extensively for immunofluorescent techniques (Riggs *et al.*, 1958). FITC continues to be the most widely used inorganic fluorochrome for labeling antibodies and owing to its long history of use has been studied extensively, particularly for labeling of antibody proteins. The conditions for optimal labeling of FITC are determined by the requirements of the isothiocyanate reactive group. In general, optimal labeling of proteins with FITC is achieved when conditions include high pH, temperature, and protein concentration. FITC is soluble in aqueous solutions but may be more completely dissolved in DMSO, resulting in more predictable results (Goding, 1976). Below pH 9.0, FITC reacts primarily with the α-amino group of the N-terminal amino acid, thus limiting the degree of substitution; however, at pH values above pH 9.0, the degree of substitution increases due to reaction with ε-amino groups of lysine (Maeda, 1969). Maximal fluorescein to protein ratios are obtainable at 37°C, but 25°C incubation provides optimal results (The and Feltkamp, 1970b). High protein concentrations allow high F/P ratios in a shorter amount of time, compared with low concentrations (The and Feltkamp, 1970b). Optimally, concentrations of 25 mg/ml are desirable, but may be impractical; proteins ranging from 1 to 10 mg/ml can be labeled with an increase in incubation time.

A modification of fluorescein, dichlorotriazinylaminofluorescein (5-DTAF), has been used to label carbohydrates and protein, including antibodies (Der-Balian *et al.*, 1988). The chlorinated *S*-triazines are highly reactive with nucleophiles and bind to α- and ε-amino groups of proteins. The conjugates appear stable, but may show changes with time, attributable to hydrolysis of the remaining, relatively inert, chloro group (Zuk *et al.*, 1979).

Another modification of fluorescein, carboxyfluorescein succinimidyl ester, known as CFSE or FAM, utilizes the reactive succinimidyl ester to form very stable protein conjugates (Zuk *et al.*, 1979). This derivative has also been modified with a seven-atom aminohexanoyl spacer group between the FAM fluorphore and the succinimidyl ester in an attempt to reduce quenching effects seen with increased F/P ratios (Haugland, 1996).

2. Biotin/Avidin

Although biotin is not a fluorochrome, its extensive use as a label in combination with fluorochrome-labeled avidin necessitates a discussion of its properties. Biotin is a small molecule (MW 244) which serves as an intermediate carrier of carbon dioxide in carboxylating enzymes. Its usefulness in immunochemistry and flow cytometry, however, is the very high affinity ($K_a = 10^{15}\ M^{-1}$) binding observed with the egg white protein avidin. Avidin is a tetramer, consisting of four subunits with a combined molecular mass of 67,000 to 68,000. Each avidin molecule contains four binding sites for biotin. Avidin is positively charged at neutral pH, having an isoelectric point of 10.5. Because of its positive charge and the presence of the oligosaccharides mannose and *N*-acetylglucosamine, avidin has been shown to bind nonspecifically to negatively charged molecules and to carbohydrate-binding proteins on cell surfaces. Avidin has also been shown to bind nonspecifically to the cytoplasmic granules of mast cells (Bussolati and Gugliotta, 1983). These and other observations have led to the more widespread use of streptavidin, particularly for fluorescence microscopy and flow cytometry.

Streptavidin, isolated from *Streptomyces avidinii*, lacks carbohydrate, reducing the potential for possible protein interaction, and it has an isoelectric point of 5–6, which lowers the overall charge of the molecule. Although it has been suggested that these characteristics of streptavidin will eliminate many of the problems associated with nonspecific binding of avidin, this may not be true in all instances. Indeed, streptavidin has been shown to contain the tripeptide sequence Arg-Tyr-Asp (RYD) that mimics the binding sequence of fibronectin Arg-Gly-Asp (RGD), a universal recognition domain of the extracellular matrix that promotes cell adhesion (Alon, 1990). Additionally, the higher isoelectric point of avidin, compared with streptavidin, may be irrelevant since it may be altered by conjugation with fluorochromes. In this regard, FITC conjugation to protein has been shown to increase their negative electrical charge (The and Feltkamp, 1970a). Another disadvantage to streptavidin is that the biotin-binding cleft is different than avidin, and may require a longer spacer arm for biotinylation to achieve optimum binding (see later). These disadvantages have prompted the introduction and use of a chemically deglycosylated form of avidin, known as variously as NeutraLite or NeutrAvidin. This form has a near neutral isoelectric point but maintains the biotin-binding affinity of native avidin (Hiller *et al.*, 1987).

The use of biotin as reagents has been reviewed (Wilchek and Bayer, 1990), as well as their use in bioassays (Wilchek and Bayer, 1988) and their methods of coupling to antibodies (Haugland and You, 1995).

Several active biotin derivatives have been produced to biotinylate proteins and glycoproteins. They are summarized in Table II. The most popular biotin derivative is the NHS ester for labeling the ε-amino groups of lysine. There are water soluble sulfo-NHS forms also available; these do not offer advantages over NHS biotin for labeling antibodies but have been used to restrict labeling to the cell surface in immunoprecipitation protocols (Lantz and Holmes, 1995). Biocytin hydrazide has been used to label carbohydrates or glycoproteins, such as the carbohydrate groups of the Fc region of antibodies. Biocytin is an adduct of biotin and lysine (N-ε-biotinyl-L-lysine), but in this application biocytin hydrazide

Table II
Biotin Derivatives

Protein functional group	Biotin derivative
Amine	N-Hydroxysuccinimide ester (NHS ester)
Tyrosyl, histidyl	p-Diazobenzoyl-biocytin
Sulfhydryl	3-(N-Maleimidopropionyl)biocytin, iodoacetyl-LC-biotin
Carboxyl	Biocytin hydrazide/carbodiimide

labeling was found to be inferior to labeling via the ε-amino groups (Diamandis and Christopoulos, 1991; Gretch *et al.,* 1987).

The biotin molecule is small and dependent on the location of the functional group on the protein to which biotin conjugates; the avidin-binding site may be inaccessible to avidin. For this reason, derivatives of biotin have been made that are known as long chain or biotin-X (or XX) which include one or two seven-atom aminohexanoic spacers attached to the carboxyl group of biotin. These separate the protein-binding site from the avidin-binding site, greatly enhancing the efficiency of formation of the avidin-binding complex by reducing steric hindrance. As noted earlier, it has been suggested that biotin-binding cleft of streptavidin is different than avidin and may require a longer spacer group for optimal binding (Haugland and You, 1995; Haugland, 1996). Haugland and You (1995) found that the use of two aminohexanoic acid spacers in biotin resulted in higher titers in enzyme-linked immunosorbent assays.

The previous discussion emphasizes the need to be knowledgeable of the reagents used when performing flow cytometric analysis or immunohistochemistry with the biotin–avidin system. The choice of use of avidin, NeutrAvidin, streptavidin, or biotin with aminohexanoic spacer groups may depend on the specific application or cell type being analyzed. When commercially prepared reagents are used, it may be difficult to determine which biotin is used for conjugations. The major antibody supply companies vary in whether they provide their antibodies coupled with biotin, its long-chain version, or both, whereas most supply streptavidin–fluorochrome conjugates.

3. Texas Red and Rhodamine

Some of the first dyes to be used as second labels in combination with FITC (or in other multicolor applications) were the sulforhodamine dyes Lissamine rhodamine B and Texas Red (sulforhodamine 101; Titus *et al.,* 1982). Both of these dyes are sulfonyl chlorides, which will form amide bonds with amino groups of the protein. They are not group specific, however, and due to a high rate of hydrolysis at the high pH (pH 9.0–9.5) required for conjugation to aliphatic amines, they require low temperature (4°C) conditions for conjugation (Lefevre *et al.,* 1996). In particular, sulfonyl chlorides will form conjugates with tyrosine, histidine, and cysteine (Wong, 1991b), which are unstable and require subsequent removal with hydroxylamine to ensure stability with storage (Brinkley, 1992). These deficiencies result in a lack of reproducibility in conjugations and an inability to successfully label some antibody classes or species (Titus *et al.,* 1982). New derivatives of Texas Red and Lissamine rhodamine B incorporate a 6-aminohexanoic spacer between the fluorophore and a NHS ester group (Lefevre *et al.,* 1996). This modification allows labeling under more mild conditions, with less precipitation of proteins and an increased fluorescence yield (Lefevre *et al.,* 1996).

D. Choice of Reactive Group

The availability of a large variety of fluorochrome derivatives provides the researcher with the opportunity to tailor their conjugations to their specific needs. The choice of the derivative is dependent on several factors: (1) the number of functional groups available on the protein. The most commonly used method of protein modification involves coupling through the aliphatic amines, the N-terminal α-amines, and the ε-amino group of lysine. This is due primarily to their high reactivity and to their abundance in proteins. Although some proteins lack N-terminal α-amines (i.e., cytochrome c and ovalbumin; Brinkley, 1992), the majority possess both α- and ε-amines. This is true for immunoglobulin G (IgG) molecules, which are composed of four polypeptide chains and approximately 90 lysine residues (Nakagawa *et al.*, 1972). (2) As previously stated, the pH of the reaction is controlled using the appropriate buffer system. However, for reactive groups requiring more alkaline pH, some proteins may be more sensitive to these conditions. For example, IgM molecules are unstable at alkaline pH and therefore necessitate the use of derivatives that can be used at neutral pH such as NHS esters (Haugland and You, 1995). (3) The most variable portion of the antibody molecule is the antigen-combining site, which implies that there will be variability in susceptibility of the antigen-combining site to be labeled during conjugation. Therefore, it may be necessary to utilize a different functional group for conjugation if it is determined that activity cannot be preserved with a particular reactive group. This is obviously of more importance when labeling monoclonal rather than polyclonal antibody preparations. For example, as stated earlier, if conjugation directed toward primary amines (i.e., NHS ester) results in loss of activity, reactive groups targeting sulfhydryls (i.e., *N*-ethylmaleimide derivatives at neutral pH) may achieve labeling with preservation of activity. In this regard, one strategy employed is to cleave the antibodies by reduction at their disulfide groups in the hinge region, using 2-mercaptoethylamine. This generates two heavy plus light chain fragments containing free sulfhydryls that are removed from the antigen-combing site and available for conjugation (Hermanson, 1996a). This may be particularly advantageous since the frequency of free thiols in proteins can be relatively low.

III. Labeling of Proteins with Phycobiliproteins

A. Background

Phycobiliproteins are water soluble fluorescent pigment proteins of the photosynthetic machinery of cyanobacteria and eukaryotic algae. They function as accessory or antenna pigments for the collection of light energy. These pigments absorb light energy of the visible spectrum in wavelengths that are poorly absorbed by chlorophyll. The phycobiliproteins in most algae are arranged in subcellular structures, known as phycobilisomes. These structures optimize the capture and transfer of light energy. All of the phycobiliproteins absorb light energy directly, and, through a series of energy transfer intermediates, transfer this light energy to the photosynthetic reaction core (phycoerythrin to phycocyanin to allophycocyanin to chlorophyll a) (Glazer, 1985). When the individual phycobiliproteins are purified and isolated, these proteins become highly fluorescent because there are no nearby molecules to act as energy acceptors. Phycobilisomes provide 30–50% of the total light-harvesting capacity of cyanobacteria and red alga cells (Glazer, 1982).

Phycobiliproteins are classified on the basis of their absorbance maxima into three major groups, the phycoerythrins, the phycocyanins, and allophycocyanin. Absorption maxima for the phycoerythrins lie between 490 and 570 nm whereas absorption maxima for the phycocyanins and allophycocyanin lie between 610 and 665 nm.

Phycobiliproteins are composed of a number of subunits, each consisting of a protein backbone to which linear tetrapyrrole chromophores are covalently bound. All phycobiliproteins contain either phycocyanobilin or phycoerythrobilin chromophores, and may also contain one of three minor bilins: phycourobilin, cryptobilin, or the 697-nm bilin.

Three phycobiliproteins are commonly used as fluorescent labels, R-phycoerythrin (R-PE), B-phycoerythrin (B-PE), and allophycocyanin (APC), due to their high quantum yields and absorbance/fluorescent properties.

The two most commonly used phycoerythrins, R-PE and B-PE, are composed of three types of subunits, α, β, and γ. The subunit structure of both R-PE and B-PE is $(\alpha\beta)_6\gamma$, producing a protein with a molecular weight of 240,000 daltons. The R- and B-prefixes refer to conventional nomenclature indicating the type of organism from which the pigment was originally isolated. These prefixes have now evolved to denote the shape of the absorbance curve of the purified phycoerythrin (PE) as more recent evidence has determined that differences in PE structure are not species specific. R-PE and B-PE are the most intensely fluorescent of the phycobiliproteins, with quantum efficiencies of 90% or more (Glazer, 1985).

Allophycocyanin is composed of two dissimilar polypeptide subunits, α and β, each containing one covalently bound phycocyanobilin chromophore (Glazer, 1982). In nature allophycocyanin exists as a trimer $(\alpha\beta)_3$ producing a protein of approximately 104,000 daltons. These allophycocyanin trimers readily dissociate into monomers $(\alpha\beta)$ on dilution to very low protein concentration (<30 ng/ml), under acidic conditions, or on exposure to chaotropic salts. Disassociated subunits typically exhibit changes in both absorbance and fluorescence spectra. For example, APC trimers typically display absorbance maxima at 650 nm, whereas the mononer has absorbance maxima of 620 nm. Techniques have been developed to cross-link APC that stabilize the trimeric structure and preserve its absorbance and fluorescence properties (Yeh et al., 1987). Dissociation of the subunit structure of the phycoerythrins has not been observed under typical laboratory conditions.

B. Principle of Coupling and the Use of Heterobifunctional Reagents

The isolation and use of phycobiliproteins from algae and cyanobacteria has revolutionized flow cytometry by providing dyes with high quantum yields and extinction coefficients, as well as large Stokes shifts, permitting their use in multicolor applications. Labeling of macromolecules with phycobiliprotein derivatives can provide absorbance coefficients 30-fold higher than with small synthetic fluorophores (Zola et al., 1990). In addition, the external environment does not easily affect these fluorophores. They are not readily quenched by conjugation to another molecule, and their fluorescence is independent of pH or ionic strength. Also, their excellent stability and solubility in aqueous solutions allows conjugation reactions under mild conditions conducive with protein integrity.

Conjugation of these fluorophores to antibodies requires the use of cross-linking reagents known as heterobifunctional cross-linkers. This is a multistep process in which reactive groups are introduced or activated on each protein before the conjugation process is begun. In addition, coupling of two proteins with heterobifunctional cross-linking agents will produce a mixture of polypeptide species, the number depending on the cross-linker chosen. Therefore, the conjugated product may require purification prior to use. This makes the bioconjugation of these dyes more complex than the inorganic dyes.

Heterobifunctional conjugation reagents contain two different reactive groups that can be used to target different functional species on proteins or other macromolecules. These reagents typically require a two- or three-step process depending on the cross-linker utilized. This allows greater control of the F/P ratio and may also be used to site-direct a conjugation reaction toward particular functional group of the target molecules. For example, as stated earlier, targeting sulfhydryls may be desirable when conjugation with ε-amino groups compromises activity of the antibody molecule. The low abundance of free sulfhydryls on antibody molecules is less problematic when performing phycobiliprotein:antibody conjugations, in contrast to organic dye:antibody conjugations. This is because the high quantum yields of the phycobiliproteins provide optimal results when an approximately 1:1 molar conjugation ratio is achieved. An added advantage of targeting sulfhydryl and carbohydrate molecules within the antibody is the spatial separation of the phycobiliprotein from the functional binding sites within the conjugate. This lessens the potential of loss of activity due to steric hindrance effects of the attached phycobiliprotein.

Several heterobifunctional agents are available for protein:protein conjugation. These have been reviewed extensively, and only the most commonly used reagents will be discussed here (Hermanson, 1996b; Wong, 1991d). Typically, the heterobifunctional reagent contains an amine-reactive moiety and a sulfhydryl-reactive group separated by a variable length spacer arm. Often the amine-reactive group is an NHS ester, whereas the sulfhydryl-reactive group can be one of several different reactive groups. Frequently, one protein is first modified with the most reactive or most labile end of the cross-linker. For example, with a NHS ester–maleimide heterobifunctional linker [i.e., succinimidyl-4-(N-maleimidomethyl)cyclohexane-1-carboxylate (SMCC)], one protein is initially bound to the cross-linker via its ε-amino groups, using the NHS ester end. Excess heterobifunctional cross-linker is removed from this reaction by gel filtration, and the maleimide reactive group is then utilized to couple the second protein to the initial reaction complex through activated sulfhydryls. The NHS ester group is utilized first because it is much more labile in aqueous solution.

1. Types of Heterobifunctional Cross-Linkers

Two groups of heterobifunctional cross-linkers are commonly used for coupling phycobiliproteins to immunoglobulins. Both classes utilize ε-amino residues of lysine for protein coupling but differ primarily in the type of sulfhydryl-reactive group presented (Figs. 4 and 5).

The first group, which includes SPDP, 4-succinimidyloxycarbonyl-α-methyl-α-(2-pyridylditio)toluene (SMPT), and SATA, introduces a reactive or activatable sulfhydryl into the protein (Fig. 4). These inserted sulfhydryl residues can then be used to form disulfide bonds with introduced or endogenous sulfhydryls of the second protein. The resulting disulfide bond linking the phycobiliprotein and the immunoglobulin when using SPDP and SATA, is, therefore, labile under reducing conditions. However, with the presence of the aromatic ring in the SPDP derivative, SMPT sterically hinders the disulfide sufficiently to increase its *in vivo* half-life (Thorpe *et al.*, 1987). Several lines of evidence suggest that short chain cross-linkers are the least immunogenic (Boeckler *et al.*, 1996). However, increasing the length of the spacer arm between conjugated proteins and antibodies results in improved antibody binding, presumably due to decreased steric hindrance (Bieniarz *et al.*, 1996). Additionally, because many sulfhydryl residues are located below the surface of the protein in more hydrophobic regions, longer spacer arms may induce a more efficient conjugation. Furthermore, the generation of disulfide bonds is intrinsically much less specific than the generation of thioether bonds. The production of phycobiliprotein:phycobiliprotein and immunoglobulin:immunoglobulin dimers is more probable, thus reducing the yield of functional product (Brinkley, 1992). Despite these shortcomings, SPDP is probably the most popular of all the amino- and sulfhydryl-directed heterobifunctional reagents. It has been used in the preparation of bispecific antibodies (Bode *et al.*, 1989), the production of immunotoxins (Bjorn *et al.*, 1986), as well as the generation of B-phycoerythrin–allophycocyanin conjugates (Glazer and Stryer, 1983).

The second group of amino- and sulfhydryl-directed heterobifunctional reagents includes SMCC, *m*-maleimidobenzoyl-*N*-hydroxysuccinimide ester (MBS), *N*-(γ-malemidobutyryloxy)succinimide ester (GMBS), and succinimidyl-4-(*p*-maleimidophenyl)butyrate (SMPB) (Fig. 5). All of these reagents contain a NHS ester on one end and a maleimide group on the other. The NHS ester can react with primary amines in macromolecules producing an amide bond, and

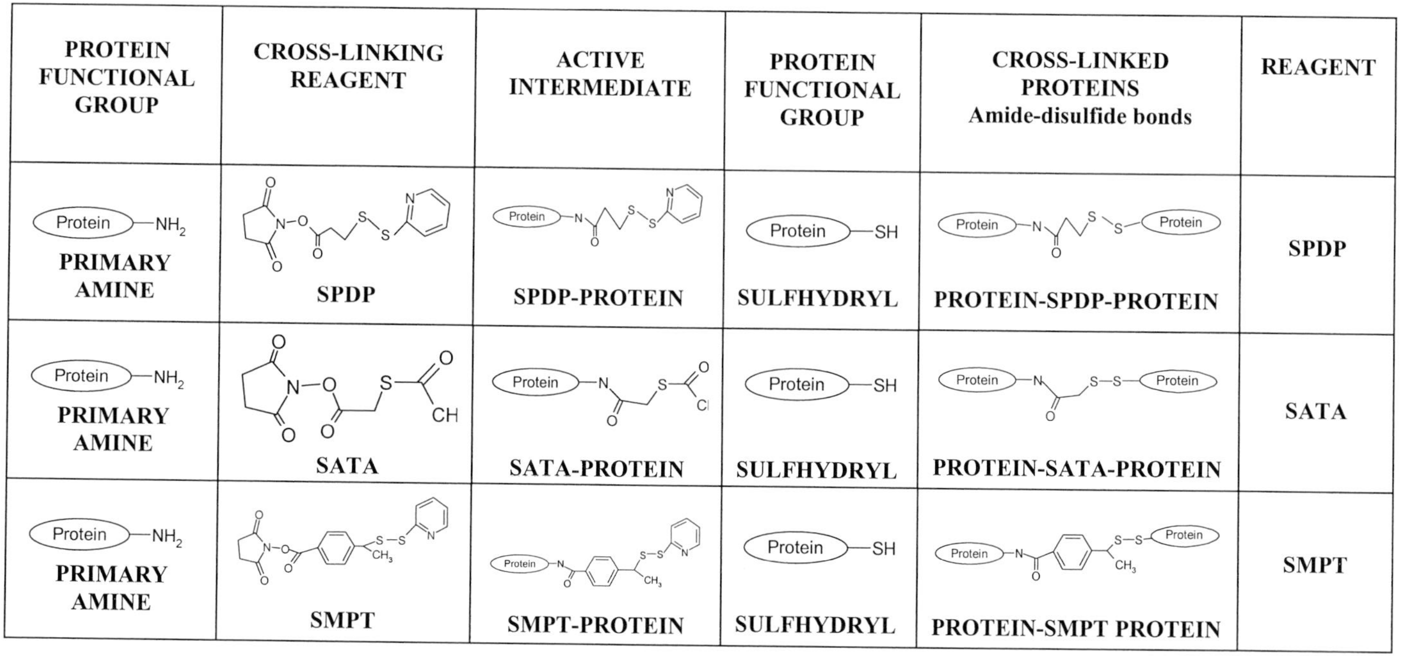

PROTEIN FUNCTIONAL GROUP	CROSS-LINKING REAGENT	ACTIVE INTERMEDIATE	PROTEIN FUNCTIONAL GROUP	CROSS-LINKED PROTEINS Amide-disulfide bonds	REAGENT
Protein—NH$_2$ PRIMARY AMINE	SPDP	SPDP-PROTEIN	Protein—SH SULFHYDRYL	PROTEIN-SPDP-PROTEIN	SPDP
Protein—NH$_2$ PRIMARY AMINE	SATA	SATA-PROTEIN	Protein—SH SULFHYDRYL	PROTEIN-SATA-PROTEIN	SATA
Protein—NH$_2$ PRIMARY AMINE	SMPT	SMPT-PROTEIN	Protein—SH SULFHYDRYL	PROTEIN-SMPT PROTEIN	SMPT

Fig. 4 Reactions of selected short chain heterobifunctional cross-linking reagents. This group of heterobifunctional cross-linking reagents introduces a reactive or activateable sulfhydryl into the protein. These inserted sulfhydryl residues can then be used to form disulfide bonds with introduced or endogenous sulfhydryls of the second protein.

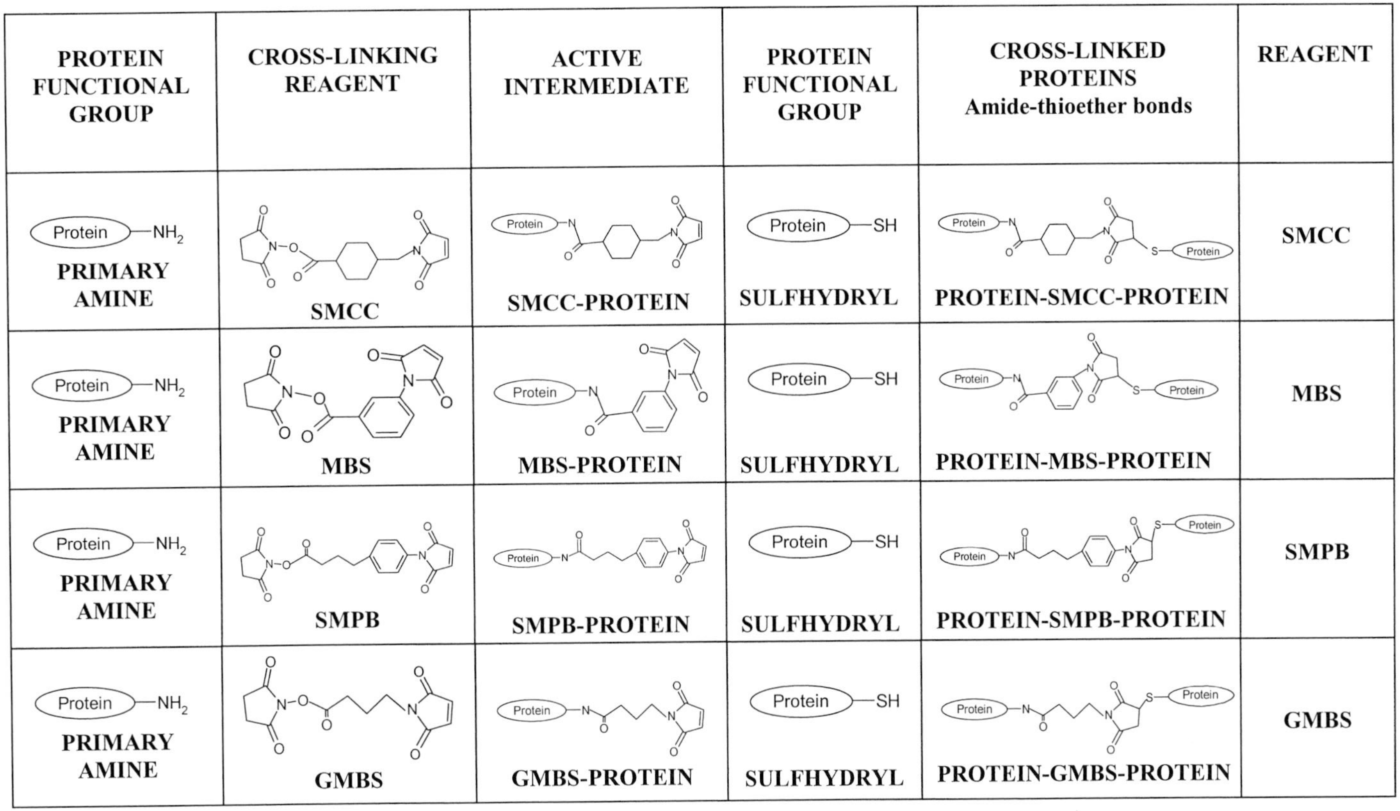

PROTEIN FUNCTIONAL GROUP	CROSS-LINKING REAGENT	ACTIVE INTERMEDIATE	PROTEIN FUNCTIONAL GROUP	CROSS-LINKED PROTEINS Amide-thioether bonds	REAGENT
Protein—NH₂ PRIMARY AMINE	SMCC	SMCC-PROTEIN	Protein—SH SULFHYDRYL	PROTEIN-SMCC-PROTEIN	SMCC
Protein—NH₂ PRIMARY AMINE	MBS	MBS-PROTEIN	Protein—SH SULFHYDRYL	PROTEIN-MBS-PROTEIN	MBS
Protein—NH₂ PRIMARY AMINE	SMPB	SMPB-PROTEIN	Protein—SH SULFHYDRYL	PROTEIN-SMPB-PROTEIN	SMPB
Protein—NH₂ PRIMARY AMINE	GMBS	GMBS-PROTEIN	Protein—SH SULFHYDRYL	PROTEIN-GMBS-PROTEIN	GMBS

Fig. 5 Reactions of selected amino- and sulfhydryl-directed heterobifunctional cross-linking reagents. This group of heterobifunctional reagents contains an NHS ester on one end and a maleimide group on the other. The NHS-ester can react with primary amines in macromolecules producing an amide bond, and the maleimide reacts with sulfhydryl groups to form thioether bonds as described above (Fig. 3).

the maleimide reacts with sulfhydryl groups to form thioether bonds as described earlier (Fig. 3). The primary difference among the members of this group of reagents is the spacer arm separating the two reactive groups. SMCC is probably the most stable compound of this group (Hermanson, 1996b). This stability is probably due to the location of the maleimide group away from the aromatic ring structure. Proteins modified with SMCC form relatively stable long-lived maleimide-activated intermediates and may be freeze-dried with minimal loss of activity (Ishikawa et al., 1983). MBS is less stable than SMCC, due to the location of the aromatic ring directly adjacent to the maleimide group. Nevertheless, MBS has enjoyed much popularity, probably because it was one of the first NHS ester–maleimide heterobifunctional reagents. SMPB is an analog of MBS containing an extended spacer arm. The maleimide group is nonetheless immediately adjacent to an aromatic ring. Both MBS and SMPB are much more labile than SMCC, and the maleimide-activated protein intermediate should be desalted and mixed with the sulfhydryl-containing molecule quickly to prevent hydrolysis of the maleimide reactive component. The maleimide group of GMBS is adjacent to an aliphatic spacer providing this heterobifunctional reagent better stability than MBS or SMPB. It is not as stable as SMCC, and maleimide-activated intermediates of GMBS should be immediately mixed with the corresponding sulfhydryl-containing molecule after purification to achieve optimal results. This class of amino- and sulfhydryl-directed heterobifunctional reagents has been utilized in cross-linking of alkaline phosphatase and human IgG F(ab′)$_2$ fragments, for generating phycobiliprotein-antibody conjugates, and for generating immunotoxin conjugates (Hardy et al., 1983; Holmes et al., 1995; Mahan et al., 1987; Myers et al., 1989).

2. Choice of Heterobifunctional Cross-Linker

The addition of the NHS ester–maleimide heterobifunctional reagents to a protein at neutral pH causes the nucleophilic attack of the NHS ester reactive group by ε-amino groups of lysine and results in the formation of an amide bond between the protein and the heterobifunctional linker. As discussed earlier (Section III,B,1), at neutral pH, the ε-amino groups of lysine are relatively unreactive toward the maleimide group of the heterobifunctional cross-linker. On addition of this protein–cross-linker complex to a second protein, which contains a sulfhydryl molecule, the maleimide group undergoes alkylation forming a thioether bond, and the two proteins are coupled. The sulfhydryl group on the second protein can either be added exogenously to the protein, or endogenous cysteines can be activated. One example of the exogenous source of sulfhydryl molecules is the first group of heterobifunctional reagents discussed earlier (i.e., SPDP, SMPT, or SATA). The activation of endogenous sulfhydryls often involves the addition of reducing agents (i.e., 2-mercaptoethanol or dithiothreitol) that reduce disulfide-bonded cysteines.

These alkylation events are random events that occur wherever lysine residues are present on the surface of the antibody. The majority of heterobifunctional cross-linker reagents, whether they add sulfhydryl (Fig. 4) or maleimide (Fig. 5) groups are directed toward lysine. As stated earlier, because these residues can be located within the antigen-combining sites of the antibody, conjugation may yield antibodies with lower activity. Because of this, site-specific methods of labeling antibodies away from the antigen-combining areas have been pursued. One method of site-directed labeling of antibodies is through the reduction of interchain disulfide bonds and the subsequent acylation with maleimide-containing heterobifunctional reagents. One important feature of the labeling of endogenous cysteine molecules is that they are generally localized away from the antigen combing sites (del Rosario *et al.*, 1990). These considerations may be of particular interest when attaching a large macromolecule such as a phycobiliprotein, in contrast to an organic dye such as FITC or biotin.

Pursuant to the previous discussion, a relatively new class of heterobifunctional cross-linking reagents is available that targets carbohydrate groups of proteins. This new class of cross-linker contains a carbonyl active group on one end and a sulfhydryl reactive group on the other end. The carbonyl reactive group is a hydrazide group that can form hydrazone bonds with aldehyde residues. Aldehyde residues are produced by the oxidation of the carbohydrate molecules with sodium periodate. The sulfhydryl reactive groups are of two types. The first type, characterized by 4-(4-*N*-maleimidophenyl)butyric acid hydrazide (MPBH) and 4-(*N*-maleimidomethyl)cyclohexane-1-carboxyl-hydrazide hydrochloride (M_2C_2H), contains a maleimide group. The principal difference between these molecules is that the maleimide group of M_2C_2H is adjacent to an aliphatic hexane ring, analogous to SMCC, and the maleimide is expected to be more stable in aqueous solutions. The maleimide group of MPBH is adjacent to an aromatic phenyl group. 3-(2-pyridyldithio)propionyl hydrazide (PDPH) contains a pyridyl disulfide group, similar to SPDP, which on reduction with dithiothreitol (DTT) forms a sulfhydryl reactive group. Because carbohydrate molecules are generally found on the Fc portion of immunoglobulins, coupling of the carbonyl reactive group to an antibody may help to preserve antigenic activity.

The coupling of phycobiliproteins to antibodies with heterobifunctional cross-linking reagents is generally not totally efficient. Unconjugated antibody and phycobiliprotein, as well as overlabeled species, must be removed to ensure optimal fluorescence signal. Unconjugated antibody will reduce the effective titer of the resulting conjugate while unconjugated phycobiliprotein and overconjugated proteins (antibodies or phycobiliproteins) may contribute to excessive background. Gel filtration is the most effective technique to separate the individual peaks. Good results are obtained using Bio-Gel A-1.5 (Bio-Rad, Hercules, CA) gel filtration medium or ion exchange chromatography using hydroxyapatite.

3. Tandem Conjugate Dyes

The demand for fluorochromes that can be used in simultaneous multicolor applications has resulted in the development of dyes that utilize fluorescence resonance energy transfer to achieve a high Stokes shift (Glazer and Stryer,

1983). These are known as tandem conjugate dyes, and the most widely used are phycoerythrin-cyanine 5 (PE–Cy5) (Lansdorp *et al.*, 1991; Shih *et al.*, 1993; Waggoner *et al.*, 1993) and phycoerythrin-Texas Red (PE–TR). Procedures for constructing tandem conjugate dyes and their coupling to antibodies basically follow the guidelines outlined in this chapter. The organic dye is first conjugated to the phycobiliprotein as described in Section II. The phycobiliprotein–dye complex is then conjugated to the antibody using a heterobifunctional cross-linking reagent as outlined earlier.

IV. Conclusion

The ability to label proteins or antibodies with fluorochrome dyes empowers the researcher with the tools to investigate a myriad of biological questions. Foremost is the ability to couple antibodies with organic dyes and phycobiliproteins for use in flow cytometry and imaging. A reflection of the usefulness of these reagents is the growth of commercially available fluorescent-labeled antibodies in more recent years. However, as new monoclonal antibodies are made or new dyes discovered, it is important to understand the relatively simple chemistries involved in fluorescent conjugation. Successful conjugations can be achieved through the use of published protocols and by following the guidelines presented here. The most important fact to remember, however, is that optimal conjugation may require slight modification of conditions for each protein labeled. As detailed earlier, this is due to the variation in amino acid sequence and three-dimensional structure of individual proteins.

References

Alon, R. (1990). Streptavidin contains an RYD sequence which mimics the RGD receptor domain of fibronectin. *Biochem. Biophys. Res. Commun.* **170**, 1236–1241.

Bieniarz, C., Husain, M., Barnes, G., King, C. A., and Welch, C. J. (1996). Extended length heterobifunctional coupling agents for protein conjugations. *Bioconjug. Chem.* **7**, 88–95.

Bjorn, M. J., Groetsema, G., and Scalapino, L. (1986). Antibody–*Pseudomonas* exotoxin A conjugates cytotoxic to human breast cancer cells in vitro. *Cancer Res.* **46**, 3262–3267.

Bode, C., Runge, M. S., Branscomb, E. E., Newell, J. B., Matsueda, G. R., and Haber, E. (1989). Antibody-directed fibrinolysis. An antibody specific for both fibrin and tissue plasminogen activator. *J. Biol. Chem.* **264**, 944–948.

Boeckler, C., Frisch, B., Muller, S., and Schuber, F. (1996). Immunogenicity of new heterobifunctional cross-linking reagents used in the conjugation of synthetic peptides to liposomes. *J. Immunol. Methods* **191**, 1–10.

Botelho, L. H., and Gurd, F. R. N. (1989). Amino acids and proteins. *In* "Practical Handbook of Biochemistry and Molecular Biology" (G. D. Fasman, ed.), pp. 359–366. CRC Press, Boca Raton, Florida.

Brinkley, M. (1992). A brief survey of methods for preparing protein conjugates with dyes, haptens, and cross-linking reagents. *Bioconjug. Chem.* **3**, 2–13.

Bussolati, G., and Gugliotta, P. (1983). Nonspecific staining of mast cells by avidin–biotin–peroxidase complexes (ABC). *J. Histochem. Cytochem.* **31**, 1419–1421.

Coons, A. H., and Kaplan, M. H. (1950). Localization of antigen in tissue cells: II. Improvements in a method for the detection of antigen by means of fluorescent antibody. *J. Exp. Med.* **91,** 1–13.

Coons, A. H., Creech, H. J., Jones, R. N., and Berliner, E. (1942). Demonstration of pneumococcal antigen in tissues by the use of fluorescent antibody. *J. Immunol.* **45,** 159–170.

del Rosario, R. B., Wahl, R. L., Brocchini, S. J., Lawton, R. G., and Smith, R. H. (1990). Sulfhydryl site-specific cross-linking and labeling of monoclonal antibodies by a fluorescent equilibrium transfer alkylation cross-link reagent. *Bioconjug. Chem.* **1,** 51–59.

Der-Balian, G. P., Kameda, N., and Rowley, G. L. (1988). Fluorescein labeling of Fab' while preserving single thiol. *Anal. Biochem.* **173,** 59–63.

Diamandis, E. P., and Christopoulos, T. K. (1991). The biotin-(strept)avidin system: Principles and applications in biotechnology. *Clin. Chem.* **37,** 625–636.

Glazer, A. N. (1982). Phycobilisomes: Structure and dynamics. *Annu. Rev. Microbiol.* **36,** 173–198.

Glazer, A. N. (1985). Light harvesting by phycobilisomes. *Annu. Rev. Biophys. Biophys. Chem.* **14,** 47–77.

Glazer, A. N., and Stryer, L. (1983). Fluorescent tandem phycobiliprotein conjugates. Emission wavelength shifting by energy transfer. *Biophys. J.* **43,** 383–386.

Goding, J. W. (1976). Conjugation of antibodies with fluorochromes: Modifications to the standard methods. *J. Immunol. Methods* **13,** 215–226.

Gretch, D. R., Suter, M., and Stinski, M. F. (1987). The use of biotinylated monoclonal antibodies and streptavidin affinity chromatography to isolate herpesvirus hydrophobic proteins or glycoproteins. *Anal. Biochem.* **163,** 270–277.

Hardy, R. R. (1986). Purification and coupling of fluorescent proteins for use in flow cytometry. *In* "Handbook of Experimental Immunology"(D. M. Weir, L. A. Herzenberg, and C. Blackwell, eds.), pp. 31.1–31.12. Blackwell, Boston.

Hardy, R. R., Hayakawa, K., Parks, D. R., and Herzenberg, L. A. (1983). Demonstration of B-cell maturation in X-linked immunodeficient mice by simultaneous three-colour immunofluorescence. *Nature* **306,** 270–272.

Haugland, R. P. (1995). Coupling of monoclonal antibodies with fluorophores. *Methods Mol. Biol.* **45,** 205–221.

Haugland, R. P. (1996). Fluorophores and their amine-reactive derivatives. *In* "Handbook of Fluorescent Probes and Research Chemicals," p. 19. Molecular Probes, Inc., Eugene, Oregon.

Haugland, R. P., and You, W. W. (1995). Coupling of monoclonal antibodies with biotin. *Methods Mol. Biol.* **45,** 223–233.

Hermanson, G. T. (1996a). Antibody modification and conjugation. *In* "Bioconjugate Techniques," p. 463. Academic Press, San Diego.

Hermanson, G. T. (1996b). Heterobifunctional cross-linkers. *In* "Bioconjugate Techniques," pp. 228–286. Academic Press, San Diego.

Hiller, Y., Gershoni, J. M., Bayer, E. A., and Wilchek, M. (1987). Biotin binding to avidin. Oligosaccharide side chain not required for ligand association. *Biochem. J.* **248,** 167–171.

Holmes, K. L., Fowlkes, B. J., Schmid, I., and Giorgi, J. V. (1995). Preparation of cells and reagents for flow cytometry. *In* "Current Protocols in Immunology"(J. E. Coligan, A. M. Kruisbeck, D. H. Margulies, E. M. Shevach, and W. Strober, eds.), pp. 5.3.1–5.3.23. Wiley, New York.

Holmes, K. L., Lantz, L. M., and Russ, W. (1997). Conjugation of fluorochromes to monoclonal antibodies. *In* "Current Protocols in Cytometry"(J. P. Robinson, Z. Darzynkiewicz, P. N. Dean, A. Orfao, P. S. Rabinovitch, C. C. Stewart, H. J. Tanke, and L. L. Wheeless, eds.), pp. 4.2.1–4.2.12. Wiley, New York.

Imam, S. A. (1979). Labelling of specific antibodies with fluorescein isothiocyanate with protection of the antigen-binding site [proceedings]. *Biochem. Soc. Trans.* **7,** 1013–1014.

Ishikawa, E., Imagawa, M., Hashida, S., Yoshitake, S., Hamaguchi, Y., and Ueno, T. (1983). Enzyme-labeling of antibodies and their fragments for enzyme immunoassay and immunohistochemical staining. *J. Immunoassay* **4,** 209–327.

Johnson, G. D., and Holborow, E. J. (1986). Preparation and use of fluorochrome conjugates. *In* "Handbook of Experimental Immunology"(D. M. Weir, L. A. Herzenberg, and C. Blackwell, eds.), pp. 28.1–28.21. Blackwell, Boston.

Lansdorp, P. M., Smith, C., Safford, M., Terstappen, L. W., and Thomas, T. E. (1991). Single laser three color immunofluorescence staining procedures based on energy transfer between phycoerythrin and cyanine 5. *Cytometry* **12,** 723–730.

Lantz, L. M., and Holmes, K. L. (1995). An improved nonradioactive cell surface labelling technique for immunoprecipitation. *BioTechniques* **18,** 56–60.

Lefevre, C., Kang, H. C., Haugland, R. P., Malekzadeh, N., and Arttamangkul, S. (1996). Texas Red-X and rhodamine Red-X, new derivatives of sulforhodamine 101 and lissamine rhodamine B with improved labeling and fluorescence properties. *Bioconjug. Chem.* **7,** 482–489.

Maeda, H. (1969). Reaction of fluorescein-isothiocyanate with proteins and amino acids. I. Covalent and non-covalent binding of fluorescein-isothiocyanate and fluorescein to proteins. *J. Biochem. (Tokyo)* **65,** 777–783.

Mahan, D. E., Morrison, L., Watson, L., and Haugneland, L. S. (1987). Phase change enzyme immunoassay. *Anal. Biochem.* **162,** 163–170.

Myers, D. E., Uckun, F. M., Swaim, S. E., and Vallera, D. A. (1989). The effects of aromatic and aliphatic maleimide crosslinkers on anti-CD5 ricin immunotoxins. *J. Immunol. Methods* **121,** 129–142.

Nakagawa, Y., Capetillo, S., and Jirgensons, B. (1972). Effect of chemical modification of lysine residues on the conformation of human immunoglobulin G. *J. Biol. Chem.* **247,** 5703–5708.

Riggs, J. L., Seiwald, R. J., Burckhalter, J. H., Downs, C. M., and Metcalf, T. G. (1958). Isothiocyanate comopounds as fluorescent labeling agents for immune serum. *Am. J. Pathol.* **34,** 1081–1091.

Shih, C. C., Bolton, G., Sehy, D., Lay, G., Campbell, D., and Huang, C. M. (1993). A novel dye that facilitates three-color analysis of PBMC by flow cytometry. *Ann. N.Y. Acad. Sci.* **677,** 389–395.

The, T. H., and Feltkamp, T. E. (1970a). Conjugation of fluorescein isothiocyanate to antibodies. II. A reproducible method. *Immunology* **18,** 875–881.

The, T. H., and Feltkamp, T. E. (1970b). Conjugation of fluorescein isothiocyanate to antibodies. I. Experiments on the conditions of conjugation. *Immunology* **18,** 865–873.

Thorpe, P. E., Wallace, P. M., Knowles, P. P., Relf, M. G., Brown, A. N., Watson, G. J., Knyba, R. E., Wawrzynczak, E. J., and Blakey, D. C. (1987). New coupling agents for the synthesis of immunotoxins containing a hindered disulfide bond with improved stability in vivo. *Cancer Res.* **47,** 5924–5931.

Titus, J. A., Haugland, R., Sharrow, S. O., and Segal, D. M. (1982). Texas Red, a hydrophilic, red-emitting fluorophore for use with fluorescein in dual parameter flow microfluorometric and fluorescence microscopic studies. *J. Immunol. Methods* **50,** 193–204.

Waggoner, A. S., Ernst, L. A., Chen, C. H., and Rechtenwald, D. J. (1993). PE-CY5. A new fluorescent antibody label for three-color flow cytometry with a single laser. *Ann. N.Y. Acad. Sci.* **677,** 185–193.

Wilchek, M., and Bayer, E. A. (1988). The avidin–biotin complex in bioanalytical applications. *Anal. Biochem.* **171,** 1–32.

Wilchek, M., and Bayer, E. A. (1990). Biotin-containing reagents. *Methods Enzymol.* **184,** 123–138.

Wong, S. S. (1991a). "Chemistry of Protein Conjugation and Cross-Linking." CRC Press, Boca Raton, Florida.

Wong, S. S. (1991b). Reactive groups of proteins and their modifying agents. *In* "Chemistry of Protein Conjugation and Cross-Linking," pp. 7–48. CRC Press, Boca Raton, Florida.

Wong, S. S. (1991c). Procedures, analysis, and complications. *In* "Chemistry of Protein Conjugation and Cross-Linking," pp. 209–220. CRC Press, Boca Raton, Florida.

Wong, S. S. (1991d). Heterobifunctional cross-linkers. *In* "Chemistry of Protein Conjugation and Cross-Linking," pp. 147–194. CRC Press, Boca Raton, Florida.

Yeh, S. W., Ong, L. J., Clark, J. H., and Glazer, A. N. (1987). Fluorescence properties of allophycocyanin and a crosslinked allophycocyanin trimer. *Cytometry* **8,** 91–95.

Zola, H., Neoh, S. H., Mantzioris, B. X., Webster, J., and Loughnan, M. S. (1990). Detection by immunofluorescence of surface molecules present in low copy numbers. High sensitivity staining and calibration of flow cytometer. *J. Immunol. Methods* **135,** 247–255.

Zuk, R. F., Rowley, G. L., and Ullman, E. F. (1979). Fluorescence protection immunoassay: A new homogeneous assay technique. *Clin. Chem.* **25,** 1554–1560.

PART II

Cell Preparation

CHAPTER 10

Preparation of Cells from Blood

J. Philip McCoy, Jr.

Cooper Hospital/UMC
Robert Wood Johnson Medical School at Camden
Camden, New Jersey 08103

I. Introduction
II. Collection, Transport, and Storage of Blood
III. Fixation and Preservation
IV. Separation of Erythrocytes from Leukocytes
V. Assessment of Cell Viability
VI. Staining
VII. Summary
    Appendix 1
    Appendix 2
    Appendix 3
    References

I. Introduction

Peripheral blood is arguably the most convenient source of *in vivo* material for flow cytometric analysis. In its native state, the cellular components of blood are both disperse and bountiful. This, for the most part, eliminates any need for processing to free individual cells from a cohesive mass. Blood is readily obtainable and, in healthy donors, is of sufficient cellularity that most experiments can be performed on 1 ml or less of peripheral blood. Table I illustrates the white blood count and cellular composition of normal human peripheral blood.

A key point in discussing the preparation of cells from blood, particularly human blood, is the need to practice universal precautions. The term universal precautions is used to define a series of steps that are to be taken to protect the technologists or technicians, as well as clinical staff and bystanders, from infectious specimens, and *every* specimen should be assumed to be infectious. There-

Table I
Cellular Constituents of Normal Adult Male Human Peripheral Blood[a]

Cellular component	Percentage of WBC	Cells ($\times 10^3$) per mm^3
Erythrocytes (RBC)	NA[b]	4760–6040
Platelets	NA	147–412
White blood cells (WBC)	100	3.5–9.8
Segmented neutrophils	50–70	2.4–7.6
Banded neutrophils	2–6	0.096–0.65
Monocytes	2–9	0.096–0.972
Lymphocytes	20–44	0.962–4.75
Eosinophils	0–4	0–0.432
Basophils	0–2	0–0.216

[a] Adapted from Kjeldsberg, C. R., ed. "Practical Diagnosis of Hematologic Disorders," 2nd Ed. ASCP Press, Chicago, Illinois, 1995.
[b] NA, not applicable.

fore, the handling of all specimens is identical. Universal precautions include the wearing of latex gloves, face shields or other eye protection, and liquid-repellant laboratory coats during all portions of specimen preparation. Thorough disinfection of all working surfaces and equipment should be performed as well as appropriate disposal of residual specimen and contaminated material in clearly labeled and leakproof biohazard containers. A more thorough description of precautions that should be used in the handling of human blood may be found in references listed at the end of this chapter (Centers for Disease Control, 1988, 1997; National Committee for Clinical Laboratory Standards, 1997; National Committee for Clinical Laboratory Standards, 1990).

II. Collection, Transport, and Storage of Blood

One of the greatest limitations to the analysis of blood by flow cytometry is the ability of blood to clot once removed from the body. A critical process in hemostasis, clot formation severely impairs analysis of the cellular components of blood, as many of these become entrapped in the clot. Fortunately, there are several chemicals, termed anticoagulants, that impede or prevent clot formation and thereby permit the retention of both red and white blood cells in single cell suspension. A list of commonly used anticoagulants is given in Table II. The choice of anticoagulant in which to collect blood is linked to which flow cytometric assay is to be performed and, to some extent, the length of time in which the specimen is to be held before processing. A more detailed description of which anticoagulant is to be used with which application is deferred to chapters pertaining to those applications.

Table II
Anticoagulants in Common Use

Color top[a]	Anticoagulant	Cytometry uses	Other anticoagulants with same color
Green	Sodium heparin	Immunophenotyping mitogens, neutrophil function, DNA, other	Lithium heparin, americium heparin
Lavender	EDTA	Immunophenotyping mitogens, neutrophil function, DNA, other	None
Yellow	Acid citrate dextrose (sol. 1)	Platelet studies immunophenotyping	Acid citrate dextrose (sol. 2) Sodium polyanetholsulfonate
Red	None	None—for serum studies only	None
Royal blue	Sodium heparin	Immunophenotyping mitogens, neutrophil function, DNA, other	None
Brown	Sodium heparin	Usually none—reserved for lead determinations	None

[a] Based on Hemoguard closures manufactured by Becton Dickinson, Inc., Franklin Lakes, NJ.

In general, the more rapidly blood is prepared for flow cytometric analyses, the more assured one is that the data are reflective of the cells *in vivo*. This fact notwithstanding, empirical and experimental data have demonstrated that blood may be held for prolonged periods of time without a pronounced effect in some assays (Centers for Disease Control, 1997; McCoy *et al.*, 1990; Thornthwaite *et al.*, 1984; Ekong *et al.*, 1993; Nicholson and Green, 1993). Many factors affect how well cells survive on storage and how well the cell-surface antigens are preserved. As mentioned earlier, the choice of anticoagulant is one factor influencing storage times. Many lymphocyte surface markers remain essentially unchanged from fresh blood when stored in sodium heparin for 72 hr or longer (Thornthwaite *et al.*, 1984; Ekong *et al.*, 1993; Nicholson and Green, 1993; Lloyd *et al.*, 1995). Not all markers, however, are equally retained, and caution must be observed, particularly with activation markers, that prolonged storage is not altering data. Temperature also exerts an influence on the suitability of blood for immunophenotyping studies (Centers for Disease Control, 1997; Thornthwaite *et al.*, 1984; Ekong *et al.*, 1993; Nicholson and Green, 1993; Lloyd *et al.*, 1995; Paxton and Bendele, 1993). In most instances blood should be stored at room temperature prior to preparation, although some data indicate that storage at 4°C may also be acceptable (Paxton and Bendele, 1993). Some investigators have suggested that the use of culture or storage media may prolong the integrity of peripheral blood cells and their markers better than anticoagulants alone (Milson *et al.*, 1986).

III. Fixation and Preservation

An alternative to determining the conditions that best keep the specimen fresh and processing cells in their native state is to preserve the cells in such a manner that the parameters to be studied are unaffected. For some assays, such as neutrophil function tests, long-term preservation of specimens is virtually, if not totally impossible, whereas for other assays, such as one-color DNA analysis, a potpourri of fixation and preservation methods exist. DNA, for example, is well preserved by cryopreservation or fixation by a number of reagents, including formalin, ethanol, methanol, and paraformaldehyde. Many of these preservation reagents, however, induce artifacts that may effect various parameters in flow cytometric analysis such as light scatter or intensity of dye staining (Overton and McCoy, 1994; Pollice et al., 1992). Preservation of cells for surface marker staining is generally more difficult. Low concentrations of paraformaldhyde have been reported to be useful for preserving blood specimens for subsequent immunophenotyping (Lal et al., 1988). A commercial reagent has become available that has been demonstrated to preserve light scatter and antigen expression (for all markers tested) of blood for up to 7 days, and of bone marrow for up to 4 days (Turpen and Collins, 1996; Saxton and Pockley, 1998).

Cryopreservation has been somewhat controversial in how well cells are preserved for immunophenotyping. Although some studies report satisfactory retention of surface marker staining as well as light scatter in cryopreserved specimens (using a variety of cryopreservation techniques), other investigators report differences between the staining of cryopreserved and fresh material (Lloyd et al., 1995; Prince and Lee, 1986; Fiebig et al., 1997; Rosillo et al., 1995).

IV. Separation of Erythrocytes from Leukocytes

Only a relatively few flow cytometric assays are performed on erythroid cells (Davis et al., 1998; Corash et al., 1988), although red cells are well suited for flow cytometric analysis. Due to the high proportion of red blood cells in blood, the assays of nonerythroid cells, such as leukocytes or platelets, generally are performed more easily after the separation of the erythroid cells from the cells of interest. Such separation may be achieved through a number of methods, including via differential staining (using CD45 and DNA dyes) (Terstappen et al., 1991), selective osmotic lysis of erythrocytes (Hoffman et al., 1980), or by density gradient separation (Boyum, 1968, 1977). The latter method, density gradient separation, was commonly used several decades ago but is now performed less frequently. In general, density gradient methods tend to be (1) more expensive, (2) more time-consuming, and (3) lower yielding than the other two approaches. Advantages of using selected density gradient separation methods include the yield of a purified mononuclear (or other) cell population and the removal of dead cells from the population of interest. A sample protocol for a common density gradient separation method, Ficoll-Hypaque separation of mononuclear cells, is given in Appendix 1 at the end of this chapter.

Today, the most commonly used approach for separation of erythrocytes is osmotic lysis. Quite simply, red blood cells lyse at an osmolarity where leukocytes do not (Hoffman *et al.*, 1980). Osmotic lysis of erythrocytes is inexpensive, quick, and provides high yields of leukocytes. The reagents used for this, often ammonium chloride solutions, do not affect most cell surface antigens, antigen–antibody binding, or fluorochrome stability. Thus erythrocyte lysis methods may be used either prior to staining or after staining. A formulation for an ammonium chloride lysing solution is given in Appendix 2, and a sample protocol for erythrocyte lysis of peripheral blood is given in Appendix 3.

The stain-then-lyse method is probably the most widely practiced, possibly due, in part, to the fact that some commercial lysing reagents incorporate small amounts of paraformaldehyde that helps to inactivate the human immunodeficiency virus in infectious specimens. There are examples where stain-then-lyse methods fail and lyse-then-stain approaches must be used. A notable example includes staining leukocytes with some lectins where these lectins will also bind sugars on erythrocytes, resulting in agglutination. Staining lymphocytes for the presence of surface immunoglobulins presents unique problems as well, since the presence of immunoglobulin-containing serum in whole blood will obscure the detection of true cell surface immunoglobulins. This problem can be overcome by washing blood cells with azide-free phosphate-buffered saline (PBS) several times prior to staining to remove all serum. Immunoglobulin bound to cells by Fc receptors may also pose a problem, and this can be obviated by incubating the washed blood in azide-free PBS for at least 30 min at 37°C followed by a final wash.

Some assays require purification of the cells of interest rather than removal or discrimination of erythroid cells. An example of this is the analysis of platelets, which is usually performed on platelet-enriched plasma or other purified, or semipurified, platelet preparations (Ault, 1988). A more detailed discussion of preparation of cells from blood involving purification of specific components of blood is deferred to other chapters in this volume relating to those specific assays.

V. Assessment of Cell Viability

Fresh peripheral blood from healthy individuals generally has an extremely high proportion of viable cells. Blood from ill donors or blood that is held for some length of time prior to processing may have an increased number of nonviable cells. Neutrophils, in particular, survive poorly *in vitro* for prolonged periods of time. Because nonviable cells can bind antibodies nonspecifically, or in some instances, contain intracellular antigens that interfere with the detection of cell surface antigens, these cells must be identified and either removed or analyzed discretely from viable cells. The most classic method of assessing cell viability is through the use of trypan blue (Philips, 1973). This is a noncytometric method in which trypan blue is mixed with blood (preferably without the red blood cells), and only the nonviable cells take up the stain. Nonviable blue-colored cells are then counted manually by light microscopy. Nonviable cells, if exceeding a predetermined threshold, must be removed by physical separation, commonly density gradient methods. The advent of cytometers capable of multicolor analysis has permitted the development of methods where viability may be assessed simultaneously with the analysis of surface marker expression using dyes such as propidium iodide, ethidium monoazide, or LDS-751 (Sasaki *et al.*, 1987; Riedy *et al.*, 1991; Terstappen *et al.*, 1988). These real-time methods have the advantage of excluding even small numbers of dead cells from analysis without the need for tedious separation procedures.

VI. Staining

It is not practical to present a brief, generic description of staining blood cells for flow cytometric analyses because a wide variety of stains, antibodies, lectins, ligands, and so forth may be used alone or in combination with each other to detect a vast array of cell surface and intracellular features or functions. Key elements to consider when staining blood cells for flow cytometric analysis are whether the antigens, or other features to be examined are cell surface or intracellular (discussed in detail in Chapter 11 of this volume), the properties of the fluorochromes to be used, including compensation issues (discussed in a previous chapter), and the specific requirements of each assay. Most contemporary cytometers are designed to permit three- or four-color analyses. In general, one is well served to take full advantage of analyzing as many parameters as possible, even when the primary goal is to study one cellular feature, to ensure that the appropriate cell population(s) in the appropriate condition is studied.

VII. Summary

Peripheral blood is a bountiful source of numerous cell types that are easily analyzed by flow cytometry for a variety of properties. The monodisperse nature of blood makes preparation of these cells relatively easy if the coagulation cascade is inhibited. For some studies the presence of serum can be a confounding factor, although this is often overcome by merely pelleting the cells from the serum and washing the remaining cells several times. Blood, particularly from humans and primates, should be considered highly infectious whether or not from a healthy donor. Therefore universal precautions should be followed at all times.

Techniques and route of collection may have a profound influence on the condition and nature of the blood being obtained. Performing venipuncture with an extremely small gauge needle may disrupt cells and prevent satisfactory analysis. Venous blood and arterial blood may yield differing data. Even blood collection at different times of day may yield diurnal variations in some assays.

The demands of each cell type and of each assay dictate specific preparation, fixation, storage, and staining protocols. However, the overall goal for preparation of cells from blood, for all of the assays, remains the same—collect and prepare the cells in such a manner that they accurately represent the *in vivo* state.

Appendix 1

A Sample Protocol for Ficoll-Hypaque Density Gradient Separation

Materials required:
- Blood anticoagulated with sodium heparin or EDTA
- Ficoll-Hypaque solution from manufacturer
- Conical centrifuge tubes, 15 ml and 50 ml
- Centrifuge, temperature controlled, capable of 400 g
- Phosphate-buffered saline (PBS)
- Disposable serological pipettes, 1 ml, 5 ml, and 10 ml (sterile)
- Pasteur pipettes
- Tissue culture medium if required

1. Using a sterile pipette, withdraw the necessary volume of Ficoll-Hypaque. This volume may vary with the brand used; however, it is usually twice the volume of blood to be used for separation. Place the Ficoll-Hypaque in a 50-ml conical centrifuge tube.
2. Mix the desired volume of blood to separate with an equal volume of PBS.
3. Slowly overlay the diluted blood onto the Ficoll-Hypaque. Do this by gently pipetting the diluted blood down the side of the tube containing the Ficoll-Hypaque.
4. Centrifuge the blood at 400 g for 40 min at 22°C.
5. Identify the mononuclear cells, which are located at the interface between the plasma (top) and the Ficoll (bottom). Remove these cells by careful aspiration with a Pasteur pipette.
6. The separated mononuclear cells can be transferred to a 15-ml conical tube for washing. Then 10 ml of PBS or tissue culture medium, if desired, should be added and mixed thoroughly with the cells. Centrifuge the cells at 400 g for 10 min at 4°C.
7. Remove the supernatant and repeat washing as needed.

Appendix 2

Formulation of Erythrocyte Lysing Solution

    1.652 g ammonium chloride (NH_4Cl)
    0.200 g potassium bicarbonate
    0.0074 g tetrasodium EDTA
    Dissolve in 200-ml sterile distilled water. Use within 24 hr.

Appendix 3

Sample Protocols for Erythrocyte Lysis of Peripheral Blood

A. Protocol for lyse-then-stain method

1. Add 200 µl of whole blood to test tube. Sodium heparin, EDTA, and ACD are all acceptable anticoagulants.
2. Add 3 ml of fresh lysing solution. (Ammonium chloride lysing solution is usually used for this. Care must be taken when using commercial formulations, as many have reagents such as paraformaldehyde that may affect later staining.)
3. Incubate 10 min at room temperature.
4. Centrifuge at 300 g for 5 min.
5. Wash cells two times in saline or isotonic buffer.
6. Stain cells as desired.

B. Protocol for stain-then-lyse method

1. Add 100 µl of whole blood to test tube. Sodium heparin, EDTA, and ACD are all acceptable anticoagulants.
2. Stain cells as desired, using minimal volumes of reagents as necessary.
3. Add 1.5 ml of fresh lysing solution. (Ammonium chloride lysing solution may be used for this, as well as a number of commercial reagents that include a poststaining fixative.)
4. Incubate 10 min at room temperature.
5. Centrifuge at 300 g for 5 min.
6. Wash cells two times in saline or isotonic buffer.
7. Fix cells as desired.

References

Ault, K. A. (1988). Flow cytometric measurement of platelet-associated immunoglobulin. *Pathol. Immunopathol. Res.* **7**, 395–408.

Boyum, A. (1968). Isolation of mononuclear cells and granulocytes from human blood. *Scand. J. Clin. Lab. Invest. Suppl.* **21**, 77–89.

Boyum, A. (1977). Separation of lymphocytes, lymphocyte subgroups and monocytes: A Review. *Lymphology* **10,** 71–76.

Centers for Disease Control (1988). Update: Universal precautions for prevention of transmission of human immunodeficiency virus, hepatitis B virus, and other blood borne pathogens in healthcare settings. *MMWR* **37,** 377–382.

Centers for Disease Control (1997). 1997 Revised guidelines for the performance of CD^{4+} T-cell determinations in person with human immunodeficiency virus infection. *MMWR* **46**(RR–2).

Corash, L., Rheinschmidt, M., Lieu, S., Meers, P. and Brew, E. (1988). Enumeration of reticulocytes using fluorescence-activated flow cytometry. *Pathol. Immunopathol. Res.* **7,** 381–394.

Davis, B. H., Olsen, S., Bigelow, N. C., and Chen, J. C. (1998). Detection of fetal red cells in fetomaternal hemorrhage using a fetal hemoglobin monoclonal antibody by flow cytometry. *Transfusion* **38,** 749–756.

Ekong, T., Kupek, E., Hill, A., Clark, C., Davies, A., and Pinching, A. (1993). Technical influences on immunophenotyping by flow cytometry. The effect of time and temperature of storage on the viability of lymphocyte subsets. *J. Immunol. Methods* **164,** 263–273.

Fiebig, E. W., Johnson, D. K., Hirschkorn, D. F., Knape, C. C., Webster, H. K., Lowder, J., and Busch, M. P. (1997). Lymphocyte subset analysis on frozen whole blood. *Cytometry* **29,** 340–350.

Hoffman, R. A., Kung, P. C., Hansen, W. P., and Goldstein, G. (1980). Simple and rapid measurement of human T lymphocytes and their subclasses in peripheral blood. *Proc. Natl. Acad. Sci. U.S.A* **77,** 4914–4917.

Lal, R. B., Edison, L. J., and Chused, T. M. (1988). Fixation and long-term storage of human lymphocytes for surface marker analysis by flow cytometry. *Cytometry* **9,** 213–219.

Lloyd, J. B., Gill, H. S., and Husband, A. J. (1995). The effect of storage on immunophenotyping of sheep peripheral blood lymphocytes by flow cytometry. *Vet. Immunol. Immunopathol.* **47,** 135–142.

McCoy, J. P., Carey, J. L., and Krause, J. R. (1990). Quality control in flow cytometry for diagnostic pathology. I. Cell surface phenotyping and general laboratory procedures. *Am. J. Clin. Pathol.* **93**(Suppl. 1), S27–S37.

Milson, T. J., Patrick, C. W., Torke, N. J., and Keller, R. H. (1986). Holding medium for cell surface phenotypic analysis. *J. Immunol. Methods* **87,** 155–159.

National Committee for Clinical Laboratory Standards. (1990). "Procedures for the Handling and Processing of Blood Specimens." Approved Guideline NCCLS Publication H18-A, Villanova, PA.

National Committee for Clinical Laboratory Standards (1997). "Protection of Laboratory Workers from Instrument Biohazards and Infectious Disease Transmitted by Blood, Body Fluids, and Tissue." Approved Guideline NCCLS Publication M29-A, Villanova, PA.

Nicholson, J. K., and Green, T. A. (1993). Selection of anticoagulants for lymphocyte immunophenotyping. Effect of specimen age on results. *J. Immunol. Methods* **165,** 31–35.

Overton, W. R., and McCoy, J. P. (1994). Reversing the effect of formalin on the binding of propidium iodide to DNA. *Cytometry* **16,** 351–356.

Paxton, H., and Bendele, T. (1993). Effect of time, temperature, and anticoagulant on flow cytometry and hematological values. *Ann. N.Y. Acad. Sci.* **667,** 440–442.

Philips, H. J. (1973). "Dye Exclusion Tests for Cell Viability, in Tissue Culture: Methods and Applications." pp. 406–408. Academic Press, New York.

Pollice, A. A., McCoy, J. P., Shackney, S. E., Smith, C. A., Agarwal, J., Burholt, D. R., Janocko, L., Hornicek, F., and Hartsock, R. J. (1992). Sequential paraformaldehyde and methanol fixation for simultaneous flow cytometric analysis of DNA, cell surface proteins, and intracellular proteins. *Cytometry* **13,** 432–444.

Prince, H. E., and Lee, C. D. (1986). Cryopreservation and short-term storage of human lymphocytes for surface marker analysis. Comparison of three methods. *J. Immunol. Methods* **93,** 15–18.

Riedy, M. C., Muirhead, K. A., Jensen, C. P., and Stewart, C. C. (1991). Use of a photolabelling technique to identify nonviable cells in fixed homologous or heterologous cell populations. *Cytometry* **12,** 133–139.

Rosillo, M. C., Ortuno, F., Rivera, J., Moraleda, J. M., and Vincente, V. (1995). Cryopreservation modifies flow-cytometric analysis of hematopoietic cells. *Vox. Sang.* **68,** 210–214.

Sasaki, D. T., Dumas, S. E., and Engelman, E. G. (1987). Discrimination of viable and non-viable cells using propidium iodide in two color immunofluorescence. *Cytometry* **8**, 413–420.

Saxton, J. M., and Pockley, A. G. (1998). Effect of ex vivo storage on human peripheral blood neutrophil expression of CD11b and the stabilizing effects of Cyto-Chex. *J. Immunol. Methods* **214**, 11–17.

Terstappen, L. W. M. M., Shah, V. O., Conrad, M. P., Recktenwald, D. and Loken, M. R. (1988). Discrimination between damaged and intact cells in fixed flow cytometric samples. *Cytometry* **9**, 477–484.

Terstappen, L. W., Johnson, D., Mickaels, R. A., Chen, J., Olds, G., Hawkins, J. T., Loken, M. R., and Levin, J. (1991). Multidimensional flow cytometric blood cell differentiation without erythrocyte lysis. *Blood Cells* **17**, 585–602.

Thornthwaite, J. T., Rosenthal, P. K., Vazquez, D. A., and Seckinger, D. (1984). The effects of anticoagulant and temperature on the measurements of helper and suppressor cells. *Diagn. Immunol.* **2**, 167–174.

Turpen, P. B., and Collins, M. (1996). "A Reagent for Stabilizing Blood Samples." *Am. Clin. Laboratory.* International Scientific Communications, Inc., Shelton, CT.

CHAPTER 11

Cell Preparation for the Identification of Leukocytes

Carleton C. Stewart and Sigrid J. Stewart

Laboratory of Flow Cytometry
Roswell Park Cancer Institute
Buffalo, New York 14263

I. Introduction
II. Antibodies
    A. Antibody Structure
    B. Antibody Binding Kinetics
    C. Influence of Epitope, Fluorochrome, and Fixation
    D. The Effect of Drugs on Expression
    E. Primary and Secondary Antibody
    F. Blocking
    G. Directly Conjugated Antibodies
    H. Wash versus No Wash
III. Tandem Fluorochromes
IV. Cell Preparation and Staining Procedures
    A. Staining Cells with One Antibody
    B. Staining Cells with Two Antibodies
    C. Staining Cells with Three Antibodies
    D. Staining Cells with Four Antibodies
    E. Intracellular Staining
    F. Cell Fixation
    G. Measuring Cell Viability
V. Titering Antibodies
    A. Processing Cells
    B. Analysis of Results
    C. Verification of Specific Antibody Binding
VI. Solutions and Reagents
    A. PBS (with Sodium Azide)
    B. Two Percent Ultrapure Formaldehyde
    C. Lysing Reagent

    D. Propidium Iodide Stock Solution
    E. Ethidium Monoazide Stock Solution
  References

I. Introduction

Immunophenotyping is the term applied to the identification of cells using antibodies to antigens expressed by these cells. These antigens are actually functional membrane proteins involved in cell communication, adhesion, or metabolism. Immunophenotyping using flow cytometry has become the method of choice in identifying and sorting cells within complex populations. Applications of this technology have occurred in both basic research and clinical laboratories. The National Committee for Clinical Laboratory Standards (now NCCLS) has prepared guidelines for flow cytometry that describe in detail the current recommendations for processing clinical samples (Landay, 1992; Borowitz, 1993).

Advances in flow cytometry instrumentation design and the availability of new fluorochromes and staining strategies have led to methods for immunophenotyping cells with two or more antibodies simultaneously (Stewart and Stewart, 1994a,b, 1995, 1997a,b, 1999; Stewart, 1994, 1996). With this progress has come the realization that the use of a single antibody is insufficient for identification of a unique cell population. Cells within one population may have proteins in common with cells of another population. By evaluating the unique repertoire of proteins using several antibodies together, each coupled with a different fluorochrome, any given cell population can be identified and their frequency in a specimen determined. The actual function of many proteins to which specific monoclonal antibodies bind are now known so that we not only know the identity of the cell but we can also determine what it is doing. For example, using antibodies to cytokines and their receptors it is now possible to identify those cells making them (Picker *et al.*, 1995; Romagnani, 1997; Prussin, 1997) and those cells responding to them (Falini *et al.*, 1995; Wognum *et al.*, 1993; Debili *et al.*, 1995; Olsson, *et al.*, 1992).

As the number of antibodies used for phenotyping increases so does the complexity caused by the overlapping spectra of the fluorochromes. Controls must also be evaluated along side the experimental samples to insure the data is collected and interpreted correctly. These process controls must be simple and inexpensive to prepare and faithfully used every time the instrument acquires data so that problems can be recognized and addressed when they occur.

Several strategies for staining cells with antibodies labeled with different colored fluorochromes are the focus of this chapter. This chapter has been divided into topics or sections to describe why the recommended procedures are performed as they are described. The behavior of antibodies toward cells is illustrated and described in the first section. Understanding the ways antibodies bind to cells using the law of mass action and the artifacts that can ensue provides a rational basis for the procedures described in the remaining sections. In Section III, we consider the tandem complexes, which are two fluorochromes chemically combined to provide unique excitation/emission spectra. These tandem fluorochromes represent a unique set of problems that must be understood to appropriately solve them. In Section IV, methods for cell preparation and staining are described in detail, and we consider dead cells, their evaluation, and their effect on data interpretation. In Section V, we describe the method for evaluating antibody quality by titering. Finally, in Section VI, the formulation of all solutions is described.

II. Antibodies

A brief review of the characteristics of antibodies germane to their use in immunophenotyping is appropriate. Antibodies bind to three-dimensional molecular structures on antigens called epitopes, and each antigen contains hundreds of different epitopes. A monoclonal antibody is specific for a single epitope, whereas polyclonal antibodies are actually the natural pool of hundreds of monoclonal antibodies produced within the animal, each one binding to its unique epitope. It is most common to use monoclonal antibodies to determine the immunophenotype of cells.

A. Antibody Structure

The basic unit of all antibodies comprises a heavy and light chain, the former defining the antibody isotype. As depicted in Fig. 1, the minimum functional antibody molecule consists of two heavy and two light chains linked together with disulfide bonds. The part of the molecule that contains only heavy chains is known as the Fc [the fragment (F) of the antibody molecule that exhibits a nonvariable or constant (c) amino acid sequence within a given isotype and subclass] portion, whereas the part that contains both heavy and light chains is the $F(ab')_2$ portion. The Fc receptor binding domain and the complement binding and activation domain (Fig. 1) are in the Fc portion. Variations in the structure of the heavy chains lead to "isotypes" and there are "subclasses" within some isotypes. These variations are of practical importance because they define the repertoire of cells to which the Fc portion of the antibody will bind.

In the $F(ab')_2$ portion, the light chains are either κ or λ, and they are associated with the heavy chains by disulfide bonding; the designation ab is "antigen binding," and the numeral 2 refers to the fact that there are two antigen binding sites in the basic functional subunit. These two can be chemically separated to produce an F(ab) using papain instead of pepsin for digestion. It is noteworthy that murine and rat antibodies almost always have κ light chains and that most *in vitro* hybridoma-produced monoclonal antibodies are of the γ [immunoglobu-

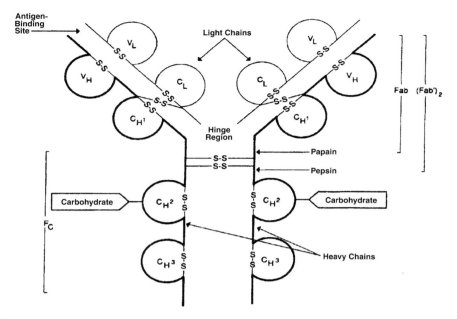

Fig. 1 Schematic representation of an IgG molecule. The Fab portion consists of the light chain and a fragment of the heavy chain. These two chains are held together by disulfide bonds. For any given IgG molecule there is a region of variable amino acid sequence (V_L) on the light chain and the heavy chain (V_H) that produces the epitope-binding site. Papain digestion produces two Fab fragments for each molecule. There are also regions where the amino acid sequence is constant for IgG molecules of similar isotype and subclass. There is one constant region for light chains (C_L) and three for heavy chains (C_{H1}, C_{H2}, and C_{H3}). The heavy chains are held together by disulfide bonds. If pepsin is used to digest the antibody, a fragment [F(ab')$_2$] is produced containing two epitope binding sites. The Fc fragment is that portion of the Ab posterior to the hinge region. From Stewart and Stewart (1994a).

lin G (IgG)] isotype. Table I shows some important properties of immunoglobulins.

B. Antibody Binding Kinetics

The single most important factor in immunophenotyping is antibody quality. This is even more important for intracellular than for membrane immunophenotyping. To better understand the basis for evaluating antibody quality, a brief review of the law of mass action may be helpful. We will illustrate the law using monoclonal antibodies (Ab) that bind to a single epitope (E) on an antigen. A more detailed description of the law for immunophenotyping cells is available (Stewart and Mayers, 2000). In Eq. (1), a single monoclonal antibody binding

Table I
Human Immunoglobulins[a]

	Immunoglobulin properties				Binding to Fc receptor class on leukocytes					
Isotype	Subclass	Serum mg/ml	Complement fixing	Protein A reaction	NK cells	B cell	Monocyte	Neutrophil	Eosinophil	Basophil
IgG	—	—	—	—	III	II	I, II	II, III	II	II, III
	IgG1	9	Yes	Yes	III	II	I, II	II, III	II	II, III
	IgG2	3	Yes	Yes	—	II	II	II	II	II
	IgG3	1	Yes	No	III	II	I, II	II, III	II	II, III
	IgG4	0.5	No	Yes	—	II	I, II	II	II	II
IgM	—	1.5	Yes	No	—	—	—	—	—	—
IgA	IgA1	3	No	No	—	—	II, III	—	—	—
	IgA2	0.5	No	No	—	—	—	II, III	—	—
IgD	—	0.3	No	No	—	—	—	—	—	—
IgE	—	Nil	No	No	—	—	—	—	II	I

[a] Modified and expanded from Roitt et al. (1989), and Stewart and Mayers (2000). FcRI is CD64, FcRII is CD32, and FcRIII is CD16.

site rapidly binds to a single epitope on the antigen with a forward rate constant k_f. As the concentration of AE increases, there is a tendency for the complex to come apart with a reverse rate constant of k_r. The affinity constant K_a is equal to the ratio of the forward and reverse rate constants. The quality of an antibody is almost solely the result of the reverse reaction

$$Ab + E \underset{k_r}{\overset{k_f}{\rightleftharpoons}} AbE. \qquad (1)$$

At equilibrium, the change in concentration of each component is zero so that

$$K_a = \frac{k_f}{k_r} = \frac{[AbE]}{[Ab][E]}. \qquad (2)$$

Antibodies are immunoglobulins (Ig), and they are naturally sticky molecules. This can be easily demonstrated by centrifuging the contents of a new vial of antibody from your favorite supplier at 15,000 g for 5 min and observing the pellet of IgG aggregates.

RULE: ALWAYS CENTRIFUGE ANTIBODIES AT 15,000 g PRIOR TO IMMUNOPHENOTYPING TO REMOVE AGGREGATES.

Immunoglobulin also binds nonspecifically because of its high electrostatic attraction to proteins. This nonspecific binding to proteins also obeys the laws of mass action, but occurs at a much lower affinity ($K_a = 10^4$ liters/M). All IgGs

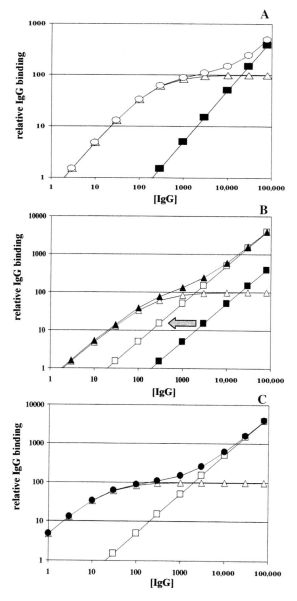

Fig. 2 Antibody binding kinetics. Antibodies are actually immunoglobulins (Ig) that can bind to cells via their antigen (epitope) binding sites or by nonspecific binding. (A) For binding to membrane antigens on viable cells, the affinity constant for the antibody describes the kinetics of specific binding to its epitope. What we measure in the flow cytometer is the sum (○) of specific (△) antibody and nonspecific (■) immunoglobulin binding. (B) If the same antibody is used for intracellular staining, the nonspecific binding increases markedly (□) due to the increased intracellular protein concentration,

bind nonspecifically in about the same concentration range at or above 30 μg/ml/10^7 cells.

Figure 2 compares the nonspecific binding of IgG ($K_a = 5 \times 10^4$) and the specific binding of an antibody with a K_a of 5×10^8 and with a K_a of 5×10^9 for a membrane epitope and an intracellular epitope. In Fig. 2A, for membrane antigen, the nonspecific binding curve is well separated for both antibodies, so it is easy to resolve epitope positive cells from nonspecific binding. In Fig. 2B, for intracellular antigens, the nonspecific binding curve has shifted to the left an order of magnitude because of the increased concentration of protein targets inside versus on the membrane of cells, that is, [E], which in this case is protein, has increased in Eq. (2). Because the concentration of protein targets has increased, there is increased nonspecific binding (Ig–protein) at equilibrium. Since nonspecific binding is so high, specific binding of the lower affinity antibody can no longer be resolved as there is no definable plateau region. This demonstrates the importance of the affinity constant in describing antibody quality.

RULE: K_a IS THE MOST IMPORTANT DESCRIPTOR OF ANTIBODY QUALITY.

Unfortunately, the value of this single most important parameter is almost never available from your favorite supplier of antibodies. We can still determine antibody quality without knowing its K_a. In Fig. 2C, increasing the K_a of the antibody to 5×10^9 results in a lower concentration to saturate the desired epitopes so that they can be resolved from nonspecific binding. Antibody binding can readily be appreciated by plotting the amount of free antibody as a function of the affinity constant as shown in Fig. 3. For a low affinity antibody ($K_a \approx 10^7$ to 10^8), 99% of it is free, that is, it is not bound to any epitope at the concentration required for epitope saturation. All of this free antibody is available to bind nonspecifically to extra- or intracellular proteins causing data misinterpretation. If, however, good antibodies are used (10^9–10^{10}), the concentration for epitope saturation is much lower and so is the amount of free antibody.

Figure 4 demonstrates the effect low affinity antibodies have on immunophenotyping and its interpretation. In Fig. 4, a preparation with 85% viable cells is stained with two antibodies having a similar low K_a of $\sim 5 \times 10^8$. As shown in Fig. 4A, Ab1 stains a mutually exclusive population from Ab2, but there is a population on the 45° line that appears to coexpress both Ab1 and Ab2. These are the permeable dead cells, which allow entry of low affinity Ab1 and Ab2 into the high concentration of intracellular protein to which they bind nonspecifi-

---

causing the curve for nonspecific binding (■) to shift to the left (arrow). There is no change in the specific binding curve so it is the same one shown in (A), but the sum (▲) measured by the flow cytometer produces a curve in which specific binding (△) can no longer be resolved. (C) The only way to correct this problem is to use a better, higher affinity antibody (●).

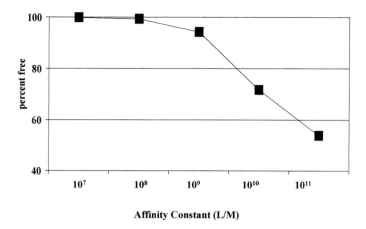

Fig. 3 Effect of affinity constant on antibody binding. This figure shows the relationship between the amount of free antibody at 50% epitope saturation as a function of the affinity constant. If the affinity constant is 10^7, 99.93% of antibody is free, but if it is 10^{11}, virtually all of it binds to the epitope.

cally. In Fig. 4B, Ab3 has an order of magnitude higher K_a so the concentration of Ig required for epitope saturation is much lower and the concentration of free antibody to nonspecifically bind at epitope saturation is much lower. However, Ab1 still binds nonspecifically producing an erroneous frequency and a heterogeneous expression as both the live and dead cells bind it. Note that in Fig. 4A the Ab1$^+$ cells were a circular cluster compared to the ellipsoid cluster in Fig. 4B.

RULE: DEAD CELLS SHOULD ALWAYS BE DETERMINED.

The presence of dead cells should always be considered in data interpretation.

C. Influence of Epitope, Fluorochrome, and Fixation

Besides antibody quality, there are two other factors to consider when choosing antibodies for multiparameter immunophenotyping, the epitope on the antigen and the fluorochrome conjugated to the antibody. Figure 5 shows an acute B-lineage lymphocytic leukemia stained with CD19 monoclonal antibody from two different clones. Because this particular leukemia exhibits aberrant CD19 expression, it is totally absent when stained with CD19, clone SJ25C1, but was found to be heterogeneously expressed when the CD19, derived from clone J4.119, was used. This is not a fluorochrome intensity effect because on normal B cells the staining intensities for both CD19 clones are virtually identical.

RULE: ANTIBODIES TO LINEAGE ANTIGENS FROM AT LEAST TWO CLONES SHOULD BE USED FOR IMMUNOPHENOTYPING MALIGNANT CELLS.

The fluorochrome can also make a profound difference on epitope detection. As shown in Fig. 6, CD69, when conjugated with phycoerythrin (PE), clearly

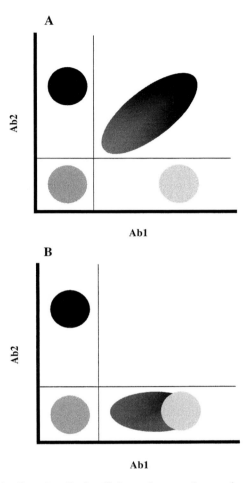

Fig. 4 Effect of dead cells and antibody affinity on immunophenotyping. In (A), Ab1 and Ab2 have a K_a of 5×10^7 liters/M, and 15% of cells are dead. The dead cells are permeable to both antibodies and bind nonspecifically to the higher concentration of intracellular protein. This produces a cluster of double positive events on a 45° angle whose frequency is equal to the dead cells in the suspension. If there were actually viable cells that are positive for both antibodies, the dead cells would obscure their frequency and the data misinterpreted. In (B), Ab3 has an affinity of 10^9, so it has no appreciable nonspecific binding, but Ab1 still binds nonspecifically as well as specifically, so the cluster of positive cells now exhibits the same heterogeneous expression as dead cells, but the viable CD3$^+$ cells cannot be resolved. This would lead to an erroneous evaluation of Ab1$^+$ cells and data misinterpretation. (See color plates.)

resolves both activated T cells and the higher epitope expressing micromegakaryocytes. When the same antibody is used, conjugated with APC, a 10-fold reduction in fluorescence intensity occurs; thus, only the brighter micromegakaryocytes are resolved, and the T cells are completely negative.

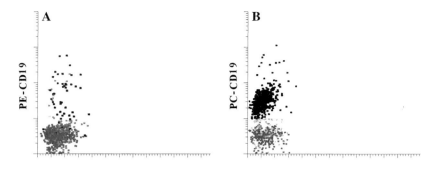

Fig. 5 Differing monoclonal antibody epitope binding. An acute B lineage lymphocytic leukemia was stained with two different clones of CD19. In (A), the CD19 antibody produced by clone SJ25C1 was used. None of the leukemic cells are positive. Leukemic cells are positive when the antibody produced by clone J4.119 was used (B).

RULE: ALWAYS SELECT ANTIBODIES CONJUGATED WITH THE BRIGHTEST FLUOROCHROMES FOR IMMUNOPHENOTYPING CELLS THAT EXPRESS ANTIGENS AT LOW DENSITY.

Choose the fluorochromes in order of brightness: PE, phycoerythrin–CY5 tandem complex (PECY5), phycoerythrin–Texas Red tandem complex (PETR), allophycocyanin (APC), fluorescein isothiocyanate (FITC), and peridinin chlorophyll protein (PerCP).

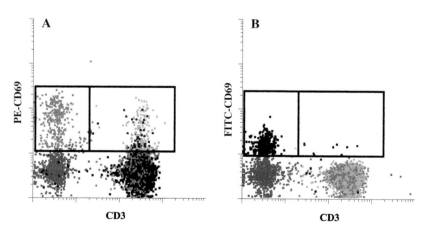

Fig. 6 Importance of fluorochrome intensity. Blood was stained with either PE-CD69 (A) or FITC-CD69 (B) and APC-CD3. In (A), CD3 negative, CD69 positive population of micromegakaryocytes exhibits a mean channel fluorescence (MCF) of 77. The CD3 positive T cell population exhibits a MCF of 44 at a resolution of 128. When APC-CD69 (B) was used, the MCF for micromegakaryocytes is only 35, and no CD69 positive T cells are found.

Fixation can also dramatically affect epitope detection, especially when intracellular staining is performed. Epitopes on fixed and permeabilized cells may no longer bind to their specific antibody because the epitope has been changed. As shown in Fig. 7, $CD19^+$ and $CD3^+$ cells are clearly resolved when viable cells are stained prior to fixation with formaldehyde. If, however, the cells are fixed first and then stained, CD3 fluorescence is not different, but CD19 fluorescence is reduced.

RULE: ALWAYS TEST THE EFFECT OF FIXATION ON EPITOPE EXPRESSION.

The epitope for the CD19 used is sufficiently altered by formaldehyde fixation that it is no longer recognized by this clone (SJ25C1) of CD19. This might be corrected by using a different clone of CD19, by using a different fixation procedure, or both. The correct interpretation of data may depend on knowledge of the behavior of the antibodies used for immunophenotyping in different settings.

D. The Effect of Drugs on Expression

Drugs can also have a profound influence on protein expression. It is customary to use the drug Brefeldin A when measuring intracellular cytokines. CD69 is an activation protein found on T cells (Testi *et al.*, 1989). Since CD69 expression is induced by stimulation, its membrane expression, shown in Fig. 8, may be completely blocked if the activation does not occur in the absence of the drug. This is caused by the inability to transport the CD69 protein to the surface membrane if it is inhibited by Brefeldin A.

RULE: ALWAYS TEST THE EFFECT OF A DRUG ON ANTIGEN EXPRESSION.

A similar phenomenon can occur for concomitantly expressed membrane proteins that have a high turnover rate.

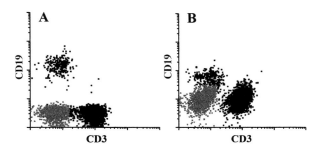

Fig. 7 Effect of fixation on antibody binding. Human blood cells were prepared by lysing the erythrocytes with lysing solution. In (A), they were first stained with the antibody combination FITC-CD3 and PE-CD19 and then fixed in 2% Ultrapure formaldehyde. In (B), cells were first fixed in 2% Ultrapure formaldehyde and then stained with the same antibody combination. CD19, but not CD3, fluorescence is significantly reduced by prefixation.

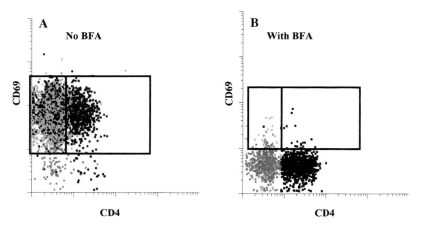

Fig. 8 Effect of the drug Brefeldin A on membrane CD69 expression. CD69 is an activation antigen on T cells that is upregulated by specific antigens and mitogens. Human blood cells were prepared by culturing them 4 hr with phorbol myristatic acid (25 ng/ml) and ionomycin (1 μg/ml) at 37°C in the absence (A) or presence (B) of Brefeldin A. The cells were put on ice and stained with the antibody combination FITC-CD4, PE-CD69, and PerCP-CD3. After gating each bivariate histogram on CD3, the CD69 expression on the CD3$^+$ T cells is shown. In addition to the complete absence of CD69 expression in the Brefeldin A treated cells (B), the CD4 expression is considerably less in the non-Brefeldin A (A) compared to the Brefeldin A (B) treated cells. Thus, drugs can affect protein expression.

E. Primary and Secondary Antibody

For immunophenotyping using a single color, it is customary to first stain cells with a primary antibody, usually a monoclonal antibody, directed to the epitope of interest. A secondary polyclonal antibody that is specific to epitopes on the primary antibody (the primary antibody, therefore, acting as an antigen in this reaction) and that is covalently coupled with a fluorochrome is then used to bind to the primary antibody, effectively coloring the cell to which the primary antibody has bound.

This method has the advantage that a single second antibody can be used for staining many different primary antibodies. Unconjugated antibodies are also less expensive than conjugated primary antibodies, and the latter may be unavailable. Using a second antibody also provides an amplification of fluorescence that may be useful for resolving cells that have a low frequency of epitopes. This advantage, however, can be negated by increased background.

The F(ab')$_2$ fragment of an affinity purified second antibody should always be used. It is of course implicit in this scheme that the second antibody does not react with the cellular epitope. Second antibodies are available with every fluorochrome conjugated to them.

RULE: ALWAYS USE F(ab) OR F(ab')$_2$ FRAGMENTS FOR SECOND ANTIBODIES.

A typical second antibody might be labeled "FITC conjugated goat F(ab')$_2$ anti-mouse IgG antibody (heavy and light chain specific, purified by affinity chromatography)." This label contains considerable information in abbreviated form. "Goat anti-mouse IgG" means the antibody was prepared by immunizing goats with mouse IgG. "Purified by affinity chromatography" means the goat serum was passed over an affinity column (usually Sepharose beads) to which mouse IgG was bound and, after elution from the column, the $F(ab')_2$ fragments were prepared and conjugated with fluorescein isothiocyanate ("FITC conjugated").

Since this goat polyclonal second antibody was prepared using a mouse IgG affinity column, this preparation contains antibodies that are specific for the heavy chain of mouse IgG. Because all isotypes found in serum, that is, IgG, IgM, and IgA, have light chains (Table I), they will also bind this second antibody because it is also "light chain specific." Therefore, this polyclonal second reagent is not specific for mouse IgG at all. In order for a second antibody to be specific for IgG it must have no light chain activity. If the antibody was only heavy chain specific the label would read "γ specific" or "$\gamma 2b$ specific," etc. A simple way to determine if a second antibody to murine IgG has light chain activity is to stain murine spleen cells with it; no cells should be positive. If there are positive cells, the reagent is defective and should not be used in applications where heavy chain specificity is required.

Any polyclonal second antibody contains all the isotypes found in the serum of the animal used to produce it (e.g., IgM, IgG, IgA). Because IgM is a pentomer and IgAs are dimers, it is very important that they all be reduced to the basic structure shown in Fig. 1. Furthermore, the FcR on polyclonal antibodies will bind to FcR on cells. These properties will increase the background binding of second antibodies, which can be effectively eliminated by using Fab fragments.

F. Blocking

Most immunophenotyping experience has been derived using hematopoietic cells in general and lymphocytes in particular. We have already described the nonspecific antibody binding, which is least for lymphoid cells and most for myeloid cells. Fc binding to Fc receptors can also lead to data misinterpretation, because each leukocyte lineage expresses a unique repertoire (Table I). This binding is not nonspecific, as so often referred to in publications, it is a saturable ligand–receptor interaction that is highly specific. Blocking the receptors with IgG prior to staining with a primary antibody can reduce this potential problem.

In the hypothetical mixture of cells shown in Fig. 9, some may have both the desired epitope and Fc receptors, FcR (cell A), some cells may have only FcR (cell B), other cells may have only epitopes (cell C), and some cells may have neither (cell D). Only the cells with the desired epitope (A and C) should be identified by the mouse monoclonal antibody (MAb) and the second antibody conjugated with a fluorochrome [fluoresceinated goat anti-mouse Ig (FGAM)]. When the MAb is added to the cells it can bind to the desired epitope via the $F(ab')_2$ portion (A and C), or to the cell via the Fc portion (A and B). If the sample contains an equal portion of each cell type, three-quarters of the cells

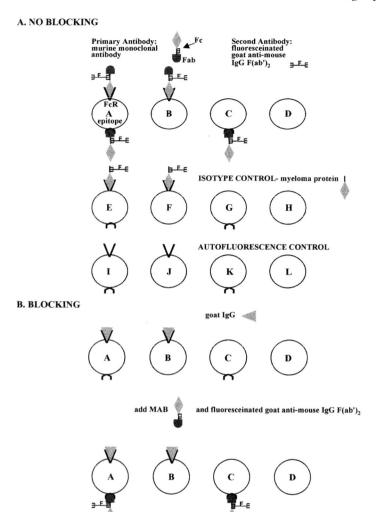

Fig. 9 Indirect immunofluorescence staining. (A) The primary antibody is shown as a stick symbol that binds to cells with epitopes shown by the half circle. The Fc portion binds to its receptor, shown by the V. The second antibody binds to the primary antibody at the m (A–D). The stick symbol for the isotype control (a myeloma protein) has the same shape for the FcR but no antibody binding site because there are no epitopes on the cells (E–H). And cells that have not been stained serve as an autofluorescence control (I–L). Epitope positive and FcR positive cells are resolved. (B) To block Fc binding, cells are incubated for 10 min with goat Ig at a concentration that will saturate FcR (200 μg/ml/10^7 cells), shown as a triangle. The murine primary MAb is then added followed after washing by the fluorochrome conjugated second antibody. The cells are washed, fixed, and analyzed. Only epitope positive cells are resolved.

present will bind the MAb. Since the second antibody is a fluorochrome conjugated F(ab')$_2$ goat anti-mouse IgG, it binds to the murine MAb (but not to any FcR). Since one-quarter of the cells have only FcR and no epitopes, these cells are inappropriately labeled and counted as epitope positive, thereby overestimating the percentage of epitope positive cells.

To account for this problem, the common practice is to use an "isotype" control. This control consists of the same cells, incubated with a myeloma protein having no epitope specificity, but having the same isotype and subclass as the specific antibody. As shown in Fig. 9A, this myeloma protein is presumed to bind to cells having FcR (cells A and B). The second antibody is added, as before, to stain cells that have bound the isotype protein. The percentage of positive cells revealed by the isotype protein is then subtracted from the percentage obtained using the specific antibody. This procedure leads to an underestimate of the epitope positive cells because one-quarter of the cells express both FcR and epitopes, and this group is subtracted from the total. Thus, isotype controls may actually lead to an erroneous conclusion.

RULE: NEVER SUBTRACT POSITIVE CELLS IN A CONTROL SAMPLE FROM POSITIVE CELLS IN AN EXPERIMENTAL SAMPLE.

The autofluorescence control should always be analyzed. It contains no antibodies and is otherwise processed the same way as the other samples. This control provides a baseline to determine the minimum fluorescence above which positive cells are identified. The isotype control and autofluorescence control should give identical results if a properly titered high quality antibody is used at exactly the same IgG concentration as the isotype control. To the extent that they differ is indicative of an improperly titered antibody, a poor quality antibody, or both.

Treating the cells with normal immunoglobulin (blocking) prior to staining them with specific antibodies can reduce the Fc binding of antibodies to cells. As shown in Fig. 9, a murine monoclonal antibody was used. Because the second antibody is derived from the goat, the blocking immunoglobulin is also derived from the goat. This is to prevent binding of the second antibody to the block. The cells are incubated for 10 min with an excess of goat IgG where the IgG binds the FcR (cells A and B). Without washing the cells, the primary murine MAb is added wherein it binds only to its epitopes (cells A and C). Because the labeled second antibody was made against murine IgG in the goat, it binds to the desired MAb but not the blocking goat IgG. To be sure that the block is effective, the "isotype control" antibody can be used in place of the primary antibody. Even if it did bind to unblocked cells, it should not bind to the blocked ones. Thus, the cells treated with an isotype control, which must be adjusted in concentration to equal that for each antibody used or adjusted to the concentration to equal the concentration of the highest antibody used, should look identical to the untreated cells that exhibit only autofluorescence.

RULE: FOR INDIRECT STAINING, BLOCK WITH THE IMMUNOGLOBULIN FRACTION FROM THE SAME SPECIES FROM WHICH THE SECOND ANTIBODY WAS MADE.

In Table I, the approximate amounts of Ig isotypes and IgG subclasses in the serum of most mammals is shown. The FcR blocking reaction is concentration dependent; there is not a high enough concentration of IgG in serum to effectively block all FcR; therefore, only purified IgG at a high enough concentration should be used for blocking.

As shown earlier in Section II,B, nonspecific binding cannot be blocked. We add 10 μl of normal IgG at 1 mg/ml to 50 μl of cells at 20×10^6/ml. This concentration provides enough IgG of each subclass to effectively block all FcR binding. The effect of blocking is shown in Fig. 10. A population of both small cells and large cells are positive in the histograms on the top of each panel. The antibody staining for the cells is shown in the histogram along the side of each panel. By projecting each histogram event into the bivariate display, the correlation of cell size with antibody staining can be visualized. In this example, directly fluoresceinated anti CD3 (FL-CD3) was used to label mononuclear cells (MNC). When cells are incubated with FL-CD3, many large cells are dimly stained (Fig. 10A). When the cells are incubated for 10 min with mouse Ig (mIg) prior to the addition of the FL-CD3, no large cells are dimly stained (Fig. 10B). Thus, blocking the cells with mIg prior to staining them had a profound effect on the type of cells that were stained by the antibody. Mouse, rather than goat IgG was used for blocking because the specific antibody was directly conjugated.

G. Directly Conjugated Antibodies

Primary antibodies are currently available conjugated with fluorescein (F), PE, PETR, or PECY5, APC, and biotin (B). Because there is no second antibody to mouse Ig, it can be used for blocking. Using a directly conjugated specific antibody the cells can be preincubated with murine IgG as the block prior to staining with a labeled murine MAb. Rat IgG would be used if the labeled specific MAb were of rat origin. When using a biotinylated antibody, labeled avidin is used as the second reagent. Avidins can be purchased conjugated with FITC, PE, PETR, PECY5, PerCP, or APC.

H. Wash versus No Wash

There is currently an ongoing discussion as to whether one should wash specimens or run them in lysing reagent without washing. Simply adding 50 μl of blood or bone marrow to a cocktail of antibodies, incubating 5 min, adding a lysing reagent for 5 min, and fixing the sample would mean that the specimen is completely ready for analysis in 10 min. For intracellular markers the incubation time is increased to 45 min to allow for permeabilization and diffusion of the antibody. Because this homogeneous assay is so simple, it is most attractive to

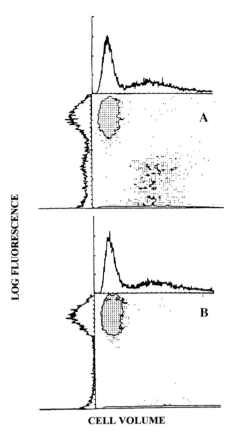

Fig. 10 Effect of blocking on MAb binding to mononuclear cells (MNC). Human cells were isolated from blood using neutrophil isolation medium (NIM, Cardinal Associates, Santa Fe, NM). Cells were adjusted to 20×10^6/ml in PBS and 50 μl were used for labeling. (A) Cells were incubated with fluoresceinated CD3 for 15 min. (B) Cells were first incubated for 10 min with 10 μg mouse IgG, and then fluoresceinated CD3 was added for 15 min. After washing, the samples were analyzed. FcR binding to monocytes of CD3 is eliminated by blocking.

suppliers and to users. The advantage is reduced processing time and a perception that cells are not lost. So why wash?

There are several reasons to consider washing the specimen. First, when measuring light chains on B cells, the specimen must be washed to remove immunoglobulins that will otherwise bind the antibodies to them instead of the cells. While healthy individuals may not have soluble free shed membrane antigens, patients with disease often have them and they will bind the antibodies in suspension thereby reducing their concentration for binding to cells. This can and does significantly bias the cellular immunophenotyping result often unbeknownst to the laboratorian. Finally, when multiple antibodies with their fluorochromes are used, the soluble fluorochrome spectra will be present in the cell stream and detected by all photomultiplier tubes (PMTs). This will cause a baseline offset to the steady state light conditions of the cell stream, reducing proportionately the signal (positive cells) to noise ratio directly affecting instrument sensitivity. Populations of cells exhibiting dim fluorescence in a washed specimen may completely disappear in an unwashed one. How often does this occur? Every time you do not wash a specimen, but you will never know if you do not wash.

Using the CAP survey results for 1998, the most commonly used lysing reagents were BDLyse (41%), Q-Prep (38%), ammonium chloride (7.6%) OrthoLyse (4.3%), and all others (9.1%). Our studies as well as several others show that ammonium chloride lysis is not different than the commercially prepared lysing reagents (Tamul et al., 1994; Carter et al., 1992; Bossuyt et al., 1997). The advantage of using an ammonium chloride solution is that it can be prepared from basic chemicals, which are an order of magnitude less expensive than the commercially obtained preparations.

III. Tandem Fluorochromes

Several tandem fluorochromes are available to provide for a third and fourth color excitable by a single or dual laser flow cytometer. The properties of these reagents have been previously described (Stewart and Stewart, 1993) and are summarized in Table II. Not included in this discussion is the new tandem fluorochrome, APC-CY7 (Beavis, 1996), which may be useful for combining with those considered here.

The PECY5 fluorochrome is the brightest tandem currently available because the CY5 absorption bandwidth is optimal for PE emission. This is reflected in

Table II
Comparison of Reagents for Three and Four Color Phenotyping

Parameter	PerCP[a]	PETR[a]	PECY5[a]	APC[b]
Relative intensity	Dim	Medium	Bright	Medium
Compensation: amount	None	High	Low	High
Batch variability	None	Some	Significant	None
Light sensitivity	Stable	Stable	Unstable	Stable
Nonspecific binding				
Lymphocytes	None	None	None	None
Monocytes	None	None	High	None
Granulocytes	None	None	Low	None
Availability of direct conjugates	Fair	Poor	Good	Good
Emission (nm)	673	613	670	670

[a] Excitation at 488 nm.
[b] Excitation at 635 nm.

the low amount of (FL2 − %FL3) compensation required (usually <3% and for some constructs, <0.5%). However, there is a great deal of variability among batches and suppliers. PerCP requires no compensation, but its emission energy is less than that found for fluorescein. It should be used for antibodies to highly expressed cellular epitopes. PerCP also rapidly degrades at high laser power density.

The PETR tandem is intermediate in its fluorescence emission intensity because the TR absorption bandwidth is not as ideally matched to PE emission as CY5. This energy mismatch results in a high degree of PE emission requiring considerable compensation (20–30%) similar to the overlap found for fluorescein in the PE channel. As shown in Fig. 11, PECY5 tandems can exhibit unacceptable variation in their compensation requirements between different suppliers of directly conjugated antibodies (Fig. 11B,D) or on different batches of the same antibody (Fig. 11C,D). A high amount of compensation is required when these

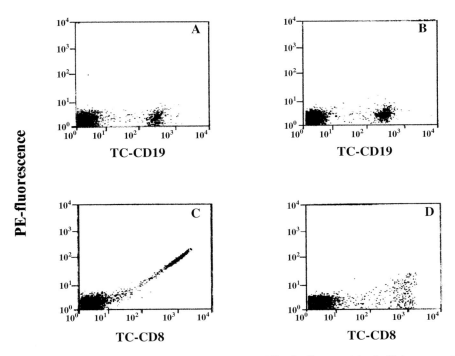

Fig. 11 Variation in compensation for PECY5 reagents. Blood cells were stained with two separate batches of PECY5-CD19 or PECY5-CD8 and compensation for FL2 − %FL3 is shown using 488 nm excitation. In (A), the PECY5-CD19 stained cells were properly compensated. When stained with a second batch (B), the same compensation is acceptable. In contrast, when the settings for PECY5-CD19 are used for PECY5-CD8, they are unacceptable (C) and new compensation settings must be found (D). If these new settings were used for PECY5-CD19, this reagent would be markedly overcompensated. (See color plates.)

reagents are combined with APC, and the amount depends on the construct and the antibody used (Stewart and Stewart, 1999).

Although otherwise an ideal fluorochrome because it is so bright, the PECY5 tandem exhibits two other problems not found for PETR, PerCP, or APC reagents. This tandem is exquisitely sensitive to light, and the CY5 molecule will become irreversibly degraded on short-term exposure (<1 hr) to ambient light. This problem can be easily recognized as an increase in PE signal like that shown in Fig. 12, requiring increasingly more compensation and a concomitant decrease in CY5 signal with light exposure. To prevent this problem, all staining should be performed in subdued light and samples stored in the dark.

RULE: KEEP PECY5 REAGENTS IN THE DARK.

The third problem is caused by the ability of the PECY5 fluorochrome to specifically bind to monocytic cells (Fig. 13). This problem is exacerbated by the tendency for some suppliers to overconjugate antibodies with this tandem. This causes an increase in both nonspecific binding to all cells and specific binding to

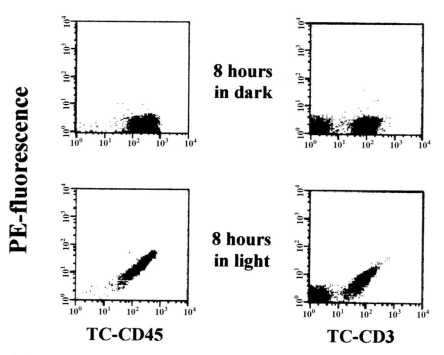

Fig. 12 Effect of light exposure on PECY5 tandem fluorescence. Blood was stained with either PECY5-CD45 or PECY5-CD3 in two sets of tubes, and one was stored 8 hr in the dark while the other was left exposed to light. Light causes the degradation of the PECY5 linkage so that efficient energy transfer is reduced, resulting in increased PE photon emission.

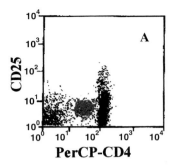 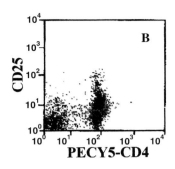 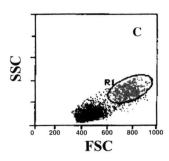

Fig. 13 PECY5 binding to monocytes. In (A), blood is stained with PerCP-CD4 and PE-CD25. Monocytes, light gray, resolved by R1 in (C) and exhibiting dim CD4 fluorescence are clearly resolved from the brighter staining T cells. In (B), blood is stained with PECY5-CD4. Because the PECY5 fluorochrome and CD4 both bind to monocytes they are much brighter and no longer can be resolved from the T cells. In (C), the forward scatter (FSC) versus side scatter (SSC) is shown.

monocytic lineage cells. A PECY5 isotype control is essential, as blocking is ineffective in eliminating this fluorescence. This control is useless if it is not conjugated in a similar manner and used at exactly the same concentration as the antibody. Although it has been suggested that CD64 is the receptor for the CY5 binding on monocytes (van Vugt *et al.*, 1996), our data do not support this claim because CD64 negative monocytic leukemias and some B myeloid lineage leukemias that are CD64 negative also bind PECY5 conjugated antibodies very strongly.

IV. Cell Preparation and Staining Procedures

The preparation of cells depends on their source. Cells contaminated with very high frequencies of erythrocytes must be lysed. Cells in tissues must be disaggregated, while cells adherent to the surface of culture containers must be removed. Cells must be uniformly monodispersed in suspension. Several chapters in this series provide procedures for dispersing cells to meet this requirement. Each of these conditions has their own set of problems, and there is no consensus on the best approach to take. The procedures described here represent 20 years of experience in immunophenotyping cells of all types. The problems associated with any procedure are cell selection, cell death, and loss of epitopes. Thus, whatever procedure is selected, all three should be determined so that the quality of the final cell suspension for immunophenotyping is known.

We recommend a whole blood (or bone marrow) lysing method to remove erythrocytes, and there are several reagents commercially available for this purpose. Although all of them work with blood, some are better than others for lysing bone marrow and fetal erythroid cells. The cost may also be a factor.

Lysing erythrocytes is not recommended before staining because the platelets and erythrocyte debris are concentrated along with the leukocytes during the subsequent washing steps, causing difficulty in the analysis of the data. Therefore, postlabeling lysis is favored.

For dispersing tissue, we recommend multiple needle sticks with a 21-gauge needle attached to a 1-ml syringe. The biopsy material is then aspirated and expelled five times into 300 μl of PBS. Several sticks throughout the tissue results in an average yield of about 5000 cells per stick (Stomper *et al.,* 1997). Our yield is in excess of 10^8 cells/g and represents the highest we have ever experienced by any method. Although it may be more tedious to perform, one must weigh the scientific benefit of high yield and less bias using this method with ease of preparation. Mechanical tissue dispersal methods produce high amounts of debris (generated by the mechanically killed cells), and enzymes may destroy the desired epitopes for immunophenotyping and kill cells.

When using cultured cells, the usual procedure for removing them is to use a surface membrane protein protease such as trypsin. This kind of treatment can also destroy the desired epitopes for phenotyping. Mechanical dispersal, such as scraping with a rubber policeman can cause severe aggregation and cell death as they are torn from the plate. We recommend treating cultures with 10 mM cold EDTA for 10 min followed by vigorous pipetting with the tip of a Pasteur pipette close to the surface of the dish. The cells are usually less adherent and quickly disperse into the rapidly flowing medium.

In the following procedures a "wash" means to add 3.5-ml phosphate buffered saline to the tube, centrifuge the cells at 1500 g for 3 min at 4°, and decant the supernatant by inverting the tubes and blotting the lip by touching its rim to an absorbent towel. Thoroughly resuspend the cells in the residual buffer that has drained back to the bottom of the tube (not to exceed 100 μl). Note the centrifugation speed is three to five times that found in most publications, that is why we do not lose cells by washing. All labeling procedures are carried out on ice. Room temperature may be fine for lymphocytes but can be problematic for other hematopoietic cells due to internalization of the antibody.

Each antibody should be titered prior to use so that the correct amount to optimally stain the cells is used (see Section V). Our general practice is to have the optimal amount of antibody in 5 or 10 μl, but no more. Commercially available antibodies that have already been titered may be diluted differently by the supplier so that more or less than 10 μl is required. In these cases, use the suppliers recommended amount (unless it is found to be incorrect by titering according to Section III).

For all the following procedures, put 50 μl of a cell suspension (do not exceed a concentration of 20×10^6 cells/ml blood, or bone marrow) into 12×75-mm tubes containing 10 μg of blocking immunoglobulin. Do not block when measuring Fc receptors unless you know the antibody you are using is not blocked by receptor occupancy.

A. Staining Cells with One Antibody

Indirect labeling with one antibody and a second antibody:

1. Add 10 μl of goat IgG (1 mg/ml) for 10 min.
2. Add primary mouse antibody for 15 min.
3. Wash.
4. Add fluoresceinated goat anti-mouse IgG F(ab′)$_2$ (FGAM) second antibody for 15 min.
5. Wash.

For the control tube, add the isotype control antibody instead of the first antibody in step 2. Also prepare a tube containing unstained cells; this will be used to measure cellular autofluorescence.

RULE: FOR INDIRECT LABELING WITH A SECOND ANTIBODY, ALWAYS BLOCK WITH IgG FROM THE SAME SPECIES AS THE SECOND ANTIBODY.

Indirect labeling with a biotinylated antibody and avidin:

1. Add 10 μl of x IgG (1 mg/ml) for 10 min.
2. Add B-Ab for 15 min.
3. Wash.
4. Add F-avidin for 15 min.
5. Wash.

Here, x is the species of IgG that is the same as the antibody. For murine antibodies block with normal mouse IgG; for rat antibodies block with normal rat IgG.

Direct labeling with one antibody:

1. Add 10 μl of x IgG (1 mg/ml) for 10 min.
2. Add F-Ab for 15 min.
3. Wash.

RULE: FOR INDIRECT LABELING WITH BIOTINYLATED ANTIBODIES OR FOR DIRECT LABELING, BLOCK WITH THE IMMUNOGLOBULIN FRACTION FROM THE SAME SPECIES AS THE ANTIBODY. NEVER BLOCK IF THE EPITOPE BEING MEASURED IS THE Fc RECEPTOR.

For example, CD16 is the FcR III receptor on granulocytes and natural killer (NK) cells. Some CD16 antibodies will not bind to this receptor if it is blocked.

B. Staining Cells with Two Antibodies

One directly conjugated antibody with one unconjugated antibody:

1. Add 10 μl of goat IgG (1 mg/ml) for 10 min.

2. Add primary mouse antibody for 15 min.
3. Wash.
4. Add phycoerythrinated goat anti-mouse Ig (PEGAM) secondary antibody for 15 min.
5. Wash.
6. Add 10 μl of x IgG (1 mg/ml) for 10 min.
7. Add F-Ab for 15 min.
8. Wash.

RULE: ALWAYS PERFORM THE INDIRECT SECOND ANTIBODY STEP FIRST AND BLOCK AFTER THIS STEP WITH IgG FROM THE SAME SPECIES AS THE PRIMARY ANTIBODY.

This is because the second antibody will have free binding sites that must be blocked so it will not bind the directly conjugated second antibody when it is added subsequently in step 7. For example, suppose that both the unconjugated and conjugated antibodies are murine MAbs. After step 5, some cells have bound MAb1 and PEGAM to them. Because free binding sites remain on the PEGAM, the F-MAb may bind to them when it is added. To prevent this binding, these free sites must be blocked by the addition of mouse IgG shown in step 6.

If a rat MAb had been used then rat IgG would be used for the block. If one MAb is rat and the other mouse then the appropriate second antibody and block would be used: Unlabeled rat, use phycoerythrinated goat anti-rat Ig (PEGAR) and block with rat IgG; unlabeled mouse, use PEGAM and block with mouse IgG. FGAM or fluoresceinated goat anti-rat Ig (FGAR) could also be used in combination with PE labeled antibodies. In summary, always perform the indirect labeling step first and block with normal IgG of the first antibody species before adding the directly conjugated antibody.

Two directly conjugated antibodies:

1. Add 10 μl of x IgG (1 mg/ml) for 10 min.
2. Add F-Ab and PE-Ab for 15 min.
3. Wash.

One antibody conjugated with fluorescein, one with biotin:

1. Add 10 μl of x IgG (1 mg/ml) for 10 min.
2. Add F-Ab and B-Ab for 15 min.
3. Wash.
4. Add PE-avidin for 15 min.
5. Wash.

Here, x IgG is derived from the species in which the conjugated antibodies were produced.

C. Staining Cells with Three Antibodies

There are several strategies that can be used for staining cells with three antibodies. For the reagents, TC (third color) refers to either PerCP, PETR, or PECY5.

Two fluorochrome conjugated antibodies and one unconjugated antibody:

1. Add 10 μl of goat IgG (1 mg/ml) for 10 min.
2. Add unlabeled mouse antibody for 15 min.
3. Wash.
4. Add B-GAM for 15 min.
5. Wash.
6. Add 10 μl of x IgG (1 mg/ml) for 10 min.
7. Add F-Ab, PE-Ab, and TC-avidin for 15 min.
8. Wash.

Here, x IgG is derived from the same species as the conjugated antibody. In this procedure a biotinylated second antibody was used followed by a TC-avidin to illustrate one of the many strategies that can be used to provide for the desired fluorochrome combination. A fluorochrome-conjugated second antibody could also have been used instead of the biotinylated one.

Three antibodies—two directly conjugated and one biotinylated antibody:

1. Block with 10 μl of x IgG (1 mg/ml) for 10 min.
2. Add FL-Ab, PE-Ab, and B-Ab.
3. Wash.
4. Add TC-avidin.
5. Wash.

Again, x IgG is derived from the same species as the three directly conjugated antibodies.

Three directly conjugated antibodies:

1. Block with 10 μl of x IgG (1 mg/ml) for 10 min.
2. Add F-Ab, PE-Ab, and TC-Ab for 15 min.
3. Wash.

D. Staining Cells with Four Antibodies

It is now possible to excite the four fluorochromes F, PE, PETR, and PECY5 or PerCP with a single laser or with two lasers where APC is used instead of PETR. This provides for the opportunity to use four antibodies simultaneously.

Two directly conjugated, one biotinylated and one unconjugated antibody:

1. Add 10 μl of goat IgG (1 mg/ml) for 10 min.

2. Add primary mouse Ab for 15 min.
3. Wash.
4. Add phycoerythrin Texas Red tandem complex conjugated to goat anti-mouse Ig (PETRGAM).
5. Wash.
6. Block with x IgG (1 mg/ml) for 10 min.
7. Add F-Ab, PE-Ab, and B-Ab for 15 min.
8. Wash.
9. Add PECY5-avidin or PerCP-avidin for 15 min.
10. Wash.

Here, x IgG is derived from the same species as the unlabeled antibody.
Three directly conjugated and one biotinylated antibody:

1. Add 10 μl of x IgG (1 mg/ml) for 10 min.
2. Add F-Ab, PE-Ab, PECY5 (or PerCP)-Ab, and B-Ab for 15 min.
3. Wash.
4. Add PETR-avidin for 15 min.
5. Wash.

Other combinations for the third color are also possible. For example, an antibody directly conjugated with PETR and a PECY5 or PerCP-avidin could have been used.

Four directly conjugated antibodies:

1. Add 10 μl of x IgG (1 mg/ml) for 10 min.
2. Add F-Ab, PE-Ab, PETR-Ab, and PECY5 (or PerCP)-Ab for 15 min.
3. Wash.

The four-color panel for immunophenotyping human cells offers the advantage of combining three antibodies for subset identification with CD45 as the fourth antibody used for resolving leukocytes from debris and other cells in general and for resolving lymphocytes in particular (Stelzer *et al.*, 1993; Mandy *et al.*, 1992). The data analysis strategy for this application will be discussed in the chapter on analysis.

E. Intracellular Staining

Unless no surface membrane staining is desired, intracellular staining should always be done after all surface membrane staining has been performed and after the cells have been fixed. Formaldehyde fixation can result in the cross-linking of epitopes already discussed in Section IV,C. There are several permeabilization reagents commercially available. They usually contain Triton X-100, Nonidet P-40 (NP-40), or saponin. Because saponin permeabilization is reversible, the wash buffer, if performed, should also contain saponin. Only antibodies with directly conjugated fluorochromes or those with biotin can be used if the cells have been stained with antibodies to membrane-bound antigens, because a second antibody will also bind to the membrane antibodies. The general procedure is as follows.

To cells in 50 μl of buffer:

1. Add the amount of permeabilizing reagent recommended by the supplier for 10 min.
2. Add the conjugated antibody(ies) to the suspension and incubate 30 min.
3. Add 3.5 ml of PBS.
4. Centrifuge cells at 1500 g for 3 min, decant supernatant, and resuspend cells in 500 μl of formaldehyde.

It is also wise to use an isotype control adjusted to the exact concentration as the antibody to the intracellular antigens. Nonspecific binding occurs at a concentration 10–50 times less than for surface phenotyping because of the increased intracellular protein concentration as discussed earlier in Section II,B.

F. Cell Fixation

Cells are often fixed after staining so data acquisition can be performed at a later time.

1. Thoroughly resuspend pelleted cells in residual PBS after staining.
2. Add 3 ml of lysing reagent and agitate cells for 3 min (rock, roll, or tumble). This step may be omitted if no erythrocytes are present.
3. Centrifuge cells at 1500 g for 3 min. Decant supernatant and resuspend pellet.
4. Wash.
5. Add 300 μl of 2% Ultrapure formaldehyde.

Paraformaldehyde is a powder that when in aqueous solution becomes Ultrapure formaldehyde. This can be purchased already prepared from PolySciences (Malvern, PA). We resuspend cells in only 300 μl of this fixative to produce a more concentrated suspension so acquisition of data will be faster. If desired, cells may be centrifuged and resuspended in less formaldehyde to have them more concentrated for data acquisition. The formaldehyde concentration should not be less than 1% for reproducible forward scatter light (FSC) versus side scatter light (SSC) characteristics. Use only Ultrapure formaldehyde because other formulations will produce increased autofluorescence. Prolonged storage of samples results in increased autofluorescence, which can compromise the resolution of dimly stained cells. Therefore, data should always be acquired within 5 days of preparation.

G. Measuring Cell Viability

As discussed in Section II, dead cells can produce errors that lead to data misinterpretation. This can be avoided by evaluating their number in the cell suspension.

1. Unfixed cell suspensions:

   a. Resuspend cell pellet in 1 ml of PBS containing 10 μl (200 μg/ml) of propidium iodide per ml.

   b. Analyze samples after 5 min of incubation.

2. Fixed cell suspensions (Riedy *et al.,* 1991):

   a. Resuspend cell pellet (prior to fixation) in residual PBS and add enough ethidium monoazide for a final concentration of 5 μg/ml. (Some batches require less EMA to reduce nonspecific fluorescence due to its binding to any protein.)

   b. Put samples 18 cm from a 40 W fluorescent light for 10 min.

   c. Wash and fix cells as described in Section IV.

V. Titering Antibodies

For trouble-free immunophenotyping it is essential to determine antibody quality and titer. This should be done for every new antibody and is recommended for new batches of previously tested antibodies. Generally, antibodies to mammals other than human cells are of poorer quality than those to human cells because of the lack of regulatory rules that govern their production. This is especially true for antibodies to intracellular antigens. By properly titering an antibody, its quality can be established according to the law of mass action. Since 3 μg/100 μl/10^6 cells exhibits the initial concentration where the nonspecific binding component for all IgGs becomes detectable, this represents a convenient starting concentration for titering.

Antibodies such as CD3 or CD19, and many more, generally stain a cluster of cells whose membrane antigen is expressed in a fairly uniform manner. This results in a distinct cell cluster. Antibodies such as CD25 or CD38, and several others, generally stain cells in a heterogeneous manner that produces a continuum of negative to positive cells. Such a continuum is often exactly like that observed for nonspecific binding, making discrimination between it and specific binding more subjective. Finally, intracellular staining requires longer incubation and wash times because antibody diffusion is restricted by the infrastructure of the cell.

A high affinity antibody at the proper antibody titer is most important if good immunophenotyping data is to be obtained. Because MAb are myeloma proteins whose antibody specificity is known, they behave both like myeloma proteins (binding by Fc and nonspecifically) and like specific antibodies (binding to epitope). For any given MAb, there may be a different degree of Fc and nonspecific binding. Proper titering can reduce nonspecific binding as well as provide the desired specific epitope binding, whereas IgG blocking eliminates the problem of FcR binding. To optimize specific binding, the mean channel fluorescence is the only parameter that is measured.

A. Processing Cells

To titer an antibody, target cells that express the epitope, target cells that do not express the epitope, the specific antibody, and an isotype control exactly matched in IgG concentration to the specific antibody are required. A target cell suspension containing a mixture of epitope positive and negative cells such as blood is best. An epitope positive and negative cell line mixed together is also appropriate.

1. Adjust the specific antibody and isotype control to 30 μg/ml in PBS and prepare five serial 1/3 dilutions of each. Also prepare a tube containing only cells (autofluorescence).
2. Make a mixture of epitope positive and epitope negative target cells if necessary, and adjust them to 2×10^7 per ml in PBS.
3. Add the antibody dilutions to five appropriately labeled 12×75-mm tubes and the isotype control dilutions to another five tubes. Prepare a tube in which only PBS is added. Then add 50 μl of cells to all tubes.
4. For extracellular epitopes, follow the staining procedure described in Section IV,A. For intracellular epitopes, follow the staining procedure described in Section IV,E.
5. Acquire data ungated.

B. Analysis of Results

Refer to Chapter 45 for data acquisition and analysis.

1. For Antibodies That Resolve Discrete Populations

a. Create a bivariate histogram. If a gated population is desired, create a bivariate plot of FSC versus SSC and establish the appropriate region for gating the univariate histogram.

b. Establish a marker to distinguish between positive and negative cells using unstained cells (autofluorescence).

c. Analyze all the files and record the mean channel fluorescence (MCF) of the positive and negative cells (see Fig. 14A).

d. For each antibody dilution, compute the ratio of the positive MCF by the negative MCF and plot these values as a function of the dilution (see Fig. 14B). Do this for both the specific antibody (Fig. 14B) and the isotype control.

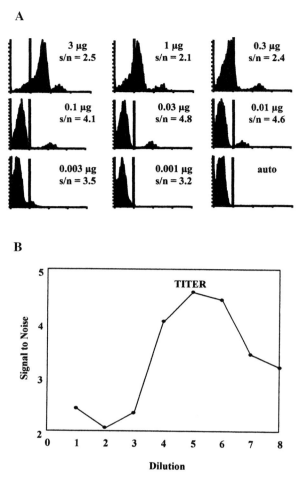

Fig. 14 Titering antibodies to epitopes with discrete expression. (A) To determine the titer, target cells are stained with various dilutions of antibody and the mean channel fluorescence of positive and negative cells determined. The marker may be set using an isotype control or unstained cells (auto). The ratio of the MCF of the positive cells to negative cells is calculated. We call this the signal-to-noise ratio (S/N). Because each antibody, when properly titered, is its own isotype control, there is no need to use a separate one. (B) By plotting S/N as a function of the concentration an inverted parabolic shaped curve is produced whose zenith is the titer. This concentration produces the best distinction between positive and negative cells. The left side of the curve represents rapid changes in nonspecific binding and very slow changes in specific binding. The right side reflects decreased saturation of epitopes by antibody.

When unblocked cells are stained with too high an antibody concentration (Fig. 14A), there is a uniform shift to the right of the negative cells so some of them are above the marker established for positive cells. As the antibody is diluted, nonspecific binding decreases so the epitope positive cells are clearly resolved. Finally, the epitope positive cells decrease in mean channel fluorescence as antibody is no longer in excess and the epitopes are no longer saturated. The kinetics can be visualized by plotting the computed ratios as shown in Fig. 14B. The optimal antibody titer is the dilution that produces the maximum signal to noise ratio, and it is often not the concentration that produces the highest percentage of positive cells.

2. For Antibodies That Resolve Nondiscrete Populations or for Intracellular Epitopes

If a discrete positive population is not found, a slightly different approach is used.

a. Create a univariate histogram, but do not establish a marker for positive versus negative cells.

b. Analyze all files and record the mean channel fluorescence of the antibody stained and isotype control stained cells.

c. For each dilution, compute the ratio of the MCF of the antibody stained to the isotype control stained cells and plot these values as a function of the dilution as shown in Fig. 15.

Many antibodies bind heterogeneously to cells and exhibit a "ski slope" appearance of nonspecific binding. To properly titer them it is necessary to use the specific antibody and the isotype control at exactly the same concentrations. No marker is used, and the MCF of the isotype control and antibody are determined. The ratios of MCF values for the antibody and isotype control are computed, and plotted as shown in Fig. 15. The titer is the concentration that produces the maximum signal to noise ratio, and this ratio must be greater than 3.0.

C. Verification of Specific Antibody Binding

Contrary to well-established dogma, the hallmark of nonspecific binding is that it cannot be blocked. No matter how much Ig is added, no matter how much other protein is added, the cells will always bind more of it. The cellular protein excess relative to the specific epitope concentration causes this unlimited binding capacity in the working range of immunophenotyping. To discriminate between specific and nonspecific binding an unconjugated specific antibody is used to block the measured specific binding of a conjugated antibody, but this block will have no affect on nonspecific binding. Only two tubes with target cells are necessary to verify that a properly titered antibody is specifically binding to cellular epitopes. If this verification fails, specific binding is not being measured. This procedure requires an appropriate target cell expressing the desired epitope, an unconjugated and a fluorochrome conjugated or biotinylated antibody.

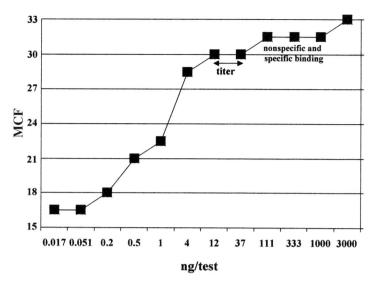

Fig. 15 Titering antibodies to epitopes with heterogeneous expression. Blood leukocytes were stained with serial 1/3 dilutions of FITC-CD38 starting with 3000 ng in a final volume of 100 μl. The mean channel fluorescence of lymphocytes was determined for the CD38 and plotted as a function of antibody concentration. Above 30 ng/test, the nonspecific binding component begins to appear. Below 10 ng/test epitope saturation no longer occurs and there is a steep slope to the titration curve. The titer is determined as the shoulder area where epitope saturation has occurred and nonspecific binding is low. If no shoulder is found, the antibody is no good. For this study, the mean channel fluorescence (MCF) for cellular autofluorescence was 5. Note that the titer of the CD38 antibody used here is 12 ng/test in a final volume of 100 μl.

1. Pipette into two tubes 10^6 target cells in 50 μl.
2. Having estimated the titer of the antibody, add an unconjugated antibody at a concentration three times the titer into the first tube. Incubate 15 min.
3. Now add to both tubes a directly conjugated antibody (with fluorochrome or biotin) at the estimated titer. Incubate 15 min for surface or 45 min for intracellular staining.
4. Acquire the data using the flow cytometer.

Figure 16 shows the expected result if the antibody is specifically binding. The binding of the fluorescent antibody is completely inhibited by the blocking unconjugated antibody(Fig. 16B). If the antibody is nonspecifically binding there is little or no effect of the block as the fluorescence is no different between the blocked and unblocked specimen (not shown). This test should always be performed when evaluating intracellular antigens.

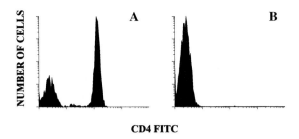

Fig. 16 Verification of antibody specificity. Human blood was incubated with either CD4-FITC (A) for 15 min or CD4 at three times titer, for 10 min, followed by CD4-FITC (B) for 15 min at the correct titer.

VI. Solutions and Reagents

A. PBS (with Sodium Azide)

Phosphate-buffered saline containing no calcium or magnesium and supplemented with 0.1% sodium azide is used for all dilutions and washes. The pH should be adjusted to 7.2. While not necessary, PBS may be supplemented with protease and nuclease free 0.5% bovine serum albumin (PAB) (Sigma Chemical, St. Louis, MO). Be sure to check osmolality of the final solution (290 ± 5 mOs*m* for human cells).

NOTE: HANDLE SODIUM AZIDE WITH EXTREME CAUTION.

B. Two Percent Ultrapure Formaldehyde

A 10% solution of Ultrapure formaldehyde can be obtained from PolySciences (Malvern, PA). This solution is diluted 1/5 with PBS to create the working solution. Do not use impure formaldehyde or formalin as these solutions will markedly increase autofluorescence.

C. Lysing Reagent

1.6520 g ammonium chloride
0.2000 g potassium bicarbonate
0.0074 g EDTA (tetra)
Make up to 200 ml with distilled water; use at room temperature.

This reagent must be prepared daily because HCO_3 combines with NH_4Cl in this strong acid weak base molecule to form CO_2, thereby rendering the solution ineffective in lysing erythrocytes. We recommend weighing the reagents and storing them as packets. The dry reagents are dissolved in water when required.

D. Propidium Iodide Stock Solution

NOTE: HANDLE PROPIDIUM IODIDE WITH EXTREME CAUTION.

Propidium iodide (MW 668), 7.4×10^{-5} M (20 mg/100 ml) (Sigma Chemical), is prepared in PBS. Store in a dark, foil-wrapped container to protect from light.

E. Ethidium Monoazide Stock Solution

Ethidium monoazide (EMA) (Molecular Probes, Eugene, OR) is very light sensitive and must be stored at $-20°C$ in a dark vial or foil-wrapped container. The stock solution is prepared in PBS at 5 mg/ml. This can then be dispersed into small aliquots of 50–100 μg/ml and stored at $-20°C$. These small aliquots can then be thawed one at a time as needed to do the assays. Discard remaining EMA after thawing.

References

Beavis, A. J., and Pennline, K. J. (1996). ALLO-7: A new fluorescent tandem dye for use in flow cytometry. *Cytometry* **24**, 390–394.

Borowitz, M. (1993). Clinical applications of flow cytometry: Immunophenotyping of leukemic cells. *NCCLS* **13**, 1–107.

Bossuyt, X., Marti, G. E., and Fleisher, T. A. (1997). Comparative analysis of whole blood lysis methods for flow cytometry. *Cytometry* **30**, 124–133.

Carter, P. H., Resto-Ruiz, S., Washington, G. C., Ethridge, S., Paline, A., Vogt, R., Waxdal, M., Fleisher, T., Noguchi, P. D., and Marti, G. E. (1992). Flow cytometric analysis of whole blood lysis, three anticoagulants and 5 cell preparations. *Cytometry* **13**, 66–74.

Debili, N., Wendling, F., Cosman, D., Titeux, M., Florinso, C., Dusanter-Fourt, I., Schooley, K., Methia, N., Charon, M., Nador, R., Bettaieb, A., and Vainchenker, W. (1995). The Mpl receptor is expressed in the megakaryocytic lineage from late progenitors to platelets. *Blood* **2**, 329–401.

Falini, B., Pileri, S., Pizzolo, G., Durkop, H., Flenghi, L., Stirpe, F., Martelli, M. F., and Stein, H. (1995). CD30 (Ki-1) molecule: A new cytokine receptor of the tumor necrosis factor receptor superfamily as a tool for diagnosis and immunotherapy. *Blood* **85**, 1–14.

Landay, A. L. (1992). Clinical applications of flow cytometry: Quality assurance and immunophenotyping of peripheral blood lymphocytes. *NCCLS* **12**, 1–76.

Mandy, F. F., Bergeron, M., Recktenwald, D., and Izaguime, C. A. (1992). A simultaneous three color T-cell subsets analysis with single laser flow cytometers using T cell gating protocol. *J. Immunol. Methods* **156**, 151–162.

Olsson, I., Gullberg, U., Lantz, M., and Richter, J. (1992). The receptors for regulatory molecules of hematopoiesis. *Eur. J. Haematol.* **48**, 1–9.

Picker, L. J., Singh, M. K., Zdraveski, Z., Treer, J. R., Waldrop, S. L., Bergstresser, P. R., and Maino, V. C. (1995). Direct demonstration of cytokine synthesis heterogeneity among human memory/effector cells by flow cytometry. *Blood* **86**, 1408–1419.

Prussin, C. (1997). Cytokine flow cytometry: Understanding cytokine biology at the single-cell level. *J. Clin. Immunol.* **17**, 195–204.

Riedy, M. C., Muirhead, K. A., Jensen, C. P., and Stewart, C. C. (1991). The use of a photolabeling technique to identify nonviable cells in fixed homologous or heterologous cell populations. *Cytometry* **12**, 133–139.

Roitt, I., Brostoff, J., and Male, D. (1989). "Immunology," Chap. 5, pp. 5.1–5.11. Mosby, St. Louis, Missouri.

Romagnani, S. (1997). The Th1/Th2 paradigm. *Immunol. Today* **18**, 263–266.

Stelzer, G. T., Shults, K. E., and Loken, M. R. (1993). CD45 gating for routine flow cytometric analysis of human bone marrow specimens. *In* "Clinical Flow Cytometry" (A. Landay, K. Ault, K. Bauer, and P. Rabinovitch, eds.), pp. 265–280.

Stewart, C. C. (1994). Multiparameter flow cytometry. *In* "Immunochemistry" (C. J. van Oss, ed.), Chap. 32, pp. 849–866. Dekker, New York.

Stewart, C. C. (1996). Clinical applications of multiparameter flow cytometry. *In* "Haematology 1996, Education Programme of the 26th Congress of the International Society of Haematology" (J. R. McArthur, S. H. Lee, J. E. Wong, and Y. W. Ong, eds.). The Int'l Society of Haematology, Singapore.

Stewart, C. C., and Mayers, G. L. (2000). Kinetics of antibody binding to cells. *In* "Immunophenotyping" (C. C. Stewart and J. K. A. Nicholson, eds.), Wiley, New York.

Stewart, C. C., and Stewart, S. J. (1993). Immunological monitoring utilizing novel probes, clinical flow cytometry. *Ann. N.Y. Acad. Sci.* **677**, 94–112.

Stewart, C. C., and Stewart, S. J. (1994a). Cell preparation for the identification of leukocytes. *In* "Methods in Cell Biology" (Z. Darzynkiewicz, J. Robinson, and H. Crissman, eds.), Vol. 41, pp. 39–60. Academic Press, New York.

Stewart, C. C., and Stewart, S. J. (1994b). Multiparameter analysis of leukocytes by flow cytometry. *In* "Methods in Cell Biology" (Z. Darzynkiewicz, J. Robinson, and H. Crissman, eds.), Vol. 41, pp. 61–79. Academic Press, New York.

Stewart, C. C., and Stewart, S. J. (1995). The use of directly and indirectly labeled monoclonal antibodies in flow cytometry. *In* "Methods in Molecular Biology" (W. C. Davis, ed.), Vol. 45., pp. 129–147. Humana Press, Totowa, New Jersey.

Stewart, C. C., and Stewart, S. J. (1997a). Titering antibodies. *In* "Current Protocols in Cytometry" (J. P. Robinson, Z. Darzynkiewicz, P. Dean, L. Dressler, P. Rabinovitch, C. Stewart, H. Tanke, and L. Wheeless, eds.), pp. 4.1.1–4.1.13. Wiley, New York.

Stewart, C. C., and Stewart, S. J. (1997b). Immunophenotyping. *In* "Current Protocols in Cytometry" (J. P. Robinson, Z. Darzynkiewicz, P. Dean, L. Dressler, P. Rabinovitch, C. Stewart, H. Tanke, and L. Wheeless, eds.), pp. 6.2.1–6.2.15. Wiley, New York.

Stewart, C. C., and Stewart, S. J. (1999). Four-color compensation. *Cytometry* **38**, 161–175.

Stomper, P. C., Nava, M. E. R., Budnick, R. M., and Stewart, C. C. (1997). Specimen mammography-guided fine-needle aspirates of clinically occult benign and malignant lesions: Analysis of cell number and type. *Invest. Radiol.* **32**, 277–281.

Tamul, K. R., O'Gorman, M. R. G., Donovan, M., Schmitz, J. L., and Folds, J. D. (1994). Comparison of a lysed whole blood method to purified cells preparations for lymphocyte immunophenotyping: Differences between healthy controls and HIV positive specimens. *J. Immunol. Methods* **167**, 237–243.

Testi, R., Philips, J. H., and Lanier, L. L. (1989). Leu-23 induction as an early marker for functional CD3/T cell antigen receptor triggering: Requirement of receptor cross-linking, prolonged elevation of intracellular *Ca++) and stimulation of protein kinase C. *J. Immunol.* **142**, 1854.

Van Vugt, M. J., van den Herik-Oudijk, I. E., and van de Winkel, J. G. J. (1996). Binding of PE-CY5 conjugates to the human high-affinity receptor for IgG (CD64). *Blood* **88**, 2358–2360.

Wognum, A. W., van Gils, F. C. J. M., and Wagemaker, G. (1993). Flow cytometric detection of receptors for interleukin-6 on bone marrow and peripheral blood cells of humans and rhesus monkeys. *Blood* **81**, 2036–2043.

CHAPTER 12

Strategies for Cell Permeabilization and Fixation in Detecting Surface and Intracellular Antigens

Steven K. Koester and Wade E. Bolton

Advanced Technology
Beckman Coulter, Inc.
Miami, Florida 33196

I. Introduction
II. Application
    A. Cell Fixation and Permeabilization
    B. Antibody and Fluorochrome Selection
    C. Antigen Properties
    D. Control Cells
    E. Specimen Handling and Processing
III. Materials and Methods
    A. Simultaneous Staining of Mitochondrial APO2.7 and Cell Surface Antigens
    B. Simultaneous Staining of T Cell Receptor ζ and Cell Surface Antigens
    C. Simultaneous Staining of PCNA, Cytokeratin, DNA, Tubulin, and Surface Antigen MC5
IV. Concluding Remarks
    References

I. Introduction

    The purpose of this chapter is to provide a logical approach for the permeabilization and fixation of cells such that surface and intracellular antigens can be preserved and quantified. Beyond the stated intent, the authors also feel compelled, whenever possible, to recommend procedures and techniques that promote and support flow cytometric observations made on unique or new antigens and techniques in a multiparametric approach to cellular investigations. Many investigations necessitate targeting a population of interest in a heterogeneous cell mixture prior to identifying a specific antigen, hence the requirement for multicolor fluorescent techniques in which surface phenotypic integrity must also be preserved. It is understood that no single fixative or permeabilization reagent will be appropriate for all antigens, and each will have its own set of advantages and limitations. However, there are ubiquitous, logical concepts from which a common framework can emerge. The level of difficulty for staining intracellular antigens is much greater than for staining of surface antigens. For this reason the authors will spend more effort in dialogue involving accessing and staining of the intracellular component. For the purpose of this chapter, it will be assumed that the cells of interest have been isolated and are ready for permeabilization and fixation.

To properly select fixation and permeabilization reagent(s), the antigen(s) of choice and their intercellular location(s) must be clearly in focus. The reagents of choice should stabilize the cell and the cytoplasmic or nuclear antigen expression over time and allow access to the antigenic site using antibody conjugates at a reasonable concentration. The logical approach we have selected will usually involve an initial fixation step, followed by cell permeabilization, followed by antibody and fluorochrome selection and antigen staining. Commercially available fixation/permeabilization reagents will not be addressed in this chapter. These reagents include a product insert indicating the intended protocol to arrive at optimum results. Not all antigen staining is compatible with these commercial products, and each one should be scrutinized individually.

II. Application

A. Cell Fixation and Permeabilization

Cell fixation will be accomplished using a cross-linking agent. This process will anchor and stabilize most antigens such that when additional permeabilization agents are added, antigen loss will be reduced to a minimum. There are several agents that fit into the category of a cross-linking agent, such as paraformaldehyde, formaldehyde, and glutaraldehyde (known collectively as noncoagulant fixatives), and to a lesser degree ethanol, methanol, and acetone (known as coagulant fixatives) (reviewed in Shapiro, 1995). From the standpoint of maintain-

Fig. 1 The effect of paraformaldehyde concentration on staining intensity in Molt-4 cells. Cells were dually stained with monoclonal anti-PCNA-fluorescein isothiocyanate (FITC) and propidium iodide following processing in 1 ml of 20 μg/ml lysophosphatidyl choline in 0.25% paraformaldehyde (a), 1.0% paraformaldehyde (b), 2.5% paraformaldehyde (c), and 4.0% paraformaldehyde (d) in phosphate-buffered saline (PBS), followed by processing in 1.0 ml of cold ($-20°$C) absolute methanol, and a wash in 1 ml of 0.1% Nonidet P-40 (NP-40) in PBS at 4°C.

12. Detection of Surface and Intracellular Antigens

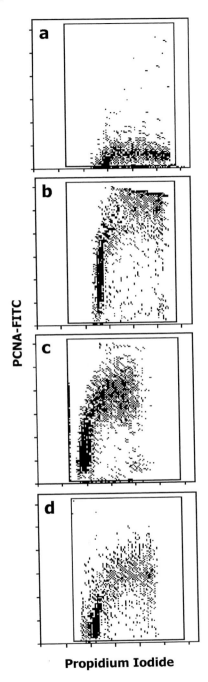

ing surface, cytoplasmic, and nuclear antigen expression, stability of reagent, and maintenance of cellular integrity, paraformaldehyde is generally the agent of choice. Paraformaldehyde has been utilized extensively in the targeting of intracellular antigens and has been found to reliably stabilize intracellular antigen expression (Bolton et al., 1992,1994a,b; Halldén et al., 1989; Healy et al., 1994; Kurki et al., 1988; Li and Darzynkiewicz, 1995; Mikulka and Bolton, 1994; O'Brien and Bolton, 1995; O'Brien et al., 1995; Pollice et al., 1975; Schmid et al., 1991; Van Bockstaele et al., 1991).

After choosing the appropriate cross-linking agent for a particular application, details of its use, such as concentration, time and temperature, need to be addressed. Paraformaldehyde has been documented for use at 0.25–4% at 4°–25°C for 2–15 min (reviewed in Clevenger and Shankey, 1993). When the cross-linking agent is too concentrated and/or the time of exposure is too long, proteins may become densely cross-linked, making accessibility for staining difficult, and extensive cell clumping may occur, making it difficult to obtain a single cell suspension. Detergents are often added with (Mikulka and Bolton, 1994; Schimenti and Jacobberger, 1992) or following (Clevenger et al., 1985; Halldén et al., 1989; Schmid et al., 1991) paraformaldehyde treatment. An example showing the effect of increasing paraformaldehyde concentrations in combination with the permeabilization agent lysolecithin for the staining of proliferating cell nuclear antigen (PCNA) can be seen in Fig. 1.

Alcoholic fixation results in the denaturation of proteins and serves as its own permeabilization agent, primarily through the extraction of phospholipids (Hopwood, 1985). Ethanol and methanol are the alcohols most often utilized for the fixation of cells. However, if used prior to a true cross-linking step, many surface and internal antigens can be irreversibly denatured. When alcohols are used after a cross-linking agent, antigen expression can be significantly altered, but reducing the temperature of the alcohol to at least −20°C can reduce cell aggregation (Jacobberger et al., 1986).

Detergents most often used for permeabilization in flow cytometry are Triton X-100, Tween 20, NP-40, lysolecithin, saponin, and digitonin. Although these detergents permeabilize effectively, and in the case of saponin (Jacob et al., 1991) and digitonin (Fig. 2) (Healy et al., 1998) (Fig. 3) have been found to be

---

Fig. 2 The response of human acute T lymphoblastic leukemia (CEM) cells to induction of apoptosis by hypoxia or etoposide treatments. Apoptosis was induced in the cells by either hypoxia by pelleting the cells in tightly capped tubes or treating continuously in 0.5 μM etoposide followed by a 5-hr incubation. Cells were stained with CD4-FITC and APO2.7-phycoerythrin (PE) following treatment or no treatment in 100 μg/ml digitonin in PBS. Histograms representing light scatter and dually stained distributions from cells without digitonin treatment are presented for noninduced control cells (a, b), 5-hr hypoxic cells (e, f), or 5-hr etoposide treated cells (i, j). Histograms representing light scatter and dually stained distributions from cells with digitonin treatment are presented for noninduced control cells (c, d), 5-hr hypoxic cells (g, h), or 5-hr etoposide treated cells (k, l). (See color plates.)

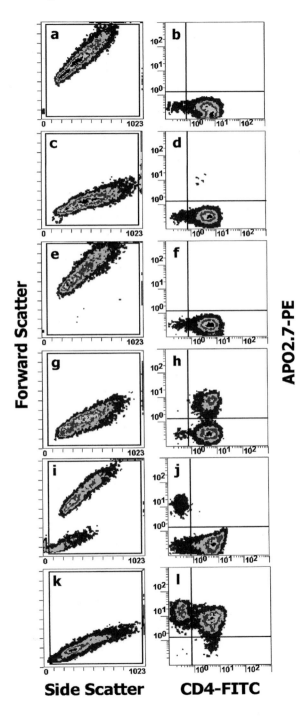

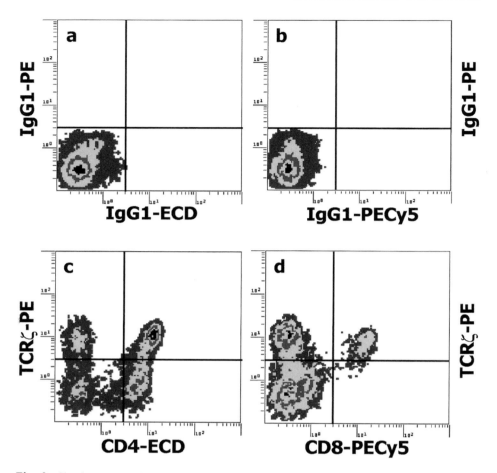

Fig. 3 Simultaneous surface and internal staining of human peripheral blood lymphocytes (PBL) with anti-T cell receptor ζ (TCRζ), anti-CD4, and anti-CD8 monoclonal antibodies. Whole blood was collected from a normal healthy donor in EDTA anticoagulant, and PBL were harvested using Ficoll-Paque. Cells were stained with CD4-ECD and CD8-PECy5, followed by treatment and staining in a solution containing 500 μg/ml digitonin in PBS and TCRζ-PE. Histograms represent IgG1-PE/IgG1-ECD isotype controls (a), IgG1-PE/IgG1-PECy5 isotype controls (b), TCRζ-PE and CD4-ECD (c), and TCRz-PE and CD8-PECy5 (d). (See color plates.)

effective in the quantitation of surface and cytoplasmic antigens, it is often at the expense of light scatter properties (Pollice *et al.*, 1975) (Fig. 2). Each lot of digitonin must be titered to obtain the desired effect for surface and intracellular antigen staining. As a rule, for effective quantitation of intracellular antigens primarily through the reduction of aggregates, detergents should only be used with or after a cross-linking agent. A summary of preferred methods used to obtain intracellular antigen measurements by flow cytometry can be observed in Table I.

B. Antibody and Fluorochrome Selection

1. Antibody Selection

Antibody specificity for the identification of an intracellular antigen will determine the outcome of an assay. The outcome of an assay will only be as good as the antibody selected to identify the intracellular antigen. Data attesting to the antibody specificity should be obtained from the antibody supplier, if not, the burden of proof rests with the investigator. Many hours of valuable research efforts have been expended in vain due to inaccurate antibody specificity. In addition to antibody specificity, if the antigen is located within the nucleus, or even further compartmentalized, such as in the nucleolus, it is advisable to use the immunoglobulin G (IgG) class of monoclonal antibodies as opposed to immunoglobulin M (IgM). The size of the pentameric IgM molecule can be restrictive depending on the antigen location and the presence of conformational epitopes.

Conventional wisdom as well as scientific dictum mandate that the negative control antibody have as many properties of the target antibody as possible. Thus, an isotype-matched negative control antibody should be used when available to determine the degree of nonspecific background staining. The isotype control should be fluorochrome-matched to the target antibody, and the fluorescein to protein (F/P) ratios of the target antibody and the isotype control antibody should be similar.

2. Fluorochrome Selection

Three things should be considered when selecting a fluorochrome for the identification of an intracellular antigen: (1) antigen location, (2) antigen density, and (3) fluorescein to protein (F/P) ratios. In general, if the antigen is located within the nucleus, the best signal is derived from those fluorochromes of a smaller molecular weight, such as fluorescein isothiocyanate (FITC). Work completed in our laboratory shows PCNA directly conjugated to phycoerytherin (PE) yields very poor signal to noise (S/N) ratios, whereas PCNA–FITC displays strong fluorescence intensities (data not shown). However, if the antigen density is low, and a greater quantum yield is required to measure a detectable signal, a larger molecular weight fluorochrome may be desirable. In these cases, empirical data may need to be collected as fluorochrome size versus quantum yield vie for contention. Fluorochrome size is not as much an issue when detecting antigens located in the cytoplasm. Larger dyes such as PE and their tandem conjugates usually work well unless the target antigen is sequestered or conformationally expressed, then some of the same problems discussed for nuclear antigens may apply.

Table I
Preferred Method to Obtain Intracellular Measurements Using Flow Cytometry

Target for staining	EtOH	MeOH	Hypotonic/NP-40	PF/TX-100	PF/EtOH	Lyso-PF/MeOH/NP-40	Digitonin	Saponin	References
DNA	X	X	X	X	X	X		X	Clevenger et al. (1985, 1987); Jacob et al. (1991); Koester et al. (1994, 1997, 1998); Krishan (1975); Kurki et al. (1988); O'Brien et al. (1995); O'Brien and Bolton (1995); Rabinovitch (1994)
Apoptotic DNA-SB[a]					X				Koester et al. (1997, 1998); Li and Darzynkiewicz (1995)
APO2.7							X		Büssing et al. (1999); Koester et al. (1997, 1998); Métivier et al. (1998); Seth et al. (1997); Zhang et al. (1996)
TCRζ							X		Healy et al. (1998)
Tubulin	X	X	X		X	X	X		Koester et al. (1997, 1998); O'Brien et al. (1995); O'Brien and Bolton (1995)
Cytokeratin						X			Bolton et al. (1992); O'Brien et al. (1995); O'Brien and Bolton (1995)
PCNA						X			Bolton et al. (1992, 1994a,b); Healy et al. (1998); Kurki et al. (1988); Mikulka and Bolton (1994); O'Brien et al. (1995)
Ki-67						X		X	Jacob et al. (1991); Healy et al. (1994); Van Bockstaele et al. (1991)
p105				X		X			Clevenger et al. (1985, 1987); Mikulka and Bolton (1994)
p120						X			Bolton et al. (1992, 1994a,b); Mikulka and Bolton (1994)
p145						X			Bolton et al. (1992, 1994); Healy et al. (1994)

[a] Abbreviations: EtOH, ethanol; MeOH, methanol; NP-40, Nonidet P-40 nonionic detergent; PF, paraformaldehyde; TX-100 Triton X-100 nonionic detergent; Lyso, lysophosphatidyl choline (lysolecithin); DNA-SB, apoptotic DNA strand breaks.

The F/P ratio is a critical feature in cell analysis that is often overlooked. The intensity of the fluorescent signal, and therefore, the expression of percent positive or mean channel fluorescent intensity (MFI), is directly related to the F/P ratio. Also, and possibly more importantly, the F/P ratio of the isotype control must be similar to that of the targeting antibody, or the S/N ratio can be erroneously elevated thus yielding false positivity. Again, the burden of responsibility rests with the investigator to know these ratios, particularly when target antibodies and isotype controls from different vendors are mixed within an experiment. The antibody F/P ratios should be available from the antibody vendors. For those individuals who directly conjugate fluorochromes to antibodies in their laboratory, it is recommended that an experiment be designed to determine the optimal F/P ratio and dosage.

C. Antigen Properties

Several properties of the target antigen may have an effect on accurate quantitation: (1) location of the antigen within the cell, (2) migration of the antigen, (3) soluble phase of the antigen, and (4) antigen configuration. The antigen location is an important consideration, particularly for intracellular antigens that are often associated with morphologically distinct organelles or structures. Fluorescent microscopy should be used to insure that the positive signal is associated with an expected location. For instance, if the fluorescent signal for PCNA or Ki67, both proliferation associated antigens located in the nucleus, appears to stain cytoplasmic components, then there is probable cause for concerns of specificity. Although this example may seem obvious to most, there are too many fluorescent microscopes within flow cytometry laboratories that collect dust from apparent nonuse.

Some antigens migrate from one location to another within the cell depending on the particular functional assignment of the antigen. For example, p70 (zap 70), a protein associated with signal transduction, migrates from the internal periphery of the plasma membrane following activation to a position within the nucleus. Following this example, the investigator could find the antigen located within different compartments of the cell due to differences in the cell activation state. This could necessitate the use of several methods for fixation and permeabilization depending where the antigen is located. Methods adequate for accessing and staining the antigen in one cellular location may destroy antigen expression in another. Just as antigens may migrate from one location to another, some may become soluble during the functional or migration phase. When the antigen enters a soluble phase, the epitope against which the antibody is directed may now have altered binding characteristics resulting in quantitatively altered expression. Since the fixation and permeabilization process renders the plasma membrane permeant to soluble factors, the degree of antigen anchoring or solubility may influence the choice of fixative.

Some antigens must alter their configuration or three-dimensional structure as part of their functional purpose. As the conformation of the antigen changes, the epitope of interest may be blocked or masked due to proximal charge changes or physical hindrance. If the epitope of the target antigen against which the antibody is directed is a conformational epitope, antibody binding, and therefore, quantitative expression, may be suppressed.

D. Control Cells

Positive and negative control cells should be considered one of the most important components of any immunological assay. Control cells are processed in the same manner as the sample and, therefore, are subject to the same conditions throughout processing. In addition, particularly for intracellular antigens, any artifacts that may be generated within the cell in the sample, such as alterations in overall charge or antigen masking, most likely will be generated in the control cell. The positive cell should be one in which the target antigen is known to be expressed in the appropriate location and can often be a cell line in which the gene encoding the antigen has been transfected for overexpression. The negative cell can be a cell line that does not contain the target antigen, or expression of the target antigen below detection level of the assay.

E. Specimen Handling and Processing

The use of proteolytic enzymes for tissue dissociation or methods used for the removal of tissue culture cells from the surface of flasks can destroy target antigens. Surface antigen proteins are therefore more susceptible to the proteolytic activity during tissue dissociation, if only due to their location or proximity to the enzyme. It is therefore important to determine whether the action of proteolytic enzymes affects the antigen of interest before proceeding with development of a staining protocol.

Time and temperature of sample incubation before and during processing can adversely affect the target antigen and the overall condition of the cell population being processed. Antigens can be internalized from the surface of cells or expressed at the surface by others following cell activation. Room temperature storage can cause antigen degradation and cell death. Time course experiments must be processed within tight time restraints and possibly at cool temperatures to halt the action of pathway enzymes. All of these adverse conditions can result in alterations of flow cytometric measurements and must be controlled.

III. Materials and Methods

Protocols for staining of surface antigens in combination with intracellular antigens are complex in nature. A general protocol would not be practical for use with all examples. For this reason, each application will be handled as an individual example.

A. Simultaneous Staining of Mitochondrial APO2.7 and Cell Surface Antigens

Intracellular staining: Koester *et al.*, 1997, 1998; Example of simultaneous staining in Fig. 2.

1. Prepare a stock solution containing 25 mg/ml digitonin (Sigma Chemical, St. Louis, MO; or WAKO BioProducts, Richmond, VA) by heating to 100°C in phosphate-buffered saline (PBS) (Beckman Coulter, Miami, FL) until completely dissolved. Remove from the heat immediately after digitonin goes into solution, and store the stock solution at 4°C for up to 1 month.
2. Prepare cell pellets containing 0.5×10^6 to 1.0×10^6 cells in 12×75 mm tubes.
3. Add recommended volume(s) of fluorescently labeled surface marker(s) of choice and incubate at room temperature for 15 min protected from light.
4. Resuspend cells in 2 ml of PBS with 2.5% fetal bovine serum (PBSF) (HyClone Laboratories, Logan, UT), centrifuge at 200 g for 5 min, and discard supernatant.
5. Add 100 μl of a 100–500 μg/ml digitonin solution diluted from the stock solution in PBS to the cell pellet, and incubate for 20 min on ice. The working concentration of digitonin should be determined by titration for each cell line or experimental condition to optimize the staining results.
6. Resuspend cells in 2 ml of PBSF, centrifuge at 200 g for 5 min, and discard supernatant.
7. Add recommended volume of APO2.7-PE or PECy5 (Immunotech, a Beckman Coulter Company, Marseille, France) and incubate at room temperature for 15 min protected from light.
8. Resuspend cells in 2 ml of PBSF, centrifuge at 200 g for 5 min, and discard supernatant.
9. Resuspend cells in 1 ml of PBSF and store on ice until analyzed on the flow cytometer. Cells may be resuspended in 1 ml of 0.5% electron microscopy (EM) grade paraformaldehyde (Electron Microscopy Sciences, Fort Washington, PA) in PBS and stored at 4°C overnight before analyzing on the flow cytometer.

B. Simultaneous Staining of T Cell Receptor ζ and Cell Surface Antigens

The T cell receptor ζ (TCRζ) chain resides on the intracellular portion of the cell membrane and appears to correlate with adequate effector cell function. Published data indicate that patients with advanced malignancy show an absence or reduced TCRζ expression in peripheral blood lymphocytes (Healy *et al.*, 1998). An example is seen in Fig. 3.

1. Prepare a stock solution containing 25 mg/ml digitonin by heating to 100°C in PBS until completely dissolved. Remove from the heat immediately after digitonin goes into solution, and store the stock solution at 4°C for up to 1 month.
2. Collect human whole blood (WB) samples in EDTA tubes (Becton Dickinson Vacutainer Systems, Franklin Lakes, NJ), store at room temperature, and process within 24 hr.
3. Harvest peripheral blood leukocytes (PBLs) by mixing 5 ml of WB with 5 ml of a complete medium and gently place over 3 ml of Ficoll-Paque (Pharmacia Biotech, Uppsala, Sweden) all at room temperature in a 15-ml conical centrifuge tube. Centrifuge for 30 min at 500 g with the centrifuge brake off.
4. Remove PBL, wash twice in 10 ml of complete medium, and resuspend to 5×10^6 cells/ml in PBS.
5. Add 200 μl of PBL (1×10^6) suspension to recommended volumes of surface marker monoclonal antibodies of choice, and incubate at room temperature for 10 min protected from light.
6. Resuspend cells in 2 ml of PBSF, centrifuge at 300 g for 5 min, and discard supernatant.
7. Resuspend cells in 200 μl of a solution consisting of a 500 μg/ml working solution of digitonin, prepared in PBS from the 25 mg/ml stock solution, plus anti-TCRζ-RD1 (Immunotech). Incubate for 10 min at room temperature protected from light.
8. Resuspend cells in 3 ml of PBSF, centrifuge at 300 g for 5 min, and discard supernatant.
9. Resuspend cells in 0.5–1 ml of PBSF and store on ice until analyzed on the flow cytometer. Cells can be resuspended in 1 ml of 0.5% EM grade paraformaldehyde in PBS and stored at 4°C overnight before analyzing on the flow cytometer.

C. Simultaneous Staining of PCNA, Cytokeratin, DNA, Tubulin, and Surface Antigen MC5

This procedure involves simultaneous surface staining, cytoplasmic staining, and nuclear staining using the following described antibodies and stains (Kurki et al., 1988; O'Brien et al., 1995; O'Brien and Bolton, 1995):

PCNA: Proliferating cell nuclear antigen
Cytokeratin: A cytoplasmic epithelial tissue-specific antigen
DAPI: DNA-specific dye
Tubulin: A ubiquitous intracellular cytoskeletal antigen
MC5: A surface membrane breast tumor-associated antigen

12. Detection of Surface and Intracellular Antigens

Cell staining is accomplished in four logical steps: (1) viability staining using antitubulin antibody prior to permeabilization; (2) fixation and permeabilization; (3) staining of PCNA, cytokeratin, and MC5; and (4) DNA staining.

1. Viability Staining

a. Prepare cell pellets containing 1.0×10^6 cells in 12×75 mm tubes.

b. Resuspend cells in 200 μl of a 1:5 dilution of antitubulin antibody (Zymed Laboratories, San Francisco, CA) and incubate at room temperature for 15 min.

c. Resuspend cells in 2 ml of PBSF, centrifuge at 200 g for 5 min, and discard supernatant.

d. Resuspend cells in 200 μl of polyclonal goat anti-mouse antibody conjugated to PECy5 (GAM-PECy5) (Immunotech) in PBSF at a concentration of 200 μg/ml, and incubate for 15 min at room temperature protected from light.

e. Resuspend cells in 2 ml of PBSF, centrifuge at 200 g for 5 min, and discard supernatant.

f. Resuspend cells in 200 μl of MsIgG1 (Immunotech) in PBSF at a concentration of 1 mg/ml, and incubate for 15 min at room temperture protected from light. Unconjugated MsIgG1 is used to block any remaining free GAM sites.

g. Resuspend cells in 2 ml of PBSF, centrifuge at 200 g for 5 min, and discard supernatant. This step should be repeated to wash out any free antibody.

2. Fixation and Permeabilization

a. Gently resuspend cells in 1 ml of 20 μg/ml lysophosphatidyl choline (Sigma Chemical) in 1% paraformaldehyde, incubate for 2 min at room temperature.

b. Resuspend cells in 2 ml of PBSF, centrifuge at 200 g for 5 min, and discard supernatant.

c. Resuspend cells in 1 ml of cold ($-20°C$) absolute MeOH (HyClone) and incubate for 10 min on ice.

d. Centrifuge at 200 g for 5 min, and discard supernatant.

e. Resuspend cells in 1 ml of cold (4°C) 0.1% NP-40 (Sigma Chemical) and incubate for 5 min on ice.

f. Centrifuge at 200 g for 5 min, and discard supernatant.

3. Staining of PCNA, Cytokeratin, and MC5

a. Resuspend cells in recommended volume of PCNA–FITC (50 μg/ml) (Immunotech) and incubate for 15 min at room temperature protected from light.

b. Resuspend cells in 2 ml of PBSF, centrifuge at 200 g for 5 min, and discard supernatant.

c. Resuspend cells in recommended volume of cytokeratin-ECD (25 µg/ml) (Immunotech) and incubate for 15 min at room temperture protected from light.

d. Resuspend cells in 2 ml of PBSF, centrifuge at 200 g for 5 min, and discard supernatant.

e. Resuspend cells in recommended volume of MC5-RD1 (7.5 µg/ml) (Immunotech) and incubate for 15 min at room temperature protected from light.

f. Resuspend cells in 2 ml of PBSF, centrifuge at 200 g for 5 min, and discard supernatant.

4. DNA Staining

a. Resuspend cells in 1 ml of 1.5 µg/ml DAPI (Sigma Chemical) in PBSF and incubate for 20 min protected from light.

b. Analyze on the flow cytometer.

IV. Concluding Remarks

Intracellular staining for flow cytometry is more of a challenge than surface staining. There are many more factors that influence the staining outcome. When developing a protocol that includes simultaneous surface and intracellular staining, several techniques should be selected as most likely approaches. Start with less complex techniques in your attempt to find fixatives and permeabilizers that will be compatible with both the target antigen and the labeling antibody–fluorochrome complex. Select fluorochrome combinations that can be excited at laser lines compatible with your flow cytometer and can be optically separated or compensated to collect valid data. And do not overlook the need for adequate controls. Use adequate combinations of isotype, positive, and negative staining controls; then set proper detector voltages, gains, and color compensations prior to analyzing test samples.

Acknowledgments

We are most grateful to Gayle Rosenthal and Dee Linder for their cell culture expertise, Cynthia Healy for kindly providing TCRζ histograms, and Julie Wilkinson for her expertise in the use of ExpO analysis applications for this work.

References

Bolton, W. E., Mikulka, W. R., Healy, C. G., Schmittling, R. J., and Kenyon, N. S. (1992). Expression of proliferation associated antigens in the cell cycle of synchronized mammalian cells. *Cytometry* **13,** 117–126.

Bolton, W. E., Freeman, J. W., Mikulka, W. R., Healy, C. G., Schmittling, R. J., and Kenyon, N. S. (1994a). Expression of proliferation-associated antigens (PCNA, p120, p145) during the reentry of G0 cells into the cell cycle. *Cytometry* **17,** 66–74.

Bolton, W. E., Zeng, X.-R., Lee, M. Y. W. T., and Mikulka, W. R. (1994b). Expression of PCNA, and DNA polymerase delta in the cell cycle of synchronized mammalian cells. *CMB* **1**, 193–197.

Büssing, A., Wagner, M., Wagner, B., Stein, G. M., Schietzel, M., Schaller, G., and Pfüller, U. (1999). Induction of mitochondrial Apo2.7 molecules and generation of reactive oxygen-intermediates in cultured lymphocytes by the toxic proteins from *Viscum album L. Cancer Lett.* **139**, 79–88.

Clevenger, C. V., and Shankey, T. V. (1993). Cytochemistry II: Immunofluorescence measurement of intracellular antigens. *In* "Clinical Flow Cytometry–Principles and Application" (K. D. Bauer, R. E. Duque, and T. V. Shankey, eds.), pp. 157–175. Williams & Wilkins, Baltimore.

Clevenger, C. V., Bauer, K. D., and Epstein, A. L. (1985). A method for simultaneous nuclear immunofluorescence and DNA content quantitation using monoclonal antibodies and flow cytometry. *Cytometry* **6**, 208–214.

Clevenger, C. V., Epstein, A. L., and Bauer, K. D. (1987). Modulation of the nuclear antigen p105 as a function of cell-cycle progression. *J. Cell. Physiol.* **130**, 336–343.

Halldén, G., Andersson, U., Hed, J., and Johansson, S. G. O. (1989). A new membrane permeabilization method for the detection of intracellular antigens by flow cytometry. *J. Immunol. Methods* **124**, 103–109.

Healy, C. G., Kenyon, N. S., and Bolton, W. E. (1994). Expression of proliferation associated antigens PCNA, p145, and Ki67 during cell cycle progression in activated peripheral blood lymphocytes: Potential utility in monitoring hematologic malignancies. *CMB* **1**, 59–70.

Healy, C. G., Simons, J. W., Carducci, M. A., DeWeese, T. L., Bartkowski, M., Tong, K. P., and Bolton, W. E. (1998). Impaired expression and function of signal-transducing ζ chains in peripheral T cells and natural killer cells in patients with prostate cancer. *Cytometry* **32**, 109–119.

Hopwood, D. (1985). Cell and tissue fixation, 1972–1982. *Histochem. J.* **17**, 389–442.

Jacob, M. C., Favre, M., and Bensa, J.-C. (1991). Membrane cell permeabilisation with saponin and multiparametric analysis by flow cytometry. *Cytometry* **12**, 550–558.

Jacobberger, J. W., Fogleman, D., and Lehman, J. M. (1986). Analysis of intracellular antigens by flow cytometry. *Cytometry* **7**, 356–364.

Koester, S. K., Maenpaa, J. U., Wiebe, V. J., Baker, W. J., Wurz, G. T., Seymour, R. C., Koehler, R. E., and DeGregorio, M. W. (1994). Flow cytometry: Potential utility in monitoring drug effects in breast cancer. *Breast Cancer Res. Treat.* **32**, 57–65.

Koester, S. K., Roth, P., Mikulka, W. R., Schlossman, S. F., Zhang, C., and Bolton, W. E. (1997). Monitoring early cellular responses in apoptosis is aided by the mitochondrial membrane protein-specific monoclonal antibody APO2.7. *Cytometry* **29**, 306–312.

Koester, S. K., Schlossman, S. F., Zhang, C., Decker, S. J., and Bolton, W. E. (1998). APO2.7 defines a shared apoptotic–necrotic pathway in a breast tumor hypoxia model. *Cytometry* **33**, 324–332.

Krishan, A. (1975). Rapid flow cytofluorometric analysis of mammalian cell cycle by propidium iodide staining. *J. Cell. Biol.* **66**, 188–195.

Kurki, P., Ogata, K., and Tan, E. M. (1988). Monoclonal antibodies to proliferation cell nuclear antigen (PCNA)/cyclin as probes for proliferating cells by immunofluorescence microscopy and flow cytometry. *J. Immunol. Methods* **109**, 49–59.

Li, X., and Darzynkiewicz, Z. (1995). Labeling DNA strand breaks with BrdUTP. Detection of apoptosis and cell proliferation. *Cell Prolif.* **29**, 571–579.

Métivier, D., Dallaporta, B., Zamzami, N., Larochette, N., Susin, S. A., Marzo, I., and Kroemer, G. (1998). Cytofluorometric detection of mitochondrial alterations in early CD95/Fas/APO-1-triggered apoptosis of Jurkat T lymphoma cells. Comparison of seven mitochondrion-specific fluorochromes. *Immunol. Lett.* **61**, 157–163.

Mikulka, W. R., and Bolton, W. E. (1994). Methodologies for the preservation of proliferation associated antigens PCNA, p120, and p105 in tumor cell lines for use in flow cytometry. *Cytometry* **17**, 246–257.

O'Brien, M. C., and Bolton, W. E. (1995). Comparison of cell viability probes compatible with fixation and permeabilization for combined surface and intracellular staining in flow cytometry. *Cytometry* **19**, 243–255.

O'Brien, M. C., Gupta, R. K., Lee, S. Y., and Bolton, W. E. (1995). Use of a multiparametric panel to target subpopulations in a heterogeneous solid tumor model for improved analytical accuracy. *Cytometry* **21,** 76–83.

O'Brien, M. C., Healy, S. F., Raney, S. R., Hurst, J. M., Avner, B., Hanly, A., Mies, C., Freeman, J. W., Snow, C., Koester, S. K., and Bolton, W. E. (1997). Discrimination of late apoptotic/necrotic cells (Type III) by flow cytometry in solid tumors. *Cytometry* **28,** 81–89.

Pollice, A. A., McCoy, J. P., Jr., Shackney, S. E., Smith, C. A., Agarwal, J., Burholt, D. R., Janocko, L. E., Hornecek, F. J., Singh, S. G., and Hartsock, R. L. (1975). Sequential paraformaldehyde and methanol fixation for simultaneous flow cytometric analysis of DNA, cell surface proteins, and intracellular proteins. *Cytometry* **13,** 432–444.

Rabinovitch, P. S. (1994). DNA content histogram and cell-cycle analysis. *In* "Methods in Cell Biology," (Z. Darzynkiewicz and J. P. Robinson, eds.), 2nd Ed., Vol. 41, pp. 263–296. Academic Press, San Diego.

Schimenti, K. J., and Jacobberger, J. W. (1992). Fixation of mammalian cells for flow cytometric evaluation of DNA content and nuclear immunofluorescence. *Cytometry* **13,** 48–59.

Schmid, I., Uittenbogaart, C. H., and Giorgi, J. V. (1991). A gentle fixation and permeabilization method for combined cell surface and intracellular staining with improved precision in DNA quantification. *Cytometry* **12,** 279–285.

Seth, A., Zhang, C., Letvin, N. L., and Schlossman, S. F. (1997). Detection of apoptotic cells from peripheral blood of HIV-infected individuals using a novel monoclonal antibody. *AIDS* **11,** 1059–1061.

Shapiro, H. M. (1995). "Practical Flow Cytometry." Wiley-Liss, New York.

Van Bockstaele, D. R., Lan, J., Snoech, H.-W., Korthout, M. L., De Bock, R. F., and Peetermans, M. E. (1991). Aberrant Ki67 expression in normal bone marrow revealed by multiparameter flow cytometric analysis. *Cytometry* **12,** 50–63.

Zhang, C., Ao, Z., Seth, A., and Schlossman, S. F. (1996). A mitochondrial membrane protein defined by a novel mitochondrial antibody is preferentially detected in apoptotic cells. *J. Immunol.* **157,** 3980–3987.

PART III

Standardization, Quality Assurance

CHAPTER 13

Stoichiometry of Immunocytochemical Staining Reactions

James W. Jacobberger

Cancer Research Center and Department of Genetics
Case Western Reserve University
Cleveland, Ohio 44106

I. Introduction
II. Structure of Immunoglobulin G
III. Cell Structure
IV. Permeabilized Cell Structure
    A. Permeabilizer Properties
    B. Fixative Properties
    C. Structure of Fixed Cells
V. Antibody–Antigen Reactions
    A. Kinetics in Solution
    B. Kinetics in Permeabilized Cells
    C. Antibody Specificity
VI. Multiparametric Analyses
VII. Summary
    References

I. Introduction

Exploitation of antibody–antigen reactions to detect and quantify intracellular proteins presents three technical problems. The first is that cells have to be rendered permeable so that the high molecular weight antibodies can diffuse through the cell. The second is that the antibody has to react specifically with its cognate epitope despite a very large number of competing binding sites that are less specific. Third, once bound, the antibody must stay in place as unbound antibody is washed away and other staining and preparatory procedural steps are enacted.

For the first problem there are three approaches. The first is to dissolve the cell membrane with detergent. This was investigated extensively by Darzynkiewicz and coinvestigators in studies using acridine orange to measure RNA and DNA content of unfixed cells (Darzynkiewicz et al., 1980, 1981). The second approach is to stabilize cells, prior to detergent treatment, by fixation with formaldehyde (F), which can be controlled by concentration, time, and temperature (Jacobberger, 1991; Clevenger and Shankey, 1993; Bauer and Jacobberger, 1994; Camplejohn, 1994; Jacobberger, 2000). The third method is to both fix and permeabilize at the same time with dehydrating-denaturing organic reagents such as acids, alcohols, or ketones (Jacobberger, 1991; Bauer and Jacobberger, 1994; Jacobberger, 2000). A fourth, but not often practiced approach is to mechanically rupture the cells by methods such as freeze–thaw (Karn et al., 1989) or electroporation (Berglund and Starkey, 1991). The subjects of fixation and permeabilization have been extensively reviewed (Jacobberger, 1991; Clevenger and Shankey, 1993; Bauer and Jacobberger, 1994; Camplejohn, 1994; Jacobberger, 2000), and the intention here is not to recapitulate. Rather, enough background will be given to present this author's viewpoint that the manner of fixation/permeabilization matters and present the author's biases about which methods do what.

The second problem relates to the first insofar as the level of the effect of competing reactions may be affected by the chemical nature of the nonspecific molecules that can be affected by the fixation/permeabilization method, and even more so, the specificity of the reaction may be affected by the chemical modification of the target molecule. The specificity of the antibody for its target is defined by the structure of the antibody and the structure of the target. If the antibody structure requires a high degree of secondary–quaternary structure of the target, then the fixation/permeabilization method may affect specificity as well as nonspecificity.

The third problem, that of high equilibrium constant, should be largely a function of the antibody structure, determined largely by the genetic evolution of the original antibody producing B cell.

Based on this introduction, this chapter will attempt to describe and list the factors that affect the stoichiometry of antibody–antigen interactions within the context of immunofluorescence staining of proteins inside cells. Fixation/permeabilization, nonspecific reactions in isolation, specific reactions in isolation, and both reactions inside cells will be addressed.

The following discussion will be limited to mouse monoclonal immunoglobulin G (IgG) antibodies, and fixation and permeabilization will be limited to information relevant to antibody staining and will largely exclude any in depth review of empirical work or technical details that have been reviewed (Clevenger and Shankey, 1993; Bauer and Jacobberger, 1994; Jacobberger, 2000).

II. Structure of Immunoglobulin G

A model of the globular IgG structure is shown in Fig. 1. The molecule is made with four peptides, two heavy (H) chains of 55 kDa and two light (L) chains of 25 kDa, covalently linked by disulfide bonds. Antigen combining regions are located at the end of the Fab fragments. The Fc region binds specific cellular receptors, especially on granulocytes and macrophages (Stewart and Mayers, 2000). Immunoglobulins are glycosylated, and IgG has carbohydrate chains attached in the Fc region. Originally described by serology, mouse IgG molecules belong to subclasses depending on the H chain (γ1, γ2a, γ2b, γ3, or γ4) and L chains (κ or λ). The average molecular mass of IgG_1 is ~154 kDa. The crystal structure for a complete monoclonal mouse IgG2a antibody has been solved (Harris et al., 1992, 1997). A model of human IgG1 has been constructed from the solved fragment structures (http://www.path.cam.ac.uk/~mrc7/mike-images.html). The model depicted in Fig. 1 emphasizes the flexibility of the molecule and the independence of the two antibody combining regions. The size of an IgG molecule ~10 nm.

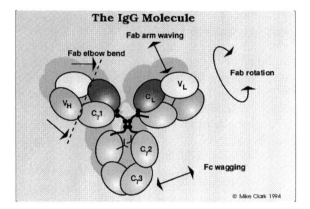

Fig. 1 Composite model of IgG structure showing two heavy chains (light gray) and two light chains (dark and lightest gray). The hinge region between heavy chain C_H1 and C_H2 globular domains has been omitted. The model emphasizes molecular flexibility. The epitope combining sites are created by sequence and length variation in six hypervariable loops (CDRs) within the V_H and V_L domains (Jelesarov et al., 1996). Antibodies digested by limited proteolysis yield fragments entitled Fab (V_H-C_H and $V_L C_L$), Fab'2 (hinged Fabs), and Fc (hinged peptides with C_H2 and C_H3 domains). The image was created by Mike Clark (Cambridge University) and was directly downloaded through the Internet from the following site: http://www.path.cam.ac.uk/~mrc7/igs/mikeimages.html#IgG. (Used with permission.)

III. Cell Structure

Eukaryotic cell models have slowly evolved from a bag segmented by membranes into sub-bags with protein and nucleic acid solutions to bags that are highly structured by intermolecular forces even in the "soluble" fraction which is thought to make up the macrotrabular network (Penman, 1995). A macroscopic visual model of the bacterial cell based on estimates of cell size, molecule size, and molecule concentration shows the cell to be a very crowded place (Fig. 2) (Goodsell, 1991). This situation has some theoretical consequences. A 160-kDa protein could travel at an average speed of 500 cm sec^{-1} at 300°K if it behaved like an ideal gas. Inside a cell, this protein diffuses slower by 1000 times. So, as a gas, the protein could move from the perimeter to the center of a 20-μm cell in 2 μsec. In solution, the time to diffuse a long distance increases as the square of the distance (Alberts *et al.*, 1989). So according to Goodsell, in the cytosol, the journey to go 10 nm would take 2 μsec, and the journey to the center of the cell would take 2 sec. The 1000-fold increase in time is due to an enormous number of interactions between the migrating protein and other molecules crowding the space. How many? Both bacterial and mammalian cells are 70% water (Alberts *et al.*, 1989). Both cell types are mostly protein (by dry weight) with mammalian cells containing 18% protein and bacteria containing 15%. An estimate of the number of proteins in log phase *Escherichia coli* growing in minimal glucose medium at 37°C is 2,350,000 (Neidhardt, 1987). Twenty-five percent of these are estimated to be in the cytoplasm (which has a volume of 0.6 μm^3), which is greater than a 4 mM solution of protein, if there were no other molecules except water and water were not bound to proteins. So, it is likely that the concentration is higher. If we apply these numbers to a 20-μm mammalian cell with a 10-μm nucleus, the cytoplasm contains something like 10^{10} peptides. If this estimate is in the ballpark, then the opportunity for the 160-kDa protein to interact with other molecules within the time period of a few minutes is enormous. This has some additional consequences. The rates of reactions and equilibria, and therefore, interactions like protein–protein binding, are very different in a crowded solvent with background interactions affecting reactions significantly (Zimmerman and Minton, 1993). In diffusion limited circumstances, crowding is predicted to lower rates of association (Zimmerman and Minton, 1993). There is a particularly good illustration of a eukaryotic cell created by Dr. Goodsell in the same manner as Fig. 2 (http://www.scripps.edu/pub/goodsell/gallery/patterson.html). This picture provides a good idea of what an antibody has to do to diffuse from the surface to the nucleus.

IV. Permeabilized Cell Structure

Cells must be made permeable to antibodies directed against intracellular antigens. Two general methods are to treat cells with nonionic detergent or

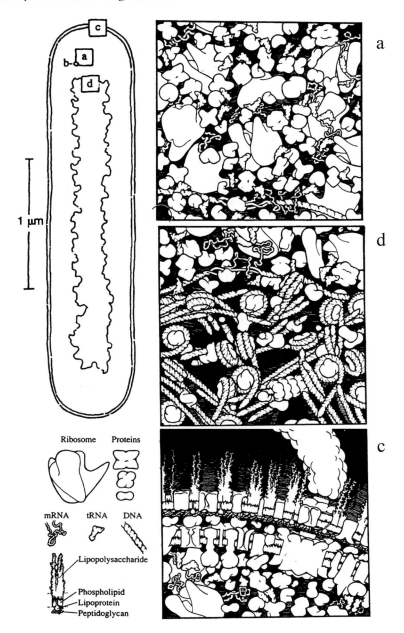

Fig. 2 Cells are crowded. A macroscopic view of small sections of an *Escherichia coli* bacterium. Molecules are drawn according to scale using hydrated space-filling models. An IgG is about 10 nm, which is the size of the largest protein depicted (lower left). For *E. coli*, Goodsell estimates that the soluble, nonribosomal fraction represents 27% of the cell protein. (Adapted from *Trends Biochem. Sci.* **16,** D. S. Goodsell. Inside a living cell, pp. 203–206. Copyright 1991, with permission from Elsevier Science.)

cholesterol binding detergents such as saponin or digitonin. These detergents permeabilize the plasma membrane but not mitochondria or nuclei. These detergents may leave a significant amount of the soluble protein inside the cell, whereas nonionic detergents essentially remove soluble proteins (Penman, 1995). Rigg *et al.* (1989) showed that several cell surface proteins were quantitatively retained after saponin permeabilization of unfixed cells. Immunologists have used a combination of formaldehyde fixation and saponin permeabilization to assay for the small molecular weight and cytoplasmic localized cytokines after enhancing expression by inhibiting Golgi function with drugs such as monensin (Maino and Picker, 1998). Darzynkiewicz and co-workers (1980) have employed nonionic detergents without fixation to measure DNA and RNA in permeabilized cells. Essentially, large structures like the cytoskeleton and ribosomes remain in cells treated with nonionic detergents. However, they are fragile and work best when suspended in serum or bovine serum albumin (BSA) solutions and at lower pH. Presumably, proteins tightly bound to large cell structures (nucleus, cytoskeleton, etc.) may be retained, and some investigators have found nonionic detergent permeabilization of unfixed cells useful for quantifying nuclear antigen positive cell fractions (e.g., Larsen *et al.,* 1991). In both the acridine orange work and the washless procedures of Larsen and co-workers, cells are not subjected to centrifugation after permeabilization. Because optimal immunostaining requires some sort of wash to remove unbound immunoglobulin, detergent permeabilization alone might result in significant cell damage. For the most part, nonionic detergents are used after formaldehyde fixation. Hallden *et al.* (1989) showed that leukocytes treated with N-octyl-β-D-glucopyranoside (OG) did not react with antivimentin antibodies (presumably this much of the cytoskeleton was lost), whereas formaldehyde fixed OG treated cells reacted quite well.

A. Permeabilizer Properties

As previously stated, the common permeabilizing reagents are nonionic detergents [Triton X-100 (TX), Tween 20, Nonidet P40, OG] and cholesterol binding detergents, saponin and digitonin. Womack *et al.* (1983) tested the ability of 50 detergents to solubilize lipid and inactivate esterase and sulfatase. Five detergents were most effective at solubilizing lipid and not inhibiting enzyme activity. These were all pure, synthetic detergents with high critical micelle concentration (CHAPS, CHAPSO, Zwittergent 310, 312, and octyl glucoside). In the study, neither TX nor Tween 20 solubilized lipid very well, and at high concentration resulted in loss of enzyme activity. Loss of enzyme activity is associated with protein denaturation. Apparently, solvents that denature proteins bind to proteins (Timasheff, 1998). Therefore, TX and Tween 20 may not completely solubilize and may weakly bind hydrophobic regions of proteins and denature them. Schmid *et al.* (1991) remark that Tween 20 is more hydrophilic than TX and therefore dissociates membranes more slowly and gently than TX. Hallden *et al.* (1989) showed results similar to Schmid *et al.* for surface marker retention with OG. From Womack *et al.*, it could be surmised that OG is "gentler" than Tween 20. However, Franek *et al.* (1994) demonstrated loss of surface but not cytoplasmic CD3 in cells treated with Tween 20 or TX compared to a mixture of saponin and digitonin. TX appeared to be "gentler" than Tween 20 in this instance. The basis for surface CD3 loss was not reported, however, it demonstrates that it can be difficult to infer staining results from chemical properties, and that while it is a good idea to design initial experiments based on fundamental chemistry, the complexity of the cell and nonpurity of many detergents may demand a more empirical comparison of the reagents commonly employed. That said, in general, there does not appear to be large differences in the choice of nonionic detergent when formaldehyde-fixed cells are permeabilized. Conversely, there is a rational reason to choose saponin/digitonin when protection of noncholesterol membranes is desired. The following general chemical properties should be kept in mind. Lipids and hydrophobic molecules are solubilized to some degree by nonionic detergents, and in general proteins are not. Further, proteins are not markedly denatured or precipitated with nonionic detergents, and one could expect the intermolecular associations to remain in cells permeabilized with these detergents.

Low molecular weight alcohols and ketones have been used as both fixatives and permeabilizers. Both exclude water and precipitate proteins as well. I am not sure whether acetone actually denatures (unfolds) proteins, but alcohols do. The temperature at which unfolding occurs is lower in aqueous alcohol solutions. The denaturization appears to be partially or in some cases wholly reversible (Fink and Painter, 1987; Hatley and Franks, 1989; Alonso and Daggett, 1995; Kamatari *et al.*, 1996, 1999). All three solvents extract lipids.

B. Fixative Properties

The fixative properties of low molecular weight organic solvents are likely to be exactly the same as their permeabilizer properties. Removal of cellular water and loss of protein secondary and tertiary structure (denaturization) reduce or eliminate enzymatic activity important for cell self-destruction (proteases, nucleases). Solubilizing lipids results in loss of ions, cofactors, and other small molecular weight compounds necessary for enzyme activity as well as some higher molecular weight proteins. An example of a high molecular weight protein that is extracted by alcohol is the mitosis marker, p105 (Clevenger *et al.*, 1985; Sramkoski *et al.*, 1999). Otherwise, cells fixed with solutions of these compounds are preserved as relatively nondegrading entities for years.

Cross-linking fixatives have been used extensively in cytometry. However, the strong cross-linkers like glutaraldehyde have been of limited utility (e.g., Davis *et al.*, 1998). The weaker, and partially reversible, formaldehyde is most often used in cytometry. A short but good review of the chemical properties of formaldehyde solutions has been provided by Clevenger and Shankey (1993). Important properties are that aqueous solutions are unstable, and formaldehyde degrades to formic acid and methanol. Therefore, solutions should be prepared often from either paraformaldehyde or electron microscope grade solutions stored under nitrogen in glass ampules. However, I have not heard a good answer to "how often" fresh solutions should be prepared, and I note that the poor fixation properties of old formaldehyde solutions were determined by poor structure in the electron microscope, not a fluorescence cytometer. Formaldehyde apparently forms reversible methylene bridges between amino, imino, sulfhydral, and hydroxyl groups of proteins, nucleic acids, and to lesser degrees lipids and carbohydrates (Clevenger and Shankey, 1993). Formaldehyde fixation is thought of as a slow process, perhaps because most of the formaldehyde in aqueous solution exists as methylene glycol which is inert. Fixation can be measured in several ways, for example, permeation of charged small molecular weight compounds, resistance to osmotic shock, or change in fluorescence properties of intercalating DNA dyes. Cell fixation by formaldehyde is a complex chemical reaction that is time, temperature, and concentration dependent. Using DNA content coefficient of variation (CV), Schimenti and Jacobberger (1992) showed that 10^6 mammalian cells were essentially 100% fixed by that criterion in 20 min at 37°C with 2% formaldehyde in phosphate-buffered saline (PBS). Table I was created from data from that paper (Schimenti and Jacobberger, 1992) and shows that, using DNA CV as a measure, cells fixed between 5 and 20 min with 1% formaldehyde at 37°C are

Table I
Degree of Fixation[a]

Formaldehyde(%)	Fixed(%)
0.1	4–15
0.5	19–78
1.0	36–88
2.0	88–100

[a] Chinese hamster cells transformed with SV40 virus were fixed with methanol or formaldehyde in PBS at the indicated concentration for 5, 10, or 20 min at 30°C. Formaldehyde-treated cells were subsequently fixed with methanol, and all samples were then washed, treated with RNase, stained with propidium diiodide, and analyzed by flow cytometry. The G_1 DNA coefficient of variation (CV) was determined, and after subtraction of the CV for methanol-fixed cells, each value was divided by that of cells fixed for 20 min in 2% formaldehyde (taken to be completely fixed) and multiplied by 100. The ranges provided are the values at 5 to 20 min. The standard assay for our laboratory is 10^6 cells in 50 μl of 0.5% formaldehyde for 10 min at 37°C. By these calculations, that corresponds to 21% fixation. Data originally published in different form in Schimenti and Jacobberger (1992).

36 to 88% fixed. The reaction is reversible, and the common practice of "antigen retrieval" relies on heat and water to reverse the extensive fixation of specimens prepared for histology. Overton and McCoy (1994) have used heat and water to reverse formaldehyde fixation and restore high quality DNA content CVs in nuclei of these overfixed cells (Overton et al., 1996). Examination of formaldehyde-fixed cells by Western blot is possible, and we assayed SV40 large T antigen (Tag) from cells that were fixed with 0.5% formaldehyde for 10 min at 37°C followed by methanol as per Schimenti and Jacobberger (1992). If these cells were solubilized with 23% sodium dodecyl sulfate (SDS) (the highest achievable concentration), all nuclei were not dissolved and Tag migrated as high molecular weight complexes. If the cells were solubilized with 23% SDS and heated for 30 min at 97°C, nuclei were dissolved and both monomer and high molecular weight Tag complexes were detected. The fraction of total Tag immunoreactivity attributed to monomers was significantly greater in unfixed or methanol-fixed cells (Frisa et al., 2000).

C. Structure of Fixed Cells

The structure of the fixed cell is an important consideration. Mann et al. (1987) showed electron micrographs of cells fixed with 70% ethanol, 0.5% formaldehyde (F) for 15 min at 4°C or 37°C, and 4% F for 15 min at 37°C. The formaldehyde-fixed cells were permeabilized with TX. The micrographs of one cell each agree with the discussion of the last paragraph. Loss of electron-dense matter was apparent in the following order 4% F at 37°C < 0.5% at 37°C ≪ 0.5% at 4°C ≪ 70% ethanol. These same investigators showed that immunoreactivity of a soluble subunit of ribonucleotide reductase increased as a function of the percentage of formaldehyde and temperature used to fix the cells. The curves describing the data were asymptotic for the highest concentrations, indicating that the reaction was complete. Similar data have been presented for soluble E. coli β-galactosidase expressed in NIH 3T3 cells (Bauer and Jacobberger, 1994) and p105 (Sramkoski et al., 1999). In the latter case, p105 was protected by formaldehyde from extraction by methanol.

From the foregoing, the following model can be developed. Cells that are fixed with formaldehyde are first protected from osmotic shock by inactivation of the plasma membrane pumps and cross-linking of proteins at the periphery of the cell. Formaldehyde may fix from the outside-in and thus create a matrix that protects cells from protein loss when subsequently subjected to wholesale removal of lipid membranes (see Jacobberger, 2000, for more discussion about this). When cells are treated with detergents, the likelihood of protein loss is increased when formaldehyde fixation is low. For example, when NIH 3T3 cells expressing E. coli β-galactosidase are fixed by the method of Clevenger, which does not completely fix cells (Clevenger et al., 1985), immunoreactivity is almost completely lost. If the same protein is directed to the nucleus by inserting a nuclear localization signal, then significant immunoreactivity is retained (K. J. Schimenti, T. L. Sladek, and J. W. Jacobberger, unpublished, 1991). For the nuclear antigen, immunoreactivity is equivalent when the cells are fixed with methanol, but the cytoplasmic immunoreactivity is increased. When 1% formaldehyde is employed and permeabilization is done with MeOH, then optimal immunoreactivity is detected for the antigen in both locations (Bauer and Jacobberger, 1994). This might mean that when alcohols dehydrate and denature proteins in the cell, they too create a matrix that protects cells from loss of high molecular weight molecules. However, loss of specific proteins depend on other constituents to which they are bound, the affinity that characterizes that binding, solubility in lipid solvents, the degree of cross-linking per se as well as specific cross-links, the degree and reversibility of the denatured state, and presumably the molecular cross section of the fixed molecule relative to the pore size of the hypothetical matrix. All this makes sense, except that for cross-links and hydrophobic intertwining, the ability of any unbound protein to diffuse out should be approximately equal to the ability of an antibody to diffuse in, which does not seem to be the case. This may mean that cross-linking and hydrophobic intertwining are the dominant mechanism of retention in the cell of what should be soluble proteins, and that any reversal of cross-linking and renaturation of peptides occurs at a rate that is slow compared to diffusion and binding of the antibody.

A comment about fixed cell structure is warranted. I have argued in the past (Jacobberger, 1991; Bauer and Jacobberger, 1994) that formaldehyde–detergent fixation leaves proteins in a more native state (and therefore intermolecular interactions are preserved), whereas formaldehyde–methanol (or any alcoholic fixation) denatures secondary and tertiary structure and therefore disrupts these interactions. From the previous discussion of the properties of cell structure, formaldehyde, and alcohol, this seems reasonable. Empirically, we observed epitope masking for SV40 Tag (Schimenti and Jacobberger, 1992) and cytokeratin but not β-galactosidase (not shown). The data were that fixation by formaldehyde–detergent produced a lower immunofluorescence signal compared to formaldehyde–methanol but formaldehyde–detergent–methanol produced a signal equal to formaldehyde–methanol. The idea still seems reasonable, and given that the degree of formaldehyde fixation can protect from antigen loss independent of the type of permeabilizer, this property seems like a major reason to choose either detergent or alcohol as a lipid solvent. There are other reasons as well, and these have been discussed at length (Bauer and Jacobberger, 1994).

A final comment about fixed cell structure. There is a least one report (Bruno *et al.*, 1992) in which the ionic strength of the formaldehyde fixative solution affected intracellular immunofluorescence. The cells were fixed according to Clevenger *et al.* (1985) (brief 0.5% formaldehyde at 4°C followed with 0.1% TX) which is a "gentle" procedure in which much of the soluble cytoplasmic protein is lost. In this case, as the NaCl concentration was increased, the immunofluorescence of proliferating cell nuclear antigen (PCNA) and Ki67 was dramatically increased while the proliferation marker, p120, was essentially unaffected. Either PCNA and/or Ki67 (both nuclear proteins) were lost at normal ionic strength, or masking was removed at high ionic strength. This should be kept in mind, and perhaps investigated, when the desire is to make statements relating intracellular immunofluorescence to the *in vivo* level of gene expression at the protein level.

V. Antibody–Antigen Reactions

What follows is a description of the chemistry of antibody–antigen reactions in nonmathematical terms. The purpose is to develop a descriptive macroscopic view of the reaction to those of us who are less mathematically intuitive. Elegant and thorough mathematical descriptions have been presented often, and the reader is directed to these works for a more complete view (e.g., Day, 1972). A good combination of mathematic language and English can be found in a chapter by Stewart and Mayers (2000). Although what follows may apply generally to any antigens of any type, the author has in mind peptides and proteins. Also, in the generalized discussion, antibody and antigen valence is taken to be one, because this is the most simple system, and is also most appropriate for discussions of staining cells with monoclonal antibodies.

A. Kinetics in Solution

In the allowed temperature ranges, reactions between antibody and antigen in solution come to equilibrium rapidly, depending on antibody affinity. Affinity is a concept that describes how well the antibody combining site (paratope) fits geometrically with the antigenic site (epitope). Specificity is related to the geometric goodness of fit in a reaction in which solvent may be displaced. The interaction appears to be largely mediated by van der Waals contacts and hydrogen bonds (Jelesarov *et al.*, 1996). Affinity can be described by the affinity constant (K_a). This value is a combined term that encapsulates rate constants for the forward and reverse reactions (antibody–antigen association and disassociation) in dilute solutions. For an antibody to have any specificity, the forward reaction rate must be greater than the reverse. The reaction proceeds until the point where the probability of binding is equal to the probability of unbinding. This is equilibrium and is a function of the concentration of the reactants. The partial rates of the reaction are concentration and temperature dependent and limited by diffusion. At any one combination of reactants, the rates of the forward and reverse reactions change as equilibrium is approached and the free concentrations of the reactants are minimized when the rates are equal.

The fraction of antibody bound at equilibrium is an illustrative measurement. A generalized description that shows the difference between antibodies of different affinities is given by the Karush equation (Day, 1972):

$$\alpha = (Kc)/(1 + Kc) \tag{1}$$

where, at equilibrium, α is the fraction of bound epitope, c is the concentration of free antibody, and K is the association constant[1]. Figure 3 shows plots of α

[1] Alternatively, α can represent the fraction of antibody bound and c can represent free antigen concentration at equilibrium.

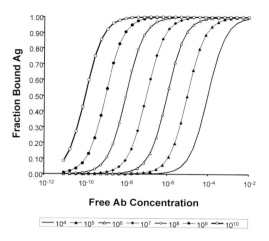

Fig. 3 Ideal titration of paratope–epitope reactions in solution. The Karush equation (see text) was used to generate curves for hypthetical antibodies with affinity constants (K_a) ranging from 10^{-4} M^{-1} to 10^{-10} M^{-1} (legend). At half-saturation of epitope the free antibody concentration numerically equals the K_a.

versus c for antibodies of different affinities. The value of this plot is that it describes the level of free antibody that will remain in solution when available epitopes are saturated. Apparently, high affinity monoclonal antibodies range from 10^{-8} to 10^{-10} M^{-1} (Stewart and Mayers, in press). The Karush equation indicates that the free antibody concentration at 99% saturation will be two orders of magnitude greater than affinity constant, and 90% saturation can be achieved at 1 order of magnitude. In useful terms, this means that if the affinity constant is 10^{-8} M^{-1}, the input antibody concentration to saturate the system has to be such that the free antibody concentration at equilibrium will be 15 μg/ml for 90% epitope saturation and 1.5 μg/ml for a K_a of 10^{-9} M^{-1}. At low concentrations of epitope the amount of bound antibody is low compared to free, so bound antibody can be ignored in these deliberations, that is, the free antibody concentration is approximately equal to the starting concentration.

In general, rates of reaction in solution are very rapid and on the order of seconds. At micromolar concentration, the time to equilibrium is on the order of ~1 min (e.g., see microcalimetry measurements in Jelesarov *et al.*, 1996), and forward rates are fractions of seconds with dissociation rates (that define affinity) substantially longer (e.g., see plasmon resonance data in MacKenzie *et al.*, 1996).

B. Kinetics in Permeabilized Cells

Three things need to be considered when extrapolating what is known about antibody–antigen reactions in solution to what is known about the antibody–antigen reactions inside a permeabilized cell. First, the epitope (antigen) is immobilized, and therefore the opportunity for bivalent antibody binding is reduced. This may simplify theoretical treatment of the data. Second, the dominant rate-limiting parameter is diffusion through the solution and cell rather than through solution alone. (From the earlier discussion, the cell is molecularly crowded, and the fixed cell may have formed a cross-linked/denatured matrix that acts as a diffusion barrier.) The extensive opportunity for binding reactions with cellular molecules other than the target should impede diffusion and increase the time to equilibrium. This appears to be the case. Flow cytometric measurements of binding kinetics for nonspecific and specific antibodies indicates that the time to reach equilibrium is approximately on the order of 15 min (Jacobberger, 1989) or longer. However, there appears to be a rapid component and a slow component for at least some specific antibodies. We have performed diffusion experiments in which fluorescien isothiocyanate (FITC)- or phycoerythrin (PE)-labeled anti-p16 antibodies (1 μg/ml) enter the methanol-fixed HeLa cells at a rate of 8% of final binding amount per minute for the first 5 min and at a rate of 1.3% per minute for the next 45 min where equilibrium levels begin to be reached (Fig. 4). In the same experiment, isotype antibodies reach equilibrium levels in 5 min at approximately the same rate. At this time, the generality of these data are not known. However, the data should serve to illustrate that some consideration of kinetics should guide formulation of staining protocols. It is especially important to allow time for weakly bound nonspecific antibody to diffuse back out of the cell during washes.

The third and final thing to consider about antibody reactions inside cells is that antibodies have the opportunity to react nonspecifically with significantly higher numbers of molecules inside the cell compared to the cell surface, that is, theoretically, the total molecular surface inside the cell is greater than that on the surface. If true, this should present a significantly greater problem for background binding. The impression is that this is true; however, I have not made a rigorous comparison, and I do not know of papers that have. It should be easy to test. Isotype control antibody could be incubated before and after fixation, and the fraction of antibody bound could be determined. Alcohol permeabilization and fixation, which likely expose hydrophobic regions of proteins, may make this condition worse.

Titration experiments, where antibody concentration is varied and cell number is kept constant, are commonly employed in a practical sense to determine the saturating concentration of antibody for flow cytometry assays. For low affinity antibodies, saturation may not be practical, and the point at which further increases in antibody concentration does not result in improved sensitivity can be determined by analysis of the signal to noise ratio (Fig. 5). In these experiments, equal dilutions of a nonreactive isotype control are often used to determine the level of nonspecific antibody binding. At saturation, this should be a good approximation of background binding since the concentration of free antibody is very high. [However, Stewart and Mayers (2000) have demonstrated that all isotype controls are not equal, and the background binding of nonspecific

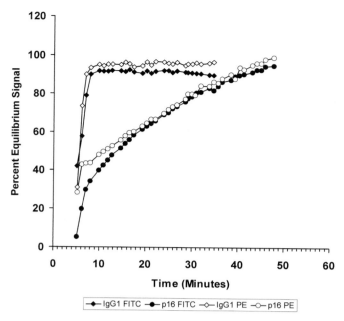

Fig. 4 Kinetics of antibody–cell reactions. HeLa cells (10^6) were analyzed by flow cytometry for autofluorescence (green and orange), then incubated with FITC-anti p16, PE-anti-p16, FITC-isotype control (IgG1), or PE-isotype control, and fluorescence was collected versus time. Antibodies were clone G175-1239 (Pharmingen) and were a generous gift from Susan Wormsley. Antibodies were used at %Fsp max (see text). Mean autofluorescence per unit time was subtracted from mean fluorescence per unit time (WinList software, Verity House, Topsham, ME) and plotted as a percentage of the final equilibrium value. (Note p16 fluorescence is higher yet at 24 hr, so the equilibrium value for the positive reactions were the last points.) The initial rate of uptake appears to be greater for both the PE-anti p16 and the PE-isotype control. The data are unpublished from Jeni Haynie and J. W. Jacobberger, 1999.

antibodies can vary significantly, so it is important to choose the "right" isotype control.] Figure 5 shows these data for three different antibodies with different apparent affinities. The use of the signal to noise ratio (sensitivity) or %Fsp (specific fluorescence/total fluorescence $\times$ 100) provides an objective staining concentration for antibodies like the HE-12 and GNS1 (Pharmingen, San Diego, CA) which have apparent affinities low enough to make determination of the saturation point difficult. One can estimate the affinity of an antibody like PAB 416 with Eq. (1), since saturation staining can be observed. In this case, the α term is taken to be the fraction of bound antigen and the term c is taken to be concentration of free antibody. Since the antibody saturates, the fraction of bound antigen (α) is equivalent to fluorescence at each antibody concentration divided by that at saturation (average of the last two data points). Since a very small amount of the input antibody is actually bound, the free antibody at

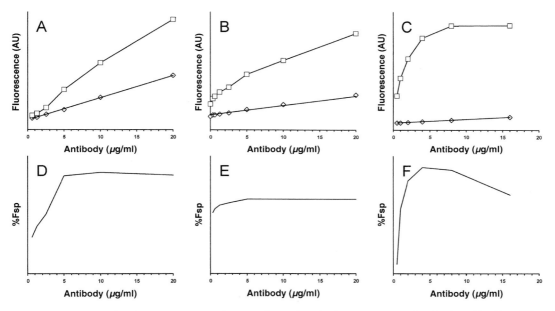

Fig. 5 Titration of antibody–cell reactions by flow cytometry. Graphs A and D, B and E, and C and F represent three different antibodies with different apparent affinities: HE-12 (A and D), GNS1 (B and E), PAB 416 (C and F). Data for HE-12 and PAB 416 were from indirect assays, whereas GNS1 data were from a direct assay. HE-12 recognizes human cyclin E, and the cells employed were DU 145. PAB 416 recognizes SV40 large T antigen, and the cells used were NIH 3T3 cells expressing large T antigen. GNS1 recognizes human cyclin B1, and the cells employed were DU 145. Graphs A, B, and C show mean fluorescence of positively stained cells (Ftot, □) for cells stained with increasing concentrations of antibody and mean fluorescence of isotype control stained cells (Fbkgd, ◇) for cells stained with antibody concentrations matched to positive stained cells. Graphs D, E, and F show sensitivity (%Fsp) that is defined by the equation %Fsp = (Ftot − Fbkgd)/Ftot × 100. The peak value of %Fsp (%Fsp max) determines the optimal staining concentration, since at that concentration the greatest distance (mode to mode) of the positive and negative fluorescence distributions is achieved. In my experience, %Fsp max for antibodies with lesser apparent affinity (identified by failure to display saturation of Ftot) identifies that concentration beyond which there is no further benefit to increasing the concentration. The progression left to right demonstrates antibodies with poor, better, and good apparent affinity. HE-12 and GNS1 were from PharMingen. PAB 416 was from Oncogene Research Products (Boston, MA).

equilibrium can be taken as the input concentration in molar units (c). The data are then fit to the Karush equation and a K_a estimated. For PAB 416, a value of 8×10^{-7}–1×10^{-8} is calculated which is close to estimates of "good" antibodies. GNS1 is a monoclonal antibody that recognizes human cyclin B1 that works reasonably well as a cytometric probe. However, the titration curve clearly shows that the apparent affinity is not very high. We have titered several monoclonal antibodies that react against cyclins D1, A, E, B1, cdk1, cdk2, p21^{WAF1}, p27^{KIP1}, p16^{INK4a}, mdm2, p53, and Rb. So far, very few show titration curves that are better than GNS1, and with the exception of DO7 (anti-p53), we have not observed titrations as good as PAB 416. I conclude from this that the current procedures for generating and testing monoclonals against these proteins do not often select for high affinity antibodies. We cannot estimate the affinity range by a flow cytometry titer as we did for PAB 416 because we cannot determine the saturation level from the data as generated. Currently we titer out to 160 mM. An open question is whether cytometry would improve if the antibodies had higher affinity.

Given the idea that the number of nonspecific sites vastly outnumbers the specific sights, it seems intuitive that the data would be much improved in terms of separating positive from negative, measuring distributions, and perhaps reproducibility if the affinity constants for the specific and nonspecific reactions were further apart. The first tenet, improved separation of negatives and positives, seems self-evident. The second tenet may not be intuitive, and ironically it is not clear (to me at least) how to determine objectively whether distribution information is getting better or worse.

Figure 6 provides an example of the features involved with this problem. In the example, HE-12 or isotype control were reacted with HeLa cells. The HE-12 antibody has a low apparent affinity as determined by measurements of the fluorescence of 5×10^5 HeLa cells reacted with 0.03–8 μg in a 50 μl volume of BSA (Fig. 6 B,C,D). Pattern information is evident even in the first sample where very few cells register as above the upper limit of background. The pattern of cyclin E immunoreactivity in normal cells rises in G_1, maximizes at the G_1/S

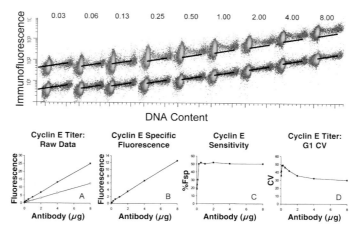

Fig. 6 Example of titration with HE-12. HeLa cells were reacted with increasing amounts (μg) of HE-12 (numbers at the top) in an indirect assay with FITC–goat anti-mouse Fab'2, then treated with RNase and stained with propidium for DNA content. The patterns displayed approximate the normal pattern expression as defined by Darzynkiewicz and co-workers (1996). The bottom graphs show analysis like that described in Fig. 5, except graph B shows Fsp, or Ftot − Fbkgd, and graph D shows the CV for the G_1 distribution as a function of antibody concentration.

interface, and decreases to background in S (Darzynkiewicz et al., 1996). HeLa cells display an almost normal pattern. The "best" pattern information that we have appears to occur between 0.25 and 1 μg of antibody. In agreement with this, the maximum sensitivity (signal to noise, %Fsp) is reached at 0.25 μg, and further additions of antibody do not provide a better separation of signal and background (Fig. 6C). However, the specific fluorescence, attributed to cell epitope content, continues upward, whereas pattern information appears to be lost. This is difficult to explain in terms of a low affinity antibody that needs to be taken to very high concentration to saturate. In that case, it is my guess that we would expect the pattern to improve as specific fluorescence increases. One possible explanation is that there are cross-reacting species of lower affinity than cyclin E and higher affinity than background that are not detected at the pattern optimum but are detected at higher concentrations. This is testable (see later). The CV for positive G_1 cells does not appear to provide information (Fig. 6D). In the good pattern range, the CVs are larger than that from less good patterns. This may mean that the biological distribution of cyclin E is broader than the distribution of autofluorescence and background binding of antibody (both are related to cell size). If true, the optimum %Fsp, pattern, and broadest CV all coincide. The CV is lower at antibody concentrations that are too low and too high because the distribution of background predominates. All this makes sense and is testable. However, how one would generalize from this is not clear. One partial approach is to titer the antibody by Western blot (see later).

The final tenet (to the idea that antibody affinity matters), that reproducibility would be improved if affinity were higher, seems intuitive as well. We have looked at reproducibility of flow cytometric measurement of immunofluorescence of PAB 416 in NIH 3T3 cells and SV40 transformed astrocytes. The CV for repeated sampling from one population is <4% (Sladek and Jacobberger, 1992a; P. S. Frisa, unpublished, 1994). We have made partial attempts to determine the reproducibility in the same way for analysis of cyclin E with antibody from clone HE-12 with mixed results. With this antibody, the CV for repeated samples was ~5% for one investigator and ~50% for another. The experiments need to done comparatively and repeated with more cell lines, but the current working hypothesis is that reproducibility with low affinity antibodies would suffer in the hands of lesser skilled technicians.

C. Antibody Specificity

Unlike nonspecific binding which presumably occurs on many surface areas of the IgG molecule, cross-reactions are epitope–paratope reactions. In this case, the epitopes are not those that enable investigators to obtain funding. Presumably, cross-reacting epitopes are those that geometrically mimic that to which the investigator wishes binding would occur. It seems reasonable to assume that the cross-reactions are lower affinity than the intended reaction, and this may be true for polyclonal antisera arising from disease. However, this does not have to be true. Hybridomas are produced by fusing myeloma cells with B cells from a responding pool of B cells. Selection may involve screening by ELISA and Western blotting. If both of these assays are done with purified antigen, there is nothing that would select against antibodies that cross-react. Figure 7 shows Western blot titrations for four monoclonal antibodies that react with cyclin D1. Two of the clones appear to react only with a peptide that has a

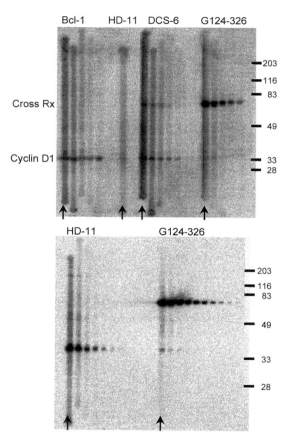

Fig. 7 Titrations of four antibodies (clone names at top) that react with human cyclin D1 by Western blot. Cell lysate was from DU 145 cells. Two bands are identified of 36 and 78 kDa. The 78-kDa band may be a cross-reacting peptide. We have observed this band in HL-60 cells as well. Each lane represents a slot on a slot blotting apparatus with dilutions of antibody concentrations starting with 10 µg/ml (arrows) and stepping down by twofold dilutions. The blotted gel had a single lane. The Western blot procedure is fully described in Jacobberger et al. (1999). HD-11 shows only one lane in the top panel. Therefore, this antibody is shown in the bottom gel, fully titered, in comparison with G124-326. Bcl-1 (Immunotech, Inc., Marseilles), HD-11 (Santa Cruz Biotechnology, Inc., Santa Cruz, CA), DCS-6 (DAKO Corp., Carpinteria, CA), and G124-326 (Pharmingen) are all monoclonal mouse antibodies. The data are unpublished from Desheng Zhang and J. W. Jacobberger.

molecular mobility, suggesting that it is cyclin D1. Two of the clones react with a second band that migrates to a position, suggesting that it has a molecular mass of 78 kDa. Of the two monoclonal antibodies that cross-react, one appears to have a higher affinity for the cross-reacting species than the specific species when compared to the other cross-reacting antibody. Because these blots were performed with epithelial cells, the data do not tell us whether the antibodies will cross-react with cyclin D2 or D3, as these are not usually expressed by epithelial cells. A flow cytometric assay with stimulated lymphocytes demonstrates significant reactivity (K. E. Shults and J. W. Jacobberger, unpublished results, 1999) with Bcl-1, therefore, it may be that this antibody cross-reacts with other D type cyclins. Because cytometric assays do not distinguish between specific (desired) and cross-reacting reactions (undesired), it is important to verify that the signal measured corresponds to the presence of the specific target molecule.

The following paragraphs describe ways to verify specificity. The value of positive and negative cell controls, microscopic localization, and functional assays seems self-evident and without controversy. Therefore, these subtopics are treated tersely with a modicum of explanation and some examples. Although Western blotting is perhaps the assay of choice for most biologists when assessing antibody specificity, it is rather rare in the cytometric literature and not without its critics. Therefore, because my laboratory has employed this method in ever-increasing amounts, some extra verbiage will be given to the reasoning behind it.

1. Positive/Negative Control Cells

Cells that express the target antigen and identical cells that do not, of the same genotype and lineage and obtained from environments producing the same growth, death, and differentiated phenotype as those that will be assayed experimentally, represent perfect controls for cytometric experiments. Almost always, this situation does not exist. Examples where it comes close are gene transfer studies, especially in virology, where the uninfected versus infected cells stained with a single specific monoclonal antibody are self-evident controls. However, keep in mind that the phenotypes are often radically different. Examples in the literature are studies of SV40 large T antigen (Lehman *et al.*, 1988, 1993; Laffin *et al.*, 1989; Kuhar and Lehman, 1991; Sladek and Jacobberger, 1992a,b, 1998). If the target epitope is coded from an endogenous gene, then the problem is less easily addressed. In some cases, cell lines can be obtained that both express and fail to express a particular gene either through aberrant (nonlineage) expression, or genetic or epigenetic (e.g., methylation) deletion. An example would be human prostate cell lines that express wild-type (LNCaP) or mutant p53 (DU 145) and cell lines in which the gene is deleted (PC-3) (Jacobberger *et al.*, 1999). In this case, the positive and negative controls are not perfect because the genotype of each cell line is different—different members of an out-bred species contributed the lines, and each has undergone significant genetic evolution as cancers within the patient and during many years of tissue culture. Even the earlier examples of viral infection do not ensure that the potential for cross-reaction is eliminated, because viruses or other gene transfer can stimulate *de novo* expression of silent genes. Therefore, additional verifying approaches will strengthen cytometric studies.

2. Western Blots, Immunoprecipitation, ELISA

Western blots are a powerful but not foolproof means of probing antibody specificity. Examples of this approach were shown in Fig. 7 for cyclin D1. However, the blots presented therein were performed as part of an ongoing study of the expression of cyclin D1 in mantle cell lymphoma (J. W. Jacobberger and K. E. Shults, unpublished, 1999). Although these blots indicate to us that two of the clones, HD-11 and Bcl-1, would be better choices than the other two, they do not ensure that we will not see cross-reactions with our target assay cell type, lymphocytes. Therefore, additional blots need to be made of this cell type. We have blotted cells that are hematopoietic tumors (myelomonocytic leukemia), but the genetically unstable nature of these cells precludes a definitive answer. They would be expected to express cyclin D2 or D3, which have molecular weights that may not be resolved from cyclin D1 on these blots. In fact, three of the antibodies react with a band of the correct molecular weight, and the results for HD-11 are not conclusive. Preliminary cytometric studies suggest that HD-11 does not react with HL-60 cells. Preliminary cytometry with stimulated peripheral blood mononuclear cells support the cross-reactivity of Bcl-1 with either cyclin D2 or D3.

Of the common immunoassay procedures—flow cytometry, immunofluorescence microscopy, immunohistology, ELISA, immunoprecipitation, and Western blotting—it is commonly stated that a particular monoclonal antibody may work well with one procedure or two procedures but not with others. At face value, this may be true, but I do not believe it arises from some unknown peculiarity of each assay. I believe it arises because of antibody affinity and cell preparation. The problem with statements about the suitability of an antibody for any particular assay is that most accounts (at least that I hear) are anecdotal. Further, reason says that any antibody that works well for one assay should work well for other assays. This is because antibodies are selected for epitopes that are colinear with the peptide chain even though the most common antibody produced does not appear to react with a colinear epitope (Van Regenmortel, 1996), epitope–paratope interaction can result in restriction of the paratope to a higher-ordered structure (Van Regenmortel, 1996), and all of the assays are conducted at relatively similar ion composition and concentration, pH, and temperature.

What are the differences then that result in anecdotal observations? The main differences between assays is preparation of antigen, presence of detergent, and the working concentration of antibody. There is good evidence for fixation affecting reactivity of epitopes (Bauer and Jacobberger, 1994; Sramkoski *et al.*,

1999). Some notable examples are p105 (Clevenger *et al.*, 1985; Sramkoski *et al.*, 1999), PCNA (Kurki *et al.*, 1988), and Ki-67 (Shibuya *et al.*, 1992; Munakata and Hendricks, 1993). In addition, lipid anchored proteins may be fixation/permeabilization sensitive. There are numerous papers describing fixation sensitivity for immunohistochemistry and many fewer that employ a more quantitative approach like flow cytometry. Because the studies exclusively test within an assay rather than between assays, it is difficult to determine whether the "sensitivity" part arises from destruction of the epitope, loss of the entire antigen (protein), failure to unmask the epitope, or use of a low affinity antibody and unfavorable staining conditions. So, without a considerable amount of evidence, I suggest that antigens that dissolve in a particular solution used to permeabilize or fix cells are likely to result in loss of that antigen to the supernatant. Examples are redistribution and loss of lipid-anchored proteins and p105 in alcohol. The formaldehyde concentrations normally employed should not dehydrate proteins to the point that the protein chains denature, and thus, the possibility that proteins that are bound to one another when the cell is alive will remain so in fixed cells and tissue seems real. This would explain epitope masking when formaldehyde is used. Alternatively, for the high concentrations and extended fixation times for tissue fixation prior to paraffin embedding, the possibility that proteins are cross-linked sufficiently to inhibit penetration of the antibody to all available epitopes seems plausible. I know of one unpublished study (Vincent Shankey, Loyola) wherein p53 immunoreactivity was compared in cells fixed in suspension with alcohol then stained and analyzed by flow cytometry to the same cells embedded in a clot then fixed with formalin, embedded in paraffin, sectioned, stained, and analyzed by microscopy. The results showed that the population was essentially 100% positive for p53 by flow cytometry and approximately 50% positive immunohistochemically. This might support epitope masking, however, like most of these studies, there are enough uncontrolled aspects that conclusions are just ideas at this point.

It seems clear that antibody affinity plays a role. For example, Western blot staining is usually done at low antibody concentration (1 μg/ml is at the high end) and in the presence of Tween 20. Low affinity antibodies do not do well under these circumstances. However, the same antibody should not do well in any other assay, but because flow cytometry is quantitative, even marginal results may show a positive reaction that can be useful. Similarly, if the epitope is not masked, immunoprecipitation may work because it is now most often carried out by incubating a cell lysate (usually a nonionic detergent mixture) with excess antibody then "precipitating" with *Staphylococcus aureus* protein A-conjugated agarose beads. This drives the reaction very far to the right even for a low affinity antibody.

In summary, if an antibody works for one assay, in my opinion, it is likely to work for all of the assays provided the chemistry of cell preparation is compatible with the antigen remaining available for reaction with the antibody. The exception to this "rule" would be the less commonly selected antibodies that recognize epitopes that are dependent on tertiary/quaternary structure of the protein or protein complex in which the epitope resides. Therefore, Western blotting and/or immunoprecipitation (either to detect the radioactive target protein on blots or to enrich the peptide in a subsequent Western blot) are an excellent method to probe the specificity of an antibody that will be subsequently used cytometrically. That said, because of the scientific uncertainties with the data that support the preceding paragraphs, this should not be the only means of validation.

3. Immunolocalization

If an antigen is known to reside in intracellular locales that can be observed microscopically, then microscopy could be employed to ensure that the antigen has remained localized in the fixed and stained cells that will be measured cytometrically. If the signal cannot be localized to the expected residence, then an alarm flag should be raised.

4. Functional Assays

One of the most effective methods to verify specificity is a functional assay. If the expression of the epitope can be modulated by known treatments that provide expected results, the confidence in the cytometric assay will improve. Examples of this are ultraviolet (UV) activated increases in p53, p53 induced expression of mdm2 (Jacobberger *et al.*, 1999), nuclear localization of nuclear factor κB (NF-κB) to the nucleus on treatment of several cell lines with tumor necrosis factor α (TNF-α) (Deptala *et al.*, 1998), and dephosphorylation of histone H3 with phosphatase (Juan *et al.*, 1998b).

VI. Multiparametric Analyses

For intracellular regulatory epitopes, which arguably are uppermost in most investigators minds, most antibodies that we have tested have an affinity that I consider to be too low for optimal cytometry. Nevertheless, they can be used effectively enough to perform studies of merit. Analysis of an epitope such as this becomes more informative in multiparametric space. The simplest and most often used additional parameter is DNA content. This provides cell cycle information. A mitotic marker can be added that provides quantification of the major cell cycle phases. Because cyclin B1 is expressed mainly in G_2 and M, the combination of a mitotic marker, cyclin B1, and DNA content provides analysis of all the major cell cycle phases of normal and endoreduplicating populations (Sramkoski *et al.*, 1999). There have been several studies that indicate that more than one antigen can be usefully integrated (e.g., Landberg *et al.*, 1990; Juan *et al.*, 1997, 1998a,b, 1999; Shackney *et al.*, 1998; Jacobberger *et al.*, 1999; Sramkoski *et al.,* 1999), however, most of the published studies have so far relied on the markers to isolate a subpopulation of cells for analysis or inferences rather than as a means to quantify two or more epitopes simultaneously. Exceptions to this are a study by Juan *et al.* (1998b), in which cyclin A and BrdUr incorporation were measured and a direct correlation established between DNA synthesis and immunoreactivity of an anti-cyclin A antibody, and a study by Jacobberger *et al.* in which p53 was introduced into cells and the correlated upregulation of the p53 activated gene, mdm2, was measured and evaluated. Because this subject is likely to be covered elsewhere in this book, in this subsection, the aspects that impact on stoichiometric binding of antibody probes will be addressed.

Multiparametric assays really depend on primary labeled probes. However, because these are scarce, investigators have made due by employing combinations of indirect and direct assays. The more easily evaluated studies have employed assays of cyclin B1 (FITC), p105 (Cy5), and DNA (Hoechst 33342); cyclin A (FITC), Histone H3 Ser 10 (PE), and DNA (7-AAD); cyclin B1 (FITC), Histone H3 Ser 10 (PE), and DNA (7-AAD); cyclin A (PE), BrdUr (FITC), and DNA (7-AAD). Each of these studies is characterized by an acceptable DNA CV (~7% or less), increased cell cycle information, and dependence on some level of the relationship (immunofluorescence intensity = antigen density) being true.

One significant problem is signal cross talk. When Hoechst dyes or DAPI are used, fluorescence emmission overlap with immunofluorescence labels is very significant, and at least with some optical designs, the interference can be significant. The system with which we have the most experience uses a collimated (defocused) light collection system. The green and red fluorescence from Hoechst 33342, presumably from a second cell residing in the isolated UV laser beam, significantly increased the immunofluorescence signal of the cell residing in the blue and colinear red beams. The only way that we found to get around this problem was to reimage the green fluorescence through an aperture that masked off any fluorescent light from the UV beam (Sramkoski *et al.,* 1999). The design and implementation of the reimaging optics was done by Coulter Electronics, Inc. (Miami, FL). However, given the overall problem with cross talk of this kind, a better system is to use 7-AAD as the DNA dye. This dye excites with 488 nm light and fluoresces above 600 nm. This is a convenient dye for analyzing cells stained with FITC and PE. However, there is significant overlap between PE and 7-AAD emission, and compensation has to be used. An additional problem is that DNA CVs are acceptable but not as good or low as they should be.

Another source of cross talk may arise in assays that couple direct and indirect assays. Theoretically, performing an indirect assay using a mouse primary and secondary from another species, then following that with a directly conjugated mouse antibody, should result in some of the second primary binding to free paratopes of the secondary reagent. In our work (Sramkoski *et al.,* 1999), we have looked for this during the course of the studies, but we have not focused on this aspect, and it deserves more attention. At present, the results we do have suggests that there are not many paratope sites free which would suggest that under optimal conditions, the secondary antibody binds bivalently. This, however, seems unlikely since we stain in a large excess of primary and secondary antibody to obtain staining that is saturated. This standard practice may need reevaluation as we gain more experience with careful measures of cross talk.

Although I would not routinely recommend it, we have successfully employed double indirect staining. This works especially well when one of the markers is not meant to be quantitative but provides a discriminating function. For example, we have stained DU 145 cells for cyclin D1 in an indirect assay, then stained with a second indirect procedure for mitotic cells with the monoclonal antibody, MPM2. This works well because MPM2 can be used at low concentration (0.13 μg/ml) and any cross talk staining (the second secondary binding to free antigenic sites on both the first and second primary antibodies) can be tolerated since the mitotic cells are still identifiable. This turns out to be useful for obtaining preliminary information about antibodies before going to the expense of conjugating them directly. Figure 8 shows a cytogram illustrating this ability. This has worked well for cell lines such as DU 145. However, when we tried to apply it to a lymphoma sample, we did not observe impressive mitotic data.

VII. Summary

This chapter reviewed and presented my biases about the factors that affect our ability to make quantitative measurements of epitope (for this chapter, equal to a protein) with monoclonal antibodies in a flow cytometric system. The

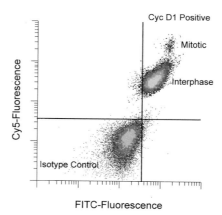

Fig. 8 DU 145 cells were fixed with methanol, stained by a double indirect method, then analyzed by flow cytometry. 10^6 fixed cells were washed with PBS, reacted with 0.5 mg Bcl-1 followed with two washes, then stained with 1.25 mg FITC goat anti-mouse Fab'2. The stained cells were washed three times, reacted with 0.125 mg MPM2, and then stained with Cy5 goat anti-mouse Fab'2, then washed three times. The data are unpublished from R. M. Sramkoski and J. W. Jacobberger, 1999.

discussion has illustrated that the chemical structure of the permeabilized cell and the affinity and specificity of the antibody are the "two" factors that are important. Pursuit of this discipline would be significantly enhanced with highly specific antibodies with a high affinity—higher than we have so far observed with the antibodies against cell cycle regulatory proteins. When moving to multiparametric space, optical systems that mask off interbeam cross talk, primary labeled antibodies, and rigorous use of fluorescence compensation is essential for the highest quality work.

Acknowledgments

I thank Phyllis Frisa, Jeni Haynie, Mike Sramkoski, and Desheng Zhang (former postdoc) from my laboratory for use of unpublished data and information; Phyllis Frisa from my laboratory for discussion and ideas on cyclin E titration and patterns; Keith Shults (Cytometry Associates, Brentwood, TN) for use of unpublished information, data, and discussion; Vince Shankey (Loyola University) for citation of unpublished data, loan of "Advanced Immunochemistry," and several in depth discussions about immunochemical reactions; Carl Stewart (Roswell Park) for use of a prepublication copy of his chapter on antibody–antigen kinetics (Stewart and Mayers, 2000), and instructive and insightful conversations about kinetics; Susan Wormsley (PharMingen) for antibodies and collaborative work; Dr. Mike Clark (Cambridge) for Fig. 1; Dr. D. S. Goodsell (Scripps Research Institute) for Fig. 2.

References

Alberts, B., Bray, D., Lewis, J., Raff, M., Roberts, K., and Watson, J. D. (1989). "Molecular Biology of the Cell." Garland, New York.

Alonso, D. O., and Daggett, V. (1995). Molecular dynamics simulations of protein unfolding and limited refolding: Characterization of partially unfolded states of ubiquitin in 60% methanol and in water. *J. Mol. Biol.* **247,** 501–520.

Bauer, K. D., and Jacobberger, J. W. (1994). Analysis of intracellular proteins. *In* "Methods in Cell Biology: Flow Cytometry" (Z. Darzynkiewicz, J. P. Robinson, and H. A. Crissman, eds.), 2nd Ed., Vol. 41, pp. 351–376. Academic Press, San Diego.

Berglund, D. L., and Starkey, J. R. (1991). Introduction of antibody into viable cells using electroporation. *Cytometry* **12,** 64–67.

Bruno, S., Gorczyca, W., and Darzynkiewicz, Z. (1992). Effect of ionic strength in immunocytochemical detection of the proliferation associated nuclear antigens p120, PCNA, and the protein reacting with Ki-67 antibody. *Cytometry* **13,** 496–501.

Camplejohn, R. S. (1994). The measurement of intracellular antigens and DNA by multiparametric flow cytometry. *J. Microsc.* **176,** 1–7.

Clevenger, C. V., and Shankey, T. V. (1993). Cytochemistry II: Immunofluorescence measurement of intracellular antigens. *In* "Clinical Flow Cytometry" (K. D. Bauer, R. E. Duque, and T. V. Shankey eds.), pp. 157–176. Williams & Wilkins, Baltimore.

Clevenger, C. V., Bauer, K. D., and Epstein, A. L. (1985). A method for simultaneous nuclear immunofluorescence and DNA content quantitation using monoclonal antibodies and flow cytometry. *Cytometry* **6,** 208–214.

Darzynkiewicz, Z., Traganos, F., and Melamed, M. R. (1980). New cell cycle compartments identified by multiparameter flow cytometry. *Cytometry* **1,** 98–108.

Darzynkiewicz, Z., Traganos, F., and Melamed, M. R. (1981). Detergent treatment as an alternative to cell fixation for flow cytometry [letter]. *J. Histochem. Cytochem.* **29,** 329–330.

Darzynkiewicz, Z., Gong, J., Juan, G., Ardelt, B., and Traganos, F. (1996). Cytometry of cyclin proteins. *Cytometry* **25,** 1–13.

Davis, B. H., Olsen, S., Bigelow, N. C., and Chen, J. C. (1998). Detection of fetal red cells in fetomaternal hemorrhage using a fetal hemoglobin monoclonal antibody by flow cytometry. *Transfusion* **38,** 749–756.

Day, E. D. (1972). "Advanced Immunochemistry." Williams and Wilkens, Baltimore.

Deptala, A., Bedner, E., Gorczyca, W., and Darzynkiewicz, Z. (1998). Activation of nuclear factor kappa B (NF-κB) assayed by laser scanning cytometry (LSC). *Cytometry* **33,** 376–382.

Fink, A. L., and Painter, B. (1987). Characterization of the unfolding of ribonuclease A in aqueous methanol solvents. *Biochemistry* **26,** 1665–1671.

Franek, K. J., Wolcott, R. M., and Chervenak, R. (1994). Reliable method for the simultaneous detection of cytoplasmic and surface CD3 epsilon expression by murine lymphoid cells. *Cytometry* **17,** 224–236.

Frisa, P. S., Lanford, R. E., and Jacobberger, J. W. (2000). Molecular quantification of cell cycle-related gene expression at the protein level. *Cytometry* **39,** 1–10.

Goodsell, D. S. (1991). Inside a living cell. *Trends Biochem. Sci.* **16,** 203–206.

Hallden, G., Andersson, U., Hed, J., and Johansson, S. G. (1989). A new membrane permeabilization method for the detection of intracellular antigens by flow cytometry. *J. Immunol. Methods* **124,** 103–109.

Harris, L. J., Larson, S. B., Hasel, K. W., Day, J., Greenwood, A., and McPherson, A. (1992). The three-dimensional structure of an intact monoclonal antibody for canine lymphoma. *Nature* **360,** 369–372.

Harris, L. J., Larson, S. B., Hasel, K. W., and McPherson, A. (1997). Refined structure of an intact IgG2a monoclonal antibody. *Biochemistry* **36,** 1581–1597.

Hatley, R. H., and Franks, F. (1989). The effect of aqueous methanol cryosolvents on the heat- and cold-induced denaturation of lactate dehydrogenase. *Eur. J. Biochem.* **184,** 237–240.

Jacobberger, J. W. (1989). Cell cycle expression of nuclear proteins. In "Flow Cytometry: Advanced Research and Applications" (A. Yen, ed.), Vol. 1, pp. 305–326. CRC Press, Boca Raton, Florida.

Jacobberger, J. W. (1991). Intracellular antigen staining: Quantitative immunofluorescence. *Methods* **2,** 207–218.

Jacobberger, J. W. (2000). Flow cytometric analysis of intracellular epitopes. In "Cytometric Cellular Analysis: Immunophenotyping" (C. C. Stewart and J. A. Nicholson, eds.), pp. 361–406. Wiley, New York.

Jacobberger, J. W., Sramkoski, R. M., Zhang, D., Zumstein, L. A., Doerksen, L. D., Merritt, J. A., Wright, S. A., and Shults, K. E. (1999). Bivariate analysis of the p53 pathway to evaluate Ad-p53 gene therapy efficacy. *Cytometry* **38,** 201–213.

Jelesarov, I., Leder, L., and Bosshard, H. R. (1996). Probing the energetics of antigen-antibody recognition by titration microcalorimetry. *Methods* **9,** 533–541.

Juan, G., Li, X., and Darzynkiewicz, Z. (1997). Correlation between DNA replication and expression of cyclins A and B1 in individual MOLT-4 cells. *Cancer Res.* **57,** 803–807.

Juan, G., Gruenwald, S., and Darzynkiewicz, Z. (1998a). Phosphorylation of retinoblastoma susceptibility gene protein assayed in individual lymphocytes during their mitogenic stimulation. *Exp. Cell Res.* **239,** 104–110.

Juan, G., Traganos, F., James, W. M., Ray, J. M., Roberge, M., Sauve, D. M., Anderson, H., and Darzynkiewicz, Z. (1998b). Histone H3 phosphorylation and expression of cyclins A and B1 measured in individual cells during their progression through G2 and mitosis. *Cytometry* **32,** 71–77.

Juan, G., Traganos, F., and Darzynkiewicz, Z. (1999). Histone H3 phosphorylation in human monocytes and during HL-60 cell differentiation. *Exp. Cell Res.* **246,** 212–220.

Kamatari, Y. O., Konno, T., Kataoka, M., and Akasaka, K. (1996). The methanol-induced globular and expanded denatured states of cytochrome *c:* A study by CD fluorescence, NMR and small-angle X-ray scattering. *J. Mol. Biol.* **259,** 512–523.

Kamatari, Y. O., Ohji, S., Konno, T., Seki, Y., Soda, K., Kataoka, M., and Akasaka, K. (1999). The compact and expanded denatured conformations of apomyoglobin in the methanol–water solvent. *Protein Sci.* **8,** 873–882.

Karn, J., Watson, J. V., Lowe, A. D., Green, S. M., and Vedeckis, W. (1989). Regulation of cell cycle duration by c-*myc* levels. *Oncogene* **4**, 773–787.

Kuhar, S. G., and Lehman, J. M. (1991). T antigen and p53 in pre- and post-crisis simian virus 40-transformed human cell lines. *Oncogene* **6**, 1499–1506.

Kurki, P., Ogata, K., and Tan, E. M. (1988). Monoclonal antibodies to proliferating cell nuclear antigen (PCNA)/cyclin as probes for proliferating cells by immunofluorescence microscopy and flow cytometry. *J. Immunol. Methods* **109**, 49–59.

Laffin, J., Fogleman, D., and Lehman, J. M. (1989). Correlation of DNA content, p53, T antigen, and V antigen in simian virus 40-infected human diploid cells. *Cytometry* **10**, 205–213.

Landberg, G., Tan, E. M., and Roos, G. (1990). Flow cytometric multiparameter analysis of proliferating cell nuclear antigen/cyclin and Ki-67 antigen: A new view of the cell cycle. *Exp. Cell Res.* **187**, 111–118.

Larsen, J. K., Christensen, I. J., Christiansen, J., and Mortensen, B. T. (1991). Washless double staining of unfixed nuclei for flow cytometric analysis of DNA and a nuclear antigen (Ki-67 or bromodeoxyuridine). *Cytometry* **12**, 429–437.

Lehman, J. M., Laffin, J., Jacobberger, J. W., and Fogleman, D. (1988). Analysis of simian virus 40 infection of CV-1 cells by quantitative two-color fluorescence with flow cytometry. *Cytometry* **9**, 52–59.

Lehman, J. M., Friedrich, T. D., and Laffin, J. (1993). Quantitation of simian virus 40 T-antigen correlated with the cell cycle of permissive and non-permissive cells. *Cytometry* **14**, 401–410.

MacKenzie, C. R., Hirama, T., Deng, S. J., Bundle, D. R., Narang, S. A., and Young, N. M. (1996). Analysis by surface plasmon resonance of the influence of valence on the ligand binding affinity and kinetics of an anti-carbohydrate antibody. *J. Biol. Chem.* **271**, 1527–1533.

Maino, V. C., and Picker, L. J. (1998). Identification of functional subsets by flow cytometry: Intracellular detection of cytokine expression. *Cytometry* **34**, 207–215.

Mann, G. J., Dyne, M., and Musgrove, E. A. (1987). Immunofluorescent quantification of ribonucleotide reductase M1 subunit and correlation with DNA content by flow cytometry. *Cytometry* **8**, 509–517.

Munakata, S., and Hendricks, J. B. (1993). Effect of fixation time and microwave oven heating time on retrieval of the Ki-67 antigen from paraffin-embedded tissue. *J. Histochem. Cytochem.* **41**, 1241–1246.

Neidhardt, F. C. (1987). Chemical composition of *Escherichia coli*. In "*Escherichia coli* and *Salmonella typhimurium* Cellular and Molecular Biology" (F. C. Neidhardt, J. L. Ingraham, K. B. Low, B. Magasanik, M. Schaechter, and H. E. Umbarger, eds.), Vol. 1, pp. 2–6. American Society for Microbiology, Washington, D. C.

Overton, W. R., and McCoy, J. P., Jr. (1994). Reversing the effect of formalin on the binding of propidium iodide to DNA. *Cytometry* **16**, 351–356.

Overton, W. R., Catalano, E., and McCoy, J. P., Jr. (1996). Method to make paraffin-embedded breast and lymph tissue mimic fresh tissue in DNA analysis. *Cytometry* **26**, 166–171.

Penman, S. (1995). Rethinking cell structure. *Proc. Natl. Acad. Sci. U.S.A.* **92**, 5251–5257.

Rigg, K. M., Shenton, B. K., Murray, I. A., Givan, A. L., Taylor, R. M., and Lennard, T. W. (1989). A flow cytometric technique for simultaneous analysis of human mononuclear cell surface antigens and DNA. *J. Immunol. Methods* **123**, 177–184.

Schimenti, K. J., and Jacobberger, J. W. (1992). Fixation of mammalian cells for flow cytometric evaluation of DNA content and nuclear immunofluorescence. *Cytometry* **13**, 48–59.

Schmid, I., Uittenbogaart, C. H., and Giorgi, J. V. (1991). A gentle fixation and permeabilization method for combined cell surface and intracellular staining with improved precision in DNA quantification. *Cytometry* **12**, 279–285.

Shackney, S. E., Pollice, A. A., Smith, C. A., Janocko, L. E., Sweeney, L., Brown, K. A., Singh, S. G., Gu, L., Yakulis, R., and Lucke, J. F. (1998). Intracellular coexpression of epidermal growth factor receptor, Her- 2/neu, and p21ras in human breast cancers: Evidence for the existence of distinctive patterns of genetic evolution that are common to tumors from different patients. *Clin. Cancer Res.* **4**, 913–928.

Shibuya, M., Miwa, T., and Hoshino, T. (1992). Embedding and fixation techniques for immunohistochemical staining with anti-DNA polymerase alpha and Ki-67 monoclonal antibodies to analyze the proliferative potential of tumors. *Biotechnol. Histochem.* **67,** 161–164.

Sladek, T. L., and Jacobberger, J. W. (1992a). Dependence of SV40 large T-antigen cell cycle regulation on T-antigen expression levels. *Oncogene* **7,** 1305–1313.

Sladek, T. L., and Jacobberger, J. W. (1992b). Simian virus 40 large T-antigen expression decreases the G1 and increases the G2 + M cell cycle phase durations in exponentially growing cells. *J. Virol.* **66,** 1059–1065.

Sladek, T. L., and Jacobberger, J. W. (1998). Cell cycle analysis of retroviral vector gene expression during early infection. *Cytometry* **31,** 235–241.

Sramkoski, R. M., Wormsley, S. W., Bolton, W. E., Crumpler, D. C., and Jacobberger, J. W. (1999). Simultaneous detection of cyclin B1, p105, and DNA content provides complete cell cycle phase fraction analysis of cells that endoreduplicate. *Cytometry* **35,** 274–283.

Stewart, C. C., and Mayers, G. L. (2000). Kinetics of antibody binding to cells. *In* "Cytometric Cellular Analysis: Immunophenotyping" (C. C. Stewart and J. A. Nicholson, eds.), pp. 1–22. Wiley, New York.

Timasheff, S. N. (1998). Control of protein stability and reactions by weakly interacting cosolvents: The simplicity of the complicated. *Adv. Protein Chem.* **51,** 355–432.

Van Regenmortel, M. H. V. (1996). Mapping epitope structure and activity: From one-dimensional prediction to four-dimensional description of antigenic specificity. *Methods* **9,** 465–472.

Womack, M. D., Kendall, D. A., and MacDonald, R. C. (1983). Detergent effects on enzyme activity and solubilization of lipid bilayer membranes. *Biochim. Biophys. Acta* **733,** 210–215.

Zimmerman, S. B., and Minton, A. P. (1993). Macromolecular crowding: Biochemical, biophysical, and physiological consequences. *Annu. Rev. Biophys. Biomol. Struct.* **22,** 27–65.

CHAPTER 14

Standardization and Quantitation in Flow Cytometry

Robert A. Hoffman
BD Biosciences
San Jose, California 95131

I. Introduction
    A. Standardization and Standards
    B. Outline of Chapter
II. General Issues
    A. What Can Be Standardized?
    B. Terminology
III. Performance Characteristics—Dynamic Range, Linearity, Resolution, and Sensitivity
    A. Dynamic Range and Logarithmic Display of Data
    B. Linearity, Log Amps, and Accuracy of Data
    C. Resolution
    D. Sensitivity and Limits of Measurements
    E. Calibration
    F. Quantitative Fluorescence Cytometry
IV. Standardization and Calibration of Common Cytometry Measurements
    A. Fluorescence
    B. Light Scatter
    C. Particle Sizing
    D. Particle Concentration
V. Examples of Applications Using Calibrated Measurements
    A. Antibodies, Antigens, Receptors, and Other Ligands Bound per Cell
    B. Bead Immunoassay
VI. Issues in Quantitation of Fluorochromes and Other Molecules
    A. Issues in Calibrated Measurements of Antibodies Bound per Cell
    B. Issues in Quantitating Antigens or Ligands per Cell
    C. Next Steps in Quantitative Fluorescence Flow Cytometry
References

I. Introduction

A. Standardization and Standards

A common sense definition of standardize is to cause to conform to a given standard or to cause to be without variation. This concept of standardization covers a wide range of flow cytometer characterization. For example, a single cytometer in an individual laboratory can be standardized so that it gives repeatable results over a long time. A single laboratory or group of laboratories can standardize multiple cytometers to give the same results on the same sample. In a quantitative mode, a universal calibration of cytometers can provide results in meaningful units such as molecules of a substance or size of cells.

The type of material used to perform the standardization determines how the analytical result from the instrument can be interpreted and used. In the most general sense a "standard" is "a 'material' against which other materials can be compared" (Henderson *et al.*, 1998). The way in which the standard allows comparison determines its usefulness. Schwartz *et al.* (1998) have proposed a classification system for flow cytometry fluorescence standards based on the degree to which the standard particle simulates a stained cell.

This chapter critically examines instrument performance standardization with a special emphasis on quantitative and calibrated measurements such as particle size and quantitative fluorescence measurements. Standardization for important physical measurements and for a variety of specific applications will be described. Sample preparation, biological factors, and general quality assurance or quality control issues are not covered here, but several excellent reviews can be recommended (Bauer, 1993; Hurley, 1997a,b; McCoy *et al.*, 1990; McCoy and Overton, 1994; Muirhead, 1993a,b; Owens and Loken, 1995).

B. Outline of Chapter

This chapter first addresses general aspects of instrument performance—dynamic range, linearity, resolution, and sensitivity. The measurement of specific sample characteristics—fluorescence, light scatter, particle size, and concentration, are considered next. A final section focuses on quantitative fluorescence cytometry (QFCM)—the measurement of fluorescent molecules or ligands labeling a particle.

II. General Issues

A. What Can Be Standardized?

In the process of sensing and measuring a property of a particle, there are several specific points at which a part of the process can be standardized. In addition the entire process, from sensing to creating the final analytical result, can be standardized at a system level. Figure 1 illustrates in block diagram form the process of sensing and analyzing a measurement of a particle in a flow cytometer. A physical property, such as fluorescent light emission, is created in a sensing zone, detected and transformed to an electrical signal, amplified, transformed (e.g., by logarithmic amplification, color compensation), quantized (e.g., by an analog to digital converter), and displayed and analyzed by a computer.

Individual subsystems involved in the process can be standardized if knowledge of the effect of the subsystem provides valuable information. For example the gain or response of a detector such as a photomultiplier (PMT) can be determined as a function of control voltage (Durand, 1994). Or the deviation from ideal logarithmic response of a log amplifier can be determined and used to generate a standardized, corrected response (Muirhead *et al.*, 1983; Schmid *et al.*, 1988).

More often, the entire system response, a defined output result for a particular input particle, is standardized. For example, a particle with defined fluorescence properties can be used to standardize a fluorescence scale defined by the mean channel in a histogram. Multiple factors may need to be adjusted and controlled to create the overall standard response.

B. Terminology

Because flow cytometry has developed through several different disciplines, the terminology used in the field is sometimes ambiguous and confusing. Precise terminology is especially important in a fundamental area such as standardization. Henderson *et al.* (1998) reviewed current usage of important terminology in quantitative flow cytometry and proposed using definitions of the National Committee on Clinical Laboratory Standards (NCCLS) as a starting point for developing standardized terminology.

Even with these definitions there is still ambiguity and overlap of the exact meaning of the terms standard, calibrator, and reference material. Henderson *et al.* take the view that: "In general, the term standards is most inclusive: all

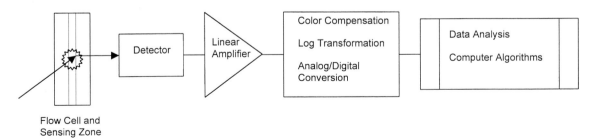

Fig. 1 Block diagram of process of generating and processing a signal. Standardization could be performed at each step in the process.

calibrators and reference materials may also be called standards. . . . Calibrators are standards with assigned numeric values that reflect some relationship between the original . . . signal and the instrument response. Reference materials are simply standards used for reference in a particular setting" (Henderson *et al.*, 1998). This chapter will use the definitions, if available, from Henderson *et al.* and, where needed for clarity, will give additional detail for some generally defined terms.

III. Performance Characteristics—Dynamic Range, Linearity, Resolution, and Sensitivity

Four fundamental performance characteristics—dynamic range, linearity, resolution and sensitivity—are inherent properties of an instrument. These performance characteristics need to be considered and standardized for all flow cytometer detection channels and measurements. General considerations that apply to all detection channels are dealt with in this section, and issues important to particular types of measurements are considered in the following section.

In many flow cytometers performance characteristics are not adjustable by the user because they require changing a physical element such as the shape and size of a focused laser spot, sheath flow rate (or equivalently particle speed), laser power, or adjusting an electronic offset in an amplifier. These performance characteristics should be distinguished from instrument characteristics such as detector and amplifier gains that can be and are routinely varied by the user to produce a standardized instrument condition.

Unless these performance characteristics are under the control of the user, they cannot be standardized directly. But they can be compared with acceptable ranges of performance. Also, if the dynamic range and deviation from linearity are known, the direct results of the measurements can be mathematically adjusted and corrected to standard dynamic ranges and a corrected linear response (Gratama *et al.*, 1998; Muirhead *et al.*, 1983; Purvis and Stelzer, 1998; Schmid *et al.*, 1988; Schwartz *et al.*, 1996). Whether to make corrections or not will depend on the accuracy required of the result. Software that allows mathematical manipulation of flow cytometry list mode data is available from some vendors. List mode data may also be read into standard spreadsheet or statistical analysis software programs for calculating corrected data.

A. Dynamic Range and Logarithmic Display of Data

The dynamic range of a measurement is, in most general terms, the range of values that can be displayed. The conventional flow cytometry measure of dynamic range for linear measurement and display of data is the number of displayed channels or "bins" from the analog to digital converter (A/D). For example, if the data is displayed in 256 channels (2^8 or 8-bit resolution), the data have a dynamic range of 256. The smallest channel measure is 0 and the largest is 255. If data is displayed in 1000 channels, then the dynamic range of linear data is 1000. When data is displayed on a logarithmic scale, however, the conventional meaning of dynamic range is the number of decades that are displayed. If the smallest measure displayed is 1 (e.g., 1 mV) and the largest displayed is 10,000 (e.g., 10,000 mV or 10 V), then the dynamic range is said to be four decades. Figure 2 illustrates linear and logarithmic displays of data with dynamic range of 256 channels for linear or log data and 10,000 (four decade, 64 channels/decade) dynamic range for logarithmic data. The term coefficient of response is used in some of the flow cytometry literature instead of the term channels per decade (Schwartz *et al.*, 1996; Schwartz and Fernandez-Repollet, 1994). The

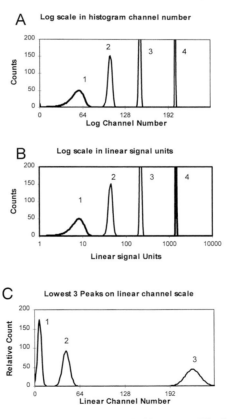

Fig. 2 Examples of linear and log displays of data in histograms. The individual populations are labeled 1–4. (A) Data displayed as log channel numbers. (B) The same data displayed in linear signal units using 64 channels per decade log conversion. (C) The data of the three lowest peaks on a linear scale.

B. Linearity, Log Amps, and Accuracy of Data

dynamic range over which physical measurements are reported by the instrument should not be confused with the dynamic range or reportable range of a biological assay. Several factors, both instrumental and assay dependent, affect the useful range of reportable results. Some of these factors are described later in this chapter.

A linear relationship between two variables x and y has the form:

$$y = mx + c. \tag{1}$$

The proportional relationship between the variables, m, is the slope of the line. The constant c is the "intercept" or the y value when $x = 0$. Figure 3 shows linear relationships displayed on a linear scale (Fig. 3A) and on a logarithmic scale (Fig. 3B). Note that on a logarithmic scale the relationship is a straight line only when the intercept, c, is zero.

Detailed discussion of the instrument factors that affect linearity are provided elsewhere (Ubezio and Andreoni, 1985; Wood, 1997). In practice, the major factor that affects linearity in commercial flow cytometers operated under typical conditions is the accuracy of the electronic amplifiers. For linear amplification, the error is usually due to a nonzero intercept in the response due to electronic offsets. Linearity can be checked by running a sample of particles that produce known relative signals in the detector. If an offset is observed, the data can be corrected using a simple linear function:

$$y_{\text{corrected}} = y + \text{constant} \tag{2}$$

where the constant may have a positive or negative value.

Uniform fluorescent particles or nuclei stained for DNA fluorescence are commonly used to check the linearity of fluorescence detection. Doublets and aggregates of the particles give simple multiples (e.g., twice, three times) of the signal of a single particle. When the integrated signal (or pulse area) is measured the mean channel for an aggregate of two particles should be twice the mean channel of a single particle, and similarly for aggregates of three and larger. This analysis does not necessarily apply accurately if pulse peak rather than pulse area is measured. Because laser illumination has a nonuniform (Gaussian) illumination profile, only particles that are small compared to the smallest dimension of the laser spot will be uniformly illuminated. Thus for laser based optical measurements two particles with the same amount of fluorochrome can give different measurement of pulse peaks. For example, if a 10 μm by 100 μm laser spot illuminates the sample stream, the peak of the fluorescence pulse from a 5-μm-diameter particle will be 90% of the peak fluorescence pulse from a 1-μm particle containing the same amount of fluorochrome.

Bagwell *et al.* (1989) describe a method for testing amplifier linearity using a mixture of two different types of particles that give slightly different fluorescent

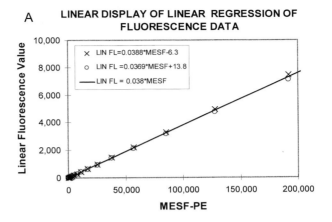

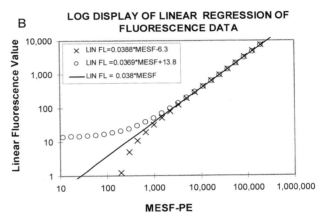

Fig. 3 Plots of the linear function $y = mx + c$ on linear (A) and log (B) scales. The plots with nonzero intercepts ($c = 0$) are best fit linear functions to data obtained on two different flow cytometers.

signals. Varying the PMT voltage causes the signals to the amplifier to vary. A plot of difference in the means of the two fluorescence populations versus the mean of one of the populations should be a straight line through the origin if the amplifier is linear with no offset.

When log amplification or transformation of signals is used, the determination of accuracy is more complicated because of the large range of data and the compression of the displayed range when log scales are used. Inaccuracies may be due to errors in the overall response (channels per decade) or due to local deviations from the nominal channels per decade response.

The log channel number is related to the signal value by

$$\text{log channel number} = (\text{channels per decade}) \times \log(\text{linear signal value}). \quad (3)$$

Conversion from log channel number to a linear value uses the following relationship:

$$\text{linear value} = 10^{C/Dc}, \quad (4)$$

where C is the channel number and Dc is the channels per decade used in the calculation.

Equation (4) is used to convert a histogram displayed with the "x"-axis in log channel numbers (as in Fig. 2A) to a histogram with the x-axis displayed as linear units on a logarithmic scale (as in Fig. 2B). Unless otherwise noted by the analysis software, the conversion will presume accurate, nominal response of the log amplifier. For example if the log amplifier is nominally four decades, and the log data are acquired into 256 channels, then the software will assume 64 channels per decade. If the data are acquired into 1024 channels, the software will assume 256 channels per decade.

The nominal dynamic range of logarithmic amplifiers (log amps) is not necessarily the true range of signal levels covered. For example, a log amp that is designed to nominally display four decades of signal over 1024 histogram channels (256 channels per decade) may in reality display 3.8 or 4.2 decades of signal (269 or 244 channels per decade) (Schwartz *et al.*, 1996). Also the fact that data is displayed over, for example, four decades dynamic range does not imply that data are accurately represented over the entire range. In fact, when log amplifier circuits are used, the data in portions of the first and last displayed decade can be more than 10% in error. At least in instruments that use analog circuits for logarithmic amplification, it is necessary to calibrate (or check factory calibration of) log amps for accurate and quantitative use of the data.

Based on surveys of laboratories doing routine immunofluorescence flow cytometry, Schwartz found the channels/decade (normalized to a scale of 256 channels for four decades) varied by $\pm 8\%$ (64 ± 5.0 channels) (Schwartz *et al.*, 1996). For routine work this was considered an acceptable range that allowed fairly uniform positioning of fluorescence results in histograms and allowed comparison across laboratories. However, such a wide variation of actual channels per decade from the nominal calibration of a log amplifier can have significant results if the linear value is computed from the log channel data.

If the wrong value of channels/decade is used to calculate linear values from Eq. (4), the histogram showing linear values on a log scale (e.g., Fig. 2B) will be in error. Thus the linear values on log histogram displays should be used with caution unless the log scale has been calibrated. Equation (5) relates the computed linear value to the input signal value:

$$\text{computed linear value} = 10^{(Dt \times \log(\text{linear signal value})/Dc)}$$
$$= (\text{linear signal value})^{Dt/Dc}, \quad (5)$$

where Dt is the true channels/decade for the amplifier.

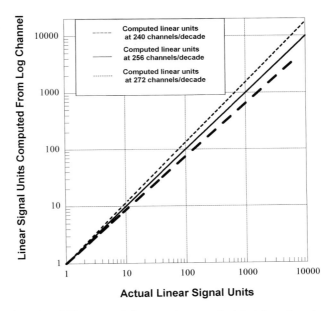

Fig. 4 Error introduced if computed linear units are calculated from log channel data using incorrect channels per decade. The plots are for a case where the calculation of linear signal unit from log channel number assumes 256 channels per decade. The calculated linear signal values for log amplifier responses of 240, 256, and 272 channels per decade are shown. Only the amplifier with 256 channels per decade gives a correct calculated value of linear signal unit.

For example if the true channels/decade for a log amplifier is 250 channels per decade and the value used for calculating the linear value from the log channel number is 256 channels per decade, then the computed linear value will be (linear signal value)$^{250/256}$ = (linear signal value)$^{0.977}$.

Unless the analysis software allows the true channels/decade to be used in the calculations, the nominal channels per decade will be used. Figure 4 shows the error that will be introduced if the nominal 256 channels per decade is used, but the instrument actually has 240 or 270 channels per decade. Some commercially available flow cytometry analysis software provides for calibration of the channels per decade and automatically rescales data displayed on the log scale.

Approaches to calibrate log amps are outlined later in this section. Good overviews of log amps are given by Shapiro (1995) and Gandler (Gandler and Shapiro, 1990).

A log amplifier that has an accurate overall response may still have large local deviations at particular values or ranges of values. This is illustrated in Fig. 5A which shows the response of one particular logarithmic amplifier from a FACSCalibur[1] (BD Biosciences, San Jose, CA). The response is quite accurate

[1] FACSCalibur, CaliBRITE, and QuantiBRITE are trademarks of Becton Dickinson.

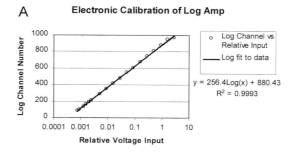

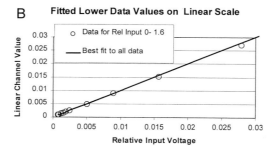

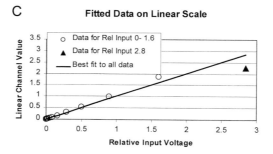

Fig. 5 Calibration of log amplifier with electronic signals of known relative value. Plots show histogram channel value versus relative pulse amplitude input to log amplifier. All the data are shown in the plot of log channel number versus log relative voltage input (A) and in the plot of linear channel values (C). The data with lower values are plotted on a linear scale in (B). Linear values in (B) and (C) are calculated from the best fit parameters for the data shown in (A).

(using the best fit line as "truth") over most of the range measured, but the last point at channel 972 is more than 20% from the best fit. Note that this data should not be taken as representative of all log amps of this type. Variability in the highest and/or lowest decades are routinely observed between amplifiers.

The data for Fig. 5 were obtained by applying electronic pulses with known relative values to the input of the signal processing electronics and recording the output with the data acquisition system of the flow cytometer. A wide range of input pulses were produced by attenuating a constant amplitude pulse with electronic attenuators whose calibrations were traceable to the U.S. National Institute for Standards and Technology. A plot of mean channel versus log of the relative input signal is shown in Fig. 5A. Over the entire range tested (channels 88–972), there is apparently a good fit to a log response with 256.4 channels/decade (solid line in Fig. 5A). Closer examination of the data on a linear scale (Fig. 5B,C) shows very good agreement from the lowest values to channel 880. Above channel 880, however, there is deviation from the 256 channel per decade fit and the signal appears to be saturating at channel 970. The deviations in the upper decade are measurable and could be corrected with a more complex fit to the data or by using a table of correction values at each channel as demonstrated by Schmid *et al.* (1988). By calibrating the electronic response of the log amplifier it is possible to extend the accuracy of measurement, but the relationship between channel number and measured value may not be simple.

Two alternative methods for calibrating a log amplifier that do not require electronic equipment have been described. The first uses a known linear amplifier as reference and compares measurements on the linear and logarithmic amplifiers (Horan *et al.*, 1990; Muirhead *et al.*, 1983). The second method uses a mixture of particles with known, different intensities. The log response is determined by measuring the separation of the peaks on the log scale as the signal level is varied by changing the PMT voltage (Schmid *et al.*, 1988).

C. Resolution

Resolution in flow cytometry has been defined as the "degree to which a flow cytometry measurement parameter can distinguish two populations in a mixture of particles that differ in mean signal intensity. Fluorescence sensitivity can be considered a special case of fluorescence resolution for which the signals are very dim" (Hoffman, 1997). Because a wide range of signal values are compressed into a small "bin" when a log scale is used, a log display of the data may not have sufficient resolution to display populations that can actually be resolved by the instrument.

The coefficient of variation, CV, is a common measure of instrument resolution based on measurement of a single population. The narrower the distribution—and smaller the CV—the better the resolution of the measurement. There are two cautions in this simple interpretation, however. The first is that the numerical value of the CV can be affected by the number of histogram channels the data are spread over (Schuette *et al.*, 1985). If the true standard deviation of the distribution is equal to one channel width in the histogram, the calculated CV will be in error by 20%. For example, if a fluorescence distribution has a true CV of 1%, the measured CV will be 1.2% if the mean channel of the histogram is at channel 100 on a 1000 channel scale. When the true standard deviation is less than the width of a histogram channel, the measured CV is unreliable. This effect is especially important when histograms on a log scale are transformed to linear data for calculation of CVs.

The second caution in using the CV of a single population as a measure of resolution, is that this CV does not necessarily predict the CV of a population with a different mean signal. This caution will be discussed in more detail later in the sections on fluorescence and light scatter standardization.

Optimal alignment of the optics with the sample stream is necessary for the smallest CV of measurement, but it is not the only factor that affects CV. On instruments that provide for user alignment of the optics and sample stream, a small CV is a good indicator of optimal alignment. The shape of the histogram distributions is also indicative of factors affecting the CV. If the histogram is skewed to lower channel values, the sample stream is probably optimally aligned, but the width of the sample stream is wider than the most intense area of illumination. This causes particles on outer portion of the sample stream to be less intensely illuminated than those in the center of the sample stream.

D. Sensitivity and Limits of Measurements

"Sensitivity" has a double meaning in flow cytometry—particularly when applied to fluorescence measurements. The two different meanings depend on whether the concept of threshold or resolution are emphasized. The first meaning has to do with the smallest amount of signal that can be detected (Schwartz *et al.*, 1996; Shapiro, 1995). This notion has also been given the name detection threshold (Schwartz *et al.*, 1996). The second meaning has to do with the ability to resolve dimly stained cells from unstained cells in a mixture (Brown *et al.*, 1986; Horan *et al.*, 1990; Shapiro, 1995). These concepts do not measure the same thing. Two instruments can have the same detection threshold but differ significantly in the ability to resolve a dimly stained population (Chase and Hoffman, 1998; Hoffman, 1997; Wood and Hoffman, 1998).

This double meaning of "sensitivity" also applies to other measurements. The difference in concepts can also be linked to what is used as a reference for "detection" and to what criterion is used for being able to "detect." The more specific term "limit" as defined by NCCLS (Henderson *et al.*, 1998) provides a clearer distinction. In the terminology used by NCCLS, the concept of detection threshold is called a "limit" and is qualified as either a detection limit ("the lowest amount of analyte that can be detected but not quantified as an exact value") or quantitation limit ("the lowest amount of analyte in a sample that can be quantitatively determined with [stated,] acceptable precision and [stated, acceptable] accuracy, under stated experimental conditions"). Examples of "analyte" in flow cytometry are (1) a subpopulation of lymphocytes whose concentration (cells/ml) is quantitated or (2) a fluorochrome on a subpopulation of cells whose amount is quantitated.

When the conditions of the measurement and the required accuracy and precision are stated, the meaning of "quantitation limit" is not ambiguous. But there is currently little consensus on what the conditions, accuracy, and precision should be. This is not surprising because there are complex instrumentation, biological, and statistical factors involved in even discussing this issue. It is worth noting that flow cytometry is not the only field to struggle with defining an objective and widely accepted criteria for a "detection limit." Scientists in the more mature field of immunoassay continue to refine and debate the meaning of minimum detectable concentration (Brown, 1996; Brown *et al.*, 1996; Buttner, 1991; Moran, 1997; Stamey, 1996).

E. Calibration

Calibration provides a "known relationship between the measurement response and the value of the substance being measured" (Henderson *et al.*, 1998). Generally, the measurement response should be determined for a range of input values. If the calibration process uses material with known amounts or values, the calibration is absolute. For example, if the flow cytometer is calibrated to measure particle diameter, the measurement response would provide a result in units of micrometers. If only the relative values of the material or process used for calibration are known, the instrument can be calibrated in relative terms. For the particle diameter example, relative calibration would adjust the measurement scale so that a 10-μm diameter particle had twice the measured value of a 5-μm diameter particle, but the exact diameter would not be read from the scale. In some cases relative calibration is used to correct for a nonlinear or nonideal response of the instrument. Examples of calibration in flow cytometry are described for important measurement parameters and for a range of applications later in this chapter.

The type of material used for calibration determines the accuracy and reliability of the calibration process. If possible, the calibration material should be traceable to a reference material from an authoritative organization. NCCLS defines a reference material as "a material or substance, one or more properties of which are sufficiently well established to be used for calibrating an apparatus. . . ." A certified reference material has the additional characteristic that "each certified value is accompanied by an uncertainty at a stated level of confidence" (Henderson *et al.*, 1998). An example of calibration material traceable to an authoritative organization are microspheres that have assigned particle diameter mean and standard deviation values traceable to the U.S. National Institute for Standards and Technology.

F. Quantitative Fluorescence Cytometry

An important special area of standardization is QFCM. The use of fluorescent probes and ligands for specific cell structures is the most powerful feature of flow cytometry. The ability to accurately quantitate the number of fluorescent molecules or the molecules to which they bind provides results in absolute terms that can be compared across instruments, between laboratories, and over time. The final section of this chapter is devoted to a review of this emerging area.

IV. Standardization and Calibration of Common Cytometry Measurements

The most common measurements made in flow cytometry are fluorescence intensity, light scatter, particle size (by several methods), and particle concentration. Each of these measurements is sample dependent and requires standardization with particles. Standardization of each of these measurements is considered in this section. An overview of each measurement and cautions on its use are intended as helpful hints and are not comprehensive.

It is helpful to distinguish between a physical measurement that is made by a flow cytometer and the properties of a particle that produce the result. Light scatter usually measures intrinsic properties of a cell or particle that do not require the addition of a probe such as a dye molecule. The intrinsic properties are useful for broadly categorizing particles. Fluorescence is normally used to detect and measure a probe that is added to a cell to give information about a particular property. The great variety of applications in flow cytometry comes from the use of fluorescent probes.

A. Fluorescence

Two different tiers of fluorescence standardization are used. The first establishes that the instrument functions within the expected limits for dynamic range, linearity, resolution, and sensitivity. The second tier of standardization establishes the operating conditions for a particular application. Standardizing fluorescence measurements in the context of specific fluorochromes requires consideration of both fluorochrome and instrumentation issues.

Fluorescence measurements in flow cytometry utilize a wide range of fluorescent probes. The wide variety of useful fluorescent molecules makes it difficult to standardize fluorescent intensity measurements for all fluorochromes by using a single fluorochrome to standardize an instrument. Some of the important properties of fluorochromes that need to be taken into account when developing a standardization method include excitation and emission spectra, sensitivity to environmental factors such as pH and other ion concentrations, saturation of emission at high illumination intensity, photobleaching, energy transfer to other fluorochrome molecules or quenchers, and fluorescence polarization. The instrument factors that affect the fluorescent measurement include excitation wavelength and intensity, light collection efficiency of the emission optics, transmission spectrum and polarization characteristics of the emission filters and mirrors, and the efficiency of the detector. The characteristics of the fluorochrome(s) and their interaction with the optical factors in the flow cytometer need to be carefully considered in choosing and using standards for fluorescence measurements. There is often a need to compromise between perfect standards for each fluorescent reagent and a few practical standards that are simple, reliable, and easy to use.

1. Measure of Fluorescence Production

Fluorescence intensity is a physical measure of light power detected, and is not easily related to biologically relevant measures such as the number of molecules producing the fluorescence. The amount of fluorescent light production from a fluorochrome molecule depends on many factors including how it is bound or incorporated into a particle. A particularly important example is the reduction in quantum yield (fluorescent photons emitted per absorbed excitation photon) when a fluorochrome such as fluorescein isothiocyanate (FITC) is conjugated to an antibody. The number of fluorochrome molecules does not necessarily predict the amount of fluorescence that will be generated. A widely accepted measure of fluorescence producing capacity is the equivalent number of molecules in solution that would produce the same amount of fluorescence as the fluorochromes bound to the particle. This has been termed molecules of equivalent soluble fluorochrome (MESF) and has been proposed as standard nomenclature for quantitative fluorescence cytometry (Henderson *et al.*, 1998).

2. Fluorescence Resolution and Sensitivity

The factors affecting fluorescence resolution are well understood and have been described by McCutcheon and others (Gaucher *et al.*, 1988; McCutcheon and Miller, 1979; Pinkel and Steen, 1982; Steen, 1992; Wood and Hoffman, 1998). At the highest fluorescence signal intensities, such as from nuclear DNA staining or highly stained microbeads, resolution is primarily limited by the uniformity of illumination of particles in the sample stream. Lasers and lamps in commercial flow cytometers routinely have less than 1% variation in output power and contribute little to the variation (noise) in illumination. Most of the variation in illumination is from spatial nonuniformity of the focused light beam, which is brightest at the center. Depending on the relative size of the sample stream and the width of the focused light spot across the stream, particles will be more or less evenly illuminated. Larger sample streams or smaller focused spots produce larger variation in the fluorescence signals and larger CVs. At lower levels of fluorescence intensity, such as from immunofluorescence, the statistical nature of light to electron conversion (photoelectron shot noise) in the detector (usually a photomultiplier) becomes the dominant factor determining the CV of a fluorescence intensity measurement. At a sufficiently low level of fluorescence intensity, background light and dark current in the detector become contributing factors to the noise and resulting CV of the measurement. These three contributing factors—illumination uniformity, photoelectron shot noise, and background—

can be independently determined, so the CV of fluorescence measurement at any fluorescence intensity level can be predicted (Chase and Hoffman, 1998; Gaucher et al., 1988; McCutcheon and Miller, 1979; Pinkel and Steen, 1982; Steen, 1992; Wood, 1998).

Some instrument design or alignment trade-offs (such as decreasing the width of the laser focus spot across the sample stream) that improve resolution of weakly fluorescent particles actually decrease resolution of strongly fluorescent particles. Standardizing the instrument setup should take this into account if the user can make optical configuration and alignment adjustments to the flow cytometer.

With strongly and uniformly fluorescent particles, standardization of resolution is usually also a check of optical alignment. Microbeads are often used for this purpose, but nuclei stained with DNA dyes also provide bright, uniform fluorescence. Commercial microbeads are available with stated maximum fluorescent CVs as measured by flow cytometry. The criterion for acceptable performance depends on the instrument and intended application. As a general rule, however, CVs of brightly fluorescent "alignment" beads should be less than 3% when the sample stream is run at its narrowest diameter (lowest sample flow rate).

Characterizing the Detection of Weakly Fluorescent Particles

Standardization of fluorescence resolution for weakly fluorescent particles requires separate consideration and is linked to the common notion of "sensitivity." Two different notions of fluorescent sensitivity have been defined by Hoffman (1997) as follows:

> "1. Degree to which a flow cytometer can measure dimly stained particles and distinguish them from a particle-free background (threshold). Threshold is important when the mean fluorescence of a dimly fluorescent population is measured. The greater the number of particles analyzed, the more accurately and precisely will the mean fluorescence be measured.
> "2. Degree to which a flow cytometer can distinguish unstained and dimly stained populations in a mixture of particles (resolution). Resolution is important for immunofluorescence analysis of subpopulations and is strongly affected by the measurement CVs for dim and unstained particles."

The commonly used "detection threshold" (Schwartz *et al.*, 1996) or equivalent "noise threshold" (Schwartz and Fernandez-Repollet, 1994) uses the position or broadness of the fluorescence histogram peak from nonfluorescent particles as a measure of instrument performance. Unfortunately this measure is incomplete and ambiguous. (For a discussion of a similar situation in the field of immunoassay see Brown, 1996; Brown *et al.*, 1996.) Because the noise in weak fluorescence detection is determined by two factors, background light and detection efficiency, the broadness or standard deviation of any fluorescence measurement is not uniquely determined. Many combinations of background and detection efficiency can produce the same measured distribution of fluorescence in a histogram. In particular, an instrument with a low amount of background signal and low detection efficiency can have the same detection threshold or noise distribution as an instrument with high background signal and high detection efficiency. But these two instruments would have a different ability to resolve dimly stained populations of cells (Chase and Hoffman, 1998; Hoffman, 1997; Wood and Hoffman, 1998). Figure 6 illustrates this with model data for detection of the same samples [assumed to be a mix of nonfluorescent particles and particles stained with phycoerythrin (PE) at levels of 500, 1000, and 2000 MESF] with two different instrument conditions: the condition for Fig. 6A,B are low background and low detection efficiency and for Fig. 6C,D high background and high detection efficiency. The detection threshold or noise distribution for the low background condition (Fig. 6A,B), is slightly lower than the noise distribution for the high background case (Fig. 6C,D). But the ability to resolve different subpopulations is dramatically different. The instrument condition with high detection efficiency is much more effective in resolving populations than the instrument with low background and low detection efficiency.

The concept of limit of quantitation defined by NCCLS as "the lowest amount of analyte in a sample that can be quantitatively determined with [stated,] acceptable precision and [stated, acceptable] accuracy, under stated experimental condi-

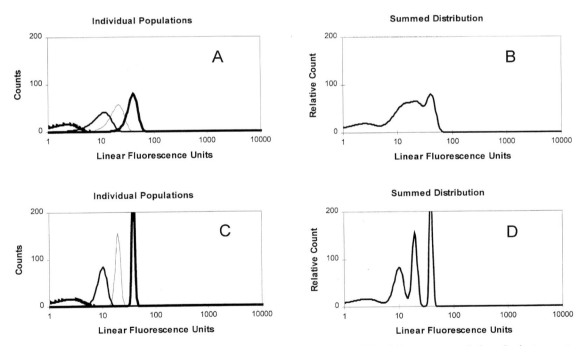

Fig. 6 Model histograms showing resolution and width of histogram populations for instruments with different amounts of background light, B, and detection efficiency, Q. In each case the histograms represent the same number of MESF. The condition in (C) and (D) represent performance with higher B and higher Q than in (A) and (B). (A) and (C) show individual populations, and (B) and (D) show the summed distribution of the individual population.

tions" (Henderson *et al.*, 1998) is a way to unambiguously characterize performance only if the conditions are stated. Different conditions can give different qualitative and quantitative limits. For example, in Fig. 6 if the criterion is discrimination of dim fluorescence from a blank, both instrument conditions give almost identical performance. If the criterion is measurement of a subpopulation where cell autofluorescence is represented by the second peak (500 MESF), then the performance represented by the instrument in Fig. 6C,D is superior. Rather than have many different criteria for performance based on narrowly defined conditions, some investigators have focused on establishing a criterion that defines and predicts performance under all conditions. Because instrument performance depends on the two factors background and detection efficiency, these two factors seem natural measures for standardization (Chase and Hoffman, 1998; Wood and Hoffman, 1998).

In 1997 and 1998 there were several workshops and meetings at which the best way to characterize instrument performance in detecting low levels of fluorescence was vigorously debated. Although there is not total agreement, a consensus seems to have been reached that performance is best characterized in terms of the background, B, and detection efficiency, Q. This allows universal characterization that can be translated into performance criteria for any particular application or defined condition. At the time of this writing, commercially available test kits to easily measure B and Q are not available. A proposed approach has been published (Chase and Hoffman, 1998; Wood and Hoffman, 1998), and a simplified protocol is in preparation. The protocol uses the standard deviation of two or more populations of uniform, dimly fluorescent beads whose mean fluorescence is calibrated in terms of MESF.

3. The Problem of Defining a Nonfluorescent (or Blank) Standard

The criterion for what is nonfluorescent depends on the sensitivity of the instrument and fluorescence channel being characterized. In general it is not sufficient to say that a particle is an adequate blank if it has less fluorescence than a lymphocyte (or some other specific cell). Many cells of interest, such as platelets, red cells, and bacteria have much less autofluorescence than lymphocytes. The least fluorescent reference measurement is of the sheath fluid with no sample or particles flowing. In instruments equipped with test pulses that simulate signals from cells, the detection electronics can be triggered to measure fluorescence when no particles are present. In Becton Dickinson flow cytometers this is achieved by pulsing a light emitting diode on the forward scatter detector while the instrument has only sheath flowing. The same effect can be achieved by running a sample of very small (less than 2 μm in diameter) plastic beads. Even if larger beads of the same material emit a small amount of fluorescence, the smaller beads have fluorescence lower by a factor dependent on the cube of the diameter. Figure 7A shows data obtained on a FACSCalibur showing green fluorescence from test pulse triggered events and several commercially available

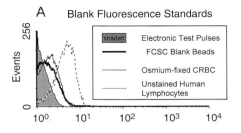

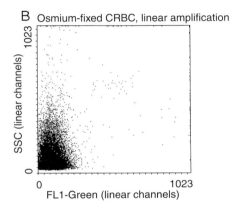

Fig. 7 (A) Histograms of green fluorescence for nominally blank particles (FCSC blank beads and osmium-fixed CRBC) and test pulse triggered fluorescence. Autofluorescence of unstained lymphocytes are shown for reference. (B) Dot plot of side scatter (SSC) versus green fluorescence for osmium-fixed CRBC showing lack of correlation between side scatter and fluorescence.

blank particles including polymer beads and osmium fixed chicken red blood cells (CRBC). Autofluorescence of human lymphocytes is shown on the same histogram for reference.

A disadvantage in using test pulse triggered fluorescence measurement is that particle-scattered light that leaks through filters or causes fluorescence in the emission optics (Loken, 1980) is not accounted for. A test for a light-scatter effect is provided by the CRBC, which are asymmetric and scatter more or less light depending on their orientation as they pass through the light beam (Loken et al., 1977a). Because fluorescence is not greatly dependent on particle orientation, a plot of fluorescence versus side scatter for the CRBC will indicate whether there is a side scatter induced signal in a fluorescence channel. If there is no correlation between side scatter and fluorescence, one can be sure there is no scatter effect on fluorescence detection—at least to the level of scatter maximally produced by the CRBC. This is shown in Fig. 7B which compares test pulse triggered fluorescence and CRBC fluorescence in a plot of SSC versus fluorescence. No correlation between SSC and fluorescence is seen in this case.

4. Linearity and Dynamic Range

The general considerations of linearity and dynamic range were discussed earlier. Commercially available fluorescent particles or fixed cells that can be stained with DNA fluorochromes are useful for testing and standardizing the fluorescence channels of the flow cytometer. Fluorescence histograms of several commercially available bead products with multiple fluorescence levels are shown in Fig. 8. Vendors are listed in Section IV, A, 9.

5. Color Compensation

When more than one fluorochrome is used to stain cells, there is often significant overlap of the emission spectra. As little as 0.1% overlap of integrated light intensity between fluorochromes can be important when measuring fluorescence intensity over four decades of dynamic range. If the measured parameter is to be expressed in units of a particular fluorochrome rather than amount of light on the detector from all fluorochromes the signals from the fluorochromes not of interest must be subtracted out (Loken *et al.*, 1977b; Bagwell and Adams, 1993).

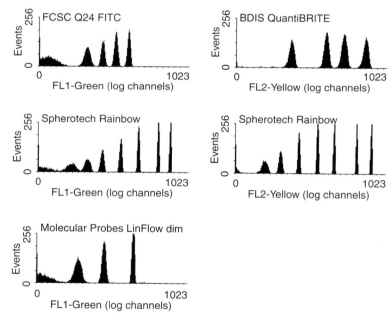

Fig. 8 Green and yellow fluorescence of beads with multiple levels of fluorescence intensity. Green fluorescence is shown for FCSC Q24, Spherotech Rainbow, and Molecular Probes low level LinFlow beads. Yellow fluorescence is shown for QuantiBRITE and Spherotech Rainbow beads. All data were were acquired with 15 mW 488 nm excitation and at constant PMT voltage settings to allow comparison of relative fluorescence of the particles. The green fluorescence filter had a bandwidth of 515–545 nm, and the yellow fluorescence filter had a bandwidth of 564–606 nm.

Correctly adjusting the compensation requires using particles with very accurately matched spectra. Stained cells or beads stained with the specific fluorochromes used to stain cells are usually used to adjust compensation. In some cases, fluorescent beads with emission spectra synthesized from multiple fluorophores provide adequate spectral matching for adjusting compensation. Setting compensation by visually observing data in a dot plot usually results in overcompensation. For quantitative fluorescence overcompensation will cause the measured fluorescence of the compensated parameter to be lower than the true value. Events in an unstained reference population that are not on scale (in channel zero) make the mean channel of an unstained population an inaccurate reference. Comparing median channels will usually give a more accurate and objective compensation setting (Zhang *et al.*, 1998). The most reliable method for adjusting compensation is to use a sample that has a double stained population whose separation from a "negative" population can be measured. Optimal compensation gives the largest separation of the double stained population from a population stained only with the interfering fluorochrome. An example of correctly compensated and overcompensated data are shown in Fig. 9.

6. Particles and Methods Used to Standardize Fluorescence Intensity

Schwartz has proposed a classification scheme for fluorescence particle standards depending on the size and uniformity of the particles and, compared to the fluorochrome intended to be standardized, the degree of spectral matching and degree to which the fluorescence responds to environmental factors such as pH (Schwartz *et al.*, 1998). Mixtures of fluorochromes embedded in polymer microbeads can very accurately reproduce the emission spectra of important tags such as FITC and PE, but will not be sensitive to environmental factors (Zhang *et al.*, 1998).

Fluorochromes such as fluorescein can have small spectral shifts depending on environment and conjugation factors, which will cause small changes in detected intensity when analyzed with a fixed detection filter (Schwartz and Fernandez-Repollet, 1993). When fluorescence emission spectra are not very closely matched, the differences in fluorescence intensity may or may not be a strong function of the wavelength range measured. Rainbow beads (Spherotech) have a broad emission spectrum due to a mixture of dyes in the particles. Figure 10 compares the emission spectra of Rainbow beads and FITC and PE CaliBRITE beads (BDIS). Typical, narrow, and broad band emission filters are represented in Fig. 10 by bars labeled G1, G2, and G3, respectively, for green fluorescence and Y1, Y2, and Y3, respectively, for yellow fluorescence. When compared to particles stained with FITC, the relative fluorescence from Rainbow beads can vary by 100% when different FITC filters are used. But when compared to PE fluorescence the relative fluorescence of Rainbow beads varies only by 15–20% when significantly different emission filters are used (Hoffman, 1997).

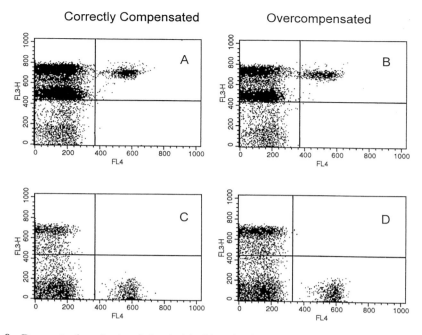

Fig. 9 Demonstration of reduced signal of doubly stained cells if fluorescence is overcompensated. Data are from human leukocytes stained with CD45-PerCP (FL3) and CD19-APC (FL4) in (A) and (B) or with CD8-PerCP (FL3) and CD19-APC (FL4) in (C) and (D). Data were acquired on a dual laser FACSCalibur (Becton Dickinson) with compensation set correctly (A and C) or overcompensated (B and D). There is essentially no autofluorescence from leukocytes in the FL4 parameter, so the reference for adjusting compensation in the FL4 channel is PMT noise. Histogram axes display fluorescence on a log scale with 256 channels per decade.

7. Making Fluorochrome-Stained Particle Standards

Fluorescent particle standards have been prepared by investigators using common biological cells or microbeads as starting material. The fluorochrome is covalently bound, adsorbed, or trapped in the particle. Table I lists a variety of particles prepared by investigators. Unlike most commercially available fluorescent particle standards, the "homemade" particles may have limited stability. In particular the fluorochrome may be released or leak from the particles over time. In addition to the fluorescent particles in Table I, a nearly nonfluorescent particle made by fixing erythrocytes with osmium tetraoxide has been reported (Loken and Herzenberg, 1975).

8. Calibrating Particles in MESF Units

The most direct method for determining the MESF of particles is to compare the fluorescence intensity of fluorochrome and particle suspensions of known

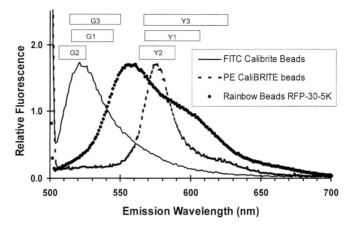

Fig. 10 Emission spectra of FITC and PE CaliBRITE beads and Rainbow beads. The transmission bands of different emission filters are represented by the bars labeled G1, G2, and G3 and Y1, Y2, and Y3.

concentration in a spectrofluorometer (Brown *et al.*, 1986; Davis *et al.*, 1998; Schwartz and Fernandez-Repollet, 1993). Concentration of the particle suspension must be measured in terms of single particles, so it may be necessary to make corrections for doublets and higher aggregates of particles. If the particles scatter light significantly in the spectrofluorometer it may also be necessary to correct for this effect on the fluorescence measurement. It is possible to dissolve some types of particles to eliminate the light scatter (Oonishi and Uyesaka, 1985). The use of erythrocyte ghosts (Doberstein *et al.*, 1995) has the advantage of very small light scatter contribution to the bulk fluorescence measurement

Table I
Methods Used to Make Fluorescent Particles

Method	Reference
Thymocyte nuclei stained with FITC or rhodamines	Brown *et al.* (1986)
Erythrocyte ghosts filled with fluorescent dextran or other fluorescently stained macromolecules such as antibodies	Doberstein *et al.* (1995)
Sephadex beads stained with FITC or rhodamine	Haaijman and Van Dalen (1974); Visser *et al.* (1978)
Stained Sephadex beads sorted for uniform size	Le Bouteiller *et al.* (1983)
Carboxylated polystyrene beads with adsorbed fluorochrome-coupled protein	Parks *et al.* (1984)
Osmium-fixed sheep erythrocytes stained with FITC	Loken and Herzberg (1975)
Osmium-fixed sheep RBC conjugated to antigen for indirect staining	Loken and Herzberg (1975)

in a spectrofluorometer and the opportunity to incorporate a wide variety of fluorescent macromolecules. Recktenwald *et al.* (1993) adapted the method of comparing a particle and fluorochrome solution fluorescence to a microscope method analyzing individual particles. The fluorescence intensity measured by a charge-coupled device (CCD) camera was calibrated by measuring the decrease in fluorescence caused by displacement of fluorochrome solution by a glass microparticle of measured diameter. The calibrated microscope and camera system was then used to measure the fluorescence from other fluorescent particles.

A simple subtractive method may also be used to measure the amount of fluorochrome bound to the particles after the staining process. In this case the fluorescence intensity of a known concentration of the staining solution is measured in a spectrofluorometer before adding the particles and measured again after the particles have been stained and removed. The difference in the fluorescence is proportional to the amount of stain taken up by the particles. By also measuring the concentration of singlet particles the MESF per particle can be determined (Brown *et al.*, 1986).

A method used by many investigators is to use an independent, quantifiable marker such as a radioisotope (de Bruin *et al.*, 1983; Dux *et al.*, 1991; Loken and Herzenber, 1975; Watson and Walport, 1986) or enzyme (Davis *et al.*, 1998) simultaneously and in known relation to the fluorochrome.

9. Commercially Available Fluorescent Particle Standards

Fluorescent particle standards are available from many commercial sources. Commercial sources for various categories of particles are listed by Schwartz (Schwartz *et al.*, 1998). Major sources of particles include Beckman Coulter (Fullerton, CA), BD Biosciences (San Jose, CA), DAKO (Carpenteria, CA), Flow Cytometry Standards Corporation/FCSC (Fishers, IN), Molecular Probes (Eugene, OR), Polysciences (Warrington, PA), Riese Enterprises (Grass Valley, CA), and Spherotech (Libertyville, IL).

10. Comparing Spectrally Matched and Unmatched Fluorescent Standards

To illustrate different types of standards, two commonly used fluorescent particles will be compared. Rainbow beads (Spherotech) are polymeric particles with a mixture of fluorochromes imbedded internally. QuantiBRITE (BD Biosciences, San Jose, CA) are polymeric particles with the fluorochrome R-PE conjugated to the surface. Both particles are reasonably stable and can be used to compare fluorescence signals from an instrument over time. But the particles do not have the same fluorescence emission spectrum—as shown in Fig. 11.

Either type of particle could provide a standard for relative fluorescence intensity because each product is a mixture of beads with different levels of fluores-

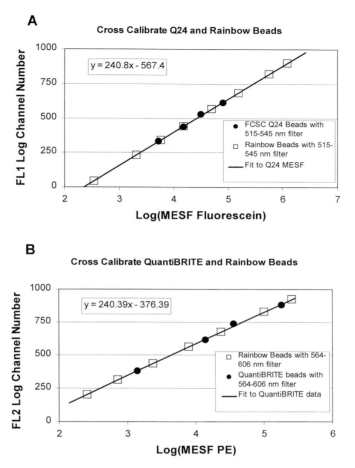

Fig. 11 (A) Calibration of log amp and cross-calibration between FCSC FITC particles and Rainbow beads with 515–545 nm filter. (B) Calibration of log amp and cross-calibration between QuantiBRITE and Rainbow beads with 564–606 nm FL2 filter. The log amplifier for green fluorescein fluorescence had 240.8 channels per decade, and the log amplifier for yellow PE fluorescence had 240.4 channels per decade.

cence intensity as shown in Fig. 9. Also over a particular wavelength range, for example, 560–600 nm, the beads might provide a consistent absolute standard. This is because both types of particles have stable values of fluorescence over this wavelength range. The fluorescence intensity of QuantiBRITE particles could be expressed in absolute terms of MESF PE for any wavelength range because the particles have the same emission spectrum (and excitation spectrum) as PE. But the emission spectra of Rainbow beads and QuantiBRITE are not the same, and the relative fluorescence of Rainbow beads compared to Quanti-BRITE will depend on the emission wavelength range, excitation wavelength, and excitation intensity.

With the Rainbow beads one can compare very accurately and quantitatively across laboratories the quantity "fluorescence from Rainbow beads." With particles conjugated to PE one can compare the quantity "PE fluorescence." Which of these standards to use depends on the analytical result one is trying to obtain from the cytometer. Generally one is more interested in knowing how many fluorescent molecules are present in a cell than in how the fluorescence relates to a broadly fluorescent particle. The PE-conjugated particles can be used directly as a secondary standard for PE fluorescence intensity by comparison with a primary standard of PE. As described later, the Rainbow beads might also be used as a tertiary standard for PE fluorescence intensity under very restrictive conditions.

11. Tertiary Fluorescence Standards and Calibrators

Since there is a constant relationship between the spectra of two different fluorescing materials, the relative intensities of particles with different spectra measured through the same optical filter is constant. Thus cross-referencing of Rainbow beads or nonspectrally matched particles to spectrally matched beads calibrated in MESF can generally be done for any *one* instrument (Schwartz *et al.*, 1996). To ensure that the secondary standard or calibrator on an instrument is reliable; however, the emission filters and the wavelength and intensity of the excitation light must remain unchanged (Schwartz *et al.*, 1996).

Figure 11A shows cross-calibration of Spherotech Rainbow beads to FCSC Q24 FITC beads with a 515- to 545-nm filter (represented by G1 in Fig. 10). The channels/decade are determined from a plot of median channel number versus MESF. For this case the log amp response was 240 channels/decade. The equation describing the relationship between channels and MESF can be used to assign MESF values to the Rainbow beads when their median channel is measured. Because Rainbow beads poorly match the FITC spectrum over most of the relevant wavelength range, very large errors can result if MESF values assigned to the Rainbow bead with one filter are used to calibrate fluorescence intensity with a different filter. For example, if MESF values were assigned to Rainbow beads with a 515–545 nm filter, they would cause a 100% error if used to calibrate fluorescein MESF with a 505–525 nm filter (represented by G2 in Fig. 10) and a 30% error if used to calibrate fluorescein fluorescence with a 512–550 nm filter (represented by G3 in Fig. 10).

Figure 11B shows cross-calibration of Spherotech Rainbow beads to PE beads (QuantiBRITE) with a 564–606 nm filter (represented by Y1 in Fig. 10). Channels per decade were determined by plotting median channel versus MESF for the QuantiBRITE beads. MESF of the Rainbow beads was determined from the equation relating channel number to MESF. If Rainbow beads were calibrated against PE beads with the 564–606 nm filter and then used as a calibration standard with the alternative filters with transmission bands of 562–588 nm or 564–627 nm (represented by Y2 and Y3 in Fig. 10), the assigned MESF values would be about 20% in error.

12. Cautions in Comparing Fluorescence from Different Types of Particles

Above a threshold excitation intensity, fluorochromes have a nonlinear increase in fluorescence intensity as excitation intensity is increased. At sufficiently high excitation intensity, the fluorescence output saturates and the fluorochrome may photobleach. The saturation and bleaching phenomena are quite variable for different fluorochromes, so relative fluorescence intensity between two fluorochromes can be expected to vary with excitation intensity (Bohmer et al., 1985; Doornbos et al., 1997). This means that a cross-calibration between two fluorescent intensity bead standards at one laser power may not be valid at a different laser power. Figure 12 shows the relative fluorescence of Rainbow beads compared to beads conjugated to either FITC or PE at various intensities of laser excitation. As excitation power is increased, the fluorescence of FITC or PE decreases relative to fluorescence from Rainbow beads. This is especially noticeable with PE.

Fluorescence resonance energy transfer (FRET) is the transfer of excited state energy from one fluorochrome to a second near-by fluorochrome molecule that subsequently fluoresces (Chan et al., 1979; Stokke et al., 1991; Szollosi et al., 1984, 1987; Tron et al., 1984). FRET is the process that produces the longer wavelength fluorescence in tandem fluorochromes such as PE-CY5 (Lansdorp et al., 1991) or PE-Texas Red. In attempting to quantitate fluorescence from particles stained with multiple fluorochromes, FRET can introduce an unexpected artifact that reduces the intensity of the fluorochrome with lower emission wavelength and increases the intensity from the fluorochrome with higher emission wavelength. At sufficiently high staining concentrations, a fluorochrome such as fluorescein can even interact with other fluorescein molecules to self-quench and shift the overall emission spectrum to the red.

Fluorescence polarization is a potentially important phenomenon that has not been widely studied for impact on standardization. The emission of fluorescence may be more or less optically polarized depending on the environment of the fluorochrome, and fluorescence polarization has been used to investigate cells in flow cytometry (Beisker and Eisert, 1989; Bock et al., 1989; Epstein et al., 1977; Fox and Delohery, 1987; Lindmo and Steen, 1977; Rolland et al., 1985; Weaver et al., 1997). Emission anisotropy is a standard measure of the degree of fluorescence polarization, with completely unpolarized fluorescence having a value of 0 and completely polarized emission a value of 1. The more rigidly bound the fluorochrome, the more the fluorescence will be polarized. Many optical components including dichroic mirrors and filters preferentially transmit or reflect one polarization component. Depending on the emission anisotropy

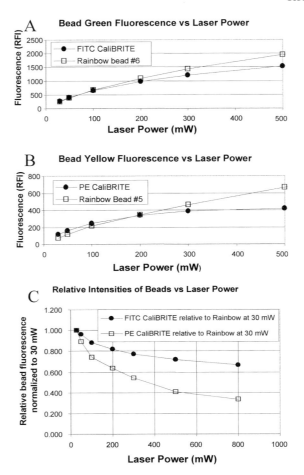

Fig. 12 Compare intensity of bead fluorescence versus laser power for beads conjugated to FITC or PE (CaliBRITE beads) and internally stained with a mixture of fluorochromes (Rainbow beads). (A and B) Relative fluorescence versus laser power for FITC CaliBRITE and Rainbow beads and for PE CaliBRITE and Rainbow beads, respectively. (C) The ratio of fluorescence of FITC or PE beads to Rainbow beads normalized to intensity at 30 mW laser power plotted as a function of laser power.

of the fluorescent particles, different flow cytometer optical designs and even different components in the same basic design can give more or less fluorescence from fluorochrome-stained particles (Beisker and Eisert, 1981, 1985; Lindmo and Steen, 1977). The relative emission anisotropy of fluorescence standardization particles and the types of fluorescently stained cells they are intended to standardize deserve more study (Dr. Ger Van den Engh, personal communication, 1998).

B. Light Scatter

Light scattered by particles provides information about size, shape, and intrinsic optical properties. There have been no attempts to calibrate or quantitate intensity of scattered light in absolute terms, but the relative intensity of light scattered in one or more angular ranges is an important discriminator in many flow cytometry applications. Although forward scattered light is roughly dependent on particle size, this measure is not simply related to size or necessarily even a monotonic function of size. For example, polystyrene beads 6 μm in diameter can have smaller forward scatter than polystyrene beads 5 μm in diameter (Doornbos *et al.*, 1994). Side scatter is a highly nonlinear function of particle diameter. This strong dependence on diameter can cause large CVs in the side scatter measurement of nearly uniform particles which have small variation in diameter (Doornbos *et al.*, 1994). The scatter signals can also depend on the orientation of nonspherical particles (Loken *et al.*, 1977a).

Figure 13 (Schwartz and Fernandez-Repollet, 1994) illustrates the variable light scatter distributions that the same sample can give on different instruments. Standardization of the angular range of light detected is the most important factor in obtaining repeatable results with light scatter (Jovin *et al.*, 1976; Kerker, 1983; Mullaney *et al.*, 1976; Sharpless *et al.*, 1977; Tycko *et al.*, 1985). This will be determined primarily by the optomechanical design and alignment of the instrument and also by the position of the sample stream within the flow cell or fluid jet (Salzman *et al.*, 1979). On commercial instruments that are not intended to require user adjustment of the optical system, the usual cause of light scatter problems is an improperly positioned sample stream. Cleaning the flow cell and/or sample injector generally returns the sample stream position and light scatter to normal conditions.

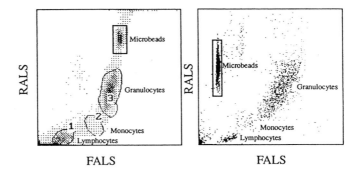

Fig. 13 Comparison of leukocyte and microbead light scatter on two different types of flow cytometer. Because the light scatter collection angles of the two cytometers are different, the relative forward scatter of beads and cells is different on the two instruments. (From Schwartz and Fernandez-Repollet, 1994, used with permission of Academic Press.)

Because designs vary among different commercial instruments, it is best to follow the manufacturers recommendations for testing and standardizing the light scatter channels. Each manufacturer will have established standardization methods and test materials that are appropriate for their design. In most applications of light scatter, a relative scale, either linear or logarithmic, is established using biological samples. Once settings for an application are determined on a particular instrument, a bead or stabilized biological sample can be analyzed under the same conditions and used to set or check the conditions for future analyses.

For lowest optical noise due to unwanted light at the detectors, the refractive index of the sample and sheath streams should be matched to minimize refraction of light at the sample/sheath interface. A common observation is that simply using deionized water as sample fluid for 2-μm beads and buffered normal saline as sheath causes sufficient refractive index mismatch so that the CV of forward scatter from the beads is degraded.

Because light scatter depends on multiple intrinsic optical properties of the particle, it is very difficult to define a simple scale that relates light scatter intensity to a specific particle property (Doornbos et al., 1994; Jovin et al., 1976; Kerker, 1983; Mullaney et al., 1976; Sharpless et al., 1977; Tycko et al., 1985). Over limited ranges and types of particles, forward light scatter has been used to measure particle size. For example, Robertson et al. calibrated forward light scatter of marine bacteria to measure size (Robertson and Button, 1989) and dry mass (Robertson et al., 1998). The ratio of light scatter intensity at two different angles can also be used to measure particle size in certain cases (Jovin et al., 1976; Tycko et al., 1985). This is discussed in the following section on particle sizing.

C. Particle Sizing

Particle size can be reliably measured as either the length (in the direction of flow) or the volume of the particle. For spherical particles, the length is equal to diameter and can be used to calculate volume. Diameter or length of a particle is determined from the width of a fluorescence or light scatter pulse (Leary et al., 1979; Schrader and Eisert, 1986; Sharpless and Melamed, 1976). The pulse width can be measured directly or calculated indirectly by the ratio of pulse area to pulse height (Sharpless and Melamed, 1976). This measurement is most precise when the light beam is tightly focused to a height (in the direction of flow) smaller than the length of the particle, but useful length measurements can be made for particles many times smaller than the beam height. Most commercial flow cytometers provide a way to make the pulse width measurement, but they have not usually been optimized for making quantitative measurements. A calibration curve can be generated with uniform microspheres of known diameter (Galbraith et al., 1988; Monroe et al., 1982; Sharpless and Melamed, 1976). Multiple sizes of calibration particles spanning the range of interest should be used.

Particle volume can be measured by either electronic impedance sensing (also called Coulter volume) (Kachel, 1990) or in certain cases by light scatter (Jovin *et al.*, 1976; Robertson and Button, 1989; Tycko *et al.*, 1985). Electronic cell volume is a routine method for sizing particles in stand-alone instruments and automated hematology analyzers, but few commercial optical flow cytometers incorporate this as an additional capability. For spherical particles, electronic volume can be calibrated with uniform microspheres of known diameter, and the diameter and surface area can also be measured once the volume measurement has been calibrated (Schwartz *et al.*, 1983). Interpretation of the results should take into account that the volume signals depend on particle shape as well as volume, and an accurate measurement depends on the particles being good electrical insulators. For cells this means the plasma membrane must be intact, or else the electronic volume measurement will be lower than the actual cell volume (Hoffman *et al.*, 1981; Kachel, 1990).

A few investigators have used the ratio of light scatted into two different angular ranges as a measure of size (Jovin *et al.*, 1976; Tycko *et al.*, 1985). The method applies strictly only to uniform spheres, a criterion met by few biological cells. Meyer and Brunsting have shown theoretically that the ratio of two angles of light scatter below 2° would be more accurate than a single angle measure to size nucleated cells (Meyer and Brunsting, 1975). Tycko *et al.* developed an elegant method for measuring the volume and hemoglobin concentration of red blood cells by isovolumetrically sphering the cells and using the correlated data from light scattered into two different angles. A unique aspect of this approach is that standardization and calibration of both the volume and hemoglobin concentration measurements can be performed with oil droplets of highly variable and unknown size but known refractive index (Tycko *et al.*, 1985).

D. Particle Concentration

In addition to measurements on individual particles, flow cytometry is also useful for measuring concentrations of particles. Sometimes this is called absolute counting. Automated hematology analyzers are specialized flow cytometers routinely used for measuring the concentrations of the major types of blood cells. The most straightforward way to measure particle concentration is to analyze a known volume of sample, and some commercially available optical flow cytometers have been designed to make volumetric measurements. Some investigators have added this capability in a homemade way (Robins and Bedo, 1994). An alternative approach is to mix a known volume of sample with a known number of reference beads or other particles (Stewart and Steinkamp, 1982). Standardization and calibration of the method usually uses reference samples whose particle concentrations are determined by reference methods such as counting chambers or calibrated electronic particle counters. It is important to prevent particles in the reference samples from aggregating because particles lost in aggregates will cause the reference concentration to be low.

High counting rates are an additional source of error due to coincidences of particles. The maximum counting rate at which accurate counting can be done depends on instrument design and counting method. The concentration range that can be accurately measured can be determined by diluting a very concentrated sample into a series of known concentrations and comparing the expected count to the measured value.

V. Examples of Applications Using Calibrated Measurements

With the exceptions of particle size and concentration described earlier, calibrated measurements in flow cytometry have used fluorescence. DNA content analysis has been widely performed as a quantitative measurement in flow cytometry, and approaches to standardize the preparation and analysis are well developed (Dressler and Seamer, 1994). Typically one or more type of reference cell is stained with the same procedure used for the test sample, and the fluorescence intensity of the test and reference cells are compared (Iversen and Laerum, 1987; Jakobsen, 1983; Vindelov *et al.*, 1983). Quantitation is usually stated as a DNA index, that is, the ratio of the mean intensity of the test sample to the reference cell type. The stainability of DNA can vary with the chromatin structure and staining procedure (Darzynkiewicz *et al.*, 1984; Evenson *et al.*, 1986), so careful attention should be given to these issues when quantitative DNA measurements are to be made (Myc *et al.*, 1992). Absolute measurement of DNA per cell is also possible if care is taken to account for stainability and if reference cells of known DNA mass are available. Tiersch *et al.* (1989) measured the DNA content in blood cells of 45 species with propidium iodide fluorescence. To convert the relative values of fluorescence to absolute values of DNA mass, the data were referenced to DNA mass of human leukocytes, which was assigned a value of 7.0 pg per cell based on a variety of independent studies. DNA mass values in the 45 species ranged from 1.5 to 110.0 pg of DNA per cell. These species could provide reference values of DNA mass for quantitative DNA flow cytometry.

Calibrated measurements of intracellular calcium (Chused *et al.*, 1987), pH (Chow *et al.*, 1996), or membrane potential (Jenssen *et al.*, 1986; Krasznai *et al.*, 1995; Shapiro, 1995) have been made using fluorescent probes whose response is calibrated against known ion concentrations or gradients. Membrane potential measurements were also calibrated against microelectrode measurements (Jenssen *et al.*, 1986; Krasznai *et al.*, 1995). The chlorophyll content of plant protoplasts has been measured by calibrating chlorophyll autofluorescence intensity to chlorophyll mass (Galbraith *et al.*, 1988).

A. Antibodies, Antigens, Receptors, and Other Ligands Bound per Cell

A wide variety of molecules on cells have been quantitated by immunofluorescence flow cytometry, and several different approaches to calibrating the measurements have been taken. (Hoffman *et al.*, 1993) At best one can expect to measure the number of antibodies or ligands bound, although the ideal immunoassay would quantitate the number of antigens on the cell. Table II lists examples of molecules quantitated by flow cytometry and the methods used for calibration. Additional discussion of this application is found in Section VI.

Table II
Examples of Molecules Quantitated by Flow Cytometry and Method Used for Calibration

Molecule quantitated	Calibration method	Reference
Molecules of FITC conjugated anti-mouse Fab antibody	Antibody double labeled with ^{125}I and FITC for cross-calibration of antibody bound to cells	Loken and Herzenberg (1975)
CD antibodies on cell lines	CD5 on cell line calibrated by radioisotope labeling and used as reference for quantitative indirect immunofluorescence (QIFI)	Poncelet and Carayon (1985)
Differentiation antigen on rat brain cells	Reference cell line whose antigen per cell was determined by radioligand binding	Dux *et al.* (1991)
GM-CSF receptor	Radioligand binding assay	Stacchini *et al.* (1996)
Surface antigens on human leukocytes	Indirect immunofluorescence and calibrated beads with known capacity for binding anti-mouse antibody (QIFI)	Bikoue *et al.* (1996)
Epidermal growth factor	Isoparametric analysis (see reference for definition of this method)	Chatelier *et al.* (1986)
Complement receptor type 1 (CR1)	Antibody double labeled with ^{125}I and FITC used for parallel radioimmunoassay and flow cytometry	Watson and Walport (1986)
Fc receptors	FITC-conjugated antibody with measured MESF/antibody molecule and calibrated MESF beads	Christensen and Leslie (1990)
CD38 molecules on lymphocytes	Compared CD4 biological standard to PE-conjugated antibody with known fluorochrome/antibody ratio and calibration beads with known number of PE molecules per bead	Iyer *et al.* (1998)
CD38 molecules on lymphocytes	CD4 on lymphocytes as a biological standard	Hultin *et al.* (1998)
Binding of fatty acid analogs to lymphoid cells	Fluorescence intensity cross-calibration of fatty acid fluorochrome tag to fluorescein calibration beads through same optical filters used on flow cytometer	Macho *et al.* (1996)
CD4 on lymphocytes	Used multiple, cross-checked methods to calibrate measurements using PE-conjugated antibody	Davis *et al.* (1998)

B. Bead Immunoassay

A wide variety of flow cytometric immunoassays using coated beads have been developed (Bishop and Davis, 1997; Collins *et al.*, 1998; Lindmo *et al.*, 1990; Lisi *et al.*, 1982; McHugh *et al.*, 1989, 1997; Renner, 1994; Saunders *et al.*, 1985). The general format uses the mean intensity of fluorescence on a capture bead as the readout. The assays are usually calibrated with reference material containing known concentrations or units of the analyte.

VI. Issues in Quantitation of Fluorochromes and Other Molecules

The interest in quantitative fluorescence measurements and their use in making quantitative measurements of specific molecules has continued to increase since the mid-1980s. Over that time many of the problems that need to be resolved have been worked on by numerous groups. Steady progress has been made, but quantitative fluorescence cytometry is not yet a mature technology. A special issue of the journal Cytometry was devoted to the topic "Quantitative Fluorescence Cytometry: An Emerging Consensus" and provides a good status report as of 1998 [Lenkei, Mandy, Marti, and Vogt (eds.), 1998]. Gratama *et al.* (1998) have given an excellent overview of the issues in immunofluorescence intensity quantitation.

There is general agreement that fluorescence should be quantitated in units of MESF. There is also general agreement that a next goal should be reliable and practical methods to accurately quantitate the number of fluorescent antibody molecules or other ligands on particles or cells. Several different methods to quantitate antibodies bound per cell (ABC) are in use, and at this time different methods often give results that do not agree (Lenkei *et al.*, 1998).

A. Issues in Calibrated Measurements of Antibodies Bound per Cell

The disagreement between methods to measure ABC is not primarily due to problems with standardizing instruments to quantitatively measure fluorescence intensity but rather due to using different methods for sample preparation and calibration. Deviation from perfect log response can be corrected, if necessary, to eliminate this source of error (Muirhead *et al.*, 1983; Purvis and Stelzer, 1998; Schmid *et al.*, 1988). Lenkei's study with three different flow cytometers showed good agreement between instruments when the same sample and same method to determine ABC was used on each instrument. But there were large differences in ABC when different methods are used to calibrate the antibodies bound.

Lenkei *et al.* (1998) reported on the performance of four different approaches to calibration standards for antibody binding capacity. The quantitative indirect immunofluorescence assay (QIFI) uses beads with calibrated amounts of mouse immunoglobulin for calibrating a second-step, anti-mouse antibody. Quantum simply cellular (QSC) beads have anti mouse antibody on the surface with calibrated binding capacity for mouse immunoglobulin G (IgG). An approach using PE antibody with presumed 1:1 ratio of PE to antibody was calibrated using QuantiBRITE beads with known amounts of PE attached to the surface. A final method used CD4 on normal lymphocytes as a biological standard. In general, the differences between these calibration methods ranged from 30 to 300%, depending on the methods and CD antibody compared.

Using the indirect QIFI method, Bikoue et al. (1996) observed that different monoclonal antibodies to a molecule (e.g., CD8) can give different ABC values on the same cells. Whether this is due to differences in binding of the primary antibody to cells or to differences in binding of the secondary antibody to the primary antibody is not clear. Altered reactivity of the second-step reagent with fluorochrome-conjugated mouse monoclonal antibody is another possible variable.

The QSC method is designed to capture quantitatively any mouse monoclonal antibody independent of fluorochrome conjugation or IgG isotype. Lenkei and Andersson (1995), however, found that conjugates of the same monoclonal antibody with different fluorophores gave different ABC values when calibrated with QSC beads. They also observed that, as with the QIFI method, different antibody clones directed against the same molecule could give differing ABC results with QSC beads (Lenkei and Andersson, 1995).

At this time, the best expectation is that one can obtain consistent results between instruments and laboratories when the same calibration method is used along with the same antibody reagents and sample preparation protocols. The method using 1:1 PE antibody conjugates and calibration of the fluorescence intensity scale in PE MESF has the fewest potential artifacts and is the easiest to interpret.

B. Issues in Quantitating Antigens or Ligands per Cell

The ultimate goal of quantitation is measuring the number of antigens or ligand binding sites on a cell. This measurement is not necessarily equal to the number of bound antibodies, because some binding sites may not be accessible or a single antibody might bind two antigens. Additional issues remain to be resolved to achieve this goal. A first issue to clarify is in what state a stained cell should be measured. Different methods of sample preparation are a major factor in determining what antigens on a cell are accessible to antibody. Different methods used to lyse erythrocytes or isolate leukocytes from whole blood cause significant differences in immunofluorescence intensity (Islam et al., 1995; Webster and Pockley, 1993).

Davis et al. (1998) provide a detailed analysis of the factors that affect quantitation of the number of antigens per cell. Measurement of antigen for CD4 was found to depend on the affinity of the antibody, whether the CD4 epitope was accessible to bind both Fab sites of an antibody, and whether the cells were fixed after staining. As is the case for quantitating ABC, consistent results can be achieved by using the same method for sample preparation and the same reagents. But standards for these do not currently exist.

C. Next Steps in Quantitative Fluorescence Flow Cytometry

The instrumentation and analysis issues for quantitative flow cytometry are well understood and can be taken into account. But the corrections to log amplifier response and potential corrections for color compensation are not easily and conveniently done with currently available systems. Automated methods for calibration and corrections to ideal response would allow accurate data to be available for all investigators.

Standardization and sample preparation methods for quantitating bound antibody (ABC) or ligand will require further comparison to understand the reasons for the different results they provide. A first step should be agreement on a reference method to define a true answer to which results of other methods can be compared. The use of radioisotope-labeled antibodies or ligands might be a useful reference method. A thorough understanding of the factors that cause differences between an alternative calibration method and the reference method will define the conditions under which the method can be expected to give accurate results.

When reference and practical methods for quantitating bound antibody or ligand are available, the quantitation of antigens or other binding sites on cells can be studied objectively and systematically. The study of CD4 antigen binding by Davis *et al.* (1998) provides a good example of how binding valency, avidity, and other factors can be taken into account. It may be necessary to develop special reagents (e.g., univalent) to provide reliable and accurate antigen quantitation. Although there are many problems to be resolved, the goal is worthwhile and should be obtainable.

References

Bagwell, C. B., and Adams, E. G. (1993). Fluorescence spectral overlap compensation for any number of flow cytometry parameters. *Ann. N.Y. Acad. Sci.* **677,** 167–184.

Bagwell, C. B., Baker, D., Whetstone, S., Munson, M., Hitchcox, S., Ault, K. A., and Lovett, E. J. (1989). A simple and rapid method for determining the linearity of a flow cytometer amplification system. *Cytometry* **10,** 689–694.

Bauer, K. D. (1993). Quality control issues in DNA content flow cytometry. *Ann. N.Y. Acad. Sci.* **677,** 59–77.

Beisker, W., and Eisert, W. G. (1981). Principles of fluorescence polarization measurements. *Anal. Quant. Cytol.* **3,** 315–322.

Beisker, W., and Eisert, W. G. (1985). Double-beam autocompensation for fluorescence polarization measurements in flow cytometry. *Biophys. J.* **47,** 607–612.

Beisker, W., and Eisert, W. G. (1989). Denaturation and condensation of intracellular nucleic acids monitored by fluorescence depolarization of intercalating dyes in individual cells. *J. Histochem. Cytochem.* **37,** 1699–1704.

Bikoue, A., George, F., Poncelet, P., Mutin, M., Janossy, G., and Sampol, J. (1996). Quantitative analysis of leukocyte membrane antigen expression: Normal adult values. *Cytometry* **26,** 137–147.

Bishop, J. E., and Davis, K. A. (1997). A flow cytometric immunoassay for β2-microglobulin in whole blood. *J. Immunol. Methods* **210,** 79–87.

Bock, G., Huber, L. A., Wick, G., and Traill, K. N. (1989). Use of a FACS III for fluorescence depolarization with DPH. *J. Histochem. Cytochem.* **37,** 1653–1658.

Bohmer, R. M., Papaioannou, J., and Ashcroft, R. G. (1985). Flow-cytometric determination of fluorescence ratios between differently stained particles is dependent on excitation intensity. *J. Histochem. Cytochem.* **33,** 974–976.

Brown, E. N. (1996). Theoretical and practical implications of a new definition of the minimal detectable concentration for immunoassays. *Proc. SPIE* **2680,** 80–91.

Brown, E. N., McDermott, T. J., Bloch, K. J., and McCollom, A. D. (1996). Defining the smallest analyte concentration an immunoassay can measure [see comments:Comment in: *Clin. Chem.* (1997) May;**43**(5), 856–857. Comment in: *Clin. Chem.* (1997) Oct.;**43**(10), 2010–2011]. *Clin. Chem.* **42,** 893–903.

Brown, M. C., Hoffman, R. A., and Kirchanski, S. J. (1986). Controls for flow cytometers in hematology and cellular immunology. *Ann. N.Y. Acad. Sci.* **468,** 93–103.

Buttner, J. (1991). Philosophy of measurement by means of immunoassays. *Scand. J. Clin. Lab. Invest. Suppl.* **205,** 11–20.

Chan, S. S., Arndt-Jovin, D. J., and Jovin, T. M. (1979). Proximity of lectin receptors on the cell surface measured by fluorescence energy transfer in a flow system. *J. Histochem. Cytochem.* **27,** 56–64.

Chase, E. S., and Hoffman, R. A. (1998). Resolution of dimly fluorescent particles: A practical measure of fluorescence sensitivity. *Cytometry* **33,** 267–279.

Chatelier, R. C., Ashcroft, R. G., Lloyd, C. J., Nice, E. C., Whitehead, R. H., Sawyer, W. H., and Burgess, A. W. (1986). Binding of fluoresceinated epidermal growth factor to A431 cell subpopulations studied using a model-independent analysis of flow cytometric fluorescence data. *EMBO J.* **5,** 1181–1186.

Chow, S., Hedley, D., and Tannock, I. (1996). Flow cytometric calibration of intracellular pH measurements in viable cells using mixtures of weak acids and bases. *Cytometry* **24,** 360–367.

Christensen, J., and Leslie, R. G. (1990). Quantitative measurement of Fc receptor activity on human peripheral blood monocytes and the monocyte-like cell line, U937, by laser flow cytometry. *J. Immunol. Methods* **132,** 211–219.

Chused, T. M., Wilson, H. A., Greenblatt, D., Ishida, Y., Edison, L. J., Tsien, R. Y., and Finkelman, F. D. (1987). Flow cytometric analysis of murine splenic B lymphocyte cytosolic free calcium response to anti-IgM and anti-IgD. *Cytometry* **8,** 396–404.

Collins, D. P., Luebering, B. J., and Shaut, D. M. (1998). T-lymphocyte functionality assessed by analysis of cytokine receptor expression, intracellular cytokine expression, and femtomolar detection of cytokine secretion by quantitative flow cytometry. *Cytometry* **33,** 249–255.

Darzynkiewicz, Z., Traganos, F., Kapuscinski, J., Staiano-Coico, L., and Melamed, M. R. (1984). Accessibility of DNA in situ to various fluorochromes: Relationship to chromatin changes during erythroid differentiation of Friend leukemia cells. *Cytometry* **5,** 355–363.

Davis, K. A., Abrams, B., Iyer, S. B., Hoffman, R. A., and Bishop, J. E. (1998). Determination of CD4 antigen density on cells: Role of antibody valency, avidity, clones, and conjugation. *Cytometry* **33,** 197–205.

de Bruin, H. G., de Leur-Ebeling, I., and Aaij, C. (1983). Quantitative determination of the number of FITC-molecules bound per cell in immunofluorescence flow cytometry. *Vox Sang.* **45,** 373–377.

Doberstein, S. K., Wiegand, G., Machesky, L. M., and Pollard, T. D. (1995). Fluorescent erythrocyte ghosts as standards for quantitative flow cytometry. *Cytometry* **20,** 14–18.

Doornbos, R. M., Hoekstra, A. G., Deurloo, K. E., De Grooth, B. G., Sloot, P. M., and Greve, J. (1994). Lissajous-like patterns in scatter plots of calibration beads. *Cytometry* **16,** 236–242.

Doornbos, R. M., de Grooth, B. G., and Greve, J. (1997). Experimental and model investigations of bleaching and saturation of fluorescence in flow cytometry. *Cytometry* **29,** 204–214.

Dressler, L. G., and Seamer, L. C. (1994). Controls, standards, and histogram interpretation in DNA flow cytometry. *Methods Cell Biol.* **41,** 241–262.

Durand, R. E. (1994). Calibration of flow cytometer detector systems. *Methods Cell Biol.* **42** Pt. B, 597–604.

Dux, R., Kindler-Rohrborn, A., Lennartz, K., and Rajewsky, M. F. (1991). Calibration of fluorescence intensities to quantify antibody binding surface determinants of cell subpopulations by flow cytometry. *Cytometry* **12,** 422–428.

Epstein, M., Norman, A., Pinkel, D., and Udkoff, R. (1977). Flow system fluorescence polarization measurements on fluorescein diacetate-stained EL4 cells. *J. Histochem. Cytochem.* **25,** 821–826.

Evenson, D., Darzynkiewicz, Z., Jost, L., Janca, F., and Ballachey, B. (1986). Changes in accessibility of DNA to various fluorochromes during spermatogenesis. *Cytometry* **7,** 45–53.

Fox, M. H., and Delohery, T. M. (1987). Membrane fluidity measured by fluorescence polarization using an EPICS V cell sorter. *Cytometry* **8,** 20–25.

Galbraith, D. W., Harkins, K. R., and Jefferson, R. A. (1988). Flow cytometric characterization of the chlorophyll contents and size distributions of plant protoplasts. *Cytometry* **9,** 75–83.

Gandler, W., and Shapiro, H. (1990). Logarithmic amplifiers [published erratum appears in *Cytometry* (1990);**11**(6), 744]. *Cytometry* **11,** 447–450.

Gaucher, J. C., Grunwald, D., and Frelat, G. (1988). Fluorescence response and sensitivity determination for ATC 3000 flow cytometer. *Cytometry* **9,** 557–565.

Gratama, J. W., D'hautcourt, J. L., Mandy, F., Rothe, G., Barnett, D., Janossy, G., Papa, S., Schmitz, G., and Lenkei, R. (1998). Flow cytometric quantitation of immunofluorescence intensity: Problems and perspectives. European Working Group on Clinical Cell Analysis. *Cytometry* **33,** 166–178.

Haaijman, J. J., and Van Dalen, J. P. (1974). Quantification in immunofluorescence microscopy. A new standard for fluorescein and rhodamine emission measurement. *J. Immunol. Methods* **5,** 359–374.

Henderson, L. O., Marti, G. E., Gaigalas, A., Hannon, W. H., and Vogt, R. F., Jr. (1998). Terminology and nomenclature for standardization in quantitative fluorescence cytometry. *Cytometry* **33,** 97–105.

Hoffman, R. A. (1997). Standardization, calibration, and control in flow cytometry. *In* "Current Protocols in Cytometry" (J. P. Robinson, ed.), Unit 1.3. Wiley, New York.

Hoffman, R. A., Johnson, T. S., and Britt, W. B. (1981). Flow cytometric electronic direct current volume and radiofrequency impedance measurements of single cells and particles. *Cytometry* **1,** 377–384.

Hoffman, R. A., Recktenwald, D. J., and Vogt, R. F., Jr. (1993). Cell-associated receptor quantitation. *In* "Clinical Flow Cytometry: Principles and Application" (K. D. Bauer, R. E. Duque, and T. V. Shankey, eds.), pp. 469–477. Williams & Wilkins, Baltimore.

Horan, P. K., Muirhead, K. A., and Slezak, S. E. (1990). Standards and controls in flow cytometry. *In* "Flow Cytometry and Sorting" (M. R. Melamed, T. Lindmo, and M. L. Mendelsohn, eds.), pp. 397–414. Wiley-Liss, New York.

Hultin, L. E., Matud, J. L., and Giorgi, J. V. (1998). Quantitation of CD38 activation antigen expression on $CD8^+$ T cells in HIV-1 infection using CD4 expression on $CD4^+$ T lymphocytes as a biological calibrator. *Cytometry* **33,** 123–132.

Hurley, A. A. (1997a). Components of quality control. *In* "Current Protocols in Cytometry" (J. P. Robinson, ed.), Unit 3.2. Wiley, New York.

Hurley, A. A. (1997b). Principles of quality control. *In* "Current Protocols in Cytometry" (J. P. Robinson, ed.), Unit 3.1. Wiley, New York.

Islam, D., Lindberg, A. A., and Christensson, B. (1995). Peripheral blood cell preparation influences the level of expression of leukocyte cell surface markers as assessed with quantitative multicolor flow cytometry. *Cytometry* **22,** 128–134.

Iversen, O. E., and Laerum, O. D. (1987). Trout and salmon erythrocytes and human leukocytes as internal standards for ploidy control in flow cytometry. *Cytometry* **8,** 190–196.

Iyer, S. B., Hultin, L. E., Zawadzki, J. A., Davis, K. A., and Giorgi, J. V. (1998). Quantitation of CD38 expression using QuantiBRITE beads. *Cytometry* **33,** 206–212.

Jakobsen, A. (1983). The use of trout erythrocytes and human lymphocytes for standardization in flow cytometry. *Cytometry* **4**, 161–165.

Jenssen, H. L., Redmann, K., and Mix, E. (1986). Flow cytometric estimation of transmembrane potential of macrophages—a comparison with microelectrode measurements. *Cytometry* **7**, 339–346.

Jovin, T. M., Morris, S. J., Striker, G., Schultens, H. A., Digweed, M., and Arndt-Jovin, D. J. (1976). Automatic sizing and separation of particles by ratios of light scattering intensities. *J. Histochem. Cytochem.* **24**, 269–283.

Kachel, V. (1990). Electrical resistance pulse sizing: Coulter sizing. In "Flow Cytometry and Sorting" (M. R. Melamed, T. Lindmo, and M. L. Mendelsohn, eds.), pp. 45–80. Wiley-Liss, New York.

Kerker, M. (1983). Elastic and inelastic light scattering in flow cytometry. *Cytometry* **4**, 1–10.

Krasznai, Z., Marian, T., Balkay, L., Emri, M., and Tron, L. (1995). Flow cytometric determination of absolute membrane potential of cells. *J. Photochem. Photobiol. B* **28**, 93–99.

Lansdorp, P. M., Smith, C., Safford, M., Terstappen, L. W., and Thomas, T. E. (1991). Single laser three color immunofluorescence staining procedures based on energy transfer between phycoerythrin and cyanine 5. *Cytometry* **12**, 723–730.

Leary, J. F., Todd, P., Wood, J. C., and Jett, J. H. (1979). Laser flow cytometric light scatter and fluorescence pulse width and pulse rise-time sizing of mammalian cells. *J. Histochem. Cytochem.* **27**, 315–320.

Le Bouteiller, P. P., Mishal, Z., Lemonnier, F. A., and Kourilsky, F. M. (1983). Quantification by flow cytofluorimetry of HLA class I molecules at the surface of murine cells transformed by cloned HLA genes. *J. Immunol. Methods* **61**, 301–315.

Lenkei, R., and Andersson, B. (1995). Determination of the antibody binding capacity of lymphocyte membrane antigens by flow cytometry in 58 blood donors. *J. Immunol. Methods* **183**, 267–277.

Lenkei, R., Gratama, J. W., Rothe, G., Schmitz, G., D'hautcourt, J. L., Arekrans, A., Mandy, F., and Marti, G. (1998). Performance of calibration standards for antigen quantitation with flow cytometry. *Cytometry* **33**, 188–196.

Lenkei, R., Mandy, F., Marti, G., and Vogt, R. (eds.) (1998). Special Issue. Quantitative fluorescence cytometry: An emerging consensus. *Cytometry* **33**, 94–287.

Lindmo, T., and Steen, H. B. (1977). Flow cytometric measurement of the polarization of fluorescence from intracellular fluorescein in mammalian cells. *Biophys. J.* **18**, 173–187.

Lindmo, T., Bormer, O., Ugelstad, J., and Nustad, K. (1990). Immunometric assay by flow cytometry using mixtures of two particle types of different affinity. *J. Immunol. Methods* **126**, 183–189.

Lisi, P. J., Huang, C. W., Hoffman, R. A., and Teipel, J. W. (1982). A fluorescence immunoassay for soluble antigens employing flow cytometric detection. *Clin. Chim. Acta* **120**, 171–179.

Loken, M. R. (1980). Evaluating optical filter efficiency in a flow cytometer. *J. Histochem. Cytochem.* **28**, 1136–1137.

Loken, M. R., and Herzenberg, L. A. (1975). Analysis of cell populations with a fluorescence-activated cell sorter. *Ann. N.Y. Acad. Sci.* **254**, 163–171.

Loken, M. R., Parks, D. R., and Herzenberg, L. A. (1977a). Identification of cell asymmetry and orientation by light scattering. *J. Histochem. Cytochem.* **25**, 790–795.

Loken, M. R., Parks, D. R., and Herzenberg, L. A. (1977b). Two-color immunofluorescence using a fluorescence-activated cell sorter. *J. Histochem. Cytochem.* **25**, 899–907.

McCoy, J. P., Jr., and Overton, W. R. (1994). Quality control in flow cytometry for diagnostic pathology: II. A conspectus of reference ranges for lymphocyte immunophenotyping. *Cytometry* **18**, 129–139.

McCoy, J. P., Jr., Carey, J. L., and Krause, J. R. (1990). Quality control in flow cytometry for diagnostic pathology. I. Cell surface phenotyping and general laboratory procedures. *Am. J. Clin. Pathol.* **93**, S27–S37.

McCutcheon, M. J., and Miller, R. G. (1979). Fluorescence intensity resolution in flow systems. *J. Histochem. Cytochem.* **27**, 246–249.

Macho, A., Mishal, Z., and Uriel, J. (1996). Molar quantification by flow cytometry of fatty acid binding to cells using dipyrromethenboron difluoride derivatives. *Cytometry* **23**, 166–173.

McHugh, T. M., Wang, Y. J., Chong, H. O., Blackwood, L. L., and Stites, D. P. (1989). Development of a microsphere-based fluorescent immunoassay and its comparison to an enzyme immunoassay for the detection of antibodies to three antigen preparations from *Candida albicans. J. Immunol. Methods* **116,** 213–219.

McHugh, T. M., Viele, M. K., Chase, E. S., and Recktenwald, D. J. (1997). The sensitive detection and quantitation of antibody to HCV by using a microsphere-based immunoassay and flow cytometry. *Cytometry* **29,** 106–112.

Meyer, R. A., and Brunsting, A. (1975). Light scattering from nucleated biological cells. *Biophys. J.* **15,** 191–203.

Monroe, J. G., Havran, W. L., and Cambier, J. C. (1982). Enrichment of viable lymphocytes in defined cycle phases by sorting on the basis of pulse width of axial light extinction. *Cytometry* **3,** 24–27.

Moran, R. F. (1997). The smallest concentration [letter; comment]. *Clin. Chem.* **43,** 856–857.

Muirhead, K. A. (1993a). Establishment of quality control procedures in clinical flow cytometry. *Ann. N.Y. Acad. Sci.* **677,** 1–20.

Muirhead, K. A. (1993b). Quality control for clinical flow cytometry. *In* "Clinical Flow Cytometry: Principles and Application" (K. D. Bauer, R. E. Duque, and T. V. Shankey, eds.), pp. 177–199. Williams & Wilkins, Baltimore.

Muirhead, K. A., Schmitt, T. C., and Muirhead, A. R. (1983). Determination of linear fluorescence intensities from flow cytometric data accumulated with logarithmic amplifiers. *Cytometry* **3,** 251–256.

Mullaney, P. F., Crowell, J. M., Salzman, G. C., Martin, J. C., Hiebert, R. D., and Goad, C. A. (1976). Pulse-height light-scatter distributions using flow-systems instrumentation. *J. Histochem. Cytochem.* **24,** 298–304.

Myc, A., Traganos, F., Lara, J., Melamed, M. R., and Darzynkiewicz, Z. (1992). DNA stainability in aneuploid breast tumors: Comparison of four DNA fluorochromes differing in binding properties. *Cytometry* **13,** 389–394.

Oonishi, T., and Uyesaka, N. (1985). A new standard fluorescence microsphere for quantitative flow cytometry. *J. Immunol. Methods* **84,** 143–154.

Owens, M. A., and Loken, M. R. (1995). "Flow Cytometry for Clinical Laboratory Practice: Quality Assurance for Quantitative Immunophenotyping." Wiley-Liss, New York.

Parks, D. R., Hardy, R. R., and Herzenberg, L. A. (1984). Three-color immunofluorescence analysis of mouse B-lymphocyte subpopulations. *Cytometry* **5,** 159–168.

Pinkel, D., and Steen, H. B. (1982). Simple methods to determine and compare the sensitivity of flow cytometers. *Cytometry* **3,** 220–223.

Poncelet, P., and Carayon, P. (1985). Cytofluorometric quantification of cell-surface antigens by indirect immunofluorescence using monoclonal antibodies. *J. Immunol. Methods* **85,** 65–74.

Purvis, N., and Stelzer, G. (1998). Multi-platform, multi-site instrumentation and reagent standardization. *Cytometry* **33,** 156–165.

Recktenwald, D., Phi-Wilson, J., and Verwer, B. (1993). Fluorescence quantitation using digital microscopy. *J. Phys. Chem.* **97,** 2868–2870.

Renner, E. D. (1994). Development and clinical evaluation of an amplified flow cytometric fluoroimmunoassay for *Clostridium difficile* toxin A. *Cytometry* **18,** 103–108.

Robertson, B. R., and Button, D. K. (1989). Characterizing aquatic bacteria according to population, cell size, and apparent DNA content by flow cytometry. *Cytometry* **10,** 70–76.

Robertson, B. R., Button, D. K., and Koch, A. L. (1998). Determination of the biomasses of small bacteria at low concentrations in a mixture of species with forward light scatter measurements by flow cytometry. *Appl. Environ. Microbiol.* **64,** 3900–3909.

Robins, D. B., and Bedo, A. W. (1994). Quantitative determination of particle concentrations in experimental and marine environmental samples. *Cytometry* **17,** 179–184.

Rolland, J. M., Dimitropoulos, K., Bishop, A., Hocking, G. R., and Nairn, R. C. (1985). Fluorescence polarization assay by flow cytometry. *J. Immunol. Methods* **76,** 1–10.

Salzman, G. C., Wilder, M. E., and Jett, J. H. (1979). Light scattering with stream-in-air flow systems. *J. Histochem. Cytochem.* **27**, 264–267.

Saunders, G. C., Jett, J. H., and Martin, J. C. (1985). Amplified flow-cytometric separation-free fluorescence immunoassays. *Clin. Chem.* **31**, 2020–2023.

Schmid, I., Schmid, P., and Giorgi, J. V. (1988). Conversion of logarithmic channel numbers into relative linear fluorescence intensity. *Cytometry* **9**, 533–538.

Schrader, H. W., and Eisert, W. G. (1986). High resolution particle sizing using the combination of time-of-flight and light-scattering measurements. *Applied Optics* **25**, 4396–4401.

Schuette, W. H., Carducci, E., Marti, G. E., Shackney, S. E., and Eden, M. (1985). The relationship between mean channel selection and the calculated coefficient of variation. *Cytometry* **6**, 487–491.

Schwartz, A., and Fernandez-Repollet, E. (1993). Development of clinical standards for flow cytometry. *Ann. N.Y. Acad. Sci.* **677**, 28–39.

Schwartz, A., and Fernandez-Repollet, E. (1994). Standardization for flow cytometry. *Methods Cell Biol.* **42** (Pt. B), 605–626.

Schwartz, A., Sugg, H., Ritter, T. W., and Fernandez-Repollet, E. (1983). Direct determination of cell diameter, surface area, and volume with an electronic volume sensing flow cytometer. *Cytometry* **3**, 456–458.

Schwartz, A., Fernandez-Repollet, E., Vogt, R., and Gratama, J. W. (1996). Standardizing flow cytometry: Construction of a standardized fluorescence calibration plot using matching spectral calibrators. *Cytometry* **26**, 22–31.

Schwartz, A., Marti, G. E., Poon, R., Gratama, J. W., and Fernandez-Repollet, E. (1998). Standardizing flow cytometry: A classification system of fluorescence standards used for flow cytometry. *Cytometry* **33**, 106–114.

Shapiro, H. M. (1995). "Practical Flow Cytometry." Wiley-Liss, New York.

Sharpless, T. K., and Melamed, M. R. (1976). Estimation of cell size from pulse shape in flow cytofluorometry. *J. Histochem. Cytochem.* **24**, 257–264.

Sharpless, T. K., Bartholdi, M., and Melamed, M. R. (1977). Size and refractive index dependence of simple forward angle scattering measurements in a flow system using sharply-focused illumination. *J. Histochem. Cytochem.* **25**, 845–856.

Stacchini, A., Fubini, L., and Aglietta, M. (1996). Flow cytometric detection and quantitative analysis of the GM-CSF receptor in human granulocytes and comparison with the radioligand binding assay. *Cytometry* **24**, 374–381.

Stamey, T. A. (1996). Lower limits of detection, biological detection limits, functional sensitivity, or residual cancer detection limit? Sensitivity reports on prostate-specific antigen assays mislead clinicians [see comments]. *Clin. Chem.* **42**, 849–852.

Steen, H. B. (1992). Noise, sensitivity, and resolution of flow cytometers. *Cytometry* **13**, 822–830.

Stewart, C. C., and Steinkamp, J. A. (1982). Quantitation of cell concentration using the flow cytometer. *Cytometry* **2**, 238–243.

Stokke, T., Holte, H., Erikstein, B., Davies, C. L., Funderud, S., and Steen, H. B. (1991). Simultaneous assessment of chromatin structure, DNA content, and antigen expression by dual wavelength excitation flow cytometry. *Cytometry* **12**, 172–178.

Szollosi, J., Tron, L., Damjanovich, S., Helliwell, S. H., Arndt-Jovin, D., and Jovin, T. M. (1984). Fluorescence energy transfer measurements on cell surfaces: A critical comparison of steady-state fluorimetric and flow cytometric methods. *Cytometry* **5**, 210–216.

Szollosi, J., Matyus, L., Tron, L., Balazs, M., Ember, I., Fulwyler, M. J., and Damjanovich, S. (1987). Flow cytometric measurements of fluorescence energy transfer using single laser excitation. *Cytometry* **8**, 120–128.

Tiersch, T. R., Chandler, R. W., Wachtel, S. S., and Elias, S. (1989). Reference standards for flow cytometry and application in comparative studies of nuclear DNA content. *Cytometry* **10**, 706–710.

Tron, L., Szollosi, J., Damjanovich, S., Helliwell, S. H., Arndt-Jovin, D. J., and Jovin, T. M. (1984). Flow cytometric measurement of fluorescence resonance energy transfer on cell surfaces. Quantitative evaluation of the transfer efficiency on a cell-by-cell basis. *Biophys. J.* **45**, 939–946.

Tycko, D. H., Metz, M. H., Epstein, E. A., and Grinbaum, A. (1985). Flow-cytometric light scattering measurement of red blood cell volume and hemoglobin concentration. *Applied Optics* **24,** 1355–1365.

Ubezio, P., and Andreoni, A. (1985). Linearity and noise sources in flow cytometry. *Cytometry* **6,** 109–115.

Vindelov, L. L., Christensen, I. J., and Nissen, N. I. (1983). Standardization of high-resolution flow cytometric DNA analysis by the simultaneous use of chicken and trout red blood cells as internal reference standards. *Cytometry* **3,** 328–331.

Visser, J., Haaijman, J., and Trask, B. (1978). Quantitative immunofluorescence in flow cytometry. *In* "Sixth International Conference on Immunofluorescence and Related Staining Techniques. Amsterdam" (W. Knapp, K. Holubar, and G. Wick, eds.), pp. 147–160. Elsevier/North-Holland, Vienna, Austria.

Watson, J. V., and Walport, M. J. (1986). Molecular calibration in flow cytometry with subattogram detection limit. *J. Immunol. Methods* **93,** 171–175.

Weaver, D. J., Jr., Durack, G., and Voss, E. W., Jr. (1997). Analysis of the intracellular processing of proteins: Application of fluorescence polarization and a novel fluorescent probe. *Cytometry* **28,** 25–35.

Webster, G. A., and Pockley, A. G. (1993). Effect of red cell lysis protocols on the expression of rat peripheral blood lymphocyte subset and activation antigens. *J. Immunol. Methods* **163,** 115–121.

Wood, J. C. S. (1997). Establishing and maintaining system linearity. *In* "Current Protocols in Cytometry" (J. P. Robinson, ed.), Unit 1.4. Wiley, New York.

Wood, J. C. (1998). Fundamental flow cytometer properties governing sensitivity and resolution. *Cytometry* **33,** 260–266.

Wood, J. C., and Hoffman, R. A. (1998). Evaluating fluorescence sensitivity on flow cytometers: An overview. *Cytometry* **33,** 256–259.

Zhang, Y. Z., Kemper, C., Bakke, A., and Haugland, R. P. (1998). Novel flow cytometry compensation standards: Internally stained fluorescent microspheres with matched emission spectra and long-term stability. *Cytometry* **33,** 244–248.

PART IV

Cell Proliferation

CHAPTER 15

Methods to Identify Mitotic Cells by Flow Cytometry

Gloria Juan, Frank Traganos, and Zbigniew Darzynkiewicz

Brander Cancer Research Institute
New York Medical College
Hawthorne, New York 10532

I. Introduction
    A. Histone H3 Phosphorylation during the Cell Cycle
    B. H3-Antibody as a Marker of Mitotic Cells
II. Materials
III. Cell Preparation and Staining
    A. Immunocytochemical Detection of H3-P in Combination with Cyclin A or B1
IV. Instruments
    A. Analysis of Cellular Fluorescence
    B. Fluorescence Microscopy
V. Critical Aspects of the Procedure
VI. Results and Discussion
    A. Immunofluorescence Microscopy
    B. Bivariate Analyses
    C. Multiparametric Analysis
    D. Identification of Mitotic Subphases
VII. Comparison of Anti-H3-P Monoclonal Antibody with Other Markers of Mitotic Cells
    References

I. Introduction

A. Histone H3 Phosphorylation during the Cell Cycle

In preparation for cell division, nuclear chromatin undergoes critical rearrangement needed for the organization of chromosomes and their separation into daughter cells. These chromatin changes are initiated during the G_2 phase of the cell cycle, and their most striking morphological manifestation is chromatin condensation, which becomes apparent during prophase and is maximal during the subsequent stages of mitosis.

One of the most characteristic molecular events that occur during mitosis is phosphorylation of histone H3 (H3). This event is highly correlated with the G_2 to M transition and appears to be essential for chromatin condensation (Ajiro *et al.*, 1983; Ajiro and Nishimoto, 1985; Gurley *et al.*, 1975; Wolffe, 1992). In mitotic cells, H3 is specifically phosphorylated at Ser-10, near its N terminus (Ajiro and Nishimoto, 1985). Dephosphorylation of H3 occurs quite rapidly after mitosis, and Ser-10 remains unphosphorylated throughout the remainder of interphase (Gurley *et al.*, 1975). Phosphorylation of H3 is also observed during premature chromosome condensation (PCC); however, when PCC in interphase nuclei of fused cells is prevented by various metabolic inhibitors, H3 remains unphosphorylated (Ajiro and Nishimoto, 1985; Hanks *et al.*, 1983).

The kinase responsible for H3 phosphorylation during the G_2 to M transition or PCC has not yet been identified. Despite the fact that its activity is inhibited by staurosporine, the kinase phosphorylating H3 is not CDC2, because CDC2 knockout cells are able to phosphorylate H3 on Ser-10 (Th'ng *et al.*, 1994). Recent studies, however, suggest that MSK1 (Thomson *et al.*, 1999) and Rsk-2 (Sassone-Corsi *et al.*, 1999) may be the H3 kinases.

B. H3-Antibody as a Marker of Mitotic Cells

Since the time of phosphorylation of H3 is restricted to mitosis, this event may be a specific marker discriminating mitotic cells. A new monoclonal antibody has been developed that specifically recognizes only H3 phosphorylated at Ser-10 (anti-H3-P MAb) (Juan *et al.*, 1998). This antibody, thus, provides the means for immunocytochemical detection of H3 phosphorylation and thereby detection of mitotic cells. Indeed, the H3 epitope is accessible *in situ* to this antibody, and mitotic cells can be identified with this approach (Juan *et al.*, 1998).

In this chapter we describe the procedure for identification of mitotic cells with the use of this antibody. There are several potential applications for this methodology. The immunocytochemical detection of H3-P provides new opportunities for studying the role of H3 phosphorylation in chromatin condensation during mitosis. In particular, it may be helpful in identification of the kinase(s) that is (are) involved in H3 phosphorylation. Since this kinase(s) is a potential target for development of new antitumor drugs designed to target the G_2 to M transition, the possibility of immunocytochemical detection of its (their) activity may be of great value in drug screening. Furthermore, this antibody, being a marker of mitotic cells, can be applied in multiparametric cytometric analysis to study H3 phosphorylation in relation to expression of other proteins critical for cell cycle progression or chromatin condensation during mitosis. It will also be used to identify mitotic cells, for example, in estimating the mitotic index (MI) in cell populations and in stathmokinetic experiments when the cells are arrested in mitosis; moreover, the rate of cell entrance to M ("cell birth rate"), emptying of the G_1 compartment, and many other kinetic parameters can be estimated (Darzynkiewicz *et al.*, 1986).

II. Materials

Cells to be analyzed

Phosphate-buffered saline (PBS)

Fixative: 1% formaldehyde in PBS and 80% ethanol, −20°C

Permeabilization solution: 0.25% (v/v) Triton X-100 in PBS, pH 7.4 (store at 4°C)

Rinsing buffer: 1% bovine serum albumin (BSA) in PBS, pH 7.4 (store at 4° C)

Antibody to phosphorylated histone H3: The anti-H3-P monoclonal antibody was developed at Oncor (Gaithersburg, MD), and the isotype of H3-P was determined to be immunoglobulin (Ig)G2a κ (Juan *et al.*, 1998); This antibody is currently available from Sigma Chemical Company (St. Louis, MO)

Cyclin antibodies: mouse monoclonal antibodies to cyclin B1 (clone GNS-1) and cyclin A (clone BF-683)

Isotypic control: mouse IgG1

1 g/liter fluorescein isothiocyanate (FITC)-conjugated goat anti-mouse IgG antibody

Propidium iodide (PI) staining buffer: 5 μg/ml PI (Molecular Probes, Eugene, OR) and 200 μg/ml DNase-free RNase A (Sigma) in PBS, pH 7.4

DAPI staining buffer: 1 μg/ml DAPI (4,6-diamidino-2-phenylindole) in PBS, pH 7.4

III. Cell Preparation and Staining

A. Immunocytochemical Detection of H3-P in Combination with Cyclin A or B1

The cells are washed with PBS and fixed in suspension at a concentration of $2-3 \times 10^6$ cells/ml in 2.0 ml of 1% formaldehyde in PBS for 15 min. After being washed in PBS, the cells are resuspended in ice-cold 80% ethanol for up to 24 hr.

After fixation, the cells are washed twice with PBS and then suspended in 1 ml of 0.25% Triton X-100 in PBS on ice for 5 min.

After further centrifugation, the cell pellet (approximately 5×10^6 cells) is suspended in 100 μl of PBS containing 0.5 μg of the anti-H3-P MAb and 1% BSA and incubated for 2 hr at room temperature.

The cells are then rinsed with PBS containing 1% BSA and incubated with the FITC-conjugated goat anti-mouse IgG antibody diluted 1:30 in PBS containing 1% BSA for 30 min at room temperature in the dark.

The cells are washed again, resuspended in 5 µg/ml of PI and 0.1% RNase A in PBS, and incubated at room temperature for 20 min prior to measurement. The isotypic control for H3-P is IgG2a κ (Oncor).

In some experiments, in addition to the detection of H3-P, cyclins A or B1 can also be immunocytochemically detected in the same cells.

In those cases, following binding of the anti-H3-P MAb, the secondary anti-mouse IgG antibody used is conjugated with phycoerythrin (PE) rather than with FITC.

After staining for H3-P, the cells are rinsed, the cell pellet suspended in 100 µl of 1% BSA in PBS containing 0.25 µg of anti-cyclin B1 or cyclin A MAb directly conjugated to FITC and incubated for 1 hr at room temperature in the dark.

The cells are then rinsed with PBS containing 1% BSA, counterstained with 1 µg/ml of DAPI in PBS, and incubated at room temperature for 20 min prior to measurement.

IV. Instruments

A. Analysis of Cellular Fluorescence

Any flow cytometer equipped with an argon ion laser tuned at 488 nm and with the capability of detecting two fluorescent signals is suitable for the detection of H3-P fluorescence and DNA content. For the detection of three or more fluorescent signals, a combination of the argon ion laser (emission at 488 nm) with the helium–cadmium laser, emitting ultraviolet (UV) light is required. In the first case, fluorescence signals are collected using the standard configuration of the flow cytometer (green fluorescence for H3-P and red fluorescence for PI). In the second case, DNA content is measured with DAPI (blue light) excited with the UV laser, while cyclins (A or B1) stained with FITC (green) and H3-P indirectly labeled with PE (orange) are excited with the blue laser. Additional details of the analysis of cellular fluorescence are presented elsewhere (Juan *et al.*, 1997; Li *et al.*, 1996).

B. Fluorescence Microscopy

The cells are initially attached to the slides by cytocentrifugation, fixed in 1% formaldehyde (as in the case of cells in suspension), rinsed, and then stained with anti-H3-P MAb on the slide. The cells are then counterstained with the appropriate DNA fluorochrome.

V. Critical Aspects of the Procedure

Cell fixation and permeabilization are critical steps for immunocytochemical detection of intracellular proteins and often must be customized for particular antigens. The fixative is expected to stabilize the antigen *in situ* and preserve its epitope in a state where it continues to remain reactive with the available antibody. The cell must be permeable to allow access of the antibody to the epitope (Clevenger *et al.,* 1987). A brief (15 min) treatment with 1% formaldehyde followed by 80% cold ethanol works well to avoid the loss of unbound or loosely bound proteins during the permeabilization process, especially for multiparameter analysis. It should be noted that extensive DNA–DNA or DNA–protein cross-linking occurs as the result of higher concentrations or longer fixation times in the presence of formaldehyde which, in turn, impair DNA stainability with intercalating dyes such as PI. Therefore, if accurate DNA distributions are required, mild formaldehyde fixation, as described earlier, is necessary.

A very critical step for the detection of cyclins is the proper choice of the antibody. Often, an antibody that is very specific and useful for Western blotting, is ineffective for the detection of the same antigen *in situ* and vice versa. Thus, while some antibodies are sensitive to protein conformation as exists *in situ,* these same antibodies may be less specific to that epitope in the denatured protein obtained during Western blotting. Alternatively, an antibody useful against a denatured protein, may not recognize the same epitope in the native. Therefore, it is critical to provide information on the reagents used in each study.

VI. Results and Discussion

A. Immunofluorescence Microscopy

Reactivity of anti-H3-P MAb, as detected immunocytochemically by fluorescence microscopy, is essentially limited to mitotic cells, from prophase to late telophase (Fig. 1). Only nuclear components are reactive with this antibody, and there is no evidence of any cytoplasmic labeling. Prophase cells show a delicate, rather diffuse, and in some instances a threadlike pattern of labeling. Intense staining of chromosomes was apparent during later stages of mitosis, with chromosomes present in metaphase and anaphase cells being the most intensely stained. It should be mentioned, however, that approximately 20% of interphase nuclei, generally of larger size, showed distinct, pointlike areas (about 1 μm in diameter), characterized by weak labeling with anti-H3-P MAb (not shown). It is difficult to capture these spots by photography due to their low fluorescence intensity, which fades during the photographic exposure. This weak "punctate" labeling of some interphase nuclei may indicate that a small fraction of H3 is phosphorylated in some areas of interphase chromatin, perhaps at the sites of initiation of chromatin condensation (centromeric domains of the chromosomes?) during late G_2, prior to prophase.

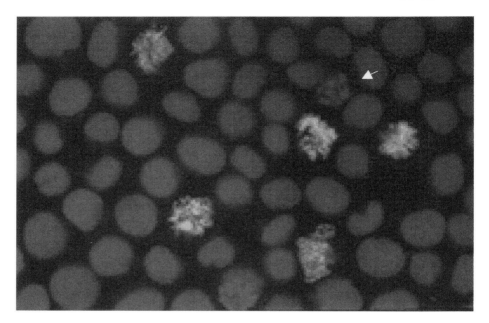

Fig. 1 Fluorescence photomicrograph of U937 cells stained with anti-H3-P MAb and counterstained with 7-aminoactinomycin D. Cells in metaphase show strong reactivity toward anti-H3-P MAb. Note that the condensed chromatin of an apoptotic cell (spontaneous apoptosis) did not react with the antibody (arrow). Photographed with a Nikon Microphot-FXA fluorescence microscope using a Plan Fluor 40× objective lens. (See color plates.)

B. Bivariate Analyses

Figure 2 illustrates flow cytometric bivariate distributions of U937 cells in which cellular DNA content is plotted versus reactivity with anti-H3-P MAb in asynchronously growing cells (left panel) and cells exposed to 50 ng/ml vinblastine for 6 hr (right panel). In untreated cell cultures, typically 1–2% of cells have high H3-P immunofluorescence which correspond to the percentage of mitotic cells in these cultures determined microscopically. Cell incubation in the presence of the microtubule poison vinblastine, which arrests cells in metaphase, increases the proportion of cells in mitosis. As is evident, there is a cell subpopulation with a G_2M DNA content that stains more intensely with anti-H3-P compared to other (e.g., S phase) cells. Taking into account that only mitotic cells show strong reactivity with anti-H3-P MAb (Fig. 1), it can be assumed that the cells with high anti-H3-P fluorescence and a G_2M DNA content are mitotic cells.

C. Multiparametric Analysis

The multiparametric analysis of H3 phosphorylation in U937 cells (with an enriched proportion of mitotic cells as a result of treatment for 6 hr with vinblas-

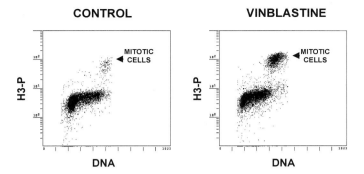

Fig. 2 Bivariate distributions of DNA content versus H3-P immunofluorescence of U937 cells growing exponentially (left, control cells) or following exposure to vinblastine for 6 hr (right). Relatively few cells in mitosis (H3-P positive) are present in asynchronously growing cultures. Treatment with vinblastine increases the percentage of cells in mitosis; those cells manifest increased stainability with the anti-H3-P MAb.

tine) is shown in Fig. 3. By simultaneous staining of cells with three different fluorochromes, one reporting the status of H3 phosphorylation (anti-H3-P MAb), another, DNA content (DAPI, cell cycle position), and still another, expression of either cyclin A or B1 (anti-cyclin MAbs), all three parameters could be correlated with each other, on a cell-by-cell basis. It is possible, therefore, to obtain information regarding the proportion of cells expressing either cyclin A or cyclin B1, which have phosphorylated H3.

D. Identification of Mitotic Subphases

During unperturbed mitosis, that is, in cultures grown in the absence of vinblastine, about 12% of the cells containing phosphorylated H3 are cyclin A negative (H3-P$^+$/A$^-$). Because cyclin A is rapidly degraded during prometaphase and is virtually absent in metaphase (Gong *et al.*, 1995; Pines and Hunter, 1991), only mitotic cells that passed the prometaphase stage are expected to be cyclin A negative. Therefore, the 88% of the cells with phosphorylated H3 also containing cyclin A (H3-P$^+$/A$^+$) have yet to enter metaphase.

Cyclin B1 is maximally expressed during mitosis (Pines and Hunter, 1991), and over 90% of cells containing phosphorylated H3 are cyclin B1 positive (H3-P$^+$/B$^+$). Expression of cyclin B1 is threefold higher among the population of cells reacting with anti-H3-P MAb (i.e., predominantly mitotic cells) than among the remaining cells with a G$_2$M DNA content, which are essentially G$_2$ cells. Since cyclin B1 is degraded very late during mitosis (Amon *et al.*, 1994), cyclin B1 negative cells that have phosphorylated H3 (H3-P$^+$/B$^-$) presumably represent cells in late telophase, just prior to cytokinesis. During prophase, when the cells are still expressing cyclin A, H3 is already phosphorylated.

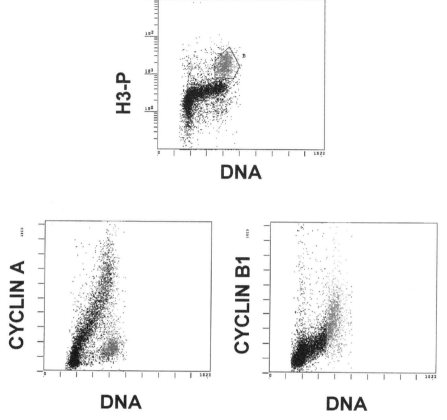

Fig. 3 Multivariate analysis of U937 cells exposed for 6 hr to vinblastine and stained for DNA content, anti-H3-P reactivity (top), and for either cyclin A (bottom left) or B1 (bottom right) expression, detected immunocytochemically. U937 cells were fixed and stained for DNA content with DAPI (blue fluorescence emission), for H3-P with secondary antibody conjugated with PE (orange fluorescence), and either for cyclin A or B1 with the respective antibody directly conjugated with FITC (green). The cells reactive with anti-H3-P MAb are selected (gated) in the top plot and using a "paint-a-gate" analysis program marked, on these scattergrams, with green color. Thus, they appear green in the bottom scattergrams, which allows one to correlate expression of the respective cyclins with H3-P immunofluorescence on a cell-by-cell basis. (See color plates.)

VII. Comparison of Anti-H3-P Monoclonal Antibody with Other Markers of Mitotic Cells

Anti-H3-P MAb is a new marker of mitotic cells which can be used immunocytochemically, in particular in flow- or laser scanning-cytometry to discern mitotic cells, for example, for their sorting, multiparametric analysis, rapid scoring of

MI, or in stathmokinetic experiments (Juan *et al.*, 1998). It offers certain advantages over other markers that have previously been identified (Table I) as discussed below:

1. Morphological features of the condensed chromatin of mitotic cells are sufficiently distinct from those of interphase cells, to alter the laser light scatter signal (Nuesse *et al.*, 1990) or generate different fluorescence intensity pulse width or fluorescence intensity peak (Gorczyca *et al.*, 1996) signals. These changes in light scatter or pulse shape were proposed as markers of mitotic cells for flow or laser scanning cytometry. These markers, however, are not unique to mitosis and do not always allow one to distinguish mitotic cells from apoptotic cells, which can also have highly condensed chromatin (Darzynkiewicz *et al.*, 1997).

2. Altered DNA stainability with different fluorochromes resulting from differences in chromatin structure between mitotic and interphase cells was also proposed to identify mitotic cells by cytometry (Darzynkiewicz *et al.*, 1977). However, these differences vary depending on the cell type, fixation, and other treatments (Darzynkiewicz *et al.*, 1977). Furthermore, DNA stainability in apoptotic cells also is altered compared to nonapoptotic cells and, based on cell stainability, apoptotic cells may be mistakenly identified as mitotic ones (Darzynkiewicz *et al.*, 1996). Chromatin of apoptotic cells, however, is not reactive with anti-H3-P MAb (Fig. 1, arrow).

3. Cell lysis followed by protein content analysis was proposed as still another means to discriminate between mitotic and interphase cells (Roti Roti *et al.*, 1982). The clusters of chromosomes from mitotic cells had distinctly lower protein content. However, because during cell lysis many constituents of chromatin leak

Table I
Characteristic Features of Chromatin Used to Distinguish Mitotic from Interphase Cells by Flow or Laser Scanning Cytometry

Markers of mitotic cells	References
Morphological features of condensed chromatin	Gorczyca *et al.* (1996), Nuesse *et al.* (1990)
Altered DNA stainability with different fluorochromes	Darzynkiewicz *et al.* (1977)
Low chromatin protein content after cell lysis	Roti Roti *et al.* (1982)
High sensitivity of DNA to denaturation	Darzynkiewicz *et al.* (1980), Juan *et al.* (1996)
Sensitivity of DNA to single strand nucleases	Juan *et al.* (1996)
Differential reactivity of chromatin with Ki-67 and PCNA antibodies	Landberg *et al.* (1990)
Absence of cyclin A and presence of cyclin B1	Gong *et al.* (1995)
Increased accessibility of the epitope reactive with AF-2 antibody	Di Vinci *et al.* (1993)
Presence of p105 antigen	Clevenger *et al.* (1987)

out, this approach cannot be used in experiments aimed at analysis of chromatin components other than DNA (e.g., cyclins, inhibitors of Cdks).

4. The procedure for detection of mitotic cells based on *in situ* DNA denaturation followed by staining with the metachromatic dye acridine orange is rapid and inexpensive (Darzynkiewicz *et al.*, 1980; Juan *et al.*, 1996). The harsh conditions required to denature DNA (acid or heat), however, destroy the native structure of many cellular components making the method incompatible with immunocytochemical detection of many other cellular constituents. Furthermore, the DNA in apoptotic cells is also sensitive to denaturation, and thus the distinction between mitotic and apoptotic cells may not be apparent (Darzynkiewicz *et al.*, 1997).

5. The method based on sensitivity of DNA *in situ* to single strand specific S1 or mung bean nucleases (Juan *et al.*, 1996) is quite specific to identify mitotic cells. It is based on cell incubation with single-strand (ss) specific endonuclease followed by labeling the resulting DNA strand breaks with labeled deoxynucleotides using exogenous terminal deoxynucleotidyl transferase (TUNEL). However, the method is rather complex and requires expensive reagents. Furthermore, because DNA strand breaks in apoptotic cells are also labeled under these conditions, the method cannot discriminate between mitotic and apoptotic cells.

6. There are methods of identification of mitotic cells that are based on the increased reactivity with Ki-67 antibody combined with low expression of proliferating cell nuclear antigen (PCNA) (Landberg *et al.*, 1990) or presence of cyclin B and absence of cyclin A (Gong *et al.*, 1995). Bivariate analysis of cellular DNA content and immunofluorescence of these markers allows one to discriminate between mitotic and interphase cells. These approaches require analysis of cell fluorescence at multiple wavelengths. This presents a limitation when one wishes to analyze additional features of mitotic versus interphase cells using a single laser instrument.

7. Two immunocytochemical methods utilizing antibodies that, as with anti-H3-P, show specificity to certain epitopes that are expressed in mitotic cells have been described. One of these methods relies on the increased accessibility of the epitope reacting with AF-2 antibody thought to occur during mitosis (Di Vinci *et al.*, 1993). Another antibody, raised against whole nuclei of mitogen-activated lymphocytes, detects a p105 protein that appears to be present only in mitotic cells (Clevenger *et al.*, 1987). These methods do not have most of the limitations discussed earlier. In the case of both antibodies, however, the nature of the target antigen in mitotic cells is poorly understood. Furthermore, the "time window" for the detection of cells with the AF-2 antibody is wider compared to anti-H3-P, since the former antibody reacts also with postmitotic cells (Di Vinci *et al.*, 1993).

In conclusion, the use of the H3-P antibody appears to provide the most advantages compared with the alternative methods of detection of mitotic cells.

It should be noted, however, that H3 appears to be phosphorylated in interphase monocytes and during monocytic differentiation of HL-60 cells (Juan *et al.*, 1999). The intensity of the immunofluorescence of monocytes stained with anti-H3-P, however, was less than that of mitotic cells. Furthermore, because mitotic cells have twice the DNA content of monocytes, these two cell types can easily be discriminated based on the bivariate analysis of DNA content versus H3-P immunofluorescence.

References

Ajiro, K., and Nishimoto, T. (1985). Specific site of histone H3 phosphorylation related to the maintenance of premature chromosome condensation. Evidence for catalytically induced interchange of the subunits. *J. Biol. Chem.* **260,** 15379–15381.

Ajiro, K., Nishimoto, T., and Takahashi, T. (1983). Histone H1 and H3 phosphorylation during premature chromosome condensation in a temperature-sensitive mutant (tsBN2) of baby hamster kidney cells. *J. Biol. Chem.* **258,** 4534–4538.

Amon, A., Irniger, S., and Nasmyth, K. (1994). Closing the cell cycle in yeast: G_2 cyclin proteolysis initiated at mitosis persists until the activation of G_1 cyclins in the next cycle. *Cell* **77,** 1037–1050.

Clevenger, C. V., Epstein, A. L., and Bauer, K. D. (1987). Quantitative analysis of a nuclear antigen in interphase and mitotic cells. *Cytometry* **8,** 280–286.

Darzynkiewicz, Z., Traganos, F., Sharpless, T., and Melamed, M. R. (1977). Interphase and metaphase chromatin. Different stainability of DNA with acridine orange after treatment at low pH. *Exp. Cell Res.* **110,** 201–214.

Darzynkiewicz, Z., Traganos, F., and Melamed, M. R. (1980). New cell compartments identified by multiparameter flow cytometry. *Cytometry* **1,** 98–108.

Darzynkiewicz, Z., Traganos, F., and Kimmel, M. (1986). Assay of cell cycle kinetics by multivariate flow cytometry. *In* "Techniques of Cell Cycle Analysis" (J. W. Gray and Z. Darzynkiewicz, eds.), pp. 291–332. Humana Press, Clifton, New Jersey.

Darzynkiewicz, Z., Gong, J., Juan, G., Ardelt, B., and Traganos, F. (1996). Cytometry of cyclin proteins. *Cytometry* **23,** 1–13.

Darzynkiewicz, Z., Juan, G., Li, X., Gorczyca, W., Murakami, T., and Traganos, F. (1997). Cytometry in cell necrobiology: Analysis of apoptosis and accidental cell death. *Cytometry* **27,** 1–20.

Di Vinci, A., Geido, E., Pfeffer, U., Vidali, G., and Giaretti, W. (1993). Quantitative analysis of mitotic and early-G1 cells using monoclonal antibodies against AF-2 protein. *Cytometry* **14,** 421–427.

Gong, J., Traganos, F., and Darzynkiewicz, Z. (1995). Discrimination of G_2 and mitotic cells by flow cytometry based on different expression of cyclins A and B1. *Exp. Cell Res.* **220,** 226–231.

Gorczyca, W., Melamed, M. R., and Darzynkiewicz, Z. (1996). Laser scanning cytometer (LSC) analysis of fraction of labeled mitoses. *Cell Prolif.* **29,** 539–547.

Gurley, L. R., Walters, R. A., and Tobey, R. A. (1975). Sequential phosphorylation of histone subfractions in the Chinese hamster cells. *J. Biol. Chem.* **250,** 3936–3944.

Hanks, S. K., Rodriguez, L. V., and Rao, P. N. (1983). Relationship between histone phosphorylation and premature chromatin condensation. *Exp. Cell Res.* **148,** 293–302.

Juan, G., Pan, W., and Darzynkiewicz, Z. (1996). DNA segments sensitive to single strand specific nucleases are present in chromatin of mitotic cells. *Exp. Cell Res.* **227,** 197–202.

Juan, G., Li, X., and Darzynkiewicz, Z. (1997). Correlation between DNA replication and expression of cyclins A and B1 in individual MOLT-4 cells. *Cancer Res.* **57,** 803–807.

Juan, G., Traganos, F., James, W. M., Ray, J. M., Roberge, M., Sauve, D. M., Anderson, H., and Darzynkiewicz, Z. (1998). Histone H3 phosphorylation and expression of cyclins A and B1 measured in individual cells during their progression through G_2 and mitosis. *Cytometry* **32,** 1–8.

Juan, G., Traganos, F., and Darzynkiewicz, Z. (1999). Histone H3 phosphorylation in human monocytes and during HL-60 cell differentiation. *Exp. Cell Res.* **246,** 212–220.

Landberg, G., Tan, E. M., and Ross, G. (1990). Flow cytometric multiparameter analysis of proliferating cell nuclear antigen/cyclin and Ki-67 antigen: A new view of the cell cycle. *Exp. Cell Res.* **187,** 111–118.

Li, X., Melamed, M. R., and Darzynkiewicz, Z. (1996). Detection of apoptosis and DNA replication by differential labeling of DNA strand breaks with fluorochromes of different color. *Exp. Cell Res.* **222,** 28–37.

Nuesse, M., Beisker, W., Hoffman, C., and Tarnok, A. (1990). Flow cytometric analysis of G1- and G2/M phase subpopulations in mammalian cells nuclei using side scatter and DNA content measurements. *Cytometry* **11,** 813–821.

Pines, J., and Hunter, T. (1991). Human cyclins A and B1 are differentially located in the cell and undergo cell cycle-dependent nuclear transport. *J. Cell Biol.* **115,** 1–17.

Roti Roti, J. L., Higashikubo, R., Blair, O. C., and Uygur, N. (1982). Cell-cycle position and nuclear protein content. *Cytometry* **3,** 91–96.

Sassone-Corsi, P., Mizzen, C. A., Cheung, P., Crario, C., Monaco, L., Jackquot, S., Hanauer, A., and Allis, C. D. (1999). Requirement of Rsk-2 for epidermal growth factor-activated phosphorylation of histone H3. *Science* **285,** 886–891.

Thomson, S., Clayton, A. L., Hazzalin, C. A., Barratt, M. J., and Mahadevan, L. C. (1999). The nucleosomal response associated with immediate-early gene induction is mediated via alternative MAP kinase cascades: MSK1 as a potential histone H3/HMG-14 kinase. *EMBO J.* **18,** 4779–4793.

Th'ng, J. P. H., Guo, X.-W., Swank, R. A., Crissman, H. A., and Bradbury, M. E. (1994). Inhibition of histone phosphorylation by staurosporine leads to chromosome decondensation. *J. Biol. Chem.* **269,** 9568–9573.

Wolffe, A. (1992). "Chromatin. Structure and Function." Academic Press, San Diego.

CHAPTER 16

Cell Cycle Kinetics Estimated by Analysis of Bromodeoxyuridine Incorporation

Nicholas H. A. Terry[*] and R. Allen White[†]

Departments of [*]Experimental Radiation Oncology and [†]Biomathematics
The University of Texas M. D. Anderson Cancer Center
Houston, Texas 77030

I. Introduction
II. Applications
III. Materials
    A. Introduction
    B. General Laboratory Reagents
IV. Methods
    A. Labeling and Fixation *in Vitro* and *in Vivo*
    B. Sample Preparation
    C. Flow Cytometry and Data Acquisition
    D. Data Analysis
V. Critical Aspects of the Procedure
    A. Labeling *in Vitro* and *in Vivo*
    B. Sample Preparation
    C. Flow Cytometry and Data Acquisition
    D. Data Analysis
References

I. Introduction

Historically, the elucidation of cell cycle kinetic parameters developed around the use of radioactively labeled thymidine, which is incorporated by cells synthesizing DNA. The fraction of labeled cells [labeling index (LI)] could, through the use of autoradiography, then be measured directly, and the percentage of labeled mitotic figures (PLM) could be counted as a function of time after labeling. The LI may be used to determine the fraction of cells synthesizing

DNA, dependent on the duration of S phase, and the changing PLM curve (or its associated cousins such as the continuous labeling curve) gave dynamic information about the progression of labeled cells through the cell cycle. The tedious techniques of multiple sampling and autoradiography are now largely supplanted by the use of halogenated pyrimidines, such as bromodeoxyuridine (BrdUrd), chlorodeoxyuridine (CldUrd), and iododeoxyuridine (IdUrd) used either singly or in combination. These agents may readily be tagged using monoclonal antibodies and visualized by fluorescent probes.

The advantages of the use of monoclonal antibodies to these thymidine analogs are not only in the increased ease and speed of analysis. The more recent methodology also offers greater precision in estimating such quantities as S-phase fraction from bivariate DNA versus thymidine-analog flow-cytometric measurements, thus avoiding the problematic assumptions inherent in single-parameter DNA histogram deconvolution. Moreover, this methodology makes it possible to estimate cell cycle kinetic parameters from measurements made at a single time after labeling, thereby making it routinely possible to evaluate clinical data concerning the relationship between cell proliferation and treatment outcome. For example, the tumor potential doubling time (T_{pot}) has been explored for utility both as a predictive assay for treatment outcome and as a selection criterion for patients for accelerated radiotherapy regimens. An understanding of the contemporary techniques of the analysis of kinetic parameters, by the use of halogenated thymidine analogs, is both essential and fundamental for current dynamic cell kinetic studies both *in vitro* and *in vivo*.

This chapter details the methods for sample preparation and staining, flow-cytometric data acquisition, and the procedures for quantitative analysis of dynamic cell-kinetic data. Whereas the PLM technology relied on labeling cells in S-phase and observing them through the window of mitosis, the two-color flow cytometric technique described here still relies on S-phase labeling but now observes the labeled cells throughout the entire cell cycle. Importantly, it utilizes the extra information discernible from the division status of the labeled population. (Parental and filial cell generational status are distinguishable one from the other.) This technology permits the rapid evaluation of the quantities required to describe many of the growth kinetic parameters of cell populations. The basis of these techniques is that cells in S phase can be selectively labeled both *in vitro* and *in vivo* by administration of a nontoxic level of BrdUrd. The cells that incorporate BrdUrd continue to progress through the cell cycle and may be sampled, or a biopsy or surgical specimen taken, at a known time later.

The sample, fixed in ethanol, may be processed to produce a suspension of single nuclei by enzymatic digestion with, for example, pepsin (Carlton *et al.*, 1991). The nuclei are analyzed simultaneously for BrdUrd and DNA content by flow cytometry (Dolbeare *et al.*, 1983). The BrdUrd-labeled nuclei are selectively stained by a monoclonal antibody to BrdUrd (Gratzner, 1982) using a fluorescein isothiocyanate-conjugated (FITC, green fluorescing) second antibody technique.

All the nuclei are also stained with propidium iodide (PI), which fluoresces red at an intensity proportional to their DNA content, thereby simultaneously defining a reference standard for relative cell ages.

At the time of labeling the BrdUrd-labeled cells are assumed to be completely and exclusively in the S-phase, with all unlabeled cells in the G_1 and G_2+M phases of the cell cycle. In the interval between the administration of BrdUrd and sampling, the cycling cells progress unperturbed to subsequent phases of the cell cycle. In particular, the BrdUrd-labeled cells progress through S, G_2+M and, subsequently, into G_1 of the next generation. These observations were the basis for the original method for calculation of the duration of DNA synthesis (T_S), and T_{pot} from a single biopsy sample (Begg et al., 1985). Subsequent modeling and experimental studies (for overviews see Terry et al., 1992; Terry and White, 1996) have considerably refined this technique.

As with all laboratory methods, sample preparation is paramount. The staining requires a balancing act between denaturing sufficient DNA for the antibodies to bind to the incorporated BrdUrd while leaving enough DNA in its normal configuration in order to derive good quality DNA histograms with low coefficients of variation (CV) on the G_1 and G_2+M peaks. Several different techniques have been developed since its origination (Dolbeare et al., 1983) to accomplish these goals. Beisker et al. (1987) and Dolbeare et al. (1990) describe a thermal denaturation method whose principal advantage is in increased sensitivity. Restriction enzyme/exonuclease III methods (Dolbeare and Gray, 1988) avoid the use of heat for denaturation, thus preserving many other antigens, and minimize cell loss. Other combined BrdUrd/DNA/cell-surface and other antigen methods have been described (Begg and Hofland, 1991; Carayon and Bord, 1992), which may allow determination of proliferation in identifiable populations of cells. Washless techniques, which require no centrifugation, also help to minimize cell loss (Larsen, 1990). The kinetic analytical methods that we describe here require sample preparative techniques that are both of high sensitivity, giving excellent visualization of incorporated BrdUrd, together with low CVs for the peaks in the DNA histograms. Such sample preparation methods generally require production of isolated nuclei, either from cells *in vitro* or disaggregated from solid tissues and tumors, by the use of digestive enzymes and then denaturation of the DNA by strong acid. The method we recommend here is developed from a Schutte et al. (1987) modification of the procedure originally described by Dolbeare et al. (1983).

II. Applications

The primary dynamic cell kinetic applications of the technique are in the quantitation of cell cycle phase durations, population doubling times, cell cycle–phase boundary transitions or stasis, together with measurement of labeling indices and growth fractions. In the clinical context the method is being used largely to measure LI, T_S, and T_{pot} (for recent reviews see Begg, 1995; Dubray *et al.*, 1995; Terry and Peters, 1995; Terry, 1996; Antognoni *et al.*, 1998).

BrdUrd-labeling is also useful in, for example, cell synchrony experiments (Bussink *et al.*, 1995), where it enables a more accurate estimation to be made of the proportion of cells in defined phases of the cell cycle. It is also helpful in studies where dynamic information about cell progression, or nonprogression such as following drug treatment, through the cycle may be informative (Sacks *et al.*, 1990; Terry *et al.*, 1997). Although beyond the scope of this chapter, extension of analytical methods following labeling with two halogenated pyrimidine analogs of thymidine (e.g., CldUrd and IdUrd), given either as two pulses or as a continuous infusion of one and a pulse label of the other, have allowed for refinement of measures of S-phase cells *in vivo* (Pollack *et al.*, 1993a, 1995). The use of different monoclonal antibodies allows specific, and simultaneous, visualization of the two thymidine analogs, together with total DNA content. These methods were developed for tumors and have been used to measure the growth fraction (GF) and T_{pot} of both DNA aneuploid tumor cells and associated DNA-diploid cells simultaneously (White *et al.*, 1994a) following continuous infusion with CldUrd. In another study, using two pulses of IdUrd and CldUrd, White *et al.* (1994b) measured three differently labeled subpopulations within S phase and the separate progression through the cell cycle of the diploid and aneuploid cells. Pollack *et al.* (1993b) used two independently visualized labels to refine calculations of T_S and T_{pot}. The proliferation kinetics of tumor cells recruited into the cycle that were previously quiescent have also been estimated (Pollack *et al.*, 1994).

III. Materials

A. Introduction

There are many ways to achieve the quality of sample labeling, tissue and tumor dissaggregation, and immunochemical visualization of incorporated BrdUrd that is needed for the analytical methods. Our general approach is to work with aseptic techniques and sterile reagents. Nonconjugated anti-BrdUrd primary antibodies, followed by a fluorochrome-conjugated second antibody, allow for greater signal enhancement than do directly conjugated antibodies. The following reagents give consistent results. All solutions are prepared, filtered through a 0.22-μm filter, and most stored at 4°C with a few exceptions:

1. Sodium borate, HCl, and Tween 20 are prepared, filtered, and stored at room temperature.

2. BrdUrd working solutions, antibodies, and PI are diluted fresh weekly and stored at 4°C in the dark.

3. Ethanol and normal goat serum are stored at −20°C.

B. General Laboratory Reagents

1. Calcium- and magnesium-free phosphate-buffered saline (PBS, Dulbecco's) (Gibco BRL, Life Technologies, Rockville, MD). Store the powder at 4°C in a desiccator.

2. 5-Bromo-2'-deoxyuridine (BrdUrd) (Sigma Chemical, St. Louis, MO). Store at −20°C in a desiccator.

   a. Stock solution for *in vitro* work is 1 mM in PBS.

   b. Working solution for injection in *in vivo* tumor/normal tissue studies is 6 mg/ml in PBS for mice and 3 mg/ml in PBS for rats.

   c. Due to the solubility limit of BrdUrd use either bromodeoxycytidine or CldUrd for continuous labeling studies using miniosmotic pumps *in vivo*. Stir to dissolve over a low heat (∼35°C). Filter and store working solutions in the dark at 4°C, discard after 1 week.

3. Ethanol. Use 200 proof, absolute, from glass or plastic containers. (The large metal drums are a source of unwanted positively charged ions which predispose for protein precipitation on fixation.) Filter and store at −20°C.

4. 2 N hydrochloric acid (HCl).

5. Digestive enzymes.

   a. Pepsin (EM Science, Cherry Hill, NJ). Working solution is 0.04% in 0.1 N HCl.

   b. Collagenase (Sigma Chemical). Working solution is 0.1% in PBS.

6. Sodium borate decahydrate (Sigma Chemical). Working solution is 0.1 M.

7. Tween 20 (polyoxyethylene-sorbitan monolaurate) (Sigma Chemical).

8. PBT = PBS + 0.5% Tween 20.

9. Bovine serum albumin (BSA) (Sigma Chemical).

10. PBTB = PBS + 0.5% Tween 20 + 0.5% BSA.

11. Normal goat serum (Sigma Chemical). Store at −20°C.

12. PBTG = PBTB + 1.0% normal goat serum.

13. Anti-BrdUrd, IU-4 nonconjugated (Caltag, South San Francisco, CA). Many other antibodies and suppliers are available. Store at −20°C. Stock aliquots are 5 μl/tube (depending on antibody activity) in PBT, stored at −20°C.

14. Second antibody (goat anti-mouse IgG-FITC conjugate) (Sigma Chemical and many other suppliers). Store at −20°C aliquoted in PBTG.

15. PI (Sigma Chemical). Store the powder at 4°C, desiccated, in the dark. A stock solution is 1.0 mg/ml PI in 70% ethanol kept at 4°C in the dark. The working solution is 10 μg/ml in PBTB. PI is a suspected carcinogen and should be handled with care.

IV. Methods

A. Labeling and Fixation *in Vitro* and *in Vivo*

1. *In Vitro* Labeling and Fixation of Cultured Cells

   a. Incubate with a final concentration of 1 μM BrdUrd (from 1 mM stock) for 20 min at 37°C.

   b. Aspirate off medium. Rinse twice with warmed serum-free medium (do this quickly).

   c. Refeed with fresh whole medium that should be warmed and pregassed. Return to incubator (quickly).

   d. At desired time interval trypsinize the cells as follows:

   1. Aspirate off medium, rinse with fresh warmed serum-free medium.
   2. Add 1.0 ml of 0.05% trypsin and incubate for 5 min at 37°C.
   3. Tap dish gently. Add 9.0 ml of whole medium (containing serum). Draw up and down and transfer to a 15-ml centrifuge tube. Draw up and down several times without bubbling.
   4. Reserve an aliquot (10–50 μl) for counting in a hemocytometer (preferred) or a Coulter counter. Record the total cell yield.
   5. Centrifuge (4 min at 350 g) remaining cells to pellet.

   e. Fixation. The final cell concentration should be 2×10^6 cells per 2 ml of solution (or any multiple of this) in 65% ethanol in PBS. Adjust the following volumes depending on the actual cell count.

   1. Aspirate off medium. Add 0.7 ml of cold PBS slowly while vortexing the pellet.
   2. Continue vortexing and trickle in 1.3 ml freezer temperature 100% ethanol. Continue to vortex for 15 sec.

   Leave to fix at 4°C overnight before staining. This fixation procedure is also suitable for cells acquired by tumor fine needle aspiration (FNA) biopsy, prepared from ascites tumors or blood or bone marrow preparations.

2. *In Vivo* Labeling and Fixation of Samples from Solid Tumors and Normal Tissues

   a. Infuse BrdUrd:

   1. 100–200 mg/m^2 BrdUrd (100 mg IdUrd) for humans as a 20-min intravenous infusion. (Note: This is given only under the authorization of the investigational drug mechanism for approved protocols.)

2. 60 mg/kg intraperitoneally for mouse (i.e., 0.10 ml/10 g body weight of a 6 mg/ml solution).

3. 30 mg/kg intraperitoneally for rats.

b. Prepare a 15-ml centrifuge tube with 65% cold ethanol (in PBS); weigh the tube containing the mixture.

c. On tissue receipt (25 mg is an operational minimum), coarsely mince tissue with scissors and place in tube.

d. Vortex for 15–30 sec to fix the tissue; reweigh (tube + ethanol + tissue chunks), store at 4°C for at least overnight.

B. Sample Preparation

1. Staining Procedure for Cultured Cells, Tumor Fine Needle Aspiration Biopsies, Ascites Tumors, and Blood or Bone Marrow Preparations

The details of step c below need to be determined by microscopic observation of the pepsin digestion.

a. Vortex fixed cells (at low speed until evenly dispersed) for 15 sec and transfer 4×10^6 cells to a 15-ml centrifuge tube. Centrifuge at 350 g for 4 min at room temperature. Note: if whole cells, rather than nuclei, are to be prepared, proceed to e below.

b. Add 5.0 ml of pepsin (0.04% in 0.1 N HCl) per 4×10^6 cells (minimum of 5 ml of pepsin even if you have fewer cells, otherwise there is a risk of cell loss due to cells sticking to the sides of the tube).

c. Incubate for 10 min on a rocker at room temperature. (These are appropriate conditions for many laboratory cell lines.) Incubate other cells + pepsin for 10–60 min on a rocker, either at room temperature or at 37°C.

Note: Optimal incubation times and temperatures vary with different cell types and tissues; therefore, a time curve should be done for each cell type. If this is not possible (i.e., one FNA from a human breast tumor), frequent observation under the microscope during pepsin digestion is strongly advised.

d. Centrifuge the tubes containing pepsin + nuclei, aspirate off the supernatant, vortex the pellet for 5–10 sec.

e. Add 3.0 ml of 2 N HCl (1.5 ml per 2×10^6 nuclei or cells) to each tube while vortexing at low speed and incubate stationary for 20 min at 37°C. Shake twice during incubation.

f. Add 6.0 ml of 0.1 M sodium borate to each tube while vortexing, continue vortexing for 10 sec, centrifuge, and aspirate off the supernatant. Vortex the pellet, then add 6.0 ml of PBTB while vortexing, and centrifuge.

g. Aspirate off the supernatant and add 0.2 ml (per 2×10^7 or fewer nuclei) of the previously aliquoted anti-BrdUrd monoclonal antibody in PBT at 1:100 dilution (dilution varies depending on vendor and lot number and needs to be established for each new batch). Incubate for 60 min, at room temperature, in the dark.

h. Add 3.0 ml of PBTB while vortexing, centrifuge, then aspirate off the supernatant.

i. While vortexing the pellet add 0.2 ml (per 2×10^7 or fewer nuclei) of second antibody (goat anti-mouse-FITC) in PBTG at 1:100 dilution (actual dilution depends on vendor and lot number), and incubate for 45 min in the dark at room temperature.

j. Add 3.0 ml of PBTB while vortexing. Save an aliquot (10–50 μl) of the suspension for counting nuclei; then centrifuge and aspirate off the supernatant.

k. While vortexing the pellet add PI (10 μg/ml in PBTB) for a final concentration of 1×10^6 nuclei per ml (based on the counts made in j above). (If there are fewer than 10^6 nuclei in total, still use a minimum PI volume of 1 ml in order to guarantee stochiometric staining.)

l. Store overnight at 4°C in the dark; run on the flow cytometer the next day. (We have stored stained cells/nuclei at 4°C for 3 months with no deterioration and, if prepared aseptically, specimens older than 5 years are still evaluable, but may need reincubating with the antibodies.)

2. Staining Procedure for Solid Tumors and Tissues

This is a general procedure and will need to be adjusted where noted for specific tissues. If aseptic techniques are used throughout then fixed tissues may be stored almost indefinitely.

a. Finely mince a portion of the ethanol-fixed tumor or tissue chunks in a preweighed 60-mm dish.

1. Air dry for approximately 5 min to evaporate surplus ethanol (the tissue should not be allowed to dry out).

2. Reweigh (dish + tissue fragments).

3. Transfer the tissue fragments to a 50-ml Erlenmeyer flask.

b. From this point on, all reagent volumes are calculated for approximately 0.1 g of tissue.

1. These volumes are minima; therefore, with less than 0.1 g of tissue still use these volumes.

2. For more than 0.1 g of tissue use appropriate multiples of each volume.

c. Assuming a potential cell yield of $1-2 \times 10^8$ cells/g of tissue, use one of the following two dissociation solutions:

1. For collagen-rich tissue: add 5.0 ml of 0.1% collagenase (in PBS) to the 50-ml Erlenmeyer flask.

a. Incubate for 15 min in a 37°C shaker water bath (cover the top of the flask with parafilm).

b. Add directly to the (collagenase + tissue) slurry 5.0 ml of pepsin (0.04% in 0.1 N HCl) and incubate further.
Proceed to d below.

2. For other solid tissues and most rodent tumors: add 5.0 ml of 0.04% pepsin (in 0.1 N HCl) to the 50-ml Erlenmeyer flask.
Incubate 20 to 90 min either in a 37°C shaker water bath (cover top of flask with parafilm), or at room temperature, in which case use a 15-ml centrifuge tube on a rocker.

d. Pepsin digestion periods. As for *in vitro* preparations this is the only step that is routinely variable. The optimum incubation time, temperature, and extent of agitation vary widely for different tumors and normal tissues and should be checked periodically (every 10 min) by microscope to monitor for clean nuclei (with little cytoplasm attached) and to obtain the maximum nuclei yield. We usually divide fixed chunks into two flasks (or centrifuge tubes) for two separate digestion times staggered apart by 10–20 min.

1. Head and neck squamous cell carcinomas: typically between 20 and 60 min.

2. Colorectal adenocarcinomas: typically between 40 and 90 min.

3. Breast and bladder tumors: 15 min collagenase + typically 20 to 60 min in pepsin.

4. Normal tissues very widely and a time course should be performed to establish optimal conditions monitored by microscopic observation of clean nuclei yield.

Note: pepsin activity reduces to zero after much longer than 60 min at 37°C. If further tissue digestion is required add a further 3–5 ml of prewarmed 0.04% pepsin.

e. Aspirate (pepsin + nuclei slurry) with a 10-cc syringe attached to an 18-gauge needle, remove the needle, and filter the slurry suspension through a 35-μm nylon mesh into 15-ml centrifuge tube.

f. Save an aliquot of the suspension for counting nuclei.

1. Store the aliquot on ice until ready to count.

2. Record the total volume of the suspension for yield calculations below, step k.

g. Centrifuge the tubes containing (collagenase), pepsin, and nuclei at 350 g for 4 min at room temperature.

h. Aspirate off the supernatant and add 1.5 ml of 2 N HCl while vortexing, incubate stationary for 20 min at 37°C. Gently shake the tubes twice during incubation.

i. Add 3.0 ml of 0.1 M sodium borate while vortexing; continue vortexing for 10 sec. Centrifuge as before.

j. Aspirate supernatant and add 3.0 ml of PBTB while vortexing, and centrifuge.

k. Count the nuclei from the reserved aliquot (f above) and calculate the total nuclei yield. [Total yield = (the average number of nuclei in one large square of a standard hemocytometer) $\times 10^4 \times$ (volume in ml of the total suspension).]

l. Add 0.2 ml per 2×10^7 nuclei of anti-BrdUrd monoclonal antibody in PBT at 1:100 dilution and incubate for 60 min at room temperature in the dark. Use a minimum volume of 0.2 ml in order to saturate the pellet. Note: antibody dilution varies depending on vendor and lot; test new batches against a standard cell line, for example, CHO cells.

m. Add 3.0 ml of PBTB to the anti-BrdUrd/nuclei suspension, mix gently, and centrifuge as before.

n. Aspirate off the supernatant and add 0.2 ml/2×10^7 nuclei of second antibody (goat anti-mouse FITC in PBTG at 1:100 dilution—dilution depends on the activity of a particular batch), and incubate for 45 min at room temperature in the dark.

o. Add 3.0 ml of PBTB to the suspension, mix gently, and centrifuge as before.

p. Aspirate off the supernatant and add PI (10 μg/ml in PBTB) so that the final concentration is 1×10^6 nuclei/ml suspension. Note, skilled personnel should lose no more than 50% of the initial number of nuclei (after pepsin) due to centrifugation and aspiration.

q. Store the nuclei suspension, for at least overnight, in PI at 4°C in the dark.

C. Flow Cytometry and Data Acquisition

While almost any flow cytometer, equipped with a single argon ion laser with two photomultiplier tubes, may be adequate, some instruments are better suited to these procedures than are others. The instrument must be equipped with log amplification to accommodate the large range of anti-BrdUrd fluorescence signal. An optimal configuration includes narrow-beam excitation optics and hardware discrimination of doublets, together with the ability to configure the optical path so that there is no cross talk between the red and green channels. The default optical configuration of most three- or more color commercial flow cytometers may readily be changed for two-color work. Most bench-top sorters, with fixed light paths and significantly defocused excitation beams are less well suited for this task.

D. Data Analysis

1. Identification of Specific Subpopulations in the Bivariate DNA versus BrdUrd Histogram (Cytogram)

The flow cytometric data that may be obtained following these procedures are shown in Figs. 1 and 2. Figure 1 illustrates results from a mouse mammary carcinoma, MCa-4, following a pulse-label of 60 mg/kg intraperitoneal BrdUrd.

16. Cell Cycle Kinetics by BrdUrd Incorporation

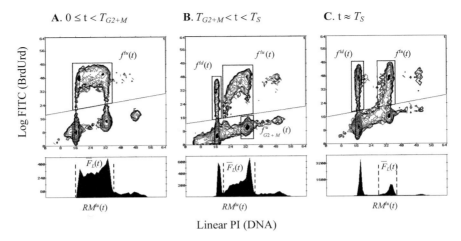

Fig. 1 Bivariate histograms of mouse mammary carcinoma MCa-4 showing DNA content (x-axis = PI fluorescence) and BrdUrd content (y-axis = log fluorescein fluorescence) following a pulse of 60 mg/kg intraperitoneal BrdUrd. Tumors were excised either shortly (20 min) after labeling (A), or 3 hr (B) and 6 hr later (C). The boxed regions indicate the subpopulations from which the fractional quantities needed for kinetic analysis may be calculated. The lower figures in each panel show the PI fluorescence distributions of the BrdUrd-labeled cells from which their relative movements may be computed.

The data are presented as DNA content (x-axis = PI fluorescence) versus BrdUrd content (y-axis = log fluorescein fluorescence). Tumors were excised either shortly (20 min) after labeling (Fig. 1A), or 3 hr (Fig. 1B) and 6 hr later (Fig. 1C). These sampling times represent periods shorter than the duration of G_2+M, longer than G_2+M but shorter than the duration of S phase, and approximately

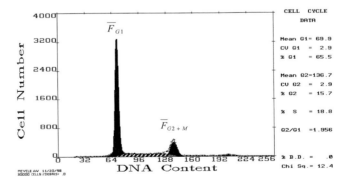

Fig. 2 The DNA content distribution (PI fluorescence) that pertains for any, and all, of the timepoints shown in Fig. 1. The mean fluorescence channel numbers and positions for the G_1 and G_2+M peaks are indicated. The total number of events is calculated from a modeled fit to the data.

the duration of S phase, respectively, for this particular model system. The boxed regions indicate the identifiable subpopulations from which the fractional quantities needed for kinetic analysis may be calculated (see Section IV,D,2). Specific calculations from the data in Fig. 1B are given in Section IV,D,2,b. The lower figures in each graph of Fig. 1 show the PI fluorescence distributions of the BrdUrd-labeled cells from which the mean fluorescence intensity of those that remain undivided at the time of sampling, $\bar{F}_L(t)$, may be computed.

Figure 2 shows the DNA content distribution (PI fluorescence) that pertains for any, and all, of the sampling times shown in Fig. 1. The important quantities that need to be measured for subsequent analysis are the mean fluorescence channel numbers for the G_1 and G_2+M peaks, together with the total number of events, which may be estimated from a modeled fit to the data.

2. Estimation of Kinetic Parameters

Kinetic parameters may be estimated by analysis of the generated bivariate halogenated thymidine analog-DNA cytograms (Fig. 1), together with the total DNA distribution (Fig. 2), obtained at a time t after labeling by the following array of procedures. The particular procedure chosen depends partly on the time following labeling and partly on the number of experimental time-points available.

As shown by several authors the following measured quantities can provide specific information on kinetic parameters. It is important to observe, as seen in Table I, that these changing, dynamic quantities are informative at different times after labeling depending on the relationship between sampling times and the underlying cell cycle parameters.

In Table I the plus signs indicate times when the measured quantity is changing and 0 and 1 are the values of the quantities when known.

The dynamic quantities listed in Table I, and selected references to their use, are

1. $f^u_{G2+M}(t)$, the fraction of unlabeled cells with G_2+M DNA content (White et al., 1990a).

Table I
The Dynamic Quantities That May Be Measured from the Cytogram, and the Cell Cycle Periods for Which They May Be Used to Obtain Information about Kinetic Parameters

Measured quantity	Minimum $t \to$ Maximum $t \to$	0 T_{G2+M}	T_{G2+M} T_S	T_S $T_S + T_{G2+M}$	$T_S + T_{G2+M}$ $T_S + T_{G2+M} + T_C$
$f^u_{G2+M}(t)$		+	0	0	0
$f^{lu}(t)$		+	+	+	0
$f^{ld}(t)$		0	+	+	+
$RM^{lu}(t)$		+	+	1	0
$RM^l(t)$		+	+	+	+

2. $f^{lu}(t)$, the fraction of labeled, undivided cells (White *et al.*, 1990a,b; Carlton *et al.*, 1991; Johansson *et al.*, 1998).
3. $f^{ld}(t)$, the fraction of labeled, divided cells (White *et al.*, 1990a,b; Carlton *et al.*, 1991).
4. $RM^{lu}(t)$, the relative movement, between a G_1 and G_2+M DNA content, of labeled undivided cells (Begg *et al.*, 1985; Carlton *et al.*, 1991; White *et al.*, 1991; Ritter *et al.*, 1992; Johansson *et al.*, 1996, 1998).
5. $RM^l(t)$, the relative movement of all labeled cells (White *et al.*, 1991; Johansson *et al.*, 1996).

The relative movements, $RM^l(t)$ and $RM^{lu}(t)$ are defined through the equation (Begg *et al.*, 1985)

$$RM(t) = \frac{\overline{F}_L(t) - \overline{F}_{G1}}{\overline{F}_{G2+M} - \overline{F}_{G1}} \qquad (1)$$

and range between 0, when all the BrdUrd-labeled cells that remain undivided have a mean red fluorescence [$\overline{F}_L(t)$, see Fig. 1] equal to the mean G_1 DNA content ($\overline{F}_{G1}$, Fig. 2), and 1 if all the undivided labeled cells have a G_2+M mean DNA content ($\overline{F}_{G2+M}$, Fig. 2).

Evidently, an examination of the values of the measured quantities can immediately provide qualitative information on the durations of G_2 and mitosis, T_{G2+M}, and S-phase, T_S, for example, $f^{ld}(t) = 0$ implies $t \leq T_{G2+M}$ while $RM^{lu}(t) = 1$ suggests $t \geq T_S$.

Quantitatively, however, different circumstances will give rise to different types of measurements being made. In the clinical situation, typically only a single time-point sample is available whereas for laboratory data, either *in vitro* or *in vivo*, multiple time-point measurements might routinely be made. In what follows the methods for analysis are grouped according to both the time after labeling and the number of different sampling times from which measurements are available. It should be noted, however, that even where multiple time-point samples exist, informative measurements can be made on different quantities such that single time-point analytical methods may be appropriate. For example, single time measurements can be compared to multiple time estimates in order to gain insight into intra- versus intersample variability. In any case careful observation of the DNA versus BrdUrd histograms should be made prior to choosing the appropriate analytical approach.

a. Single Time-Point Measurements with $f^{ld}(t) = 0$

The procedure used here is based on methods described by Ritter *et al.* (1992). The first step computes T_S from $RM^{lu}(t)$, and the second step combines T_S and information about the fraction of labeled cells contained in $f^{lu}(t)$ and $f^u{}_{G2+M}(t)$, to determine the potential doubling time, T_{pot}. Define

$$\nu \equiv \ln\left[\frac{1 + f^u_{G2+M}(t) + f^{lu}(t)}{1 + f^u_{G2+M}(t)}\right], \qquad (2)$$

for which it has been shown that

$$\nu = cT_S, \qquad (3)$$

where $c = \ln(2)/T_{pot}$ and ln is the natural logarithm. Equations (2) and (3) may be derived from equations in White *et al.* (1990a). These equations are similar to those in Ritter *et al.* (1992) but are more accurate for short sampling times.

Using the value of ν just obtained for cT_S, compute the initial intercept of the relative movement curve (Carlton *et al.*, 1991; White *et al.*, 1991):

$$RM^I \equiv \frac{1 - e^{-\nu}(1 + \nu)}{\nu(1 - e^{-\nu})}. \qquad (4)$$

Using the relative movement of the BrdUrd-labeled undivided cells, we denote

$$RM\Delta t = RM^{lu}(t) - RM^I. \qquad (5)$$

T_S can now be estimated from

$$T_S = \frac{t}{2RM\Delta t(1 + \sqrt{1 - 2RM\Delta t})} \qquad (6)$$

with T_{pot} given by

$$T_{pot} = \frac{\ln(2)T_S}{\nu}. \qquad (7)$$

Such short time estimates should be expected to be unreliable.

b. Single Time-Point Measurements with $RM^{lu}(t) < 1$ and $f^{ld}(t) > 0$

The calculations are similar to, but are more likely to result in more precise parameter estimates than, those employed for shorter times. Define (White *et al.*, 1990a,b)

$$\nu \equiv \ln\left[\frac{1 + f^{lu}(t)}{1 - f^{ld}(t)/2}\right]. \qquad (8)$$

The initial intercept of the relative movement curve, RM^{II} under these circumstances, is given by (Carlton *et al.*, 1991)

$$RM^{II} \equiv \frac{1 - e^{-\nu} - \nu e^{-1.3\nu}}{\nu(1 - e^{-1.3\nu})}, \qquad (9)$$

and now the changing relative movement of the BrdUrd-labeled undivided cells is

$$RM\Delta t = RM^{lu}(t) - RM^{II}. \qquad (10)$$

T_S and T_{pot} are computed from

$$T_S = \frac{t}{2RM\Delta t} \qquad (11)$$

$$T_{pot} = \frac{\ln(2)T_S}{\nu}. \qquad (12)$$

As an example, the data from Fig. 1B resulted in the following: $RM^{lu}(t) = 0.763$, $f^{lu}(t) = 0.166$, and $f^{ld}(t) = 0.045$, where $t = 3$ hr. From these quantities values of 8.9 hr for T_S and 35 hr for T_{pot} were computed (see also Section V,D).

c. Multiple Time-Point Measurements

There are several possible methods for fitting kinetic parameters from measurements made at multiple time-points after labeling (White *et al.*, 1990a,b, 1991). All of these provide, in contrast to the previously described methods, direct estimates of T_{G2+M}, T_S, and T_{pot}. The most complete approach is to fit $RM^{lu}(t)$ or $RM^l(t)$, $f^{lu}(t)$ and either $f^u{}_{G2+M}(t)$ or $f^{ld}(t)$, using nonlinear methods such as the Marquardt (1963) algorithm (Press *et al.*, 1992). For simplicity, however, the following linear functions may be used as a substitute.

1. Multiple time-point measurements with $f^u_{G2+M}(t) > 0$

Since

$$\ln[1 + f^u_{G2+M}(t)] = cT_{G2+M} - ct, \tag{13}$$

standard linear regression packages may be used to compute c and T_{pot} as well as T_{G2+M}. Further T_S may be computed from Equation (3).

2. Multiple time-point measurements with both $f^{ld}(t)$ and $f^{lu}(t) > 0$

In this case, we write

$$\ln[1 + f^{lu}(t)] = c(T_S + T_{G2+M}) - ct \tag{14}$$

$$\ln[1 - f^{ld}(t)/2] = cT_{G2+M} - ct, \tag{15}$$

and again linear regression may be used to obtain the kinetic parameters. Note that these equations hold for times up to $T_S + T_{G2+M}$ in contrast to single time-point methods which are limited to times shorter than T_S.

V. Critical Aspects of the Procedure

A. Labeling *in Vitro* and *in Vivo*

When labeling cells *in vitro* it is important to minimize perturbation of the cultures. The parasynchrony induced by such insults as leaving the dishes out of the incubator for more than 1 min or so, or refeeding without using warmed, pregassed medium, will be readily discernible in the data. For *in vivo* work do not refill syringes with needles attached that have been used for injections; otherwise contamination of the stock BrdUrd solution will result.

B. Sample Preparation

Adequate sample preparation techniques have been identified as probably the most important requirement for production of the accurate flow cytometric histograms that these studies need (Terry, 1996). This is particularly true in the case of preparations from solid tumors. Different pepsin digestion times of tumors can produce strikingly different flow cytometric profiles despite their preparation from a homogenate of the same specimen. For example, tumors that under optimal digestion conditions would contain an aneuploid population might, if prepared inadequately, be misclassified as diploid. Because no pepsin digestion procedure can be considered "standard," even for tumors of similar histologies from similar sites, many problems may be obviated by making multiple digests, optimized for nucleus yield, from a homogenate of the same sample. In a comprehensive assessment of the sources of error in interlaboratory comparisons, Haustermans *et al.* (1995) also demonstrated that variations in the sample preparation and staining steps were the largest contributors to the overall variance.

The pepsin digestion step to give nuclei is the only part of the procedure that we routinely adjust. Depending on the sample properties, any or all of digestion time, temperature, and degree of agitation may be manipulated to optimize tissue disaggregation and nuclei yield. Frequent microscopic observation is the key to success. It should be noted that not all commercial pepsins are the same and, if bought in small quantities, may differ in their activity from batch to batch.

All centrifugation is at 350 g for 4 min at room temperature. This is a relatively gentle centrifugation and care should be taken not to disturb the pellet before aspirating off the supernatant.

Our approach to sample staining is very standardized. It is always based on knowledge of cell numbers, and the use of defined minimal volumes of reagents when counts are low. The goal is to approach a plateau of FITC staining, together with stochiometric PI staining, 50- to 500-fold levels of FITC/background staining, and low CVs around the G_1 and G_{2+M} peaks of the DNA histogram. CVs of 2–3% about the G_1 peak are readily attainable for *in vitro* sample preparations. The 2.5–4% G_1 CVs are achievable goals for solid tumors and normal tissues. CVs in excess of 5% are generally unacceptable for subsequent quantitative analyses.

C. Flow Cytometry and Data Acquistion

High quality flow cytometry is required for these analytical procedures. The samples must be adequately stained and, if nuclei are to be used (preferred), no undigested cells should remain in the sample. Flow should be stable as long run times may sometimes be required. The instrument should be in good optical alignment with no spectral overlap of the red and green channels. [Fluorescent signal compensation if required, but performed imprecisely, will compromise estimation of quantities such as $RM^{lu}(t)$.] For the same reason there must be good linearity of analog-to-digital converters and amplifiers. Depending on the time of sampling after labeling some of the fractional quantities required for analysis might be of low frequency. Hence, it is important to collect sufficient events, after hardware gating for doublet discrimination. For *in vitro* samples, 30,000 total nuclei usually suffice, DNA-aneuploid tumors usually require 50,000 gated events, and normal tissues, with low labeling indices, may need 100,000 or more nuclei to be acquired.

Furthermore, Haustermans *et al.* (1995) have implicated placement of analytical regions on the flow cytometric histograms as a significant source of potential error. Exploration of the data, by adjusting the regions of interest, usually gives sufficient feedback regarding the stability of estimates of $RM^{lu}(t)$, $f^{lu}(t)$, and $f^{ld}(t)$. There are objective criteria (White and Terry, 1992) that may be used to help distinguish BrdUrd-labeled from unlabeled cells, and aid in analytical region placement, in instances when this distinction is not absolute.

D. Data Analysis

Before embarking on any numerical analysis it is important to look carefully at the DNA versus BrdUrd histograms to ensure that the data are of sufficient quality that estimation of kinetic parameters is worthwhile. For example in DNA-aneuploid samples multiple overlapping populations may hinder or preclude analysis. A special case is that of DNA-diploid tumors as we show in Figs. 1 and 2. In these circumstances two completely overlapped populations (tumor parenchyma together with normal stromal and infiltrating cell populations) are present but indistinguishable, one from the other, based only on DNA contents. If tumor cells are in the majority then a minimum value for LI may be approximated and, while a reasonable value for T_S may often be obtained, the tumor T_{pot} has to be shorter than that value which results from analysis of the measured quantities. The use of whole cells, rather than nuclei, together with three-color flow cytometry, may be helpful in the case of DNA-diploid tumors if markers exist to discriminate tumor from normal cells (Begg and Hofland, 1991).

An important concept to appreciate is whether or not the time interval between BrdUrd labeling and sampling is greater, or less, than the duration of G_2+M (T_{G2+M}). Depending on this timing, the calculation of the duration of S phase (T_S) differs by a factor of two (White and Meistrich, 1986). This can usually be readily deduced by inspection of the bivariate DNA versus BrdUrd flow cytometric histogram and checking for the presence or absence of BrdUrd-labeled cells that have divided [$f^{ld}(t)$] in the period since labeling.

Although the kinetic parameters estimated by the analysis of the bivariate data are often treated as equivalent, it should be observed that there are subtle differences among them. Whereas T_S and T_{G2+M} are direct measures of the duration of cell cycle phases, based on the observed progression of cells, the potential doubling time is, in fact, a derived quantity depending on $RM^{lu}(t)$, $f^{lu}(t)$, and $f^{ld}(t)$. The interpretation of T_{pot} thus depends on the homogeneity of the populations making up the fractions of labeled cells. Thus it may be necessary to fit overlapping DNA populations before computing $f^{lu}(t)$ and $f^{ld}(t)$ in order to obtain a T_{pot} value for a tumor subpopulation. Moreover, as also pointed out by Bertuzzi *et al.* (1997), changes in the pattern of cell loss can strongly influence the computed value of T_{pot}. Thus, caution is required in interpreting the relationships between T_{pot} values obtained from different tumors.

Acknowledgments

The authors thank Mrs. Nalini Patel for her expert technical assistance. This work was supported by the National Institutes of Health Grant No. CA 06294 and the State of Texas Higher Education Coordinating Board Advanced Technology Program.

References

Antognoni, P., Terry, N. H. A., Richetti, A., Luraghi, R., Tordiglione, M., and Danova, M. (1998). The predictive role of flow cytometry-derived tumor potential doubling time (T_{pot}) in radiotherapy: Open questions and future perspectives (review). *Int. J. Oncol.* **12,** 245–256.

Begg, A. C. (1995). The clinical status of T_{pot} as a predictor? Or why no tempest in the T_{pot}! *Int. J. Radiat. Oncol. Biol. Phys.* **32,** 1539–1541.

Begg, A. C., and Hofland, I. (1991). Cell kinetic analysis of mixed populations using three-color fluorescence flow cytometry. *Cytometry* **12,** 445–454.

Begg, A. C., McNally, N. J., Shrieve, D. C., and Kärcher, H. (1985). A method to measure duration of DNA synthesis and the potential doubling time from a single sample. *Cytometry* **6,** 620–626.

Beisker, W., Dolbeare, F., and Gray, J. W. (1987). An improved immunocytochemical procedure for high-sensitivity detection of incorporated bromodeoxyuridine. *Cytometry* **8,** 235–239.

Bertuzzi, A., Gandolfi, A., Sinisgalli, C., Starace, G., and Ubezio, P. (1997). Cell loss and the concept of potential doubling time. *Cytometry* **29,** 34–40.

Bussink, J., Terry, N. H. A., and Brock, W. A. (1995). Cell cycle analysis of synchronized Chinese hamster cells using bromodeoxyuridine labeling and flow cytometry. *In Vitro Cell. Dev. Biol.* **31,** 547–552.

Carayon, P., and Bord, A. (1992). Identification of DNA-replicating lymphocyte subsets using a new method to label the bromo-deoxyuridine incorporated into the DNA. *J. Immunol. Methods* **147,** 225–230.

Carlton, J. C., Terry, N. H. A., and White, R. A. (1991). Measuring potential doubling times of murine tumors using flow cytometry. *Cytometry* **12,** 645–650.

Dolbeare, F., and Gray, J. W. (1988). Use of restriction endonucleases and exonuclease III to expose halogenated pyrimidines for immunochemical staining. *Cytometry* **9,** 631–635.

Dolbeare, F. A., Gratzner, H. G., Pallavicini, M. G., and Gray, J. W. (1983). Flow cytometric measurement of total DNA content and incorporated bromodeoxyuridine. *Proc. Natl. Acad. Sci. U.S.A.* **80,** 5573–5577.

Dolbeare, F., Kuo, W. L., Beisker, W., Vanderlaan, M., and Gray, J. W. (1990). Using monoclonal antibodies in bromodeoxyuridine-DNA analysis. *In* "Methods in Cell Biology" (Z. Darzynkiewicz and H. Crissman, eds.), Vol. 33, p. 207. Academic Press, San Diego.

Dubray, B., Maciorowski, Z., Cosset, J.-M., and Terry N. H. A. (1995). Le point sur le temps de doublement potentiel. *Bull. Cancer/Radiother.* **82,** 331–338.

Gratzner, H. (1982). Monoclonal antibody to 5-bromo and 5-iododeoxyuridine: A new reagent for detection of DNA replication. *Science* **218,** 474–475.

Haustermans, K., Hofland, I., Pottie, G., Ramaekers, M., and Begg, A. C. (1995). Can measurements of potential doubling time (T_{pot}) be compared between laboratories? A quality control study. *Cytometry* **19,** 154–163.

Johansson, M. C., Baldetorp, B., Bendahl, P. O., Fadeel, I. A., and Oredsson, S. M. (1996). Comparison of mathematical formulas used for estimation of DNA synthesis time of bromodeoxyuridine-labelled cell populations with different proliferative characteristics. *Cell Prolif.* **29,** 525–538.

Johansson, M. C., Johansson, R., Baldetorp, B., and Oredsson, S. M. (1998). Comparison of different labelling index formulae used on bromodeoxyuridine-flow cytometry data. *Cytometry* **32,** 233–240.

Larsen, J. K. (1990). Washless double staining of a nuclear antigen (Ki-67 or bromodeoxyuridine) and DNA in unfixed nuclei. *In* "Methods in Cell Biology" (Z. Darzynkiewicz and H. Crissman, eds.), Vol. 33, p. 227. Academic Press, San Diego.

Marquardt, D. W. (1963). An algorithm for least-squares estimation of nonlinear parameters. *J. Soc. Indust. Appl. Math.* **11,** 431–441.

Pollack, A., Terry, N. H. A., Van, N. T., and Meistrich, M. L. (1993a). Flow cytometric analysis of two incorporated halogenated thymidine analogues and DNA in a mouse mammary tumor grown in vivo. *Cytometry* **14,** 168–172.

Pollack, A., White, R. A., Cao, S., Meistrich, M. L., and Terry, N. H. A. (1993b). Calculating potential doubling time (T_{pot}) using monoclonal antibodies specific for two halogenated thymidine analogues. *Int. J. Radiat. Oncol. Biol. Phys.* **27,** 1131–1139.

Pollack, A., Terry, N. H. A., White, R. A., Cao, S., Meistrich, M. L., and Milas, L. (1994). Proliferation kinetics of recruited cells in a mouse mammary carcinoma. *Cancer Res.* **54,** 811–817.

Pollack, A., Terry, N. H. A., Wu, C. S., Wise, B. M., White, R. A., and Meistrich, M. L. (1995). Specific standing of iododeoxyuridine and bromodeoxyuridine in tumors double-labelled in vivo: A cell kinetic analysis. *Cytometry* **20,** 53–61.

Press, W. H., Teukolsky, S. A., Vetterling, W. T., and Flannery, B. P. (1992). "Numerical Recipes in FORTRAN. The Art of Scientific Computing," 2nd Ed. Cambridge Univ. Press, Cambridge.

Ritter, M. A., Fowler, J. F., Kim, Y., Lindstrom, M. J., and Kinsella, T. J. (1992). Single biopsy, tumor kinetic analyses: A comparison of methods and an extension to shorter sampling intervals. *Int. J. Radiat. Oncol. Biol. Phys.* **23,** 811–820.

Sacks, P. G., Oke, V., Calkins, D. P., Vasey, T., and Terry, N. H. A. (1990). Effects of β-all-trans retinoic acid on growth, proliferation and cell death in a multicellular tumor spheroid model for squamous carcinomas. *J. Cell. Physiol.* **144,** 237–243.

Schutte, B., Reynders, M. M. J., van Assche, C. L. M. V. J., Hupperets, P. S. G. J., Bosman, F. T., and Blijham, G. H. (1987). An improved method for the immunocytochemical detection of bromodeoxyuridine labeled nuclei using flow cytometry. *Cytometry* **8,** 372–376.

Terry, N. H. A. (1996). Predictive assays for radiotherapy: The role of tumor proliferation (T_{pot}) measurements. *Onkologie* **19,** 322–327.

Terry, N. H. A., and Peters, L. J. (1995). The predictive value of tumor-cell kinetic parameters in radiotherapy: Considerations regarding data production and analysis. *J. Clin. Oncol.* **13,** 1833–1836.

Terry, N. H. A., and White, R. A. (1996). Lessons from multiparameter thymidine analogue-DNA flow cytometry for one parameter DNA cytometry. *Clin. Immunol. Newslett.* **16,** 46–50.

Terry, N. H. A., White, R. A., and Meistrich, M. L. (1992). Cell kinetics: From tritiated thymidine to flow cytometry. *Br. J. Radiol. Suppl.* **24,** 153–157.

Terry, N. H. A., Milross, C. G., Patel, N., Mason, K. A., White, R. A., and Milas, L. (1997). The effect of paclitaxel on the cell cycle kinetics of a murine adenocarcinoma in vivo. *The Breast Journal* **3,** 99–105.

White, R. A., and Meistrich, M. L. (1986). A comment on "A method to measure the duration of DNA synthesis and the potential doubling time from a single sample." *Cytometry,* **7,** 486–490.

White, R. A., and Terry, N. H. A. (1992). A quantitative method for evaluating bivariate flow cytometric data obtained using monoclonal antibodies to bromodeoxyuridine. *Cytometry* **13,** 490–495.

White, R. A., Terry, N. H. A., and Meistrich, M. L. (1990a). New methods for calculating kinetic in vitro properties using pulse labeling with bromodeoxyuridine. *Cell Tissue Kinet.* **23,** 561–573.

White, R. A., Terry, N. H. A., Meistrich, M. L., and Calkins, D. P. (1990b). Improved method for computing potential doubling time from flow cytometric data. *Cytometry* **11,** 314–317.

White, R. A., Terry N. H. A., Baggerly, K. A., and Meistrich, M. L. (1991). Measuring cell proliferation by relative movement. I. Introduction and in vitro studies. *Cell Prolif.* **24,** 257–270.

White, R. A., Fallon, J. F., and Savage, M. P. (1992). On the measurement of cytokinetics by continuous labeling with bromodeoxyuridine with applications to chick wing buds. *Cytometry* **13,** 553–556.

White, R. A., Pollack, A., and Terry, N. H. A. (1994a). Simultaneous cytokinetic measurement of aneuploid tumors and associated diploid cells following continuous labelling with chlorodeoxyuridine. *Cytometry* **15,** 311–319.

White, R. A., Pollack, A., Terry., N. H. A., Meistrich, M. L., and Cao, S. (1994b). Double labeling to obtain S-phase subpopulations; Application to determine cell kinetics of diploid cells in an aneuploid tumor. *Cell Prolif.* **27,** 123–127.

CHAPTER 17

Flow Cytometric Analysis of Cell Division History Using Dilution of Carboxyfluorescein Diacetate Succinimidyl Ester, a Stably Integrated Fluorescent Probe

A. Bruce Lyons,[*] Jhagvaral Hasbold,[†] and Philip D. Hodgkin[*,†,‡]

[*]Discipline of Pathology
Faculty of Health Science
The University of Tasmania
Hobart, Australia

[†]The Centenary Institute of Cancer Medicine and Cell Biology
Sydney, Australia

[‡]Medical Foundation
University of Sydney
Sydney, Australia

    I. Introduction and Background
   II. Reagents and Solutions
       A. Carboxyfluorescein Diacetate Succinimidyl Ester Stock
       B. Buffers and Culture Medium
  III. Preparation and Labeling of Cells
       A. Labeling of Cells with CFSE (Standard Protocol)
       B. Determining Appropriate Levels of CFSE Staining Intensity
       C. Improving Resolution by Presorting
  IV. Gathering of Information Concurrent with Division
       A. Cell Surface Marker Staining
       B. Intracellular Staining
       C. Bromodeoxyuridine and DNA Staining
       D. The Source of Division Asynchrony in Cultures
       E. Cell Sorting on CFSE Peaks for Functional and Analytical Studies
       F. Tracking the Fate of Dead Cells

V. Analysis of Data
   A. Monitoring the Position of the Undivided Group
   B. Finding the Position of Each Peak
   C. Estimating the Number of Cells per Division
   D. The Method of Slicing Data from Asynchronous Cultures for Differentiation Analysis
VI. Application of Carboxyfluorescein Diacetate Succinimidyl Ester to *in Vitro* Culture of Lymphocytes
   A. B Lymphocyte Cultures
   B. T Lymphocyte Cultures
VII. Monitoring Lymphocyte Responses *in Vivo*
   A. Intravenous Transfer of Lymphocytes
   B. Preparation of Cell Suspensions and Strategies for Analysis
VIII. Antigen Receptor Transgenic Models
   References

I. Introduction and Background

The cells of the immune system undergo significant expansion and differentiation as a result of immune stimulation, as well as during the production of the formed hemopoietic elements. In contrast to tissues with defined microanatomy, where associations between cells and structures are relatively simple to determine, the mobile nature of lymphohemopoietic cells makes it more difficult to define lineage relationships.

Here we review the use of a technique based on the serial dilution of a stably binding intracellular fluorochrome, carboxyfluorescein diacetate succinimidyl ester (CFSE), which allows eight to ten sequential cell divisions to be analyzed by flow cytometry (Fig. 1). When incubated with cells, the fluorescein-based CFSE crosses the cell membrane and attaches to free amine groups of cytoplasmic cell proteins. After enzymatic removal of carboxyl groups by endogenous intracellular esterases, CFSE acquires identical spectral characteristics to fluorescein, with optimal excitation by 488 nm argon laser light, emitting strongly at 519 nm, and as such is compatible with almost all single and multiple laser flow cytometers. On cell division, CFSE is distributed equally between progeny, allowing the division history of a cell population to be determined. This technique can be used to investigate the behavior of cells *in vitro*, as well as division of transferred cells *in vivo*.

Other competing techniques for monitoring cell proliferation, such as the use of tritiated thymidine incorporation, can quantify overall division behavior of a population but give no information on the division history of individual cells. Furthermore, appropriately conjugated monoclonal antibodies can be employed to identify the cells undergoing division and whether their phenotypic properties change with division number. A major advantage of the CFSE based technique is

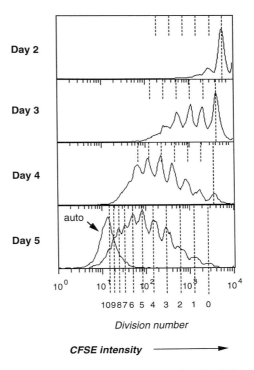

Fig. 1 Tracking division of B lymphocytes using CFSE dilution. Small dense murine B cells labeled with CFSE were cultured for up to 5 days with supernatants containing interleukin (IL)-5 (31 U/ml) and IL-4 (140 U/ml) and a source of CD40L in the form of membranes from an activated T helper cell clone (H66.61). At intervals of 2, 3, 4, and 5 days, cells were harvested and analyzed by flow cytometry. Division is characterized by sequential twofold reduction in CFSE fluorescence, resulting in equally spaced peaks on a logarithmic scale. Dashed lines indicate the division cycle number for each panel. Note the slow decay in intensity of peaks independent of division. The arrow indicates the autofluorescence level of stimulated but unstained cells. (Reproduced from *The Journal of Experimental Medicine*, 1996, Vol. 184, pp. 277–281, by copyright permission of The Rockefeller University Press.)

that viable cells from defined division cycles can be recovered, allowing functional characteristics to be related to differentiation stage.

Since the introduction of the CFSE based technique in 1994, it has become the method of choice for investigating the division related differentiation of lymphohemopoietic cells, as well as the kinetics of cellular expansion during an immune response. Among the cell types that have been investigated using the technique are hemopoietic stem cells, T and B lymphocytes, natural killer (NK) cells, and a number of cell lines.

II. Reagents and Solutions

A. Carboxyfluorescein Diacetate Succinimidyl Ester Stock

A stock solution of 5 mM CFSE [5(and 6)-carboxyfluorescein diacetate succinimidyl ester] (MW 557, Molecular Probes, Eugene, OR) is made by dissolving CFSE in DMSO at a concentration of 2.785 mg/ml. Gentle pipetting is sometimes necessary to achieve this. The stock solution is aliquoted into convenient volumes (e.g., 50 μl) and stored frozen at $-20°C$ under dessicating conditions. Stock solutions can be kept for over 1 year under these conditions. To ensure reproducibility of staining intensity, aliquots are thawed a maximum of three times and then discarded.

B. Buffers and Culture Medium

Standard isotonic phosphate-buffered saline (PBS), pH 7, is usually employed in the staining procedure, however, culture medium such as RPMI with no added serum has also been successfully used. The addition of a small amount of protein such as bovine serum albumin (BSA) in the staining procedure can improve poststaining viability (see later).

For culture of cells after staining, standard culture medium such as RPMI or DMEM supplemented with 5–10% serum is routinely used.

III. Preparation and Labeling of Cells

A. Labeling of Cells with CFSE (Standard Protocol)

This standard protocol has been successfully employed to stain B and T lymphocytes of both mouse and human origin. There have been a number of "in house" modifications of the original staining procedure, including staining on ice or at room temperature, which can sometimes improve viability of sensitive cells. However, the standard protocol will be suitable for most applications.

Cells to be labeled are suspended at 5×10^7 ml in PBS/0.1% BSA, ensuring that cells are well suspended with no aggregates. The resolution of cell division is critically dependent on uniformity of staining. The inclusion of a small amount of BSA in the staining step does not markedly affect the intensity of staining achieved, but can improve the viability of cells, especially when sensitive cells such as B lymphocytes are stained. Alternatively, cells can be stained in a culture medium such as RPMI without added serum, containing 0.1% BSA which can also improve poststaining viability.

To each milliliter of cell suspension, 2 μl of CFSE stock solution is added and immediately mixed to ensure uniform staining, resulting in a final concentration of 10 μM CFSE. The cells are incubated for 10 min at 37°C, and the labeling is quenched by adding five volumes of ice-cold RPMI/10% fetal calf serum

(FCS). Resolution of division related fluorescence peaks is critically dependent on achieving uniform staining of the starting population, so it is essential to mix cell suspension well at the point of addition of CFSE. Another way to ensure even staining is to resuspend cells at 10^8 ml and add an equal volume of CFSE prediluted to 20 μM. Due to the labile nature of CFSE, it is important to predilute immediately before adding to the cell suspension. More detailed discussion of the methodology can be found in Lyons and Doherty (1998).

B. Determining Appropriate Levels of CFSE Staining Intensity

The standard protocol is suitable for tracking cell division from day 2 to 14 after staining, both *in vitro* (Fig. 1) and *in vivo*. For time points earlier than 2 days CFSE staining intensity may need to be reduced. This is because within the first few days after staining, there is a sharp drop in CFSE fluorescence as rapidly turned over components are catabolized. This period is then followed by a slower loss of fluorescence (Fig. 2). Note that this fluorescence loss occurs in the absence of cell division, and also occurs in divided cells, so that the relationship between fluorescence intensity between cells undergoing different

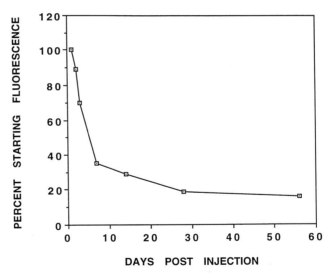

Fig. 2 Decay of CFSE fluorescence intensity of nondivided cells. Murine splenic lymphocytes 2×10^7, labeled with CFSE were injected intravenously into syngeneic hosts. At various time intervals, spleens were removed and analyzed by flow cytometry. The fluorescence intensity of nondividing cells was determined, and expressed as a percentage of starting CFSE fluorescence intensity. (From Flow cytometric analysis of cell division by dye dilution. A. B. Lyons and K. V. Doherty, J. P. Robinson *et al.*, Eds. "Current Protocols in Cytometry." Copyright 1998 John Wiley & Sons. Reprinted by permission of Wiley-Liss, Inc., a subsidiary of John Wiley & Sons, Inc.)

numbers of cell divisions remains constant (see Fig. 1). In addition, when a CFSE stained population of cells is both cultured *in vitro* and transferred *in vivo,* the rate of loss of CFSE in undivided cells is the same under both conditions (A. B. Lyons, 1999). As the relationship between fluorescence intensity obtained is essentially linear with respect to CFSE concentration (Fig. 3), the fluorescence intensity can be manipulated to suit the experiment. For example, lower concentrations can be used for short experiments, whereas for extremely long-term experiments, such as transfer and tracking of cells *in vivo,* a high initial staining intensity may be required. However, it is important to consider that CFSE labeling may be toxic to some cell types at high concentrations. For this reason the experimenter will need to determine appropriate levels of staining for each application. Note that very intensely CFSE-stained cells may make flow cytometric compensation difficult or impossible due to very bright fluorescence spilling over into orange and red channels. Conversely, understaining will limit the number of cycles of division that can be resolved.

C. Improving Resolution by Presorting

If the starting population of cells being stained is heterogenous with respect to size, this will result in a broad range of starting fluorescence intensities. As a

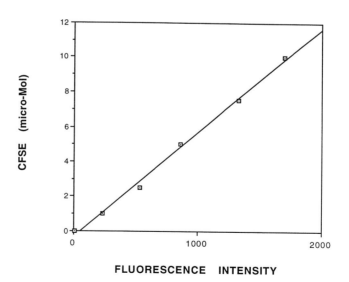

Fig. 3 Linearity of staining with respect to concentration of CFSE. Murine splenic lymphocytes were stained using the standard protocol, except differing final concentrations of CFSE were used. Fluorescence intensity (arbitrary units) was plotted against CFSE concentration, showing staining is linear with respect to CFSE concentration. (From Flow cytometric analysis of cell division by dye dilution. A. B. Lyons and K. V. Doherty, J. P. Robinson *et al.,* Eds. "Current Protocols in Cytometry." Copyright 1998 John Wiley & Sons. Reprinted by permission of Wiley-Liss, Inc., a subsidiary of John Wiley & Sons, Inc.)

result, the separation of division related peaks will not resolve as well as a more uniformly labeled starting population. Some researchers have overcome this limitation by presorting the stained cell population such that their fluorescence intensity is over a 40-channel interval on a 1024-channel scale (Nordon *et al.*, 1997), ensuring high resolution of division. The only caveat for this approach is to ensure that sorting for a narrow range of fluorescent intensity does not select for a distinct subset of the starting population, as this approach will tend to select cells on the basis of size.

IV. Gathering of Information Concurrent with Division

A. Cell Surface Marker Staining

The spectral characteristics of CFSE are essentially identical to fluorescein making it possible to monitor cell phenotype and division history simultaneously by the use of specific antibodies coupled with compatible fluorochromes.

Cells are labeled with CFSE by the previously mentioned standard method, and then cultured appropriately to induce proliferation and differentiation. Cells are harvested at various time points after cell culture initiation and incubated with surface marker specific antibodies for 45 min followed by secondary conjugate on ice. Cells are washed two times with ice-cold PBS/0.1% BSA/0.1% sodium azide between incubations. Stained cells then are analyzed on a flow cytometer. Appropriate fluorochromes for use in conjunction with CFSE for a single laser (488 nm) are phycoerythrin (PE) and tandem dyes such as PE/Texas Red or PE/Cy5, or alternatively peridinium chlorophyll protein (PerCP). More elaborate multilaser cytometers will support the use of other fluorochromes such as allophycocyanin (APC), allowing more complex analyses to be performed. Electronic compensation between detecting channels should be performed according to the instructions of the instrument manufacturer. However, very bright staining with CFSE may result in compensation being difficult to achieve, in which case it may be necessary to adjust labeling to suit the experiment.

It is recommended to have *in vitro* culture controls for flow cytometry analysis. The first control is use of CFSE labeled cells either unstimulated in culture or stimulated in such a way as to maintain viability without inducing division. For example, interleukin (IL)-4 will keep resting B cells alive for 3–5 days of *in vitro* culture, without inducing proliferation. This control is important for marking the starting point for calibrating division number (i.e., division 0). The second control is use of cells that have not been CFSE labeled but stimulated to proliferate. This control is for the estimation of autofluorescence intensity, which provides information on the number of divisions that can be tracked. As the CFSE labeling can vary between experiments, in some cases the intensity of CFSE level may interfere with the other channels in flow cytometry analysis. Therefore, it is important to use single color CFSE labeled cells to check and appropriately adjust compensation before analysis.

In response to various stimuli immune cells proliferate and differentiate. The differentiation of activated cells is usually associated with the appearance or downregulation of cell surface markers. As the stimulated cells progress through the differentiation pathway, they display a distinct composition of surface molecules that help determine the cells functional capabilities. A number of variables can affect the differentiation profile of cultured cells such as the time of culture or the rate of proliferation. The CFSE labeling method provides a unique opportunity to separate these two variables and examine the changes of cell surface molecule expression in relation to cell division number. An example is shown in Fig. 4, which demonstrates the division related changes in immunoglobulin (Ig) isotype expressed on CFSE labeled murine B cells. This method is based on small dense B cells purified from murine spleen; however, differentiation changes can be successfully followed in *in vitro* stimulated murine T cells (Gett and Hodgkin, 1998), as well as in human peripheral B and T lymphocytes (Fig. 6) (Kindler and Zubler, 1997).

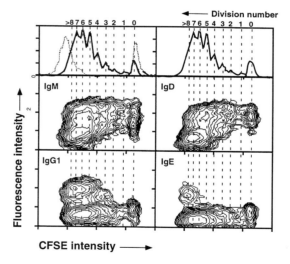

Fig. 4 Isotype switching to IgE requires more rounds of cell division than switching to IgG1. Small dense murine B lymphocytes were cultured in the presence of IL-4 (1000 U/ml) and an optimal concentration of plasma membrane of a CD40L transfected insect cell line for 5 days. The culture was harvested and stained for surface expression of different immunoglobulin isotypes. The top panels show histograms of CFSE fluorescence, with the dotted histogram on the left representing autofluorescence of unstained stimulated cells, and the dotted histogram on the right representing control unstimulated CFSE stained cells. Dashed lines represent positions of the cell division peaks. Note that switching to IgG1 occurs after three cell divisions, but acquisition of surface IgE requires at least six rounds of cell division. (This figure is reproduced from Hasbold *et al.*, 1998, and is used with kind permission of the publisher.)

B. Intracellular Staining

As described earlier, double staining of CFSE labeled cells for cell surface markers provides unique information about cell division regulated differentiation. It has also proved possible to couple the CFSE method with the detection of intracellular components. Potentially, the intracellular staining method can be used to detect any cytoplasmic or nuclear molecule, depending on the sensitivity of antibody used. This method has also proved to be more sensitive than surface staining for some rare markers, especially when they are difficult to detect by a standard surface staining techniques, such as IgE.

Many different procedures have been described for fixation and permeabilization of cells for intracellular staining. The primary concern in fixing and permeabilizing CFSE labeled cells is to retain the CFSE and, at the same time, to preserve the cell morphology and interior for antibody recognition. Most fixation and permeabilization methods described in the literature were not suitable for this purpose. However, an adaptation of a method using paraformaldehyde and Tween 20 detergent proved successful.

Appropriately stimulated CFSE labeled cells ($\sim 10^6$ per sample) are fixed with 0.25 ml 2% paraformaldehyde (PFA) for 10 min at room temperature followed by the addition of 0.25 ml PBS and 0.5 ml 0.2% Tween 20 solution. The cells are then mixed on a vortex and left overnight at 4°C to permeabilize. Next day cells are centrifuged and the pellets resuspended in PBS containing 0.1% BSA and 0.1% sodium azide, and left on ice for a further 30 min to block any free PFA residues. Note that after overnight incubation with PFA and Tween 20, plastic tubes usually become highly hydrophobic, and it may be difficult to remove supernatants after centrifugation. Changing tubes after overnight incubation and/or the addition of 0.05% Tween 20 to the washing buffer (PBS/BSA/0.05% Tween 20) for subsequent washes will help reduce cell losses. For intracellular staining of fixed and permeabilized cells any standard staining protocol for flow cytometry is suitable. However, if streptavidin reagents are used higher backgrounds may occur due to the nonspecific binding to cytoplasmic components.

C. Bromodeoxyuridine and DNA Staining

Commonly used methods to assess overall cell proliferation, such as tritiated thymidine incorporation and bromodeoxyuridine (BrdU) incorporation provide a comparative cumulative assessment of cell proliferation. However, BrdU staining in combination with the CFSE labeling technique also allows the examination of the division-related proliferation rate and cell cycle profile as well as division history. The following method is designed to detect simultaneously cell division history, cell cycle position, and proliferation rate in a population of dividing cells by flow cytometry.

CFSE labeled cells are stimulated and pulsed for 3 hr before harvest with BrdU (100 μg/ml final concentration) on various days of culture. The harvested cells are fixed and permeabilized with PFA and Tween 20 as described earlier.

The method of Tough and Sprent (1994) is then used to detect the level of BrdU incorporation. Cells are treated with DNase I (20 μg/ml final concentration diluted in 50 mM Tris-HCl buffer, pH 7.4, with 10 mM MgCl$_2$) for 30 min at 37°C in a water bath. Subsequently, cells are stained with an appropriate BrdU specific antibody conjugate. For obvious reasons, this antibody should be conjugated to fluorochromes other than fluorescein isothiocyanate (FITC). If BrdU specific antibodies are not conjugated to fluorochromes and secondary antibody or avidin conjugate is used, the following staining controls are suggested: (1) fixed and permeabilized cells stained with secondary conjugate only and (2) cells not pulsed with BrdU but treated and stained in same way as BrdU pulsed samples. In some instances, especially when cells are proliferating at a high rate, a dramatic reduction of BrdU incorporation occurs after 72 hr of *in vitro* culture due to cell overgrowth and medium depletion. Therefore, it is advisable to keep the initial starting cell density low for day 4 and 5 cultures. CFSE labeled and BrdU stained cells run on the flow cytometer allow an estimation of the rate of cell division and how it might differ at each division (see Fig. 5). If BrdU is being detected with a PE-conjugated antibody it is also possible to monitor simultaneously the position of each cell within the cell cycle. The DNA intercalating molecule 7-aminoactinomycin D (7-AAD) fluoresces at 647 nm and is compatible with both CFSE and PE in a single laser (488 nm) flow cytometer (Rabinovitch *et al.*, 1986). After detection of BrdU, CFSE labeled fixed cells are incubated with 7-AAD at 1 μg/ml for 30 min on ice, before washing and analysis by flow cytometry. The 7-AAD channel should be collected in linear acquisition mode. Figure 5C indicates the simultaneous staining with CFSE and 7-AAD.

D. The Source of Division Asynchrony in Cultures

One of the most curious features of CFSE profiles is the tremendous variation in the number of divisions that otherwise identical cells will undergo at a given time point after stimulation. This level of asynchrony is affected by time and the strength of stimulation (Hodgkin *et al.*, 1997). BrdU pulsing can be used to determine whether the source of asynchony is either due to differing rates of cell division, or variation in the time a cell responds to activation signals eliciting division. Figure 5B shows that in the case of B cells stimulated with CD40L, the rate of division is constant once cells enter their first division cycle. Therefore, at the population level, there is a difference in time taken to enter first division among apparently homogeneous cells. This procedure can also be followed *in vivo* as mice carrying transferred CFSE labeled cells can be fed BrdU to reveal information about division asynchrony.

E. Cell Sorting on CFSE Peaks for Functional and Analytical Studies

A major advantage of the CFSE dye dilution technique is the ability to recover viable cells that have undergone defined numbers of rounds of cell division.

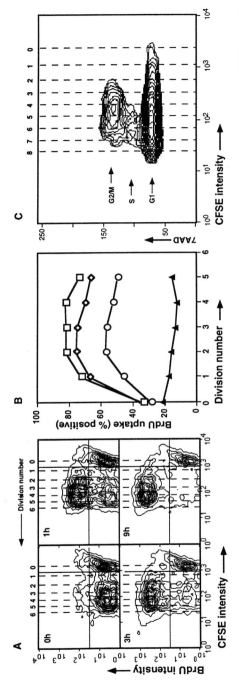

Fig. 5 Simultaneous use of bromodeoxyuridine incorporation and CFSE dilution enables rate of cell division to be compared. CFSE stained B cells were cultured for 3 days under the same conditions as in Fig 1. For the final 1, 3, or 9 hr of incubation, cultures were pulsed with 50 μg BrdU. Cells were harvested, permeabilized, and stained with a biotinylated anti-BrdU antibody detected with streptavidin tricolor. (A) Contour plots of CFSE versus BrdU incorporation. From this data, the percentage of BrdU positive cells in each division cycle was determined, and represented graphically in (B), showing 9-hr pulse (open squares), 3-hr pulse (open diamonds), 1-hr pulse (open circles), and 0-hr pulse (background control, closed triangles). (C) A contour plot of 7-AAD stained cells. CFSE labeled cells were stimulated with CD40L and IL-4. Four days later cells were fixed and permeabilized, and stained with 10 ng/ml 7-AAD. Dashed lines represent positions of the cell divisions. (A and B are reproduced from Hasbold *et al.*, 1998, and are used with kind permission of the publisher.)

Sorting profiles can be further refined by expression of one or more surface markers, in combination with division history.

Cells recovered in this way can be analyzed in functional assays, such as secretion of cytokines or antibodies (Gett and Hodgkin, 1998), or ability to lyse targets in the case of cytotoxic T cells. Cells can also be restained with CFSE to assess further proliferative potential. Sorted cells can be used as a source of RNA or DNA to assess levels of specific message for cytokines and other products, or to determine genomic recombination events in antigen receptors of B and T cells, or changes in DNA methylation patterns associated with differentiation.

F. Tracking the Fate of Dead Cells

Immune cells in culture exhibit varying amounts of cell death, usually by the process of apoptosis. In many instances, it is helpful to determine what contribution cell death makes to the overall proliferation behavior of cells at the population level. To investigate cell death, the membrane integrity of harvested CFSE labeled cells can be probed with nucleic acid binding fluorochromes. Again, 7-AAD is especially useful, as it is compatible with fluorochromes such as PE, allowing surface marker expression to be simultaneously analyzed along with division and cell death. Unfixed, CFSE stained cells are stained on ice for 30 min with 1 μg/ml 7-AAD, and washed prior to flow cytometric analysis. Alternatively, CFSE division analysis can be combined with staining for annexin V to simultaneously detect apoptotic cells (Warren and Kinnear, 1999).

On dying, CFSE stained cells retain their fluorescence, allowing the division cycle in which cell death occurred to be determined (Wells *et al.*, 1997; Hasbold *et al.*, 1999).

V. Analysis of Data

A. Monitoring the Position of the Undivided Group

After labeling with CFSE the small resting population runs as a log-normal distribution with respect to fluorescence intensity, as shown in Fig. 1. As mentioned previously, the mean intensity of labeled cells progressively reduces with time in culture even without any cell division (Figs. 1 and 2). Therefore, for any quantitative analysis of division history it is important to determine the mean fluorescence intensity of an undivided control group of cells. This control is usually provided by cells cultured under similar conditions but without any proliferative stimulation. A potential problem with this procedure would result if unstimulated cells lost CFSE intensity at a different rate to activated cells. Remarkably, this does not occur. Figure 6 shows fluorescence histograms and dot plots of forward scatter against CFSE fluorescence of small resting and larger activated cells at day 3 after stimulation. Clearly, the mean CFSE staining

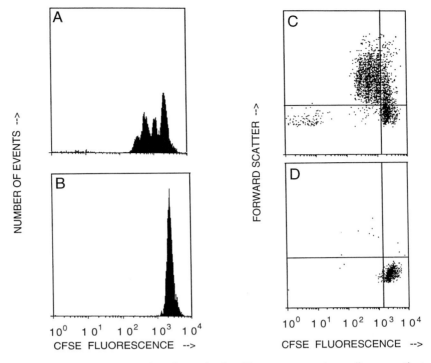

Fig. 6 Fidelity of CFSE partitioning after activation. Human mononuclear cells were activated *in vitro* with PHA and sheep red blood cells. After 3 days, control (B and D) and activated (A and C) cells were harvested and analyzed by flow cytometry. (A and B) Histograms of CFSE fluorescence intensity (C and D), CFSE fluorescence intensity versus forward scatter as an indication of size, therefore of activation status, with the lower right quadrant indicting the position of undivided, unactivated cells. This figure demonstrates that the fidelity of CFSE inheritence is not affected by blastogenesis.

intensity of the small cells in both unstimulated and stimulated cultures is identical to that of the large activated cells, and the intensity predicted from twofold dilution of subsequent divisions is faithfully preserved. The consistent, proportional lessening of CFSE intensity across all divisions is also apparent from the time course data shown in Fig. 1. For this reason, unstimulated control cultures can be used to indicate the position of the undivided peak at the time of harvesting an *in vitro* stimulation experiment.

B. Finding the Position of Each Peak

When a cell divides the retained CFSE label is distributed to each daughter cell. Experience with lymphocytes indicates that this partitioning is essentially even and that the mean intensity of the cells found in division 1 is accurately half that of the undivided peak recorded at the same time. A geometric twofold reduction in intensity is faithfully followed with each successive cell division, allowing the position of the mean of cells in each division to be calculated from the starting fluorescence. When plotted as log CFSE intensity as is usually done, the peaks are evenly separated (Fig. 1). This diminution in fluorescence, however, has a natural limitation. Unlabeled but stimulated cells increase in size and autofluorescence intensity (arrowed in the lower panel of Fig. 1). This autofluorescence is essentially insignificant when compared to the CFSE intensity of cells in early divisions, but contributes increasingly to the total fluorescence as cells undergo further divisions. The mean intensity of each peak can be determined by the following equation, which requires only the experimental values for D_0 and A—the mean fluorescence intensity of the undivided cells and unlabeled cells respectively.

$$D_i = [(D_0 - A)/2^{(i)}] + A \qquad (1)$$

Where i is the division number.

C. Estimating the Number of Cells per Division

For quantitative analysis of cell proliferation it is often necessary to estimate the total cell number found in each division. As the cells in adjacent divisions always overlap to some extent, then the numbers must be estimated from the curve. This can be achieved by using an interval analysis, or a more accurate computer based curve fitting procedure.

1. Interval Analysis

This is the simplest procedure, and requires the use of software to prepare intervals around the predicted mean. An example is shown in Fig. 7B. Intervals are assigned around the peak channel, and the number of events in that interval is divided by the total event number (all gated live cells) to determine the proportion of cells in that division. Clearly this procedure will become less accurate at later divisions as autofluorescence becomes significant and the overlap between divisions increases.

2. Fitting by Computer

As mentioned previously, the distribution of fluorescence intensity in the starting population is log normal. Remarkably, for T and B cells the standard deviation of this distribution is preserved with each division, indicating that the redistribution of CFSE between daughter cells at mitosis is accurately twofold through at least seven generations. As a consequence, the standard CFSE profile of asynchronously dividing cell populations is made up of a series of log-normal distributions with the mean intensity of each predicted by Equation (1) and with

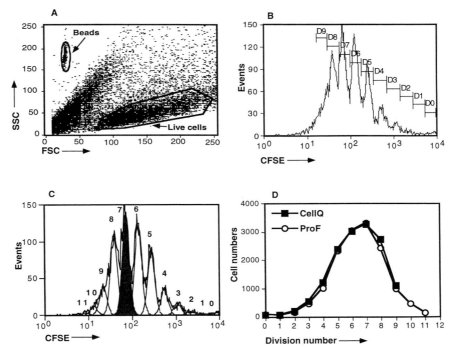

Fig. 7 Quantitative data analysis. CFSE labeled B cells were stimulated with CD40L and IL-4 for 4 days. (A) Dot plot indicates gated live cells and the position of Calibrite beads on forward and side scatter profiles. (B) Interval analysis of the CFSE histogram using CellQuest software. Markers above the histogram represent gates that correspond to the different division numbers. (C) A series of log-normal Gaussian curves fitted to the same data by Pro Fit. Division numbers are shown above each peak. (D) This panel represents the number of live cells in each CFSE peak based on a CellQuest interval analysis (closed square) or Pro Fit curve fitting analysis (open circles) following reference to the number of beads as described in the text. Data represents mean and SE of triplicate cultures (error bars smaller than symbol).

a constant standard deviation. Many computer-based curve fitting programs are available that can be adapted to take gated flow cytometric data and fit a series of overlapping log-normal curves and return the value of the area of each. Pro Fit for Macintosh (Quantumsoft, Zurich) or Peakfit for PC (Jandel, CA) are suitable for this purpose. Figure 7D compares an interval analysis using Cellquest software (Becton Dickinson, Palo Alto, CA) with a computer based Pro Fit analysis, and shows good correlation over the first nine divisions.

3. Converting to Total Cell Numbers

Both interval and curve-fitting analyses return the proportion of cells found within each division. It is often of more interest to know the absolute number of cells found within a cell culture. Estimates of total cell count can be made by reference to bead numbers run at the same time as the cells. Calibrite beads (Becton Dickinson, Palo Alto, CA) are suitable for this application. Prior to harvesting cells a known number of beads are added to the culture. Subsequently, when the culture is harvested and run on the flow cytometer the cells and beads appear with the beads easily distinguished from lymphocytes by their distinct forward and side scatter profile (Fig. 7A). A ratio of live cell events versus the number of beads acquired in the same sample is then used to determine the number of cells in culture. For example, if on acquisition 3000 live cells were counted together with 2000 beads then there were 1.5 times as many cells as beads. Thus, if 10,000 beads had been added to culture then 15,000 total cells were present. The total number of cells within each division can then be determined from the proportions calculated earlier.

D. The Method of Slicing Data from Asynchronous Cultures for Differentiation Analysis

Many lymphocyte differentiation events follow a division based "map" that is not affected by the time in culture or time spent in a single division. Thus, CFSE can be used to separate the two variables of time and division cycle number to prepare such maps. In order to process data for a differentiation map analysis, flow cytometry data showing phenotype changes with division need to be converted to a quantitative estimate of the percentage of cells of interest in each division. Figure 8 illustrates an example of the "slicing" method. CFSE labeled B cells were stimulated *in vitro*, and after 4 days of culture cells were stained with anti-IgG1 antibodies and analyzed by flow cytometry. The viable cell population was gated using side and forward scatter parameters. Individual peaks on the CFSE histogram profile were gated and the percentage of cells within the peak expressing surface IgG1 calculated. This proportion is then plotted against division number to reveal the division related surface immunoglobulin expression (Fig. 8D).

VI. Application of Carboxyfluorescein Diacetate Succinimidyl Ester to *in Vitro* Culture of Lymphocytes

A. B Lymphocyte Cultures

Lymphocytes have provided a useful subject for CFSE applications, as they can be isolated in a resting state and then activated *in vitro* to undergo differentiation changes equivalent to that observed *in vivo*. B lymphocytes have proved particularly useful for exploring the effect of signals on, and the relationship between, cell division and phenotypic changes associated with differentiation mentioned previously. For these studies it is important to begin with a homogeneous small resting population that retains differentiation potential and will label with CFSE

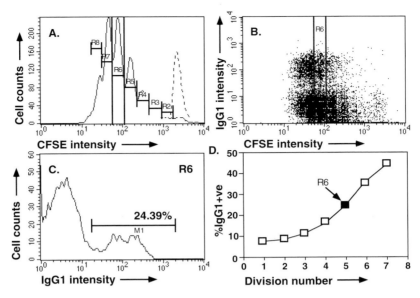

Fig. 8 Quantitative analysis of division linked differentiation by "slicing" data from asynchronous cultures. CFSE-labeled murine B cells were stimulated for 4 days in the presence of CD40L and IL-4, then stained for surface IgG expression. (A) Peaks on the CFSE profile were gated individually. (B) A representative peak was gated as R6, with IgG1 associated fluorescence on the vertical axis. This was used to generate a histogram of IgG1 expression (C), allowing quantitation of IgG1 expression. This data was generated for division cycles 0 through 7, allowing changes in expression at each cycle to be represented in graphical form (D). (This figure is reproduced from Hasbold *et al.*, 1998, and is used with kind permission of the publisher.)

as a tight, homogeneous population thereby giving good resolution of division peaks. Such cells can be prepared from murine spleen by teasing organs apart through a steel mesh, followed by hypotonic lysis of red cells. After removal of adherent cells by incubation of the resulting suspension in plastic tissue culture vessels, T cells are depleted using antibodies against CD4, CD8, and CD90 (Thy-1) and complement mediated lysis. Small, dense B cells are then recovered from Percoll gradients at the 65–72% interface. Detailed description of this procedure can be obtained from Hodgkin and Kehry (1995).

To induce B cell proliferation, differentiation, and isotype switching, cells are cultured with cytokines (e.g., IL-4, IL-5, or IL-6, depending on the experiment) as well as with a stimulus for proliferation (e.g., lipopolysaccharide or anti-Ig antibodies) or a source of signaling through the CD40 molecule. This latter stimulation can be delivered by an agonistic anti-CD40 antibody, purified membranes from activated T lymphocyte clones, or a CD40L transfected insect cell line.

The initial starting density of B lymphocytes should be kept low, preferably below 10^5 cells per ml, as the intense burst of cell proliferation seen in maximally stimulated cultures after around 3 to 5 days can result in exhaustion of medium and failure to complete the threshold number of division cycles required for differentiation to production of downstream Ig isotypes such as IgE. See Hodgkin and Kehry (1995) and Hodgkin *et al.* (1997) for a discussion of these issues. Typically B cell proliferation will begin at around 36–48 hr and proceed at a rate of about one division per 8–12 hr thereafter. Figures 4 and 8 demonstrate the information which can be gained using the CFSE technique with respect to division related isotype switching to IgG1 and IgE. See Hodgkin *et al.* (1996), Kindler and Zubler (1997), and Hasbold *et al.* (1998) for more detailed discussions of B cell differentiation studies and the effects of time, stimulation dose, and division number.

B. T Lymphocyte Cultures

T lymphocytes can be purified from peripheral lymphoid tissues or blood using flow cytometry or depletion by complement mediated lysis. Alternatively, contaminating B and other cells can be excluded from analysis by immunophenotypic staining, providing they do not interfere in a functional sense.

The CFSE division monitoring technique has been used to explore the relation between cytokine production and division number (Gett and Hodgkin, 1998; Bird *et al.*, 1998), the kinetics of T cell proliferative responses (Wells *et al.*, 1997; Gett and Hodgkin, 1998), and the acquisition of characteristics of memory cells (Lee and Pelletier, 1998). These studies also have clinical applications, such as determining antigen specific T cell responses to *Candida* (Angulo and Fulcher, 1998). This approach could be extended to other pathogens, as well as for monitoring efficacy of vaccination.

Another application of the CFSE technique is in the determination of the frequency of responding T cells in a mixed population. Figure 9 shows the response of CFSE labeled BALB/c splenocytes, gated for T cells, cultured with irradiated allogeneic C57/Bl6 splenocytes. The proportion of events in each peak can be determined, allowing the frequency of responders in the starting population to be calculated. Table I shows how the data generated from Fig. 9 can be used to calculate this frequency.

In Fig. 10, a comparison of CD4 and CD8 T cell responses to immobilized anti-CD3 and soluble anti-CD28 reveals a more vigorous response by CD8 T cells.

VII. Monitoring Lymphocyte Responses *in Vivo*

In addition to *in vitro* culture to investigate division linked differentiation, it is possible to use animal models to gain information on the division and differentiation of lymphocytes *in vivo*. CFSE labeled lymphocytes can be reintroduced into syngeneic hosts, and their division behavior monitored by flow cytometry of cell suspensions derived from secondary lymphoid or other tissues. The

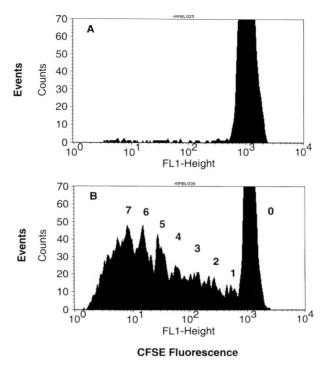

Fig. 9 *In vitro* alloresponse (mixed lymphocyte response). Murine splenic lymphocytes from BALB/c mice were stained with CFSE and cultured alone (A) or at a 16:1 ratio with C57/Bl6 irradiated, nonlabeled splenocytes (B). After 7 days culture, cells were stained with CD45R-PE, and T cells (CD45R-) were analyzed by flow cytometry. Note division of a proportion of T lymphocytes cultured with allogeneic stimulators, and no division when cultured alone. The number of events in each peak can be determined, allowing the frequency of responders in the starting population to be calculated, which in this case is approximately 6.5%. (From Flow cytometric analysis of cell division by dye dilution. A. B. Lyons and K. V. Doherty, J. P. Robinson *et al.*, Eds. "Current Protocols in Cytometry." Copyright 1998 John Wiley & Sons. Reprinted by permission of Wiley-Liss, Inc., a subsidiary of John Wiley & Sons, Inc.)

exquisite sensitivity of flow cytometry allows tracking of labeled lymphocytes in mice for up to 6 months (Lyons and Parish, 1994; Lyons, 1997).

A. Intravenous Transfer of Lymphocytes

The usual sources of lymphocytes in sufficient numbers for transfer is the spleen or alternatively lymph nodes. From a single murine spleen, approximately $5-10 \times 10^7$ lymphocytes can be obtained, with roughly equal numbers of B and T cells, depending on the strain. Typically, lymph nodes yield around 10-fold

Table I
In Vitro **Alloresponse (Mixed Lymphocyte Response)**[a]

Division number (D_n)	Mean fluorescence	Events	Divisor	Undivided cohort number
0	1124	3533	2^0	3533
1	513	158	2^1	79
2	248	281	2^2	70.25
3	116	358	2^3	44.75
4	59	320	2^4	20
5	32	436	2^5	13.62
6	15	718	2^6	11.22
7	8	628	2^7	4.90

[a] Data from Fig. 9B, were used to generate this table. Note that the fluorescence intensity of peaks closely follows the predicted serial dilution with each division. The number of cells ("Events") in a given division number (D_n) are divided by 2^{Dn} to calculate the number of original undivided cells they derived from. This number is referred to as the undivided cohort number and is shown in the final column. The sum of this column is the total undivided cohort number (= 3776). The sum of cohorts from division 1 to 7 represent the number of these precursors that have been activated to proliferate within the time of the assay. In this experiment, this number is 244. Therefore, the proportion of the starting population induced into division is: 244/3776 or 6.5%, which is within published estimates of alloresponding precursor percentages obtained using limiting dilution analysis. (From Flow cytometric analysis of cell division by dye dilution. A. B. Lyons and K. V. Doherty, J. P. Robinson *et al.,* Eds. "Current Protocols in Cytometry." Copyright 1998 John Wiley & Sons. Reprinted by permission of Wiley-Liss, Inc., a subsidiary of John Wiley & Sons, Inc.)

fewer cells. Specific subsets of lymphocytes can be purified from cell suspensions, or alternatively, the division behavior of subsets within a mixed transferred population can be monitored using appropriately congugated antibodies (see Section IV). The usual route for transfer to recipients is via the lateral tail vein, with a maximum volume of 0.2 ml injected. The usual injecting medium is PBS. Transfer of increasing numbers of lymphocytes shows a linear relationship to the percentage of donor cells found in the spleen, up to 5×10^7 lymphocytes transferred per recipient mouse, before apparent saturation occurs. At this point, donor cells will represent about 1–2% of splenic lymphocytes.

B. Preparation of Cell Suspensions and Strategies for Analysis

Cell suspensions from both lymphoid organs and other tissues can be prepared using standard techniques. In the majority of cases, where cell division occurs, it will be within secondary lymphoid organs such as spleen and nodes. It will usually be necessary to determine how many cells in total need to be analyzed

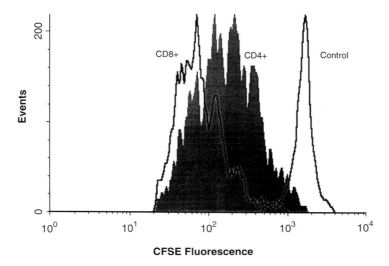

Fig. 10 *In vitro* proliferation of subsets of T cells. Murine splenic T cells purified by flow cytometry were stained with CFSE and cultured alone, or in the presence of immobilized anti-CD3 antibody and anti-CD28. After 3 days culture, cells were labeled with PE conjugated anti-CD4 or anti-CD8 antibodies before analysis. Histograms show CFSE fluorescence of undivided control cells, CD4 T cells (shaded histogram) and CD8 T cells (open histogram). Note that CFSE staining in conjunction with immunophenotyping allows the kinetics of proliferation in different populations to be compared. In this example, the response of CD8 T cells is more vigorous than that of CD4 T cells. (From Flow cytometric analysis of cell division by dye dilution. A. B. Lyons and K. V. Doherty, J. P. Robinson *et al.*, Eds. "Current Protocols in Cytometry." Copyright 1998 John Wiley & Sons. Reprinted by permission of Wiley-Liss, Inc., a subsidiary of John Wiley & Sons, Inc.)

to obtain sufficient events for meaningful cell division information to be obtained from the transferred cells, especially if only a subpopulation responds. Initially, it will be important to determine the percentage of CFSE labeled cells in a given population. From this information, the total number of cells needed for meaningful analysis can be determined. Where CFSE positive events are rare, gating will be required to exclude the recipient's own cells, resulting in a data file of CFSE positive events. See Fig. 11 and Lyons and Parish (1994), Lyons (1997), and Fulcher *et al.* (1996) for examples.

VIII. Antigen Receptor Transgenic Models

One limitation in the study of immune responses at the individual cell level is the diversity of the immune repertoire, which dictates that the proportion of cells of specificity for a given epitope will be low. The development of antigen receptor transgenic animals can circumvent such limitations. Mice with B cells

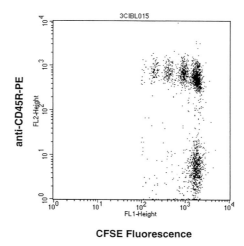

Fig. 11 *In vivo* B cell division in absence of T division after injection of splenic lymphocytes. Murine splenic lymphocytes (2×10^7) labeled with CFSE were injected intravenously. After 14 days, a cell suspension of the recipient's spleen was labeled with a CD45R antibody conjugated with PE, to enable discrimination between B and T cells. Events with green fluorescence above the autofluorescence background were collected, ensuring that only CFSE positive events were analyzed, which in this experiment represented approximately 0.5% of total events. 5000 events were collected, and anti-CD45R staining revealed cell division in the B, but not T, cell compartment. (From Flow cytometric analysis of cell division by dye dilution. A. B. Lyons and K. V. Doherty, J. P. Robinson *et al.*, Eds. "Current Protocols in Cytometry." Copyright 1998 John Wiley & Sons. Reprinted by permission of Wiley-Liss, Inc., a subsidiary of John Wiley & Sons, Inc.)

expressing a single immunoglobulin molecule recognizing a protein antigen such as hen egg lysozyme are available, as are mice in which defined T cell receptors are expressed, which are restricted for peptide presentation on either class I or class II major histocompatibility complex (MHC). Such mice allow the dissection of *in vivo* responses to antigen. These models are not without their shortcomings, as they are by nature nonphysiological. For example, currently available B cell receptor transgenic animals cannot be used to study immunoglobulin class switching. Also, immunization with cognate antigen can result in massive expansion of immune cells followed by equally dramatic apoptotic cell death, due to the inability of such population increases to be sustained. However, by adoptively transferring transgenic lymphocytes to recipient animals, such responses can more closely resemble physiological ones while still ensuring that the proportion of responding cells is large enough to study without exceeding the capacity of the animal to support the expanded population. Used with caution, such models combined with the CFSE cell division technique can yield much useful information on the kinetics of immune responses as well as on differentiation of lymphocytes (Fulcher *et al.*, 1996; Kurts *et al.*, 1997).

Note Added in Proof It has recently been demonstrated that cell nuclei retain CFSE stain, allowing levels of nuclear transcription factors to be measured simultaneously with cell division (Hasbold and Hodgkin, 2000).

References

Angulo, R., and Fulcher, D. A. (1998). Measurement of *Candida*-specific blastogenesis: Comparison of carboxyfluorescein succinimidyl ester labelling of T cells, thymidine incorporation, and CD69 expression. *Cytometry* **34,** 143–151.

Bird, J. J., Brown, D. R., Mullen, A. C., Moskowitz, N. H., Mahowald, M. A., Sider, J. R., Gajewski, T. F., Wang, C.-R., and Reiner, S. L. (1998). Helper T cell differentiation is controlled by the cell cycle. *Immunity* **9,** 229–237.

Fulcher, D. A., Lyons, A. B., Korn, S., Cook, M. C., Koleda, C., Parish, C., Fazekas de St. Groth, B., and Basten, A. (1996). The fate of self-reactive B-cells depends primarily on the degree of antigen receptor engagement and availability of T-cell help. *J. Exp. Med.* **183,** 2313–2328.

Gett, A. V., and Hodgkin, P. D. (1998). Cell division regulates the T cell cytokine repertoire, revealing a mechanism underlying immune class regulation. *Proc. Natl. Acad. Sci. U.S.A.* **95,** 9488–9493.

Hasbold, J., Lyons, A. B., Kehry, M. R., and Hodgkin, P. D. (1998). Cell division number regulates IgG1 and IgE switching of B cells following CD40L and IL-4 stimulation. *Eur. J. Immunol.* **28,** 1040–1051.

Hasbold, J., Hong, J. S., Kehry, M. R., and Hodgkin, P. D. (1999). Integrating signals from IFN-gamma and IL-4 by B cells: positive and negative effects on CD40 ligand-induced proliferation, survival, and division-linked isotype switching to IgG1, IgE, and IgG2a. *J. Immunol.* **163,** 4175–4181.

Hasbold, J., and Hodgkin, P. D. (2000). Flow cytometric cell division tracking using nuclei. *Cytometry* **40,** 230–237.

Hodgkin, P. D., and Kehry, M. R. (1995). Methods for polyclonal B lymphocyte activation and proliferation and Ig secretion in vitro. *In* "Handbook of Experimental Immunology" (D. M. Weir, C. Blackwell, L. A. Herzenberg, and L. A. Herzenberg, eds.) Blackwell, Oxford.

Hodgkin, P. D., Lee, J. H., and Lyons, A. B. (1996). B cell differentiation and isotype switching is related to division cycle number. *J. Exp. Med.* **184,** 277–281.

Hodgkin, P. D., Bartell, G., Mamchak, A., Doherty, K. V., Lyons, A. B., and Hasbold, J. (1997). The importance of efficacy and partial agonism in evaluating models of B lymphocyte activation. *Int. Rev. Immunol.* **15,** 101–127.

Kindler, V., and Zubler, R. H. (1997). Memory, but not naive, peripheral blood B lymphocytes differentiate into Ig-secreting cells after CD40 ligation and costimulation with IL-4 and the differentiation factors IL-2, IL-10, and IL-3. *J. Immunol.* **159,** 2085–2090.

Kurts, C., Carbone, F. R., Barnden, M., Blanas, E., Allison, J., Heath, W. R., and Miller, J. F. (1997). $CD4^+$ T cell help impairs $CD8^+$ T cell deletion induced by cross-presentation of self-antigens and favors autoimmunity. *J. Exp. Med.* **186,** 2057–2062.

Lee, W. T., and Pelletier, W. J. (1998). Visualizing memory phenotype development after in vitro stimulation of $CD4^+$ T cells. *Cell. Immunol.* **188,** 1–11.

Lyons, A. B. (1997). Pertussis toxin pretreatment alters the *in vivo* division behaviour and survival of B lymphocytes, after intravenous transfer. *Immunol. Cell Biol.* **75,** 7–12.

Lyons, A. B. (1999). Divided we stand: Tracking cell proliferation with carboxyfluorescein diacetate succinimidyl ester. *Immunol. Cell Biol.* **77,** 509–515.

Lyons, A. B., and Doherty, K. V. (1998). Flow cytometric analysis of cell division by dye dilution. *In* "Current Protocols in Cytometry" (J. P. Robinson, *et al.,* eds.), Unit 9.11. Wiley, New York.

Lyons, A. B., and Parish, C. R. (1994). Determination of lymphocyte division by flow cytometry. *J. Immunol. Methods* **171,** 131–137.

Nordon, R. E., Ginsberg, S. S., and Eaves, C. J. (1997). High resolution cell division tracking demonstrates the Flt3 ligand dependence of human marrow $CD34^+CD38^-$ cell production in vitro. *Br. J. Haematol.* **98,** 528–539.

Rabinovitch P. S., Torres R. M., and Engel, D. (1986). Simultaneous cell cycle analysis and two-color surface immunofluorescence using 7-amino-actinomycin D and single laser excitation: Applications to study of cell activation and the cell cycle of murine Ly-1 B cells. *J. Immunol.* **136,** 2769–2775.

Tough, D. F., and Sprent, J. (1994). Turnover of naive- and memory-phenotype T cells. *J. Exp. Med.* **179,** 1127–1135.

Warren, H. S., and Kinnear, B. F. (1999). Quantitative analysis of the effect of CD16 ligation on human NK cell proliferation. *J. Immunol.* **162,** 735–742.

Wells, A. D., Gudmundsdottir, H., and Turka, L. A. (1997). Following the fate of individual T cells throughout activation and clonal expansion. Signals from T cell receptor and CD28 differentially regulate the induction and duration of a proliferative response. *J. Clin. Invest.* **100,** 3173–3183.

CHAPTER 18

Antibodies against the Ki-67 Protein: Assessment of the Growth Fraction and Tools for Cell Cycle Analysis

Elmar Endl, Christiane Hollmann, and Johannes Gerdes

Division of Molecular Immunology
Research Center Borstel
D-23845 Borstel, Germany

I. Introduction
II. Application
III. Materials and Methods
    A. Preparation of Paraformaldehyde in Phosphate-Buffered Saline
    B. Paraformaldehyde Fixation
    C. Acetone Fixation of Cell Suspensions
    D. Paraformaldehyde/Methanol Fixation of Cell Suspensions
    E. Cryopreservation of Cells
    F. Enzymatic Cell Isolation from Fresh Solid Tissue
    G. Simultaneous Staining of Ki-67 and Cell Surface Antigens
IV. Critical Aspects
V. Controls and Standards
VI. Examples of Results
    References

I. Introduction

The monoclonal antibody Ki-67 was found by intensive studies at the University of Kiel (Ki), Germany, aimed at the production of monoclonal antibodies (MAb) to Hodgkin and Sternberg Reed cells (Gerdes *et al.,* 1983). The clone found in the sixty-seventh well of a 96-well microtiter plate (Ki-67) produced an antibody recognizing a human nuclear antigen that is exclusively expressed in proliferating cells. Detailed analysis of the cell cycle distribution revealed that this proliferation associated antigen, later designated as the Ki-67 protein, is present in all active parts of the cell cycle, that is, G_1, S, G_2, and mitosis, but not in resting cells, that is, G_0 (Gerdes, *et al.*, 1984a; Baisch and Gerdes, 1987; Endl *et al.*, 1997). Immunohistological methods based on antibodies against the Ki-67 protein provided reliable and reproducible means for the determination of the growth fraction of a given human cell population, well within the scope of a routine histological laboratory (Gerdes, 1985). Since then Ki-67 has become one of the most cited markers for cell proliferation (Brown and Gatter, 1990; Gerdes, 1990; Schwarting, 1993; Scholzen and Gerdes, 2000).

Clinical studies prove that the Ki-67 protein is an independent prognostic marker in human neoplasms, including mammary carcinoma (Gerdes *et al.*, 1987a; Pavelic *et al.*, 1992; Jansen *et al.*, 1998), soft tissue sarcoma (Heslin *et al.*, 1998), bladder carcinoma (Popov *et al.*, 1997; Vollmer *et al.*, 1998), meningnomas (Perry *et al.*, 1998), and non-Hodgkin's lymphoma (Gerdes *et al.*, 1984b, 1987b; Grogan *et al.*, 1988; Hall *et al.*, 1988). Besides the ability to characterize the growth fraction in histopathology and cancer research, the Ki-67 protein has some additional properties that fulfill the criteria for a "robust" proliferation marker. No invalid expression of the Ki-67 protein has been reported for precancerous genetic alterations in human tumors that already lead to the unscheduled expression of cell cycle regulating proteins like the cyclins (Jares *et al.*, 1996). Furthermore, the expression of the Ki-67 protein is not related to any DNA repair processes in cells following DNA damaging treatment such as ultraviolet (UV) exposure (Hall *et al.*, 1993), and there is also no evidence that excessive expression of the Ki-67 protein is involved in apoptosis (Pittman *et al.*, 1994; Coates *et al.*, 1996).

The characterization of the Ki-67 protein in molecular terms was thought to help to explain its role in the system of proteins regulating the cell cycle (Gerdes *et al.*, 1991; Duchrow *et al.*, 1994). The cloning of the cDNA identified two differentially spliced mRNAs encoding for the two isoforms of the Ki-67 protein with predicted molecular masses of 359 and 320 kDa (Schlüter *et al.*, 1993). Computer aided analysis revealed nuclear targeting sequences, potential phosphorylation sites, and several PEST sequences (Duchrow *et al.*, 1994) known to be present in proteins that show a high turnover and are susceptible to degradation by proteases (Rechsteiner and Rogers, 1996). There is evidence now that the Ki-67 protein is posttranslationally modified during mitosis and that this modification results from phosphorylations that are regulated by the cyclin B/CDC2 kinase pathway (MacCallum and Hall, 1999; Endl and Gerdes, 2000). Homologues to the human Ki-67 protein have been described in rodents (Starborg *et al.*, 1996; Gerlach *et al.*, 1997; Scholzen and Gerdes, 2000) and in rat kangaroo PtK2 cells (Takagi *et al.*, 1999). Whereas no significant homology to other cell cycle regulated proteins has been elucidated. The Ki-67 protein sequence therefore remains unique and its relationship to other proteins unknown. Bacterially expressed parts of the Ki-67 cDNA, however, enabled the generation of new monoclonal antibodies (Cattoretti *et al.*, 1992; Key *et al.*, 1993; Gerlach *et al.*, 1997; Scholzen and Gerdes, 2000) that are better suited for further investigations using cytometric techniques than the prototype antibody Ki-67.

Microscopy showed that the regulation of the Ki-67 protein expression is accompanied by characteristic spatial distribution patterns within the different cell cycle compartments (Braun *et al.*, 1988; van Dierendonck *et al.*, 1989; Endl and Gerdes, 2000). Expression is predominantly localized in the nucleus during interphase with an intensive staining of the nucleoli (van Dierendonck *et al.*, 1989, 1991; Kill, 1996). The Ki-67 protein is associated with the dense fibrillar components of the nucleoli, but the domains are distinct from domains containing two of the major nucleolar antigens fibrillarin and RNA polymerase I (Kill, 1996). During mitosis it is associated with the chromosomes, where it covers the chromatin (Gerdes *et al.*, 1984a; Bading *et al.*, 1989; Verheijen *et al.*, 1989a,b; Kill, 1996). Redistribution of the Ki-67 protein during postmitotic formation of the nucleus is thought to be relevant for nucleolar reorganization (Bridger *et al.*, 1998).

In addition, flow cytometric analysis revealed that the level of the Ki-67 protein expression is related to the different cell cycle compartments. It is upregulated in lymphocytes stimulated with phytohemagglutinin A just before the cells enter the very first S phase (Gerdes *et al.*, 1984a). The expression of Ki-67 protein is necessary for the transit from G_1 into S phase, as was demonstrated by a decreased incorporation of [^{3}H]thymidine in IM-9 cells incubated with antisense deoxynucleotides complementary to the deduced translation start site of the Ki-67 mRNA (Schlüter *et al.*, 1993). Staining for the Ki-67 protein increases during S phase and displays a high intensity during mitosis (Lopez *et al.*, 1991; Bruno and Darzynkiewicz, 1992; Endl *et al.*, 1997). Cells reenter the subsequent G_1 phase with a high expression of the Ki-67 protein (Du Manoir *et al.*, 1991; Lopez *et al.*, 1991; Bruno and Darzynkiewicz, 1992). Two ways of regulation are supposed for the protein expression during the following G_1 phase. Ki-67 detection can decline in cells during early G_1 with an estimated half-life of less than 1 hr (Bruno and Darzynkiewicz, 1992) as calculated for HL-60 cells after vinblastine block of mitosis and pulse–chase experiments using [^{35}S]methionine in L428 cells (Heidebrecht *et al.*, 1996), whereas the level of protein is still detectable in cells that continue to grow until the cells reenter a new S phase (Du Manoir *et al.*, 1991; Lopez *et al.*, 1991).

Despite the accumulating knowledge about the structure, localization, and regulation of the Ki-67 protein, thus far there is no clue to the function of the protein. Its role in the cell cycle regulating network awaits further characterization. Considering the unique and complex character of the Ki-67 protein, a multidisciplinary and multiparameter approach to elucidate the function of the protein may therefore be not only advantageous but necessary.

II. Application

The methods to retrieve the protein in formalin fixed paraffin embedded tissue are standardized (Cattoretti *et al.*, 1992, 1993; Gerdes *et al.*, 1992; McCormick *et al.*, 1993a; Shi *et al.*, 1995) and can be controlled in a standardized fashion (Ruby and McNally, 1996). However, the diversity of protocols for Ki-67 measurement in flow cytometry is sometimes confusing (Table I). The attempt of this chapter is therefore to give a summary of methods already used in cell biology and some examples of applications on different cell types.

Antibodies against the Ki-67 protein can be used for cell kinetic investigations of cell cultures (Schwarting *et al.*, 1986; Littleton *et al.*, 1991; Tsurusawa and Fujimoto, 1995), peripheral blood mononuclear cells (Palutke *et al.*, 1987; Campana *et al.*, 1988; Lopez *et al.*, 1991; Cavanagh *et al.*, 1998), and solid tumors (Camplejohn, 1994; Schutte *et al.*, 1995). Counterstaining with propidium iodide enables a correlation of Ki-67 expression with the position of cells in the cell cycle, on a single cell basis (Landberg *et al.*, 1990). This information can be used to investigate antigens with yet unknown regulation within proliferating and resting cells for their usefulness as proliferation markers (Neubauer *et al.*, 1989; Pellicciari *et al.*, 1995; Endl *et al.*, 1997).

Staining with monoclonal antibodies against the Ki-67 protein is also an easy method to estimate the growth fraction of a given cell population. This makes it superior to autoradiographic methods (Scott *et al.*, 1991) and detection of incorporated nucleotide analogs like bromodeoxyuridine (Wilson *et al.*, 1996). Additional information about the proliferation characteristics within a given cell population can be obtained by counterstaining with specific antibodies for a particular subpopulation of cells. Cytokeratin was used to precisely estimate the amount of proliferating tumor cells in solid tumor preparations (Schutte *et al.*, 1995) and T cell turnover in both healthy and HIV-1 infected adults by combining CD4 and CD8 assessment and staining with monoclonal antibodies against the Ki-67 protein (Sachsenberg *et al.*, 1998).

Monoclonal antibodies against the Ki-67 protein in mouse and rat (MIB-5, TEC-3, Dianova, Hamburg, Germany) have been described for proliferation assessment in rodents (Gerlach *et al.*, 1997; Scholzen and Gerdes, 2000) and have to be evaluated for their use in flow cytometry. Progress in the development of fluorochromes that can easily be conjugated to the preferred antibodies for proliferation associated antigens may promote multiparameter analysis in proliferation studies as an alternative to the commonly used counterstaining with DNA specific dyes.

III. Materials and Methods

A. Preparation of Paraformaldehyde in Phosphate-Buffered Saline

Fixation and preservation of the antigen during immunochemical staining procedures are the most critical steps in the analysis of intracellular antigens (Clevenger and Shankey, 1993; Bauer and Jacobberger, 1994). Formaldehyde is a common fixative for immunohistochemistry and has become more and more popular for preparation of single cell suspensions in flow cytometry. It both preserves the morphology of the cells and protects the antigen from leaking

Table I
Protocols for Ki-67 Measurement in Flow Cytometry

Fixative	%	Time	Temperature	Detergent	%	Time	Temperature	Cells	Reference
PFA in PBS	0.5	5 min	4°C	Triton X-100	0.1	3 min	4°C	MCF7, K562	Kute and Quadri (1991)
PFA in PBS[a]	0.5	10 min	On ice	Triton X-100	0.1	4 min	On ice	HL-60	Bruno et al. (1992)
PFA in PBS	1	10 min	On ice	Triton X-100	0.25	5 min	On ice	J82	Endl et al. (1997)
Acetone	95	10 min	Room temp.	—	—	—	—	U937	Schwarting et al. (1986)
Acetone	100	30 min	−20°C	—	—	—	—	HeLa S3	Sasaki et al. (1987)
Acetone	80	30 min	Room temp.	—	—	—	—	PB-MNC	Lopez et al. (1991)
Methanol	70	30 min	−18°C	—	—	—	—	HL-60	Wersto et al. (1988)
Methanol[b]	70	10 min	−20°C	Triton X-100	0.5	15 min	On ice	MR65	Schutte et al. (1995)
Methanol	100	1 hr	4°C	—	—	—	—	Molt-4	Tsurusawa and Fujimoto (1995)

[a] With varying concentrations of NaCl.
[b] First perform detergent treatment then fixation by adding pure methanol.

by forming covalent cross-links between uncharged amino groups. Subsequent treatment with detergents enables the antibody to penetrate the cells and recognize its specific antigen.

A fresh, purified, and methanol free solution of formaldehyde is recommended and is commercially available from vendors for electron microscopy (e.g., Polysciences, Warrington, PA). It should be kept in mind that methanol is sometimes added as a stabilizer and may therefore introduce an undesired variable in the fixation procedure.

Fresh formaldehyde/phosphate-buffered saline (PBS) solutions can be prepared from paraformaldehyde (PFA) and can be used for 1 week when stored in dark glass bottles. Stock solutions can be frozen at $-70°C$ and are suitable for several months. Avoid any inhalation during preparation. Formaldehyde is harmful if inhaled or absorbed through skin.

Classic Preparation of Paraformaldehyde in Phosphate-Buffered Saline

1. To prepare a solution of 2% PFA in 10 ml PBS, add 0.2 mg PFA to 1 ml distilled water and warm to 70°C in a water bath.
2. Add NaOH (2 M) dropwise while stirring until solution becomes clear.
3. Incubate another 30 min at 70°C while stirring.
4. Add 5 ml of PBS and adjust buffer to pH 7.4 by adding HCl dropwise.
5. Adjust volume to 10 ml with PBS.
6. Filter through a sterile filter (0.2 μm) to remove aggregates.

Rapid Preparation of Paraformaldehyde in Phosphate-Buffered Saline

1. Add 10 ml PBS to 0.2 mg PFA in a glass bottle.
2. Put the bottle in a microwave oven (600 W) and heat at medium power settings. Avoid exposure to vapor. Reduce vapor by plugging the bottle neck with cotton wool.
3. Watch the solution for bubble formation and stop heating when the first bubbles appear.
4. Remove the bottle from the microwave oven and dissolve remaining PFA by shaking.
5. If there is some undissolved PFA left, put the bottle back into the microwave oven and start the procedure again.
6. If necessary adjust to pH 7.4.
7. Filter through a sterile filter and store in dark glass bottles or freeze at $-70°C$.

B. Paraformaldehyde Fixation

1. Spin 10^6 cells (600 g, 5 min), decant, and resuspend pellet in 500 μl ice-cold PBS.

2. Transfer the 500 µl of cell suspension into 500 µl of ice-cold PBS containing 2% PFA (final concentration 10^6 cells in 1 ml PBS containing 1% PFA). Incubate 10 min on ice.
3. Add 2 ml of ice-cold PBS containing 0.5% bovine serum albumin (BSA), spin, and decant.
4. Resuspend in 500 µl ice-cold PBS containing 0.25% Triton X-100. Incubate 5 min on ice.
5. Add 2 ml of ice-cold PBS containing 0.5% BSA, spin, and decant.
6. Resuspend in 100 µl of primary Ki-67 antibody in PBS containing 0.5% BSA. The following antibodies may be used:

    MIB-1 for human cells, dilute to 2 µg/ml (Dianova)
    Anti-Ki-67 polyclonal rabbit serum for human cells (DAKO, Hamburg, Germany)
    MIB-5 for rat cells, dilute to 5 µg/ml (Dianova)
    TEC-3 for mouse cells, dilution 1:10 (Dianova)

7. Add 2 ml of ice-cold PBS containing 0.5% BSA, spin, and decant.
8. Resuspend in 100 µl of secondary antibody. The secondary antibody may be fluorescein-conjugated goat anti-mouse, goat anti-rat, or goat anti-rabbit antibodies, according to the primary antibody.
9. Incubate 30 min at 4°C in the dark.
10. Add 2 ml of ice-cold PBS containing 0.5% BSA, spin and decant.
11. Resuspend in 500 µl PBS containing 50 µg/ml propidium iodide (PI) and 50 µg/ml RNase. Incubate 20 min at 37°C in the dark.
12. Analyze cells by flow cytometry.

C. Acetone Fixation of Cell Suspensions

The protocol is based on one of the first publications on the use of antibodies against the Ki-67 protein in flow cytometry (Gerdes and Baisch, 1984a; Baisch and Gerdes 1987, 1990).

1. Spin 1×10^6 cells (600 g for 5 min) and decant.
2. Resuspend cells in 200 µl of 150 mM NaCl at 4°C. Transfer cell suspension slowly into 800 µl of pure acetone ($-20°C$) while gently shaking.
3. Incubate at $-20°C$ for at least 30 min.
4. Centrifuge and decant.
5. Resuspend cells in 2 ml ice-cold PBS containing 0.5% BSA, centrifuge, and decant.
6. For antibody and DNA staining, proceed as described for paraformaldehyde fixation (steps 6–11).

D. Paraformaldehyde/Methanol Fixation of Cell Suspensions

Methanol fixation is hypotonic and coagulant and may therefore alter the cellular morphology and solubilize nucleoproteins. An interesting variation is fixation with PFA prior to methanol treatment (Jacobberger, 1991; Schimenti and Jacobberger, 1992). This preserves the antigens of interest and may unmask additional epitopes recognized by the primary antibody, owing to the denaturing properties of alcohol fixation (Clevenger and Shankey, 1993).

1. Spin 10^6 cells (600 g, 5 min), decant, and resuspend pellet in 500 μl ice-cold PBS.
2. Transfer the 500 μl of cell suspension into 500 μl of ice-cold PBS containing 2% PFA (final concentration 10^6 cells in 1 ml PBS containing 1% PFA).
3. Incubate 10 min on ice.
4. Add 2 ml of ice-cold PBS containing 0.5% BSA, spin, and decant.
5. Resuspend in 100 μl ice-cold PBS.
6. Transfer cells to 900 μl methanol ($-20°C$) while gently shaking.
7. Incubate 30 min at $-20°C$.
8. Add 2 ml of ice-cold PBS containing 0.5% BSA, spin, and decant.
9. For antibody and DNA staining, proceed as described for paraformaldehyde fixation (steps 6–11).

E. Cryopreservation of Cells

Experiments may require a longer storage of cells. The following procedure describes the freezing of single cell suspensions and should preserve the integrity of the Ki-67 protein for several months.

Determine the concentration and volume of the suspension, calculate the total number of available cells, and perform viability assessment. Cells are frozen at 5×10^6 cells per milliliter of freezing medium. Freezing medium contains 90% fetal calf serum (FCS)/0.05% azide/10% DMSO. Pellet the cells and, after removing the supernatant, gently resuspend with 1 ml volume of freezing medium. Avoid any mechanical stress. Divide into freezing vials in 1-ml aliquots.

Freezing

1. Wrap the vials in cotton wool and place them in a Styrofoam box.
2. Store the racks at $-20°C$ for at least 24 hr, then transfer to $-70°C$ freezer; after 1 week transfer the vials to liquid nitrogen storage.
3. Best results are obtained by freezing rates of approximately 1°C per hour. Because only a few laboratories own apparatus for controlled freezing rates, this is a simple and easy to use method to imitate the procedure.

Thawing

1. Each vial thawed will require 15 ml of defrosting medium (60% RPMI and 40% FCS). Warm the defrosting medium to 37°C before thawing the sample. Put aside 1 ml of this medium for use in final resuspension of thawed cells.

2. Remove a vial from the freezer and immediately place in a 37°C water bath. Use constant but gently swirling until just a small ice crystal remains. Do not immerse the cap of the vial into the water bath during thawing, as the sealing ring on the vial may deform when it warms and may not be able to protect the vial from water leakage.
3. Open the vial and add 500 μl of defrosting medium. Transfer the cells immediately to the remaining defrosting medium by slowly adding the cell suspension to the medium. Avoid any mechanical stress, that is, bubble formation, by rinsing the pipette. Gently invert the tube to mix the cells.
4. Centrifuge, decant, and resuspend the pellet in the 1 ml of defrosting medium previously set aside.

F. Enzymatic Cell Isolation from Fresh Solid Tissue

The following protocol describes the disaggregation and staining of peripheral human lymphoid tissue. The different subsets of cells were characterized by the antibody CD4, by the B cell specific antibodies CD10 and surface immunoglobulin D (IgD), and by the follicular dendritic cell specific monoclonal antibody R4/23. Proliferative activity of the different subsets was monitored by counterstaining with the DNA binding dye Hoechst 33258 and fluorescein isothiocyanate (FITC) conjugated MIB-1 antibody. Figure 4 represents the corresponding cytograms.

1. Cut tissue into 2 mm cubes with a scalpel in a petri dish.
2. Incubate cubes in 10 ml PBS containing 350 U/ml collagenase type IV and 500 Kunitz units/ml DNase I for 20 min at 37°C while gently shaking.
3. Let the tissue fragments sediment, and transfer supernatant to FCS to stop enzymatic activity (final concentration 10% FCS in PBS).
4. Centrifuge, decant, and resuspend cells in cold RPMI supplemented with 10% FCS.
5. Resuspend remaining tissue fragments in PBS containing 350 U/ml collagenase type IV and 500 Kunitz units/ml DNase I for 20 min at 37°C.
6. Repeat steps 2 to 4. The procedure can be repeated three times, and cells can be pooled for further proceedings.
7. Pass cells through a 53-μm filter before starting the fixation and staining procedure.

G. Simultaneous Staining of Ki-67 and Cell Surface Antigens

1. Prepare single cell suspensions from solid tissue as described in Section III,F.
2. Separate cells by centrifugation on a Ficoll-Hypaque density gradient.

3. Incubate cells with monoclonal antibody against surface antigen. The following may be used:

    CD4, IgG1 (Leu 3a, Becton Dickinson, Franklin Lakes, NJ)
    CD10, IgG2a (B-E3, ImmunoQuality Products, Gromingen, Netherlands)
    Anti-IgD, IgG1 (IgD26, DAKO)
    CD21L, IgM (R4/23, Naiem et al., 1983)

4. Wash in ice-cold PBS containing 1% FCS, spin, and decant.
5. Incubate cells with secondary antibodies conjugated with fluorochromes suitable for multiparameter analysis. The secondary antibodies may be goat anti-mouse IgG PE (excitation at 488 nm, emission at 578 nm) and goat anti-mouse IgM biotin.
6. Wash in ice-cold PBS containing 1% FCS, spin, and decant.
7. Incubate with Streptavidin-Red 670 (excitation at 488 nm, emission at 670 nm).
8. Wash in ice-cold PBS containing 1% FCS, spin, and decant.
9. Resuspend in 500 μl ice-cold PBS and transfer cell suspension to 500 μl PBS containing 2% PFA (final concentration 10^6 cells in 1 ml PBS containing 1% PFA).
10. Incubate 10 min on ice, centrifuge, and decant.
11. Resuspend cells in PBS containing 0.1% Nonidet P-40.
12. Incubate 10 min on ice, centrifuge, and decant.
13. Add FITC conjugated MIB-1 antibody (Dianova) in PBS supplemented with 1% FCS in the appropriate concentration determined by titration experiments. Incubate 30 min at 4°C.
14. Wash in ice-cold PBS containing 1% FCS, spin, and decant.
15. If your flow cytometer is equipped with a UV excitation source, you can use Hoechst 33258 for a simultaneous assessment of DNA content. Resuspend in 500 μl PBS and add Hoechst 33258 (stock solution 1 mg in distilled water) to a final concentration of 2 μg/ml. Incubate for 15 min.
16. Analyze by flow cytometry.

IV. Critical Aspects

Cell fixation should be tested for different cell types. Antibodies should be titrated, and staining efficiency should be controlled by fluorescence microscopy (Jacobberger, 1991; Bauer and Jacobberger, 1994). Staining of the Ki-67 protein is restricted to the nucleus during interphase, with a high density of protein located in the nucleoli. This staining pattern serves as a control for proper staining procedure.

We suggest that one try the formaldehyde fixation first, because in our hands it yields the best results with regard to staining intensity and reproducibility. Formaldehyde forms cross-links with protein end groups and is therefore well suited for preservation of the antigen, but it usually produces DNA histograms of poor quality (Clevenger and Shankey, 1993). Detergent treatment provides improved accessibility for DNA binding dyes and enhances antibody staining. Formaldehyde concentrations can be varied from 0.5 to 1%. Higher concentrations may disturb the precise measurement of DNA histograms, due to a limited binding of PI to cross-linked DNA. Triton X-100 concentrations can vary from 0.1 to 0.25%. Concentrations should be checked for loss of binding sites in an experimental setup using multiparameter analysis of membrane bound antigens. Other modifications of the protocol include variations in time and temperature (see Table I).

The use of milder detergents like Nonidet P-40 may help in preserving membrane bound antigens when precise measurement of cell cycle distribution by DNA intercalating dyes is not required. Because cross-linking agents like formaldehyde can mask epitopes in a way that they are no longer recognizable by the antibody, alcohol and acetone fixation represent reasonable alternatives when monoclonal antibodies against the Ki-67 protein are combined with other antibodies.

The procedure for isolation of cells from solid tissue may be modified. However, it should be kept in mind that the Ki-67 protein is unstable and can easily be destroyed by minimal quantities of trypsin or pepsin (Gerdes et al., 1991). Another alternative is the isolation of bare nuclei because the Ki-67 protein is restricted to the nucleus (Larsen et al., 1991; Landberg and Roos, 1992). A critical aspect of this procedure is that cells in mitosis will be lost during the preparation. Isolated nuclei can be fixed and processed as described for whole cells (Schutte et al., 1995). Loss of protein by detergents and buffers used during isolation procedure of cell nuclei should be controlled by other methods, for example, Western blotting.

V. Controls and Standards

Unspecific binding should always be monitored by running an appropriate isotype control. The MIB-1 antibody shows a rapid saturation curve (McCormick et al., 1993b), whereas unspecific binding sites for IgG follow different staining kinetics. The required amount of antibody should therefore be titrated to achieve a maximum signal to noise ratio (Jacobberger, 1991; Bauer and Jacobberger, 1994). The principal components of background fluorescence are autofluorescence of the cells (Aubin, 1979), unspecific binding of the primary and secondary antibodies, and spectral overlap from propidium iodide staining. Unspecific binding can be minimized by using monoclonal primary antibodies and F(ab')$_2$ fragments of the fluorochrome conjugated secondary antibody absorbed against species specific immunoglobulin. Autofluorescence can be reduced by using alternative fluorochromes for antibody labeling such as the indodicyanine dye Cy5 (Amersham, Buckinghamshire, England). The excitation wavelength of Cy5 is in the range of 625–650 nm where autofluorescence is almost negligible (Benson *et al.,* 1979). This both increases the sensitivity of antigen detection and bypasses the problem of spectral overlap from propidium iodide staining.

An appropriate isotype control is required for an accurate determination of the percentage of cells that stain positive for the Ki-67 protein. A precise separation of positive/negative antibody staining can be achieved by normalized subtraction of the isotype control. This alternative software approach runs by subtracting the negative control fluorescence histogram at each interval from the corresponding axis interval of the positive staining distribution (Overton, 1988).

VI. Examples of Results

Figure 1 represents the typical staining pattern of the MIB-1 antibody on PFA/Triton X-100 treated HeLa cells during interphase. The staining is characterized by a strong intensity of nucleolar staining. Numerous small, discrete foci of staining in the nucleoplasm can also be recognized. The staining intensity of the nucleoplasm and the number of nucleoli and nucleoplasmic speckles may vary depending on the cell type and the position of the cells in the cell cycle (Braun *et al.,* 1988; Kill, 1996). The cytoplasm is completely negative.

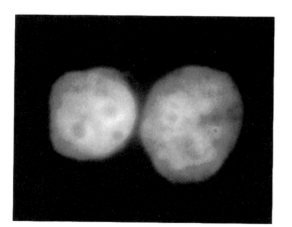
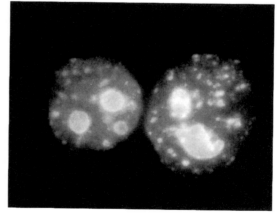

Fig. 1 Cytospin preparations of HeLa cells in interphase. Cells were fixed as single cell suspension with PFA/Triton X-100 stained with Hoechst 33258 for the representation of the nucleus (left) and MIB-1 antibody staining of the Ki-67 protein followed by FITC conjugated secondary antibody (right).

Figure 2 represents two parameter displays of DNA versus Ki-67 expression in cell lines derived from various species. Figure 2a,b demonstrate the variation of Ki-67 staining in human tumor cell lines with different *in vitro* growth characteristics. Figure 2a is derived from the nontumorigenic urothelial human cell line HCV29, which is sensitive to variations in the concentration of growth factors and is growth inhibited by cell–cell contact *in vitro*. Cells that are in the G_1 phase exhibit a broad variation in staining intensity that ranges from high intensity in postmitotic cells to no expression in resting cells. It should also be noted that cells enter S phase with an intermediate staining intensity. Figure 2b represents a two parameter display DNA versus Ki-67 of HeLa cells. These cells are less sensitive to varying growth conditions *in vitro*, and all of these cells display a detectable level of Ki-67. The rat monoclonal antibody TEC-3 raised against the murine Ki-67 equivalent gives a similar profile for the mouse bladder carcinoma cell line MB49, as shown in Fig. 2c. MIB-5 staining in YB210 rat myeloma cells

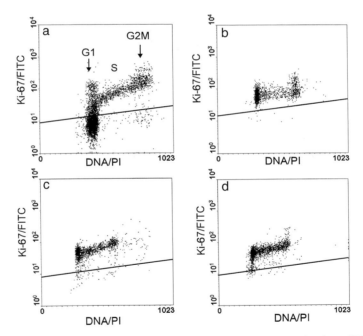

Fig. 2 Two parameter display of Ki-67 expression versus DNA distribution for cell lines derived from different species. Cells were fixed with PFA followed by Triton X-100 permeabilization and stained for the Ki-67 protein with monoclonal antibodies MIB-1 for human cells, TEC-3 for mouse cells, and MIB-5 for rat cells. Cells were counterstained with PI to display the cell cycle distribution. Cytograms (a) and (b) represent the nontumorigenic human urothelial cell line HCV29 (a) and IM-9 cells derived from human lymphoma (b). MB49 cells isolated from mouse bladder carcinoma and the YB210 rat myeloma cell line are shown in (c) and (d), respectively. The solid line indicates the upper limit of the fluorescence intensity of the corresponding isotype control.

show virtually the same staining characteristics. The solid line indicates the upper limit of the fluorescence intensity of the corresponding isotype control.

Figure 3 compares the effect of different fixation and permeabilization procedures on the detection of the Ki-67 protein using the MIB-1 antibody on HeLa cells. PFA/Triton X-100, acetone, and PFA/methanol yield the same staining

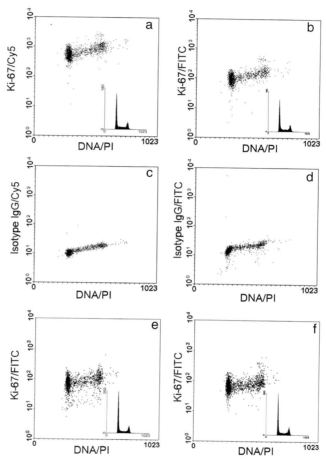

Fig. 3 Two parameter displays of Ki-67 expression versus DNA content of HeLa cells prepared with different fixation protocols and stained with different fluorochrome conjugated secondary antibodies. (a, b) HeLa cells fixed with PFA followed by Triton X-100 treatment. Cells were stained with MIB-1 antibody against the Ki-67 protein and Cy5 conjugated or FITC conjugated secondary antibodies in (a) and (b), respectively. (c, d) Corresponding cytograms for isotype IgG as primary antibody. (e, f) Display cytograms of HeLa cells stained analogously but fixed in acetone (e) and PFA/methanol (f). Histograms of the DNA distribution are included in the corresponding two parameter display.

pattern for HeLa cells stained with MIB-1 and goat anti-mouse F(ab')$_2$ fragments. Sensitivity can be improved by using Cy5 (Amersham) conjugated goat anti-mouse F(ab')$_2$ as secondary antibodies as is shown in Fig. 3a. The coefficients of variation of the G$_1$ fraction of cells for the different fixation protocols are approximately 3% and therefore sufficient for a calculation of the cell cycle distribution for all protocols.

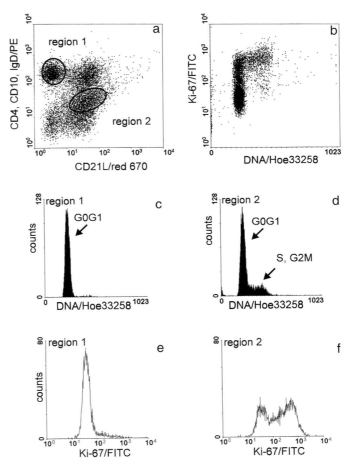

Fig. 4 Characterization of isolated lymphoid human tissue by multiparameter analysis. (a) Cells were stained with CD4, CD10, and anti-IgD followed by phycoerythrin (PE) conjugated secondary antibody and CD21L followed by Red 670 conjugated secondary antibody. (b) Corresponding simultaneous measurement of cell cycle characteristics by antibody staining of the Ki-67 protein and DNA counterstaining with Hoechst 33258 for all cells displayed in (a). (c, e) Expression of the Ki-67 protein and percentage of cells in the different cell cycle compartments for cells corresponding to region 1. Data for cells from region 2 are displayed in (d) and (f).

Figure 4 illustrates the possibility of multiparameter analysis and cell proliferation assessment using FITC conjugated MIB-1 antibody and detection of cell surface antigens that are specific for subpopulations of cells in human lymphoid tissue. Several populations of cells can be identified in a two parameter display of CD4, CD10, and anti-IgD versus CD21L (Fig. 4a). Region 1 consists of cells that show a strong reactivity to antibodies for CD4. These cells represent nonproliferating T cells, monocytes, and dendritic cells (Sprenger *et al.*, 1995). The corresponding DNA histogram confirms that nearly all of these cells are in same state of the cell cycle according to their DNA content (Fig. 4c). Assessment of the Ki-67 protein by MIB-1 antibody staining reveals that all of these cells are in G_0, since they are all negative (Fig. 4e). Cells that are positive for CD21L and stain for CD10 and IgD (region 2), however, involve both proliferating and resting cells according to their Ki-67 protein expression (Fig. 4f). More than 60% are in an active state of the cell cycle. The proliferative activity is further confirmed by the DNA staining, which reveals cells in S and G_2M phase (Fig. 4d). These cells represent germinal center B lymphocytes, and the results obtained by multiparameter analysis are in concordance with the known biology of these cells.

References

Aubin, J. E. (1979). Autofluorescence of viable cultured mammalian cells. *J. Histochem. Cytochem.* **27**, 36–43.

Bading, H., Rauterberg, E. W., and Moelling, K. (1989). Distribution of c-myc, c-myb, and Ki-67 antigens in interphase and mitotic human cells evidenced by immunofluorescence staining technique. *Exp. Cell Res.* **185**, 50–59.

Baisch, H., and Gerdes, J. (1987). Simultaneous staining of exponentially growing versus plateau phase cells with the proliferation-associated antibody Ki-67 and propidium iodide: Analysis by flow cytometry. *Cell Tissue Kinet.* **20**, 387–391.

Baisch, H., and Gerdes, J. (1990). Identification of proliferating cells by Ki-67 antibody. *In* "Methods in Cell Biology" (Z. Darzynkiewicz, J. P. Robinson, and H. A. Crissman, eds.), Vol. 41, pp. 217–226. Academic Press, New York.

Bauer, K. D., and Jacobberger, J. W. (1994). Analysis of intracellular proteins. *In* "Methods in Cell Biology" (Z. Darzynkiewicz, J. P. Robinson, and H. A. Crissman, eds.), Vol. 41, pp. 351–376. Academic Press, New York.

Benson, H. C., Meyer, R. A., and Zaruba M. E. (1979). Cellular autofluorescence—Is it due to flavins? *J. Histochem. Cytochem.* **27**, 44–48.

Braun, N., Papadopoulos, T., and Muller-Hermelink, H. K. (1988). Cell cycle dependent distribution of the proliferation-associated Ki-67 antigen in human embryonic lung cells. *Virch. Arch. B Cell Pathol. Incl. Mol. Pathol.* **56**, 25–33.

Bridger, J. M., Kill, I. R., and Lichter, P. (1998). Association of pKi-67 with satellite DNA of the human genome in early G1 cells. *Chromosome Res.* **6**, 13–24.

Brown, D. C., and Gatter, K. C. (1990). Monoclonal antibody Ki-67: Its use in histopathology. *Histopathology* **17**, 489–503.

Bruno, S., and Darzynkiewicz, Z. (1992). Cell cycle dependent expression and stability of the nuclear protein detected by Ki-67 antibody in HL-60 cells. *Cell Prolif.* **25**, 31–40.

Bruno, S., Gorczyca, W., and Darzynkiewicz, Z. (1992). Effect of ionic strength in immunocytochemical detection of the proliferation associated nuclear antigens p120, PCNA, and the protein reacting with Ki-67 antibody. *Cytometry* **13**, 496–501.

Campana, D., Coustan-Smith, E., and Janossy, G. (1988). Double and triple staining methods for studying the proliferative activity of human B and T lymphoid cells. *J. Immunol. Methods* **107,** 79–88.

Camplejohn, R. S. (1994). The measurement of intracellular antigens and DNA by multiparametric flow cytometry. *J. Microsc.* **176,** 1–7.

Cattoretti, G., Becker, M. H., Key, G., Duchrow, M., Schluter, C., Galle, J., and Gerdes, J. (1992). Monoclonal antibodies against recombinant parts of the Ki-67 antigen (MIB 1 and MIB 3) detect proliferating cells in microwave-processed formalin-fixed paraffin sections. *J. Pathol.* **168,** 357–363.

Cattoretti, G., Pileri, S., Parravicini, C., Becker, M. H., Poggi, S., Bifulco, C., Key, G., D'Amato, L., Sabattini, E., and Feudale, E. (1993). Antigen unmasking on formalin-fixed, paraffin-embedded tissue sections. *J. Pathol.* **171,** 83–98.

Cavanagh, L. L., Saal, R. J., Grimmett, K. L., and Thomas, R. (1998). Proliferation in monocyte-derived dendritic cell cultures is caused by progenitor cells capable of myeloid differentiation. *Blood* **92,** 1598–1607.

Clevenger, C. V., and Shankey, T. V. (1993). Cytochemistry II: Immunofluorescence measurement of intracellular antigens. *In* "Clinical Flow Cytometry" (K. D. Bauer, R. E. Duque, and T. V. Shankey, eds.), pp. 157–176. Williams & Wilkins, Baltimore.

Coates, P. J., Hales, S. A., and Hall, P. A. (1996). The association between cell proliferation and apoptosis: Studies using the cell cycle-associated proteins Ki67 and DNA polymerase α. *J. Pathol.* **178,** 71–77.

Duchrow, M., Schluter, C., Wohlenberg, C., Flad, H. D., and Gerdes, J. (1994). Molecular characterization of the gene locus of the human cell proliferation-associated nuclear protein defined by monoclonal antibody Ki-67. *Cell Prolif.* **29,** 1–12.

Du Manoir, M. S., Guillaud, P., Camus, E., Seigneurin, D., and Brugal, G. (1991). Ki-67 labeling in postmitotic cells defines different Ki-67 pathways within the 2c compartment. *Cytometry* **12,** 455–463.

Endl, E., and Gerdes, J. (2000). Post-translational modifications of the Ki-67 protein coincide with two major check points during mitosis. *J. Cell. Physiol.* **182,** 371–380.

Endl, E., Steinbach, P., Knuchel, R., and Hofstadter, F. (1997). Analysis of cell cycle-related Ki-67 and p120 expression by flow cytometric BrdUrd-Hoechst/7AAD and immunolabeling technique. *Cytometry* **29,** 233–241.

Gerdes, J. (1985). An immunohistological method for estimating cell growth fractions in rapid histopathological diagnosis during surgery. *Int. J. Cancer* **35,** 169–171.

Gerdes, J. (1990). Ki-67 and other proliferation markers useful for immunohistological diagnostic and prognostic evaluations in human malignancies. *Semin. Cancer Biol.* **1,** 199–206.

Gerdes, J., Schwab, U., Lemke, H., and Stein, H. (1983). Production of a mouse monoclonal antibody reactive with a human nuclear antigen associated with cell proliferation. *Int. J. Cancer* **31,** 13–20.

Gerdes, J., Lemke, H., Baisch, H., Wacker, H. H., Schwab, U., and Stein, H. (1984a). Cell cycle analysis of a cell proliferation-associated human nuclear antigen defined by the monoclonal antibody Ki-67. *J. Immunol.* **133,** 1710–1715.

Gerdes, J., Dallenbach, F., Lennert, K., Lemke, H., and Stein, H. (1984b). Growth fractions in malignant non-Hodgkin's lymphomas (NHL) as determined in situ with the monoclonal antibody Ki-67. *Hematol. Oncol.* **2,** 365–371.

Gerdes, J., Pickartz, H., Brotherton, J., Hammerstein, J., Weitzel, H., and Stein, H. (1987a). Growth fractions and estrogen receptors in human breast cancers as determined in situ with monoclonal antibodies. *Am. J. Pathol.* **129,** 486–492.

Gerdes, J., Van Baarlen, J., Pileri, S., Schwarting, R., Van Unnik, J. A., and Stein, H. (1987b). Tumor cell growth fraction in Hodgkin's disease. *Am. J. Pathol.* **128,** 390–393.

Gerdes, J., Li, L., Schlueter, C., Duchrow, M., Wohlenberg, C., Gerlach, C., Stahmer, I., Kloth, S., Brandt, E., and Flad, H. D. (1991). Immunobiochemical and molecular biologic characterization of the cell proliferation-associated nuclear antigen that is defined by monoclonal antibody Ki-67. *Am. J. Pathol.* **138,** 867–873.

Gerdes, J., Becker, M. H., Key, G., and Cattoretti, G. (1992). Immunohistological detection of tumour growth fraction (Ki-67 antigen) in formalin-fixed and routinely processed tissues. *J. Pathol.* **168,** 85–86.

Gerlach, C., Golding, M., Larue, L., Alison, M. R., and Gerdes, J. (1997). Ki-67 immunoexpression is a robust marker of proliferative cells in the rat. *Lab. Invest.* **77,** 697–698.

Grogan, T. M., Lippman, S. M., Spier, C. M., Slymen, D. J., Rybski, J. A., Rangel, C. S., Richter, L. C., Miller, T. P. (1988). Independent prognostic significance of a nuclear proliferation antigen in diffuse large cell lymphomas as determined by the monoclonal antibody Ki-67. *Blood* **71,** 1157–1160.

Hall, P. A., Richards, M. A., Gregory, W. M., d'Ardenne, A. J., Lister, T. A., and Stansfeld, A. G. (1988). The prognostic value of Ki67 immunostaining in non-Hodgkin's lymphoma. *J. Pathol.* **154,** 223–235.

Hall, P. A., McKee, P. H., Menage, H. D., Dover, R., and Lane, D. P. (1993). High levels of p53 protein in UV-irradiated normal human skin. *Oncogene* **8,** 203–207.

Heidebrecht, H. J., Buck, F., Haas, K., Wacker, H. H., and Pawaresch R. (1996). Monoclonal antibodies Ki-S3 and Ki-S5 yield new data on the "Ki-67" proteins. *Cell Prolif.* **29,** 413–425.

Heslin, M. J., Cordon-Cardo, C., Lewis, J. J., Woodruff, J. M., and Brennan, M. F. (1998). Ki-67 detected by MIB-1 predicts distant metastasis and tumor mortality in primary high grade extremity soft tissue sarcoma. *Cancer* **83,** 490–497.

Jacobberger, J. W. (1991). Intracellular antigen staining: Quantitative immunofluorescence. *Methods* **2,** 207–218.

Jansen, R. L., Hupperets, P. S., Arends, J. W., Joosten-Achjanie, S. R., Volovics, A., Schouten, H. C., and Hillen, H. F. (1998). MIB-1 labelling index is an independent prognostic marker in primary breast cancer. *Br. J. Cancer* **78,** 460–465.

Jares, P., Campo, E., Pinyol, M., Bosch, F., Miquel, R., Fernandez, P. L., Sanchez-Beato, M., Soler, F., Perez-Losada, A., Nayach, I., Mallofre, C., Piris, M. A., Montserrat, E., and Cardesa, A. (1996). Expression of retinoblastoma gene product (pRb) in mantle cell lymphomas. Correlation with cyclin D1 (PRAD1/CCND1) mRNA levels and proliferative activity. *Am. J. Pathol.* **148,** 1591–1600.

Key, G., Becker, M. H. G., Baron, B., Duchrow, M., Schlueter, C., Flad, H. D., and Gerdes, J. (1993). New Ki-67 equivilant murine monoclonal antibodies (MIB 1–3) generated against bacterially expressed parts of the Ki-67 cDNA containing three 66bp repetitive elements encoding for the Ki-67 epitope. *Lab. Invest.* **68,** 629–635.

Kill, I. R. (1996). Localisation of the Ki-67 antigen within the nucleolus. Evidence for a fibrillarin-deficient region of the dense fibrillar component. *J. Cell Sci.* **109,** 1253–1263.

Kute, T. E., and Quadri, Y. (1991). Measurement of proliferation nuclear and membrane markers in tumor cells by flow cytometry. *J. Histochem. Cytochem.* **39,** 1125–1130.

Landberg, G., and Roos, G. (1992). Flow cytometric analysis of proliferation associated nuclear antigens using washless staining of unfixed cells. *Cytometry* **13,** 230–240.

Landberg, G., Tan, E. M., and Roos, G. (1990). Flow cytometric multiparameter analysis of proliferating cell nuclear antigen/cyclin and Ki-67 antigen: A new view of the cell cycle. *Exp. Cell Res.* **187,** 111–118.

Larsen, J. K., Christensen, I. J., Christiansen, J., and Mortensen, B. J. (1991). Washless double staining of unfixed nuclei for flow cytometric analysis of DNA and nuclear antigen (Ki-67 or bromodeoxyuridine). *Cytometry* **12,** 429–437.

Littleton, R. J., Baker, G. M., Soomro, I. N., Adams, R. L., and Whimster, W. F. (1991). Kinetic aspects of Ki-67 antigen expression in a normal cell line. *Virch. Arch. B Cell Pathol. Incl. Mol. Pathol.* **60,** 15–19.

Lopez, F., Belloc, F., Lacombe, F., Dumain, P., Reiffers, J., Bernard, P., and Boisseau, M. R. (1991). Modalities of synthesis of Ki-67 antigen during the stimulation of lymphocytes. *Cytometry* **12,** 42–49.

MacCallum, D. E., and Hall, P. A. (1999). Biochemical characterization of pKi67 with the identification of a mitotic-specific form associated with hyperphosphorylation and altered DNA binding. *Exp. Cell Res.* **252,** 186–198.

McCormick, D., Chong, H., Hobbs, C., Datta, C., and Hall, P. A. (1993a). Detection of the Ki-67 antigen in fixed and wax-embedded sections with the monoclonal antibody MIB1. *Histopathology* **22,** 355–360.

McCormick, D., Yu, C., Hobbs, C., and Hall, P. A. (1993b). The relevance of antibody concentration to the immunohistological quantification of cell proliferation-associated antigens. *Histopathology* **22,** 543–547.

Naiem, M., Gerdes, J., Abdulaziz, Z., Stein, H., and Mason, D. Y. (1983). Production of a monoclonal antibody reactive with human dendritic reticulum cells and analysis of human lymphoid tissue. *J. Clin. Pathol.* **36,** 167–175.

Neubauer, A., Serke, S., Siegert, W., Kroll, W., Musch, R., and Huhn, D. (1989). A flow cytometric assay for the determination of cell proliferation with a monoclonal antibody directed against DNA-methyltransferase. *Br. J. Haematol.* **72,** 492–496.

Overton, W. (1988). Modified histogram subtraction technique for analysis of flow cytometry data. *Cytometry* **9,** 619–626.

Palutke, M., KuKuruga, D., and Tabaczka, P. (1987). A flow cytometric method for measuring lymphocyte proliferation directly from tissue culture plates using Ki-67 and propidium iodide. *J. Immunol. Methods* **105,** 97–105.

Pavelic, Z. P., Pavelic, L., Lower, E. E., Gapany, M., Gapany, S., Barker, E. A., and Preisler, H. D. (1992). c-myc, c-erbB-2, and Ki-67 expression in normal breast tissue and in invasive and noninvasive breast carcinoma. *Cancer Res.* **52,** 2597–2602.

Pellicciari, C., Mangiarotti, R., Bottone, M. G., Danova, M., and Wang, E. (1995). Identification of resting cells by dual-parameter flow cytometry of statin expression and DNA content. *Cytometry* **21,** 329–337.

Perry, A., Stafford, S. L., Scheithauer, B. W., Suman, V. J., and Lohse, C. M. (1998). The prognostic significance of MIB-1, p53, and DNA flow cytometry in completely resected primary meningiomas. *Cancer* **82,** 2262–2269.

Pittman, S. M., Strickland, D., and Ireland, C. M. (1994). Polymerization of tubulin in apoptotic cells is not cell cycle dependent. *Exp. Cell Res.* **215,** 263–272.

Popov, Z., Hoznek, A., Colombel, M., Bastuji-Garin, S., Lefrere-Belda, M. A., Bellot, J., Abboh, C. C., Mazerolles, C., and Chopin, D. K. (1997). The prognostic value of p53 nuclear overexpression and MIB-1 as a proliferative marker in transitional cell carcinoma of the bladder. *Cancer* **80,** 1472–1481.

Rechsteiner, M., and Rogers, S. W. (1996). PEST sequences and regulation by proteolysis. *Trends Biochem. Sci.* **21,** 267–271.

Ruby, S. G., and McNally, A. C. (1996). Quality control of proliferation marker (MIB-1) in image analysis systems utilizing cell culture based control materials. *Am. J. Clin. Pathol.* **106,** 634–639.

Sachsenberg, N., Perelson, A. S., Yerly, S., Schockmel, G. A., Leduc, D., Hirschel, B., and Perrin, L. (1998). Turnover of CD4+ and CD8+ T lymphocytes in HIV-1 infection as measured by Ki-67 antigen. *J. Exp. Med.* **187,** 1295–1303.

Sasaki, K., Murakami, T., Kawasaki, M., and Takahashi, M. (1987) The cell cycle associated change of the Ki-67 reactive nuclear antigen expression. *J. Cell. Physiol.* **133,** 579–584.

Schimenti, K. J., and Jacobberger, J. W. (1992). Fixation of mammalian cells for flow cytometric evaluation of DNA content and nuclear immunofluorescence. *Cytometry* **13,** 48–59.

Schlüter, C., Duchrow, M., Wohlenberg, C., Becker, M. H., Key, G., Flad, H. D., and Gerdes, J. (1993). The cell proliferation-associated antigen of antibody Ki-67: A very large, ubiquitous nuclear protein with numerous repeated elements, representing a new kind of cell cycle-maintaining proteins. *J. Cell Biol.* **123,** 513–522.

Scholzen, T., and Gerdes, J. (2000). The Ki-67 protein: From the known and the unknown. *J. Cell. Physiol.* **182,** 311–322.

Schutte, B., Tinnemans, M. M., Pijpers, G. F., Lenders, M. H., and Ramaekers, F. C. (1995). Three parameter flow cytometric analysis for simultaneous detection of cytokeratin, proliferation associated antigens and DNA content. *Cytometry* **21,** 177–186.

Schwarting, R. (1993). Little missed markers and Ki-67. *Lab. Invest.* **68,** 597–599.

Schwarting, R., Gerdes, J., Niehus, J., Jaeschke, L., and Stein, H. (1986). Determination of the growth fraction in cell suspensions by flow cytometry using the monoclonal antibody Ki-67. *J. Immunol. Methods* **90,** 65–70.

Scott, R. J., Hall, P. A., Haldane, J. S., van Noorden, S., Price, Y., Lane, D. P., and Wright, N. A. (1991). A comparison of immunohistochemical markers of cell proliferation with experimentally determined growth fraction. *J. Pathol.* **165,** 173–178.

Shi, S. R., Imam, S. A., Young, L., Cote, R. J., and Taylor, C. R. (1995). Antigen retrieval immunohistochemistry under the influence of pH using monoclonal antibodies. *J. Histochem. Cytochem.* **43,** 193–201.

Sprenger, R., Toellner, K. M., Schmetz, C., Lüke, W., Stahl-Henning. C., Ernst, M., Hunsmann, G., Schmitz, H., Flad, H. D., Gerdes, J., and Zimmer, J. P. (1995). Folicular dendritic cells productively infected with immunodeficiency viruses transmit infection to T cells. *Med. Microbiol. Immunol.* **184,** 129–134.

Starborg, M., Gell, K., Brundell, E., and Höög, C. (1996). The murine Ki-67 cell proliferation antigen accumulates in the nucleolar and heterochromatic regions of interphase cells and at the periphery of the mitotic chromosomes in a process essential for cell cycle progression. *J. Cell Sci.* **109,** 143–153.

Takagi, M., Matsuoka, Y., Kurihara, T., and Yoneda, Y. (1999). Chmadrin: A novel Ki-67 antigen-related perichromosomal protein possibly implicated in higher order chromatin structure. *J. Cell Sci.* **112,** 2463–2472.

Tsurusawa, M., and Fujimoto, T. (1995). Cell cycle progression and phenotypic modification of Ki67 antigen-negative G1- and G2-phase cells in phorbol ester-treated Molt-4 human leukemia cells. *Cytometry* **20,** 146–153.

Van Dierendonck, J. H., Keijzer, R., van de Velde, C. J., and Cornelisse, C. J. (1989). Nuclear distribution of the Ki-67 antigen during the cell cycle: Comparison with growth fraction in human breast cancer cells. *Cancer Res.* **49,** 2999–3006.

Van Dierendonck, J. H., Wijsman, J. H., Keijzer, R., van de Velde, C. J., and Cornelisse, C. J. (1991). Cell-cycle-related staining patterns of anti-proliferating cell nuclear antigen monoclonal antibodies. Comparison with BrdUrd labeling and Ki-67 staining. *Am. J. Pathol.* **138,** 1165–1172.

Verheijen, R., Kuijpers, H. J., van Driel, R., Beck, J. L., van Dierendonck, J. H., Brakenhoff, G. J., and Ramaekers, F. C. (1989a). Ki-67 detects a nuclear matrix-associated proliferation-related antigen. II. Localization in mitotic cells and association with chromosomes. *J. Cell Sci.* **92,** 531–540.

Verheijen, R., Kuijpers, H. J., Schlingemann, R. O., Boehmer, A. L., van Driel, R., Brakenhoff, G. J., and Ramaekers, F. C. (1989b). Ki-67 detects a nuclear matrix-associated proliferation-related antigen. I. Intracellular localization during interphase. *J. Cell Sci.* **92,** 123–130.

Vollmer, R. T., Humphrey, P. A., Swanson, P. E., Wick, M. R., and Hudson, M. L. (1998). Invasion of the bladder by transitional cell carcinoma: Its relation to histologic grade and expression of p53, MIB-1, c-erb B-2, epidermal growth factor receptor, and bcl-2. *Cancer* **82,** 715–723.

Wersto, R. P., Herz, F., Gallagher, R. E., and Koss, L. G. (1988). Cell cycle-dependent reactivity with the monoclonal antibody Ki-67 during myeloid cell differentiation. *Exp. Cell Res.* **179,** 79–88.

Wilson, G. D., Saunders, M. I., Dische, S., Daley, F. M., Robinson, B. M., Martindale, C. A., Joiner, B., and Richman, P. I. (1996). Direct comparison of bromodeoxyuridine and Ki-67 labelling indices in human tumours. *Cell Prolif.* **29,** 141–152.

CHAPTER 19

Detection of Proliferating Cell Nuclear Antigen

Jørgen K. Larsen,* Göran Landberg,† and Göran Roos†

*Finsen Laboratory
Finsen Center, Rigshospitalet
Copenhagen University Hospital
DK-2100 Copenhagen, Denmark

†Department of Pathology
University of Umeå
S-90187 Umeå, Sweden

I. Introduction
II. Molecular Biology of Proliferating Cell Nuclear Antigen
III. Methods for Immunochemical Detection and Quantification of Proliferating Cell Nuclear Antigen
    A. Anti-PCNA Antibodies
    B. Cell Preparation and Immunochemical Staining
IV. Results of Cytometric Analysis of Proliferating Cell Nuclear Antigen Expression
    A. Flow Cytometry
    B. Image Cytometry
V. Applications in Toxicology, Pathology, and Oncology
    A. Analysis of DNA Repair
    B. Analysis of Cell Proliferation
References

I. Introduction

The expression of the proliferating cell nuclear antigen (PCNA) molecule is associated with cell proliferation, being relatively low in quiescent cells and higher in cycling cells, especially during S phase. In addition, elevated PCNA expression is observed in cells under DNA repair. The possibility of utilizing PCNA expression as an empirical marker of cell proliferation, assessed by quantitative immunofluorescence cytometry without dependency on labeling of cells with DNA precursors, has motivated numerous investigations due to its potential value in toxicological, pathological, and oncological applications.

II. Molecular Biology of Proliferating Cell Nuclear Antigen

PCNA was originally discovered by Miyachi *et al.* (1978), who in sera from patients with systemic lupus erythematosus (SLE) found autoantibodies which reacted with nuclear antigens in proliferating cells. The use of such antisera as markers for the study of proliferating cells was recognized (Tan *et al.*, 1987). Using immunoprecipitation with anti-PCNA antibodies from SLE sera, it was demonstrated by Mathews *et al.* (1984) that PCNA was identical to the acidic 36 kDa protein "cyclin" that was defined with two-dimensional gel electrophoresis by Bravo *et al.* (1981). Since the term cyclin became ambiguous owing to confusion with the types of cyclin molecules that are described in Chapter 27, *Cytometry, 2nd Ed.* (*Methods in Cell Biology, Vol. 41*), the term PCNA was proposed.

PCNA is a phylogenetically highly conserved molecule. Considerable homology is found between mammalian PCNA and yeast, plant, and viral forms, implying that a primordial gene for PCNA evolved more than one billion years ago and must play an essential role in the cell cycle and in the maintenance of species (Dietrich, 1993; Kelman and Hurwitz, 1998). PCNA was early recognized as an auxiliary factor of DNA polymerase δ (Bravo *et al.*, 1987). The human PCNA gene has been localized to chromosome 20, with two pseudogenes on chromosome X and 6 (Ku *et al.*, 1989), and it has been molecularly cloned and completely sequenced (Travali *et al.*, 1989). The PCNA protein is in *Escherichia coli* structured as a homodimer, and in yeast and humans as a homotrimer, with a closed circular structure that can encircle DNA with a minimum of specific interactions. Measurements by small-angle neutron scattering support the trimeric ringlike structure of functionally active human PCNA in solution, in good agreement with model calculations based on the crystal structure from yeast PCNA (Schurtenberger *et al.*, 1998).

PCNA functions as a part of the DNA polymerase holoenzyme at the DNA replication forks for semidiscontinuous DNA synthesis (Brush and Kelly, 1996). The current model for the function of PCNA, mainly derived from studies of the cell-free SV40 system, suggests that PCNA acts as a sliding clamp that allows DNA polymerases (pol-δ/ε) to move rapidly along the DNA while remaining topologically bound to it. It has been shown that PCNA forms a complex together with the replication factor C (RF-C, clamp loader) at the primer in an ATP-dependent manner. This primer recognition complex can efficiently recruit pol-δ or pol-ε, but not pol-α, to the primer terminus (Hübscher *et al.*, 1996). The dynamic movement of PCNA on and off the DNA renders this protein an ideal communicator for a variety of proteins that are essential for DNA metabolic events in eukaryotic cells (Jonsson and Hubscher, 1997). In eukaryotic cells DNA replication is highly compartmentalized into replication centers, each containing many replication forks, and the PCNA has been detected at these replication centers by immunofluorescence microscopy (Bravo and Macdonald-Bravo, 1987; Cardoso et al., 1993).

The expression of PCNA declines in quiescent cells and is strongly induced by mitogens (Mathews et al., 1984). Although PCNA is synthesized shortly before the onset of cellular DNA replication and its abundance in the nucleus is highest during S phase, it is a stable protein with a half-life of about 20 hr that is present throughout the cell cycle (Hassell and Brinton, 1996). Results have indicated that PCNA can interact with the cell cycle regulatory machinery.

Cell cycle progression is controlled by the cyclin-dependent protein kinases (CDKs) and their regulatory partners, the cyclins. In experiments with isolation of kinase complexes containing cyclins A, B, D, and E, it has been shown that a complex of PCNA and the CDK inhibitor p21 associates with each cyclin/CDK dimer to form quaternary complexes (Jonsson and Hubscher, 1997; Prosperi et al., 1994; Xiong et al., 1992). On transformation with tumor viruses, such as SV40, PCNA and p21 are displaced from most of these complexes, implying that PCNA may play a role in cell cycle control through these interactions (Xiong et al., 1993). It seems likely that p21 inhibits DNA replication by acting as a competitor for other PCNA-binding proteins. It has been shown that PCNA is a target for binding through a conserved motif (PIP-box) of several regulating proteins that are not part of the DNA polymerase apparatus: p21, replication endonuclease Fen1, DNA (cytosine)-5 methyltransferase, and DNA repair endonuclease XPG (Warbrick, 1998).

PCNA has been shown to participate in nucleotide excision DNA repair and to be required for the gap-filling step (Celis et al., 1986; Shivji et al., 1992). Exposure to DNA damaging agents results in transcriptional activation of p21 and Gadd45 through a p53-dependent pathway. Gadd45 is known to interact with both PCNA and p21 (Warbrick, 1998). PCNA interacts physically with the protein Gadd45 that stimulates DNA excision repair and inhibits entry into S phase (Smith et al., 1994).

In proliferating cells, PCNA cycles between a chromatin-bound detergent-insoluble state in S phase and a diffuse soluble state when DNA is not being replicated or repaired (Bravo and Macdonald-Bravo, 1987; Celis et al., 1986). The solubility is associated with phosphorylation of PCNA and/or RF-C (Prosperi et al., 1993, 1994; Weisshart and Fanning, 1996). The loading of PCNA onto DNA seems sufficient to explain the detergent insolubility and association with replication sites.

III. Methods for Immunochemical Detection and Quantification of Proliferating Cell Nuclear Antigen

According to the choice of anti-PCNA antibody and the procedure for permeabilization, fixation, and immunofluorescent staining, it is possible in cell suspensions or tissue sections to identify the subpopulation of S-phase cells or of cycling cells. The level of expression of PCNA can be quantitatively analyzed by flow cytometry; however, image cytometry must be applied for a quantitative analysis of the subcellular distribution of PCNA.

A. Anti-PCNA Antibodies

In addition to the originally applied human autoantisera (Miyachi et al., 1978; Tan et al., 1987), a series of mouse monoclonal antibodies have been developed that bind to different epitopes in the PCNA molecule, and some of these are commercially available. The first two monoclonals were 19A2 [immunoglobulin M (IgM)] and 19F4 (IgG) (Kurki et al., 1988; Ogata et al., 1987) (available from, e.g., Coulter-Immunotech and Roche). From a series of 11 monoclonal antibodies developed by cloning cDNA for rat PCNA into a series of bacterial expression vectors and immunizing mice with the resulting protein (Waseem and Lane, 1990), the PC-10 antibody in particular has shown useful for cytometric analysis (available from, e.g., DAKO and Roche). Since then, using synthetic overlapping peptides of PCNA, the epitopes of a variety of monoclonal anti-PCNA antibodies have been mapped (Roos et al., 1993). As an example, the antibody 74B1 that effectively inhibits DNA replication *in vitro* was mapped to amino acids 121–135, a region of the PCNA protein containing the interdomain connector implicated in intermolecular interactions (Roos et al., 1996).

B. Cell Preparation and Immunochemical Staining

Because two distinct populations of PCNA molecules are simultaneously present in the cell and both are confined to the nucleus, namely, one that is localized diffusely in the nucleoplasm and is dominating in quiescent cells and another that is bound to nuclear structures resembling replicon clusters and is seen in the S-phase nucleus (Bravo and Macdonald-Bravo, 1987), the procedure for cell preparation and staining must be designed accordingly. It was an early experience from staining with the human PCNA autoantibodies that selective immunofluorescence staining of the replicon clusters required fixation with methanol, whereas both types of PCNA were fixed by formaldehyde. The diffusely localized PCNA could be extracted with Triton X-100 before fixation, without removal of the replicon-bound PCNA (Bravo and Macdonald-Bravo, 1987; Madsen and Celis, 1985).

This experience from immunofluorescence microscopy has been utilized in the flow cytometric approach to quantitative analysis of PCNA expression on a cell by cell basis. Fixation of cell suspensions with alcohols and/or formaldehyde before immunostaining generally results in PCNA-positive staining of all cycling cells. With the monoclonal 19A2 and 19F4 antibodies, an increased staining of S-phase cells has been obtained using permeabilization and fixation with a mixture of formaldehyde and lysolecithin followed by fixation with cold methanol (Kurki *et al.*, 1988). Fixation with formaldehyde at a higher ionic strength resulted in increased PCNA immunofluorescence (Bruno *et al.*, 1992). With the monoclonal PC-10 antibody selective staining of S-phase cells has been obtained using (1) extraction of unfixed cells with nonionic detergents such as Triton X-100 or Nonidet P-40 in ice-cold buffers, followed by fixation with methanol at −20°C before immunostaining (Landberg and Roos, 1991a; Wilson *et al.*, 1992), (2) fixation with 96% ethanol followed by incubation at 37°C with pepsin/HCl before immunostaining (Zölzer *et al.*, 1994), or (3) sequential treatment with acetone and methanol at −20°C (Beppu *et al.*, 1994). With human PCNA autoantibody, selective PCNA staining of S-phase cells is possible by addition of a buffer with a mixture of nonionic detergent and human PCNA autoantibody to the unfixed cells and performing the flow cytometric analysis without further washings; however, this simple method was not feasible with the monoclonal PCNA antibodies tested, including PC-10 (Landberg and Roos, 1992).

There are many critical aspects of the methodology for flow cytometric analysis of the PCNA expression. The preparation and staining procedures have to be adapted to each individual cell type. Most of the reported procedures have only been applied to cultured cell lines or hematological tissues. The choice and the dosage of antibodies for direct or indirect immunostaining protocols have to be optimized by careful titration experiments. Additional samples of cells with known positive and negative expression of PCNA must be included as biological controls, and if possible measured as internal references. It may be useful to calibrate the measurement system using standard fluorescencent beads with known loads of fluorochrome. Comparison of the PCNA measurements with those of other cell cycle related markers set up in parallel experiments may be helpful in validating the PCNA data. Thus, comparison with DNA content, bromodeoxyuridine (BrdUrd) incorporation, Ki-67 expression, or tritiated thymidine incorporation might be helpful. Critical aspects of general importance for flow cytometric, quantitative immunofluorescence analysis of intracellular markers are reviewed by Jacobberger (1991) as well as in Chapters 13, 18, and 22 of this book.

IV. Results of Cytometric Analysis of Proliferating Cell Nuclear Antigen Expression

A. Flow Cytometry

Bivariate analysis of PCNA expression and DNA content seems to be the most widely used methodology in flow cytometric applications, and several methods may be used, although with different selectivity with regard to the PCNA staining of S-phase cells, as mentioned earlier (Beppu *et al.*, 1994; Kurki *et al.*, 1988; Landberg and Roos, 1991a; Wilson *et al.*, 1992; Zölzer *et al.*, 1994). For the data shown in Fig. 1, indicating increased expression of PCNA induced by phytohemagglutinin in human lymphocytes, and in Fig. 2, indicating decreased expression induced by dimethylsulfoxide in HL-60 cells, a procedure was used that was based on cold extraction of unfixed cells with Triton X-100 and subsequent fixation with methanol, followed by staining with fluorescein isothiocyanate (FITC)-conjugated PC-10 antibody (Landberg and Roos, 1991a). The PCNA staining was validated by parallel staining for incorporated BrdUrd.

Bivariate analysis of PCNA and Ki-67 expression reveals distinct clusters of G_0, G_1, S, and G_2M cells without any staining of DNA (Landberg et al., 1990).

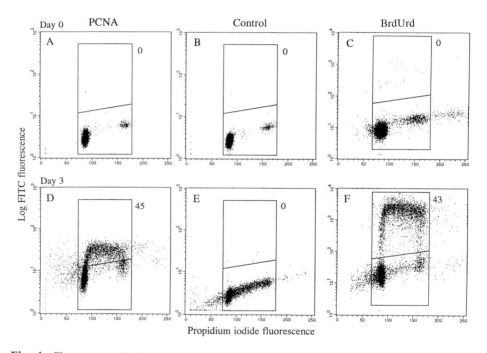

Fig. 1 Flow cytometric bivariate analysis of PCNA expression and DNA content, indicating an increased PCNA expression of S-phase cells in phytohemagglutinin stimulated normal human blood lymphocytes for 3 days. PCNA was stained selectively in S-phase cells, using cold extraction of unfixed cells with Triton X-100 and subsequent fixation with methanol, followed by staining with fluorescein isothiocyanate (FITC)-conjugated PC-10 antibody (DAKO M-879), as described by Landberg and Roos (1991a) (A, day 0; D, day 3). The PCNA staining was validated by staining with irrelevant isotype control antibody (DAKO X-933) (B, day 0; E, day 3), and by staining of DNA-synthesizing cells according to their incorporation of bromodeoxyuridine (BrdUrd) after pulse-labeling for 1 hr with 10 μM BrdUrd, using the method described by Jensen et al. (1993) (C, day 0; F, day 3). The subpopulation of S-phase cells was estimated by the upper part of the window region in the dot plots, and it is indicated by its percentage. The cells were analyzed in a Becton Dickinson FACS IV (excitation at 488 nm, FITC fluorescence at 515–540 nm, propidium iodide fluorescence at >620 nm), using gating in forward and side scatter. (Reproduced with permission from Henrik Flyger and Peter Østrup Jensen.)

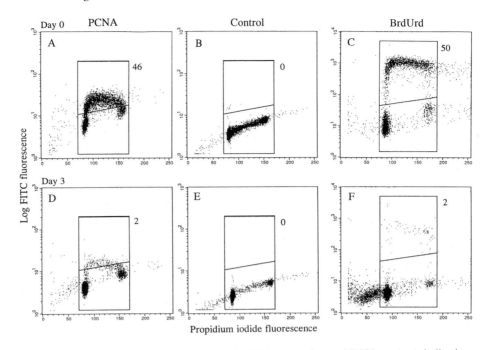

Fig. 2 Flow cytometric bivariate analysis of PCNA expression and DNA content, indicating a decreased PCNA expression in HL-60 cells treated with 1.5% dimethylsulfoxide for 3 days, using the same procedure for staining and analysis as in Fig. 1 (A, day 0; D, day 3). The staining for PCNA expression was validated by staining with irrelevant isotype control antibody (B, day 0; E, day 3), and by staining for BrdUrd incorporation (C, day 0; F, day 3). (Reproduced with permission from Henrik Flyger and Peter Østrup Jensen.)

This is illustrated in Fig. 3, showing two cases of malignant lymphoma, based on indirect immunostaining of unfixed cells with human PCNA autoantibody and monoclonal Ki-67 antibody (Landberg and Roos, 1992). The flow cytometric investigation may be extended from a "snap shot" of the cell cycle distribution into a cell kinetic analysis, when multivariate analysis is applied to a series of sequentially sampled BrdUrd-labeled cell populations, and based on simultaneous fluorescence measurements of PCNA, Ki-67, DNA by Hoechst 33258 (quenched by BrdUrd), and DNA by 7-aminoactinomycin D (Landberg and Roos, 1993).

B. Image Cytometry

Immunofluorescence microscopy enables observation and quantification of the distribution of the PCNA expression between cells in culture or in tissue sections, and the subcellular PCNA staining pattern may be characterized in comparison

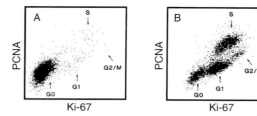

Fig. 3 Flow cytometric bivariate analysis of PCNA and Ki-67 expression of two human malignant lymphomas, indicating the cell cycle distribution by distinct clusters of G_0, G_1, S, and G_2M cells, based on indirect immunostaining of unfixed and detergent treated cells with a human PCNA antibody according to the method described by Landberg and Roos (1992). The expression of PCNA and Ki-67 is presented in log scale. (A) Small cell lymphoma with low proliferation. (B) Large cell lymphoma with high proliferation. [Reproduced from Landberg and Roos (1991b, Fig. 2, p. 918) with permission from the publisher.]

with the staining patterns of, for example, chromatin, DNA, BrdUrd, and Ki-67 (Bravo and Macdonald-Bravo, 1987; Celis *et al.*, 1987; Hyde-Dunn and Jones, 1997; van Dierendonck *et al.*, 1991). Image cytometric analysis based on the simultaneous staining of PCNA, Ki-67, AT-DNA, and CG-DNA in comparison with chromatin texture parameters has resulted in a detailed map of the cell cycle with several subphases of $G_{0/1}$ and S (Santisteban and Brugal, 1994, 1995). For image cytometric analysis it has also been possible to stain simultaneously for PCNA and BrdUrd (Humbert *et al.*, 1992).

V. Applications in Toxicology, Pathology, and Oncology

A. Analysis of DNA Repair

PCNA is a valuable DNA repair associated marker for cytometric investigations. An increased PCNA expression was found by immunofluorescence microscopy in transformed human amnion cells that had been exposed to ultraviolet light under conditions that induced nucleotide excision DNA repair, and it was found that the PCNA expression correlated well with tritiated thymidine labeling (Celis *et al.*, 1986). Examinations of histological sections of human skin exposed to ultraviolet irradiation that induced mild sunburn showed that the expression of PCNA and p53, but not of Ki-67 or DNA polymerase α, increased within 2 hr in keratinocytes and superficial dermal cells, with the same spatial and temporal distribution as that of *in situ* end labeling (ISEL) of damaged DNA (Coates *et al.*, 1995; Hall *et al.*, 1993). By flow cytometry a dose-dependent increase of the nuclear bound PCNA was found in the G_1 and G_2M compartments, but not in the S phase, of cultured human fibroblasts after treatment with DNA repair inducing alkylating agents and compounds that induce oxidative DNA damage (Savio *et al.*, 1998).

B. Analysis of Cell Proliferation

As an empirical marker of S-phase cells or cycling cells in studies of cell proliferation, and in particular as a potential prognostic indicator for several types of cancer, immunochemical detection of PCNA has become a valuable substitute for the labeling of cells with tritiated thymidine or halogenated deoxyuridines. For pathological applications it is a great advantage that detection of PCNA is applicable to microscopical examination of formalin-fixed paraffin-embedded tissue, as reviewed by Yu and Filipe (1993). Among the more recent reports based on microscopical examination, those demonstrating a multivariate independent prognostic value of PCNA, for example, for breast cancer (Aaltomaa et al., 1993), colorectal cancer (Paradiso et al., 1996), gastric cancer (Matturri et al., 1998), bladder cancer (Lipponen and Eskelinen, 1992), renal cell cancer (Morell-Quadreny et al., 1998), and non-Hodgkin's lymphoma (Korkolopoulou et al., 1998), may be opposed to a similar number of reports concluding only a univariate prognostic value of PCNA, not to mention the even larger number of reports concluding no prognostic value.

Future investigations with more refined and standardized methodologies with regard to the cell preparation and staining, sampling strategy, and image cytometrical instrumentation may help to clarify this ambiguous situation. Obviously, the value of a multivariate survival analysis for independent prognostic significance of PCNA expression is rather limited if all other relevant histopathological and clinical parameters are not included for comparison. In scoring the fraction of PCNA positive cells (PCNA index) of a heterogeneous tissue biopsy, practical decisions have to be taken with respect to sampling of the microscopical fields in which to count the number of PCNA positive and negative cells. Some of the diverging results on the correlation between PC-10 immunoreactivity and S phase in pathological specimens may be explained by the rather long half-life of PCNA (~ 20 hr), resulting in staining persisting in cells that may have recently left the cell cycle (Bravo and Macdonald-Bravo, 1987; Scott et al., 1991). Although levels of PCNA expression as assessed by immunofluorescence microscopy are normally very low in noncycling tissues, high levels of PCNA have been observed in the normal tissues surrounding human breast and pancreatic tumors, suggesting that tumors may elaborate growth factors that induce PCNA expression in nearby normal cells (Hall et al., 1994). Furthermore, the PCNA upregulation seen after DNA damage can in certain cases explain a poor correlation between PCNA expression and other proliferation markers.

Using flow cytometry, human malignant hematopoietic cells have been studied by bivariate analysis of PCNA and Ki-67 expression (Landberg and Roos, 1991b) (Fig. 3). In samples from patients with hematological disorders, the S-phase fraction and the cycling cell fraction according to PCNA expression were analyzed by flow cytometry (Erlanson et al., 1997). For a set of non-Hodgkin's lymphomas the PCNA data were consistent with estimates of S-phase fraction, S-phase duration, and potential doubling time calculated from parallel measurements on iododeoxyuridine-labeled cells. On fresh biopsies from colorectal tumors, Sawtell *et al.* (1995) performed flow cytometric bivariate analysis of PCNA expression and DNA content, using cycling cell specific staining of PCNA, and these data, showing a large variation in the PCNA positive fraction of cells, were compared with data from immunofluorescence microscopy of PCNA expression and BrdUrd-incorporation in a similar set of archival BrdUrd-labeled specimens. Prognostic investigations based on flow cytometrical measurement of PCNA expression in human tumor cells are quite sparse, if any.

Perspectives for future investigations are opened by the steadily increasing knowledge about the biology of the PCNA molecule and the increasing number of antibodies developed against specific, biologically important peptide sequences in the PCNA molecule. With respect to the oncological applications of PCNA detection it is interesting that a unique form of PCNA has been found in malignant breast cells (Bechtel *et al.*, 1998).

References

Aaltomaa, S., Lipponen, P., and Syrjanen, K. (1993). Proliferating cell nuclear antigen (PCNA) immunolabeling as a prognostic factor in axillary lymph node negative breast cancer. *Anticancer Res.* **13,** 533–538.

Bechtel, P. E., Hickey, R. J., Schnaper, L., Sekowski, J. W., Long, B. J., Freund, R., Liu, N., Rodriguez-Valenzuela, C., and Malkas, L. H. (1998). A unique form of proliferating cell nuclear antigen is present in malignant breast cells. *Cancer Res.* **58,** 3264–3269.

Beppu, T., Ishida, Y., Arai, H., Wada, T., Uesugi, N., and Sasaki, K. (1994). Identification of S-phase cells with PC10 antibody to proliferating cell nuclear antigen (PCNA) by flow cytometric analysis. *J. Histochem. Cytochem.* **42,** 1177–1182.

Bravo, R., and Macdonald-Bravo, H. (1987). Existence of two populations of cyclin/proliferating cell nuclear antigen during the cell cycle: Association with DNA replication sites. *J. Cell Biol.* **105,** 1549–1554.

Bravo, R., Fey, S. J., Bellatin, J., Larsen, P. M., Arevalo, J., and Celis, J. E. (1981). Identification of a nuclear and of a cytoplasmic polypeptide whose relative proportions are sensitive to changes in the rate of cell proliferation. *Exp. Cell Res.* **136,** 311–319.

Bravo, R., Frank, R., Blundell, P. A., and Macdonald-Bravo, H. (1987). Cyclin/PCNA is the auxiliary protein of DNA polymerase-δ. *Nature* **326,** 515–517.

Bruno, S., Gorczyca, W., and Darzynkiewicz, Z. (1992). Effect of ionic strength in immunocytochemical detection of the proliferation associated nuclear antigens p120, PCNA, and the protein reacting with Ki-67 antibody. *Cytometry* **13,** 496–501.

Brush, G. S., and Kelly, T. J. (1996). Mechanisms for replicating DNA. *In* "DNA Replication in Eukaryotic Cells" (M. L. DePamphilis, ed.), pp. 1–43. Cold Spring Harbor Laboratory, New York.

Cardoso, M. C., Leonhardt, H., and Nadal-Ginard, B. (1993). Reversal of terminal differentiation and control of DNA replication: Cyclin A and Cdk2 specifically localize at subnuclear sites of DNA replication. *Cell* **74,** 979–992.

Celis, J. E., Madsen, P., Nielsen, S., and Celis, A. (1986). Nuclear patterns of cyclin (PCNA) antigen distribution subdivide S-phase in cultured cells—some applications of PCNA antibodies. *Leuk. Res.* **10,** 237–249.

Celis, J. E., Madsen, P., Nielsen, S., and Rasmussen, H. H. (1987). Expression of cyclin (PCNA) and the phosphoprotein dividin are late obligatory events in the mitogenic response. *Anticancer Res.* **7,** 605–616.

Coates, P. J., Save, V., Ansari, B., and Hall, P. A. (1995). Demonstration of DNA damage/repair in individual cells using in situ end labelling: Association of p53 with sites of DNA damage. *J. Pathol.* **176,** 19–26.

Dietrich, D. R. (1993). Toxicological and pathological applications of proliferating cell nuclear antigen (PCNA), a novel endogenous marker for cell proliferation. *Crit. Rev. Toxicol.* **23,** 77–109.

Erlanson, M., Landberg, G., Lindh, J., and Roos, G. (1997). Flow cytometric evaluation of proliferating cell nuclear antigen expression in human hematopoietic malignancies. *Acta Oncol.* **36,** 17–22.

Hall, P. A., McKee, P. H., du Menage, H., Dover, R., and Lane, D. P. (1993). High levels of p53 protein in UV-irradiated normal human skin. *Oncogene* **8,** 203–207.

Hall, P. A., Coates, P. J., Goodlad, R. A., Hart, I. R., and Lane, D. P. (1994). Proliferating cell nuclear antigen expression in non-cycling cells may be induced by growth factors in vivo. *Br. J. Cancer* **70,** 244–247.

Hassell, J. A., and Brinton, B. T. (1996). SV40 and polyomavirus DNA replication. In "DNA Replication in Eukaryotic Cells" (M. L. DePamphilis, ed.), pp. 639–677. Cold Spring Harbor Laboratory Press, New York.

Hübscher, U., Maga, G., and Podust, V. (1996). DNA replication accessory proteins. In "DNA Replication in Eukaryotic cells" (M. L. DePamphilis, ed.), pp. 525-543. Cold Spring Harbor Laboratory, New York.

Humbert, C., Santisteban, M. S., Usson, Y., and Robert-Nicoud, M. (1992). Intranuclear co-location of newly replicated DNA and PCNA by simultaneous immunofluorescent labelling and confocal microscopy in MCF- 7 cells. *J. Cell Sci.* **103,** 97–103.

Hyde-Dunn, J., and Jones, G. E. (1997). Visualization of cell replication using antibody to proliferating cell nuclear antigen. *Methods Mol. Biol.* **75,** 341–347.

Jacobberger, J. W. (1991). Intracellular antigen staining: quantitative immunofluorescence. *Methods* **2,** 207–218.

Jensen, P. O., Larsen, J. K., Christensen, I. J., and van Erp, P. E. J. (1993). Discrimination of bromodeoxyuridine labelled and unlabelled mitotic cells in flow cytometric bromodeoxyuridine/DNA analysis. *Cytometry* **15,** 154–161.

Jonsson, Z. O., and Hubscher, U. (1997). Proliferating cell nuclear antigen: More than a clamp for DNA polymerases. *BioEssays* **19,** 967–975.

Kelman, Z., and Hurwitz, J. (1998). Protein–PCNA interactions: A DNA–scanning mechanism? *Trends Biochem. Sci.* **23,** 236–238.

Korkolopoulou, P., Angelopoulou, M. K., Kontopidou, F., Tsengas, A., Patsouris, E., Kittas, C., and Pangalis, G. A. (1998). Prognostic implications of proliferating cell nuclear antigen (PCNA), AgNORs and P53 in non-Hodgkin's lymphomas. *Leuk. Lymphoma* **30,** 625–636.

Ku, D. H., Travali, S., Calabretta, B., Huebner, K., and Baserga, R. (1989). Human gene for proliferating cell nuclear antigen has pseudogenes and localizes to chromosome 20. *Somat. Cell Mol. Genet.* **15,** 297–307.

Kurki, P., Ogata, K., and Tan, E. M. (1988). Monoclonal antibodies to proliferating cell nuclear antigen (PCNA)/cyclin as probes for proliferating cells by immunofluorescence microscopy and flow cytometry. *J. Immunol. Methods* **109,** 49–59.

Landberg, G., and Roos, G. (1991a). Antibodies to proliferating cell nuclear antigen as S-phase probes in flow cytometric cell cycle analysis. *Cancer Res.* **51,** 4570–4574.

Landberg, G., and Roos, G. (1991b). Expression of proliferating cell nuclear antigen (PCNA) and Ki-67 antigen in human malignant hematopoietic cells. *Acta Oncol.* **30,** 917–921.

Landberg, G., and Roos, G. (1992). Flow cytometric analysis of proliferation associated nuclear antigens using washless staining of unfixed cells. *Cytometry* **13,** 230–240.

Landberg, G., and Roos, G. (1993). Proliferating cell nuclear antigen and Ki-67 antigen expression in human haematopoietic cells during growth stimulation and differentiation. *Cell Prolif.* **26,** 427–437.

Landberg, G., Tan, E. M., and Roos, G. (1990). Flow cytometric multiparameter analysis of proliferating cell nuclear antigen/cyclin and Ki-67 antigen: A new view of the cell cycle. *Exp. Cell Res.* **187,** 111–118.

Lipponen, P. K., and Eskelinen, M. J. (1992). Cell proliferation of transitional cell bladder tumours determined by PCNA/cyclin immunostaining and its prognostic value. *Br. J. Cancer* **66,** 171–176.

Madsen, P., and Celis, J. E. (1985). S-phase patterns of cyclin (PCNA) antigen staining resemble topographical patterns of DNA synthesis. A role for cyclin in DNA replication? *FEBS Lett.* **193,** 5–11.

Mathews, M. B., Bernstein, R. M., Franza, B. R. J., and Garrels, J. I. (1984). Identity of the proliferating cell nuclear antigen and cyclin. *Nature* **309,** 374–376.

Matturri, L., Biondo, B., Cazzullo, A., Colombo, B., Giordano, F., Guarino, M., Pallotti, F., Turconi, P., and Lavezzi, A. M. (1998). Prognostic significance of different biological markers (DNA index, PCNA index, apoptosis, p53, karyotype) in 126 adenocarcinoma gastric biopsies. *Anticancer Res.* **18,** 2819–2825.

Miyachi, K., Fritzler, M. J., and Tan, E. M. (1978). Autoantibody to a nuclear antigen in proliferating cells. *J. Immunol.* **121,** 2228–2234.

Morell-Quadreny, L., Clar-Blanch, F., Fenollosa-Enterna, B., Perez-Bacete, M., Martinez-Lorente, A., and Llombart-Bosch, A. (1998). Proliferating cell nuclear antigen (PCNA) as a prognostic factor in renal cell carcinoma. *Anticancer Res.* **18,** 677–682.

Ogata, K., Kurki, P., Celis, J. E., Nakamura, R. M., and Tan, E. M. (1987). Monoclonal antibodies to a nuclear protein (PCNA/cyclin) associated with DNA replication. *Exp. Cell Res.* **168,** 475–486.

Paradiso, A., Rabinovich, M., Vallejo, C., Machiavelli, M., Romero, A., Perez, J., Lacava, J., Cuevas, M. A., Rodriquez, R., Leone, B., Sapia, M. G., Simone, G., and De, L. M. (1996). p53 and PCNA expression in advanced colorectal cancer: Response to chemotherapy and long-term prognosis. *Int. J. Cancer* **69,** 437–441.

Prosperi, E., Stivala, L. A., Sala, E., Scovassi, A. I., and Bianchi, L. (1993). Proliferating cell nuclear antigen complex formation induced by ultraviolet irradiation in human quiescent fibroblasts as detected by immunostaining and flow cytometry. *Exp. Cell Res.* **205,** 320–325.

Prosperi, E., Scovassi, A. I., Stivala, L. A., and Bianchi, L. (1994). Proliferating cell nuclear antigen bound to DNA synthesis sites: Phosphorylation and association with cyclin D1 and cyclin A. *Exp. Cell Res.* **215,** 257–262.

Roos, G., Jiang, Y., Landberg, G., Nielsen, N. H., Zhang, P., and Lee, M. Y. (1996). Determination of the epitope of an inhibitory antibody to proliferating cell nuclear antigen. *Exp. Cell Res.* **226,** 208–213.

Roos, G., Landberg, G., Huff, J. P., Houghten, R., Takasaki, Y., and Tan, E. M. (1993). Analysis of the epitopes of proliferating cell nuclear antigen recognized by monoclonal antibodies. *Lab. Invest.* **68,** 204–210.

Santisteban, M. S., and Brugal, G. (1994). Image analysis of *in situ* cell cycle related changes of PCNA and Ki-67 proliferating antigen expression. *Cell Prolif.* **27,** 435–453.

Santisteban, M. S., and Brugal, G. (1995). Fluorescence image analysis of the MCF-7 cycle related changes in chromatin texture. Differences between AT- and GC-rich chromatin. *Anal. Cell Pathol.* **9,** 13–28.

Savio, M., Stivala, L. A., Bianchi, L., Vannini, V., and Prosperi, E. (1998). Involvement of the proliferating cell nuclear antigen (PCNA) in DNA repair induced by alkylating agents and oxidative damage in human fibroblasts. *Carcinogenesis* **19,** 591–596.

Sawtell, R. M., Rew, D. A., Stradling, R. N., and Wilson, G. D. (1995). Pan cycle expression of proliferating cell nuclear antigen in human colorectal cancer and its proliferative correlations. *Cytometry* **22,** 190–199.

Schurtenberger, P., Egelhaaf, S. U., Hindges, R., Maga, G., Jonsson, Z. O., May, R. P., Glatter, O., and Hubscher, U. (1998). The solution structure of functionally active human proliferating cell nuclear antigen determined by small-angle neutron scattering. *J. Mol. Biol.* **275,** 123–132.

Scott, R. J., Hall, P. A., Haldane, J. S., van Noorden, S., Price, Y., Lane, D. P., and Wright, N. A. (1991). A comparison of immunohistochemical markers of cell proliferation with experimentally determined growth fraction. *J. Pathol.* **165,** 173–178.

Shivji, K. K., Kenny, M. K., and Wood, R. D. (1992). Proliferating cell nuclear antigen is required for DNA excision repair. *Cell* **69,** 367–374.

Smith, M. L., Chen, I. T., Zhan, Q., Bae, I., Chen, C. Y., Gilmer, T. M., Kastan, M. B., O'Connor, P. M., and Fornace, A. J. J. (1994). Interaction of the p53-regulated protein Gadd45 with proliferating cell nuclear antigen. *Science* **266,** 1376–1380.

Tan, E. M., Ogata, K., and Takasaki, Y. (1987). PCNA/cyclin: A lupus antigen connected with DNA replication. *J. Rheumatol.* **14** (Suppl. 13), 89–96.

Travali, S., Ku, D. H., Rizzo, M. G., Ottavio, L., Baserga, R., and Calabretta, B. (1989). Structure of the human gene for the proliferating cell nuclear antigen. *J. Biol. Chem.* **264,** 7466–7472.

van Dierendonck, J. H., Wijsman, J. H., Keijzer, R., van de Velde, C. J., and Cornelisse, C. J. (1991). Cell-cycle-related staining patterns of anti-proliferating cell nuclear antigen monoclonal antibodies. Comparison with BrdUrd labeling and Ki- 67 staining. *Am. J. Pathol.* **138,** 1165–1172.

Warbrick, E. (1998). PCNA binding through a conserved motif. *BioEssays* **20,** 195–199.

Waseem, N. H., and Lane, D. P. (1990). Monoclonal antibody analysis of the proliferating cell nuclear antigen (PCNA). Structural conservation and the detection of a nucleolar form. *J. Cell Sci.* **96,** 121–129.

Weisshart, K., and Fanning, E. (1996). Roles of phosphorylation in DNA replication. *In* "DNA Replication in Eukaryotic Cells" (M. L. DePamphilis, ed.), pp. 295–330. Cold Spring Harbor Laboratory, New York.

Wilson, G. D., Camplejohn, R. S., Martindale, C. A., Brock, A., Lane, D. P., and Barnes, D. M. (1992). Flow cytometric characterisation of proliferating cell nuclear antigen using the monoclonal antibody PC10. *Eur. J. Cancer* **28A,** 2010–2017.

Xiong, Y., Zhang, H., and Beach, D. (1992). D type cyclins associate with multiple protein kinases and the DNA replication and repair factor PCNA. *Cell* **71,** 505–514.

Xiong, Y., Zhang, H., and Beach, D. (1993). Subunit rearrangement of the cyclin-dependent kinases is associated with cellular transformation. *Genes Dev.* **7,** 1572–1583.

Yu, C. C., and Filipe, M. I. (1993). Update on proliferation-associated antibodies applicable to formalin- fixed paraffin-embedded tissue and their clinical applications. *Histochem. J.* **25,** 843–853.

Zölzer, F., Streffer, C., and Pelzer, T. (1994). A comparison of different methods to determine cell proliferation by flow cytometry. *Cell Prolif.* **27,** 685–694.

CHAPTER 20

Lymphocyte Activation Associated Antigens

Andrea Fattorossi, Alessandra Battaglia, and Cristiano Ferlini

Institute of Obstetrics and Gynecology
Università Cattolica del Sacro Cuore
00136 Rome, Italy

I. Introduction
II. Methodological Aspects
    A. General Considerations
    B. More Details
III. To Flow or Not to Flow for Assessing Lymphocyte Activation/Proliferation? And If Yes, How Reliable Is Immunophenotyping?
    A. Some Advantages
    B. Some Shortcomings
IV. Additional Approaches
V. Concluding Remarks
    References

I. Introduction

Lymphocyte activation and proliferation represent an essential step in the immune response. Lymphocytes are activated and then proliferate, expanding specific clones of cells that are reactive against foreign antigens. A disregulation of the response, either diminished or enhanced, is responsible for a variety of pathological states. Lack of response leads to immunodeficiency, and the series of severe clinical symptoms following human immunodeficiency virus (HIV) infection is a typical, albeit not exclusive, example of the consequences of an impaired immune response. In contrast, the lack of inhibitory control pathways of the immune response is responsible for the development of autoimmune diseases and, possibly, allergies. Because balanced lymphocyte activation and proliferation play such an important role, studies regarding this topic have always attracted the attention of investigators. The measurement of [^{3}H]thymidine incorporation into DNA performed with bulk lymphocyte cultures has been used for decades, and it probably remains the most commonly used method to quantify lymphocyte proliferation, although some doubts about its validity have been raised in some circumstances (Weir *et al.*, 1993).

More recently, the problems inherent to the handling of radionuclides have led to the development of nonradioactive alternatives to evaluate proliferation and reproductive integrity (Roehm *et al.*, 1991; Cory *et al.*, 1991; Ahmed *et al.*, 1994). Soon after the beginning of such studies, it became clear that lymphocyte activation and proliferation *in vitro* reflected the complexity of the immune response *in vivo* and was not a simple consequence of the interaction, in isolation, of an individual cell with a stimulus. Rather, the process involved the interaction between a number of different cell types, each playing a specific role and often changing its characteristics with time. This notion raised interest in assessing the peculiar role played by the various lymphocyte subsets during activation and proliferation, in identifying them, and in selectively measuring the features that were most related to the activation/proliferation process.

Multiparameter flow cytometry analysis, made possible by the very successful marriage between monoclonal antibody (MAb) technology and the increasing availability of fluorochromes excitable by relatively inexpensive light sources, has facilitated the discovery that lymphocyte activation and proliferation involve a finely tuned modulation of several preexisting molecules and/or a *de novo* expression of molecules on the outer cellular membrane. Activation and proliferation are distinct phenomena and, as will be exemplified later, the latter does not necessarily follow the former. Therefore, the more general definition "activation/proliferation" will be used throughout to define the molecules that are somehow modulated on stimulation.

Although a modulation of membrane antigens on activation can be observed on all lymphoid cell populations, namely, natural killer (NK) cells (Wang *et al.*, 1992; Reyburn *et al.*, 1997; Craston *et al.*, 1998), B lymphocytes (Banchereau and Rousset 1992; Buhling *et al.*, 1995; Dorfman *et al.*, 1997; Agematsu *et al.*, 1998), and T lymphocytes, T lymphocytes have received by far the highest attention because of the major influence they exert on the regulation of the immune response. We will therefore focus on the phenotypic changes shown by T lymphocytes undergoing stimulation.

Classically, the term "activation antigens" refers to molecules that either are not expressed or are expressed at a very low level by resting T lymphocytes and are augmented on cellular activation. CD25, CD38, CD69, and CD71 (Neckers and Cossman, 1983; Beverley, 1987; Triebel *et al.*, 1986; Nakamura *et al.*, 1989; Biselli *et al.*, 1992) are examples of molecules that are rapidly upmodulated (4–6 hr for CD69, formerly referred to as very early activation antigen, and around 24 hr for the others), whereas CD49a–e (formerly referred to as very late activation antigens) appear or are modulated after 3–4 weeks of *in vitro* activation (Hemler, 1990). Class II major histocompatibility complex (MHC) antigens can be defined as activation antigens, as they are expressed at a high level by activated

T lymphocytes (Triebel et al., 1986; Gansbacher and Zier, 1988; Fattorossi et al., 1992).

A more comprehensive view of the concept of activation related antigens should include a number of additional membrane molecules, indicating that they are modulated on exposure to some stimuli. Damle and Doyle (1989) showed that simultaneous binding of CD3 and CD28 on quiescent T lymphocytes induced the expression of functional receptors for interleukin (IL)-4, normally absent on resting lymphocytes. Insulin-like growth factor-1 receptor expression on human T lymphocytes is modulated on activation and correlates with CD25 expression in some models (Schillaci et al., 1998). CD2 is modulated on stimulated T lymphocytes and is probably involved in cell-to-cell adhesion (Redelman, 1987; Makgoba et al., 1989; Biselli et al., 1992). The expression of CD6, a protein highly regulated during T lymphocyte ontogeny, can be increased on exposure to anti-CD2 MAb or phorbol ester (Cosman, 1994). CD30 and LAG-3 have been reported to be expressed differently by polarized human type 1 and type 2 helper lymphocytes (Annunziato et al., 1997). CD96 (Tactile) is upregulated by T lymphocytes after 6–9 days of culture *in vitro* and is thought to be involved in cell adhesion (Wang et al., 1992). CD152 is transiently expressed on activated T lymphocytes (Castan et al., 1997) and is a ligand for CD80 and CD86 present on the membrane of activated B lymphocytes (Vallé et al., 1990; Dorfman et al., 1997). CD154 (CD40 ligand) has been demonstrated to increase during T lymphocyte activation and can play a role in functional responses of these cells during physiological interactions with other immunocompetent cells (Noelle et al., 1992; Roy et al., 1993; Brugnoni et al., 1996). CD45RO, CD45RA, and CD62L represent further examples of molecules that are not essential for the cell to proliferate but play a critical role in the regulation of cell function (Kishimoto et al., 1990; Biselli et al., 1992; Clement, 1992; Johannisson and Festin, 1995). Sandilands et al. (1997) showed that human T lymphocytes stimulated in a mixed leukocyte culture assay modulated CD32 expression.

Paradoxically, the increasing knowledge of the intricate relationship between cell proliferation and cell death deems it necessary to include among activation antigens additional molecules, such as the mitochondrial protein Bcl-2 that exerts a considerable regulatory activity on the process of programmed cell death (Akbar et al., 1993; Veis et al., 1993). CD95, also an apoptosis-related molecule, is upregulated on activated T lymphocytes (Yoshino et al., 1994). Figure 1 illustrates the modulation of several "classic" and "nonclassic" activation antigens occurring in a $CD4^+$ T cell clone following phytohemagglutinin (PHA) restimulation.

A close examination of all possible molecules playing a role in lymphocyte activation/proliferation and the numerous ways of inducing T lymphocyte activation *in vitro* (polyclonal mitogens such as plant lectins and others, superantigens, antigens, and cross-linking of some membrane molecules) or *in vivo* goes beyond the scope of this discussion. Instead, we mainly focus on the problems the flow cytometrist faces when attempting to measure the variations in T lymphocyte

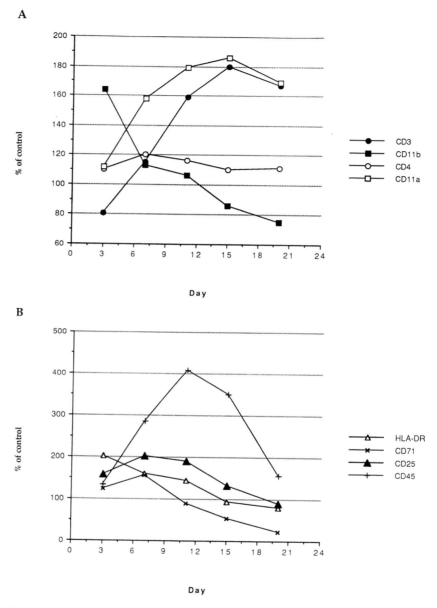

Fig. 1 A CD4⁺ T cell clone was left unstimulated for 15 days and then stimulated with 1 mg/ml PHA in the presence of autologous antigen presenting cells (day 0). The modulation of various membrane antigens was assessed by flow cytometry using phycoerythrin (PE)- or fluorescein isothiocyonate (FITC)-labeled MAb. Data are expressed as percent variation of the median fluorescence intensity calculated assuming 100% fluorescence at day 0. The scale of (A) and (B) are different to better visualize the variations occurring in the various antigens. All antigens were consistently up-

phenotypes during activation/proliferation. As will be discussed and exemplified later, the immunophenotyping of activated/proliferating lymphocytes is the quintessential example of difficulties that can be encountered in flow cytometry.

II. Methodological Aspects

A. General Considerations

Detection of activated/proliferating lymphocytes *ex vivo* in experimental animal models (e.g., Baldinelli *et al.*, 1992; Aiuti *et al.*, 1995; Roos *et al.*, 1996), or in patients suffering from a variety of diseases, is very popular and has produced an impressive amount of data. A very brief list includes the following: autoimmune disorders (Infante *et al.*, 1996; Chan *et al.*, 1996; Jason *et al.*, 1997), tumors (Reyes *et al.*, 1997), parasitic diseases (Sztein *et al.*, 1990; Worku *et al.*, 1997), allergies (Majori *et al.*, 1997; Muller-Suur *et al.*, 1997; Hallstrand *et al.*, 1998), grafts (Reinke *et al.*, 1994; Beik *et al.*, 1998), and viral infections (Tomkinson *et al.*, 1987). Activated lymphocytes have been found also in human milk and are thought to be important for the defense of mucosal sites (Wirt *et al.*, 1992). Human immunodeficiency virus (HIV) infection has been the object of intensive investigation, producing an overwhelming body of data on the expression of activation/proliferation antigens (e.g., Ziegler-Heitbrock *et al.*, 1988; Hofmann *et al.*, 1991; Levacher *et al.*, 1992; Callebaut *et al.*, 1993; Lees *et al.*, 1993; Borthwick *et al.*, 1994; Plaeger-Marshall *et al.*, 1993; Kestens *et al.*, 1994; Bouscarat *et al.*, 1996; Gougeon *et al.*, 1996). Albeit informative, this kind of evaluation of the immune system status may be misleading, and there is always the possibility that lymphocytes expressing an activated phenotype may be deficient in the ability to actually generate a satisfactory proliferative response. This has rendered it necessary, in most situations, to assess directly the activation/proliferation capacity of these cells in the presence of appropriate stimulation *in vitro*.

In principle, the care and basic technical procedures for measuring activation/proliferation antigens by flow cytometry on the surface of stimulated lymphocytes does not intrinsically differ from those common to the phenotyping of normal resting lymphocytes (other chapters of this book are dedicated to the general aspects of immunophenotyping, which will not be examined in close detail here). However, some aspects are worth emphasizing. The appropriate choice and usage

---

or downmodulated on stimulation (with the exception of CD4), indicating their involvement in the process of lymphocyte activation/proliferation. Assessing the modulation of the phenotypic profile during the various restimulation cycles is of practical relevance in studies involving T cell clones, because the proliferative response to various stimuli (polyclonal activators, antigens, interleukin-2, etc.) correlates with the phenotype (Ferlini *et al.*, 1996b).

of reagents are of paramount importance. For example, MAb conjugated with the more efficient fluorochromes, such as phycoerythrin (PE), is preferable when staining for the activation markers (or all markers usually expressed in a continuous manner), whereas use of MAb to identify lymphocyte subsets, usually expressed in an all-or-none manner, gives satisfactory results when conjugated with relatively less efficient fluorochromes, such as fluorescein isothiocyanate (FITC). The importance of this issue is highlighted by the fact that by using MAb labeled with the more efficient fluorochrome labeling of MAb, some activation antigens, once thought not to be expressed by resting T lymphocytes, are now well documented, albeit at low density, on the cell surface (Zola *et al.*, 1989; Jackson *et al.*, 1990; Jackson and Bell, 1990). It is also advisable to titrate any new batch of MAb to minimize nonspecific binding of the MAb and establish the possible presence of free fluorochrome molecules. Free fluorochrome molecules adsorb onto cells and contribute to an elevated background. In our personal experience their presence, due to nonoptimal labeling and/or storage conditions, is most likely to occur with PE and tandem conjugates, such as PE-Cy5.

Some of the major problems encountered when immunophenotyping activated/proliferating lymphocytes are as follows:

1. Heterogeneity in autofluorescence signal due to the simultaneous existence of various cell subsets.
2. Variations in autofluorescence signal intensity due to changes occurring in a number of intrinsic cellular factors caused by activation.
3. Modifications in lymphocyte size on stimulation with subsequent changes in intrinsic components (autofluorescence) and surface area (MAb binding).
4. Unimodal expression of most antigens on the cell membrane, which generates histograms with poor resolution between positive and negative events.
5. Variable but consistent cell mortality and cell clumping.

Finally, all of the problems are complicated by the fact that commitment of lymphocytes to activation/proliferation is a dynamic process, and, consequently, the expression of activation/proliferation markers occurs not only progressively in each given lymphocyte but, to make things worse, with different kinetics depending on the lymphocyte subset and/or the quality/quantity of the stimulus.

A cell suspension containing activated/proliferating lymphocytes therefore appears as a mix of viable and nonviable cells heterogeneous in size and also in fluorescence intensity, from quite low to very high, and because of Shapiro's second law of flow cytometry ("What you see is what you get"), there is the serious risk of getting "garbage out in response to garbage in" (Shapiro, 1985). We will approach the problem starting from the point of view of a flow cytometrist sitting in front of a single argon laser instrument (by far the most widely used) and dealing with acquisition and analysis of a peripheral blood sample cultured in the presence of a suitable stimulus and double stained with MAb against classic T cell and activation markers, for example, CD3 and CD25, respectively (three-color or multicolor phenotyping have the same kinds of difficulties, only amplified; see later).

B. More Details

1. The Gating Procedures

Usually performed on the forward scatter (FSC) versus side scatter (SSC) dot plot, the acquisition gating must take into account the relatively poor viability of the sample and the large heterogeneity in size. The natural tendency is to gate out whatever is not viable, that is, whatever has low FSC and high SSC (Fig. 2) or whatever looks like contaminating cells or aggregates. This procedure is usually satisfactory to exclude debris and dead cells. However, it is advisable to compare data from scatter signals with the measurement of the uptake of some membrane impermeant nucleic acid dyes, such as propidium iodide (PI), ethidium bromide (EB), or 7-aminoactinomycin D (7-AAD). An appropriate combination of these dyes with the MAb of interest allows one to ameliorate the discrimination between live and dead cells with relative ease (Fig. 3). It may be worth remembering that early apoptotic lymphocytes may exhibit marginally altered scatter signals and admit measurable amounts of the diverse nucleic acid dyes (Swat *et al.*, 1991; Ferlini *et al.*, 1996a). Gating these cells out, one may lose valuable information. Gating procedures are further complicated by the fact that these kinds of samples usually also contain variable proportions of cells other than lymphocytes (e.g., monocytes) whose light scatter characteristics can resemble those of lymphocytes and whose relative proportion varies on stimulation (Figs. 2 and 4).

2. The Fluorescence Background Assessment

Fluorescence background assessment is a point of major importance when dealing with activation/proliferation antigens. It is often overlooked and is the source of pitfalls and misinterpretation when the specific fluorescent signal to be measured is dim. (See also Section II,B,5.) Indeed, background fluorescence is a common obstacle in immunophenotyping, making the detection of low level specific fluorescence difficult. Total background fluorescence depends on intrinsic autofluorescence and the nonspecific binding capacity of MAb and/or free fluorochrome molecules sometimes present in preparations of conjugated MAb (storage!). A complete description on how to deal with background fluorescence and related problems is presented in other chapters of this book.

For the sake of the present discussion, it is important to highlight that autofluorescence depends on cell type (cells of the mononuclear phagocyte lineage and polymorphonuclear leukocytes, typical contaminants of lymphocyte preparations, are more autofluorescent than resting lymphocytes, possibly reflecting a diverse content of pyridine and flavin nucleotides, Aubin, 1979; Figs. 4 and 5)

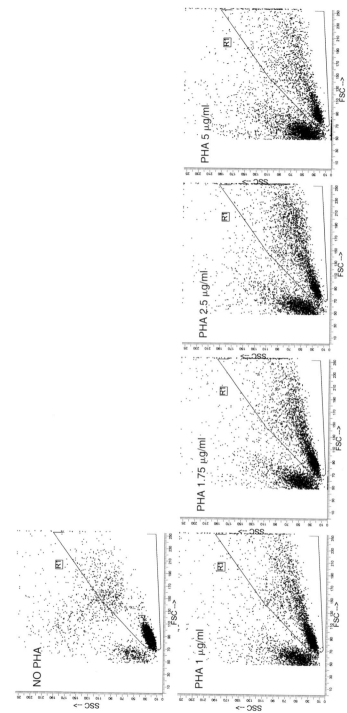

Fig. 2 FSC versus SSC dot plots of human peripheral blood mononuclear cells prepared by density gradient centrifugation and cultured for 3 days in the presence of various concentrations of PHA. Gate R1 excludes dead cells, as confirmed by 7-AAD stain, and thus restricts phenotypic assessment to viable cells only. The number of cells with increased FSC and SSC augments along with PHA concentration and was used to calculate the "lymphocyte performance" according to Janossy et al. (1993) (see Table 1). The variations in size and internal complexity contribute to the enhanced background value depicted in Fig. 4. The gate was set in stimulated samples so as to include all viable lymphocyte blasts and was not changed in the unstimulated sample for the sake of comparison. FSC/SSC makes resting lymphocytes easy to distinguish from monocytes in unstimulated cultures. In PHA stimulated cultures, blasts change their scatter properties, rendering it difficult to distinguish them from monocytes (see also Fig. 5). Restricting the gate so as to completely exclude monocytes would lead to an error by excluding viable lymphoblasts.

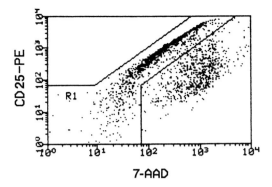

Fig. 3 7-AAD versus CD25-PE dot plot. Dead cells are identified by their increased uptake of 7-AAD. The plot was not compensated, intending to show that the emission of 7-AAD does not significantly interfere with the evaluation of PE generated fluorescence. The spillover of PE fluorescence into the "red" channel is uninfluential in interpreting the CD25 staining.

and cell size (a typical change is an augmentation in cell size that occurs as the lymphocyte becomes activated). The latter cooperates in determining autofluorescence in at least two ways. First, a large cell will obviously contain a higher amount of autofluorescent constituents than a smaller one of the same origin (the possibility also exists of a selective increase of autofluorescent constituents). Second, a large cell, due to its increased surface area, will be more prone to have a greater amount of nonspecific binding of the MAb antibody and/or free fluorochrome molecules. Note that large differences in fluorescence background between subsets may exceed the dynamic range (Fig. 5). Overlooking size changes can also bias the intepretation of a specific immunofluorescence signal. If a surface unit accommodates the same amount of MAb, the resulting specific fluorescent signal of a large cell will be higher than that of a small cell despite the fact that the density of the antigen has not changed. Conversely, an unchanged specific total fluorescence signal may indicate changes in antigen density if detected on cells differing in surface area. Measuring cell size by Coulter volume or FSC can serve acceptably. However, it would be better if the two ways for measuring size can be performed simultaneously. Measuring the Coulter volume helped to determine that it was the density of CD2 that was modulated on activation (Biselli *et al.*, 1992).

In our experience, the best way to obtain a dependable measure of total background fluorescence for a "difficult" conjugated MAb preparation is by specifically blocking the access of conjugated MAb to the epitope by a 5 to 10 molar excess of unconjugated MAb (Fig. 6). This approach takes into account all variables described earlier, especially those related to poor quality reagents, which sometimes may be the only ones available on the market for specific purposes. Preincubating samples with mouse serum for 15 min and then adding

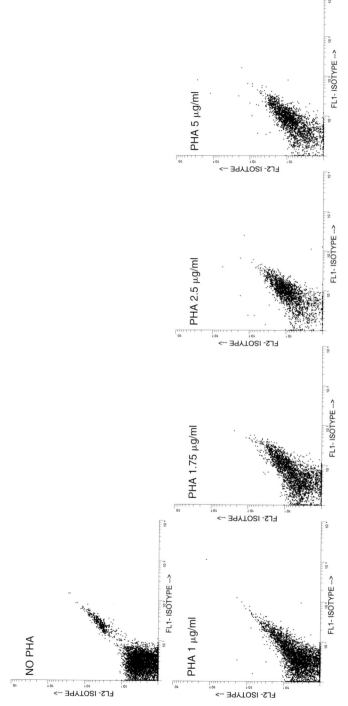

Fig. 4 "Green" versus "orange" (FL1 versus FL2) background fluorescence dot plots generated from cells included in gate R1 of Fig. 2. Background fluorescence augments in step with PHA concentration, reflecting variations in size and autofluorescent compounds. The cluster with the high background value observed in the sample cultured in the absence of PHA is made up of monocytes (see Fig. 5).

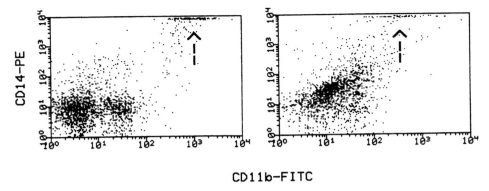

Fig. 5 CD11b-FITC versus CD14-PE dot plots generated from cells included in gate R1 of Fig. 2. The cluster with the high background visible in unstimulated cultures stains intensely with both CD11b and CD14, indicating its monocytic nature. Monocytes in the gate diminish, although they never disappear, in PHA cultures, possibly because of adherence to the culture plate or clumping. Diminution of monocytes in the high background region in PHA samples was also confirmed by morphological examination of cells relocated by laser scanning cytometer (CompuCyte). Arrows have been set to indicate the position of monocytes, which is off scale in the PE channel due to their high background and specific stain.

the conjugate helps preventing nonspecific staining via Fc receptors and perhaps also contributes to lowering cell adhesiveness.

3. Compensation

Compensation is not an issue restricted to the immunophenotyping of activated/proliferating lymphocytes, but it is certainly of major relevance here because this sort of sample typically contains cells with antigens expressed in an all-or-none fashion, antigens expressed in a continuous manner from dim to bright, and antigens whose intensity varies depending on the activation status of each lymphocyte subset. Compensation values, which should always be established on single-stained samples, remain valid only if the intensity of the stain remains relatively constant. This leads to the horrifying conclusion that it would be necessary to have single-stained samples for each pair of MAb and for each condition of stimulation. Because it usually is impractical and quite expensive to run such a large number of single-stained samples, an acceptable alternative is to run, at least, samples single-stained for the more brilliant markers. An appropriate choice of conjugates can help, however. If activation/proliferation markers are PE-labeled, their variations in intensity are less likely to influence the FITC conjugate signal (usually, the spillover of PE into the "green channel" of most commercial flow cytometers is low). Unfortunately, PE conjugates do influence peridinium chlorophyll protein (PerCP) or tandem conjugate signal

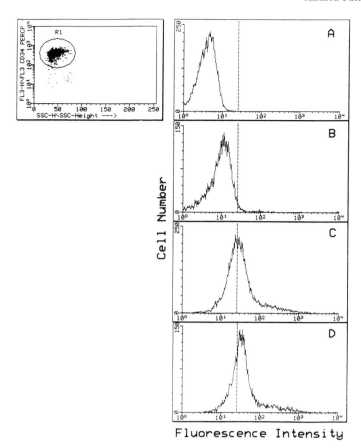

Fig. 6 Example of a poor quality MAb. CD34+ hematopoietic progenitors were purified by immunomagnetic beads. The dot plot on the left shows the gate used in the analysis to exclude the minimal remnants of contaminating non-CD34+ cells. Histograms on the right depict the fluorescence intensity of cells incubated in the presence of (A) nothing, (B) irrelevant MAb, (C) 10 molar excess unconjugated CD105 followed by FITC-CD105, and (D) FITC-CD105. The fluorescence in (C) is clearly due to the nonspecific binding of the MAb and/or to the presence of free flurochrome molecules and is consistently higher than that generated by the irrelevant MAb (B). Relying on the latter to establish background would lead to the false conclusion of a strong specific staining. In fact, cells do react specifically with FITC-CD105, but the magnitude of the staining is to be assessed by comparing (C) and (D). Median values of fluorescence are 4.1, 10, 28.4, and 34 for (A), (B), (C), and (D) histograms, respectively.

(>600 nm emitting fluorochomes). Cytometrists with the possibility to perform color compensation using interactive/automatic compensation system on list mode files have a distinct advantage when performing three-color (or more) analysis.

4. The Number of Cells of Interest to Be Acquired

One of the most powerful capabilities of flow cytometry consists of its ability to identify subset-specific differences in activation/proliferation antigens among lymphocytes subsets. When gating on a small subset, a list file of 10,000 events may be dramatically reduced to a dozen or so cells of interest. As underlined by Parks *et al.* (1989), cell sample size greatly affects the significance of population frequency measurements and makes it dangerous to state the outcome of a culture on the basis of measures made on a sample containing a limited number of cells carrying the antigen of interest.

5. Data Interpretation

The percentage of positive cells is usually used to summarize measures of lymphocyte subsets. The percentage of positive cells refers to the proportion of cells that exceed the staining intensity of cells labeled with an isotype control reagent. Often, this is not easily feasible when measuring activation/proliferation antigens due to the usually poor resolution between negative and positive cells. One should therefore wonder whether it is always correct to force data to look for "percent positive." There are cases in immunophenotyping, activation/proliferation markers being typical in this respect, in which the underlying biology and/or instrument resolution does not produce a resolved bimodal distribution but a unimodally skewed one (Figs. 6 and 7). In these cases, any boundary chosen to define positive cells is arbitrary and likely to miscalculate the real frequency. It may be more reasonable and informative to resort to the mean or median value of the whole histogram. The median (50th percentile point) offers a distinct advantage over the mean when distribution is skewed: including or excluding a few events falling outside the main population has in fact less effect on the median than on the mean. As shown in Fig. 7, however, the possibility exists that a unimodal fluorescence distribution is masking an unresolved bimodal one. The effort presently made to render flow cytometry quantitative and to allow comparison among different laboratories is a possible way around the problem (Lenkei and Andersson, 1995), and measurement of the fluorescence intensity of CD38 in HIV infection has been proposed as a marker of prognosis (Liu *et al.*, 1996; Zhiyuan *et al.*, 1996). The vast literature on the relevance of quantitating antigen expression in hematological malignancies will not be mentioned here (see appropriate chapters).

Incidentally, there is no general consensus on the actual utility of isotype controls, as the right isotype control should match exactly the characteristics, that is, class and subclass of immunoglobulin, fluorochrome labeling procedure, storage conditions, and so on, of the specific MAb.

6. Antigen Uptake

Some membrane antigens are released on activation (Parish and McKenzie, 1977; Ziegler-Heitbrock and Ulevitch, 1993; Spronk *et al.*, 1994; Suda *et al.*, 1996)

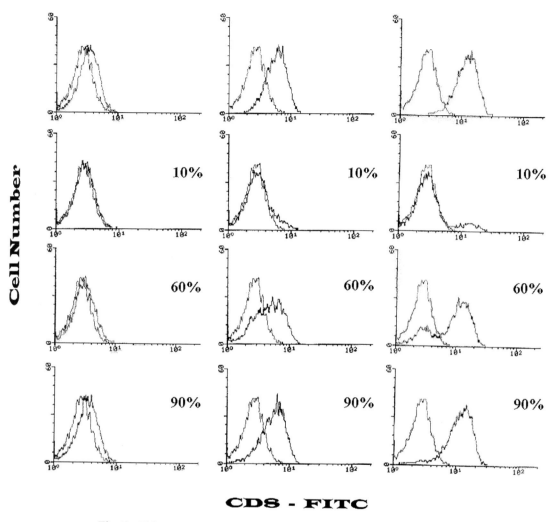

Fig. 7 This experiment was designed to show the relevance of both intensity of specific fluorescence and proportion of positive cells in evaluating flow data. A CD8+ T cell clone was stained with FITC-CD8. Mixtures containing variable proportions (shown are 10, 60, and 90%) of stained and nonstained cells were then prepared and 5000 cells acquired using a FACScan (Becton Dickinson, San Jose, CA). To reduce the intensity of the stain to simulate low density antigen expression, cells were preincubated with a molar excess of unconjugated OKT8. This figure reports data from mixtures generated using samples with three different fluorescence intensities where the lowest, intermediate, and highest are in the left, the middle, and the right column, respectively. The upper row of panels has been generated by superimposing data from unstained, left histograms, and stained sample, shifted histograms, to show the maximum fluorescence above background of the three samples. Numbers in other panels represent the proportion of stained cells contained in the sample. When the fluorescence over background is high (right column) there is no difficulty in determining the boundary between negative and positive, even when the proportion of positive cells is low. When

and, under certain circumstances, are transferred to the membrane of neighboring cells, resulting in false positivity. Class II MHC molecules are typical in this respect, as was demonstrated *in vitro* in a murine CD4+ T cell clone model (Le Moli *et al.*, 1989). This phenomenon was correlated with the magnitude of the stimulus and required close cell-to-cell contact. Remarkably, murine spleen CD4+ lymphocytes also adsorb class II MHC molecules *in vivo* (A. Fattorossi, unpublished data, 1989).

III. To Flow or Not to Flow for Assessing Lymphocyte Activation/Proliferation? And, If Yes, How Reliable Is Immunophenotyping?

A. Some Advantages

There is little doubt that flow cytometry is virtually the only choice if lymphocyte activation/proliferation has to be evaluated *ex vivo*. The case is different when one wants to document the response to stimulation in *in vitro* models. As mentioned earlier, radiolabeled and nonradiolabeld probes to evaluate lymphocyte response *in vitro* in bulk culture have been successfully used for years. These methods are quite well standardized, and their widespread use facilitates comparison among laboratories. Before deciding to resort to flow cytometry based methods, it is important to evaluate the information one wants. Bulk methods are generally carried out in microplates. Microplate technology has distinct advantages. It usually requires a relatively low number of cells, thus permitting the establishment of replicate samples in a single culture plate, thereby overcoming the inherent inaccuracy of bioassays based on cell cultures. Also it allows automated devices for evaluating the response so that multiple wells can be examined in minutes with a limited intervention by the operator. The main drawback consists of the fact that it is impossible to establish which specific cell subset is responding. Cell sorting, panning, or microbeads have been used to establish cultures containing only selected subsets; well suited for specific purposes, these approaches are cumbersome to perform and suffer from the potential

---

the specific fluorescence intensity decreases (middle column), establishing the boundary becomes increasingly difficult in step with the diminution of the proportion of positive cells. No meaningful boundary can be established when the specific fluorescence intensity is very low (left column), even when the proportion of positive cells is quite high. In this case, increasing the proportion of positive cells generates progressively shifted histograms, apparently giving the false impression of a homogeneous increase of antigen on the whole population (courtesy of A. Kunkl and P. Samoggia, unpublished, 1992). A mathematical approach to the problem of samples with poor positive/negative resolution can be found in papers by Overton (1988) and Lampariello and Aiello (1998). However, these methods are also likely to fall short with extremely dim specific fluorescence signals, for example, the left column.

disadvantage that preculture steps may introduce artifacts and disrupt the physiological cooperation between the various cell components of the immune response (Akbar et al., 1991).

Multivariate flow cytometry easily averts this problem: it allows examination of activation/proliferation antigens in discrete subsets and helps establish whether a response observed in a bulk culture reflects a modification in the recruitment of distinct subsets or a modulation of the response occurring in all subsets. A simple dual color assay will give at least two kinds of information. First, it gives the proportion of cells exhibiting each of the possible combinations generated by the two probes. Second, it provides the proportion of cells carrying the activation/proliferation antigen within a given subset (Fig. 8). The latter information is often overlooked in spite of its relevance. In fact, calculating the percentage of responding cells within the potential responder population allows solving the bias inherent in bulk assays caused by a scarcity of responder cells. In some situations (typical in this respect are blood specimens from seriously ill patients) the percentage of T lymphocytes in density gradient preparations used to set up cultures can be greatly reduced by the presence of other cell populations, namely,

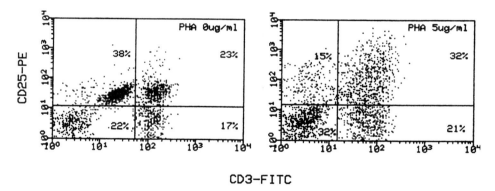

Fig. 8 CD3-FITC versus CD25-PE dot plots generated from peripheral blood mononuclear cells prepared by density gradient centrifugation from a patient 20 days after being transplanted with CD34$^+$ hematopoietic progenitors following high dose chemotherapy. Numbers represent the percentage of cells in each quadrant. The left-hand plot shows data from a nonstimulated culture in which a consistent proportion of T lymphocytes expresses CD25, indicating a status of activation *in vivo*. In nonstimulated cultures, the large amount of monocytes with high background fluorescence, often particularly abundant in density gradient preparations from this kind of patient, made it necessary to shift the negative/positive boundary to specifically evaluate the proportion of CD25$^+$ cells within the cell subset of interest, namely, CD3$^+$ T lymphocytes. Fifty-eight percent of CD3$^+$ cells were positive. Repeating the operation after PHA stimulation showed that 60% of T lymphocytes expressed CD25. Thus, PHA was not able to induce a *de novo* expression; of CD25 on T lymphocytes, although it did augment the intensity of the expression, and this change was related to actual proliferation (see Fig. 9). Note that in this kind of measurement, the presence of a double negative population helps establish the negative/positive boundary.

NK cells, monocytes, and granulocytes, that will obviously not respond or respond less to stimulation. Thus, a low [^{3}H]thymidine uptake will reflect the scarcity of responding cells rather than an actually impaired response.

An additional drawback of bulk cultures is the lack of information on cell loss, because methods relying on DNA synthesis or cell metabolism do not easily lend themselves to discrimination between viable and nonviable cells. A variable but consistent proportion of lymphocytes stimulated *in vitro* dies in culture, essentially by apoptosis (Martin, 1993; Kabelitz *et al.*, 1993; Ferlini *et al.*, 1996a) (Fig. 9). These considerations become of major concern if one has to examine samples from certain diseased subjects, for example, HIV and tumor patients receiving chemotherapy, often containing lymphoid cells already committed to apoptosis (Carbonari *et al.*, 1994). Flow cytometric approaches for the measuring of apoptosis rate can be found in excellent reviews (Darzynkiewicz *et al.*, 1997) and are also detailed in other chapters of this book. Interestingly, Medina *et al.* (1994) put forward the concept of activation-associated cell death to account for a consistent cell loss occurring in the absence of clear signs of apoptosis.

B. Some Shortcomings

Several pros and cons about relying on flow cytometry to assess lymphocyte activation/proliferation are summarized in the literature on CD69. CD69 is currently considered as the earliest activation antigen expressed by human circulating T lymphocytes on stimulation, and it is virtually undetectable or expressed at a very low level by resting lymphocytes (Testi *et al.*, 1989a; Biselli *et al.*, 1992; Mardiney *et al.*, 1996). Measuring CD69 modulation after a short incubation *in vitro* has been proposed as a quick indicator of the capacity of T lymphocytes to respond to a variety of agents (Maino *et al.*, 1995; Simms and Ellis, 1996). It soon became evident that the quality and quantity of the stimulus conditioned the reliability of the test. CD69 was able to identify activated lymphocytes as early as 4–6 hr when polyclonal mitogens, but not specific antigens, were used. In the latter cases, a measurable CD69 expression could only be detected after 18–24 hr, thereby negating the advantage of eliminating the need of a tissue culture facility (Mardiney *et al.*, 1996; Gibbons and Evans, 1996). Perfetto *et al.* (1997) showed that measuring CD69 on peripheral T lymphocytes from HIV patients on exposure to soluble CD2 or immobilized CD3 allowed the identification of subjects more at risk of progression. Similar findings were also reported by Krowka *et al.* (1996). However, probably reflecting methodological differences, the proportion of CD69$^+$ lymphocytes reported by the two groups differed. CD69 is not obviously related to cell proliferation, as suggested by the fact that stimuli whch induce a dramatic upregulation of CD69, such as phorbol ester, do not induce proliferation in the absence of an additional stimulating agent, such as calcium ionophore (Testi *et al.*, 1989b). To add to the complexity, CD69 may also be upregulated by death stimuli in the absence of detectable proliferation. As reported more recently, γ-radiation initiates cell death by apoptosis concomi-

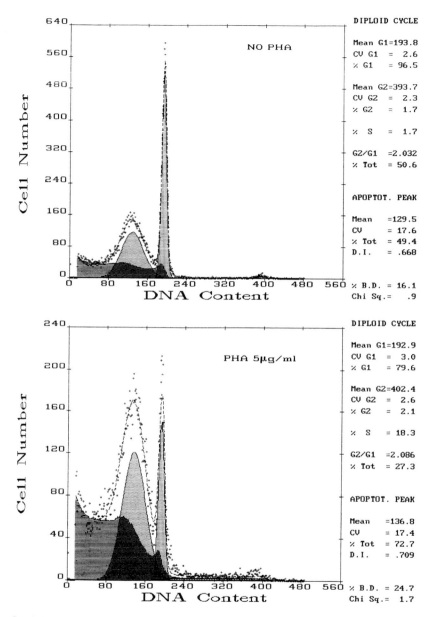

Fig. 9 Cell cycle analysis and apoptosis evaluation by MultiCycle of ethanol-fixed PI stained cultures from the same experiment in Fig. 8. PHA stimulation drove lymphocytes to proliferate (cells in S–G_2/M rose from 3.4 to 20.4%) but also to undergo apoptosis (cells in the apoptotic peak rose from 49.4 to 72.7%), suggesting that lymphocytes from this patient were particularly susceptible to activation-induced apoptosis.

tantly with an enhanced CD69 expression, suggesting that CD69 serves a variety of, as yet not well understood, functions and does not necessarily reflect an activation leading to proliferation (Chen *et al.,* 1997).

The complexity of the issue of regulation and significance of activation/proliferation antigen expression is also exemplified by the observation that class II MHC expression on T lymphocytes can be independent of the regulation of CD25 and proliferation induction (Salvadori *et al.,* 1991). Similarly, T cell clones specific for certain superantigens such as staphylococcal enterotoxin (SE) comodulate CD25 and class II MHC molecules in different manners depending on whether PHA or specific SE is used, and this reflects a different capacity to proliferate in response to a further stimulation (Fig. 10; see legend for details). A further example drawn from the lymphocyte clone–SE model is reported in Fig. 11,

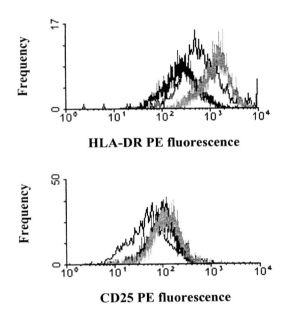

Fig. 10 Fluorescence histograms generated by HLA-DR-PE and CD25-PE MAb in a T cell clone responding to staphylococcal enterotoxin E (SEE) and restimulated with SEE or PHA. The black line represents no stimulus, the dark gray line is for SEE, and the light gray line shows PHA. The experiment was set to establish the effect of equipotent concentrations of PHA and SEE (defined as the amount of stimulus producing the same [^{3}H]thymidine uptake) on activation/proliferation antigen modulation and induction of anergy. CD25 expression was upregulated in the same manner by superantigen and PHA (the two histograms are superimposed, and this renders difficult their distinction), whereas only the latter enhanced HLA-DR expression. The lack of class II MHC molecule modulation was related to anergy induction because the PHA-treated clone, but not the SEE-treated clone, proliferated following a subsequent stimulation with PHA. The experiment was performed 7 days after the last restimulation, and this explains the high expression of HLA-DR and CD25 in the nonstimulated sample.

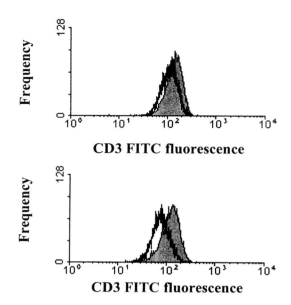

Fig. 11 Fluorescence histograms generated by CD3-FITC MAb in a T cell clone responding to SED, but not SEA, incubated with the two superantigens at 4°C or 37°C for 1 hr, washed, and assayed for CD3 expression. Shaded and empty histograms are from SEA and SED treated samples, respectively. (Top) CD3 expression on cells incubated with either SE at 4°C remains essentially unchanged. (Bottom) CD3 expression is reduced on contact, at 37°C, with specific SE, indicating a rapid downmodulation of the antigen that precedes cell proliferation.

depicting the effect of 1 hr of contact of a T cell clone with the appropriate SE on CD3 expression (see legend for detail).

A lymphocyte carrying an activation/proliferation antigen is not necessarily engaged in cell proliferation (Salvadori et al., 1991; Nisini et al., 1992, 1996; LaSalle et al., 1992; June et al., 1994). This issue has been addressed by Rutella et al. (1998) in an elegant series of experiments describing the effect of serum from healthy subjects receiving granulocyte colony-stimulating factor on the T lymphocyte response to PHA. Phenotypic markers of activation/proliferation, namely, CD69, CD25, CD71, and class II MHC molecules, were upmodulated; cells underwent blast transformation, but the entry into S phase was greatly reduced. There was no measurable apoptosis, and lymphocytes promptly started S phase on addition of exogenous IL-2. This suggested a partial activation status similar to that observed in T cell clone models (Sloan-Lancaster et al., 1994; Nisini et al., 1996).

Lastly, flow cytometry falls short in measuring a lymphocyte response when the cell population of interest is small owing to a limited number of reacting cells and/or to an asynchronous response so that the amount of cells at the same stage of activation at a given time point is too low to be detected (it may be interesting to note that such a situation is reminiscent of the concept of the "time window" typical of apoptosis; Darzynkiewicz *et al.*, 1997). This explains the relative difficulty of measuring antigen specific responses, typically carried out by a minor proportion of specific lymphocytes, as compared to responses to polyclonal activators such as plant lectins and superantigens.

IV. Additional Approaches

The flow cytometry approach to the issue of lymphocyte activation/proliferation is not restricted to immunophenotyping. Other probes can be efficiently used. As early as 1981, Darzynkiewicz *et al.* showed that the uptake of a mitochondrial fluorescent probe, rhodamine 123 (Rh123), was enhanced in stimulated lymphocytes. Confirming and extending that observation, we later demonstrated the suitability of this probe in detecting lymphocyte stimulation in various *in vitro* models (Ferlini *et al.*, 1995). It was also possible to double stain cells with MAb so as to compare mitochondrial behavior on stimulation with the appearance of activation/proliferation antigens on the cell membrane. It was demonstrated that lymphocytes not expressing detectable amounts of CD25, and thus not definably stimulated, were in fact metabolically activated, though to a lesser extent than CD25$^+$ lymphocytes. Lyons and Parish (1994) first proposed using the fluorescein derived carboxyfluorescein diacetate succinimidyl ester (CFDA-SE, formerly CFSE) for the identification of cell progeny and analysis of the division history of individual cells in an animal model. We have adopted the method for tracking division of human lymphocytes (Figs. 12 and 13), cell lines (Viora *et al.*, 1997;

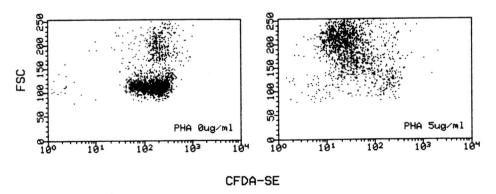

Fig. 12 CFDA-SE versus FSC dot plots generated from peripheral blood mononuclear cells prepared by density gradient centrifugation and cultured in the presence or absence of PHA. In the stimulated sample, CFDA-SE fluorescence is reduced in cells with increased FSC, indicating a fair correlation between acquisition of a blastlike morphology and actual proliferation.

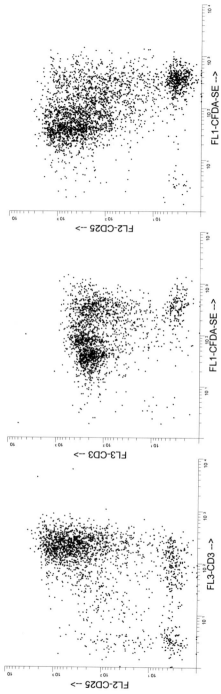

Fig. 13 Three-color analysis of peripheral blood mononuclear cells prepared by density gradient centrifugation and cultured for 3 days in the presence of PHA. CFDA-SE loaded cells were costained with CD3-peridinium Chlorophyll protein (PerCP) and CD25-PE. Although the great majority of T lymphocytes expresses CD25 (left), a consistent amount of these cells did not half CFDA-SE, indicating lack of proliferation (middle and right). This confirms that the presence of CD25 is not sufficient for defining a cell as proliferating. The present three-color assay requires some skill, as the CFDA-SE emission spills over into the "orange" and "red" channels of most commercial flow cytometers and renders compensation somewhat difficult. Here, compensation established during acquisition on a FACScan (Becton Dickinson) was touched up by WinList software (Verity Software House, Topsham, ME). However, this method for monitoring lymphocyte activation *in vitro*, possibly coupled with cell cycle and apoptosis analysis, and additional or alternative activation/proliferation or subset markers, probably represents the most comprehensive manner to effectively assess the actual capacity of lymphocytes to respond to stimuli by flow cytometry.

Ercoli *et al.*, 1998), and hematopoietic progenitors (Fattorossi *et al.*, 1996; Pierelli *et al.*, 1998a,b; and this book) and have found that coupling CFDA-SE with immunophenotyping is a promising tool to approach several issues related to hematopoietic progenitor differentiation and lymphocyte subset response to various stimuli. A similar dye, PKH26, has been also used to assess cell division of different subsets of cells (Horan and Slezak, 1989). The probe is also amenable to three-color staining (Allsopp *et al.*, 1998). However, its "orange" emission prevents exploiting the favorable spectral characteristics of PE conjugates for the best detection of activation/proliferation antigens.

Other flow cytometric methods for detecting lymphocyte activation/proliferation have been described, for example, ionized calcium measurement and cell cycle monitoring (Braunstein *et al.*, 1975; Adriaansen *et al.*, 1990; Biselli *et al.*, 1992; Bussa *et al.*, 1993; Nisini *et al.*, 1992; Bartoleschi *et al.*, 1994; Kohler *et al.*, 1997), each with their own pros and cons (see appropriate chapters in this book). In recent years the possibility of assessing intracellular cytokines produced by stimulated lymphocytes by flow cytometry has received increasing attention because of the sensitivity of the method and possibility of dual staining of cells with membrane antigens so as to establish which subset is producing which cytokine (Maino and Picker, 1998).

V. Concluding Remarks

At this point, the reader may wonder which is to be considered the best way of approaching the issue of assessing lymphocyte activation/proliferation by flow cytometry. As is often the case in science, there is no definitive answer, and the choice of the best suited approach remains the responsibility of the investigator. Surely, each method could be used interchangeably with no significant loss of information when the study involves samples from normal subjects. In these cases, the method routinely used in the laboratory, or the one the investigator feels more familiar with, is probably the best choice. In contrast, failing to take into account the possible pitfalls inherent in the various assays that have been underlined earlier with samples from diseased individuals can become a source of misinterpretation. Consequently, the general conclusion is that to get a complete picture of the events taking place during lymphocyte activation/proliferation a combination of methods, possibly flow cytometry and not flow cytometry, is to be preferred (Fattorossi *et al.*, 1991). To exemplify the concept, data obtained by using a choice of the several possible approaches in assessing a standard response of normal human lymphocytes to scalar amounts of PHA have been summarized in Table I. As is immediately evident, no method can be defined superior in absolute terms; each provides a relevant piece of information, but all contribute to delineate the complete picture of events taking place in lymphocytes on PHA contact.

Table I
Parallel Assessment of CD25+ Cells[a]

Treatment	CD25+ %[b]	CD25+ %[c]	Proliferation index[d]	S–G$_2$/M Phase (%)[e]	Lymphocyte performance (%)[f]	[^{3}H]Thymidine (cpm)	Apoptosis (%)[g]
No PHA	15	24	1.21	1	5	1,000	0
PHA 1.0 μg/ml	37	53	1.47	11	28	45,790	38
PHA 1.75 μg/ml	40	57	1.68	14	34	52,900	38
PHA 2.5 μg/ml	47	80	2.05	33	56	71,640	40
PHA 5.0 μg/ml	51	81	1.46	26	57	90,675	33

[a] Proliferation index was calculated by ModFit (Verity)/Cell Proliferation Model (Sigma) software, percent cells in the proliferative compartments of the cell cycle was computed by MultiCycle software (Phoenix), lymphocyte performance was expressed as a percentage of viable CD3+ cells with increased FSC compared with unstimulated cells, and [^{3}H]thymidine uptake was measured in normal lymphocytes cultured in the presence of various concentrations of PHA for 72 hr. Apoptosis evaluated by MultiCycle computation is also shown. The amount of CD25+ cells is reported in two ways: the column on the left reports CD3+CD25 cells evaluated as percent cells on the whole population of viable cells (gated on a FSC/SSC dot plot); the column on the right reports the percentage of CD25+ cells within the population of CD3+ cells. This kind of computation is recommended when dealing with samples containing low proportion of CD3+ cells (or other subset of interest). In these cases, a low CD3+CD25+ value would not indicate a reduced response but merely reflects a low amount of responders. There was a fair correlation between the various indicators of activation/proliferation and also between these and [^{3}H]thymidine uptake, although flow-based assays failed to detect the difference between 2.5 and 5 mg/ml PHA, or even suggested a decrease (proliferation index). The apparent major sensitivity of the isotopic method is readily explained, since [^{3}H]thymidine measures DNA synthesis that occurred over a long period, typically 18 hr, whereas in the flow methods measures are snapshots representative of events taking place at that given time. It should also be noted that [^{3}H]thymidine remaining in the nuclei of dead cells adds to the total radioactivity, whereas data from flow-based methods refer to viable cells only.
[b] Calculated as CD3+CD25+ cells within the viable cell population.
[c] Calculated as CD25+ cells within CD3+ viable cell population.
[d] By ModFit/Cell Proliferation Model computation.
[e] By MultiCycle computation.
[f] According to Janossy et al. (1993).
[g] By MultiCycle computation of apoptotic peak.

Acknowledgments

We gratefully acknowledge Leila Andreocci's invaluable technical contribution, and we are indebted to Paolo Malinconico for assistance in the analysis of flow data.

References

Adriaansen, H. J., Osman, C., Van Dongen, J. J. M., Wijdenes-De Bresser, J. H. F. M., Kappetijn-Van Tilborg, C. M. J. M., and Hooijkaas, H. (1990). Immunological marker analysis of mitogen-induced proliferating lymphocytes using BrdU incorporation or screening of metaphase. Staphylococcal protein A is a potent mitogen for CD4$^+$ lymphocytes. *Scand. J. Immunol.* **32,** 687–694.

Agematsu, K., Nagumo, H., Shinozaki, K., Hokibara, S., Yasui, K., Terada, K., Kawamura, N., Toba, T., Nonoyama, S., Ochs, H. D., and Komiyama, A. (1998). Absence of IgD-CD27($^+$) memory B cell population in X-linked hyper-IgM syndrome. *J. Clin. Invest.* **102,** 853–860.

Ahmed, S. A., Gogal, R. M., and Walsh, J. E. (1994). A new rapid and simple non-radioactive assay to monitor and determine the proliferation of lymphocytes: An alternative to [^{3}H]thymidine incorporation assay. *J. Immunol. Methods* **170,** 211–224.

Aiuti, A., Forte, P., Simeoni, L., Lino, M., Pozzi, L., Fattorossi, A., Giacomini, P., Ginelli, E., Beretta, A., Siccardi, A., and Fantoni, A. (1995). Membrane expression of HLA-Cw4 free chains in activated T cells of transgenic mice. *Immunogenetics* **42,** 368–375.

Akbar, A. N., Salmon, M., and Janossy, G. (1991). The synergy between naive and memory T cells during activation. *Immunol. Today* **12,** 184–188.

Akbar, A. N., Borthwick, N., Salmon, M., Gombert, W., Bofill, M., Shamsadeen, N., Pilling, D., Pett, S., Grundy, J. E., and Janossy, G. (1993). The significance of low bcl-2 expression by CD45RO T cells in normal individuals and patients with acute viral infections. The role of apoptosis in T cell memory. *J. Exp. Med.* **178,** 427–438.

Allsopp, C. E. M., Nicholls, S. J., and Langhorne, J. (1998). A flow cytometric method to assess antigen-specific proliferative responses of different subpopulations of fresh and cryopreserved human peripheral blood mononuclear cells. *J. Immunol. Methods* **214,** 175–186.

Annunziato, F., Manetti, R., Cosmi, L., Galli, G., Hesser, C. H., Romagnani, S., and Maggi, E. (1997). Opposite role for interleukin-4 and interferon-γ on CD30 and lymphocyte activation gene-3 (LAG-3) expression by activated naive T cells. *Eur. J. Immunol.* **27,** 2239–2244.

Aubin, J. E. (1979). Autofluorescence of viable cultured mammalian cells. *J. Histochem. Cytochem.* **27,** 36–43.

Baldinelli, L., Biselli, R., and Fattorossi, A. (1992). Influence of ST 789 on murine thymocytes: A flow cytometry study of thymocyte subset distribution and of intracellular free Ca^{++} increase upon activation. *Thymus* **19,** S53–S61.

Banchereau, J., and Rousset, F. (1992). Human B lymphocytes: Phenotype, proliferation, and differentiation. *Adv. Immunol.* **52,** 125–262.

Bartoleschi, C., Nisini, R., Biselli, R., Casciaro, A., and Fattorossi, A. (1994). Flow cytometric analysis of cell function: Cytosolic Ca^{++} modulation in human polymorphonuclear leukocytes and T lymphocytes. *Eur. J. Histochem.* **38**(Suppl. 1), 45–72.

Beik, A. I., Morris, A. G., Higgins, R. M., and Lam, F. T. (1998). Serial flow cytometric analysis of T-cell surface markers can be useful in differential diagnosis of renal allograft dysfunction. *Clin. Transplant.* **12,** 24–29.

Beverley, P. C. L. (1987). Activation antigens: New and previously defined clusters. *In* "Leucocyte Typing III" (A. J. McMichael, ed.), pp. 516–528. Oxford Univ. Press, Oxford.

Biselli, R., Matricardi, P. M., D'Amelio, R., and Fattorossi, A. (1992). Multiparametric flow cytometric analysis of the kinetics of surface molecule expression after polyclonal activation of human peripheral blood T lymphocytes. *Scand. J. Immunol.* **35,** 439–447.

Borthwick, N. J., Bofill, M., Gombert, W. M., Akbar, A. N., Medina, E., Sagawa, K., Lipman, M. C., Johnson, M. A., and Janossy, G. (1994). Lymphocyte activation in HIV-1 infection. II. Functional defects of CD28$^-$ T cells. *AIDS* **8,** 431–441.

Bouscarat, F., Levacher-Clergeot, M., Dazza, M. C., Strauss, K. W., Girard, P. M., Ruggeri, C., and Sinet, M. (1996). Correlation of CD8 lymphocyte activation with cellular viremia and plasma HIV RNA levels in asymptomatic patients infected by human immunodeficiency virus type 1. *AIDS Res. Hum. Retroviruses* **12,** 17–24.

Braunstein, J. D., Melamed, M. R., Sharpless, T. K., Hansen, J. A., Dupont, B., and Good, R. A. (1975). Quantitation of transformed lymphocytes by flow cytofluorometry. I. Phytohaemagglutinin response. *Clin. Immunol.* **4,** 209–215.

Brugnoni, D., Airo, P., Graf, D., Marconi, M., Molinari, C., Braga, D., Malacarne, F., Soresina, A., Ugazio, A. G., Cattaneo, R., Kroczek, R. A., and Notarangelo, L. D. (1996). Ontogeny of CD40L [corrected] expression by activated peripheral blood lymphocytes in humans. *Immunol. Lett.* **49,** 27–30.

Buhling, F., Junker, U., Reinhold, D., Neubert, K., Jager, L., and Ansorge, S. (1995). Functional role of CD26 on human B lymphocytes. *Immunol. Lett.* **45,** 47–51.

Bussa, S., Rumi, C., Leone, G., and Bizzi, B. (1993). Evaluation of a new whole blood cytometric lymphocyte transformation test for immunological screening. *J. Clin. Lab. Immunol.* **40,** 39–46.

Callebaut, C., Krust, B., Jacotot, E., and Hovanessian, A. G. (1993). T-cell activation antigen, CD26, as a cofactor for entry of HIV in $CD4^+$ cells. *Science* **262,** 2045–2050.

Carbonari, M., Cibati, M., Cherchi, M., Sbarigia, D., Pesce, A. M., Dell'Anna, L., Modica, A., and Fiorilli, M. (1994). Detection and characterization of apoptotic peripheral blood lymphocytes in human immunodeficiency virus infection and cancer chemotherapy by a novel flow immunocytometric method. *Blood* **5,** 1268–1277.

Castan, J., Tenner-Racz, K., Racz, P., Fleischer, B., and Broker, B. M. (1997). Accumulation of CTLA-4 expressing T lymphocytes in the germinal centres of human lymphoid tissues. *Immunology* **90,** 265–271.

Chan, E. Y., Lau, C. S., and Zola, H. (1996). Expression of IL-2R, IL-4R, IL-6R on peripheral blood lymphocytes in systemic lupus erythematosus and correlation with disease activity: A prospective study. *J. Clin. Pathol.* **49,** 660–663.

Chen, J. C., Davis, B. D., Leone, M. A., and Leong, L. C. (1997). Gamma radiation induces CD69 expression on lymphocytes. *Cytometry* **30,** 304–312.

Clement, L. T. (1992). Isoforms of the CD45 common leukocyte antigen family—markers for human T-cell differentiation. *J. Clin. Immunol.* **12,** 1–10.

Cory, A. H., Owen, T. C., Barltrop, J. A., and Cory, J. G. (1991). Use of an aqueous soluble tetrazolium/formazan assay for cell growth in culture. *Cancer Commun.* **3,** 207–212.

Cosman, D. (1994). A family of ligands for the TNF receptor superfamily. *Stem Cells* **12,** 440–455.

Craston, R., Koh, M., Mc-Dermott, A., Ray, N., Prentice, H. G., and Lowdell, M. W. (1998). Temporal dynamics of CD69 expression on lymphoid cells. *J. Immunol. Methods* **209,** 37–45.

Damle, N. K., and Doyle, L. V. (1989). Stimulation via the CD3 and CD28 molecules induces responsiveness to IL-4 in $CD4^+CD29^+CD45R^-$ memory T lymphocytes. *J. Immunol.* **143,** 1761–1767.

Darzynkiewicz, Z., Staiano-Coco, L., and Melamed, M. R. (1981). Increased mitochondrial uptake of rhodamine 123 during lymphocyte stimulation. *Proc. Natl. Acad. Sci. U.S.A.* **78,** 2383–2387.

Darzynkiewicz, Z., Juan, G., Li, X., Gorczyca, W., Murakami, T., and Traganos, F. (1997). Cytometry in cell necrobiology: Analysis of apoptosis and accidental cell death (necrosis). *Cytometry* **27,** 1–20.

Dorfman, D. M., Schultze, J. L., Shahsafei, A., Michalak, S., Gribben, J. G., Freeman, G. J., Pinkus, G. S., and Nadler L. M. (1997). In vivo expression of B7-1 and B7-2 by follicular lymphoma cells can prevent induction of T-cell anergy but is insufficient to induce significant T-cell proliferation. *Blood* **90,** 4297–4306.

Ercoli, A., Scambia, G., Fattorossi, A., Raspaglio, G., Battaglia, A., Cicchillitti, L., Malorni, W., Rainaldi, G., Benedetti Panici, P., and Mancuso, S. (1998). Comparative study on the induction of cytostasis and apoptosis by ICI 182,780 and Tamoxifen in an estrogen receptor-negative ovarian cancer cell line. *Int. J. Cancer* **76,** 47–54.

Fattorossi, A., Matricardi, P. M., Pizzolo, J. C., Le Moli, S., Antonelli, G., and D'Amelio, R. (1991). Lack of specificity in the mechanisms involved in the enhancement of the Concanavalin A driven human T lymphocyte stimulation by β-endorphin: Studies on activation marker expression, cell cycle and interleukin release. *J. Biol. Reg. Homeost. Agents* **5,** 91–97.

Fattorossi, A., Biselli, R., Casciaro, A., Trinchieri, V., and De Simone, C. (1992). Oral polyvalent vaccine (Buccalin Berna™) administration activates selected T-cell subsets and regulates the expression of polymorphonuclear leukocyte membrane molecules. *J. Clin. Lab. Immunol.* **38,** 95–101.

Fattorossi, A., Pierelli, L., Scambia, G., Ciarli, M., Bonanno, G., Battaglia, A., Menichella, G., Benedetti-Panici, P., Bizzi, B., and Mancuso, S. (1996). A multiparametric flow cytometric approach to hematopoietic differentiation *in vitro. Cytometry* **S8,** 79.

Ferlini, C., Biselli, R., Nisini, R., and Fattorossi, A. (1995). Rhodamine 123: A useful probe for monitoring T cell activation. *Cytometry* **21,** 284–293.

Ferlini, C., Di Cesare, S., Rainaldi, G., Malorni, W., Samoggia, P., Biselli, R., and Fattorossi, A. (1996a). Flow cytometric analysis of the early phases of apoptosis by cellular and nuclear techniques. *Cytometry* **24,** 106–115.

Ferlini, C., Nisini, R., and Fattorossi, A. (1996b). Membrane antigen modulation in superantigen driven human T cell activation. *Cytometry* **S8,** 97.

Gansbacher, B., and Zier, K. S. (1988). Regulation of HLA-DR, DP, and DQ expression in activated T cells. *Cell. Immunol.* **117,** 22–34.

Gibbons, D. C., and Evans, T. G. (1996). CD69 expression after antigenic stimulation. *Cytometry* **23,** 260–261.

Gougeon, M. L., Lecoeur, H., Dulioust, A., Enouf, M. G., Crouvoiser, M., Goujard, C., Debord, T., and Montagnier, L. (1996). Programmed cell death in peripheral lymphocytes from HIV-infected person: Increased susceptibility to apoptosis of CD4 and CD8 T cells correlates with lymphocyte activation and with disease progression. *J. Immunol.* **156,** 3509–3520.

Hallstrand, T. S., Ault, K. A., Bates, P. W., Mitchell, J., and Schoene, R. B. (1998). Peripheral blood manifestations of T(H)2 lymphocyte activation in stable atopic asthma and during exercise induced bronchospasm. *Ann. Allergy Asthma Immunol.* **80,** 424–432.

Hemler, M. E. (1990). VLA proteins in the integrin family: Structures, functions, and their role on leukocytes. *Annu. Rev. Immunol.* **8,** 365–400.

Hofmann, B., Nishanian, P., Fahey, J. L., Esmail, I., Jackson, A. L., Detels, R., and Cumberland, W. (1991). Serum increases and lymphoid cell surface losses of IL-2 receptor CD25 in HIV infection—distinctive parameters of HIV-induced change. *Clin. Immunol. Immunopathol.* **61,** 212–224.

Horan, P. K., and Slezak, S. E. (1989). Stable cell membrane labelling. *Nature* **430,** 167–170.

Infante, A. J., Infante, P. D., Jackson, C. E., Barohn, R. J., Tami, J., Itturiaga, E., Talib, S., Kraig, E., Clarkin, K. Z., and Krolick, K. A. (1996). Evidence against chronic antigen-specific T lymphocyte activation in myasthenia gravis. *J. Neurosci. Res.* **45,** 492–499.

Jackson, A. L., Matsumoto, H., Janzen, M., Maino, V., Blidy, A., and Shye, S. (1990). Restricted expression of p55 interleukin 2 receptor (CD25) on normal T cells. *Clin. Immunol. Immunopathol.* **54,** 126–133.

Jackson, D. G., and Bell, J. I. (1990). Isolation of a cDNA encoding the human CD38 (T10) molecule, a cell surface glycoprotein with an unusual discontinuous pattern of expression during lymphocyte differentiation. *J. Immunol.* **144,** 2811–2815.

Janossy, G., Borthwick, N., Lomnitzer, R., Medina, E., Berthel Squire, S., Phillips, A. N., Lipman, M., Johnson, M. A., Lee, C., and Bofill, M. (1993). Lymphocyte activation in HIV-1 infection. I. Predominant proliferative defects among CD45RO[+] cells of the CD4 and CD8 lineages. *AIDS* **7,** 613–624.

Jason, J., Gregg, L., Han, A., Hu, A., Inge, K. L., Eich, A., Tham, I., and Campbell, R. (1997). Immunoregulatory changes in Kawasaki disease. *Clin. Immunol. Immunopathol.* **84,** 296–306.

Johannisson, A. and Festin, R. (1995). Phenotype transition of CD4[+] cells from CD45RA to CD45RO is accompanied by cell activation and proliferation. *Cytometry* **19,** 343–352.

June, C. H., Bluestone, J. A., Nadler, L. M., and Thompson, C. B. (1994). The B7 and CD28 receptor families. *Immunol. Today* **15,** 321–331.

Kabelitz, D., Pohl, T., and Pechhold, K. (1993). Activation-induced cell death (apoptosis) of mature peripheral T lymphocytes. *Immunol. Today* **14,** 338–344.

Kestens, L., Vanham, G., Vereecken, C., Vandenbruaene, M., Vercauteren, G., Colebunders, R. L., and Gigase, P. L. (1994). Selective increase of activation antigens HLA-DR and CD38 and CD4$^+$CD45RO$^+$ T-lymphocytes during HIV-1 infection. *Clin. Exp. Immunol.* **95,** 436–441.

Kishimoto, T. K., Jutila, M. A., Berg, E. L., and Butcher, E. C. (1990). Identification of a human peripheral lymph node homing receptor: A rapidly down-regulated adhesion molecule. *Proc. Natl. Acad. Sci U.S.A.* **87,** 2244–2248.

Kohler, C., Kolopp-Sarda, M. N., De March-Kennel, A., Barbaud, A., Béné, M. C., and Faure, G. C. (1997). Sequential assessment of cell cycle S phase in flow cytometry: A nonisotopic method to measure lymphocyte activation *in vitro*. *Anal. Cell. Pathol.* **14,** 51–59.

Krowka J. F., Cuevas, B., Maron, D. C., Steimer, K. S., Ascher, M. S., and Sheppard, H. W. (1996). Expression of CD69 after *in vitro* stimulation: A rapid method for quantitating impaired lymphocyte response in HIV-infected individuals. *J. Acquir. Immune Defic. Syndr.* **11,** 95–104.

Lampariello, F., and Aiello, A. (1998). Complete mathematical modeling method for the analysis of immunofluorescence distributions composed of negative and weakly positive cells. *Cytometry* **32,** 241–254.

LaSalle, J. M., Tolentino, P. J., Freeman, G. J., Nadler, L. M., and Hafler, D. A. (1992). Early signaling defects in human T cells anergized by T cell presentation of autoantigen. *J. Exp. Med.* **176,** 177–186.

Lees, O., Ramzaoui, S., Gilbert, D., Borsa, F., Humbert, G., Leblanc, D., Lagarde, M., and Tron, F. (1993). The impaired *in vitro* production of interleukin-2 in HIV infection is negatively correlated to the number of circulating CD4$^+$DR$^+$ T-cells and is reversed by allowing T-cells to rest in culture—Arguments for *in vivo* CD4$^+$ T-cell activation. *Clin. Immunol. Immunopathol.* **67,** 185–191.

LeMoli, S., Fattorossi, A., Corradin, G., and D'Amelio, R. (1989). Class II MHC antigens on stimulated L3T4$^+$ murine T cell clones. *Basic Appl. Histochem.* **33,** 149.

Lenkei, R., and Andersson, B. (1995). Determination of antibody binding capacity of lymphocyte membrane antigens by flow cytometry. *J. Immunol. Methods* **183,** 267–277.

Levacher, M., Hulstaert, F., Tallet, S., Ullery, S., Pocidalo, J. J., and Bach, B. A. (1992). The significance of activation markers on lymphocytes-CD8 in human immunodeficiency syndrome—Staging and prognostic value. *Clin. Exp. Immunol.* **90,** 376–382.

Liu, Z., Cumberland, W. G., Hultin, L. E. Prince, H. E., Detels, R., and Giorgi, J. V. (1996). Elevated CD38$^+$ T cells is a stronger marker for the risk of chronic HIV disease progression to AIDS and death in the Multicenter AIDS Cohort Study than CD4$^+$ cell count, soluble immune activation markers, or combinations of HLA-DR and CD38 expression. *J. Acquir. Immune Defic. Syndr. Hum. Retrovirol.* **16,** 83–92.

Lyons, A. B., and Parish, C. R. (1994). Determination of lymphocyte division by flow cytometry. *J. Immunol. Methods* **171,** 131–137.

Maino, V. C., and Picker, L. J. (1998). Identification of functional subset by flow cytometry: Intracellular detection of cytokine expression. *Cytometry* **34,** 207–215.

Maino, V. C., Suni, M. A., and Ruitenberg, J. J. (1995). Rapid flow cytometric method for measuring lymphocyte subset activation. *Cytometry* **20,** 127–133.

Majori, M., Piccoli, M. L., Melej, R., Pileggi, V., and Pesci, A. (1997). Lymphocyte activation markers in peripheral blood before and after natural exposure to allergen in asthmatic patients. *Respiration* **64,** 45–49.

Makgoba, M. W., Sanders, M. E., and Shaw, S. (1989). The CD2-LFA-3 and LFA-1-ICAM patways: Relevance to T cell recognition. *Immunol. Today* **10,** 417–422.

Mardiney III, M., Brown, M. R., and Fleisher, T. A. (1996). Measurement of T-cell CD69 expression: A rapid and efficient means to assess mitogen- or antigen-induced proliferative capacity in normals. *Cytometry* **26,** 305–310.

Martin, S. J. (1993). Protein or RNA synthesis inhibition induces apoptosis of mature human CD4$^+$ T cell blasts. Immunol. *Lett.* **35,** 125–134.

Medina, E., Borthwick, N., Johnson, M. A., Milleri, S., and Bofill, M. (1994). Flow cytometric of the stimulatory response of T cell subsets from normal and HIV-1$^+$ individuals to various mitogenic stimuli *in vitro*. *Clin. Exp. Immunol.* **97,** 266–272.

Muller-Suur, C., Larsson, K., Malmberg, P., and Larsson, P. H. (1997). Increased number of activated lymphocytes in human lung following swine dust inhalation. *Eur. Respir. J.* **10,** 376–380.

Nakamura, S., Sung, S. S. J., Bjorndahl, J. M., and Fu, S. M. (1989). Human T-cell activation IV. T-cell activation and proliferation via the early activation antigen. *J. Exp. Med.* **169,** 677–689.

Neckers, L. M., and Cossman, J. (1983). Transferrin receptor induction in mitogen-stimulated human T lymphocytes is required for DNA synthesis and cell division and is regulated by interleukin-2. *Proc. Natl. Acad. Sci. U.S.A.* **80,** 3494–3498.

Nisini, R., Matricardi, P. M., Fattorossi, A., Biselli, R., and D'Amelio, R. (1992). Presentation of superantigen by human T-cell clones: A model of T–T cell interaction. *Eur. J. Immunol.* **22,** 2033–2039.

Nisini, R., Fattorossi, A., Ferlini, C., and D'Amelio, R. (1996). One cause for the apparent inability of human T cell clones to function as professional super-antigen presenting cells is autoactivation. *Eur. J. Immunol.* **26,** 797–803.

Noelle, R. J., Roy, M., Shepard, D. M., Stamenkovic, I., Ledbetter, J. A., and Aruffo, A. (1992). A 39-kDa protein on activated helper T cells binds CD40 and transduces the signal for cognate activation of B cells. *Proc. Natl. Acad. Sci U.S.A.* **89,** 6550–6554.

Overton, W. R. (1988). Modified histogram subtraction technique for analysis of flow cytometry data. *Cytometry* **9,** 619–626.

Parish, C. R., and McKenzie, I. F. C. (1977). Mitogens and T-independent antigens stimulate T lymphocytes to secrete Ia antigens. *Cell. Immunol.* **33,** 134–144.

Parks, D. R., Herzenberg, L. A., and Herzenberg, L. A. (1989). Flow cytometry and fluorescence-activated cell sorting. *In* "Fundamental Immunology" (W. E. Paul, ed.), pp 781–802. Raven, New York.

Perfetto, S. P., Hickey, T. E., Blair, P. J., Maino, V. C., Wagner, K. F., Zhou, S., Mayers, D. L., St. Louis, D., June, C. H., and Siegel, J. N. (1997). Measurement of CD69 induction in the assessment of immune function in asymptomatic HIV-infected individuals. *Cytometry* **30,** 1–9.

Pierelli, L., Scambia, G., Fattorossi, A., Bonanno, G., Battaglia, A., Perillo, A., Menichella, G., Panici, P. B., Leone, G., and Mancuso, S. (1998a). In vitro effect of amifostine on haematopoietic progenitors exposed to carboplatin and non-alkylating antineoplastic drugs: Haemoprotection acts as a drug-specific progenitor rescue. *Br. J. Cancer* **78,** 1024–1029.

Pierelli, L., Scambia, G., Fattorossi, A., Bonanno, G., Battaglia, A., Rumi, C., Marone, M., Mozzetti, S., Rutella, S., Menichella, G., Romeo, V., Mancuso, S., and Leone, G. (1998b). Functional, phenotypic and molecular characterization of cytokine low-responding circulating CD34$^+$ haemopoietic progenitors. *Br. J. Haematol.* **102,** 1139–1150.

Plaeger-Marshall, S., Hultin, P., Bertolli, J., O'Rourke, S., Kobayashi, R., Kobayashi, A. L., Giorgi, J. V., Bryson, Y., and Stiehm, E. R. (1993). Activation and differentiation antigens on T cells of healthy at-risk and HIV-infected children. *J. Acquir. Immune Defic. Syndr.* **6,** 984–993.

Redelman, D. (1987). Simultaneous increased expression of E-rosette receptor (CD2, T11) and T cell growth factor receptor on human T lymphocyte during activation. *Cytometry* **8,** 170–183.

Reinke, P., Fietze, E., Docke, W. D., Kern, F., Ewert, R., and Volk, H. D. (1994). Late acute rejection in long-term renal allograft recipients. Diagnostic and predictive value of circulating activated T cells. *Transplantation* **58,** 35–41.

Reyburn, H., Mandelboim, O., Valés-Goméz, M., Sheu, E. G., Pazmany, L., and Strominger, J. L. (1997). Human NK cells, their ligands, receptors and functions. *Immunol. Rev.* **155,** 119–125.

Reyes, E., Carballido, J., Prieto, A., Molto, L., Manzano, L., and Alvarez-Mon, M. (1997). T lymphocytes infiltrating the bladder wall of patients with carcinoma of urinary bladder are *in vivo* activated. *Eur. J. Urol.* **31,** 472–477.

Roehm, N. W., Rodgers, G. H., Hatfield, S. M., and Glasebrook, A. L. (1991). An improved colorimetric assay for cell proliferation and viability utilizing the tetrazolium salt XTT. *J. Immunol. Methods* **142,** 257–265.

Roos, A., Claessen, N., Weening, J. J., and Aten, J. (1996). Enhanced T lymphocyte expression of LFA-1, ICAM-1, and the TNF receptor family member OX40 in HgCl2 induced systemic autoimmunity. *Scand. J. Immunol.* **43**, 507–518.

Roy, M., Waldschmidt, T., Aruffo, A., Ledbetter, J. A., and Noelle, R. J. (1993). The regulation of the expression of gp39, the CD40 ligand, on normal and cloned CD4$^+$ T cells. *J. Immunol.* **151**, 2497–2510.

Rutella, S., Rumi, C., Lucia, M. B., Sica, S., Cauda, R., and Leone, G. (1998). Serum of healthy donors receiving granulocyte colony-stimulating factor induces T cell unresponsiveness. *Exp. Hematol.* **26**, 1024–1033.

Salvadori, S., Pizzimenti, A., Cohen, S., and Zier, K. S. (1991). The control of class II expression on T cells is independent of the regulation of Tac and the induction of proliferation. *Clin. Exp. Immunol.* **86**, 544–549.

Sandilands, G. P., MacPherson, S. A., Burnett, E. R., Russell, A. J., Downie, I., and MacSween, R. N. (1997). Differential expression of CD32 isoforms following alloactivation of human T cells. *Immunology* **91**, 204–211.

Schillaci, R., Brocardo, M. G., Galeano, A., and Roldan, A. (1998). Downregulation of insulin-like growth factor-1 receptor (IGF-1R) expression in human T lymphocyte activation. *Cell. Immunol.* **183**, 157–161.

Shapiro, H. M. (1985). "Practical Flow Cytometry," 1st Ed. Alan R. Liss, New York.

Simms, P. E., and Ellis, T. M. (1996). Utility of flow cytometric detection of CD69 expression as a rapid method for determining poly- and oligoclonal lymphocyte activation. *Clin. Diagn. Lab. Immunol.* **3**, 301–304.

Sloan-Lancaster, J., Evavold, B. D., and Allen, P. M. (1994). Th2 cell clonal anergy as a consequence of partial activation. *J. Exp. Med.* **190**, 1195–1202.

Spronk, P. E., Ter Borg, E. J., Huitema, M. G., Limburg, P. C., and Kallenberg, C. G. M. (1994). Changes in levels of soluble T-cell activation markers, sIL-2R, sCD4 and sCD8, in relation to disease exacerbations in patients with systemic lupus erythematosus. A prospective study. *Ann. Rheum. Dis.* **53**, 235–239.

Suda, T., Tanaka, M., Miwa, K., and Nagata, S. (1996). Apoptosis of mouse naive T cell induced by recombinant soluble Fas ligand and ativation-induced resistance to Fas ligand. *J. Immunol.* **157**, 3918–3924.

Swat, W., Ignatowicz, L., and Kisielow, P. (1991). Detection of apoptosis of immature CD4$^+$8$^+$ thymocytes by flow cytometry. *J. Immunol. Methods* **137**, 79–87.

Sztein, M. B., Cuna, W. R., and Kierszenbaum, F. (1990). *Trypanosoma cruzi* inhibits the expression of CD3, CD4, CD8, and IL-2R by mitogen-activated helper and cytotoxic human lymphocytes. *J. Immunol.* **144**, 3558–3562.

Testi, R., Phillips, J. H., and Lanier, L. L. (1989a). Leu 23 induction as an early step of functional CD3/T cell antigen receptor triggering. Requirement for receptor cross-linking, prolonged elevation of intracellular (Ca^{++}) and stimulation of protein kinase C. *J. Immunol.* **142**, 1854–1860.

Testi, R., Phillips, J. H., and Lanier, L. L. (1989b). T cell activation via Leu-23 (CD69). *J. Immunol.* **143**, 1123–1128.

Tomkinson, B. E., Wagner, D. K., Nelson, D. L., and Sullivan, J. L. (1987). Activated lymphocytes during acute Epstein-Barr virus infection. *J. Immunol.* **139**, 3802–3807.

Triebel, F., De Roquefeuil, S., Blanc, C., Charron, D. J., and Debre, P. (1986). Expression of MHC class II and Tac antigens on IL-2 activated human T-cell clones that can stimulate in MLR, AMLR, PLT, and can present antigen. *Hum. Immunol.* **15**, 302–308.

Vallé, A., Aubry, J. P. Durand, I., and Banchereau, J. (1990). MAb 104, a new monoclonal antibody, recognizes the B7 antigen that is expressed on activated B cells and HTLV-1-trasformed T cells. *Immunology* **69**, 531–535.

Veis, D. J., Sentman, C. L., Bach, E. A., and Korsmeyer, S. J. (1993). Expression of the bcl-2 protein in murine and human thymocytes and in peripheral T lymphocytes. *J. Immunol.* **5**, 2546–2554.

Viora, M., Di Genova, G., Rivabene, R., Malorni, W., and Fattorossi, A. (1997). Interference with cell cycle progression and induction of apoptosis by dideoxinucleoside analogs. *Int. J. Immunopharmacol.* **19**, 311–321.

Wang, P. L., O'Farrell, S., Clayberger, C., and Krensky, A. M. (1992). Identification and molecular cloning of tactile. A novel human T cell activation antigen that is a member of the Ig gene superfamily. *J. Immunol.* **148,** 2600–2608.

Weir, M. R., Peppler, R., Gomolka, D., and Handwerger, B. S. (1993). Calcium channel blockers inhibit cellular uptake of thymidine, uridine and leucine: The incorporation of these molecules into DNA, RNA and protein in the presence of calcium channel blockers is not a valid measure of lymphocyte activation. *Immunopharmacology* **25,** 75–82.

Wirt, D. P., Adkins, L. T., Palkowetz, K. H., Schmalstieg, F. C., and Goldman, A. S. (1992). *Cytometry* **13,** 282–290.

Worku, S., Bjorkman, A., Troye-Blomberg, M., Jemaneh, L., Farnet, A., and Christensson, B. (1997). Lymphocyte activation and subset redistribution in the peripheral blood in acute malaria illness: Distinct $\gamma\delta^+$ T cell patterns in *P. falciparum* and *P. vivax* infections. *Clin. Exp. Immunol.* **108,** 34–41.

Yoshino, T., Kondo, E., Cao, L. Takahashi, K., Nomura, S., and Akagi, T. (1994). Inverse expression of Bcl-2 protein and FAS antigen in lymphoblasts in peripheral lymph nodes and activated peripheral blood T and B lymphocytes. *Blood* **83,** 1856–1861.

Zhiyuan, L., Hultin, E. H., Cumberland, W. G., Hultin, P., Schmid, I., Matud, J. L., Detels, R., and Giorgi, J. V. (1996). Elevated relative fluorescence intensity of CD38 antigen expression on CD8$^+$ T cells is a marker of poor prognosis in HIV infection. Results of 6 years of follow-up. *Cytometry* **26,** 1–7.

Ziegler-Heitbrock, H. W. L., and Ulevitch, R. J. (1993). CD14: Cell surface receptor and differentiation marker. *Immunol. Today* **14,** 121–124.

Ziegler-Heitbrock, H. W., Stachel, D., Schlunk, T., Gurtler, L., Scharamm, W., Froschi, M., Bogner, J. F., and Riethmuller, G. (1988). Class-II (DR) antigen expression on CD8$^+$ lymphocyte subsets in acquired immune deficiency syndrome (AIDS). *J. Clin. Immunol.* **8,** 473–478.

Zola, H., Mantzioris, B. X., Webster, J., and Kette, F. E. (1989). Circulating human T and B lymphocytes express the p55 interleukin-2 receptor molecule (TAC, CD25). *Immunol. Cell. Biol.* **67,** 233–237.

PART V

Cell Death/Apoptosis

CHAPTER 21

Analysis of Mitochondria during Cell Death

Andrea Cossarizza* and Stefano Salvioli[†]

*Chair of Immunology
Department of Biomedical Sciences
University of Modena and Reggio
Emilia School of Medicine
41100 Modena, Italy

[†]Department of Experimental Pathology
University of Bologna
40126 Bologna, Italy

I. Introduction
II. Scientific Background
III. Apoptosis and Mitochondria
IV. Method
V. Results
VI. Pitfalls and Misinterpretation of the Data
VII. Comparison with Other Methods
VIII. Reviews of the Applications
IX. Biological and Biomedical Information
X. Future Directions
References

I. Introduction

Cell death can be schematically divided into two different types, necrosis or apoptosis. It is now well known that necrosis is the consequence of a degenerative process or the passive consequence of gross injury to the cell, whereas apoptosis represents the end point of a process that actively involves several cellular structures and organelles, including mitochondria. Classic studies on apoptosis have clarified that this type of death is first decided, then driven by the cell in a finely tuned fashion. Apoptosis is a phenomenon of crucial importance in all multicellular, living organisms, being involved in several physiological and pathological phenomena, including, among others, embryogenesis, differentiation, development of the immune system, and control of cell growth, carcinogenesis, and viral production (Wyllie et al., 1980). Apoptosis can be triggered by a wide variety of stimuli, such as deprivation of growth factors, γ-irradiation, oxygen free radical (OFR) production, receptor–ligand interaction, and inhibition of protein kinases, among others. Each stimulus activates its own "private" pathway, which is thought to converge into one or more central mechanism(s), the "common pathway," leading to the classic morphological and biochemical changes of the cell dying by apoptosis (Golstein, 1997; Hetts, 1998).

The assumption that apoptosis is an active process led to the idea that energy supply, and thus mitochondrial functionality, must be well preserved. This concept has been questioned by much experimental data showing that mitochondria can be involved in the induction of apoptosis in several manners. Indeed, they can produce OFR (Hockenbery et al., 1993; Buttke and Sandstrom, 1994), alter the redox state of the cell (Marchetti et al., 1997), or cause cycling of Ca^{2+} ions (Richter et al., 1995). It has been hypothesized that the apoptotic elimination of cells with high amounts of OFR, which are mostly produced in mitochondria, can be considered a defense mechanism that favors the homeostasis of the entire organism (Skulachev, 1996). Nevertheless, the complex process of apoptosis is supposed to have a high energy demand, as the coordinated activity of several enzymes is required. Indeed, some apoptotic pathways need RNA and protein synthesis (Cohen and Duke, 1984; Wyllie et al., 1984; McConkey et al., 1989), and the active formation of blebs at the plasma membrane level is a common feature of late apoptosis (Straface et al., 1995). Most intracellular energy derives from the hydrolysis of ATP, which is mainly produced into mitochondria. Accordingly, it is reasonable to hypothesize that, at least in early apoptotic phases, cells have to maintain a good driving force for ATP synthesis, and, consequently, a high mitochondrial membrane potential ($\Delta\Psi$). On the other hand, it has been demonstrated that the extrusion of some molecules usually sequestrated into mitochondria, such as cytochrome c (Liu et al., 1996), or the so called apoptosis inducing factor, AIF (Susin et al., 1996), can activate the caspase cascade and nuclear apoptosis (see later).

Thus, the analysis of $\Delta\Psi$, which is a reliable mirror of the capacity to synthesize ATP and of mitochondrial membrane impermeability, can be an important parameter to be considered when studying apoptosis.

II. Scientific Background

In the mitochondrial respiratory chain, the energy released during oxidation reactions is stored as an electrochemical gradient generated by the active extrusion of protons from the mitochondrial matrix to the cytosol. Such a gradient consists of two components: a transmembrane electrical potential, negative inside, of about 180–200 mV, and a proton gradient of about 1 pH unit. This energy drives the synthesis of ATP. In order to analyze the mechanisms that regulate changes in membrane potential of organelles with a negative interior, several membrane-permeable, lipophilic cations have been used, due to their ability to accumulate in such organelles and/or liposomes. Such probes include those that exhibit optical and fluorescence activity after accumulation into energized systems [3,3'-dihexyloxadicarbocyanine iodide, nonyl acridine orange (NAO), safranine O, rhodamine-123 (Rh123), etc.], radiolabeled probes ([^{3}H]methyltriphenylphosphonium, etc.), and unlabeled probes used with specific electrodes [tetraphenylphosphonium ion (TPP$^+$), etc.]. These systems have several possible disadvantages:

1. time required to achieve equilibrium distribution of a mitochondrial membrane probe;
2. degree of passive (unspecific) binding of probes to a membrane component, such as in the case of NAO, which detects mitochondrial mass as it binds to cardiolipin (Maftah *et al.*, 1989), or Rh123, which has several energy-independent binding sites (Lopez-Mediavilla *et al.*, 1989);
3. toxic effects of probes on mitochondrial functional integrity;
4. sampling procedures;
5. interference from light scattering changes and from absorption changes of mitochondrial components;
6. requirement of large amounts of biological materials.

The TPP electrode affords an easy and precise tool to measure $\Delta\Psi$ not only due to the low interferences between bound TPP$^+$ and the membrane, but also because of the lack of responses of the electrode to species other than TPP$^+$. However, this method requires discrete amounts of biological samples, and, in contrast to uptake by isolated mitochondria, the uptake of this lipophilic cation by several intact mammalian cells is a slow process.

By using the lipophilic cation 5,5',6,6'-tetrachloro-1,1',3,3'-tetraethylbenzimidazolcarbocyanine iodide (JC-1), we developed a cytofluorimetric technique to detect variations in $\Delta\Psi$ at the single cell level (Cossarizza *et al.*, 1993), which has been validated at the single mitochondrion level (Cossarizza *et al.*, 1996). This probe is allowing us, as well as many colleagues, to analyze $\Delta\Psi$ by using not only flow cytometry, but also confocal microscopy [see, e.g., the marvelous images by Drs. I. Nicoletti and R. Mannucci with this probe used on NIH-3T3 cells (Microscopy CD-ROM Vol. 1, Purdue University, 1998)]. The use of JC-1 is more advantageous over the use of rhodamines and other carbocyanines that are capable of entering selectively into mitochondria, since JC-1 changes reversibly its color from green to orange as membrane potential increases. This property is due to the reversible formation of JC-1 aggregates on membrane polarization that causes shifts in emitted light from 530 nm (i.e., emission of JC-1 monomeric form) to 590 nm (i.e., emission of J aggregate), when excited at 490 nm (Hada

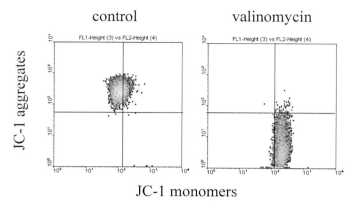

Fig. 1 Valinomycin-induced depolarization of mitochondria in U937 cells as revealed by staining with JC-1. Note that after treatment with 100 nM valinomycin cells go from the middle of the quadrant (high FL1, high FL2) to the lower right corner (higher FL1, lower FL2).

et al., 1977; Reers et al., 1991). Thus, the color of the dye changes reversibly from green to orange as the mitochondrial membrane becomes more polarized (Smiley et al., 1991). Both colors can be detected using the filters commonly mounted in flow cytometers or confocal microscopes, so that green emission can be analyzed in fluorescence channel 1 (FL1) and greenish orange emission in channel 2 (FL2). Results with JC-1 are mostly qualitative, considering the shift from green to orange fluorescence emission, but they can also give quantitative information, considering the pure fluorescence intensity, which can be detected in both FL1 and FL2 channels. Figure 1 shows the basic fluorescent changes of the probe after loading of cell line U937 with JC-1 and after treatment with the depolarizing agent valinomycin.

III. Apoptosis and Mitochondria

There is a growing interest in the role of mitochondria during apoptosis. A crucial role for this organelle has been postulated since the observation that the antiapoptotic protein Bcl-2 is located in the outer mitochondrial membrane (Monaghan et al., 1992; de Jong et al., 1994). Indeed, it seems that Bcl-2 family proteins are located preferentially at the contact sites where the inner and outer membranes come into close apposition (de Jong et al., 1994), near proteinaceous structures that form pores in the inner mitochondrial membrane (Reed et al., 1998). Such pores are referred to as mitochondrial "megachannels" or permeability transition pores (Zoratti and Szabò, 1995). The opening of these pores allows for the free distribution of solutes <1500 Da and is thought to cause a decrease in $\Delta\Psi$ (Zoratti and Szabò, 1995; Zamzami et al., 1996), swelling of the organelle, rupture of the outer membrane, and release of cytochrome c (Liu *et al.*, 1996) and/or AIF (Susin *et al.*, 1996) from the intermembrane space. Both of the latter compounds are able to induce nuclear apoptosis, that is, apoptotic changes in isolated nuclei, that can be inhibited by Bcl-2 protein activity (Susin *et al.*, 1996) and by protease inhibitors (Marchetti *et al.*, 1996). AIF appears to directly activate some caspases, resulting in the activation of a proteolytic cascade leading to nuclear apoptosis (Susin *et al.*, 1996), whereas cytochrome c activates caspases through its effects on Apaf-1 (apoptic protease activation factor 1), cytosolic factor (Zou *et al.*, 1997). On binding of cytochrome c, Apaf-1 becomes competent at binding and activating procaspase-9 (Li *et al.*, 1997). The antiapoptotic or proapoptotic effect of Bcl-2 family members may reside in the regulation of the release of such apoptogenic molecules into cytosol. However, the precise mechanism by which they promote or prevent cell death is far from clear.

On the whole, the data indicate that mitochondria are deeply involved in apoptosis, and it has been proposed that they could act as the central executioner of apoptosis (Susin *et al.*, 1997). It has also been suggested that the fall in $\Delta\Psi$ could be a common feature of the program. On this concern, contrasting results have been obtained by different groups, but they were likely due, at least in part, to methodological problems. In fact, in many studies, $\Delta\Psi$ has been measured by flow cytometry with fluorescent probes that are not fully satisfactory, such as $DiOC_6(3)$, which is not completely specific for mitochondria as it binds also to endoplasmic reticulum (ER) (Terasaki *et al.*, 1984) and is very sensitive to plasma membrane depolarization (Salvioli *et al.*, 1997; Fricker *et al.*, 1997), or such as Rh123, which has a low specificity for changes in $\Delta\Psi$ (Lopez-Mediavilla *et al.*, 1989; Salvioli *et al.*, 1997).

It has been demonstrated that the release of cytochrome c can occur without a simultaneous $\Delta\Psi$ drop (Kluck *et al.*, 1997; Yang *et al.*, 1997). It is not clear how a protein such as cytochrome c can transit from the intermembrane space into the cytosol without disruption of $\Delta\Psi$. It has been proposed that pores are created *ex novo* in the outer membrane, allowing some cytochrome c to escape, while leaving the permeability of the inner membrane intact (Van der Heiden *et al.*, 1997). Therefore, it has been postulated that the release of apoptogenic molecules, and not the collapse of $\Delta\Psi$, is the true link between mitochondria and apoptosis (Lemasters *et al.*, 1998). On the other hand, significant portions of the Bcl-2 protein are also integrated either into the membranes of the ER or into the nuclear envelope. ER-targeted Bcl-2 retained the ability to block apoptosis in some experimental models (Zhu *et al.*, 1996), thus suggesting that Bcl-2 could possess some nonmitochondrial mechanisms for promoting cell survival. In this regard, the finding that some fluorescent probes which accumulate mostly on ER membranes rapidly lose their fluorescence intensity during early phases of apoptosis may suggest that these organelles, and not only mitochondria, are precociously involved in cell death regulation. Finally, a more recent study provided evidence that none of the following events represent a point of no return in the apoptotic process: loss of $\Delta\Psi$, changes in mitochondrial permeability transition (MPT), or mitochondrial swelling (Minamikawa et al., 1999). Indeed, intact human osteosarcoma cells may undergo MPT and mitochondrial swelling in a fully reversible manner, without inducing cell death.

Our group has focused on the study of mitochondrial involvement in apoptosis for many years, and we could demonstrate that in many (but not in all) experimental models the drop of $\Delta\Psi$ is a late event in the process (see later), suggesting that $\Delta\Psi$ collapse should not always be considered an early and common feature of apoptotic cells. This has been confirmed by the fact that low doses of a $\Delta\Psi$ dissipator molecule such as valinomycin failed to induce apoptosis in the same experimental models (Salvioli et al., 2000a). It has been reported that valinomycin can induce apoptosis (Ojcius et al., 1991; Deckers et al., 1993; Furlong et al., 1998), but at much higher doses, suggesting that this is likely due to another pharmacological action(s), since the dose capable of inducing a complete and prolonged mitochondrial depolarization does not affect cell viability nor activation of caspases, at least in the first 8 hr of treatment. In any case, a $\Delta\Psi$ drop during cell death is a complex and heterogeneous phenomenon either at the single cell or single organelle level (Salvioli et al., 2000b).

IV. Method

The staining with JC-1 is, in principle, quite simple. Adjust a cell suspension to a density of 0.5–1 million cells/ml and incubate it in complete medium [RPMI 1640 with 10% fetal calf serum (FCS), but many other culture media work as well] for at least 10–15 min at room temperature in the dark with 5–10 μg/ml JC-1. Incubation can also be performed at 37°C, for 10 min. The temperature, time of incubation, and dose of the probe depend on the cell type. JC-1 has to be dissolved and stored according to the manufacturer's instruction, that is, in dimethyl sulfoxide (DMSO) or dimethyl formamide (DMF). We prefer the latter, as it does not freeze at -20°C, always avoiding freeze–thaw cycles. During the addition of JC-1, or immediately after, it is important to vortex the mixture vigorously, as the solubility of the probe is low in water; with vortexing, aggregates of the probe should not be present or should disappear. At the end of the incubation period cells have to be washed twice in phosphate-buffered saline (PBS), resuspended in PBS, and analyzed.

We suggest that one should always prepare a functional, "negative" control, treating a parallel sample with drugs able to collapse $\Delta\Psi$, such as the K^+ ionophore valinomycin (100 nM or more) or the proton translocator carbonyl cyanide p-(trifluoromethoxy)phenylhydrazone (FCCP, 250 nM). A dramatic change of the fluorescence distribution has to be occur (Fig. 1) and is required to set the quadrants. In other words, this is the best system to see where cells with depolarized mitochondria go, and to adjust the compensation either in control or treated samples.

Another useful control that should be performed is that concerning mitochondrial mass. This parameter can be measured by using different fluorescent molecules, such as NAO, which binds stoichiometrically to cardiolipin, (Septinus *et al.*, 1985; Maftah *et al.*, 1989), or the new probes MitoFluor Green or MitoTracker Green (from Molecular Probes, Eugene, OR). A debate actually exists on the sensitivity of such probes. Indeed, an interesting paper showed that $\Delta\Psi$ is able to affect the staining of mitochondria by these three probes, as their fluorescence can change after treatment with $\Delta\Psi$-altering drugs (Keij *et al.*, 2000). In our hands, NAO gave only marginal problems and has been used as a control for mitochondrial mass during different processes, including apoptosis (even if in the models we studied $\Delta\Psi$ collapse is a late event and NAO-stained cells still had a high $\Delta\Psi$) (Cossarizza *et al.*, 1994). In any case, NAO fluorescence can be detected either in FL1 or in FL3 (Petit *et al.*, 1994), and it can provide an idea not only of the mass of mitochondria present within a cell, but also of the cardiolipin distribution in the leaflets of the inner mitochondrial membrane (Garcia Fernandez *et al.*, 2001). In general, the mitochondrial mass has to constant; if a given cell population has not only a low $\Delta\Psi$, but also a reduced mass, the changes of $\Delta\Psi$ that are seen with a potential sensitive probe could be simply due to the disruption or loss of organelles caused, for example, by cell death of the necrotic type. This topic is clearly of crucial importance, and further studies are required to identify sensitive systems that allow a precise analysis of mitochondrial mass.

If the cells under investigation are adherent to a substrate or plastic, such as fibroblasts, and cannot be shaken, one can simply dissolve JC-1 at the final concentration in complete medium and then cover the cell layer with such a solution. After 10–15 min of incubation, rinse the plate with PBS or other adequate saline solution two to four times, each time leaving the saline solution for 5 min in order to equilibrate, and subsequently dilute, the fluorescent probe. This allows the analysis of JC-1 fluorescence in morphologically and metabolically intact, adherent cells with microscopy techniques (Fig. 2). If flow cytometry analysis is required, first detach cells from the substrate as gently as possible, and then stain them following the previously described procedure.

V. Results

After staining with JC-1, cells with high $\Delta\Psi$ display green-orange fluorescence, owing to J aggregates. On the other hand, in cells with low $\Delta\Psi$, JC-1 maintains its monomeric form and thus shows only green fluorescence. Because JC-1 can also bind to membranes other than mitochondrial ones, a non-$\Delta\Psi$-related green fluorescence is also always present in cells with well-polarized organelles. However, usually the green fluorescence signal of depolarized cells is higher than that of polarized cells, simply because of the presence of higher amounts of monomers.

When the sample contains an heterogeneic cell population, it is possible to discriminate the different subpopulations not only by physical parameters [i.e., forward scatter light (FSC) and side scatter light (SSC)], but also by the different

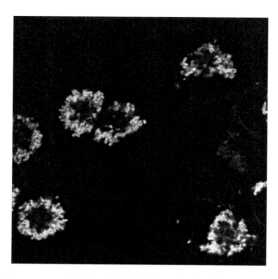

Fig. 2 Confocal microscopy analysis of U937 cells stained with JC-1. The red emission from JC-1 aggregates is shown. Note that this cell line is extremely rich in mitochondria, which appear as brilliant, linear spots. Courtesy of Dr. Jurek Dobrucki, Jagellonian University, Krakow, Poland.

JC-1 fluorescence patterns, owing to the variable content in intracellular membranes and mitochondria. This is the typical case of peripheral blood mononuclear cells (PBMC), which include lymphocytes and monocytes, the first being smaller and with less mitochondrial content. Accordingly, the fluorescence pattern of JC-1 in a sample containing PBMC shows two distinct peaks, one corresponding to lymphocytes, and the second, brighter in both FL1 and FL2, corresponding to monocytes. When the staining is performed at 37°C, it is possible to further discriminate another subpopulation, intermediate between lymphocytes and monocytes not only for the fluorescence intensity but also for its physical parameters, which likely corresponds to large lymphocytes (Fig. 3). Finally, when cells are bigger than PBMC (e.g., hepatocytes, keratinocytes, fibroblasts), the increased intensity of green fluorescence in samples with depolarized organelles sometimes is not well evident at all, because the ratio of the monomeric and aggregate forms of JC-1 is only slightly affected by the release of the aliquot of the aggregate dye contained in mitochondria.

JC-1 is a sensitive probe, but it has some disadvantages. The main disadvantage is that the emission spectrum of the aggregate form is quite large, and in the most common single-laser flow cytometers it tends to occupy not only the second, but also the third channel, that is, that of FL3. This inhibits de facto the use of other probes, unless there is the possibility of exciting them in the ultraviolet (UV) area.

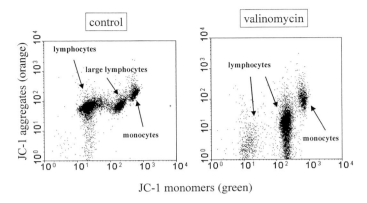

Fig. 3 Flow cytometry analysis of peripheral blood mononuclear cells stained with JC-1. Note that monoytes have a higher green emission (due to JC-1 monomers) than lymphocytes. Treating cells with 100 nM valinomycin for a few minutes results in a consistent depolarization of mitochondria present in both cell types, with the loss of JC-1 aggregates.

A family of new fluorescent probes, called MitoTracker, has been developed for the detection of $\Delta\Psi$. Such molecules are rosamine derivatives and present a thiol-reactive chloromethyl moiety, which allows them to bind irreversibly and selectively to protein components of mitochondria. They display a single fluorescence emission, thus overcoming the problem of the dual emission of JC-1 (Macho *et al.*, 1996; Poot and Singer, 1997). As the accumulation is strictly dependent on $\Delta\Psi$, they are more reliable than other single-emission probes such as Rh123 and $DiOC_6(3)$ (see later), and they are also suitable for samples that have to be fixed. In this regard, such probes can be used for the simultaneous detection of $\Delta\Psi$ and other parameters related to, for example, gene expression, apoptosis, and cell cycle (Salvioli *et al.*, 2000b). Although MitoTracker dyes present many advantages in the detection of $\Delta\Psi$, there are some limitations. Indeed, the fluorescence emission of red MitoTracker has a peak at 599 nm with a broad spectrum, occupying both FL2 and FL3 channels (exactly like JC-1), with FL3 being the more sensitive to $\Delta\Psi$ changes. The simultaneous use of such a probe together with a fluorescein isothiocyanate (FITC)-labeled probe can create some problems in compensation of the fluorescences. We suggest that one should detect the fluorescence of MitoTracker in the FL2 channel and compensate it with FITC fluorescence in the FL1 channel. This will give a partial loss of sensitivity to $\Delta\Psi$ changes, but it will produce better cytograms. Another disadvantage of MitoTracker is intrinsic in its single fluorescence emission. HL-60 cells treated with staurosporine, a powerful inducer of apoptosis, and stained with Mito-Tracker show, as expected, a decrease in fluorescence intensity, but the staining of the same cells with JC-1 reveals that such a decrease can be further divided

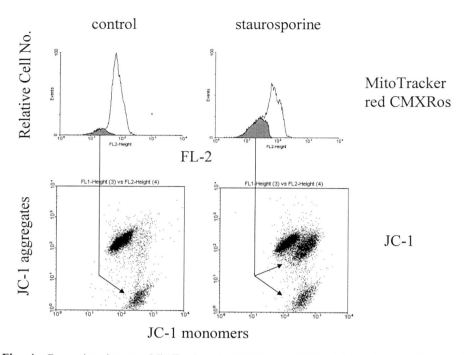

Fig. 4 Comparison between MitoTracker red CMXRos and JC-1 staining. HL-60 cells treated with 5 μM staurosporine show a marked decrease in both MitoTracker and JC-1 fluorescence. The aliquot of cells with low $\Delta\Psi$ is indicated in gray in MitoTracker-stained samples (upper part), but, according to JC-1 staining, are formed by two groups, one with low JC-1 FL2 fluorescence and another with intermediate FL2 intensity (lower part).

into two parts, providing more detailed information concerning the heterogeneity of the cell population undergoing the apoptotic process (Fig. 4).

VI. Pitfalls and Misinterpretation of the Data

One of the main problems with JC-1 is to understand which is which. Since it is very difficult to simultaneously label cells with other probes, it would be hard to discriminate different subpopulations only for their JC-1 fluorescence. If the experimental samples contain heterogeneous cell populations, it is recommended to perform multiparametric analysis using JC-1 green and orange fluorescence together with FSC and SSC parameters, and using back gates as well. Moreover, since JC-1 binds to all membranes simply because of their lipophilic environment, it is important to be sure that the events passing through the laser beam are viable cells and not dead or broken cells, which still possess membranes and can thus bind JC-1. Indeed, this would make the results of the analysis

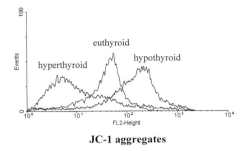

Fig. 5 Changes of the distribution of cells within a broad range of $\Delta\Psi$. Rat hepatocytes from animals with different thyroid functionality stained with JC-1 show different intensity in FL2 fluorescence, suggesting that a different energy coupling efficiency exists.

unclear and not reliable. For example, when analyzing the depolarizing effect of a cytotoxic agent, one must be sure that the observed drop of JC-1 orange fluorescence and the rise of JC-1 green fluorescence are due to a fall in $\Delta\Psi$ and not simply to the death of the cells caused by that agent. For this purpose, only cells that maintain normal physical parameters (checked through a back gate) should be considered for analysis of JC-1 fluorescences.

Sometimes it happens that samples display a broad, uninterrupted JC-1 orange fluorescence intensity. How does one decide where to put the threshold between polarized and depolarized cells? If even a control, untreated sample shows a similar fluorescence pattern, it is first important to ascertain whether green fluorescent cells are viable or not, by using multiparametric analysis, as mentioned earlier. Second, it must be clear that the JC-1 staining technique is mostly qualitative rather than quantitative, and thus one can decide on a marker threshold in order to compare different samples using the same marker (obviously a sample of cells treated with a depolarizing agent is of great help). This will give an idea of the changes of the distribution of cells within a given range of $\Delta\Psi$. An example of such a situation is shown in Fig. 5.

VII. Comparison with Other Methods

The technique of JC-1 staining has been developed with the intent to detect $\Delta\Psi$ in intact, viable cells. In this regard, the probe acts as a reliable marker of mitochondrial activity. This method has been compared to the classic TPP^+ electrode measurement on isolated mitochondria (Cossarizza *et al.*, 1996), and the results were impressively superimposed, with a R^2 of about 0.9. Cytofluorimetric detection of $\Delta\Psi$ by JC-1 is now a well-assessed method, but, although a growing number of publications show the benefit of the capabilities of such a probe (see later), other less sensitive molecules are still employed. Rh123 and other carbocyanines are widely used, but they are not completely reliable. We have performed accurate, methodological studies to compare the JC-1 staining technique with some of these probes, such as Rh123 and $DiOC_6(3)$ (Salvioli et al., 1997). JC-1 always showed a much higher specificity and sensitivity for $\Delta\Psi$ and its changes. On the contrary, $DiOC_6(3)$ showed a significant sensitivity to changes in plasma membrane potential, and its fluorescence was not affected by the collapse in $\Delta\Psi$ provoked by low doses of valinomycin or FCCP. Rh123, even if sensitive to changes in plasma membrane potentials, was less sensitive than JC-1 to $\Delta\Psi$ changes, as already observed (Lopez-Mediavilla et al., 1989).

VIII. Reviews of the Applications

The JC-1 probe has been extensively used by several groups, allowing one to demonstrate the relationship between mitochondrial membrane potential and cell function in different experimental models, such as rat cardiac myocytes (Di Lisa et al., 1995), rat hippocampal neurons (Overly et al., 1996), and thymocytes exposed to oxidants (Virag et al., 1998). The possible involvement of mitochondria in apoptosis induced by different stimuli, such as 2-deoxy-D-ribose, cladribine, cycloheximide plus tumor necrosis factor α (TNF-α), concanavalin A, and oxidative stress, has been studied in human PBMC (Monti et al., 1997; Barbieri et al., 1998) and in human fibroblasts (Kletsas et al., 1998; Kulkarni et al., 1998; Roberg et al., 1999). Used with fluorescence or confocal microscopy, spectrometry, and flow cytometry, JC-1 has helped many scientists in demonstrating the following: the characteristics of Ca^{2+} waves provoked by metacholine treatment on cortical oligodendrocytes are dependent on mitochondrial location and function (Simpson and Russell, 1996); accumulation of Ca^{2+} and subsequent MPT may be involved in the N-methyl-D-aspartate (NMDA) receptor-mediated excitotoxic cascade (White and Reynolds, 1996; Hoyt et al., 1997); MPT in brain mitochondria is consequent to activation of Ca^{2+} cycling-dependent and independent pathways (Kristal and Dubinsky, 1997); pyruvate dehydrogenase-deficient mitochondria have a decreased ability to take up cytosolic Ca^{2+} (Padua et al., 1998); and both ionotropic and metabotropic stimulation in oligodendrocytes progenitor cells evoke changes in $\Delta\Psi$ and Ca^{2+} levels (Simpson and Russell, 1998).

Further applications of the JC-1 staining technique allowed the analysis of the organization and functionality of mitochondria in living sensory neurons (Dedov and Roufogalis, 1999) or in intact, isolated nerve terminals (Chinopoulos et al., 1999), the evaluation of $\Delta\Psi$ changes in neurons after exposure to alkaline pH (Hoyt and Reynolds, 1998), in osteoblasts treated with hormones that regulate their metabolism (Troyan et al., 1997), in a colon carcinoma cell line treated with herbimycin A (Mancini et al., 1997), in breast and prostate cancer cell lines treated with epoxide-containing piperazine, a new class of anticancer agents (Eilon et al., 2000), in acute myeloblastic leukemia cells undergoing apoptosis after treatment with all-*trans*-retinoic acid (Zheng et al., 1999), during FCCP-induced apoptosis in PC12 neuroblastoma cells (Dispersyn et al., 1999), and in rat lung cells after ischemic injury (Al-Mehdi et al., 1998).

Because JC-1 acts as a marker of mitochondrial activity, it has been used for the assessment of viability and functional status of animal sperm cells after cryopreservation (Garner et al., 1997; Thomas et al., 1998; Garner and Thomas, 1999) and for studies in human sperm cells from asthenozoospermic subjects (Troiano et al., 1998).

It has also been demonstrated that JC-1 accumulation into cells is inversely correlated with the expression of permeability glycoprotein Pgp170, which is responsible for multidrug resistance (Kuhnel et al., 1997). Other studies showed that changes in $\Delta\Psi$ induced by different drugs or production of oxygen free radicals can be proposed as useful tools for the assessment of *in vitro* chemosensitivity of breast cancer (Mancini et al., 1998). These research approaches suggest novel applications for JC-1, which can be used as either a marker for drug resistance or a marker for revealing the sensitivity of tumor cells.

Finally, it has been shown that the $\Delta\Psi$ of human neuroblastoma cells deficient in mitochondrial DNA and repopulated with mitochondria from patients with Alzheimer's disease or Parkinson's diseases (cybrids) was lower than that present in cybrids obtained with mitochondria from healthy donors (Trimmer *et al.*, 2000).

IX. Biological and Biomedical Information

Using the JC-1 staining technique to measure $\Delta\Psi$, we can demonstrate the following:

1. In the early hours of a classic apoptotic process, such as that occurring in dexamethasone-treated rat thymocytes, mitochondria functionality is maintained (Cossarizza et al., 1994).
2. Maintenance of $\Delta\Psi$ by N-acetylcysteine is crucial for protection of human cells from TNF-α-induced apoptosis (Cossarizza et al., 1995a). Interestingly, using different experimental models, other authors found no early alterations of $\Delta\Psi$ (Gorman et al., 1997; Quillet et al., 1997), reinforcing the hypothesis we put forward that membrane potential, and thus ATP/ADP ratio, is crucial for the decision of cell fate (Richter et al., 1996; Leist et al., 1997). Indeed, it seems that when the ADP/ATP ratio falls below 0.2, cells undergo apoptosis (Smets et al., 1994), whereas in the presence of a severe drop of ATP, cells die by necrosis.
3. Mitochondria play a role in the cytotoxic damage induced by L-histidine, whose action is enhanced by inorganic or organic hydroperoxide (Guidarelli et al., 1995).
4. Mitochondria are a target for the protective effects of heat shock proteins after an oxidative stress (Polla et al., 1996a).

5. $\Delta\Psi$ can be modified by the action of cytokines such as TNF-α, but TNF-α-resistant cell clones do not change $\Delta\Psi$ after treatment with this cytokine (Polla et al., 1996b).
6. Mitochondria functionality is well preserved in healthy centenarians (Cossarizza et al., 1997a), that is, in individuals able to reach the extreme limit of human life with an intact immune system (Franceschi et al., 1995).
7. Mitochondrial functional alterations and a dramatic tendency to undergo apoptosis are characteristics of lymphocytes obtained from patients experiencing acute, primary HIV syndrome (AHS) (Cossarizza et al., 1997b, 2000), where a variety of immunological alterations take place (Cossarizza et al., 1995b).
8. Human peripheral blood granulocytes (PBG) and monocytes (PBM) from patients with AHS express markers related to apoptosis without disruption of $\Delta\Psi$, and treatment with highly active antiretroviral therapy decreases the tendency to undergo spontaneous apoptosis in PBG and PBM from such patients (Cossarizza et al., 1999).
9. Sendai virus or herpes simplex virus type 1, two potent inducers of interferon-α, caused cell death in a consistent number of human peripheral blood mononuclear cells as well as a decrease in $\Delta\Psi$, which is concomitant with the appearance of the hypodiploid peak (Tropea et al., 1995).

Beside the applications listed earlier, mainly related to studies on apoptosis, we applied the JC-1 staining technique to other purposes. In particular, we were able to

1. demonstrate that small and large coelomocytes from the earthworm *Eisenia foetida* have a different composition in mitochondrial mass and a different sensitivity to depolarizing agents (Cossarizza et al., 1995c);
2. measure the mitochondrial toxicity of the doses of chromium that are regularly used by most researchers in studies on natural killer (NK) cell functions (Borella et al., 1995);
3. evaluate $\Delta\Psi$ in intact isolated rat hepatocytes by flow cytometry and study the effects of thyroid hormones on mitochondrial energy coupling in such cells (Bobyleva et al., 1998; Salvioli et al., 1998).

X. Future Directions

The growing development of new fluorescent probes for the detection of biological parameters related to mitochondrial activity allows researchers to continuously set up new methods and techniques. Moreover, the use of multiple lasers and of even more sophisticated activated cell sorters or confocal microscopes dramatically increases the number of possible investigations. It is obvious that in the very near future it will be possible to detect more and more parameters at the single cell level. With flow cytometry, microscopy systems of high efficiency, and new specific probes, it will be extremely interesting to combine the advantages of the cytometric approach with those of molecular biology, to go deeper into the details of the biological processes and mechanisms that regulate cell death.

Acknowledgments

Dr. Jurek Dobrucki (Jagellonian University, Krakow, Poland) is acknowledged for precious help in confocal microscopy; Professor Valentina Bobyleva, Dr. Leonarda Troiano, Dr. Laura Moretti (University of Modena and Reggio Emilia, Italy), Professor Edwin L. Cooper (University of California, Los Angeles), Professor Claudio Franceschi and Dr. Miriam Capri (University of Bologna), Professor Daniela Monti (University of Florence), Professor Vladimir P. Skulachev (University of Moscow, Russia), and Dr. Maria Garcia Fernandez (University of Navarra, Pamplona, Spain) are acknowledged for helpful discussions.

References

Al-Mehdi, A. B., Zhao, G., and Fisher, A. B. (1998). ATP-independent membrane depolarization with ischemia in the oxygen-ventilated isolated rat lung. *Am J. Respir. Cell Mol. Biol.* **18,** 653–661.

Barbieri, D., Abbracchio, M. P., Salvioli, S., Monti, D., Cossarizza, A., Ceruti, S., Brambilla, R., Cattabeni, F., Jacobson, K. A., and Franceschi, C. (1998). Apoptosis by 2-chloro-2'-deoxyadenosine and 2-chloro-adenosine in human peripheral blood mononuclear cells. *Neurochem. Int.* **32,** 493–504.

Bobyleva, V., Pazienza, T. L., Maseroli, R., Tomasi, A., Salvioli, S., Cossarizza, A., and Franceschi, C. (1998). Decrease in mitochondrial energy coupling by thyroid hormones: A physiological effect rather than a pathological hyperthyroidism consequence. *FEBS Lett.* **430,** 409–413.

Borella, P., Bargellini, A., Salvioli, S., Incerti Medici, C., and Cossarizza, A. (1995). The use of non-radioactive chromium as an alternative to ^{51}Cr in NK assays. *J. Immunol. Methods* **186,** 101–110.

Buttke, T. M., and Sandstrom, P. A. (1994). Oxidative stress as a mediator of apoptosis. *Immunol. Today* **15,** 7–10.

Chinopoulos, C., Tretter, L., and Adam-Vizi, V. (1999). Depolarization of in situ mitochondria due to hydrogen peroxide-induced oxidative stress in nerve terminals: Inhibition of α-ketoglutarate dehydrogenase. *J. Neurochem.* **73,** 220–228.

Cohen, J. J., and Duke, R. D. (1984). Glucocorticoid activation of a calcium-dependent endonuclease in thymocyte nuclei leads to cell death. *J. Immunol.* **132,** 38–42.

Cossarizza, A., Baccarani Contri, M., Kalashnikova, G., and Franceschi, C. (1993). A new method for the cytofluorimetric analysis of mitochondrial membrane potential using the J-aggregate forming lipophilic cation 5,5',6,6'-tetrachloro-1,1',3,3'-tetraethylbenzimidazolcarbocyanine iodide (JC-1). *Biochem. Biophys. Res. Commun.* **197,** 40–45.

Cossarizza, A., Kalashnikova, G., Grassilli, E., Chiappelli, F., Salvioli, S., Capri, M., Barbieri, D., Troiano, L., Monti, D., and Franceschi, C. (1994). Mitochondrial modifications during rat thymocyte apoptosis: A study at the single cell level. *Exp. Cell Res.* **214,** 323–330.

Cossarizza, A., Franceschi, C., Monti, D., Salvioli, S., Bellesia, E., Rivabene, R., Biondo, L., Rainaldi, G., Tinari, A., and Malorni, W. (1995a). Protective effect of N-acetylcysteine in tumor necrosis factor α-induced apoptosis in U937 cells: The role of mitochondria. *Exp. Cell Res.* **220,** 232–240.

Cossarizza, A., Ortolani, C., Mussini, C., Borghi, V., Guaraldi, G., Mongiardo, N., Bellesia, E., Franceschini, M. G., De Rienzo, B., and Franceschi, C. (1995b). Massive activation of immune cells with an intact T cell repertoire in acute HIV syndrome. *J. Infect. Dis.* **172,** 105–112.

Cossarizza, A., Cooper, E. L., Quaglino, D., Salvioli, S., Kalachnikova, G., and Franceschi, C. (1995c). Mitochondrial mass and membrane potential in coelomocytes from the earthworm *Eisenia foetida*: Studies with fluorescent probes in single intact cells. *Biochem. Biophys. Res. Commun.* **214,** 503–510.

Cossarizza, A., Ceccarelli, D., and Masini, A. (1996). Functional heterogeneity of isolated mitochondrial population revealed by cytofluorimetric analysis at the single organelle level. *Exp. Cell Res.* **222,** 84–94.

Cossarizza, A., Ortolani, C., Monti, D., and Franceschi, C. (1997a). Cytometric analysis of immunosenescence. *Cytometry* **270,** 297–314.

Cossarizza, A., Mussini, C., Mongiardo, N., Borghi, V., Sabbatini, A., De Rienzo, B., and Franceschi, C. (1997b). Mitochondria alterations and dramatic tendency to apoptosis in peripheral blood lymphocytes during acute HIV syndrome. *AIDS* **11,** 19–26.

Cossarizza, A., Mussini, C., Borghi, V., Mongiardo, N., Nuzzo, C., Pedrazzi, J., Benatti, F., Pinti, M., Paganelli, R., Franceschi, C., and De Rienzo, B. (1999). Apoptotic features of peripheral blood granulocytes and monocytes during primary, acute HIV infection. *Exp. Cell Res.* **247,** 304–311.

Cossarizza, A., Stent, G., Mussini, C., Paganelli, R., Borghi, V., Nuzzo, C., Pinti, M., Pedrazzi, J., Benatti, F., Esposito, R., Røsok, B., Nagata, S., Franceschi, C., and De Rienzo, B. (2000). Deregulation of the CD95/CD95L system in lymphocytes from patients with primary, acute HIV infection. *AIDS* **14,** 345–355.

Deckers, C. L. P., Lyons, A. B., Damuel, K., Sanderson, A., and Maddy, A. H. (1993). Alternative pathways of apoptosis induced by methylprednisolone and valinomycin analyzed by flow cytometry. *Exp. Cell Res.* **208,** 362–370.

Dedov, V. N., and Roufogalis, B. D. (1999). Organization of mitochondria in living neurons. *FEBS Lett.* **456,** 171–174.

de Jong, D., Prins, F. A., Mason, D. Y., Reed, J. C., van Ommen, G. B., and Kluin, P. M. (1994). Subcellular localization of bcl-2 protein in malignant and normal lymphoid cells. *Cancer Res.* **54,** 256–260.

Di Lisa, F., Blank, P. S., Colonna, P. S., Gambassi, G., Silverman, H. S., Stern, M. D., and Hansford, R. G. (1995). Mitochondrial membrane potential in single living adult rat cardiac myocytes exposed to anoxia or metabolic inhibition. *J. Physiol.* **486,** 1–13.

Dispersyn, G., Nuydens, R., Connors, R., Borgers, M., and Geerts, H. (1999). Bcl-2 protects against FCCP-induced apoptosis and mitochondrial membrane potential depolarization in PC12 cells. *Biochim. Biophys. Acta* **1428,** 357–371.

Eilon, G. F., Gu, J., Slater, L. M., Hara, K., and Jacobs, J. W. (2000). Tumor apoptosis induced by etoposide-containing piperazines, a new class of anti-cancer agents. *Cancer Chemother. Pharmacol.* **45,** 183–191.

Franceschi, C., Monti, D., Sansoni, P., and Cossarizza, A. (1995). The immunology of exceptional individuals: The lesson of centenarians. *Immunol. Today* **16,** 12–16.

Fricker, M., Hollinshead, M., White, N., and Vaux, D. (1997). Interphase nuclei of many mammalian cell types contain deep, dynamic tubular membrane-bound invaginations of the nuclear envelope. *J. Cell Biol.* **136,** 531–544.

Furlong, I. J., Lopez-Mediavilla, C., Ascaso, R., Lopez-Rias, A., and Collins, M. K. L. (1998). Induction of apoptosis by valinomycin: Mitochondrial permeability transition causes intracellular acidification. *Cell Death Differ.* **5,** 214–221.

Garcia Fernandez, M., Troiano, L., Moretti, L., Pedrazzi, J., Salvioli, S., Castilla-Cortazar, I., and Cossarizza, A. (2001). Change in intramitochondrial cardiolipin distribution in apoptosis-resistant HCW-2 cells, derived from the human promyelocytic leukemia HL-60. Submitted for publication.

Garner, D. L., and Thomas, C. A. (1999). Organelle-specific probe JC-1 identifies membrane potential differences in the mitochondrial function of bovine sperm. *Mol. Reprod. Dev.* **53,** 222–229.

Garner, D. L., Thomas, C. A., Joerg, H. W., DeJarnette, J. M., and Marshall, C. E. (1997). Fluorometric assessment of mitochondrial function and viability in cryopreserved bovine spermatozoa. *Biol. Reprod.* **57,** 1401–1406.

Golstein, P. (1997). Controlling cell death. *Science* **275,** 1081–1082.

Gorman, A. M., Samali, A., McGowan, A. J., and Cotter, T. G. (1997). Use of flow cytometry techniques in studying mechanisms of apoptosis in leukemic cells. *Cytometry* **29**, 97–105.

Guidarelli, A., Sestili, P., Cossarizza, A., Franceschi, C., Cattabeni, F., and Cantoni, O. (1995). Evidence for dissimilar mechanisms of enhancement of inorganic and organic hydroperoxide cytotoxicity by L-histidine. *J. Pharmacol. Exp. Ther.* **275**, 1575–1582.

Hada, H., Honda, C., and Tanemura, H. (1977). Spectroscopic study on the J-aggregate of cyanine dyes. I. Spectral changes of UV bands concerned with J-aggregate formation. *Photogr. Sci. Eng.* **21**, 83–91.

Hetts, S. W. (1998). To die or not to die—An overview of apoptosis and its role in disease. *JAMA* **279**, 300–307.

Hockenbery, D. M., Oltvai, Z. M., Yin, X. M., Milliman, C. L., and Korsmeyer, S. J. (1993). Bcl-2 functions in an antioxidant pathway to prevent apoptosis. *Cell* **75**, 241–251.

Hoyt, K. R., and Reynolds, I. J. (1998). Alkalinization prolongs recovery from glutamate-induced increases in intracellular Ca^{2+} concentration by enhancing Ca^{2+} efflux through the mitochondrial Na^+/Ca^{2+} exchanger in cultured rat forebrain neurons. *J. Neurochem.* **71**, 1051–1058.

Hoyt, K. R., Sharma, T. A., and Reynolds, I. J. (1997). Trifluoroperazine and dibucaine-induced inhibition of glutamate-induced mitochondrial depolarization in rat cultured forebrain neurones. *Br. J. Pharmacol.* **122**, 803–808.

Keij, J. F., Bell-Prince, C., and Steinkamp, J. A. (2000). Staining of mitochondrial membrane with 10-nonyl acridine orange, MitoFluor Green, and MitoTracker Green is affected by mitochondrial membrane potential-altering drugs. *Cytometry* **39**, 203–210.

Kletsas, D., Barbieri, D., Stathakos, D., Botti, B., Bergamini, S., Tomasi, A., Monti, D., Malorni, W., and Franceschi, C. (1998). The highly reducing sugar 2-deoxy-D-ribose induces apoptosis in human fibroblasts by reduced glutathione depletion and cytoskeletal disruption. *Biochem. Biophys. Res. Commun.* **243**, 416–425.

Kluck, R. M., Bossy-Wetzel, E., Green, D. R., and Newmeyer, D. D. (1997). The release of cytochrome c from mitochondria: A primary site for bcl-2 regulation of apoptosis. *Science* **275**, 1132–1136.

Kristal, B. S., and Dubinsky, J. M. (1997). Mitochondrial permeability transition in the central nervous system: Induction by calcium cycling-dependent and -independent pathways. *J. Neurochem.* **69**, 524–538.

Kuhnel, J. M., Perrot, J. Y., Faussat, A. M., Marie, J. P., and Schwaller, M. A. (1997). Functional assay of multidrug resistant cells using JC-1, a carbocyanine fluorescent probe. *Leukemia* **11**, 1147–1155.

Kulkarni, G. V., Lee, W., Seth, A., and McCulloch, C. A. G. (1998). Role of mitochondrial membrane potential in concanavalin A-induced apoptosis in human fibroblasts *Exp. Cell Res.* **245**, 170–178.

Leist, M., Single, B., Castoldi, A. F., Kuhnle, S., and Nicotera, P. (1997). Intracellular adenosine triphosphate (ATP) concentration: A switch in the decision between apoptosis and necrosis. *J. Exp. Med.* **185**, 1481–1486.

Lemasters, J. J., Nieminen, A. L., Qian, T., Trost, L. C., Elmore, S. P., Nishimura, Y., Crowe, R. A., Cascio, W. E., Bradham, C. A., Brenner, D. A., and Herman, B. (1998). The mitochondrial permeability transition in cell death: A common mechanism in necrosis, apoptosis and autophagy. *Biochim. Biophys. Acta* **1366**, 177–196.

Li, P., Nijhawan, D., Budihardjo, I., Srinivasula, S. M., Ahmad, M., Alnemri, E. S., and Wang, X. (1997). Cytochrome c and dATP-dependent formation of Apaf-1/caspase-9 complex initiates an apoptotic protease cascade. *Cell* **91**, 479–489.

Liu, X., Kim, C. N., Yang, J., Jemmerson, R., and Wang, X. (1996). Induction of apoptotic program in cell-free extracts: Requirement for dATP and cytochrome c. *Cell* **86**, 147–157.

Lopez-Mediavilla, C., Orfao, A., Gonzales, M., and Medina, J. M. (1989). Identification by flow cytometry of two distinct rhodamine-123-stained mitochondrial populations in rat liver. *FEBS Lett.* **254**, 115–120.

McConkey, D. J., Nicotera, P., Hartzell, P., Bellomo, G., Wyllie, A. H., and Orrenius, S. (1989). Glucocorticoids activate a suicide process in thymocytes through an elevation of cytosolic Ca^{2+} concentration. *Arch. Biochem. Biophys.* **269**, 365–370.

Macho, A., Decaudin, D., Castedo, M., Hirsch, T., Susin, S. A., Zamzami, N., and Kroemer, G. (1996). Chloromethyl-X-rosamine is an aldehyde-fixable potential-sensitive fluorochrome for the detection of early apoptosis. *Cytometry* **25**, 333–340.

Maftah, A., Petit, J. M., Ratinaud, M. H., and Julien, R. (1989). 10-*N* nonyl-acridine orange: A fluorescent probe which stains mitochondria independently of their energetic state. *Biochem. Biophys. Res. Commun.* **164**, 185–190.

Mancini, M., Anderson, B. O., Caldwell, E., Sedghinasab, M., Paty, P. B., and Hockenbery, D. M. (1997). Mitochondrial proliferation and paradoxical membrane depolarization during terminal differentiation and apoptosis in a human colon carcinoma cell line. *J. Cell Biol.* **138**, 449–469.

Mancini, M., Sedghinasab, M., Knowlton, K., Tam, A., Hockenbery, D., and Anderson, B. O. (1998). Flow cytometric measurement of mitochondrial mass and function: A novel method for assessing chemoresistance. *Ann. Surg. Oncol.* **5**, 287–295.

Marchetti, P., Hirsch T., Zamzami, N., Castedo, M., Decaudin D., Susin S. A., Masse, B., and Kroemer, G. (1996). Mitochondrial permeability transition triggers lymphocytes apoptosis. *J. Immunol.* **157**, 4830–4836.

Marchetti, P., Decaudin, D., Macho, A., Zamzami, N., Hirsch, T., Susin, S. A., and Kroemer, G. (1997). Redox regulation of apoptosis: Impact of thiol oxidation status on mitochondrial function. *Eur. J. Immunol.* **27**, 289–296.

Minamikawa, T., Williams, D. A., Bowser, D. N., and Nagley, P. (1999). Mitochondrial permeability transition and swelling can occur reversibly without inducing cell death in intact human cells. *Exp. Cell Res.* **246**, 26–37.

Monaghan, P., Robertson, D., Amos, T. A., Dyer, M. J., Mason, D. Y., and Greaves, M. F. (1992). Ultrastructural localization of Bcl-2 protein. *J. Histochem. Cytochem.* **40**, 1819–1825.

Monti, D., Macchioni, S., Guido, M., Pagano, G., Zatterale, A., Calzone, R., Cossarizza, A., Straface, E., Malorni, W., and Franceschi, C. (1997). Resistance to apoptosis in Fanconi's anaemia. An *ex vivo* study in peripheral blood mononuclear cells. *FEBS Lett.* **409**, 365–369.

Ojcius, D. M., Zychlinsky, A., Zheng, L. M., and Young, J. D.-E. (1991). Ionophore-induced apoptosis: Role of DNA fragmentation and calcium fluxes. *Exp. Cell Res.* **197**, 43–49.

Overly, C. C., Rieff, H. I., and Hollenbeck, P. J. (1996). Organelle motility and metabolism in axons vs dendrites of cultured hippocampal neurons. *J. Cell Sci.* **109**, 971–980.

Padua, R. A., Baron, K. T., Thyagarajan, B., Campbell, C., and Thayer, S. A. (1998). Reduced Ca^{2+} uptake by mitochondria in pyruvate dehydrogenase-deficient human diploid fibroblasts. *Am. J. Physiol.* **274**, C615–C622.

Petit, J. M., Huet, O., Gallet, P. F., Maftah, A., Ratinaud, H. M., and Julien, R. (1994). Direct analysis and significance of cardiolipin transversal distribution in mitochondrial inner membranes. *Eur. J. Biochem.* **220**, 871–879.

Polla, B. S., Kantengwa, S., François, D., Salvioli, S., Franceschi, C., Marsac, C., and Cossarizza, A. (1996a). Mitochondria as targets for the protective effects of heat shock against oxidative injury. *Proc. Natl. Acad. Sci. U.S.A.* **93**, 6458–6463.

Polla, B. S., Kantengwa, S., Mariéthoz, E., Jacquier-Sarlin, M., Hennet, T., Russo-Marie, F., and Cossarizza, A. (1996b). TNF-α alters mitochondrial membrane potential in L929 but not in TNF-α-resistant L929.12 cells: Relationship with the synthesis of heat shock proteins and superoxide dismutase activity. *Free Radical Res.* **25**, 125–131.

Poot, M., Gibson, L. L., and Singer, V. L. (1997). Detection of apoptosis in live cells by MitoTracker red CMXRos and SYTO dye flow cytometry. *Cytometry* **27**, 358–364.

Quillet, M. A., Jaffrezou, J. P., Mansat, V., Bordier, C., Naval, J., and Laurent, G. (1997). Implication of mitochondrial hydrogen peroxide generation in ceramide-induced apoptosis. *J. Biol. Chem.* **272**, 21388–21395.

Reed, J. C., Jurgensmeier, J. M., and Matsuyama, S. (1998). Bcl-2 family proteins and mitochondria. *Biochim. Biophys. Acta* **1366**, 127–137.

Reers, M., Smith, T. W., and Chen, L. B. (1991). J-aggregate formation of a carbocyanine as a quantitative fluorescent indicator of membrane potential. *Biochemistry* **30**, 4480–4486.

Richter, C. H. T., Gogvadze, V., Laffranchi, R., Schlapbach, R., Schweizer, M., Sutter, M., Walter, P., and Yaffee, M. (1995). Oxidants in mitochondria: From physiology to diseases. *Biochim. Biophys. Acta* **1271,** 67–74.

Richter, C., Schweizer, M., Cossarizza, A., and Franceschi, C. (1996). Control of apoptosis by the cellular ATP level. *FEBS Lett.* **378,** 107–110.

Roberg, K., Johansson, U., and Ollinger, K. (1999). Lysosomial release of cathepsin D precedes relocation of cytochrome *c* and loss of mitochondrial transmembrane potential during apoptosis induced by oxidative stress. *Free Radicals Biol. Med.* **27,** 1228–1237.

Salvioli, S., Ardizzoni, A., Franceschi, C., and Cossarizza, A. (1997). JC-1, but not $DiOC_6(3)$ or rhodamine 123, is a reliable fluorescent probe to assess $\Delta\Psi$ in intact cells. Implications for studies on mitochondrial functionality during apoptosis. *FEBS Lett.* **411,** 77–82.

Salvioli, S., Maseroli, R., Pazienza, T. L., Bobyleva, V., and Cossarizza, A. (1998). Use of flow cytometry as a tool to study mitochondrial membrane potential in isolated, living hepatocytes. *Biochemistry (Moscow)* **63,** 235–238.

Salvioli, S., Barbi, C., Dobrucki, J., Troiano, L., Moretti, L., Pinti, M., Pedrazzi, J., Pazienza, T. L., Bobyleva, V., Franceschi, C., and Cossarizza, A. (2000a). Opposite role of changes in mitochondrial membrane potential ($\Delta\Psi$) in different apoptotic processes. *FEBS Lett.* **469,** 186–190.

Salvioli, S., Dobrucki, J., Moretti, L., Troiano, L., Garcia Fernandez, M., Pinti, M., Pedrazzi, J., Franceschi, C., and Cossarizza A. (2000b). Mitochondrial heterogeneity during staurosporine-induced apoptosis in HL60 cells: Analysis at the single cells and single organelle level. *Cytometry* **40,** 189–197.

Septinus, M., Berthold, T., Naujok, A., and Zimmerman, H. W. (1985). Hydrophobic acridine dyes for fluorescent staining of mitochondria in living cells. 3. Specific accumulation of the fluorescent dye NAO on the mitochondrial membranes in HeLa cells by hydrophobic interaction. Depression of respiratory activity, changes in the ultrastructure of mitochondria due to NAO. Increase of fluorescence in vital stained mitochondria in situ by irradiation. *Histochemistry* **82,** 51–66.

Simpson, P. B., and Russell, J. T. (1996). Mitochondria support inositol 1,4,5-triphosphate-mediated Ca^{2+} waves in cultured oligodendrocytes. *J. Biol. Chem.* **271,** 33493–33501.

Simpson, P. B., and Russell, J. T. (1998). Mitochondrial Ca^{2+} uptake and release influence metabotropic and ionotropic cytosolic Ca^{2+} responses in rat oligodendrocytes progenitors. *J. Physiol.* **508,** 413–426.

Skulachev, V. P. (1996). Why are mitochondria involved in apoptosis? Permeability transition pores and apoptosis as selective mechanisms to eliminate superoxide-producing mitochondria and cell. *FEBS Lett.* **397,** 7–10.

Smets, L. A., Van den Berg, J., Acton, D., Top, B., Van Rooij, H., and Verwijs-Janssen, M. (1994). BCL-2 expression and mitochondrial activity in leukemic cells with different sensitivity to glucocorticoid-induced apoptosis. *Blood* **84,** 1613–1619.

Smiley, S. T., Reers, M., Mottola-Hartshorn, C., Lin, M., Chen, A., Smith, T. W., Steele, G. D., and Chen, L. B. (1991). Intracellular heterogeneity in mitochondrial membrane potential revealed by a J-aggregate-forming lipophilic cation JC-1. *Proc. Natl. Acad. Sci. U.S.A.* **88,** 3671–3675.

Straface, E., Santini, M. T., Donelli, G., Giacomoni, P. U., and Malorni, W. (1995). Vitamin E prevents UVB-induced cell blebbing and cell death in A431 epidermoid cells. *Int. J. Radiat. Biol.* **68,** 579–587.

Susin, S. A., Zamzami, N., Castedo, M., Hirsch, T., Marchetti, P., Macho, A., Daugas, E., Geuskens, M., and Kroemer, G. (1996). Bcl-2 inhibits the mitochondrial release of an apoptogenic protease. *J. Exp. Med.* **184,** 1331–1341.

Susin, S. A., Zamzami, N., Castedo, M., Daugas, E., Wang, H. G., Geley, S., Fassy, F., Reed, J. C., and Kroemer, G. (1997). The central executioner of apoptosis: Multiple connections between protease activation and mitochondria in Fas/APO-1/CD95- and ceramide-induced apoptosis. *J. Exp. Med.* **186,** 25–37.

Terasaki, M., Song, J., Wong, J. R., Weiss, M. J., and Chen, B. L. (1984). Localization of endoplasmic reticulum in living and glutaraldehyde-fixed cells with fluorescent dyes. *Cell* **38,** 101–108.

Thomas, C. A., Garner, D. L., DeJarnette, J. M., and Marshall, C. E. (1998). Effect of cryopreservation of sperm organelle function and viability as determined by flow cytometry. *Biol. Reprod.* **58,** 786–793.

Trimmer, P. A., Swerdlow, R. H., Parks, J. K., Kenney, P., Bennet, J. P., Jr., Miller, S. W., Davis, R. E., and Parker, W. D., Jr. (2000). Abnormal mitochondrial morphology in sporadic Parkinson's and Alzheimer's disease cybrids cell lines. *Exp. Neurol.* **162,** 37–50.

Troiano, L., Granata, A. R., Cossarizza, A., Kalashnikova, G., Bianchi, R., Pini, G., Tropea, F., Carani, C., and Franceschi, C. (1998). Mitochondrial membrane potential and DNA stainability in human sperm cells: A flow cytometry analysis with implications for male infertility. *Exp. Cell Res.* **241,** 384–393.

Tropea, F., Troiano, L., Monti, D., Lovato, E., Malorni, W., Rainaldi, G., Mattana, P., Viscomi, G., Portolani, M., Cermelli, C., Cossarizza, A., and Franceschi, C. (1995). Sendai virus and Herpes virus type I induce apoptosis in human peripheral blood mononuclear cells. *Exp. Cell Res.* **218,** 63–70.

Troyan, M. B., Gilamn, V. R., and Gay, C. V. (1997). Mitochondrial membrane potential changes in osteoblasts treated with parathyroid hormone and estradiol. *Exp. Cell Res.* **233,** 274–280.

Van der Heiden, M. G., Chandel, N. S., Williamson, E. K., Schumaker, P. T., and Thompson, C. B. (1997). Bcl-xL regulates the membrane potential and volume homeostasis of mitochondria. *Cell* **91,** 627–637.

Virag, L., Salzman, A. L., and Szabo, C. (1998). Poly(ADP-ribose) synthetase activation mediates mitochondrial injury during oxidant-induced cell death. *J. Immunol.* **161,** 3753–3759.

White, R. J., and Reynolds, I. J. (1996). Mitochondrial depolarization in glutamate-stimulated neurons: An early signal specific to excitotoxin exposure. *J. Neurosci.* **16,** 5688–5697.

Wyllie, A. H., Kerr, J. F. R., and Currie, A. R. (1980). Cell death: The significance of apoptosis. *Int. Rev. Cytol.* **68,** 251–255.

Wyllie, A. H., Morris, R. G., Smith, A. L., and Dunlop, D. (1984). Chromatin cleavage in apoptosis: Association with condensed chromatin morphology and dependence on macromolecular synthesis. *J. Pathol.* **142,** 67–77.

Yang, J., Liu, X., Bhalla, K., Kim, C. N., Ibrado, A. M., Cai, J., Peng, T. I., Jones, D. P., and Wang, X. (1997). Prevention of apoptosis by Bcl-2: Release of cytochrome c from mitochondria blocked. *Science* **275,** 1129–1132.

Zamzami, N., Marchetti, P., Castedo, M., Hirsch, T., Susin, S. A., Masse, B., and Kroemer, G. (1996). Inhibitors of permeability transition interfere with the disruption of the mitochondrial transmembrane potential during apoptosis. *FEBS Lett.* **384,** 53–57.

Zheng, A., Mantymaa, P., Saily, M., Siitonen, T., Savolainen, E. R., and Koistinen, P. (1999). An association between mitochondrial function and all-trans retonoic acid-incuded apoptosis in acute myeloblastic leukemia cells. *Br. J. Haematol.* **105,** 215–224.

Zhu, W., Cowie, A., Wasfy, G. W., Penn, L. Z., Leber, B., and Andrews D. S. (1996). Bcl-2 mutants with restricted subcellular location reveal spatially distinct pathways for apoptosis in different cell types. *EMBO J.* **15,** 4130–4141.

Zoratti, M., and Szabò, I. (1995). The mitochondrial permeability transition. *Biochim. Biophys. Acta* **1241,** 139–176.

Zou, H., Henzel, W. J., Liu, X., Lutschg, A., and Wang, X. (1997). Apaf-1, a human protein homologous to *C. elegans* CED-4, participates in cytochrome c-dependent activation on caspase-3. *Cell* **90,** 405–413.

CHAPTER 22

Cytometry of Caspases

Steven K. Koester and Wade E. Bolton

Advanced Technology
Beckman Coulter, Inc.
Miami, Florida 33196

I. Introduction
II. Materials and Methods: Caspase Peptide Inhibitors and Methods to Monitor Responses
    A. Caspase Inhibition Followed by CD95 Induction of Jurkat Cells
    B. Staining of Cells for APO2.7 Expression
    C. Staining of Cells for Viability Status with Antitubulin Antibody
    D. Method for Distinguishing Viable, Early Apoptotic, Late Apoptotic, and Necrotic Cells by Flow Cytometry
    E. Annexin V Staining for Phosphatidylserine Exposure
    F. Light Scatter Measurements by Flow Cytometry
    G. Detection of DNA Fragments
    H. Staining of Cells for Viability Status with Trypan Blue
    I. Ultrastructural Analysis with Transmission Electron Microscopy
    J. Fluorogenic Substrates for Flow Cytometric Identification of Caspase Activation
III. Results and Discussion
    References

I. Introduction

The study of caspases and their role in apoptosis has grown out of genetic studies related to the development of the nematode *Caenorhabditis elegans* (Cohen, 1997; Golstein, 1997; Yuan *et al.*, 1993). Exciting is the fact that the complete DNA sequence of this small invertebrate animal is now complete (Hodgkin *et al.*, 1998). The only other eukaryote with a sequenced genome is the budding yeast, *Saccharomyces cerevisiae*. The genetically programmed cell death in *C. elegans* appears similar to the type of cell death we term apoptosis. Studies indicate that two *C. elegans* genes, *ced-3* and *ced-4,* are required for programmed cell death. The *ced-3* gene encodes a cysteine protease that is very similar to a known human enzyme, interleukin-1β-converting enzyme (ICE) (Xue and Horvitz, 1995; Yuan *et al.*, 1993), or more specifically, caspase-3 (Xue *et al.*, 1996), and *ced-4* is now thought to be homologous to the gene for mitochondrial apoptotic protease activation factor, Apaf-1 (Jacobson, 1997; Zou *et al.*, 1997). Initial studies showed that the activation of caspases (cysteine-containing, aspartate-specific proteases) was directly, or indirectly through a caspase cascade, responsible for activation of caspase-3 (Muzio *et al.*, 1996; Schlegel *et al.*, 1996). Most investigators now agree that apoptotic signaling involves a cascade of events in which caspases play an important role. Caspases include proteases that contain the amino acid cysteine in their active sites and cleave their targets after the amino acid aspartate (Alnemri *et al.*, 1996). Thirteen ICE-related proteins, or caspases, have now been identified (Thornberry and Lazebnik, 1998).

Caspases are initially found as inactive proteins called zymogens and must be cleaved to become active effectors in the death cascade (Barinaga, 1998; Orth *et al.*, 1996). Active cleaved products can now specifically cleave other zymogens to create a cascade of enzymatic reactions. Much effort is presently directed toward unraveling the sequence of events in cell death signaling. The APO-1/Fas death signaling cascade has received much attention in these efforts, and it should be considered an impressive model for caspase studies (Koester *et al.*, 1997, reviewed in Koester and Bolton, 1998, 1999) (Fig. 1).

Nuclear and cytoplasmic membrane changes are a hallmark of the apoptotic process. Time related morphological cellular changes in CD95 induced Jurkat cells have been shown to be brought about directly or indirectly though the activity of caspases (Koester *et al.*, 1997, Koester and Bolton, 1998, 1999). Normally, no single method is sufficient to track the changes that take place during the cell death cascade.

Flow cytometric applications have been developed to aid in identifying the sequence of caspase activation and for monitoring cell death responses to the effects of specific caspases. One method employs the use of caspase peptide inhibitors that pass through the cell membrane of living cells to block caspase activity. Questions can now be asked about death signaling and the effect of specific caspases or the position of a caspase in the signaling cascade (Enari *et al.*, 1996; Hirata *et al.*, 1998; Koester and Bolton, 1999; Lincz, 1998) using monitoring methods mentioned earlier. It should be noted, however, that caspase homologies exist that can cause cross-reactions with caspases showing the homologous regions (Lincz, 1998; Margolin *et al.*, 1997; Villa *et al.*, 1997). Therefore, experiments should be designed with this possibility in mind. A second method utilizes unique fluorogenic substrates to identify activation of specific caspases. Synthetic substrates having specificity for a single enzyme will pass through the cell membrane of living cells, and once inside the cell, enzymatic hydrolysis of the substrate frees a fluorescent dye that accumulates in the cell, with the intensity of fluorescence proportional to the activity of the specific enzyme (Lucas *et al.*, 1996; Hamilton *et al.*, 1995; Ruez *et al.*, 1996; Woodard *et al.*, 1993).

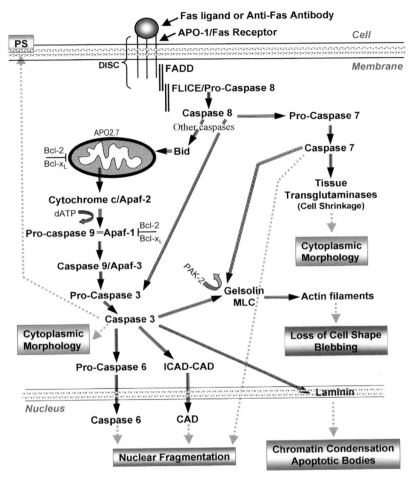

Fig. 1 Caspase signaling cascade represented by ligation of APO-1/Fas cell surface receptor. Apaf, Apoptotic protease activation factor; CAD, caspase-activated deoxyribonuclease; caspase, cysteine-containing, aspartate-specific protease; DISC, death-inducing signaling complex; FADD, Fas associated death domain; Fas, FS7-associated cell surface antigen; FasL, Fas ligand; FLICE, FADD like ICE; ICAD, inhibitor of caspase-activated deoxyribonuclease; ICE, interleukin-1β-converting enzyme; MLC, myosin regulatory light chains; PAK-2, p21 activated kinase-2; PS, phosphatidylserine. Reprinted from *Clinical Immunol. Newsletter,* Vol. 18, S. K. Koester and W. E. Bolton, The APO-1/Fas death signaling pathway: A life and death balance, pp. 97–102. Copyright 1998, with permission from Elsevier Science.

Several indirect flow cytometric methods can be utilized in concert with direct methods to map caspase activation in apoptosis. These methods include detection of DNA fragmentation (Gorczyca *et al.,* 1992; Koester *et al.,* 1997, 1998, 1999; Li and Darzynkiewicz, 1995), which can be compared to DNA laddering by gel electrophoresis (Duke *et al.*, 1983; Koester *et al.*, 1997, Koester and Bolton, 1999; Wyllie, 1980; Zhang *et al.*, 1995), detection of cell surface expression of phosphatidylserine (PS) using annexin V (Koester *et al.*, 1997, 1998; Koester and Bolton, 1999; Koopman *et al.*, 1994; Vermes *et al.*, 1995), light scattering changes (Koester *et al.*, 1997, 1998; Koester and Bolton, 1999; Ormerod *et al.*, 1994), detection of mitochondrial membrane specific APO2.7 expression (Berthou *et al.*, 1997; Koester *et al.*, 1997, 1998; Koester and Bolton, 1998, 1999; Metivier *et al.*, 1998; Seth *et al.*, 1997; Zhang *et al.*, 1996), detection of cell viability/permeability with antitubulin antibody (Koester *et al.*, 1997, 1998; Koester and Bolton, 1998, 1999; O'Brien and Bolton, 1995; O'Brien *et al.*, 1995, 1997), which can be compared to trypan blue staining by light microscopy (Koester *et al.*, 1997, 1998; Koester and Bolton, 1998, 1999). And finally, comparing all of the methods to ultrastructural characteristics seen by employing transmission electron microscopy is informative (Glauert, 1995; Koester *et al.*, 1998, Koester and Bolton, 1998, 1999; Wyllie *et al.*, 1980). Questions can now be directed toward identifying caspase activity during death signaling, by a combination of the methods mentioned earlier.

II. Materials and Methods: Caspase Peptide Inhibitors and Methods to Monitor Responses

A. Caspase Inhibition Followed by CD95 Induction of Jurkat Cells

1. Prepare 1.0×10^6 Jurkat cells/ml in AIM V lymphocyte medium (GIBCO BRL-Life Technologies, Grand Island, NY). Incubate at 37°C, 5% CO_2, and 85% humidity.
2. Treat cells for 1 hr with 300 μM final concentration of peptide inhibitor Ac-YVAD-cmk or Ac-DEVD-cho (BACHEM, Torrance, CA). Follow published reports for recommended concentrations of other caspase peptide inhibitors.
3. Add anti-CD95 antibody [immunoglobulin M (IgM)] (Immunotech, a Beckman Coulter Company, Marseille, France). Add 1 μg of anti-CD95 antibody/ml per 1.0×10^6 cells.
4. Sample cells for testing at designated time points.

B. Staining of Cells for APO2.7 Expression

Staining without Digitonin Processing to Permeabilize

1. Prepare a cell pellet containing 0.5×10^6 to 1.0×10^6 cells in a 12×75 mm tube.

2. Add APO2.7-phycoerythrin (PE) or APO2.7-PECy5 (Immunotech) per product insert, and incubate at room temperature protected from light.
3. Resuspend cells in 2 ml of phosphate-buffered saline (PBS) (Coulter, a Beckman Coulter Company, Miami, FL) with 2.5% fetal bovine serum (FBS, HyClone Laboratories, Logan, UT) (PBSF, PBS, and FBS). Centrifuge cell suspension at 200 g for 5 min and discard supernatant.
4. Resuspend cells in 1 ml of PBSF and store on ice until analyzed on the flow cytometer. Cells can be resuspended in 1 ml of 0.5% electron microscopy (EM) grade paraformaldehyde (Electron Microscopy Sciences, Fort Washington, PA) in PBS and stored at 4°C overnight before analyzing on the flow cytometer.

Staining after Digitonin Processing to Permeabilize

1. Prepare a stock solution containing 25 mg/ml digitonin (Sigma, St. Louis, MO; or WAKO BioProducts, Richmond, VA) by heating in PBS to 100°C until completely dissolved. Remove from the heat immediately after digitonin goes into solution, and store the stock solution at 4°C for up to 1 month.
2. Prepare a cell pellet containing 0.5×10^6 to 1.0×10^6 cells in a 12×75 mm tube.
3. Add 100 μl of a 100–500 μg/ml digitonin solution to the cell pellet, diluted from the stock solution in PBS, and incubate for 20 min on ice. The concentration of digitonin should be determined for each cell line or experimental condition to optimize the staining results.
4. Resuspend cells in 2 ml of PBSF, centrifuge cell suspension at 200 g for 5 min, and discard supernatant.
5. Add APO2.7-PE or APO2.7-PECy5 and incubate at room temperature for 15 min protected from light.
6. Resuspend cells in 2 ml of PBSF, centrifuge cell suspension at 200 g for 5 min, and discard the supernatant.
7. Resuspend cells in 1 ml of PBSF and store on ice until analyzed on the flow cytometer. Cells can be resuspended in 1 ml of 0.5% EM grade paraformaldehyde in PBS and stored at 4°C overnight before analyzing on the flow cytometer.

C. Staining of Cells for Viability Status with Antitubulin Antibody

Indirect Staining Method

1. Resuspend a pellet of cells containing 0.5×10^5 to 1.0×10^6 cells in 100 μl of a 1:10 dilution of antitubulin antibody in PBSF (Zymed Laboratories, San Francisco, CA) and incubate for 15 min at room temperature.

2. Resuspend cells in 2 ml of PBSF, centrifuge cell suspension at 200 g for 5 min, and discard supernatant.
3. Add 100 μl of 1:40 sheep anti-mouse–fluorescein isothiocyanate (SAM-FITC) (Silenus Laboratories, Victoria, Australia) in PBSF, and incubate at room temperature for 15 min protected from light.
4. Resuspend cells in 2 ml of PBSF, centrifuge cell suspension at 200 g for 5 min, and discard supernatant.
5. Resuspend cells in 1 ml of PBSF and store on ice until analyzed on the flow cytometer. Cells can be resuspended in 1 ml of 0.5% EM grade paraformaldehyde in PBS and stored at 4°C overnight before analyzing on the flow cytometer.

Direct Staining Method

1. Resuspend a pellet of cells containing 0.5×10^5 to 1.0×10^6 cells in 100 μl of a predetermined concentration of antitubulin-FITC (Immunotech, in development) and incubate for 15 min at room temperature protected from light.
2. Resuspend cells in 2 ml of PBSF, centrifuge cell suspension at 200 g for 5 min, and discard the supernatant.
3. Resuspend cells in 1 ml of PBSF and store on ice until analyzed on the flow cytometer. Cells can be resuspended in 1 ml of 0.5% EM grade paraformaldehyde in PBS and stored at 4°C overnight before analyzing on the flow cytometer.

D. Method for Distinguishing Viable, Early Apoptotic, Late Apoptotic, and Necrotic Cells by Flow Cytometry

1. Add predetermined volumes of antitubulin-FITC and APO2.7-PE conjugated antibodies to 0.5×10^6 to 1.0×10^6 cells in a 12×75-mm tube, gently mix, and incubate for 15 min at room temperature protected from light.
2. Resuspend cells in 2 ml of PBSF, centrifuge cell suspension at 200 g for 5 min, and discard the supernatant.
3. Gently resuspend cells in 100 μl of 100 μg/ml digitonin in PBS and incubate for 20 min on ice, protected from light.
4. Resuspend cells in 2 ml of PBSF, centrifuge cell suspension at 200 g for 5 min, and discard the supernatant.
5. Add predetermined volume of APO2.7-PECy5 conjugated antibody to the cell pellet, gently mix, and incubate for 15 min at room temperature, protected from the light.
6. Resuspend cells in 2 ml of PBSF, centrifuge cell suspension at 200 g for 5 min, and discard the supernatant.

7. Resuspend cells in 1 ml of PBSF and store on ice until analyzed on the flow cytometer. Cells can be resuspended in 1 ml of 0.5% EM grade paraformaldehyde in PBS and stored at 4°C overnight before analyzing on the flow cytometer.

E. Annexin V Staining for Phosphatidylserine Exposure

The method is as described in the product insert for the annexin V-FITC kit (Immunotech).

1. For the 20-test kit, dilute the 10× concentrated binding buffer 10-fold with distilled water. For the 200-test kit, dilute the annexin V-FITC stock solution 10-fold with ice-cold (4°C) diluted buffer. Place the diluted buffer on ice.
2. Suspend the cell samples in 2 ml of ice-cold culture medium or PBS and centrifuge at 500 g for 5 min at 4°C. Discard the supernatant and resuspend the cell pellet in ice-cold, diluted binding buffer to 10^5 to 10^6 cells/ml. Store on ice.
3. Dissolve the 250 μg propidium iodide (PI) in 1 ml of diluted binding buffer.
4. To the cells stored on ice, add 5 μl of diluted annexin V-FITC or ready-to-use annexin V-FITC solution for the 200-test kit and the 20-test kit, respectively; and 5 μl dissolved PI to 490 μl of the cell suspension as prepared earlier. Mix contents gently.
5. Store the sample on ice, protected from light, for 10 min prior to analyzing on the flow cytometer.

F. Light Scatter Measurements by Flow Cytometry

Light scatter measurements should be collected for each flow cytometric analysis. A gate should be set on this histogram to eliminate debris and instrument noise from the cellular events collected for analysis.

1. Create a dual parameter histogram representing forward light scatter (FSC) and side scatter (SSC).
2. Set detector voltages and gains to present both live and dead cell distributions on the same histogram.

G. Detection of DNA Fragments

DNA Strand Break Analysis with Flow Cytometry: Cell Fixation

1. Resuspend cells in 1% paraformaldehyde prepared in PBS and incubate at 4°C for 15 min.

2. Centrifuge cell suspension at 200 g for 5 min in a refrigerated centrifuge and discard the supernatant.
3. Resuspend cells in 1 ml of cold (4°C) PBS and, using a small bore pipette, deliver the cell suspension into 2.5 ml of −20°C absolute ethyl alcohol (~70% ethanol final dilution).
4. Store cell suspension overnight at −20°C.

DNA Strand Break Analysis with Flow Cytometry: Cell Staining

1. Dispense 0.5×10^6 to 1.0×10^6 cells into 12×75 mm tubes and resuspend in 2 ml of PBS.
2. Centrifuge the cell suspension at 200 g for 5 min and discard the supernatant.
3. Add 50 μl of terminal deoxynucleotidyltransferase (TdT) reaction mixture [containing 5 μl of 10× cobalt chloride, 10 μl of 5× reaction buffer, 0.25 μl of biotin-16-dUTP (50 nmol/vial), with 0.5 μl of TdT (25 units/μl, 500 unit vial) for test or without TdT for the control (Boehringer Mannheim, Indianapolis, IN), and 34.25 μl distilled water] to each test, gently resuspend cells, and incubate for 30 min in a 37°C water bath.
4. Resuspend cells in 2 ml of PBS, centrifuge the cell suspension at 200 g for 5 min, and discard the supernatant.
5. Add 100 μl of avidin-FITC reaction mixture [containing 0.25 μg avidin-FITC (Vector Laboratories, Burlingame, CA) and 0.1% Triton X-100 in 4× saline sodium citrate (SSC)] to each test and control, gently resuspend cells, and incubate for 30 min in a 37°C water bath.
6. Resuspend cells in 2 ml of PBS, centrifuge the cell suspension at 200 g for 5 min, and discard the supernatant.
7. Resuspend cells in 1 ml of 5 μg/ml PI in PBS.
8. Collect the FITC and PI signals on a two-parameter flow cytometric histogram. Split the fluorescence from the FITC and PI emission using a 550-nm dichroic filter with the FITC emission collected through a 525-nm/30-nm bandwidth filter and the PI emission collected through a 675-nm long-pass filter.

DNA Ladder with Gel Electrophoresis

1. Cell pellets are lysed in 0.3 ml 10 mM Tris (pH 8.0), 100 mM NaCl, 25 mM EDTA (pH 8.0), 0.5% sodium dodecyl sulfate (SDS), and 0.1 μg/ml protease.
2. Extract DNA by adding an equal volume of phenol/chloroform/isoamyl alcohol (25:24:1 ratio) to the supernatants.
3. Transfer the aqueous phase supernatants to new tubes and combine with 7.5 M ammonium acetate and 100% ethanol.

4. Recover the DNA by centrifugation, dry the pellets, resuspend in Tris/EDTA buffer, and electrophorese in a 1% agarose gel containing 0.05% ethidium bromide.
   5. Visualize gels under ultraviolet light and photograph.

H. Staining of Cells for Viability Status with Trypan Blue

   1. Combine 100 μl of trypan blue (GIBCO, BRL-Life Technologies) with 1.0×10^5 cells in 100 μl of medium, gently mix, and load on a hemacytometer.
   2. Count cells within 5 min.

I. Ultrastructural Analysis with Transmission Electron Microscopy

   1. Centrifuge 2×10^6 to 10×10^6 cells and discard supernatant.
   2. Fix in 2.5% glutaraldehyde with 3% glucose in 0.1 M cacodylate buffer, pH 7.2, for 24 hr at 4°C.
   3. Discard the above fixative and post fix in 1% osmium tetroxide for 1 hr at room temperature.
   4. Dehydrate stepwise through increasing concentrations of ethanol terminating in absolute ethanol. Use propylene oxide as a transition solvent, and embed the cells in Spurr's resin.
   5. Cut ultrathin sections using an ultramicrotome, mount sections on support grids, stain in uranyl acetate and lead citrate, and examine/photograph using the transmission electron microscope.

J. Fluorogenic Substrates for Flow Cytometric Identification of Caspase Activation

Flow cytometric identification of caspase activation makes use of CellProbe-flow cytoenzymology (Beckman Coulter, Miami, FL). First, induce apoptosis in the cell culture model of choice following determination of the optimal drug concentration. Additional flasks of cells should be maintained and used for assay controls (e.g., no drug treatment, drug solvent effects, caspase inhibitor peptides to block caspase activity, and enzyme blockers for interfering enzymes). Prior to apoptosis induction in cell culture, cells should be maintained in log phase growth.

Second, sample cells over a time course to identify activation sequence for caspase of interest. The following sampling protocol is recommended.

   1. Wash the cells, prepare the appropriate single cell concentration for the application, and determine cell viability.
   2. Add 50 μl of the washed cell sample carefully into the bottom of the tube.

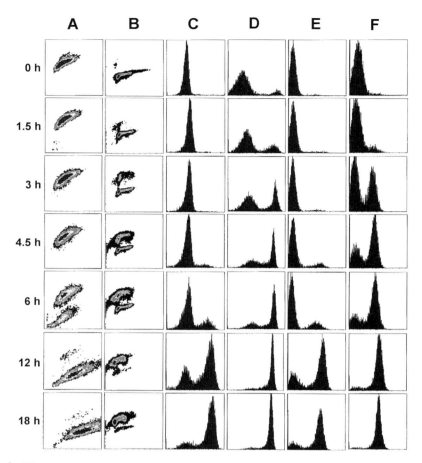

Fig. 2 Histograms representing data points collected over 18 hr to monitor responses to CD95 induction in Jurkat cells. Data points are represented by light scatter changes where the abscissa and the ordinate represent linear side scatter and linear forward scatter, respectively (A); DNA strand break analysis where linear DNA-PI content is represented on the abscissa and log DNA strand break-FITC on the ordinate (B); staining of intracellular tubulin represented by log antitubulin-FITC (C); staining of phosphatidylserine represented by log Annexin V-FITC (D); staining of APO2.7 on the mitochondria of apoptotic cells represented by log anti-APO2.7-PECy5 without digitonin processing (no permeabilization) (E); and log anti-APO2.7-PECy5 following digitonin processing (permeabilization) (F). Reprinted from Monitoring early cellular responses in apoptosis is aided by the mitochondrial membrane protein-specific monoclonal antibody APO2.7. S. K. Koester, P. Roth, W. R. Mikulka, S. F. Schlossman, C. Zhang, and W. E. Bolton, *Cytometry* **29**, 306–312. Copyright © 1997, Wiley-Liss, Inc., a subsidiary of John Wiley & Sons, Inc.

3. Prewarm the sample to 37°C in a water bath for 10 min.
4. Add 25 µl of the specific CellProbe reagent to the tube, gently mix by hand, and return immediately to the water bath. Incubate for 1, 5, or 10 min depending on the enzyme activity for the CellProbe reagent used. Follow reagent package insert protocol.
5. Place in crushed ice for 3–20 min to stop kinetic reaction and to restore cellular membrane integrity.
6. Lyse whole blood samples in Q-Prep or resuspend other cell samples in 1 ml of ice-cold (4°C) PBS.
7. Hold prepared samples on ice and analyze on the flow cytometer within 30 min following completion of initial CellProbe incubation.

III. Results and Discussion

Several methods have been outlined in this chapter for tracking the responses of cells to the cascade of events leading to cell death by apoptosis. Figure 2 represents flow cytometric histograms of the time-related behavior of several apoptosis related parameters in CD95 induced Jurkat cells. These histograms represent the apoptosis specific, mitochondrial membrane-specific APO2.7 expression in digitonin processed cells (Fig. 2F) and in nondigitonin processed

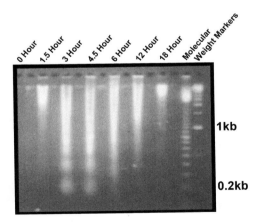

Fig. 3 Gel electrophoresis of DNA extracted from Jurkat cells, following CD95 induction, over an 18 hr time course. Reprinted from Monitoring early cellular responses in apoptosis is aided by the mitochondrial membrane protein-specific monoclonal antibody APO2.7. S. K. Koester, P. Roth, W. R. Mikulka, S. F. Schlossman, C. Zhang, and W. E. Bolton, *Cytometry* **29**, 306–312. Copyright © 1997, Wiley-Liss, Inc., a subsidiary of John Wiley & Sons, Inc.

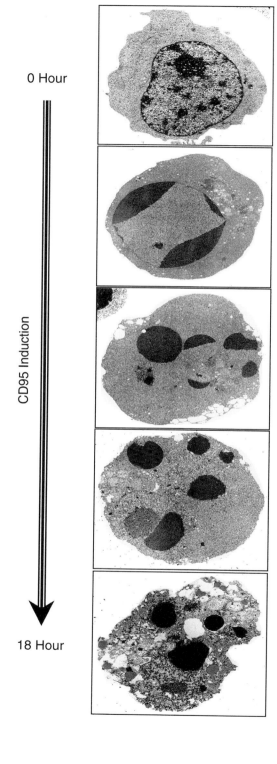

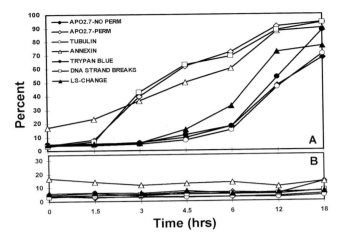

Fig. 5 Summary of time-related responses in CD95 induced Jurkat cells, represented by several documented methods for identifying cell death (A), and in noninduced Jurkat cell controls (B). All data represented were derived from flow cytometric analysis with the exception of trypan blue counts, which were derived from light microscopic hemacytometer counts. Reprinted from Monitoring early cellular responses in apoptosis is aided by the mitochondrial membrane protein-specific monoclonal antibody APO2.7. S. K. Koester, P. Roth, W. R. Mikulka, S. F. Schlossman, C. Zhang, and W. E. Bolton, *Cytometry* **29,** 306–312. Copyright © 1997, Wiley-Liss, Inc., a subsidiary of John Wiley & Sons, Inc.

cells (Fig. 2E), light scatter measurements (Fig. 2A), DNA strand break measurements in combination with DNA cell cycle (Fig. 2B), cell membrane integrity measurements to detect the ubiquitous intracellular tubulin (Fig. 2C), and detection of phosphatidylserine using annexin V (Fig. 2D). Non-flow cytometric measurements were also utilized to aid in detecting apoptosis. DNA fragment laddering can be detected by gel electrophoresis (Fig. 3) and by the gold standard, ultrastructural characteristics as seen by transmission electron microscopy (Fig. 4).

It is important that time course evaluations are accomplished when evaluating a cell death model. This will eliminate the possibility of overlooking information. The graphs in Fig. 5 represent a time course of several assays for tracking apoptosis in CD95 induced Jurkat cells using methods outlined in this chapter.

Fig. 4 Transmission electron micrographs representing classic ultrastructural characteristics in Jurkat cells, following CD95 induction, through an 18 hr time course (Electron Microscopy kindly provided by Susan Decker, University of Miami, Department of Anatomy and Cell Biology). Reprinted from *Clinical Immunol. Newsletter*, Vol. 18, S. K. Koester and W. E. Bolton, The APO-1/Fas death signaling pathway: A life and death balance, pp. 97–102. Copyright 1998, with permission from Elsevier Science.

Figures 6A and 6B represent time-related APO2.7 expression and DNA strand break detection, respectively, in CD95 induced Jurkat cells following a 1 hr pretreatment with caspase peptide inhibitors Ac-DEVD-cho or Ac-YVAD-cmk.

Finally, a life and death assessment can be made using two fluorochrome conjugates of anti-APO2.7, in combination with the cell membrane integrity marker antitubulin in a third fluorochrome, as shown in example histograms for

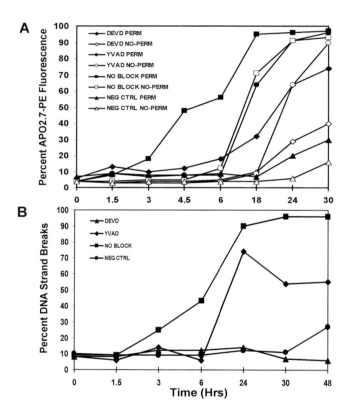

Fig. 6 Caspase inhibitor data over the indicated time course showing the percentage of cells expressing APO2.7 (A) or the percentage of cells showing DNA fragments as detected by flow cytometric DNA strand break assay (B). Cells were treated for 1 hr with 300 μM, final concentration, of either DEVD or YVAD peptides, followed by CD95 induction using 1 μg/ml anti-CD95 antibody. Cells were stained with APO2.7-PE prior to or after permeabilization using 100 μg/ml digitonin (A); alternatively, cells were labeled with biotinylated dUTP and avidin-FITC, in the reaction catalyzed by exogenous terminal deoxynucleotidyltransferase, and counterstained with PI to correlate cell cycle position to DNA strand breaks (B). Reprinted from *Clinical Immunol. Newsletter*, Vol. 18, S. K. Koester and W. E. Bolton, The APO-1/Fas death signaling pathway: A life and death balance, pp. 97–102. Copyright 1998, with permission from Elsevier Science.

the CD95 induced Jurkat cell model (Fig. 7A, B) and the MDA-MB-175-VII breast tumor hypoxia model (Fig. 7C).

The methods outlined in this chapter aid in the study of the cell death cascade and provide information to help understand the caspase cascade. The purpose

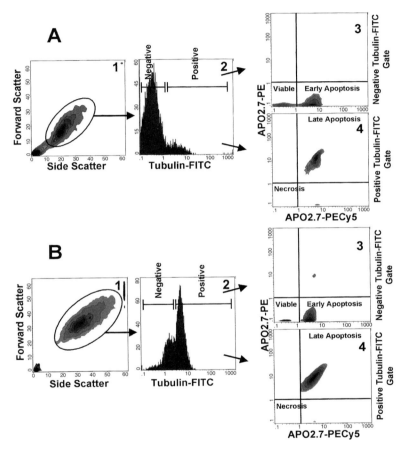

Fig. 7 Examples of a three-color assay to identify viable, early apoptotic, late apoptotic, and necrotic cells in the CD95 induced Jurkat cell model (A, B) and in the hypoxia induced MDA-MB-175-VII cell model (C). An antitubulin-FITC histogram (A2, B2, C2), gated on a 90° light scatter versus forward light scatter histogram (A1, B1, C1), is collected. APO2.7-PECy5 versus APO2.7-PE histograms are collected, one gated on the antitubulin-FITC negative population (A3, B3, C3) and one gated on the antitubulin-FITC positive population (A4, B4, and C4). Histograms A3, B3, and C3 represent viable and early apoptotic cells, and histograms A4, B4, and C4 represent late apoptotic and necrotic cells.

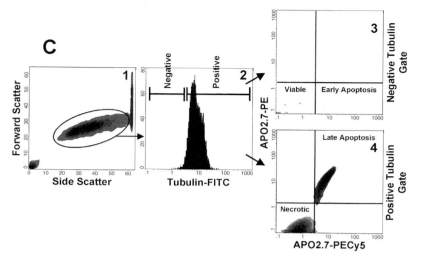

Fig. 7 (*continued*).

of this chapter was to introduce methods that will provide the investigator the means to track cell death responses, possibly augmenting presently used methods.

References

Alnemri, E. S., Livingston, D. J., Nicholson, D. W., Salvesen, G., Thornberry, N. A., Wong, W. W., and Yuan, J. (1996). Human ICE/CED-3 protease nomenclature. *Cell* **87,** 171.

Barinaga, M. (1998). Death by dozens of cuts. *Science* **280,** 32–34.

Berthou, C., Michel, L., Soulie, A., Jean-Louis, F., Flageu, I. B., Duvertret, L., Sigaux, F., Zhang, Y., and Sasportes, M. (1997). Acquisition of granzyme B and Fas ligand proteins by human keratinocytes contributes to epidermal cell defense. *J. Immunol.* **159,** 5293–5300.

Cohen, G. M. (1997). Caspases: The executioners of apoptosis. *Biochem. J.* **326,** 1–16.

Duke, R. C., Chervenak, R., and Cohen, J. J. (1983). Endogenous endonuclease-induced DNA fragmentation: An early event in cell-mediated cytolysis. *Proc. Natl. Acad. Sci. U.S.A.* **80,** 6361–6365.

Enari, M., Talanian, R. V., Wong, W. W., and Nagata, S. (1996). Sequential activation of ICE-like and CPP32-like proteases during Fas-mediated apoptosis. *Nature* **380,** 723–726.

Glauert, A. M. (1995). Embedding. *In* "Fixation, Dehydrating and Embedding of Biological Specimens. Practical Methods in Electron Microscopy" (A. M. Glauert, ed.), Vol. 3. Part 1, pp. 123–176. North-Holland Publ., Amsterdam.

Golstein, P. (1997). Controling cell death. *Science* **275,** 1081–1082.

Gorczyca, W., Bruno, S., Darzynkiewicz, R., Gong, J., and Darzynkiewicz, Z. (1992). DNA strand breaks occurring during apoptosis: Their early *in situ* detection by the terminal deoxynucleotidyl transferase and nick translation assays and prevention by serine protease inhibitors. *Int. J. Oncol.* **1,** 639–648.

Hamilton, M., Steele, B., Kuylen, N., Lucas, F., and Carter, J. (1995). The use of novel fluorogenic enzyme substrates to detect nucleated red blood cells. *Clin. Chem.* **41,** S239.

Hirata, H., Takahashi, A., Kobayashi, S., Yonehara, S., Sawai, H., Okazaki, T., Yammamoto, K., and Sasada, M. (1998). Caspases are activated in a branched protease cascade and control distinct downstream processes in Fas-induced apoptosis. *J. Exp. Med.* **187,** 587–600.

Hodgkin, J., Horvitz, R. H., Jasny, B. R., and Kimble, J. (1998). *C. elegans:* Sequence to biology. *Science* **282,** 2011.

Jacobson, M. D. (1997). Programmed cell death: A missing link found. *Cell Biol.* **7,** 467–469.

Koester, S. K., and Bolton, W. E. (1998). The APO-1/Fas death signaling pathway: A life and death balance. *Clin. Immunol. Newslett.* **18,** 97–102.

Koester, S. K., and Bolton, W. E. (1999). Differentiation and assessment of cell death. *J. Clin. Chem. Lab. Med.* **37,** 311–317.

Koester, S. K., Roth, P., Mikulka, W. R., Schlossman, S. F., Zhang, C., and Bolton, W. E. (1997). Monitoring early cellular responses in apoptosis is aided by the mitochondrial membrane protein-specific monoclonal antibody APO2.7. *Cytometry* **29,** 306–312.

Koester, S. K., Schlossman, S. F., Zhang, C., Decker, S., and Bolton, W. E. (1998). APO2.7 defines a shared apoptotic–necrotic pathway in a breast tumor hypoxia model. *Cytometry* **33,** 324–332.

Koopman, G., Reutelingsperger, C. P. M., Kuijten, G. A. M., Keehnen, R. M. J., Pals, S. T., and van Oers, M. H. J. (1994). Annexin V for flow cytometric detection of phosphatidylserine expression on B cells undergoing apoptosis. *Blood* **84,** 1415–1420.

Li, X., and Darzynkiewicz, Z. (1995). Labeling DNA strand breaks with BrdUTP. Detection of apoptosis and cell proliferation. *Cell. Prolif.* **28,** 571–579.

Lincz, L. F. (1998). Deciphering the apoptotic pathway: All roads lead to death. *Immunol. Cell. Biol.* **76,** 1–19.

Lucas, F., Kuylen, N., Barcelon, M., and Carter, J. (1996). Identification of a cell type based on a profile of enzymatic activity in a flow cytoenzymological assay. Eleventh Annual Meeting of Clinical Applications of Cytometry, Charleston, South Carolina.

Margolin, N., Raybuck, S. A., Wilson, K. P., Chen, W., Fox, T., Gu, Y., and Livingston, D. J. (1997). Substrate and inhibitor specificity of interleukin-1β-converting enzyme and related caspases. *J. Biol. Chem.* **272,** 7223–7228.

Metivier, D., Dallaporta, B., Zamzami, N., Larochette, N., Susin, S. A., Marzo, I., and Kroemer, G. (1998). Cytofluorometric detection of mitochondrial alteration in early CD95/Fas/APO-1-triggered apoptosis of Jurkat T lymphoma cells. Comparison of seven mitochondrion-specific fluorochromes. *Immunol. Lett.* **61,** 157–163.

Muzio, M., Chinnaiyan, A. M., Kischkel, F. C., O'Rourke, K., Shevchenko, A., Ni, J., Gentz, R., Mann, M., Krammer, P. H., Peter, M. E., and Dixit, V. M. (1996). FLICE, a novel FADD-homologous ICE-CED-3-like protease, is recruited to the CD95 (Fas/APO-1) death-inducing signaling complex. *Cell* **85,** 817–827.

O'Brien, M. C., and Bolton, W. E. (1995). Comparison of cell viability probes compatible with fixation and permeabilization for combined surface and intracellular staining in flow cytometry. *Cytometry* **19,** 243–255.

O'Brien, M. C., Gupta, R. K., Lee, S. Y., and Bolton, W. E. (1995). Use of a multiparametric panel to target subpopulations in a heterogeneous solid tumor model for improved analytical accuracy. *Cytometry* **21,** 76–83.

O'Brien, M. C., Healy, S. F., Jr., Raney, S. R., Hurst, J. M., Avner, B., Hanly, A., Mies, C., Freeman, J. W., Snow, C., Koester, S. K., and Bolton, W. E. (1997). Discrimination of late apoptotic/necrotic cells (type III) by flow cytometry in solid tumors. *Cytometry* **28,** 81–89.

Ormerod, M. G., O'Neal, C. F. O., Robertson, D., and Harrap, K. R. (1994). Cisplatin induces apoptosis in human ovarian carcinoma cell line without a concomitant internucleosomal degradation of DNA. *Exp. Cell Res.* **211,** 231–237.

Orth, K., O'Rourke, K., Salvesen, G. S., and Dixit, V. M. (1996). Molecular ordering of apoptotic mammalian CED-3/ICE-like proteases. *J. Biol. Chem.* **271,** 20977–20980.

Ruez, P., Nassiri, M., Steele, B., and Viciana, A. (1996). Cytofluorographic evidence that thymocyte DPP IV (CD26) activity is altered with stage of ontogeny and apoptotic status. *Cytometry* **23,** 322–329.

Schlegel, J., Peters, I., Orrenius, S., Miller, D. K., Thornberry, N. A., Yamin, T.-T., and Nicholson, D. W. (1996). CPP32/Apopain is a key interleukin 1B converting enzyme-like protease involved in Fas-mediated apoptosis. *J. Biol. Chem.* **271,** 1841–1844.

Seth, A., Zhang, C., Letvin, N., and Schlossman, S. F. (1997). Detection of apoptotic cells from peripheral blood of HIV-infected individuals using a novel monoclonal antibody. *AIDS* **11,** 1059–1061.

Thornberry, N. A., and Lazebnik, Y. (1998). Caspases: Enemies within. *Science* **281,** 1312–1316.

Vermes, I., Haanen, C., Steffens-Nakken, H., and Reutelingsperger, C. (1995). A novel assay for apoptosis flow cytometric detection of phosphatidylserine expression on early apoptotic cells using fluorescein labelled Annexin V. *J. Immunol. Methods* **184,** 39–51.

Villa, P., Kaufmann, S. H., and Earnshaw, W. C. (1997). Caspases and caspase inhibitors. *Trends Biochem. Sci.* **22,** 388–393.

Woodard, M. D., Steele, B. W., Shukla, R. S., Lucas, F. J., and Carter, J. H. (1993). Study of neoplastic enzyme patterns by flow cytometry. *Clin. Chem.* **39,** 1195.

Wyllie, A. H. (1980). Glucocorticoid-induced thymocyte apoptosis is associated with endogenous endonuclease activation. *Nature* **284,** 555–556.

Wyllie, A. H., Kerr, J. F. R., and Currie, A. R. (1980). Cell death: The significance of apoptosis. *Int. Rev. Cytol.* **68,** 251–306.

Xue, D., and Horvitz, H. R. (1995). Inhibition of the *Caenorhabditis elegans* cell-death protease CED-3 by a CED-3 cleavage site in baculovirus p35 protein. *Nature* **377,** 248–251.

Xue, D., Shaham, S., and Horvitz, H. R. (1996). The *Caenorhabditis elegans* cell-death protein CED-3 is a cysteine protease with substrate specificities similar to those of human cpp32 protease. *Genes Dev.* **10,** 1073–1083.

Yuan, J.-Y., Shaham, S., Ledoux, S., Ellis, M. H., and Horvitz, R. H. (1993). The *C. elegans* cell death gene ced-3 encodes a protein similar to mammalian interleukin-1β converting enzyme. *Cell* **75,** 641–652.

Zhang, C., Robertson, M. J., and Schlossman, S. F. (1995). A triplet of nuclease proteins (NP$^{42-50}$) is activated in human Jurkat cells undergoing apoptosis. *Cell. Immunol.* **165,** 161–167.

Zhang, C., Ao, Z., Seth, A., and Schlossman, S. F. (1996). A mitochondrial membrane protein defined by a novel monoclonal antibody is preferentially detected in apoptotic cells. *J. Immunol.* **157,** 3980–3987.

Zou, H., Henzel, W. J., Liu, X., Lutschg, A., and Wang, X. (1997). Apaf-1, a human protein homologous to *C. elegans* ced-4, participates in cytochrome *c*-dependent activation of caspase-3. *Cell* **90,** 405–413.

CHAPTER 23

Analysis of Apoptosis in Plant Cells

Iona E. Weir

Horticulture and Food Research Institute of New Zealand, Ltd.
Auckland, New Zealand

I. Introduction
II. Apoptosis in Plants
III. Problems Associated with Analyzing Plant Cells Using Flow Cytometry
    A. Analyzing Plant Cells with a Flow Cytometer
    B. Autofluorescence
    C. Probe Loading
    D. Fluorescence Energy Transfer
IV. Morphological Changes of Plant Cells
    A. Changes in Scattering Properties
    B. Measurement of the Degree of Chromatin Condensation
    C. Techniques to Measure Nucleosomal DNA Fragmentation
V. Physiological Changes during Apoptosis
    A. Plasma Membrane Integrity
    B. Calcium Measurements
    C. Oxidative Stress
    D. Mitochondrial Membrane Potential
    E. Phosphatidylserine Migration
VI. Conclusion
    References

I. Introduction

Death of cells in both plants and animals can be either fortuitous, as in necrosis caused by external agents or traumatic environmental events, or under genetic control. In the latter case death occurs as a response to outside signals activating biochemical pathways. Programmed cell death (PCD) can be defined as the death of cells by a genetically controlled process. Apoptosis is programmed cell death characterized by specific transformations such as membrane-enclosed vesicles, cellular shrinkage, chromatin condensation, and fragmentation of nuclear DNA (Earnshaw, 1995).

Apoptotic cells were first recognized (Kerr et al., 1972) by their distinct morphological features. These morphological changes can be differentiated from what occurs in necrosis. The distinctive characteristics of necrosis in mammalian cells center around collapse of membrane function and cell metabolism, mitochondrial swelling, cell enlargement, a loss of compartmentation, and an early loss of plasma membrane integrity (Dive et al., 1992). In contrast to apoptotic cells, necrotic cells have no obvious early changes in DNA (Wyllie et al., 1980; Darzynkiewicz et al., 1992; Dive et al., 1992). However, at the later stage of necrosis, the chromatin loses its structure and the nucleus appears amorphous.

Apoptosis, by comparison, involves an ordered set of events. The morphological properties of apoptotic cells are (1) a rapid shrinkage of the cell as it becomes dehydrated, followed by a detachment from its neighboring cells, (2) condensation of the chromatin, (3) systematic DNA cleavage with disintegration of the nucleus and fragmentation into discrete apoptotic bodies, and (4) a late loss of plasma membrane integrity (Martin et al., 1995).

II. Apoptosis in Plants

In plants, apoptosis is involved in a variety of situations. These include responses to environmental stresses, the hypersensitive response to pathogen attack, and plant senescence and fruit ripening. In addition, genetically controlled cell death is also likely to be involved in pollen tube growth down the style of flowers, in xylem development, and in other processes where cells die as part of normal development and differentiation processes (Pennell and Lamb, 1997; O'Brien et al., 1998a). It is only more recently that there has been any evidence provided in plant cells for the presence of morphological features of apoptosis characteristic of mammalian cells. The evidence to date centers on DNA fragmentation detected by DNA laddering and terminal deoxynucleotidyltransferase (TdT)-mediated dUTP labeling (TUNEL) in plant cells responding to fungal infection (Ryerson and Heath, 1996) and to phytotoxin exposure (Wang et al., 1996). Apoptosis is also likely to be involved in xylem development. For instance, differentiation of vascular elements has been shown by Mittler and Lam (1995) and Groover et al. (1997) to involve nucleosomal DNA (nDNA) fragmentation. Apoptosis has also been shown to play a role in plant senescence; nDNA fragmentation has been shown to occur during the synthesis of ethylene inducing carpel senescence (Orzaez and Granell, 1997).

In most cases, this evidence is from *in situ* TUNEL assays. However, this technique has also been adapted for flow cytometry and used to determine the extent of nDNA fragmentation at each stage of apoptosis in plant cells (O'Brien et al., 1997, 1998a). Flow cytometry has been further used to investigate chromatin condensation and nDNA fragmentation both in senescing cells and in those exposed to cytotoxic agents. In the latter case, this has involved establishing that calcium and oxidative signaling pathways are associated with apoptosis in plants, both in its induction and progression (O'Brien *et al.*, 1998a). Furthermore, annexin V binding, indicative of apoptotic changes in phosphatidylserine exposure, has been identified as an early indicator of apoptosis in plants (O'Brien *et al.*, 1997).

A distinctive feature of plant apoptosis is the amplitude of the change in intercalation of DNA by propidium iodide (PI), followed by a sub-G_1 peak known as the apoptotic peak (O'Brien *et al.*, 1997, 1998a). This peak is indicative of the start of DNA loss from the nucleus and is a specific characteristic of apoptosis. These large shifts in fluorescence have provided a model system that has been used to distinguish between different stages of apoptosis and to identify that the early stage is reversible (O'Brien *et al.*, 1998a). It may be that in mammalian cells the early stages of chromatin condensation are reversible, but they are not observed owing to the differences in chromatin structure resulting in smaller shifts in PI fluorescence.

A variety of techniques can be used to study apoptosis. These include microscopy, flow cytometry, and a range of biochemical tests. Fluorescence microscopy has been used to examine chromatin structure by using fluorescent dyes such as acridine orange, PI, ethidium bromide, DAPI (4,6-diamidino-2-phenylindole), or Hoechst 33342. Some of these fluorescent probes have been used to look at changes within the nucleus during the progression of apoptosis in plant tissue (Groover *et al.*, 1997). Ethidium bromide and PI have been used to characterize the changes within the nucleus of protoplasts undergoing senescence (O'Brien *et al.*, 1998a,b). As apoptosis progressed, increasing chromatin condensation was observed as shown by marginalization of the chromatin onto the nuclear membrane, followed by swelling of the nucleolus and a loss of chromatin (Fig. 1). Confocal microscopy has been used to further examine chromatin condensa-

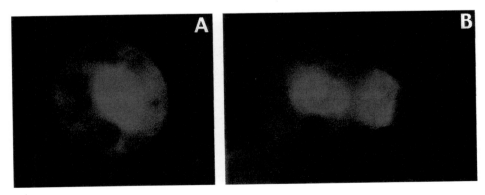

Fig. 1 Apoptotic protoplasts when stained with ethidium bromide (B) show nuclear marginalization onto the nuclear membrane with chromatin loss, resulting in the enlarged nucleolus becoming prominent. Control protoplasts show dense nuclear staining (A). Micrographs are at 1000× magnification. From O'Brien *et al.* (1998). Early stages of the apoptotic pathway in plant cells are reversible. *Plant Journal* **13**(6), 803–814, with permission from Blackwell Science Ltd. (See color plates.)

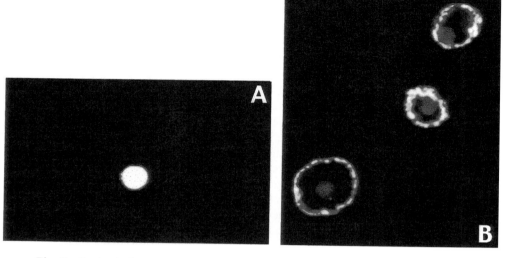

Fig. 2 Confocal microscopy scanning of nuclei stained with PI shows a dense even nuclear structure in noncondensed nuclei (A), in contrast to a loss of DNA and marginalization of chromatin on the nuclear membrane in apoptotic nuclei (B). All micrographs are at 1000× magnification. (See color plates.)

tion and nDNA fragmentation in protoplasts using PI (O'Brien *et al.*, 1998b), with similar results obtained (Fig. 2). This condensing of the chromatin onto the nuclear membrane accompanied by nucleolar swelling can also often be observed in protoplasts by light microscopy (Fig. 3).

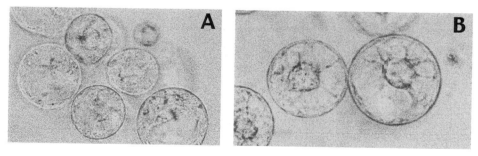

Fig. 3 Micrographs of healthy protoplasts (A) in contrast to apoptotic protoplasts (B) as seen under light microscopy. The normal protoplasts are cytoplasmically dense, in contrast to the vacuolated apoptotic protoplasts in which the nucleus has become more distinct as the cytoplasm becomes more vacuolated. Micrographs are at 1000× magnification. From O'Brien *et al.* (1998). Early stages of the apoptotic pathway in plant cells are reversible. *Plant Journal* **13**(6), 803–814, with permission from Blackwell Science Ltd. (See color plates.)

Examination of changes in morphological and physiological properties in plant cells has resulted in a distinction between necrosis, as demonstrated by the collapse of the DNA, and the loss of membrane integrity and apoptosis, as characterized by chromatin condensation, nDNA fragmentation, and membrane integrity (Groover et al., 1997; O'Brien et al., 1998a). The only mammalian characteristic for apoptosis that has not been observed to date in plant cells is plasma membrane blebbing.

III. Problems Associated with Analyzing Plant Cells Using Flow Cytometry

A. Analyzing Plant Cells with a Flow Cytometer

In most cases there are unique problems associated with the use of flow cytometry to study apoptosis and plant signaling pathways. Plant cells are surrounded by a rigid cell wall, which must first be removed to allow the cells to become single and capable of passing through the flow cell. The plant cell wall is removed by cell wall-degrading enzymes such as cellulases and pectinases in combination with an osmoticum source and antioxidants, to produce protoplasts (Tables I and II).

In Table I the first three ingredients are dissolved one at a time and then gently stirred for 30 min. The remaining ingredients are then sequentially added and fully dissolved, before the next ingredient is added. This approach is essential to ensure enzyme activity is retained. The solution is adjusted to pH 5.8 with 1 M KOH, centrifuged at 10,000 g for 10 min, and the supernatant decanted into a beaker. The solution is filter sterilized and divided into aliquots stored at −20°C.

To isolate protoplasts from whole tissue, shoots are excised, surface sterilized if required in 1.5% hypochlorite solution for 20 min, and rinsed with sterile

Table I
Protoplast Isolation Medium[a]

Ingredient	Amount (g/liter)
Cellulase[b]	10
Macerozyme[c]	3
PY-23[d]	0.5
Sorbitol	91
Calcium nitrate	1.18
Ascorbic acid	1

[a] O'Brien and Lindsay (1993).
[b] Worthington, Freehold, NJ.
[c] Yakult Honsha, Tokyo, Japan.
[d] Seishin Pharmaceuticals, Tokyo, Japan.

Table II
Protoplast Wash

Ingredient	Amount (per liter)
$CaCl_2 \cdot 2H_2O$	73.5 mg
KH_2PO_4	7 mg
KNO_3	146.6 mg
$MgSO_4 \cdot 7H_2O$	55.6 mg
NH_4NO_3	28 mg
$Na_2EDTA\ FeSO_4 \cdot 7H_2O$	6.4 mg
NaCl	15.78 g

[a] Binding (1974). Adjust to pH 5.8.

distilled water. The plant tissue is placed into several droplets of the enzyme solution (Table I) on the glass lid of a glass petri dish. This tissue is finely sliced using a fine edged razor blade dragged through a pair of fine tipped forceps. The sliced tissue is placed into 10 ml of the enzyme solution and digested overnight at 24°C. Protoplasts are sieved through 35-μm steel mesh and pelleted by centrifugation (100 g for 4 min). They are then resuspended in the wash solution (Table II) and repelleted at 80 g for 4 min. This wash step is repeated before the protoplasts are resuspended in fresh wash solution (Table II) to a final concentration of 10^6 cells/ml. For suspension-cultured cells, the cells are harvested by pelleting at 300 g for 5 min, resuspended in the above enzyme solution, and incubated in the dark with gentle rotation (45 rpm) for 2–3 hr. The isolated protoplasts are then treated as described earlier.

This process of removing the cell wall may also induce stress if conditions are not optimized for the cells in question, resulting in the release of phenolics and nucleases from stressed or burst cells, making subsequent analysis futile. The presence of phenolics presents another problem; they are highly fluorescent, often resulting in quenching of the desired fluorescent signal, and should therefore be avoided by the use of antioxidants and careful handling of the cells. The presence of nucleases during protoplast isolation can also be a problem. These can result either from the release of nucleases from dying cells or from the use of nonpurified cellulases.

Owing to the extreme sensitivity of protoplasts to changes in osmotica, pressure, and speed, several of the assays described here cannot be easily performed on analyzer type instruments. A flow cytometer in which sheath pressures and flow rate can be adjusted, combined with a large flow cell tip, is required. Otherwise, the result will be a collapse of the protoplasts as they are aspirated, combined with blockages of the flow cell, requiring extreme measures for their removal. Therefore, protoplasts need to be analyzed using a large flow cell tip combined with a slow flow rate and low sheath pressure.

Another option we have explored to avoid the problem of having to produce protoplasts is to analyze plant apoptosis using macerated leaf or callus tissue that releases intact cells. The technique of macerating leaf tissue to release individual plant cells was first developed by Fukuda and Komamine (1980) using *Zinnia elegans* leaf tissue. These cells can be analyzed on the flow cytometer in the same manner as protoplasts. We have used this model system for flow cytometric analysis of tracheary element development in plant cells. Some species of unicellular algae have also been examined using flow cytometry; in general, however, flow cytometric analysis of plant cells is limited to protoplasts.

B. Autofluorescence

Autofluorescence of plant cells is a perpetual problem. This is often observed under fluorescence microscopy as blue fluorescence from the vacuole and as red fluorescence from the chloroplasts. This is compounded by the incidence of increasing autofluorescence as the plant cell becomes stressed, most likely due to the increase in flavonoids, phenolics, and NADH as defense mechanisms. Autofluorescence consequently causes two different complications in flow cytometric analysis of plant cells. First, quenching caused by this autofluorescence reduces the magnitude of any signal increase. Second, there is a possibility of false positives, as the cells become more autofluorescent with increasing stress.

To overcome this background autofluorescence, it is best to compare unlabeled protoplasts at different stages of apoptosis and analyze them for their level of autofluorescence. Any fluorescence of cells loaded with fluorescent probes above this level can therefore be considered positive. Furthermore, it is best to ensure that the negative control used has undergone the same stress conditions; otherwise, the level of autofluorescence can be different, giving spurious results, for example, in TUNEL overlays.

C. Probe Loading

Another problematic feature of plant cells is that they usually contain a large vacuole into which many organic compounds are rapidly sequestered; unfortunately, these often include the fluorescent compound of interest. Compartmentalization of the probe in the vacuole also causes problems in analysis, as only a short window of time must be examined, instead of allowing the analysis of the same sample for a longer period. Internal sequestration and secretion of agents have been reported as occurring in mammalian cells, but to a lesser extent (Malgaroli *et al.,* 1987; Steinberg *et al.,* 1988). Several compounds have been suggested to inhibit this activity, and sequestration has previously been inhibited by probenecid, an inhibitor of anion transport (Steinberg *et al.,* 1988). Owing to the problem of compartmentalization, it is often critical that a fluorescent microscope is used as a check for correct loading by the probe.

The use of fluorescent probes for the measurement of physiological responses has met with limited success in plant cells (Elliott and Petkoff, 1990). The essential problem in using these probes for plant cells is the loading into the cell. Loading mammalian cells is often achieved using acetomethoxy esters, but the permeability of these species is different for plant cells. Various techniques have been tested for loading plant cells with fluorescent probes, including microinjection (Bush and Jones, 1990; Ayling et al., 1994), electroporation (Bush and Jones, 1990), acid loading involving a decrease in pH of the loading solution (Bush and Jones, 1990; Elliott and Petkoff, 1990), and cold treatment for 2 hr (Zhang et al., 1998). Loading of the probe into the cell becomes even more difficult when the cell wall is present, inasmuch as the rate of loading is the rate at which the extracellular esterases cleave the probe versus the rate of passive loading (Zhang et al., 1998). Subsequently this makes single cell fluorescence imaging of plant cells complicated. We have also observed that the rate and the extent of probe loading varies widely between plant species and cell type, and therefore it is prudent to test loading with each new species studied.

D. Fluorescence Energy Transfer

Fluorescence energy transfer often appears to occur when using multiple fluorescent probes for simultaneous analysis in plant cells. In protoplasts containing high numbers of chloroplasts, we have observed that the fluorescence signal of one probe can falsely enhance the signal of another. An example of this is shown in Fig. 4 where *Zinnia* cells are analyzed for their viability [fluorescein diacetate (FDA)] versus plasma membrane integrity (PI). First levels of autofluorescence were determined, and then FDA was added to the cell population. However, along with the expected increase in green fluorescence there was also an increase in the red fluorescence signal. On the addition of PI, a second population appears that is higher in red fluorescence (therefore PI positive) and negative for FDA uptake. No amount of color compensation or discrimination can be used to remove this artifactual result. This increase in fluorescence of another signal was not observed in our albino cell culture, suggesting that this phenomenon is a result of fluorescence energy transfer from chlorophyll within the chloroplasts. It is therefore important to always test for enhancement of one emission wavelength by either a second probe or chlorophyll when using multiple probes in plant cells. If fluorescence energy transfer is occurring, it is best to identify a baseline to determine whether there is a real change in the fluorescence of your desired probe.

IV. Morphological Changes of Plant Cells

A. Changes in Scattering Properties

During cell death, the capability of the cell to scatter light alters as a result of morphological changes such as cell shrinkage, chromatin condensation, and

GALLERY

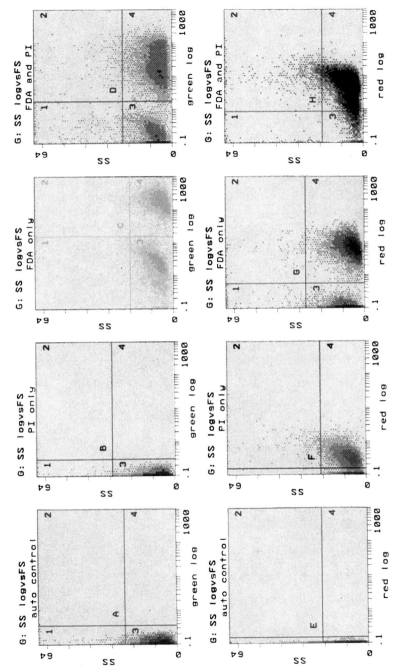

Fig. 4 The eight two-dimensional histograms presented show changes in the fluorescence intensity of fluorescent probes on interaction with chlorophyll in plant cells. The cells used are *Zinnia elegans* cells 2 days after isolation by maceration. The top four two-dimensional histograms display 525 nm emission (green log), while the bottom four display 610 emission (red log). The autocontrol samples demonstrate the level of autofluorescence from the cells for each of these wavelengths. The PI only two-dimensional histograms show the effect of adding 10 μg/ml of PI to the cells; there is an increase in red fluorescence of some of the cells but not of green fluorescence. In contrast for the FDA only two-dimensional histograms, there is not only the expected increase in green fluorescence when 1 μg/ml FDA is added to the cells, but there is also an increase in red fluorescence. The effect of adding both FDA and PI to the same cells is shown in the last two two-dimensional histograms, where both green and red fluorescence appear differently than when the probes are added separately.

nucleosomal fragmentation (Givan, 1992). Generally, during apoptosis there is a decrease in cell size as indicated by a decrease in forward scatter with an increase or no change in side scatter, a measure of cell granularity. In the later stages of apoptosis both forward and side scatter decreases (Darzynkiewicz *et al.,* 1994). This is also observed with plant cells and is a useful way to compare subpopulations by back gating on these differences within the heterogeneous cellular population (Fig. 5). In contrast, necrotic plant cells will dramatically increase in forward scatter just prior to bursting. This can often be observed during drug interactions studies where on the addition of the cytotoxic agent the cells respond by an increase in the parameter of interest; however, on bursting there is a sudden decline. Therefore, the use of monitoring changes in forward versus side scatter is highly recommended to identify differences between an apoptotic versus a necrotic response. Changes in scatter can also result from broken protoplasts, chloroplasts, or necrotic protoplasts as they plasmolyze, and therefore, although an indicator of apoptosis, light scatter changes may not be specific for apoptosis.

B. Measurement of the Degree of Chromatin Condensation

The stability of DNA in chromatin is based on its interactions with histones, other nuclear proteins, and the nuclear matrix (Darzynkiewicz, 1990, 1994).

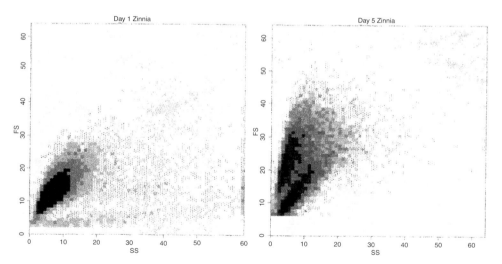

Fig. 5 Changes in morphology of *Zinnia elegans* leaf cells, either immediately following isolation (day 1) or 5 days after isolation (day 5). Changes in the forward scatter versus side scatter were compared using a Coulter EPICS Elite flow cytometer with the 488 nm line of a 15 mW air-cooled argon laser. On day 1 only a single population was observed; however, by day 5, two distinct populations are present. The second population, observed on day 5, has increased forward scatter and decreased side scatter; these cells are developing into tracheary elements and undergoing changes in size and shape. These cells are in the later stages of apoptosis.

Chromatin condensation is a result of increased coiling around the histones. This results in reduced fluorescent labeling with intercalatory dyes such as PI, ethidium bromide, and acridine orange, with binding sites becoming blocked as coiling increases. As the chromatin condenses it therefore also becomes less sensitive to DNA denaturation by acid, heat, DNase, or salt, again as access to the activity sites becomes limited by the tighter coiling (Hotz *et al.,* 1992, 1994). This response is often measured by flow cytometry using changes in fluorescent staining as an indicator of DNA sensitivity to denaturation *in situ*. This correlates with the degree of chromatin condensation (Darzynkiewicz, 1994).

As the chromatin condenses further during the latter stages of apoptosis, it changes from a state of reduced to increased sensitivity to DNA denaturation during acid, DNase I, heat, or salt treatment. This change is probably a result of proteolysis of histone I, followed by proteolysis of other nuclear proteins prior to nDNA fragmentation by endonucleases (Hotz *et al.,* 1994). This phenomenon has been used to detect apoptotic cells on the basis of their increased sensitivity to DNA denaturation by acid treatment (Hotz *et al.,* 1992, 1994; Darzynkiewicz *et al.,* 1993; Gorczyca *et al.,* 1993). We have utilized this to compare for plant nuclear responses at different stages of senescence and have identified that the same three stages described for mammalian cells happen in plant nuclei undergoing apoptosis (O'Brien *et al.,* 1998b).

In nuclei from tobacco cells, a large decrease in PI fluorescence intensity (100 μg/ml) was observed, and this correlated with the age of the cell culture used. This is associated with progressive chromatin condensation, which was confirmed by treatment of nuclei from cells of different ages with 0.1 M HCl for 1 min, washing in Galbraith's buffer (Galbraith *et al.,* 1983), and resuspending in PI (100 μg/ml). Log and quiescent phase cells responded to mild HCl treatment by an increase in PI fluorescence intensity. This indicated that the chromatin was capable of relaxation on disassociation of the histones, with a consequent increase in fluorescence as the intercalating dye PI would have greater access to binding sites. This response was not found in older cells, where nuclei with greater condensation showed little change in fluorescence, most likely a result of the acid treatment not being effective in disassociating the histones due to the increased coiling of the DNA around the histones. In the late stages of the growth cycle, there was a collapse of DNA in the presence of acid, indicating that these nuclei were probably in the later stages of apoptosis as defined in mammalian cells (Hotz *et al.,* 1992, 1994; Darzynkiewicz *et al.,* 1994). This method can also be used to verify whether DNA content analysis results are being affected by chromatin condensation.

In other plant species the concentration of HCl required to return the condensed nuclei to the normal level of PI fluorescence varies, suggesting that the degree of sensitivity of accessibility of chromatin differs from species to species. With each species, a concentration curve should be carried out prior to using this method to determine the DNA content. It is also critical to ensure that following HCl treatment the nuclei are returned to a neutral pH environment, as otherwise there will be a collapse in the DNA.

DNase I digestion of nuclear DNA has also been widely used as an indicator of chromatin condensation (Darzynkiewicz, 1990; Sgorbati *et al.*, 1993), because the extent of condensation determines the rate at which DNase I digestion occurs. Noncondensed nuclei undergo DNA cleavage more rapidly as cleavage sites on the DNA are easily accessible to the DNase. In comparison, the condensed nuclei with their increased coiling around the histones limit the access of the DNase to cleavage sites, resulting in a reduced rate of digestion. Apoptotic cells exhibit increased sensitivity to DNA denaturation (Darzynkiewicz *et al.*, 1993) as demonstrated by a collapse of the apoptotic nuclei.

These three different responses were identified in isolated plant nuclei, with rapid DNA degradation in uncondensed nuclei in contrast to slow DNA degradation in highly condensed nuclei. Nuclei in the final stages of apoptosis collapsed within 10 min in the presence of DNase I, again suggesting, as with HCl treatment, that these cells were in the late stages of apoptosis. These results correlate with the identification of three different responses of *in situ* DNA sensitivity to denaturation during the progression of apoptosis as described by others (Hotz *et al.*, 1992, 1994; Darzynkiewicz *et al.*, 1993; Gorczyca *et al.*, 1993).

C. Techniques to Measure Nucleosomal DNA Fragmentation

Systematic DNA cleavage into 30–50 kbp fragments representative of the size of chromatin loops, followed by random DNA cleavage at links between the nucleosomes resulting in ~200 bp fragments, is unique to apoptosis (Hotz *et al.*, 1994). Several techniques have been developed to use this phenomenon as an apoptotic marker.

1. Apoptotic Peak Detection

Ethanol fixation of cells, followed by aqueous washes, results in the partial extraction of the low molecular weight DNA (~200 bp) fragments produced during the endonucleolysis of DNA. By removing this fraction and subsequently staining with a DNA fluorochrome such as PI, cells will have lower DNA fluorescence since there will be less DNA within the nucleus. With flow cytometry, this smaller DNA subpopulation appears as a sub-G_1 or hypodiploid population, commonly known as the apoptotic peak (Darzynkiewicz *et al.*, 1992; Dive *et al.*, 1992). This apoptotic peak was detected by flow cytometry in plant cells with induced apoptosis or in the late stages of senescence (O'Brien *et al.*, 1997, 1998b). The apoptotic peak (Fig. 6) was obtained by the modification of the nuclear isolation buffer (Galbraith *et al.*, 1983) by the addition of an extra 15 mM citric acid to the buffer to improve the clarity of the sub-G_1 peak (Table III).

2. DNA Laddering

The second technique used to identify nDNA fragmentation in the later stages of apoptosis involves the use of pulse field gel electrophoresis or field inversion

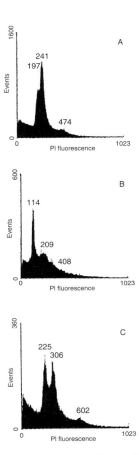

Fig. 6 Appearance of the subdiploid peak of nuclei isolated from apoptotic (as determined by TUNEL) protoplasts using different buffers. The subdiploid peak appears as a shoulder in nuclei isolated with Galbraith's buffer (Galbraith *et al.*, 1983) (A). A distinct subdiploid population is apparent (B) with phosphate–citrate buffer (Hotz *et al.*, 1994) at 45 mM citric acid, but the normal population has collapsed, with a decrease in fluorescence of both populations. The two populations are distinct using modified Galbraith's buffer containing 45 mM sodium citrate (Table III) with little change in fluorescence (C). From Annexin V and TUNEL use in monitoring the progression of apoptosis in plants; I. E. W. O'Brien, C. P. M. Reutelingsperger, and K. M. Holdaway; *Cytometry*. Copyright © 1997 John Wiley & Sons, Inc. Reprinted by permission of Wiley-Liss, Inc., a subsidiary of John Wiley & Sons, Inc.

gel electrophoresis. This results in a DNA "ladder" of different sized DNA fragments corresponding to the apoptotic peak identified by flow cytometry (Dive *et al.*, 1992). Necrotic or nonapoptotic cells, in contrast, produce a DNA "smear" with gel electrophoresis. Although this assay can be used to differentiate between apoptotic and necrotic populations, it is rather limited in its applicability: not all cell types produce a DNA ladder, it conceals heterogeneity, and it does not provide quantification of the number of apoptotic cells or the degree of

Table III
Modified Galbraith's Chopping Buffer[a]

Ingredient	Amount (g/liter)
$MgCl_2 \cdot 6H_2O$	9.15
MOPS	4.19
Sodium citrate	13.23
0.1% Triton X-100	1.0

[a] O'Brien et al. (1997). Adjust to pH 7.0.

DNA breakage (Dive et al., 1992). Nevertheless, it is an important and useful indicator of fragmentation. DNA laddering has been routinely observed in plant cells undergoing apoptosis using equivalent DNA extraction techniques (Levine et al., 1996; Ryerson and Heath, 1996; McCabe et al., 1997).

3. TUNEL

The third technique, using enzymatic labeling, allows earlier detection of DNA breaks (in contrast to the other techniques), while also providing quantification of apoptotic cell numbers and the degree of DNA breakage, using fluorescence labeling detected by flow cytometry. Two forms of enzymatic labeling are available. These are nick translation using DNA polymerase and end labeling using terminal deoxynucleotidyltransferase (TdT).

Nick translation is based on exogenous DNA polymerase catalyzing the addition of modified nucleotides when one strand of the double stranded (ds) DNA has been cut. Hence, this technique is template-dependent and not specific to apoptotic DNA breaks (Gold et al., 1994; Hotz et al., 1994). Alternatively, end-labeling or TUNEL (terminal deoxynucleotidyltransferase *in situ* end labeling) is template-independent, by labeling blunt ends of dsDNA breaks, and therefore is the preferred technique for measuring apoptotic DNA breaks.

a. TUNEL by Flow Cytometry

To compare chromatin condensation versus nDNA fragmentation, a two-step fixation procedure is used involving first cross-linking the proteins and then dehydrating the cell. This enables TUNEL to be used in combination with PI staining. In mammalian cells this is routinely done by using paraformaldehyde followed by methanol fixation (O'Brien et al., 1997). However, in plant cells this results in reduced fluorescence of TUNEL-positive nuclei (Fig. 7). The use of methanol often results in a browning of the solution, indicating a release of phenolics from the vacuole. Oxidized phenolics probably result in the quenching of the TUNEL assay. Methyl Cellosolve is often used in protoplast microscopy as it dehydrates the protoplast at a slower, constant rate and therefore does not result in a sudden osmotic

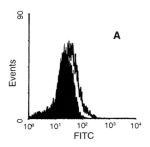

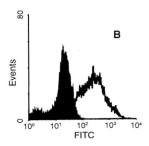

Fig. 7 A population of apoptotic protoplasts split and dehydrated with different fixatives resulted in variation in the degree of TUNEL detected in apoptotic protoplasts. Minimal TUNEL activity was detected in protoplasts dehydrated with methanol (A) in contrast to the high levels of DNA fragmentation indicated when protoplasts were dehydrated with Methyl Cellosolve (B). The negative control is displayed as the shaded population, with the overlaid population as the positive population. From Annexin V and TUNEL use in monitoring the progression of apoptosis in plants; I. E. W. O'Brien, C. P. M. Reutelingsperger, and K. M. Holdaway; *Cytometry*. Copyright © 1997 John Wiley & Sons, Inc. Reprinted by permission of Wiley-Liss, Inc., a subsidiary of John Wiley & Sons, Inc.

change, which causes the vacuole to lyse. Apart from these changes in fixation of the cells, the only other variance for plant cells in regard to TUNEL is to ensure that the increase in fluorescence observed is not due to increasing levels of autofluorescence as the cells become increasingly stressed. Spurious results can readily be obtained with this technique if not all precautions are taken. Therefore, always have both a positive and a negative control.

b. *In Situ* TUNEL

In situ TUNEL is becoming a routine technique in many plant apoptosis studies (Groover *et al.,* 1997; Mittler and Lam, 1995); however, there are many sources of error. It is critical to ensure complete fixation of the tissue, while preventing the release of DNases from burst cells. Autofluorescence can also play a major role in false positives, and it is beneficial to conduct dual labeling with probes that are observed at different wavelengths, for example, DAPI and fluorescein isothiocyanate (FITC), to provide better visualization of changes within the nucleus. Including a negative (deletion of the TdT) and a positive (DNase treated) control helps to ensure validity of the assay.

V. Physiological Changes during Apoptosis

In addition to morphological and nuclear changes, apoptosis can further be characterized by cell signaling events. These include an increase in free cytosolic calcium concentration, an oxidative burst manifest as an increase in free oxygen radical activity, and consequent protective measures such as an increase in cellular glutathione, a collapse in mitochondrial membrane potential, and exposure of the acidic phospholipid phosphatidylserine (PS) on the outer surface of the plasma membrane (Castedo et al., 1996; Hale et al., 1996; Petit et al., 1996; Van Engeland et al., 1996).

All of these physiological changes can be measured by flow cytometry or confocal microscopy using fluorescent probes that, depending on the type used, result either in an increase in fluorescence intensity or in a fluorescent wavelength emission shift. This allows for physiological and biochemical responses to be aligned with morphological changes within the cell while avoiding problems of population heterogeneity. Ultimately this should lead to identification of the cell-signaling pathway for apoptosis.

A. Plasma Membrane Integrity

To measure the loss of plasma membrane function, charged dyes such as PI, which are excluded by the intact membrane of live cells, are used. This provides a simple assay able to distinguish between on the one hand normal and apoptotic cells, which exclude the dye, and on the other necrotic cells, which readily label with PI (Givan, 1992). To analyze further membrane integrity, fluorescein diacetate (FDA), which is nonfluorescent until taken up and hydrolyzed within live cells, is widely used in flow cytometry. By dual staining with FDA and PI, live cells fluoresce green (fluorescein) and dead cells fluoresce red (PI), providing rapid distinction between the two populations (Fig. 8).

B. Calcium Measurements

Indo-1 is preferred to other fluorescent agents such as Fura-red or Fluo-3, since with Indo-1 ratiometric measurements of intracellular calcium levels can be performed, therefore producing a quantification independent loaded probe concentration. Second, in plant cells Fura-red and Fluo-3 are rapidly sequestered into the vacuole within 3 to 10 min, thereby making time analysis futile. Confocal microscopic observation to check loading of Indo-1 showed that after a period of 30 min, Indo-1 was sequestered into the vacuole, and accordingly, protoplasts should not be used after periods of longer than 30 min. To load protoplasts with

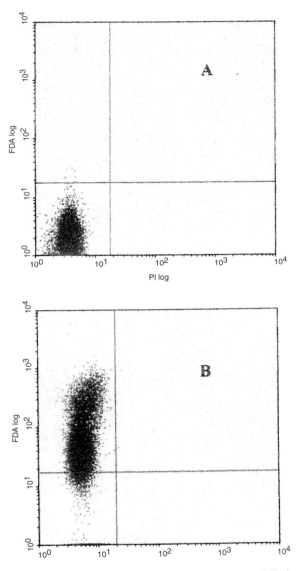

Fig. 8 Analysis of the viability of *Nicotiana plumbaginifolia* (albino cell line) protoplasts labeled with FDA and PI. The autofluorescent (no stain added) control (A) was positioned in the lower left quadrant of the histogram. Normal (or apoptotic) protoplasts showed FDA uptake (as demonstrated by an increase in FDA fluorescence) indicative of cell viability but no PI labeling (no shift to the right) and therefore integrity of the plasma membrane.

Indo-1, warming of the protoplasts at 28°C for 10 min at a pH of 5.8 has been found to be essential (O'Brien *et al.*, 1998a). If too high a temperature is used, heat shock can occur that could create artifacts. The use of Indo-1-AM (acetoxymethyl ester) to label plant cells resulted in a slower rate of sequestration of the probe into the vacuole. This was also observed by Bush and Jones (1990). However, the use of Indo-1-AM resulted in decreased fluorescence in comparison to Indo-1, suggesting incomplete loading, and it appeared to be slightly toxic to the cell. This toxicity has been noted by Bush and Jones (1990) and is possibly due to an accumulation of harmful by-products from hydrolysis of the ester. In some plant species, no loading with Indo-1-AM has been observed.

C. Oxidative Stress

Dichlorofluorescein diacetate (DCFDA) can be loaded into cells for 10 min at 25°C and used to measure reactive oxygen species in protoplasts (Fig. 9).

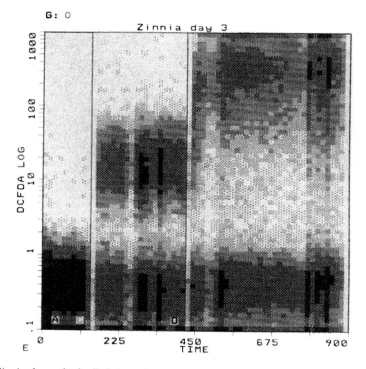

Fig. 9 *Zinnia elegans* leaf cells 3 days after isolation measured for their level of oxidative activity. Cells were split into two tubes; the first was used as an autofluorescent control and analyzed for the first 0–200 sec. Concurrently, the second tube of cells was loaded with DCFDA for 10 min and analyzed for the next 200–450 sec to obtain the base level of oxidative activity within the cells as demonstrated by an increase in fluorescence (*y*-axis). At 450 sec, 130 μM of H_2O_2 was added to the sample with a rapid increase in oxidative activity as shown by an increase in fluorescence (*y*-axis).

Monochlorobimane (MBCl) is used as the main fluorescent probe to measure reduced glutathione (GSH). Loading by warming the protoplasts was also found to be the key to successful labeling with MBCl. Autofluorescence from the protoplasts is very high at the emission wavelength (450–500 nm) at which MBCl is measured, with this autofluorescence often appearing as a second population. At certain stages of oxidative stress the level of MBCl labeling within the protoplasts may not rise above the level of autofluorescence; however, by biochemical assay GSH is present. Therefore, often only dramatic changes in GSH levels are detectable using MBCl due to the high levels of background interference due to the autofluorescence.

D. Mitochondrial Membrane Potential

Measurement of mitochondrial membrane potential in plant cells is problematic. We have tested a wide range of probes [rhodamine 123 (Rh-123), 5,5',6,6'-tetrachloro-1,1'3,3'-tetraethylbenzimidazolcarbocyanine iodide (JC-1), $DiOC_6$] and found with all that there was rapid sequestration of the probe into mitochondria and other organelles such as the chloroplasts, followed by movement to the nuclear membrane and cytoplasm in general. This can occur in less than 30 sec. The rate and extent of this sequestration vary widely between plant species, cell type, and cell condition. It is best to load plant cells starting with a 0.1 μM concentration for 5 min, and to use fluorescent microscopy to follow the probe loading to identify possible concentration and loading times. Loading to the mitochondria can be verified using responses to cytotoxic agents for mitochondria such as carbonyl cyanide m-chlorophenyl-hydrazone (CCCP). Plant mitochondria have been analyzed using Rh-123 (Petit, 1992), but this has been on isolated pure populations of mitochondria.

E. Phosphatidylserine Migration

Cell surface exposure of phosphatidylserine (PS) has been identified as an early indicator of apoptosis in a wide range of mammalian cell types under the action of a variety of apoptosis-inducing signals (Fadok *et al.*, 1992; Koopman *et al.*, 1994; Vermes *et al.*, 1995; Martin *et al.*, 1995; Van Engeland *et al.*, 1996; O'Brien *et al.*, 1997). Using a fluorescent conjugate of annexin V, which binds preferentially to negatively charged phospholipids such as PS (Tait *et al.*, 1989; Andree *et al.*, 1990), it has been shown that the phenomenon of PS exposure precedes nuclear changes and DNA fragmentation (Martin *et al.*, 1995). Again, annexin V binding can be measured by flow cytometry. Protoplasts should be resuspended in a protoplast wash containing either 3 mM calcium chloride or calcium nitrate (calcium is essential for annexin V binding) and labeled with the annexin V for 10 min prior to analysis (O'Brien *et al.*, 1997). The buffer provided in many of the kits available for annexin V are toxic to plant cells, and preferably an alternative should be used that is optimal for the plant cells being used.

VI. Conclusion

When we first started to compare the features of mammalian and plant apoptosis we expected that there would be very few features in common. Instead, we found that the similarities are striking. There are a number of identical morphological and physiological features occurring during apoptosis in both plant and mammalian cells, including nDNA fragmentation, chromatin condensation, retention of plasma membrane integrity, cell shrinkage, oxidative stress, changes in intracellular calcium, and increased exposure of phosphatidylserine on the plasma membrane.

The major difference observed so far has been the extent of the changes in chromatin condensation: in plants there is far greater chromatin condensation during apoptosis than occurs in mammalian cells (O'Brien *et al.*, 1998b). This greater condensation of chromatin in plant cells suggests that chromatin in plants is more dynamic in that there is more coiling of DNA around the histones over a longer period, suggesting that this may be a defense mechanism valuable for an organism vulnerable to static exposure.

A major problem facing plant researchers is the lack of any evidence for the key genes involved in mammalian apoptosis including *p53, Bcl-2,* and *Bax* families. Currently, no one has reported isolating any gene from plants homologous to any of the major genes involved in the mammalian apoptotic pathway. Whether they are present or whether there are in fact other similar regulatory genes present remains a matter of great interest.

The study of plant apoptosis using flow cytometry and/or confocal microscopy has its own set of unique technical problems. However, with persistence these technical problems can be overcome, providing insight into the signal transduction pathway(s) of apoptosis in plants.

References

Andree, H. A. M., Reutelingsperger, C. P. M., Hauptmann, R., Hemker, H. C., Hermens, W. T., and Willems, G. M. (1990). Binding of vascular anticoagulant α (VAC α) to planar phospholipid bilayers. *J. Biol. Chem.* **265,** 4923–4928.

Ayling, S. M., Brownlee, C., and Clarkson, D. T. (1994). The cytoplasmic streaming response of tomato root hairs to auxin; observations of cytosolic calcium levels. *J. Plant Physiol.* **143,** 184–188.

Binding, H. (1974). Regeneration of haploid and diploid plants from protoplasts of *Petunia hybrida* L. *Z. Pflanzenphysiol.* **74,** 327–356.

Bush, D. S., and Jones, R. L. (1990). Measuring intracellular Ca^{2+} levels in plant cells using the fluorescent probes, Indo-1 and Fura-2. *Plant Physiol.* **93,** 841–845.

Castedo, M., Hirsch, T., Susin, S. A., Zamzami, N., Marchetti, P., Macho, A., and Kroemer, G. (1996). Sequential acquisition of mitochondrial and plasma membrane alterations during early lymphocyte apoptosis. *J. Immunol.* **157,** 512–521.

Darzynkiewicz, Z. (1990). Probing nuclear chromatin by flow cytometry. *In* "Flow Cytometry and Sorting" (M. R. Melamed, T. Lindmo, and M. L. Mendelsohn, eds.), 2nd Ed., pp. 315–340. Wiley-Liss, New York.

Darzynkiewicz, Z. (1994). Acid-induced denaturation of DNA *in situ* as a probe of chromatin structure. *In* "Methods in Cell Biology" (Z. Darzynkiewicz, J. P. Robinson, and H. A. Crissman, eds.), Vol. 41, pp. 527–541. Academic Press, San Diego.

Darzynkiewicz, Z., Bruno, S., Del Bino, G., Gorczyca, W., Hotz, M. A., Lassota, P., and Traganos, F. (1992). Features of apoptotic cells measured by flow cytometry. *Cytometry* **13,** 795–808.

Darzynkiewicz, Z., Gorczyca, W., Lassota, P., and Traganos, F. (1993). Altered sensitivity of DNA *in situ* to denaturation in apoptotic cells. *Ann. N.Y. Acad. Sci.* **677,** 334–340.

Darzynkiewicz, Z., Li, X., and Gong, J. (1994). Assays of cell viability: Discrimination of cells dying by apoptosis. *In* "Methods in Cell Biology" (Z. Darzynkiewicz, J. P. Robinson, and H. A. Crissman, eds.), Vol. 41, pp. 15–38. Academic Press, San Diego.

Dive, C., Gregory, C. D., Phipps, D. J., Evans, D. L., Milner, A. E., and Wyllie, A. H. (1992). Analysis and discrimination of necrosis and apoptosis (programmed cell death) by multiparameter flow cytometry. *Biochim. Biophys. Acta* **1133,** 275–285.

Earnshaw, W. C. (1995). Apoptosis: Lessons from *in vitro* systems. *Trends Cell Biol.* **5,** 217–220.

Elliott, D. C., and Petkoff, H. S. (1990). Measurement of cytoplasmic free calcium in plant protoplasts. *Plant Sci.* **67,** 125–131.

Fadok, V. A., Voelker, D. R., Campbell, P. A., Cohen, J. J., Bratton, D. L., and Henson, P. M. (1992). Exposure of phosphatidylserine on the surface of apoptotic lymphocytes triggers specific recognition and removal by macrophages. *J. Immunol.* **148,** 2207–2216.

Fukuda, H., and Komamine, A. (1980). Establishment of an experimental system for the study of tracheary element differentiation from single cells isolated from the mesophyll of *Zinnia elegans*. *Plant Physiol.* **65,** 57–60.

Galbraith, D. W., Harkins, K. R., Maddox, J. M., Ayres, N. M., Sharma, D. P., and Firoozabady, E. (1983). Rapid flow cytometric analysis of the cell cycle in intact plant tissues. *Science* **220,** 1049–1051.

Givan, A. L. (1992). "Flow Cytometry First Principles." Wiley-Liss, New York.

Gold, R., Schmied, M., Giegerich, G., Breitschopf, H., Hartung, H. P., Toyka, K. V., and Lassman, H. (1994). Differentiation between cellular apoptosis and necrosis by the combined use of *in situ* tailing and nick translation techniques. *Lab. Invest.* **71,** 219–225.

Gorczyca, W., Gong, J., and Darzynkiewicz, Z. (1993). Detection of DNA strand breaks in individual apoptotic cells by the *in situ* terminal deoxynucleotidyl transferase and nick translation assays. *Cancer Res.* **53,** 1945–1951.

Groover, A., DeWitt, N., Heidel, A., and Jones, A. (1997). Programmed cell death of plant tracheary elements differentiating *in vitro*. *Protoplasma* **196,** 197–211.

Hale, A. J., Smith, C. A., Sutherland, L. C., Stoneman, V. E. A., Longthorne, V. L., Culhane, A. C., and Williams, G. T. (1996). Apoptosis: Molecular regulation of cell death. *Eur. J. Biochem.* **236,** 1–26.

Hotz, M. A., Traganos, F., and Darzynkiewicz, Z. (1992). Changes in nuclear chromatin related to apoptosis or necrosis induced by the DNA topoisomerase II inhibitor fostriecin in MOLT-4 and HL-60 cells are revealed by altered DNA sensitivity to denaturation. *Exp. Cell Res.* **201,** 184–191.

Hotz, M. A., Gong, J., Traganos, F., and Darzynkiewicz, Z. (1994). Flow cytometric detection of apoptosis: Comparison of the assays of *in situ* DNA degradation and chromatin changes. *Cytometry* **15,** 237–244.

Kerr, J. F. R., Wyllie, A. H., and Currie, A. R. (1972). Apoptosis: A basic biological phenomenon with wide-ranging implications in tissue kinetics. *Br. J. Cancer* **26,** 239–257.

Koopman, G., Reutelingsperger, C. P. M., Kuijten, G. A. M., Keehnen, R. M. J., Pals, S. T., and van Oers, M. H. J. (1994). Annexin V for flow cytometric detection of phosphatidylserine expression on B cells undergoing apoptosis. *Blood* **84,** 1415–1420.

Levine, A., Pennell, R. I., Alvarez, M. E., Palmer, R., and Lamb, C. (1996). Calcium-mediated apoptosis in a plant hypersensitive disease resistance response. *Curr. Biol.* **6,** 427–437.

McCabe, P. F., Levine, A., Meijer, P. J., Tapon, N. A., and Pennell, R. I. (1997). A programmed cell death pathway activated in carrot cells cultured at low cell density. *Plant J.* **12,** 267–280.

Malgaroli, A., Milani, D., Meldolesi, J., and Pozzan, T. (1987). Fura-2 measurement of cytosolic free Ca^{2+} in monolayers and suspensions of various types of animal cells. *J. Cell Biol.* **105,** 2145–2155.

Martin, S. J., Reutelingsperger, C. P. M., McGahon, A. J., Rader, J. A., van Schie, R. C. A. A., LaFace, D. M., and Green, D. R. (1995). Early redistribution of plasma membrane phospatidylserine is a general feature of apoptosis regardless of the initiating stimulus: Inhibition by overexpression of Bcl-2 and Abl. *J. Exp. Med.* **182**, 1545–1556.

Mittler, R., and Lam, E. (1995). *In situ* detection of nDNA fragmentation during the differentiation of tracheary elements in higher plants. *Plant Physiol.* **108**, 489–493.

O'Brien, I. E. W., and Lindsay, G. C. (1993). Protoplasts to plants in Gentianaceae. Regeneration of lisianthus (*Eustoma grandiflorum*) is affected by calcium ion preconditioning, osmolality and pH of the culture media. *Plant Cell Tissue Organ Cult.* **33**, 31–37.

O'Brien, I. E. W., Reutelingsperger, C. P. M., and Holdaway, K. M. (1997). The use of annexin V and TUNEL to monitor the progression of apoptosis in plants. *Cytometry* **29**, 28–33.

O'Brien, I. E. W., Baguley, B. C., Murray, B. G., Morris, B. A. M., and Ferguson, I. B. (1998a). The early stages in the pathway for apoptosis in plant cells are reversible. *Plant J.* **13**, 803–814.

O'Brien, I. E. W., Murray, B. G., Baguley, B. C., Morris, B. A. M., and Ferguson, I. B. (1998b). Major changes in chromatin condensation suggest the presence of an apoptotic pathway in plant cells. *Exp. Cell Res.* **241**, 46–54.

Orzaez, D., and Granell, A. (1997). DNA fragmentation is regulated by ethylene during carpel senescence in *Pisum sativum*. *Plant J.* **11**, 137–144.

Pennell, R. I., and Lamb, C. (1997). Programmed cell death in plants. *Plant Cell* **9**, 1157–1168.

Petit, P. X. (1992). Flow cytometric analysis of rhodamine 123 fluorescence during modulation of the membrane potential in plant mitochondria. *Plant Physiol.* **98**, 279–286.

Petit, P. X., Susin, S.-A., Zamzami, N., Mignotte, B., and Kroemer, G. (1996). Mitochondria and programmed cell death: Back to the future. *FEBS Lett.* **396**, 7–13.

Ryerson, D. E., and Heath, M. C. (1996). Cleavage of nuclear DNA into oligonucleosomal fragments during cell death induced by fungal infection or by abiotic treatments. *Plant Cell* **8**, 393–402.

Sgorbati, S., Berta, G., Trotta, A., Schellenbaum, L., Citterio, S., Dela Pierre, M., Gianinazzi-Pearson, V., and Scannerini, S. (1993). Chromatin structure variation in successful and unsuccessful arbuscular mycorrhizas of pea. *Protoplasma* **175**, 1–8.

Steinberg, T. H., Swanson, J. A., and Silverstein, S. C. (1988). A prelysosomal compartment sequesters membrane-impermeant fluorescent dyes from the cytoplasmic matrix of J774 macrophages. *J. Cell Biol.* **107**, 887–896.

Tait, J. F., Gibson, D., and Fujikawa, K. (1989). Phospholipid binding properties of human placental anticoagulant protein-I, a member of the lipocortin family. *J. Biol. Chem.* **264**, 7944–7949.

Van Engeland, M., Ramaekers, F. C. S., Schutte, B., and Reutelingsperger, C. P. M. (1996). A novel assay to measure loss of plasma membrane asymmetry during apoptosis of adherent cells in culture. *Cytometry* **24**, 131–139.

Vermes, I., Haanen, C., Steffens-Nakken, H., and Reutelingsperger, C. P. M. (1995). A novel assay for apoptosis. Flow cytometric detection of phosphatidylserine expression on early apoptotic cells using fluorescein labelled annexin V. *J. Immunol. Methods* **184**, 39–51.

Wang, H., Li, J., Bostock, R. M., and Gilchrist, D. G. (1996). Apoptosis: A functional paradigm for programmed cell death induced by a host-selective phytotoxin and invoked during development. *Plant Cell* **8**, 375–391.

Wyllie, A. H., Kerr, J. F. R., and Currie, A. R. (1980). Cell death: The significance of apoptosis. *Int. Rev. Cytol.* **68**, 251–306.

Zhang, W. H., Rengel, Z., and Kuo, J. (1998). Determination of intracellular Ca^{2+} in cells of intact wheat roots: Loading of acetoxymethyl ester of Fluo-3 under low temperature. *Plant J.* **15**, 147–151.

CHAPTER 24

Difficulties and Pitfalls in Analysis of Apoptosis

Zbigniew Darzynkiewicz,[*] Elżbieta Bedner,[†] and Frank Traganos[*]

[*]Brander Cancer Research Institute
New York Medical College
Hawthorne, New York 10532

[†]Department of Pathology
Pomeranian School of Medicine
Szczecin, Poland

I. Introduction
II. Apoptotic Index May Not Be Correlated with Incidence of Cell Death
III. Difficulties in Estimating Frequency of Apoptosis by Analysis of DNA Fragmentation
IV. The Lack of Evidence Is Not Evidence for the Lack of Apoptosis
V. Misclassification of Apoptotic Bodies or Nuclear Fragments as Single Apoptotic Cells
VI. Apoptosis versus Necrosis versus "Necrotic Stage" of Apoptosis
VII. Selective Loss of Apoptotic Cells during Sample Preparation
VIII. Live Cells Engulfing Apoptotic Bodies Masquerade as Apoptotic Cells
IX. The Problems with Commerical Kits and Reagents
X. Cell Morphology Is Still the Gold Standard for Identification of Apoptotic Cells
XI. Laser Scanning Cytometry: Have Your Cake and Eat It Too
References

I. Introduction

There has been an explosive growth of interest in mechanisms associated with cell death, in particular cell death by apoptosis (for reviews, see Dragovich *et al.*, 1998; Kerr *et al.*, 1994; Kroemer, 1998; Meier and Evan, 1998; Nuñez *et al.*, 1998; Reed, 1998; Vaux and Korsmeyer, 1999). Analysis of cell death is now commonplace not only in field of cell and molecular biology but also in oncology, immunology, embryology, endocrinology, hematology, neurology, and other disciplines. Flow cytometry has become the preferred methodology for analysis of apoptosis both in the research environment as well as in clinical settings (for reviews, see Darzynkiewicz *et al.*, 1992, 1997a,b; Ormerod, 1998). It offers the possibility of rapid, accurate, and unbiased analysis of large populations of individual cells. The most attractive feature of cytometry is that it provides the possibility of measuring several cell attributes simultaneously in a large number of individual cells. Such a measurement directly reveals correlations between the attributes measured within the same cells. This feature is of particular value in studies of the mechanisms of cell death and regulatory pathways predisposing to, or protecting the cell from, death. In this application cytometry is primarily used to measure immunocytochemically labeled cell constituents that play a role in the regulation of apoptosis. Members of Bax/Bcl-2 protein family, caspases, the protooncogenes c-*myc* and *ras,* or products of tumor suppressor genes p53 and pRB are most frequently measured (Meier and Evan, 1998; Reed, 1998). By virtue of their close association with the early (and perhaps still reversible) stages of apoptosis, changes in mitochondrial metabolism, particularly in mitochondrial electrochemical transmembrane potential ($\Delta\Psi_m$), and the presence of reactive oxygen intermediates (ROIs) are also frequently studied by cytometry (Kroemer, 1998; Zamzani *et al.*, 1998).

Another, even more frequent application of cytometry is in the identification and quantitation of apoptotic cells. Their identification is generally based on a particular cytochemical or molecular change that is characteristic, and hopefully unique, for apoptosis. Numerous methods have been developed for identification of apoptotic cells. Some methods rely on the detection of apoptosis-associated changes in the distribution of plasma membrane phospholipids or altered transport function of the membrane (Fadok *et al.*, 1992; Koopman *et al.*, 1994). Others are based on endonucleolytic DNA degradation that results in loss of fragmented DNA from the cell; apoptotic cells are then recognized by their fractional DNA content (Nicoletti *et al.*, 1991; Umansky *et al.*, 1981). When DNA extraction is prevented by fixation with formaldehyde, the *in situ* presence of DNA strand breaks in apoptotic cells can be detected by their labeling with fluorochrome-conjugated nucleotides in a reaction utilizing exogenous terminal deoxynucleotidyltransferase (TdT) (Gorczyca *et al.*, 1992, 1993a,b; Li and Darzynkiewicz, 1995; Li *et al.*, 1996). The apoptosis associated changes in cell size and granularity can be detected by analysis of laser light scattered by the cell in the forward and side directions (Ormerod *et al.*, 1995). Activation of the apoptosis-associated proteases, caspases (Nuñez *et al.*, 1998; see Chapter 22 of this volume), the increased sensitivity of DNA *in situ* to denaturation (single-strandedness) (Hotz *et al.*, 1992), or the appearance of apoptosis-associated antigens (Koester *et al.*, 1998) provide still other markers of apoptotic cells used in cytometry. A variety of reagent kits to detect apoptosis based on markers listed above are being offered by many vendors.

Several method chapters and protocols on the use of cytometry to identify apoptotic cells have been published (Darzynkiewicz and Li, 1996; Darzynkiewicz *et al.,* 1994, 1997a,b; see Chapter 22, this volume). Many of the commercially available reagent kits include detailed protocols and instructions for use. However, most of these chapters and protocols fail to adequately address certain problems and difficulties that are often encountered in the analysis of apoptosis. The common pitfalls and inappropriate uses of the methodology are apparent from reviewing the literature. Some of these problems are generic to most of the methods: they generally pertain to data interpretation, in particular how the frequency of apoptotic cells (apoptotic index) relates to the incidence of cell death in cultures or in tissue. Other problems are specific to particular methods or cell systems. Certain issues associated with the inappropriate use of flow cytometry in the analysis of apoptosis were described previously (Darzynkiewicz *et al.,* 1998). The most common errors in measurement of apoptosis as well as the frequent mistakes in the analysis and interpretation of the data have been updated and are discussed in this chapter.

II. Apoptotic Index May Not Be Correlated with Incidence of Cell Death

It is often assumed that the frequency of apoptotic cells (apoptotic index, AI), *in vivo* or in cultures, is a reflection of "how many" cells underwent apoptosis, for example, as a result of the treatment or over a given time interval. Thus, for example, when a particular drug is administered to a culture and several hours later the percentage of apoptotic cells is higher in this culture than in a culture treated with another drug, an assumption is often made that the first drug was more effective in *inducing* apoptosis or in *cell kill*. Likewise, the increased AI *in vivo,* in the tissue, is often interpreted as reflecting the increased *cell death incidence* in this tissue. This is not always a correct assumption because apoptosis is a kinetic event. The entire apoptotic process, from the initiation to the total disintegration of the cell, is of short and variable duration. The time window during which individual apoptotic cells demonstrate their characteristic features (markers) that allow them to be recognizable varies depending on (1) the method used, (2) the cell type, and/or (3) the nature of the inducer of apoptosis. Thus, for example, variable estimates of the AI in the same cell population are expected when different methods, differing in the width of the time window through which they recognize apoptosis, are used. Furthermore, some inducers may slow down or accelerate the apoptotic process. This may occur if the rate of either formation and/or shedding of apoptotic bodies, endonucleolysis, or proteolysis is affected by the inducer or by the growth conditions (e.g., temperature, pH). Thus, when the duration of apoptosis is shortened the frequency of apoptotic cells (AI) is diminished even if the incidence of cell death remains the same. Conversely, prolongation of apoptosis in absence of any change in incidence of cell death manifests by the increase in AI. Protease inhibitors, for example, including inhibitors of serine proteases, delay nuclear fragmentation and prolong the process of apoptosis (Hara *et al.*, 1996). Induction of apoptosis in their presence, therefore, is expected to be reflected by the increased AI compared to parallel cultures with the same incidence of cell death but where apoptosis is of shorter duration.

The duration of apoptosis is also different in different cell types and tissues, as well as *in vivo* and *in vitro*. Cells of hematopoietic tumor lines (e.g., such as HL-60 cells) *in vitro*, when triggered by DNA damage, for example, by DNA topoisomerase inhibitors, progress through the entire process of apoptosis rapidly; the cells disintegrate totally within 4 to 6 hr after the treatment. The same treatment of MCF-7 breast carcinoma cells triggers apoptosis after a 24-hr delay and leads to an apoptotic process that is of much longer duration (Del Bino *et al.*, 1999). Apoptosis *in vivo*, within tissues, appears to be rapid, with the remains of apoptotic cells completely removed from the tissue in a short time. This is evidenced by the fact that under conditions of homeostasis the AI is often similar to the mitotic index (e.g., 1–2%). Because the duration of mitosis is approximately 1 hr, the duration of apoptosis must be similarly short. It is quite possible, however, that the rate of removal of the remains of apoptotic cells by neighboring cells and by macrophages also varies depending on the tissue type.

In conclusion, an observed increase in AI, for example, as a result of a particular treatment, may indicate that indeed the incidence of cells dying by apoptosis was increased by the treatment. As discussed, however, it may also indicate that the same number of cells were dying but that the duration of apoptosis was prolonged. A combination of both, namely, increased incidence of apoptosis and prolongation of this process, could occur as well. Unfortunately, no methods yet exist to arrest cells in apoptosis and therefore to obtain a cumulative estimate of the rate of cell entrance to apoptosis. Such an approach is available, for example, for mitosis, which can be arrested by microtubule poisons, and the rate of cell entrance to mitosis can be calculated to obtain a quantitative estimate of the *cell birth rate* (Darzynkiewicz *et al.*, 1986). Perhaps the apoptotic process can be arrested by some inhibitors of caspases, which would then allow one to measure the rate of cell entrance to apoptosis or the *cell death rate*. Nevertheless, the percentage of apoptotic cells in a cell population as presently estimated by any given method (AI) is not a measure of the incidence of cells dying by apoptosis.

To estimate the incidence of cell death, for example, as a result of treatment with chemical or physical agent, the absolute number (not the percentage) of live cells should be measured in the treated culture and compared with the appropriate untreated control. A correction should also be made to account for the rate of cell proliferation. The latter may be obtained from the classic cell growth curves, when the number of live cells is plotted against the time in culture. Alternatively, it may be obtained from rate of cell entrance to mitosis in a stathmokinetic experiment by arresting cells in mitosis (Darzynkiewicz *et al.*, 1986). The observed deficit in the actual number of live cells from the expected number of live cells estimated based on the rate of cell birth provides an estimate of the cumulative cell loss (death) during the measured time interval. Indirectly, the cell proliferation rate can be inferred from the percentage of cells incorporating bromodeoxyuridine (BrdU) or from the mitotic index, under the assumption that the treatment which induces apoptosis does not affect the duration of any particular phase of the cell cycle. The estimate of apoptosis (by detection of cells with DNA strand breaks) may be combined with analysis of BrdU incorporation and cell cycle position, by flow or laser scanning cytometry (Li et al., 1996). This methodology, which reveals AI and the fraction of cells replicating DNA in the same sample, may be particularly useful in evaluating the proliferative potential of tumors.

III. Difficulties in Estimating Frequency of Apoptosis by Analysis of DNA Fragmentation

A common misconception in analysis of apoptosis is that the amount of fragmented (low MW, "extractable") DNA detected in cultures, tissue, or cell extracts, etc., is proportional to the frequency of apoptosis. Many methods were developed to estimate the amount of fragmented DNA, and numerous reagent kits are being sold for that purpose. They include direct quantitative colorimetric analysis of "soluble" DNA, densitometry of "DNA ladders" on gels, and immunochemical assessment of nucleosomes. Some of these approaches are advertised by the vendors as quantitative, in the sense that they are intended to provide information regarding the frequency of apoptosis in cell populations. Such claims are grossly incorrect. Namely, the amount of fragmented (low MW) DNA that can be extracted from a single apoptotic cell varies over a wide range depending on the stage of apoptosis. Although early during apoptosis only a small fraction of DNA is degraded, when apoptosis is more advanced nearly all DNA is fragmented. Thus, the amount of low MW DNA that is extracted from a single apoptotic cell varies manyfold depending on the stage of apoptosis. As a result, the total content of low MW DNA extracted from the cell population, or the ratio of the low to high MW fraction, does not provide information about the frequency of apoptotic cells (apoptotic index), even in relative terms, for example, for comparison of cell populations. For this reason biochemical methods based on analysis of fragmented DNA cannot be used to quantitatively estimate the frequency of apoptosis.

DNA "laddering" observed during electrophoresis provides evidence of internucleosomal DNA cleavage that is considered one of the hallmarks of apoptosis (Arends et al., 1990). Analysis of DNA fragmentation by gel electrophoresis to detect such laddering is thus a valuable method to demonstrate the apoptotic mode of cell death. It should not be used, however, as a means to quantitate the frequency of apoptosis.

In some cell systems apoptosis may occur without internucleosomal DNA cleavage; the products of DNA fragmentation are large DNA sections that cannot be easily extracted from the cell (Oberhammer et al., 1993). Obviously, in these systems apoptosis cannot be revealed by the presence of DNA laddering on gels or by analysis of low MW products. These instances are discussed in the next section of this chapter.

IV. The Lack of Evidence Is Not Evidence for the Lack of Apoptosis

There are numerous publications describing cell death that resembles apoptosis which lacks, however, one or more characteristic apoptotic features ("atypical apoptosis") (e.g., Cohen et al., 1992; Collins et al., 1992; Ormerod et al., 1994; Zakeri et al., 1993; Zamai et al., 1996). Thus, for example, apoptosis-associated DNA endonucleolysis frequently terminates after generating 50- to 300-kb breaks and does not proceed to generate internucleosomal-sized DNA fragments (Oberhammer et al., 1993). Such cells contain relatively few in situ DNA strand breaks compared with classic apoptosis. Methods based on detection of DNA laddering on gels will fail to identify apoptosis in such situations. [It should be noted, however, that the 50- to 300-kb fragments can be detected by pulsed field electrophoresis (Oberhammer et al., 1993).] Because of the paucity of DNA strand breaks under such circumstances, it is also difficult to identify such cells by the DNA strand break TdT-mediated dUTP-biotin nick end labeling (TUNEL) assay (Del Bino et al., 1999).

Apoptosis is often induced by agents that are inhibitors of a particular enzyme or metabolic pathway that is associated with apoptosis. Identification of apoptotic cells based on activity of this enzyme or analysis of the pathway involved will be unsuccessful. For example, apoptosis can be induced by certain protease inhibitors (Hara et al., 1996). Because these inhibitors (perhaps by inhibiting proteolysis of nuclear lamin) prevent nuclear fragmentation, this feature (nuclear fragmentation) cannot be used as a marker distinguishing apoptotic cells.

Application of more than one method, each based on a different principle (i.e., detecting a different cellular feature of apoptosis), offers a better chance of detecting atypical apoptosis than does any single method. As mentioned, if DNA in apoptotic cells is fragmented to 50- to 300-kb sections it is not extractable, and such cells cannot be identified as apoptotic either by the method based on analysis of DNA content or DNA laddering during electrophoresis. It is likely, however, that such apoptotic cells can be recognized based on their reduced F-actin stainability with fluorescein isothiocyanate (FITC)-phalloidin (Endersen et al., 1995), by their reactivity with a fluorochromed annexin V (Koopman et al., 1994), by the drop in mitochondrial transmembrane potential detected by transmembrane potential-sensing flourochrome probes (Cossarizza et al., 1994; Zamzani et al., 1998), or by other markers.

V. Misclassification of Apoptotic Bodies or Nuclear Fragments as Single Apoptotic Cells

Identification of apoptotic cells often relies on cellular DNA content measurements by flow cytometry. It was initially observed that following cell fixation in ethanol and staining of their DNA, apoptotic cells were recognized by virtue of their lower stainability compared to G_1 cells (Umansky et al., 1981). On the DNA content frequency histograms they occupied a position between the origin of the DNA coordinate and the G_1 peak ("sub-G_1" cells). Their decreased DNA stainability was explained as partially due to the extraction of fragmented, low MW DNA during the staining procedure (Darzynkiewicz et al., 1992; Gong et al., 1994). To optimize the distinction between intact G_1 and apoptotic cells, the extent of DNA extraction can be enhanced by using buffers of higher molar strength prior to fixation (Gong et al., 1994). The apoptotic (sub-G_1) cells can thus be distinguished based on their fractional DNA content, which may partly be due to DNA extraction during the staining procedure but may also reflect DNA loss due to shedding of apoptotic bodies, or even to changes in chromatin structure (condensation) that make DNA less accessible to the fluorochrome.

The major problem with this methodology stems from the fact that, commonly, the analysis is performed on cells that were subjected to treatment with a detergent or hypotonic solution instead of fixation. Such treatment lyses the plasma membrane and gives rise to the following artifacts. (1) Because the nucleus of an apoptotic cell is fragmented, numerous individual chromatin fragments are present in a single cell. On cell lysis, each individual chromatin fragment, having a fractional DNA content, is separately released and, when measured, is erroneously identified as a single apoptotic cell. Therefore, the percentage of objects represented by the sub-G_1 peak significantly overestimates the actual AI. (2) Similar problems arise when, for instance, chromosomes are released from lysed mitotic cells. Both individual chromosomes as well as chromosome aggregates having a fractional DNA content may mistakenly be identified as apoptotic cells. This problem is exacerbated when apoptosis is induced by agents that increase the proportion of mitotic cells, for example, taxol or other microtubule poisons. (3) Often, following cell irradiation or treatment of cells with DNA damaging drugs, micronuclei are formed. Their number depends on the degree of DNA damage and duration of treatment. Having a fractional DNA content, micronuclei may also be erroneously identified as apoptotic cells.

A gentle permeabilization of the cell with a detergent but in the presence of exogenous proteins such as serum or serum albumin prevents lysis of the plasma membrane. It was shown that the presence of 1% (w/v) albumin or 10% (v/v) serum protects cells from lysis (e.g., induced by 0.1% Triton X-100) without affecting their permeabilization by detergent. In fact, this method is used for simultaneous analysis of DNA and RNA as well as for detection of apoptotic cells characterized by reduced DNA content (Darzynkiewicz, 1994). However, apoptotic or nonapoptotic cells suspended in saline containing detergent and serum proteins remain very fragile, and pipetting, vortexing, or even shaking the tube containing the suspension causes their lysis and release of the cell constituents into solution.

Logarithmic amplification of the fluorescence signal is frequently used to measure and display cellular DNA content in the methods that employ detergents to quantitate apoptosis. A logarithmic scale allows one to measure and record events with 1 or even 0.1% of the DNA content of intact, nonapoptotic cells. The majority of such objects cannot be individual apoptotic cells. In the case of cell lysis by detergents, as discussed earlier, these objects represent nuclear fragments, individual apoptotic bodies, individual chromosomes, chromosome aggregates, micronuclei, or contaminating bacteria.

To exclude objects with a minimal DNA content that may not be apoptotic cells from analysis, it is advisable during fluorescence measurement to set the threshold of DNA detection at a constant level, for example, at 1/10th or 1/20th the fluorescence value of intact G_1 cells. This would eliminate all particles with a DNA content less than 10 or 5% of that of G_1 cells from the analysis. Although the AI may then be underestimated, the underestimate is constant and introduces less error than would occur if all objects with a fractional DNA content were counted. In essence, there is little reason to use a logarithmic scale to measure DNA content because a linear scale provides better assurance that objects with a minimal DNA content are not included in the analysis.

Difficulties in identification of apoptotic cells by this methodology may also occur when G_2/M or late S phase cells undergo apoptosis, and the extraction of low MW DNA from them is insufficient to shift them to a position below the G_1 peak. On DNA frequency histograms, such cells may overlap with the nonapoptotic G_1 cells. Likewise, when cells grow at two DNA ploidy levels, the apoptotic cells of a higher DNA ploidy despite the loss of DNA may still overlap in DNA content with the nonapoptotic cells of lower DNA ploidy. As mentioned, a more extensive DNA extraction, which is provided using high molarity buffers and improves separation of apoptotic cells on DNA frequency histograms (Gong et al., 1994), may be useful in such situations.

The method of identification of apoptotic cells on the basis of their fractional DNA content, even if it is based on analysis of fixed cells (i.e., is devoid of artifacts of cell lysis discussed earlier), is not very specific. Mechanically damaged cells, in particular cell fragments that may remain in suspension, for example, after isolation of cells from the tissue, will have fractional DNA content and be indistinguishable from apoptoptic cells. Likewise, late necrotic cells are also characterized by loss of DNA and therefore may have similar DNA stainability as apoptotic cells on DNA histograms.

VI. Apoptosis versus Necrosis versus "Necrotic Stage" of Apoptosis

There are many differences between typical apoptotic and necrotic cells (Arends *et al.*, 1990; Kerr *et al.*, 1972; Majno and Joris, 1995), and they have provided a basis for development of numerous markers and methods that can discriminate between these two modes of cell death (Darzynkiewicz *et al.*, 1992, 1997a,b). The major difference stems from the early loss of integrity of the plasma membrane during necrosis. This event results in a loss of the ability of the cell to exclude many fluorochromes. In contrast, the plasma mambrane and membrane transport functions remain, to a large extent, preserved during the early stages of apoptosis. The permeability of a cell to prodidium iodide (PI), or its ability to retain some fluorescent probes such as products of enzyme activity, is the most common marker distinguishing apoptosis from necrosis. Thus, for example, a combination of fluorochrome-conjugated annexin V with PI was proposed to distinguish live cells (unstainable with both dyes) from apoptotic cells (stainable with annexin V but unstainable with PI) from necrotic cells (stainable with both dyes) (Koopman *et al.*, 1994) (Fig. 1). The fluorochrome-conjugated annexin V binds to phosphatidylserine, the phospholipid of the plasma membrane that is inaccessible to this conjugate in live cells but becomes exposed on the outer leaflet of the plasma membrane and, therefore, accessible

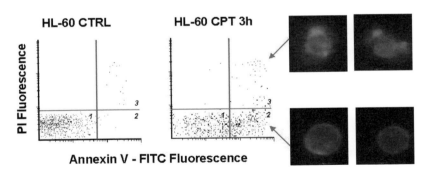

Fig. 1 Discrimination between nonapoptotic, early apoptotic, and late apoptotic ("necrotic stage" of apoptosis) or necrotic cells following staining with annexin V-FITC conjugate and PI, and fluorescence measurement by laser scanning cytometry (LSC). The cells in quadrant 1 are live; they exclude PI and do not bind annexin V-FITC. They predominate in HL-60 untreated (control) cultures. The cells in quadrant 2 are early apoptotic; they bind annexin V-FITC but still exclude PI. The cells in quadrant 3 are late apoptotic (or necrotic) and they bind both PI and annexin V-FITC. The quadrant 2 and 3 cells are more frequent in the camptothecin-treated sample (CPT). The possibility of observing the morphology of the cells selected based on their fluorescence, as offered by LSC, helps to confirm the identity of apoptotic cells. (See color plates.)

during apoptosis (Fadok *et al.*, 1992; Koopman *et al.*, 1994). Although this approach works well in many instances, it has limitations and possible pitfalls, as follows:

1. Late stage apoptotic cells resemble necrotic cells to such an extent that, to define them, the term "apoptotic necrosis" was proposed (Majno and Joris, 1995). This is a consequence of the fact that the integrity of the plasma membrane of late apoptotic cells is compromised, which makes the membrane leaky and permeable to charged cationic dyes such as PI. Thus, since the ability of such cells to exclude these dyes is lost, the discrimination between late apoptosis and necrosis cannot be accomplished by methods based on the use of plasma membrane permeability probes or annexin V.

2. The permeability and asymmetry of plasma membrane phospholipids (accessibility of phosphatidylserine) may change, for example, as a result of prolonged treatment with proteolytic enzymes (trypsinization), mechanical damage (e.g., cell removal from flasks by a rubber policeman, cell isolation from solid tumors, or even repeated centrifugations), electroporation, or treatment with some drugs.

3. Many flow cytometric methods designed to quantify the frequency of apoptotic or necrotic cells are based on the differences between live versus apoptotic versus necrotic cells in the permeability of their plasma membrane to different fluorochromes such as PI, 7-aminoactinomycin D (7-AAD), or Hoechst dyes. It should be stressed, however, that plasma membrane permeability may vary depending on the cell type and on many other factors, unrelated to apoptosis or necrosis. The assumption, therefore, that live cells maximally exclude a particular dye, early apoptotic cells are somewhat leaky, while late apoptotic or necrotic cells are totally permeable to the dye, and that these differences are large enough to identify these cells, is not universally applicable.

It is particularly difficult to discriminate between apoptotic and necrotic cells in suspensions from solid tumors. Necrotic areas form in tumors as a result of massive local cell death, for example, due to poor accessibility to oxygen and growth factors when tumors grow in size and their local vascularization becomes inadequate. Needle aspirate samples or cell suspensions from the resected tumors may contain many cells from the necrotic areas. Such cells are indistinguishable from late apoptotic cells by many markers. Because the AI in solid tumors, representing spontaneous or treatment induced apoptosis, should not include cells from the necrotic areas, one has to carefully eliminate such cells from analysis. Because incubation of cells with trypsin and DNase I selectively and totally digests all cells whose plasma membrane integrity is compromised, that is, primarily necrotic cells (Darzynkiewicz *et al.*, 1994), such a procedure may be used to remove necrotic cells from suspensions. It should be noted, however, that late apoptotic cells have somewhat permeable plasma membranes and are expected to also be sensitive to this treatment.

VII. Selective Loss of Apoptotic Cells during Sample Preparation

Relatively early during apoptosis cells will detach from the surface of culture flasks and float in the medium. Thus, the standard procedure of discarding the medium, followed by trypsinization or EDTA treatment of the attached cells and their collection, results in selective loss of apoptotic cells that are discarded with the medium. Such loss may vary from flask to flask depending on how the culture is handled, for example, the degree of mixing or shaking, efficiency in discarding the old medium. Surprisingly, cell trypsinization and discarding the medium is still occasionally reported by some authors. Needless to say, such an approach cannot be used for quantitative analysis of apoptosis. To estimate the frequency of apoptotic cells in adherent cultures, it is essential to collect floating cells, pool them with the trypsinized ones, and measure them as a single sample. It should be stressed that trypsinization, especially if prolonged, results in digestion of cells with a compromised plasma membrane. Thus, collection of cells from cultures by trypsinization is expected to cause selective loss of late apoptotic and necrotic cells.

Similarly, density gradient separation of cells (e.g., using Ficoll-Hypaque or Percoll solutions) may result in selective loss of dying and dead cells. This is due to the fact that early during apoptosis the cells become dehydrated, have condensed nuclei and cytoplasm, and therefore have a higher density compared to nonapoptotic cells. Knowledge of any selective loss of dead cells in cell populations purified by such an approach is essential when one is studying apoptosis.

Repeated centrifugations lead to cell loss by at least two mechanisms. One involves electrostatic cell attachment to the tubes and may be selective to a particular cell type. Thus, for example, preferential loss of monocytes and granulocytes was observed during repeated centrifugation of white blood cells, while lymphocytes remained in suspension (Bedner *et al.*, 1997). Cell loss is of particular concern when hypocellular samples ($<5 \times 10^4$ cells) are processed. In such a situation carrier cells in excess (e.g., chick erythrocytes) may be added to preclude disappearance of the cells of interest through centrifugations. The second mechanism of cell loss involves preferential disintegration of fragile cells. Because apoptotic cells, especially at late stages of apoptosis, are very fragile, they may selectively be lost from samples that require centrifugation or are repeatedly vortexed, pipetted, etc. Addition of serum or bovine serum albumin to cell suspensions, shortened centrifugation time, and decreased gravity force all may have a protective effect against cell breakage by mechanical factors. Apoptotic cells may also preferentially disintegrate in biomass cultures that require constant cell mixing.

VIII. Live Cells Engulfing Apoptotic Bodies Masquerade as Apoptotic Cells

Exposure of phosphatidylserine on the outer leaflet of the plasma membrane that occurs during apoptosis (Fadok *et al.*, 1992) makes such cells and their fragments (apoptotic bodies) attractive to neighboring cells, which phagocytize them. In addition to professional phagocytes, cells of fibroblast or epithelial lineage also have the ability to engulf apoptotic bodies. It is frequently observed, especially in solid tumors, that the cytoplasm of both nontumor as well as tumor cells located in the neighborhood of apoptotic cells contains inclusions typical of apoptotic bodies. The remains of apoptotic cells engulfed by neighboring cells contain altered plasma membranes, fragmented DNA, and other constituents with attributes characteristic of apoptosis. Thus, if the distinction is based on any of these attributes, the live, nonapoptotic cells that phagocytized apoptotic bodies cannot be distinguished from genuine apoptotic cells by flow cytometry (Fig. 2).

IX. The Problems with Commercial Kits and Reagents

A large number of commercial kits designed to detect apoptosis have become available, and reagent companies are racing to introduce new kits, often advertising them as "unique apoptosis detection kits." Some of these kits have solid experimental foundations and have been repeatedly tested on a variety of cell

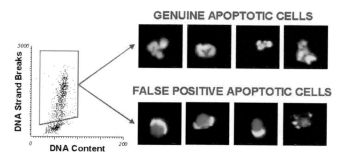

Fig. 2 Discrimination between genuine and false positive apoptotic cells by LSC. Peripheral blood mononuclear cells were obtained from a leukemia patient during chemotherapy (Gorczyca *et al.*, 1993a). DNA strand breaks were labeled with fluoresceinated dUTP using exogenous terminal deoxynucleotidyltransferase as described (Gorczyca *et al.*, 1992; 1993a). Genuine apoptotic cells are characterized by morphology typical of apoptosis. However, the nonapoptotic cells resembling monocytes, which contain cytoplasmic inclusions stainable with PI and having DNA strand breaks, also are present in this sample. These cells (false positive) most likely represent monocytes that phagocytized apoptotic bodies. By flow cytometry, which does not allow morphological identification, such cells are mistakenly classified as apoptotic. (See color plates.)

systems. Other kits, however, especially those advertised by vendors who do not fully explain the principle of detection of apoptosis on which the kit is based, and do not list its chemical composition, may not be universally applicable. It is not uncommon for a kit to be introduced after being tested on one or two cell lines using a single agent to trigger apoptosis (generally, classic apoptosis of either a T cell leukemia line treated with the Fas ligand or HL-60 cells treated with a DNA topoisomerase inhibitor).

Before application of any new kit, it is advisable to confirm that at least three to four independent laboratories have already successfully used it on different cell types. Furthermore, it is good practice to initially use the new kit in parallel with a well-established methodology, in a few experiments. This would allow one, by comparison of the apoptotic indices, to estimate the time window of detection of apoptosis by the new method and its sensitivity, compared to the one that is already established and accepted in the field.

One potential pitfall in using commercial kits has been described by Bedner *et al.* (1999a). The problem pertains an erroneous identification of live nonapoptotic eosinophils as apoptotic cells. This misclassification was due to the fact that there are trace amounts of unconjugated FITC in most commercially available reagents and kits (e.g., FITC-conjugated dUTP, FITC-conjugated avidin, FITC-conjugated primary or secondary antibodies). This FITC reacts avidly with granule proteins of eosinophils, labeling them very strongly. Thus, all the methods of identification of apoptotic cells that rely on the use of FITC-conjugated reagents and permeabilized cells are expected to nonspecifically label eosinophils, which then can be erroneously classified as apoptotic cells by flow cytometry (Bedner *et al.*, 1999a).

X. Cell Morphology Is Still the Gold Standard for Identification of Apoptotic Cells

Apoptosis was originally defined as a specific mode of cell death based on very characteristic changes in cell morphology (Kerr *et al.*, 1972) (Fig. 3). The characteristic morphological features of apoptosis and necrosis are listed in Table I. Although individual features of apoptosis may serve as markers for detection and analysis of the proportion of apoptotic cells in the cell populations studied by flow cytometry or other quantitative methods, the mode of cell death should *always* be identified by inspection of cells by light or electron microscopy. Therefore, when quantitative analysis is done by flow cytometry, it is essential to *confirm* the mode of cell death on the basis of morphological criteria. Furthermore, if there is any ambiguity regarding the mechanism of cell death, the morphological changes should be the deciding attribute in resolving the uncertainty.

It should be stressed that optimal preparations for light microscopy require cytospining of live cells following by their fixation and staining on slides. The cells are then flat and their morphology is easy to assess. On the other hand,

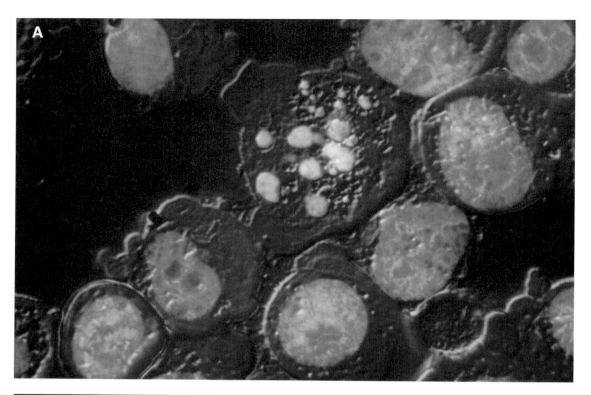

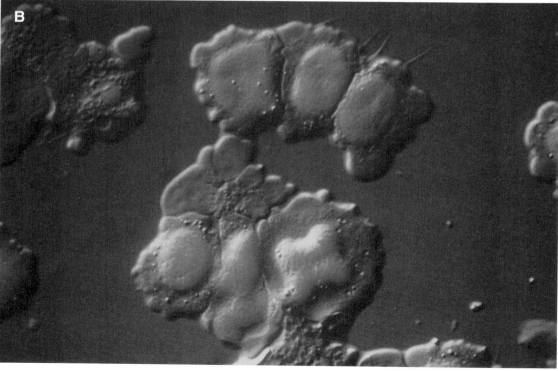

Table I
Morphological Criteria for Identification of Apoptosis or Necrosis

Apoptosis	Necrosis
Reduced cell size, convoluted cell shape	Cell and nuclear swelling
Plasma membrane undulations (blebbing/budding)	Patchy chromatin condensation
Chromatin condensation (DNA hyperchromicity)	Swelling of mitochondria
Loss of the structural features of the nucleus (smooth appearance)	Vacuolization in cytoplasm
	Plasma membrane rupture (ghostlike cells)
Nuclear fragmentation (karyorrhexis)	Dissolution of nuclear chromatin (karyolysis)
Presence of apoptotic bodies	
Dilatation of the endoplasmic reticulum	Attraction of inflammatory cells
Relatively unchanged cell organelles	
Shedding of apoptotic bodies	
Phagocytosis of the cell remnants	
Cell detachment from tissue culture flasks	

when the cells are initially fixed and stained in suspension, then transferred to slides and analyzed under the microscope, their morphology is obscured by the unfavorable geometry: the cells are spherical and thick and require confocal microscopy to reveal details such as early signs of apoptotic chromatin condensation.

Differential staining of cellular DNA and protein of cells on slides with 4,6-diamidino-2-phenylindole (DAPI) and sulforhodamine 101, respectively, is rapid and simple and provides very good morphological resolution of apoptosis and necrosis (Darzynkiewicz et al., 1997a). A combined cell illumination with ultraviolet (UV) light (to excite the DNA fluorochrome, e.g., DAPI) and light transmission microscopy utilizing interference contrast (Nomarski illumination) is our favorite method of cell visualization to identify apoptotic cells (Fig. 3). Other DNA fluorochromes, such as PI, 7-AAD, or acridine orange (AO), can be used as well.

XI. Laser Scanning Cytometry: Have Your Cake and Eat It Too

As mentioned earlier, characteristic changes in cell morphology provided the deciding criteria for identification of apoptotic cells. Quantitation of AI by micros-

Fig. 3 Morphology of apoptotic cells from U937 cultures treated with tumor necrosis factor α (TNF-α), cytocentrifuged, fixed in formaldehyde, and stained with 4, 6-diamidino-2-phenylindole (DAPI) (A) or 7-AAD (B), as revealed by their examination by fluorescence microscopy combined with interference contrast (Nomarski) illumination. Photographs were taken with a Nikon Microphot-FXA, with a 40× objective. (See color plates.)

copy or by classic image analysis techniques, however, was cumbersome and slow, and selection of cells for visual inspection was biased. Flow cytometry provided rapid and unbiased analyses but did not allow for morphological identification of the measured cell. Although the cells could be electronically sorted for their identification, cell sorters are expensive, the procedure is time-consuming and cumbersome, and there are many technical problems involved in recovering apoptotic cells for morphological analysis following sorting.

A new instrument, the laser scanning cytometer (LSC), satisfies both requirements for analysis of apoptosis: rapidity of the measurements and the possibility of morphological examination of the selected cells. The LSC is a microscope-based cytofluorometer that combines advantages of flow and image cytometry (Kamentsky and Kamentsky, 1991; see Chapter 3 of this volume). The fluorescence of individual cells can be measured rapidly, with sensitivity and accuracy comparable to those of flow cytometry. Cell staining and measurement on slides eliminates cell loss, which inevitably occurs during the repeated centrifugations necessary for sample preparation for flow cytometry. Since the spatial $x-y$ coordinates of each cell on the slide are recorded, these cells can be relocated after the initial LSC measurement, for example, for visual microscopy after staining with another dye to carry out additional image analysis. Furthermore, because the geometry of the cells cytocentrifuged or smeared on the slide is more favorable for morphometric analysis than is the case for cells in suspension, more information on cell morphology can be obtained by laser scanning than by flow cytometry. Advantages and limitations of the LSC for analysis of apoptosis have been reviewed (Bedner *et al.*, 1999b; Li and Darzynkiewicz, 1999; Darzynkiewicz *et al.*, 1999).

Two attributes of LSC make it an instrument of choice for analysis of apoptosis. As mentioned, the first attribute is the possibility of morphological examination of the cells of interest. Thus, several thousand cells can be measured per sample, with rates approaching 100 cells/sec, to quantify the frequency of apoptotic cells identified on the basis of a particular marker. Morphology of the presumed apoptotic cells can then be discerned following their relocation and microscopic examination, as shown in Fig. 4. This attribute of the LSC made it possible, for example, to identify the false positive apoptotic cells in bone marrow of leukemia patients undergoing chemotherapy. The latter, showing the presence of DNA strand breaks (TUNEL positive), were actually nonapoptotic monocytes and macrophages that engulfed apoptotic bodies, the products of disintegration of apoptotic cells (Bedner *et al.*, 1998, 1999b) (Fig. 2). Flow cytometry, of course, was unable to discriminate between the genuine apoptotic cells and the false positive ones. The LSC thus allows one to quantify the frequency of apoptotic cells at rates approaching those of flow cytometry, and it also allows one to confirm the accuracy of classification based on the gold standard of apoptosis, morphology.

The second attribute that is unique to the LSC and of great value in analysis of apoptosis relates to the possibility of repeated measurements of the same set of cells and integration (merging) of the results of all sequential measurements into a single file (Kamentsky and Kamentsky, 1991). Multivariate analysis of

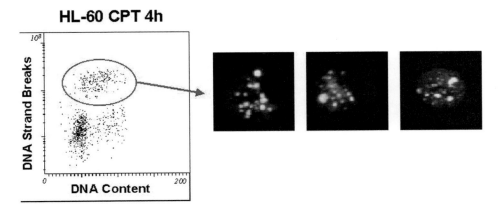

Fig. 4 Morphological identification of apoptotic cells detected based on the presence of DNA strand breaks, after relocation, by LSC. HL-60 cells treated with 0.15 μM CPT for 4 hr were subjected to DNA strand break labeling with FITC-conjugated digoxigenin antibody (Gorczyca *et al.*, 1992) and counterstained with PI. Note nuclear fragmentation and yellow fluorescence (due to colocalization and spectral overlap of FITC and PI) of the nuclear fragments. (See color plates.)

the integrated data reveals correlations between measured cell features. This capability of the LSC was employed to combine analysis of functional features of live cells with cell attributes that can be probed only after fixation (Li and Darzynkiewicz, 1999; Li *et al.*, 2000). Specifically, the functional changes that occur during apoptosis, dissipation of the mitochondrial transmembrane potential ($\Delta\Psi_m$; Petit *et al.*, 1995; Zamzani *et al.*, 1998), and oxidative stress (increase in ROIs; Hedley and McCulloch, 1996; Sheng-Tanner *et al.*, 1998) were correlated with the attributes measured in fixed cells, the cell cycle position, and the presence of DNA strand breaks. The cells were first measured when alive to assess their $\Delta\Psi_m$ or ROIs, and then the cells were fixed and subjected to analysis of DNA content and/or presence of DNA strand breaks. The results of both measurements were integrated into a single file for multivariate analysis. It was possible, therefore, to reveal the status of DNA fragmentation or cell cycle position of the same cell whose $\Delta\Psi_m$ or ROI level was measured. This approach appears to be of particular utility in mapping sequences of intracellular events that include both functional and structural changes. It is currently used in our laboratory to map the sequences and correlate the following: the decrease in intracellular pH; rise in Ca^{2+}; ROIs; changes in reduced glutathione; exposure of phosphatidylserine on the outer leaflet of the plasma membrane; activation of caspases; dissipation of $\Delta\Psi_m$; translocation of Bax to, and leakage of cytochrome *c* from, mitochondria; nuclear translocation of nuclear factor (NF-κB); chromatin condensation; and DNA fragmentation, events that occur during apoptosis (Li *et al.*, 2000).

Acknowledgments

Work was supported by the National Cancer Institute (Grant RO1 CA 28 704), by the Chemotherapy Foundation, and by "This Close" for Cancer Research Foundation.

References

Arends, M. J., Morris, R. G., and Wyllie, A. H. (1990). Apoptosis: The role of endonuclease. *Am. J. Pathol.* **136,** 593–608.

Bedner, E., Burfeind, P., Gorczyca, W., Melamed, M. R., and Darzynkiewicz, Z. (1997). Laser scanning cytometry distinguishes lymphocytes, monocytes and granulocytes by differences in their chromatin structure. *Cytometry* **29,** 191–196.

Bedner, A., Burfeind, P., Hsieh, T-C., Wu, J. M., Augero-Rosenfeld, M. E., Melamed, M. R., Horowitz, H. W., Wormser, G. P., and Darzynkiewicz, Z. (1998). Cell cycle effects and induction of apoptosis caused by infection with human granulocytic Ehrlichiosis pathogen measured by laser scanning cytometry. *Cytometry* **33,** 47–55.

Bedner, E., Halicka, H. D., Cheng, W., Salomon, T., Deptala, A., Gorczyca, W., Melamed, M. R., and Darzynkiewicz, Z. (1999a). High affinity of binding of fluorescein isothiocynate to eosinophils detected by laser scanning cytometry: A potential source of error in analysis of blood samples utilizing fluorescein-conjugated reagents in flow cytometry. *Cytometry* **36,** 77–82.

Bedner, E., Li, X., Gorczyca, W., Melamed, M. R., and Darzynkiewicz, Z. (1999b). Analysis of apoptosis by laser scanning cytometry. *Cytometry* **35,** 181–195.

Cohen, G. M., Su, X.-M., Snowden, R. T., Dinsdale, D., and Skilleter, D. N. (1992). Key morphological features of apoptosis may occur in the absence of internucleosomal DNA fragmentation. *Biochem. J.* **286,** 331–334.

Collins, R. J., Harmon, B. V., Gobe, G. C., and Kerr, J. F. R. (1992). Internucleosomal DNA cleavage should not be the sole criterion for identifying apoptosis. *Int. J. Radiat. Biol.* **61,** 451–453.

Cossarizza, A., Kalashnikova, G., Grassilli, E., Chiappelli, F., Salvioli, S., Capri, M., Barbieri, D., Troiano, L., Monti, D., and Franceschi, C. (1994). Mitochondrial modifications during rat thymocyte apoptosis: A study at a single cell level. *Exp. Cell Res.* **214,** 323–330.

Darzynkiewicz, Z. (1994). Simultaneous analysis of cellular RNA and DNA content. *In* "Methods in Cell Biology" (Z. Darzynkiewicz, J. P. Robinson, and H. A. Crissman, eds.), Vol. 41, pp. 401–420. Academic Press, San Diego.

Darzynkiewicz, Z., and Li, X. (1996). Measurements of cell death by flow cytometry. *In* "Techniques in Apoptosis. A User's Guide" (T. G. Cotter and S. J. Martin. eds.) pp. 71–106. Portland Press, London.

Darzynkiewicz, Z., Traganos, F., and Kimmel, M. (1986). Assay of cell cycle kinetics by multivariate flow cytometry. *In* "Techniques of Cell Cycle Analysis" (J. W. Gray and Z. Darzynkiewicz, eds.), pp. 291–332. Humana Press, Clifton, New Jersey.

Darzynkiewicz, Z., Bruno, S., Del Bino, G., Gorczyca, W., Hotz, M. A., Lassota, P., and Traganos, F. (1992). Features of apoptotic cells measured by flow cytometry. *Cytometry* **13,** 795–808.

Darzynkiewicz, Z., Li, X., and Gong, J. (1994). Assays of cell viability. Discrimination of cells dying by apoptosis. *In* "Methods in Cell Biology" (Z. Darzynkiewicz, J. P. Robinson, and H. A. Crissman, eds.), Vol. 41, pp. 16–39. Academic Press, San Drego.

Darzynkiewicz, Z., Juan, G., Li, X., Murakami, T., and Traganos, F. (1997a). Cytometry in cell necrobiology: Analysis of apoptosis and accidental cell death (necrosis). *Cytometry* **27,** 1–20.

Darzynkiewicz, Z., Bedner, E., Traganos, F., and Murakami, T. (1998). Critical aspects in the analysis of apoptosis. *Hum. Cell* **11,** 3–12.

Darzynkiewicz, Z., Li, X., Gong, J., and Traganos, F. (1997b). Methods for analysis of apoptosis by flow cytometry. *In* "Manual of Clinical Laboratory Immunology" (N. R. Rose, E. C. de Macario, J. D. Folds, H. C. Lane, and R. Nakamura eds.), pp. 334–344. ASM Press, Washington, D. C.

Darzynkiewicz, Z., Bedner, E., Li, X., Gorczyca, W., and Melamed, M. R (1999). Laser scanning cytometry. A new instrumentation with many applications. *Exp. Cell Res.* **249,** 1–12.

Del Bino, G., Darzynkiewicz, Z., Degraef, C., Mosselmans, R., Fokan, D., and Galand, P. (1999). Comparison of methods based on annexin V binding, DNA content or TUNEL for evaluating cell death in HL-60 and adherent MCF-7 cells. *Cell Prolif.* **32,** 25–37.

Dragovich, T., Rudin, C. M., and Thompson, C. B. (1998). Signal transduction pathways that regulate cell survival and cell death. *Oncogene* **17,** 3207–3213.

Endersen, P. C., Prytz, P. S., and Aarbakke, J. (1995). A new flow cytometric method for discrimination of apoptotic cells and detection of their cell cycle specificity through staining of F-actin and DNA. *Cytometry* **20,** 162–171.

Fadok, V. A., Voelker, D. R., Cammpbell, P. A., Cohen, J. J., Bratton, D. L., and Henson, P. M. (1992). Exposure of phosphatidylserine on the surface of apoptotic lymphocytes triggers specific recognition and removal by macrophages. *J. Immunol.* **148,** 2207–2216.

Gong, J., Traganos, F., and Darzynkiewicz, Z. (1994). A selective procedure for DNA extraction from apoptotic cells applicable for gel electrophoresis and flow cytometry. *Anal. Biochem.* **218,** 314–319.

Gorczyca, W., Bruno, S., Darzynkiewicz, R., Gong, J., and Darzynkiewicz, Z. (1992). DNA strand breaks occurring during apoptosis: Their early in situ detection by the terminal deoxynucleotidyl transferase and nick translation assays and prevention by serine protease inhibitors. *Int. J. Oncol.* **1,** 639–648.

Gorczyca, W., Bigman, K., Mittelman, A., Ahmed, T., Gong, J., Melamed, M. R., and Darzynkiewicz, Z. (1993a). Induction of DNA strand breaks associated with apoptosis during treatment of leukemias. *Leukemia* **7,** 659–670.

Gorczyca, W., Gong, J., and Darzynkiewicz, Z. (1993b). Detection of DNA strand breaks in individual apoptotic cells by the in situ terminal deoxynucleotidyl transferase and nick translation assays. *Cancer Res.* **52,** 1945–1951.

Hara, S., Halicka, H. D., Bruno, S., Gong, J., Traganos, F., and Darzynkiewicz, Z. (1996). Effect of protease inhibitors on early events of apoptosis. *Exp. Cell Res.* **232,** 372–384.

Hedley, D. W., and McCulloch, E. A. (1996). Generation of oxygen intermediates after treatment of blasts of acute myeloblastic leukemia with cytosine arabinoside: Role of bcl-2. *Leukemia* **10,** 1143–1149.

Hotz, M. A., Traganos, F., and Darzynkiewicz, Z. (1992). Changes in nuclear chromatin related to apoptosis or necrosis induced by the DNA topoisomerase II inhibitor fostriecin in MOLT-4 and HL-60 cells are revealed by altered DNA sensitivity to denaturation. *Exp. Cell Res.* **201,** 184–191.

Kamentsky, L. A., and Kamentsky, L. D. (1991). Microscope-based multiparameter laser scanning cytometer yielding data comparable to flow cytometry. *Cytometry* **12,** 381–387.

Kerr, J. F. R., Wyllie, A. H., and Curie, A. R. (1972). Apoptosis: A basic biological phenomenon with wide-ranging implications in tissue kinetics. *Br. J. Cancer* **26,** 239–257.

Kerr, J. F. R., Winterford, C. M., and Harmon, V. (1994). Apoptosis, its significance in cancer and cancer therapy. *Cancer* **73,** 2013–2026.

Koopman, G., Reutelingsperger, C. P. M., Kuijten, G. A. M., Keehnen, R. M. J., Pals, S. T., and van Oers, M. H. J. (1994). Annexin V for flow cytometric detection of phosphatidylserine expression of B cells undergoing apoptosis. *Blood* **84,** 1415–1420.

Koester, S. K., Schlossman, S. F., Zhang C., Decker, S. J., and Bolton, W. E. (1998). APO2.7 defines a shared apoptotic–necrotic pathway in a breast tumor hypoxia model. *Cytometry* **33,** 324–332.

Kroemer, G. (1998). The mitochondrion as an integrator/coordinator of cell death pathways. *Cell Death Differ* **5,** 547–548.

Li, X., and Darzynkiewicz, Z. (1995). Labeling DNA strand breaks with BdrUTP. Detection of apoptosis and cell proliferation. *Cell Prolif.* **28,** 571–579.

Li, X., and Darzynkiewicz, Z. (1999). The Schrödinger's cat quandary in cell biology: Integration of live cell funtional assays with measurements of fixed cells in analysis of apoptosis. *Exp. Cell Res.* **249,** 404–412.

Li, X., Melamed, M. R., and Darzynkiewicz, Z. (1996). Detection of apoptosis and DNA replication by differential labeling of DNA strand breaks with fluorochromes of different color. *Exp. Cell Res.* **222,** 28–37.

Li, X., Du, L., and Darzynkiewicz, Z. (2000). During apoptosis of HL-60 and U-937 cells caspases are activated independently of dissipation of mitochondrial electrochemical potential. *Exp. Cell Res.* **257,** 290–297.

Majno, G., and Joris, I. (1995). Apoptosis, oncosis, and necrosis. An overview of cell death. *Am. J. Pathol.* **146,** 3–16.

Meier, P., and Evan, G. (1998). Dying like flies. *Cell* **95,** 295–298.

Nicoletti, I., Migliorati, G., Pagliacci, M. C., Grignani, F., and Riccardi, C. (1991). A rapid and simple method for measuring thymocyte apoptosis by propidium iodide staining and flow cytometry. *J. Immunol. Methods* **139,** 271–280.

Nuñez, G., Benerdict, M. A., Hu, Y., and Inohara, N. (1998). Caspases: The proteases of the apoptotic pathway. *Oncogene* **17,** 3237–3245.

Oberhammer, F., Wilson, J. M., Dive, C., Morris, I. D., Hickman, J. A., Wakeling, A. E., Walker, P. R., and Sikorska, M. (1993). Apoptotic death in epithelial cells: Cleavage of DNA to 300 and/or 50 kb fragments prior to or in the absence of internucleosomal fragmentation. *EMBO J.* **12,** 3679–3684.

Ormerod, M. G. (1998). The study of apoptotic cells by flow cytometry. *Leukemia* **12,** 1013–1025.

Ormerod, M. G., O'Neill, C. F., Robertson, D., and Harrap, K. R. (1994). Cisplatin induced apoptosis in a human ovarian carcinoma cell line without a concomitant internucleosomal degradation of DNA. *Exp. Cell Res.* **211,** 231–237.

Ormerod, M. G., Cheetham, F. P. M., and Sun, X.-M. (1995). Discrimination of apoptotic thymocytes by forward light scatter. *Cytometry* **21,** 300–304.

Petit, P. X., LeCoeur, H., Zorn, E., Dauguet, C., Mignotte, B., and Gougeon. M. L. (1995). Alterations of mitochondrial structure and function are early events of dexamethasone-induced thymocyte apoptosis. *J. Cell Biol.* **130,** 157–165.

Reed, J. C. (1998). Bcl-2 proteins. *Oncogene* **17,** 3225–3236.

Umansky, S. R., Korol', B. R., and Nelipovich, P. A. (1981). In vivo DNA degradation in the thymocytes of gamma-irradiated or hydrocortisone-treated rats. *Biochim. Biophys. Acta* **655,** 281–290.

Sheng-Tanner, X., Bump, E. A., and Hedley, D. A. (1998). An oxidative stress-mediated death pathway in irradiated human leukemia cells mapped using multilaser flow cytometry. *Radiat. Res.* **150,** 636–647.

Vaux, D. L., and Korsmeyer, S. J. (1999). Cell death in development. *Cell* **96,** 245–254.

Zakeri, Z. F., Quaglino, D., Latham, T and Lockshin, R. A. (1993). Delayed internucleosomal DNA fragmentation in programmed cell death. *FASEB J.* **7,** 470–478.

Zamai, L., Falcieri, E., Marhefka, G., and Vitale, M. (1996). Supravital exposure to propidium iodide identifies apoptotic cells in the absence of nucleosomal DNA fragmentation. *Cytometry* **23,** 303–311.

Zamzani, N., Brenner, C., Marzo. I., Susin, S. A., and Kroemer, G. (1998). Subcellular and submitochondrial mode of action of Bcl-2-like oncoproteins. *Oncogene* **16,** 2265–2282.

PART VI

Cell–Cell, Cell–Environment Interactions

CHAPTER 25

Analysis of Cell Migration

Nicole Dodge Zantek and Michael S. Kinch

Department of Basic Medical Sciences
Purdue University
West Lafayette, Indiana 47907

I. Introduction and Application
II. General Strategies to Measure Cell Migration
   A. Time-Lapse Microscopy
   B. Boyden Chamber Assays
References

I. Introduction and Application

Cell migration is a key event in a number of biological processes, including tumor invasion and metastasis, angiogenesis, wound healing, and development. By gaining a deeper understanding into how and why cells migrate, we also gain great insight into these processes. Assays designed to study migration are also perfectly suited for testing modulators of migration that may be useful clinically, such as antimetastatic, angiogenic, and anti-inflammatory agents.

Cell migration is a complex multistep process. It includes polarization of the cell, extension of the membrane, formation of cell adhesion molecule attachment, stabilization of cell adhesion molecules to the cytoskeleton, generation of force, contraction of the cell, release of adhesions at the trailing end of the cell, and recycling of adhesion proteins (Sheetz *et al.*, 1998; Lauffenburger and Horwitz, 1996). Although a number of cell adhesion (e.g., integrins, cadherins, immunoglobulins), cytoskeletal (e.g., actin), and signaling proteins (e.g., focal adhesion kinase and β-catenin) have been implicated in this process, the intricate regulation of many of these stages remains to be clearly understood.

II. General Strategies to Measure Cell Migration

A. Time-Lapse Microscopy

1. Basic Technique

Microscopic examination of adherent cells by time-lapse videography allows for direct measurement of the extent of cell migration. The most basic technique measures the movement of cells plated onto an appropriate substrate. In brief, cell displacement is measured at physiological temperature (using a heated stage or heater) on a calibrated grid over time. Microscopic images are generally captured at fixed intervals using a time-lapsed videographer (VCR) or digital camera attached to the camera port of the microscope. The distances of cell displacement are measured either with specially designed software (e.g., 2D-DIAS, Solltech, Iowa City, IA; Soll, 1995) or by hand with a ruler to measure cell displacement on a calibrated grid (Khosravi-Far et al., 1995).

When cell migration is measured via time-lapse microscopy, a number of variables (e.g., temperature, density, substrate, media) must be optimized. In addition, a number of modifications of this approach have been developed to address different questions, as discussed later.

The choice of cells must be dictated by the questions being addressed. A highly migratory cell line, such as some fibroblast lines, may be useful for studying the regulation and mechanics of locomotion. In other circumstances, the migration of different cell populations (e.g., normal and neoplastic cells) may be compared as a measure of metastatic potential. However, when such comparisons are performed, it is critical to employ equivalent conditions (temperature, media, density, substrates) to minimize variables resulting from environmental factors.

Cell density can also impact the outcome of migration assays. At high cell density, epithelial cells may form colonies that preclude cell migration. In addition, cell–cell contact can deliver signals that dramatically alter subsequent migratory behavior (e.g., via cadherins; Chen et al., 1997). These types of events can dramatically alter the interpretation of migration assays. Conversely, cells at very low densities often do not migrate, perhaps due to a requirement for juxtacrine factors secreted by neighboring cells. One distinct advantage of time-lapsed videography is the ability to select the appropriate cell density for study. However, a proper sampling must be obtained (generally at least 30 different cells) to ensure one will obtain a proper sampling of the population of the cells to be studied.

The substrate on which cells are allowed to migrate can dramatically alter the resultant migration. Cells on tissue culture plastic are certainly the easiest samples to generate. However, it is our experience that cells often do not adhere or migrate as rapidly on tissue culture plastic. Cells plated onto glass coverslips can easily be used for migration assays, and this is also amenable for histochemistry. Another advantage is that the coverslip can be placed into a larger plate of cells. After removing the coverslip for analysis of migration, the remaining cells in the plate can then be subjected to relevant biochemical assays. However, we find that the migration capacity can vary when comparing the same cell on glass or plastic. To obviate this, it is sometimes necessary to treat the coverslips with purified matrix materials (e.g., fibronectin, vitronectin) prior to the addition of cells in order to promote cell migration. Also, it must be considered that serum in the culture medium can provide important matrix and signaling factors that can change the rates of cell migration.

As mentioned, the migration of some cells is selectively enhanced when the cells are attached to certain substrates (e.g., laminin, collagen, fibronectin). For these assays, culture dishes (or coverslips) are coated by incubation overnight at 4°C with a 50 μg/ml solution of purified matrix material. After washing away unbound material, the cells are then added to the dish and allowed to spread prior to the migration assay. A preferential migration on specific substrates may also identify critical cell adhesion molecules (e.g., integrins or selectins) involved in the migration. One important consideration for the analyses of migration assays on purified materials is that many cell types retain the capacity to secrete and remodel underlying matrices on which they migrate (Fath *et al.*, 1989). This secretion of new matrices can be prevented by treating cells with 50 μg/ml cycloheximide for at least 2 hr prior to the assay to block the synthesis of new proteins (Burridge *et al.*, 1992).

One potential problem with these assays concerns the duration in which cell migration can be reliably measured. Over extended times, cells often decrease their migration as a consequence of changes in environment (usually pH or medium depletion). If the experiment is to be conducted outside of a CO_2 environment, a steadily increasing pH can alter migratory behaviors. This limitation can be overcome by using either a carbon dioxide chamber fitted to the microscope or by including buffers such as 15 mM HEPES [N-(2-hydroxyethyl)-piperazine-N'-(2-ethanesulfonic acid)] in the assay buffer.

Chronic exposure to intense heat or light (especially fluorescent light and UV) can desiccate or fix the sample. In an effort to minimize light exposure, we installed a relatively inexpensive intervelometer to the light source that opens the light path only so long as necessary to capture the images. Maintenance of correct temperature can be achieved using a heated stage or, as a less expensive alternative, an appropriately calibrated heater (e.g., a hair dryer). However, the use of air heaters can often generate vibrations that impede image capture or analysis. Thus, the installation of a heated stage insert onto the microscope is usually preferable.

2. Modifications of Basic Assay

a. Wound Healing Assay

Much investigation has focused on how cells migrate, fill, and heal a wound. This has application for studies of angiogenesis, metastasis, and development. In addition to simple migration, wound healing requires a cell to be able to break away from its contact with neighboring cells, and it sometimes involves cell division and growth as well. Thus, there has been considerable development of experimental systems to study wound healing *in vitro*. Many of these techniques use time-lapse microscopy. For these assays, a wound is generated by scraping an adherent monolayer with a rubber plate scraper, plastic pipette tip, or a razor blade. The medium and suspended cells are removed and the wounded culture washed. This is an important step, as detached cells in the medium could otherwise reattach to the denuded area and begin to fill the wound. The average width of the wound, as measured microscopically, is then measured at specific time points following the initial wound (usually 6–36 hr). Wounding in such a manner allows for a synchronous movement of a number of cells along a front. Notably, this technique can allow one to study changes in both cell migration and cell–cell adhesion.

There are two considerations that can potentially confound the interpretations of wound healing assays. First, the creation of a wound can release a compressive effect, especially in the study of tightly packed cells, and this can allow for a wave of expansion that fills the wound (rather than true cell migration). This can be minimized by employing newly confluent samples and by delaying the initial measurement of the wound until after expansion has been allowed to occur (usually within the first hour or two). A second consideration is that the long time periods employed for wound healing assays (sometimes days) may not allow the investigator to distinguish between the contributions imparted by cell migration versus proliferation in the healing of the wound. This is further compounded by the fact that the loss of cell–cell contact after wounding removes contact-inhibitory regulation of cell proliferation. To minimize the contributions of cell proliferation, the time of data collection for wound healing assays should be kept at a minimum. Also, one can determine the relative contribution of cell proliferation by including 3 μg/ml bromodeoxyuridine (BrdU) in the culture medium. BrdU is incorporated into newly synthesized DNA and can be detected using immunohistochemistry with commercially available reagents.

b. Chemotactic Assays

It is well established that cell migration during development and disease is controlled by a gradient of chemical factors that promote and direct migration. This behavior can be mimicked *in vitro* using time-lapse videomicroscopy. A number of techniques have been developed to generate gradients. The simplest methods introduce a test sample at one end of the field of view in the microscope or the tissue culture dish and measure migration toward the area. Soluble samples can be introduced via a pipette tip. They can also be embedded in a solid matrix such as agarose or Sepharose beads, from which the samples will slowly diffuse. Drawing from development studies in which regions of the development embryo are explanted into new sites, similar assays can be performed *in vitro* by introducing pieces of tissue into the migrating culture of cells. Further purification can be done to identify specific factors contained in the sample that regulate migration.

c. Other Assays

In addition to the common basic techniques, a number of more specialized protocols have been developed to assess migration by tracking cell movement over time using videomicroscopy (although a detailed discussion of these assays is beyond the scope of this chapter). A number of assays examine the cell migration away from aggregates, such as spheroid (Lund-Johansen *et al.*, 1990) and microcarrier bead assays (Jozaki *et al.*, 1990; Marucha *et al.*, 1987; Rosen *et al.*, 1990), or from explants, such as pieces of tumor or embryonic tissues. Leukocyte chemotaxis is frequently measured by an under agarose assay which measures cell migration toward a well of test or control chemoattractant (Nelson *et al.*, 1978). In addition there is a great interest in understanding how cells react and migrate in response to changes in topography. For example, how does a cell respond to a bump or a change in the underlying matrix in its pathway? A number of treatments to the surface of migration have been developed in the field of topography (Curtis and Wilkinson, 1998).

3. Data Analysis

Analysis of time-lapse videography data can be formidable and highly subjective. Here we will review some basic factors that must be considered when collecting time-lapse data. For a complete discussion of this, the reader should refer to Soll and Voss (1998).

Cells typically migrate in a random direction and change directions frequently if there are no attempts made to encourage chemotactic migration in a specific direction. Over long time periods (minutes to hours), a cell may migrate a greater distance than the net vector indicates (especially if direction changes are frequent). To obtain a more accurate measurement of migration rates, vector analysis of the data is usually required.

The choice of a point of reference is critical, and care should be taken to avoid reference points that may change during the course of an experiment. For example, the leading edge of migration may be useful for studying such processes as axonal growth cone migration, wound healing, or chemotaxis when the migration is directional. However, when frequent changes in direction are encountered, the leading edge can change in shape and direction very rapidly. In general, changes in lamella structure can obscure accurate measurements of cell displacement. It is thus more appropriate to select a center of mass (centroid translocation), such as the nucleus, as the reference point. This is often more useful for tracking the random two-dimensional migration vectors often encountered when comparing cells not moving toward a chemoattractant. Establishing a proper time interval between data collection points is also crucial. In many cell types, migration occurs in spurts and cycles. Cells may remain rather static and then undergo a periodic burst in migration. Frequent changes in direction can greatly underestimate cell migration rates.

Other changes in cell behavior, most notably mitosis and cell death, can dramatically alter the rates of cell migration. To minimize these types of problems, it is often helpful to employ shorter time intervals than are to be analyzed (e.g., collect at 2-min intervals despite the fact that 10-min intervals may be measured). This allows one to test for changes in cell direction, bursts of migration, or entry into mitosis. With shorter intervals, the data can be reviewed and edited to preclude analysis of specific populations (e.g., mitotic cells) that might otherwise prevent an accurate assessment of migration rates.

4. Applications

The advantage of time-lapse microscopy is the ability to directly monitor the effects of specific intervention (drug treatment, etc.). The microscopic identification of specific cells, for example, cells expressing a transfected or microinjected marker, can be singled out and compared with control cells within the sample. Time-lapse microscopy also can allow one to document dynamic cellular processes linked with cell migration. For example, we found that migration rates correlate with the dynamics of membrane ruffling in transformed cells (Khosravi-Far *et al.*, 1995). Similarly, the extent of membrane ruffling and pseudopodal extension has been used to develop a grading system to predict the metastatic potential of neoplastic cells (Mohler *et al.*, 1987a,b).

B. Boyden Chamber Assays

Boyden introduced the Boyden chamber for the analysis of leukocyte chemotaxis in 1962 (Boyden, 1998). In brief, the basic principle behind the Boyden chamber is to measure the migration of cells through a porous membrane. Cells are added to an upper chamber and allowed to migrate through a filter over time and into a lower chamber. Cell migration is then determined either by counting the cells on the bottom side of the filter or by counting the number of cells that have migrated through the filter into the lower chamber. Often, biological stains [Diff-Quick (Scientific Products, Columbia, MD), etc.] are employed to aid in the accurate determination of cells that are attached to the filter. Extensive modifications to this basic principle have increased the ease of use and widened the application to studies of many different cell types and conditions.

1. Apparatus

There are typically two types of Boyden chambers: so-called open-well and blind-well chambers. The open-well type consists of a cylinder with a membrane on the bottom that is inserted into a larger well. The membrane separates the two chambers, such that the upper chamber is bounded by the cylinder and the membrane and the lower chamber is bound by the well wall and the membrane. In this arrangement the lower chamber is open to the environment and can thus be accessed for the addition of media, chemotaxic agents, etc. In the closed well apparatus, the membrane is laid over the well of the lower chamber, and then the upper chamber is stacked on top of the lower chamber. Once the apparatus is assembled, the lower chamber is not accessible. An important technical consideration is to minimize forces (e.g., gravity, osmolarity) that may otherwise promote a gradient flow between chambers and to minimize hydrostatic pressure to prevent curving of the membrane. For complete collection of the cells, a catcher membrane may also be included in the lower chamber to "catch" cells that have migrated through the membrane but do not adhere to it.

There are many different membranes that can be used in Boyden chamber assays. Three important factors to consider in choosing a membrane are membrane material (e.g., polycarbonate and polyester), wetting agents, and pore size. Attachment-dependent cells frequently require treatment of the membrane with a purified extracellular matrix protein (e.g., collagen, fibronectin, laminin). The exact protocol varies with each substrate, though typically this involves diluting the matrix protein, adding the solution to the inside of the insert, and allowing it to dry onto the filter.

2. Chemotaxis and Chemokinesis Assays

Boyden chambers are classically used to determine migration in response to chemotaxis or chemokinesis. Chemotaxis is the migration of a cell toward a substance. In this assay, a chemoattractant is placed in the lower chamber or on the lower side of the membrane to promote migration from the upper side of the membrane to the lower. Common chemoattractants include extracellular matrix components (purified fibronectin) and tissue or tumor extracts (e.g., fibroblast conditioned medium). Chemokinesis is a term used to describe general increases in migration that occur in response to a specific chemoattractant. In these assays, the chemokinetic agent is placed in both the upper and lower chambers at an equal concentration; therefore, no gradient is established (as opposed to a chemotaxis assay). Growth factors and serum are examples of chemokinetic agents.

3. Matrigel Invasion Assays

In 1987 Albini *et al.* modified Boyden chamber assays to measure tumor cell invasion. Since then, the technique has been improved and modified, and today these types of assays have become a standard measure of tumor cell invasiveness. However, although Matrigel invasion assays are frequently used to determine the metastatic phenotype of cells, Noel *et al.* (1991) reported that the degree of migration may not always correlate with the extent of metastatic phenotype.

A typical invasion assay measures the migration of cells through a layer of Matrigel (Collaborative Biomedical Products, Bedford, MA), though other materials may be employed as well. Matrigel is a semipurified, natural basement membrane produced by Engelbreth–Holm–Swarm (EHS) mouse sarcoma cells (Kleinman *et al.*, 1982, 1986). It contains extracellular matrix proteins (laminin, collagen, and heparan sulfate proteoglycans) (Kleinman *et al.*, 1982, 1986) and a variety of growth factors (McGuire and Seeds, 1989). Conveniently, Matrigel is a liquid at 4°C but solidifies at 37°C. This facilitates the coating of cold Matrigel liquid onto membranes and the subsequent polymerization into a gel by incubation at 22–35°C.

An important technical consideration for invasion assays concerns the thickness of the Matrigel layer. The number of cells that can invade through the layer decreases with increasing Matrigel thickness. Inconsistencies in seeding density within a sample can thus substantially alter the results. This is especially important if cell migration is to be reported as the average of random fields within the sample. The surface tension of Matrigel on the sides of the Boyden chamber can be a major cause for unevenness in Matrigel deposition. This can create a meniscus that results in the center of the layer being thinner than the edges. To minimize this, the sides of the Boyden chamber can be treated with Parafilm prior to incubation with Matrigel (Imamura *et al.*, 1994).

4. Membrane Invasion Assays

An alternative approach to the Matrigel invasion assay is invasion through a biological membrane. A number of different membranes have been used, including bladder wall, amnion, lens capsule, and chicken chorioallantonic membrane (Hart and Fidler, 1978; Poste *et al.*, 1980; Liotta *et al.*, 1980; Hendrix *et al.*, 1985; Magnatti *et al.*, 1986; Starkey *et al.*, 1984; Marcel *et al.*, 1981). In these assays, the membrane is stretched across an apparatus similar to a Boyden chamber. Suspended cells are placed on one side of the membrane, and migration through the membrane is assessed. An advantage gained by the use of biological membranes is that they often more closely model surfaces encountered during invasion *in vivo*. However, these membrane preparations are often difficult to obtain and prepare, and they generally have more intrinsic variability than Matrigel.

5. Data Collection and Analysis

Boyden chamber assay data can be collected by visually counting the cells attached to the membrane or by using an indicator assay. Visual collection requires the cells to be fixed (usually with formaldehyde) and stained with biological dyes. Generally, a nuclear stain, such as hematoxylin, is employed as a marker. The membrane is then mounted on a glass slide under a coverslip, and cell numbers are counted using light microscopy. A complete set of data includes an assessment of cell numbers at the upper membrane surface (cells that attached but did not marginate), the lower membrane surface (cells currently migrating through the membrane), and the catcher membrane (cells that have previously migrated through the membrane). For ease of data collection, a specified number of cells is applied to the chamber, and usually only the cells on the lower membrane surface are reported. In this case, the upper cells are removed with a cotton applicator prior to fixation and staining to preclude the possibility that these might be inadvertently counted.

Typically the number of cells on the entire surface of the membrane is not counted. A fixed number of random fields are selected for data collection. As mentioned earlier, this can strongly bias the data as the degree of migration can vary on the membrane. For example, meniscus formation with Matrigel can result in an uneven distribution and thereby impact migration locally. Visual collection is also tedious and time-consuming and thus introduces the possibility of operator error.

Because of the disadvantages of visual counting, a number of indicator assays have been developed to enumerate migration in a Boyden chamber. Cells can be labeled prior to the experiment with a radiolabel such a [^{3}H]thymidine or [^{125}I]iododeoxyuridine and the membranes measured using liquid scintillation (Repesh, 1998). However, this approach does not distinguish live from dead cells and generates considerable radioactive waste.

A number of colorimetric approaches developed to label viable cells have been applied to the Boyden chamber assay. Variations of the MTT [3-(4,5-dimethylthiazol-2-yl)-2,5-diphenyltetrazolium bromide dye] assay are increasingly popular. MTT is metabolized by mitochondrial hydrolases to produce a formazan dye that is detected using spectrophotometry. As this conversion only occurs in living cells, MTT can distinguish live and dead cells (Mosmann, 1983). In brief, cells are labeled with 0.5 mg/ml MTT for 4 hr at 37°C to allow for fomazan dye development. Acid–isopropanol (0.04 N HCl in isopropanol, 100 μl per 100 μl growth medium) or 5% sodium dodecyl sulfate and 0.05 N HCl is added to the well to dissolve the crystals and the optical density read at 570 nm on an ELISA plate reader (Mosmann, 1983). Boyden chamber filters can be also be stained with MTT and visualized directly as described above or scored qualitatively without a microscope (Sasaki and Passaniti, 1998).

Combining these approaches, Imamura et al. (1994) developed a colorimetric assay for measuring migration in a Boyden chamber. The apparatus used in this assay has an upper chamber that inserts into a lower chamber. After the migration assay is completed, the upper chamber containing the membrane is removed and the cells on the upper surface removed with a cotton applicator. MTT is added to the lower chamber, the upper chamber is returned, and the apparatus is incubated for 4 hr. The upper chamber is then moved to another chamber containing dimethyl sulfoxide that dissolves the crystals. The resulting solution can then be read on an ELISA plate reader. A similar assay using crystal violet has since been developed (Saito et al., 1997). Both the radiolabeled and the colorimetric assays require preparations of standard curves for each cell line. The colorimetric assays are advantageous as they are easy to perform and more fully examine the membrane than do visual techniques.

References

Albini, A., Iwamoto, Y., Kleinman, H. K., Martin, G. R., Aaronson, S. A., Kozlowski, J. M., and McEwan, R. N. (1987). A rapid in vitro assay for quantitating the invasive potential of tumor cells. *Cancer Res.* **47,** 3239–3245.

Boyden, S. V. (1998). The chemotactic effect of mixtures of antibody and antigen on polymorphonuclear leucocytes. *J. Exp. Med.* **1115,** 453–466.

Burridge, K., Turner, C. E., and Romer, L. H. (1992). Tyrosine phosphorylation of paxillin and pp125FAK accompanies cell adhesion to extracellular matrix: A role in cytoskeletal assembly. *J. Cell Biol.* **119,** 893–903.

Chen, H., Paradies, N. E., Fedor-Chaiken, M., and Brackenbury, R. (1997). E-cadherin mediates adhesion and suppresses cell motility via distinct mechanisms. *J. Cell Sci.* **110,** 345–356.

Curtis, A. and Wilkinson, C. (1998). Topographical control of cell migration. *In* "Motion Analysis of Living Cells" (D. R. Soll and D. Wessels, eds.), pp. 141–156. Wiley-Liss, New York.

Fath, K. R., Edgell, C. J., and Burridge, K. (1989). The distribution of distinct integrins in focal contacts is determined by the substratum composition. *J. Cell Sci.* **92,** 67–75.

Hart, I. R., and Fidler, I. J. (1978). An in vitro quantitative assay for tumor cell invasion. *Cancer Res.* **38,** 3218–3224.

Hendrix, M. J. C., Gehlsen, K. R., Wagner, H. N., Rodney, S. R., Misiorowski, R. L., and Meyskens, J. R. (1985). In vitro quantification of melanoma tumor cell invasion. *Clin. Exp. Metastasis* **3,** 221–233.

Imamura, H., Takao, S., and Takashi, A. (1994). A modified invasion-3-(4,5-dimethylthiazole-2-yl)-2,5-diphenyltetrazolium bromide assay for quantitating tumor cell invasion. *Cancer Res.* **54,** 3620–3624.

Jozaki, K., Marucha, P. T., Despins, A. W., and Kreutzer, D. L. (1990). An in vitro model of cell migration: Evaluation of vascular endothelial cell migration. *Anal. Biochem.* **190,** 39–47.

Khosravi-Far, R., Solski, P. A., Clark, G. J., Kinch, M. S., and Der, C. J. (1995). Activation of Rac1, RhoA, and mitogen-activated protein kinases is required for Ras transformation. *Mol. Cell. Biol.* **15,** 6443–6453.

Kleinman, H. K., McGarvey, M. L., Liotta, L. A., Robey, P. G., Tyrggvason, K., and Martin, G. (1982). Isolation and characterization of Type IV procollagen, laminin, and heparan sulfate proteoglycan from the EHS sarcoma. *Biochemistry.* **21,** 6188–6193.

Kleinman, H. K., McGarvey, M. L., Hassell, J. R., Star, V. L., Cannon, F. B., Laurie, G. W., and Martin, G. R. (1986). Basement membrane complexes with biological activity. *Biochemistry* **25,** 312–318.

Lauffenburger, D. A., and Horwitz, A. F. (1996). Cell migration: A physically integrated molecular process. *Cell* **84,** 359–369.

Liotta, L. A., Lee, W. C., and Morakis, D. J. (1980). New method for preparing large surface of intact basement membrane for tumor invasion studies. *Cancer Lett.* **11,** 141–147.

Lund-Johansen, M., Bjerkvig, R., Humphrey, P. A., Bigner, S. H., Bigner, D. D., and Laerum, O. D. (1990). Effect of epidermal growth factor on glioma cell growth, migration, and invasion in vitro. *Cancer Res.* **50,** 6039–6044.

McGuire, P. G., and Seeds, N. W. (1989). The interaction of plasminogen activator with a reconstituted basement membrane matrix and extracellular macromolecules produced by cultured epithelial cells. *J. Cell. Biochem.* **40,** 215–227.

Magnatti, P., Robbins, E., and Rifkin, D. (1986). Tumor invasion through the human amniotic membrane: Requirement for a proteinase cascade. *Cell* **47,** 487–498.

Marcel, M. M., DeBruyre, G. K., Vandesande, F., and Dragonetti, C. (1981). Immunohistochemical study of embryonic chick heart invaded by malignant cells in three dimensional culture. *Invasion Metastasis* **1,** 195–204.

Marucha, P. T., Jozaki, K., Despins, A. W., and Kreutzer, D. L. (1987). A new in vitro method for the quantitation of endothelial cell migration. *Fed. Proc. Fed. Am. Soc. Exp. Biol.* **46,** 978.

Mohler, J. L., Partin, A. W., and Coffey, D. S. (1987a). Prediction of metastatic potential by a new grading system of cell motility: Validation in the Dunning R-3327 prostatic adenocarcinoma model. *J. Urol.* **138,** 168–170.

Mohler, J. L., Partin, A. W., Isaacs, W. B., and Coffet, D. S. (1987b). Time lapse videomicroscopic identification of Dunning R-3327 adenocarcinoma and normal rat prostate cells. *J. Urol.* **137,** 544–547.

Mosmann, T. (1983). Rapid colorimetric assay for cellular growth and survival: Application to proliferation and cytotoxicity assays. *J. Immunol. Methods.* **65,** 55–63.

Nelson, R. D., McCormack, R. T., and Fiegel, V. D. (1978). Chemotaxis of human leukocytes under agarose. *In* "Leukocyte Chemotaxis: Methods, Physiology, and Clinical Implications" (J. I. Gallin and P. G. Quie, eds.), pp. 25–42. Raven, New York.

Noel, A. C., Calle, A., Emonard, H. P., Nusgens, B. V., Simar, L., Foidart, J., Lapiere, C. M., and Foidart, J.-M. (1991). Invasion of reconstituted basement membrane matrix is not correlated to the malignant metastatic cell phenotype. *Cancer Res.* **51,** 405–414.

Poste, G., Doll, J., Hart, I. R., and Fidler, I. J. (1980). In vitro selection of murine B16 melanoma variants with enhanced tissue invasive properties. *Cancer Res.* **40,** 1636–1644.

Repesh, L. A. (1998). A new in vitro assay for quantitating tumor cell invasion. *Invasion Metastasis* **9,** 192–208.

Rosen, E. M., Meromsky, L., Setter, E., Vinter, D. W., and Goldberg, I. D. (1990). Quantitation of cytokine-stimulated migration of endothelium and epithelium by a new assay using microcarrier beads. *Exp. Cell Res.* **186,** 22–31.

Saito, K., Oku, T., Ata, N., Miyashiro, H., Hattori, M., and Saiki, I. (1997). A modified and convenient method for assessing tumor cell invasion and migration and its application to screening for inhibitors. *Biol. Pharm. Bull.* **20,** 345–348.

Sasaki, C. Y., and Passaniti, A. (1998). Identification of anti-invasive but noncytoxic chemotherapeutic agents using tetrazolium dye MTT to quantitate viable cells in matrigel. *BioTechniques* **24,** 1038–1043.

Sheetz, M. P., Felsenfeld, D. P., and Galbraith, C. G. (1998). Cell migration: Regulation of force on extracellular-matrix–integrin complexes. *Trends Cell Biol.* **8,** 51–54.

Soll, D. R. (1995). The use of computers in understanding how cells crawl. *Int. Rev. Cytol.* **163,** 43–104.

Soll, D. R., and Voss, E. (1998). Two- and three-dimensional computer systems for analyzing how animal cells crawl. *In* "Motion Analysis of Living Cells" (D. R. Soll and D. Wessels, eds.) pp. 25–52. Wiley-Liss, New York.

Starkey, J. R., Hosick, H. L., Stanford, D. R., and Liggitt, H. D. (1984). Interaction of metastatic tumor cells with bovine lens. *Cancer Res.* **44,** 1585–1594.

CHAPTER 26

Three-Dimensional Extracellular Matrix Substrates for Cell Culture

Sherry L. Voytik-Harbin

Department of Basic Medical Sciences
School of Veterinary Medicine and
Department of Biomedical Engineering
Purdue University
West Lafayette, Indiana 47907

I. Introduction
    A. Importance of Extracellular Matrix to Cell Biology
    B. Classification of Extracellular Matrix
II. Application
    A. Cell Culture: A New Dimension
III. Methods
    A. Intact Extracellular Matrices
    B. Reconstituted Extracellular Matrices
IV. Application of Intestinal Submucosa as a Three-Dimensional Extracellular Matrix Substrate
    A. Introduction
    B. Materials and Methods
    C. Results
V. Summary
References

I. Introduction

A. Importance of Extracellular Matrix to Cell Biology

    The importance of understanding cell–extracellular matrix (ECM) interaction is now evident as biologists, tissue engineers, and physicians search for novel scaffolds that support and maintain tissue-specific cellular growth and function both *in vivo* and *in vitro*. In the past, cellular and molecular biology research involved the propagation and growth of cells primarily in two-dimensional

(2-D) culture systems. Despite the limitation of 2-D culture environments, they have formed the cornerstone on which the fundamentals of cell biology are built. Regardless, it is difficult to ignore the fact that, *in vivo*, cells exist within 3-D supramolecular structures with biophysical and biochemical properties that influence cell behavior.

Within tissues and organs, cells communicate with other cells within a 3-D molecular framework known as the ECM to collectively give rise to differentiated form and function. A major breakthrough in our perception of the information network that regulates cell behavior was made, however, by the realization that the ECM not only provides a mechanical framework for tissue architecture but also plays an active role in regulating the signaling process. The cell and its ECM coexist in a state of "dynamic reciprocity" (Bissell *et al.*, 1982). Both biochemical and biophysical signals from the ECM modulate fundamental cellular activities, including adhesion, migration, proliferation, differential gene expression, and programmed cell death (Adams and Watt, 1993; Lelievre *et al.*, 1996). In turn, the cell can modify its ECM environment by modulating synthesis, degradation, and organization of specific matrix components.

Many comprehensive reviews have detailed the evidence supporting the role for ECM in mediating critical processes such as acquisition and maintenance of differentiated phenotypes during embryogenesis, the development of form (morphogenesis), angiogenesis, tissue repair, and even tumor metastasis (Stetler-Stevenson *et al.*, 1993; Lin and Bissell, 1993; Vernon and Sage, 1995; Adams and Watt, 1993; Ingber, 1992; End and Engel, 1991; Hay, 1993). Much less information is available describing the mechanisms by which the ECM can control these complex phenomena. Therefore, there now exists a renewed interest in characterizing ECM constituents and the basic mechanisms of cell–ECM interaction, particularly in three dimensions.

B. Classification of Extracellular Matrix

The ECM represents a complex macromolecular assembly of proteins and polysaccharides that are produced, processed, and assembled by resident cells. Although the composition of the ECM is often tissue- and organ-specific, there are several major components that are common to nearly all, and they are the members of the following molecular families: collagens, glycosaminoglycans, proteoglycans, and glycoproteins. It is the variation in the relative amounts of these different types of matrix molecules and the way they are organized in the ECM that gives rise to a diversity of forms, each adapted to the functional requirements of a particular tissue. For example, ECM may have a high collagen to carbohydrate ratio and be ropelike in organization, as observed in tendons and ligaments, to provide tremendous tensile strength. Alternatively, the matrix may become calcified to form the rock-hard structure characteristic of bone or teeth. The diversity in the molecular composition of ECM also contributes to the overwhelming versatility in ECM signaling capacity. In fact, the ECM serves as a depot for potent regulatory molecules such as growth factors, matrix degrading enzymes, and their inhibitors.

Extracellular matrices found throughout the body can be divided into two major categories: basement membranes (BM) and interstitial connective tissues (ICT). These two types of ECM can be distinguished based on composition, location, and function. BMs are thin, acellular, sheetlike structures that are present practically everywhere in the body, where they serve to separate organ cells, epithelia, and endothelia from the ICT. In addition to compartmentalization, BMs provide orientative function for proliferating and differentiating cells as required during development and tissue remodeling processes. Likewise, BMs may serve as selective filters for macromolecules, as is exhibited by the glomerular basement membrane of the kidney. The major biochemical components of BM are type IV collagen, laminin, entactin/nidogen, and heparan sulfate and chondroitin sulfate proteoglycans. Type IV collagen forms a 3-D network structure that is further stabilized by laminin–entactin/nidogen complexes.

In contrast, ICT is a plentiful fibrillar matrix with major biomechanical and supportive functions in the body. ICT is distinct from BM in that it houses a cast of supporting cells (e.g., fibroblasts, mast cells, macrophages) and specialized cells (e.g., osteocytes, chondrocytes). It is the specialized cells along with the specific biochemical composition that distinguishes the different forms of ICT found throughout the body, including bone, tendon, cartilage, ligament, fasciae, and stroma. ICT, when compared to BM, contains more collagen by weight with significant representation from the fibrillar collagen family. Type I collagen is the predominant collagen present in most ICT, with the exception of cartilage which contains primarily type II. Several other less abundant collagen types have been identified in interstitial extracellular matrices, including III, V, IV, VII, VIII, IX, and X. These and other noncollagenous proteins of ICT are embedded in a highly hydrated gel-like "ground substance" composed of glycosaminoglycan (e.g., hyaluronic acid) and proteoglycan molecules. The glycosaminoglycan and proteoglycan components of both ICT and BM play a significant role in determining the hydrodynamic properties of the matrix. Likewise, glycosaminoglycans and proteoglycans contribute to the ability of ECMs to serve as a selective barrier to cells and macromolecules as well as a reservoir for potent regulatory molecules such as growth factors, proteinases, and their inhibitors. For reviews on BM and ICT, see Tryggvason *et al.* (1987), Paulsson (1992), and Engvall (1995).

II. Application

A. Cell Culture: A New Dimension

Much of our knowledge of form and function at the cell, tissue, and organ levels has come from the isolation of mammalian cells and their culture *in vitro*. To date the majority of cell culture experimentation has been performed in a 2-D format on artificial (synthetic) substrata consisting of glass or plastic. Although polystyrene is the most commonly used artificial substrate, cells have also been successfully grown on polyvinylchloride (PVC), polycarbonate, polytetrafluoroethylene (PTFE), Melinex, and Thermanox (TPX) (Freshney, 1994). To provide an *in vitro* culture environment that would more closely mimic cell–ECM interaction *in vivo,* purified ECM components such as collagen, fibronectin, laminin, and glycosaminoglycans (e.g., hyaluronic acid, heparan sulfate) have been used to derivatize (coat) artificial substrata for augmentation of cell adhesion, growth, and morphology. While it has long been realized that these 2-D culture systems provide a convenient way of preparing and observing a culture and allow a high rate of proliferation, they lack the 3-D microenvironment characteristic of whole tissues *in vivo*. Therefore, when cells are isolated from their natural ECM, cultured, and propagated under these conditions, the resultant cell phenotype is often different from that observed in the tissue from which it is derived.

The first attempts to culture animal tissues actually used 3-D gel systems formed of clotted lymph or plasma on glass (Harrison, 1907; Carrel, 1912). Unfortunately, these systems were plagued by rapid substrate dissolution. In 1956, Ehrmann and Gey reported that a viscous, soft-gel preparation of purified rat tail collagen could be used to study cell behavior *in vitro*. Elsdale and Bard (1972) later modified this technique to form a "hydrated collagen lattice" or moldable collagen gel for use as a cell culture substrate. Investigations with these matrices demonstrated the importance of 3-D architecture in the establishment of a tissue-like histology (Emerman *et al.,* 1977, 1979; Chambard *et al.,* 1981; Bell *et al.,* 1979). More complex 3-D scaffolds representing combinations of ECM components in a natural or processed form have also been studied. Intact ECM substrates representing blood vessels, lens capsule, amnion, chorioallantoic membrane, and small intestinal submucosa have been fashioned from allogeneic and xenogeneic tissue sources for use *in vitro* (Dehm and Kefalides, 1978; Armstrong and Quigley, 1982; Meezan *et al.,* 1975; Liotta *et al.,* 1980; Voytik-Harbin *et al.,* 1998; Mareel, 1983; Starkey, 1990). These techniques involve processing the extracellular matrix derived from various tissues and organs such that viable cells are eliminated but the natural architecture and composition of the ECM are minimally perturbed. Although intact ECM substrates can provide the most natural *in vitro* environment for cells, methods of isolation and processing must be carefully considered. For example, techniques employed for elimination of viable cells and cellular debris as well as for disinfection/sterilization may have deleterious effects on the biological activity and organization of matrix molecules.

Complex mixtures of soluble ECM molecules prepared by extraction or other biochemical techniques have been used to develop reconstituted 3-D gel-like scaffolds. With these 3-D ECM systems, one can readily alter substrate composition and structure to determine how it affects cell shape, phenotype, and function. However, because cells secrete and modify (proteolytically processed, sulfated, oxidized, and cross-linked) matrix molecules before and after formation of functional polymeric assemblies, reconstituted matrices likely do not have all the properties they have when they are assembled by cells *in vivo* (Olsen, 1997).

Herein, a listing of 3-D ECM substrates, both intact and reconstituted, is provided with details on preparation and applications in cell culture. These substrates and their associated ECM classification, biochemical composition, and references are summarized in Table I.

III. Methods

A. Intact Extracellular Matrices

1. Preparation of Intact Basement Membranes from Various Tissue Sources

Meezan and co-workers (1975) described a simple procedure for the isolation of intact BMs from a number of different tissues, including bovine retinal and brain blood vessels, rabbit renal tubules, and rat renal glomeruli. In brief, isolated organ subfractions (Meezan *et al.*, 1974; Brendel and Meezan, 1973) are placed in a large volume (100:1) of water and stirred for 1–2 hr to lyse the cells and release intracellular contents. The tissue suspension is centrifuged at 10,000 rpm for 10 min and the supernatant discarded. The pellet is suspended in a solution of 1 M NaCl containing 2000 units (U) of DNase and stirred for 2–4 hr at room temperature. Following centrifugation at 10,000 rpm for 10 min the pellet is treated with 4% deoxycholate to solubilize cellular membranes. The suspension is again centrifuged at 10,000 rpm and the pellet washed extensively with water. Structural and chemical characterization studies performed on BMs prepared by this technique suggest that they are ultrastructurally indistinguishable from their *in vivo* counterparts and that the chemical composition is not altered (Meezan *et al.*, 1975; Kefalides, 1971; Carlson *et al.*, 1978).

2. Preparation of Intact Amniotic Membranes from Human Placentas

In 1980, Liotta and co-workers used the Meezan technique with modifications to prepare human amniotic BM for *in vitro* studies. This method involves fresh human placentas obtained at the time of normal term delivery and kept on ice prior to dissection. The inner amniotic membrane is teased away from the chorion by gentle blunt dissection to obtain an intact membrane with a surface area of at least 200 cm^2. Under sterile conditions, the membrane is washed in ice-cold phosphate-buffered saline (PBS) containing penicillin/streptomycin (100 U/ml) and Fungizone (100 U/ml) followed by Dulbecco's minimum essential medium and stored in the same medium at 4°C. Portions of the amniotic membrane are then clamped in ring-shaped plastic holders to create a surface area of 3.8 cm^2. The amnion in the holder is first rinsed in distilled water followed by water containing 1 mM N-ethylmaleimide (NEM) for 1 hr at 4°C. The epithelial side of the membrane is exposed to 4% deoxycholate for 2 hr at 20°C. The epithelium

Table I
Three-Dimensional Extracellular Matrix Substrates

Extracellular matrix substrate	ECM classification	Components identified to date	References
Intact extracellular matrices			
Intact basement membrane from bovine retinal blood vessels, bovine brain blood vessels, rabbit renal tubules, and rat renal glomeruli	BM	ND[a]	Meezan et al. (1975)
Intact amniotic membrane from human placenta	BM/ICT	ND	Liotta et al. (1980)
Small intestinal submucosa from porcine intestine	ICT	Collagen types: I, III, V Proteins: fibronectin Glycosaminoglycans: hyaluronic acid, heparin, heparan sulfate, chondroitin sulfate A, chondroitin sulfate B, dermatan sulfate Growth Factors: FGF-2, TGF-β	Voytik-Harbin et al. (1998)
Reconstituted extracellular matrices			
Purified collagen gel (Vitrogen)	ICT	Collagen types: I, III	Ehrmann and Gey (1956), Elsdale and Bard (1972)
EHS matrix (Matrigel)	BM	Collagen types: IV Proteins: laminin, entactin/nidogen Proteoglycans: heparan sulfate proteoglycan Growth Factors: TGF-β, EGF, FGF-2, IGF-1, and PDGF Miscellaneous: plasminogen, tPA, MMP-2, MMP-9	Kleinman et al. (1983), Kleinman et al. (1986)
Amgel from human amniotic membranes	BM/ICT	Collagen types: I, IV Proteins: entactin, laminin, tenascin Proteoglycans: heparan sulfate proteoglycan	Siegal et al. (1993)
Humatrix from human myoepithelial tumors	BM/ICT	Collagen types: I, IV Proteins: fibronectin, laminin, nidogen fragment, BM90/fibulin, bamin, and BM40/osteonectin/SPARC Glycosaminoglycans: chondroitin sulfate, hyaluronic acid, heparan sulfate Growth factors: EGF, IGF-1 Miscellaneous: protease nexin II, α_1-antitrypsin, thrombospondin-1, maspin	Kedeshian et al. (1998)
Reconstituted interstitial ECM from Intestinal submucosa	ICT	Collagen types: I, III, V	Voytik-Harbin et al. (1998)

[a] ND, not determined.

is denuded by gentle agitation with a rubber policeman, leaving the underlying BM surface intact. Prior to tissue culture studies, the denuded amnion is soaked overnight at 4°C in complete medium containing antibiotics and Fungizone. This technique yields a continuous BM overlying a thin avascular stroma composed of interstitial collagen and elastin. This intact BM substrate has been used primarily as a model system for the study of tumor invasion *in vitro* (Liotta *et al.*, 1980; Mignatti *et al.*, 1986; Thorgeirsson *et al.*, 1982).

3. Preparation of Intact Interstitial Extracellular Matrix (Intestinal Submucosa) from Porcine Intestine

Our laboratory has investigated an intact interstitial extracellular matrix known as intestinal submucosa for use as a cell culture substrate (Voytik-Harbin *et al.*, 1998). This biomaterial can be purchased (Cook Biotech, West Lafayette, IN) or prepared from the small intestines of market-weight pigs obtained from a local meat-processing plant. The tube of intestinal material is rinsed free of contents, everted, and the superficial layers of the mucosa removed by mechanical delamination with blunt-ended instruments such as the handle end of a hemostat. The intestinal tissue is reverted to its original orientation, and the external muscle layer is removed. The prepared intestinal submucosa tube is split open longitudinally and rinsed extensively in water to lyse any cells associated with the matrix as well as to eliminate cell degradation products. Immediately, after rinsing, intestinal submucosa can be disinfected with 0.1% peracetic acid for cell culture applications or stored at $-80°C$ for preparation of reconstituted interstitial ECM (refer to Section III,B,5). Disinfection is achieved by soaking intestinal submucosa in a solution of 4% ethanol containing 0.1% peracetic acid for 2 hr at room temperature using a ratio of 10:1 (milliliters peracetic acid solution:grams intestinal submucosa) or greater. The intestinal submucosa then is rinsed extensively in sterile water or PBS and stored at 4°C or colder.

Microscopic studies demonstrate that intestinal submucosa consists of the submucosa, muscularis mucosa, and remnant lamina propria layers of the intestine. The topographical features of the mucosal and serosal surfaces of the material are distinct (Voytik-Harbin *et al.*, 1998). Ultrastructurally, the mucosal surface is characterized by more densely packed fibers, which form discontinuous layers varying in orientation. Alternatively, the serosal side exhibits a fine network of loosely organized fibers, most of which are <1 μm in diameter. Preliminary biochemical characterization studies indicate that intestinal submucosa contains at least three different collagen types. Collagen types I and III are predominant within intestinal submucosa, with lesser quantities of type V (S. L. Voytik-Harbin, unpublished data). A number of glycoaminoglycans (Hodde *et al.*, 1996), fibronectin (McPherson and Badylak, 1998), and the growth factors TGF-β (transforming growth factor β) and FGF-2 (fibroblast growth factor 2) (Voytik-Harbin *et al.*, 1997) have also been identified in intestinal submucosa.

To date, at least 18 different cell types have been evaluated on intestinal submucosa, including established cell lines and primary cells derived from normal and neoplastic tissues of human and other mammalian sources. In fact, the different cell types survived and developed a tissue-like morphology when grown on intestinal submucosa *in vitro* (Voytik-Harbin *et al.*, 1998; Sturgis *et al.*, 1998). For example, when cultured on intestinal submucosa, mesenchymal cell types (e.g., fibroblasts) appear spindle-shaped and readily penetrate and integrate within the ECM components. Alternatively, intestinal submucosa induces polarization, stratification, and acini formation of glandular epithelial cells derived from canine prostate adenocarcinoma. Studies by other laboratories have shown that articular chondrocytes grown on intestinal submucosa maintain their phenotype, synthesize type II collagen and large proteoglycans, and form a cartilaginous tissue reminiscent of hyaline cartilage (Peel *et al.*, 1998). Intestinal submucosa has also been used as an *in vitro* culture substrate for studying and comparing the proliferation, attachment, and migration properties of tumor cells varying in metastatic capacity (S. L. Voytik-Harbin, unpublished data). For additional details regarding application of intestinal submucosa as a 3-D ECM substrate, see Section IV.

B. Reconstituted Extracellular Matrices

1. Preparation of Purified Collagen Gel

Collagen represents the predominant component of the extracellular matrix, so it is not surprising that 3-D substrates prepared from this macromolecule have been extensively applied to study cell behavior *in vitro* (Ehrmann and Gey, 1956; Elsdale and Bard, 1972). Type I collagen is most widely used because of its relative abundance and ease of purification. This collagen type can be readily prepared from virtually any tissue or organ, with the most convenient sources being tendon, dermis, and bone.

Prior to the isolation of collagen, soluble noncollagenous substances are removed from source tissues by treatment with water or neutral salt solvents (0.15–1 M NaCl) in the presence of protease inhibitors [e.g., 20 mM (ethylenedinitrilo)tetraacetic acid (EDTA), 1 mM phenylmethanesulfonyl fluoride (PMSF), 1 μg/ml pepstatin, and 2 mM NEM]. The collagen is solublilized with 0.5 M acetic acid alone or in the presence of pepsin (1:10 weight of pepsin to dry weight of tissue) for 1–2 days at 4°C (Miller and Rhodes, 1982). Tissue particulate is removed by filtering the suspension through several layers of cheesecloth followed by centrifugation at 40,000 g. The clarified collagen solution then is dialyzed against dilute acid (0.01 M acetic acid or hydrochloric acid). Chloroform may be added to the reservoir (0.5%, v/v) during dialysis to achieve a sterile collagen preparation. For polymerization into a 3-D substrate, the collagen solution is adjusted to physiological pH and ionic strength by the addition of 10× concentrated medium or PBS and dilute NaOH and HCl. The neutralized solution is divided into aliquots in cultureware and incubated at 37°C for 30–60 min. A purified collagen preparation from bovine dermis (Vitrogen) is available from Collagen Aesthetics (Palo Alto, CA).

Three-dimensional collagen substrates have received widespread application in cell culture. More specifically, collagen substrata have been used as a model system to facilitate the study of cellular proliferation, differentiation, and metabolism as well as the more complex processes of morphogenesis and metastasis (Ehrmann and Gey, 1956; Elsdale and Bard, 1972; Chambard *et al.*, 1981; Michalopoulos and Pitot, 1975; Grinnell, 1982; Schor *et al.*, 1983; Lee *et al.*, 1984; Schor *et al.*, 1982; Bissell *et al.*, 1987; Bell *et al.*, 1979).

2. Preparation of Engelbreth-Holm-Swarm Matrix

In the 1950s, the Engelbreth-Holm-Swarm (EHS) tumor appeared spontaneously in ST/Eh mice and was found to produce copious amounts of BM components (Swarm, 1963). Since that time, Kleinman *et al.* (1982, 1986) described an extract of this tumor that forms a 3-D gel-like BM *in vitro*. The EHS tumor is propagated in C57/Bl mice made lathyritic by inclusion of 0.1% β-aminopropionitrile fumarate (which blocks the cross-linking of collagen) in their drinking water as previously described (Orkin *et al.*, 1977). Four weeks following subcutaneous injection (0.2 ml) of tumor homogenate (2 g/ml) in lathyritic mice, the tumor is harvested. Tumor tissue is suspended (1 g/ml) and homogenized in a buffer containing 3.4 M NaCl, 4 mM sodium EDTA, and 2 mM NEM in 50 mM Tris-HCl, pH 7.4. Following centrifugation at 10,000 rpm for 30 min at 4°C, the recovered pellet is extracted in a buffer containing 2 M urea and 0.15 M NaCl in 50 mM Tris-HCl, pH 7.4, for 24 hr at 4°C. The mixture is clarified by centrifugation at 14,000 rpm and the supernatant dialyzed against 0.15 M NaCl in 50 mM Tris-HCl, pH 7.4. The extract may be sterilized by inclusion of 0.5% (v/v) chloroform in the dialysis buffer followed by dialysis against sterile serum-free medium. The EHS matrix extract is stored frozen at −20°C. For application as a 3-D culture substrate, the EHS matrix is thawed slowly on ice or at 4°C. Polymerization of the EHS matrix is achieved by placing aliquots of the viscous solution into culture ware followed by incubation at 37°C. Because this ECM-based substrate is temperature sensitive, it is recommended that all culture ware and pipette tips be chilled prior to use. The EHS matrix is commercially available as Matrigel from Collaborative Research (Bedford, MA).

Compositional analyses have revealed that the EHS matrix has laminin as its predominant component, with lesser amounts of type IV collagen, heparan sulfate proteoglycan, and entactin (Kleinman *et al.*, 1982). The potent growth factors TGF-β, FGF-2, epidermal growth factor (EGF), insulin-like growth factor-1 (IGF-1) and platelet-derived growth factor (PDGF), (Vukicevic *et al.*, 1992) as well as significant quantities of plasminogen (Farina *et al.*, 1996), tissue plasminogen activator (tPA) (McGuire and Seeds, 1989), matrix metalloproteinase-2 (MMP-2) (Mackay *et al.*, 1993), and matrix metalloproteinase-9 (MMP-9) (Mackay *et al.*, 1993) have also been identified in the matrix. The EHS matrix has been utilized as a model system for the study of cellular differentiation as well as the complex processes of angiogenesis and morphogenesis (Nicosia and Ottinetti, 1990; Hadley et al., 1985; Bendayan et al., 1986; Kleinman et al., 1986; Li et al., 1987; Bissell et al., 1987). Albini and others developed a chemoinvasion assay system with EHS matrix that has been used extensively for studying the mechanisms involved in tumor and endothelial cell invasion of BM (Albini et al., 1987; Albini, 1998). Likewise, the EHS matrix-based assay has been used for screening of anti-invasive agents (Albini, 1998).

3. Preparation of Amgel from Human Amniotic Membranes

A reconstituted ECM also has been prepared from human amnions (Siegal et al., 1993). Amniotic membranes are harvested from normal, full-term, human placentas. The epithelial layer is denuded from the amnion by a brief exposure to 0.1 N ammonium hydroxide, which lyses any contaminating cells, followed by gentle scraping and extensive washing in PBS. Membranes then are homogenized in 0.5 N acetic acid and the homogenate adjusted to pH 2. Pepsin (8 mg/g starting tissue) is added to the homogenate and incubated at 4°C overnight. The pepsin-treated tissue is centrifuged at 3000 g for 15 min and the recovered supernatant adjusted to pH 7.8 to inactivate pepsin. Solid Tris and NaCl are slowly added to the supernatant to achieve final concentrations of 50 mM Tris and 4 M NaCl, respectively. After stirring 18–20 hr, the gel-like solution is centrifuged at 17,500 g for 30 min and the pellet resuspended in 0.5 N acetic acid. Again the solution is adjusted to 50 mM Tris and 4 M NaCl and then centrifuged. The pelleted product is dialyzed extensively against 5 mM acetic acid containing 0.14 M NaCl and 5 mM KCl. The resultant ECM preparation can be further concentrated by ultrafiltration with a molecular weight cutoff of 12,000 and stored at 4°C for immediate use or at −20°C for later studies.

Biochemical analyses of Amgel have shown that collagen is a major component, with both type I and type IV being present. The BM components laminin, entactin, tenascin, and heparan sulfate proteoglycan have also been identified within Amgel. Amgel has been used to coat polycarbonate filters for creation of a tumor invasion assay system. The proclaimed advantages of this ECM substrate include its derivation from normal physiological rather than pathological tissue and increased human clinical relevance (Siegal et al., 1993).

4. Preparation of Humatrix from Human Myoepithelial Tumors

Similar to Kleinman et al. (1983, 1986), Kedeshian and co-workers (1998) identified human myoepithelial tumors that produced and secreted an abundant extracellular matrix, from which to prepare a reconstituted ECM. Human myoepithelial tumors, harvested from human salivary gland and breast tissue, are propagated in mice rendered lathyritic by feeding β-aminoproprionitrile fumarate. After initial homogenization in high salt buffer (3.4 M NaCl, 20 mM EDTA, and 10 mM NEM in 50 mM Tris-HCl, pH 7.4), the homogenate was centrifuged for 15 min at 12,000 g at 4°C. The supernatant was discarded and the pellet extracted (0.5 ml/g starting material) overnight at 4°C with a solution containing 6 M urea, 2 M guanidinium hydrochloride, 20 mM EDTA, 10 mM NEM, and 2 mM dithiothreitol (DTT) in 50 mM Tris-HCl, pH 7.4. Following centrifugation at 24,000 g at 4°C, the supernatant was dialyzed against several changes of 0.15 M NaCl, 20 mM EDTA, and 10 mM NEM in 50 mM Tris-HCl, pH 7.4, followed by sequential dialyses against 0.5% (v/v) chloroform and sterile cell culture medium. This ECM solution may be concentrated by ultrafiltration and is stored at 4°C. To prepare a 3-D ECM substrate, the solution is divided into aliquots in culture ware and allowed to polymerize at 25°–37°C.

Because of its myoepithelial origin, Humatrix is said to represent a more natural source of extracellular matrix molecules and bound factors that carcinoma cells encounter *in vivo* (Kedeshian *et al.*, 1998). In addition to the BM components type IV collagen, laminin, and entactin/nidogen, Humatrix consists of type I collagen and fibronectin, which are characteristic of ICT. Humatrix also contains an array of bound growth factors and proteinase and angiogenic inhibitors that reflect the constitutive gene expression profile of the myoepithelial cell. For more specific compositional information, see Table I. When used as a culture substrate, Humatrix has been shown to selectively stimulate the growth and tumorigenicity of human myoepithelial cell lines, but it inhibits invasion, angiogenesis, and metastasis of other nonmyoepithelial malignant cell lines.

5. Preparation of Reconstituted Interstitial Extracellular Matrix from Intestinal Submucosa

As described previously (refer to Section III,A,3), intestinal submucosa is an interstitial ECM derived from porcine small intestine that is suitable as an intact 3-D cell culture substrate. Our laboratory also has developed a solubilized form of this biomaterial that forms a 3-D gel of ICT components *in vitro* (Voytik-Harbin *et al.*, 1998). For preparation of reconstituted interstitial ECM, frozen intestinal submucosa (refer to Section III,A,3) is pulverized under liquid nitrogen with an industrial blender. Intestinal submucosa powder is suspended (5%, w/v) in 0.5 M acetic acid containing 0.1% (w/v) pepsin and vigorously stirred for 72 hr at 4°C. The mixture then is centrifuged at 12,000 rpm for 20 min at 4°C to remove tissue particulate. The supernatant is dialyzed extensively against 0.01 M acetic acid at 4°C in Spectrapor tubing with a molecular weight cutoff of 3500 (Spectrum Medical Industries, Laguna Hills, CA). To obtain a sterile preparation, the solution is dialyzed against 0.01 M acetic acid containing chloroform (0.5%, v/v), followed by several changes of sterile 0.01 M acetic acid. To polymerize the intestinal submucosa solution, 10× PBS (1/10 final volume) is added and the pH adjusted between 8.0 and 9.0 with NaOH. This solution then is brought to pH 7.4 ± 0.2 with HCl, divided into aliquots in culture ware, and gelled at 37°C for 30–60 min. The substrate is equilibrated to culture conditions by incubation with PBS or serum-free medium.

Reconstituted interstitial ECM consists of collagen types I, III, and V as well as significant levels of glycosaminoglycans and glycoproteins that have yet to be fully characterized (S. L. Voytik-Harbin, unpublished data). When compared to routinely used reconstituted ECM substrates, Matrigel and Vitrogen, reconstituted interstitial ECM possesses unique architectural and biological properties. In fact, many cell types proliferated and developed a tissue-like morphology when grown on reconstituted interstitial ECM *in vitro* (Voytik-Harbin et al., 1998).

IV. Application of Intestinal Submucosa as a Three-Dimensional Extracellular Matrix Substrate

A. Introduction

The majority of the 3-D ECM substrates described herein were developed and utilized by a single research group for a specific research application. Only two of the scaffolds, namely, collagen gel and reconstituted basement membrane, have received more widespread use and are now commercially available as Vitrogen and Matrigel, respectively. Unfortunately, neither of these two substrates represents the complex mixture of components and structure of the interstitial ECM. With an interest in 3-D culture systems, we have evaluated intact intestinal submucosa and a reconstituted interstitial ECM derived from intestinal submucosa for their ability to support cellular growth and behavior *in vitro*. As stated previously, intestinal submucosa represents a tissue-derived interstitial ECM processed such that viable cells are eliminated but the natural architecture and composition of the matrix are minimally perturbed. In addition, many of the component macromolecules of this intact interstitial ECM have been solubilized and used to form a reconstituted (gel-like) interstitial ECM.

In preliminary studies, we compared the morphology and behavior of several different cell types (mesenchymal and epithelial) on different 3-D scaffolds, including intact intestinal submucosa, intestinal submucosa-derived gel, type I collagen, and reconstituted basement membrane. Two of the more than 18 specific cell types evaluated included an endothelial cell line derived from rat pulmonary arteries and low passage smooth muscle (stromal) cells derived from human urinary bladders. Interestingly, intestinal submucosa supported distinct morphological responses of these two cell types. In fact, individual cultures of smooth muscle cells and endothelial cells demonstrated a cell-specific response to intestinal submucosa that more closely approximated those observed *in vivo*, especially when compared to plastic, reconstituted basement membrane, and type I collagen (Voytik-Harbin *et al.*, 1998). These results emphasize the importance of substrate composition and structure on cell behavior (Figs. 1 and 2). On intestinal submucosa, rat pulmonary artery endothelial cells grew primarily along the surface of the scaffold as cuboidal shaped cells, creating a "cobblestone" pattern. After just 4 days in culture, the endothelial cells formed a cellular layer one to two cells thick, reminiscent of the endothelial layer common to blood vessels. In

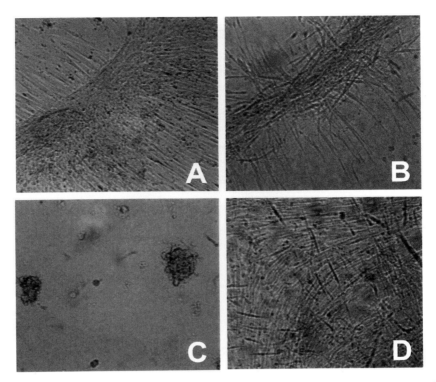

Fig. 1 Morphology of urinary bladder smooth muscle (stromal) cells (day 7) on plastic (A), Vitrogen (B), Matrigel (C), and small intestinal submucosa-derived gel (D). Magnification ×10.

contrast, urinary bladder smooth muscle cells maintained their spindle shape with a centrally located prominent nucleus. Within 7 days, the cells proliferated and migrated throughout the matrix, organizing into thick bands or multilayers of parallel aligned cells.

With the overall goal of creating functional tissue equivalents (e.g., blood vessels) for scientific studies *in vitro* or tissue replacement *in vivo*, the next generation studies involved smooth muscle cells and endothelial cells derived from human coronary arteries. For these studies intact intestinal submucosa was applied as the 3-D culture substrate, and the maintenance of cell-specific phenotype and function was explored.

B. Materials and Methods

1. Propagation of Cells

Tertiary cultures of vascular smooth muscle cells or vascular endothelial cells isolated from normal human coronary arteries were obtained from Clonetics

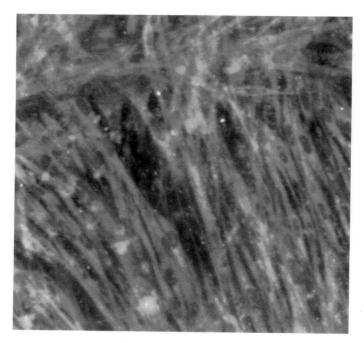

Fig. 2 Morphology of PKH26-labeled urinary bladder smooth muscle (stromal) cells on intact intestinal submucosa, day 7. Magnification ×20.

(San Diego, CA). The smooth muscle cells were propagated on standard tissue culture plastic using Smooth Muscle Cell Basal Medium (Clonetics) supplemented with 10 ng/ml human recombinant epidermal growth factor, 5 µg/ml insulin, 22 ng/ml human recombinant fibroblast growth factor, 50 µg/ml gentamicin, 50 ng/ml amphotericin B, and 5% fetal bovine serum. Endothelial cells were propagated in Endothelial Cell Growth Medium (Clonetics) supplemented with 10 ng/ml human recombinant epidermal growth factor, 1 µg/ml hydrocortisone, 4 µg/ml bovine brain extract, 20 µg/ml gentamicin, 50 ng/ml amphotericin B, and 5% fetal bovine serum. Cell populations representing limited passage numbers (three to eight) were used in the experiments.

2. Establishment of Three-Dimensional Culture Systems

The intestinal submucosa material was affixed in polypropylene frames with either the mucosal or serosal surface facing upward to create a surface area of 0.5 cm². Each scaffold was equilibrated with sterile PBS, pH 7.4, prior to the application of cells. Cells were seeded on the mucosal side of intestinal submucosa at densities ranging from 1×10^3 to 1×10^6 cells/cm². Medium contents were varied to determine the effect of the presence and absence of serum or specific mitogens on cell phenotype and behavior. Culture plates were incubated at 37°C in a humidified atmosphere of 5% CO_2 in air and fed 2–3 times weekly. After incubation for up to 14 days, the cells and associated substrate were processed for qualitative and quantitative evaluation.

3. Morphological Evaluation

General morphological evaluation was conducted using fluorescence and confocal microscopy on cells dually labeled with stain specific for F-actin and DNA for simultaneous visualization of the cytoplasm and nucleus. Whole mount specimens were fixed in neutral buffered formalin, permeabilized in 0.1% Triton X-100, stained with a dye mixture containing 30 μM Hoechst 33342 (Molecular Probes, Eugene, OR) and 0.13 μM Oregon Green-conjugated phalloidin (Molecular Probes), and mounted.

4. Evaluation of Cell-Specific Phenotype and Function

Endothelial cells express a number of specific markers that are helpful in identifying these cells both *in vivo* and in culture. These include expression of *Ulex europaeus* I (UEI) agglutinin binding antigen and the functional internalization and degradation of 1,1'-dioctadecyl-3,3,3',3'-tetramethylindocarbocyanine perchlorate-labeled acetylated low density lipoprotein (DiI-Ac-LDL). Uptake of DiI-Ac-LDL was examined according to Voyta *et al.* (1984) and the manufacturer's specifications.

In brief, cultures were incubated in complete medium containing 10 μg/ml DiI-Ac-LDL (Biomedial Technologies) for 6 hr at 37°C, washed extensively in probe-free medium and images collected. Endothelial cells were also identified by staining with rhodamine-conjugated UEI lectin (20 μg/ml) (Vector Laboratories, Burlingame, CA). This lectin binds α-L-fucose containing glycocompounds specific to endothelial cells. Vascular smooth muscle cells were evaluated as a negative control and did not bind the UEI lectin. The specificity of the lectin was demonstrated by competitive inhibition of binding by 0.2 M α-L-fucose. Expression of a smooth muscle-specific contractile regulating protein, calponin, was used as an indicator of the smooth muscle phenotype. Whole mount specimens were fixed in neutral buffered formalin and rinsed in PBS, pH 7.4. Monoclonal anticalponin (Sigma, St. Louis, MO) and a rhodamine-conjugated donkey anti-mouse antibody (Chemicon, Temecula, CA) were applied as the primary and secondary antibodies, respectively.

5. Imaging

Two-dimensional and 3-D images were collected using a Nikon Labophot fluorescence microscope and a Bio-Rad MRC 1024 UV/vis confocal microscope, respectively.

C. Results

Consistent with our previous observations in which intestinal submucosa was used as a 3-D cell culture substrate, cell-specific behavior was observed using tertiary cultures of endothelial cells and smooth muscle cells derived from human coronary arteries (Sturgis *et al.*, 1998). When cultured on intact intestinal submucosa, human vascular endothelial cells formed a continuous monolayer of polygonal-shaped cells along the surface of substrate and continued to express endothelial phenotype (e.g., *Ulex europaeus* I agglutinin binding antigen) and function (e.g., uptake of DiI-Ac-LDL) for up to 14 days (Fig. 3). This particular growth pattern along the surface of the relatively opaque intestinal submucosa matrix allowed morphological, phenotypic, and functional properties to be visualized by standard fluorescence microscopy. Unlike endothelial cells, vascular smooth muscle cells demonstrated a spindle-shaped morphology (Fig. 4A) as would be observed *in vivo*. In addition, smooth muscle cells not only grew along the surface of intestinal submucosa but readily migrated into the matrix. This 3-D growth pattern was most apparent using confocal microscopy in conjunction with computer-aided 3-D reconstruction (Fig. 5). An intense, diffuse calponin staining pattern was observed at all time points up to 14 days on intestinal submucosa, suggesting maintenance of phenotype (Fig. 4B).

V. Summary

In summary, the understanding of cell biology will be furthered as cell culture expands from 2-D to 3-D systems. In choosing which substrate, synthetic or

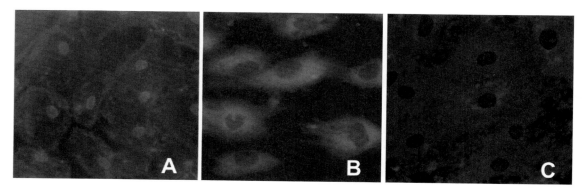

Fig. 3 Endothelial cells cultured on intestinal submucosa formed a monolayer of polygonal-shaped cells along the surface of the substrate and continued to express endothelial cell specific phenotype and function. (A) For morphological evaluation cells were dually labeled with Oregon Green-conjugated phalloidin (cytoplasmic stain) and Hoechst 33342 (nuclear stain). (B) Cells continued to express endothelial cell function as indicated by their ability to uptake DiI-Ac-LDL. (C) Cells stained positively with rhodamine-conjugated *Ulex europaeus* I lectin, indicating the presence of an endothelial cell-specific surface marker. Magnification ×40. (See color plates.)

26. Three-Dimensional ECM Substrates

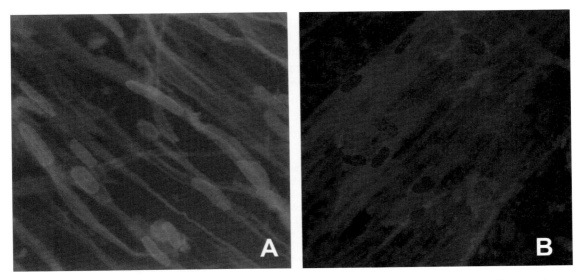

Fig. 4 Smooth muscle cells cultured on intestinal submucosa demonstrate a spindle-shaped morphology and continue to express smooth muscle specific phenotype. (A) For morphological evaluation cells were dually labeled with Oregon Green-conjugated phalloidin (cytoplasmic stain) and Hoechst 33342 (nuclear stain). (B) Cells continued to express the contractile regulating protein calponin (red fluorescence). Magnification ×40. (See color plates.)

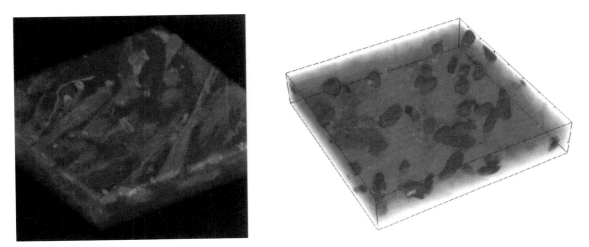

Fig. 5 3-D reconstructed confocal images of human coronary artery smooth muscle cells grown on intestinal submucosa. Cells were dually labeled with Oregon Green-conjugated phalloidin and Hoechst 33342. The left image shows the cytoplasmic and nuclear aspects of the cells within the intestinal submucosa matrix, whereas the right image highlights the nucleus of each cell. (See color plates.)

biologically derived, is most well suited for a specific application, substrate composition and structure as well as cell type(s) must be carefully considered. In addition, optimization of seeding densities, medium conditions, growth factor supplements, and other culture parameters may be necessary. Finally, cytometric analyses of such 3-D culture systems will require concurrent innovations in 3-D imaging and methods for quantitating cell morphology, phenotype, and function.

References

Adams, J. C. and Watt, F. M. (1993). Regulation of development and differentiation by the extracellular matrix. *Development* **117**, 1183–1198.

Albini, A. (1998). Tumor and endothelial cell invasion of basement membranes: The matrigel chemoinvasion assay as a tool for dissecting molecular mechanisms. *Pathol. Oncol. Res.* **4**, 230–241.

Albini, A., Iwamoto, Y., Kleinman, H. K., Martin, G. R., Aaronson, S. A., Kozlowski, J. M., and McEwan, R. N. (1987). A rapid *in vitro* assay for quantitating the invasive potential of tumor cells. *Cancer Res.* **47**, 3239–3245.

Armstrong, P. B., and Quigley, J. B. (1982). Transepithelial invasion and intramesenchymal infiltration of the chick embryo chorioallantois by tumor cell lines. *Cancer Res.* **42**, 1826.

Bell, E., Ivarsson, B., and Merrill, C. (1979). Production of a tissue-like structure by contraction of collagen lattices by human fibroblasts of different proliferative potential *in vitro*. *Proc. Natl. Acad. Sci. U.S.A.* **76**, 1274–1278.

Bendayan, M., Duhr, M. A., and Gingras, D. (1986). Studies on pancreatic acinar cells in tissue culture: Basal lamina (basement membrane) matrix promotes three-dimensional reorganization. *Eur. J. Cell Biol.* **42**, 60–67.

Bissell, D. M., Arenson, D. M., Maher, J. J., and Roll, F. J. (1987). Support of cultured hepatocytes by a laminin-rich gel. *J. Clin. Invest.* **79**, 801–812.

Bissell, M. J., Hall, H. G., and Parry, G. (1982). How does the extracellular matrix direct gene expression? *J. Theor. Biol.* **99**, 31–68.

Brendel, K., and Meezan, E. (1973). Properties of a pure metabolically active glomerular preparation from rat kidneys. II. Metabolism. *J. Pharmacol. Exp. Ther.* **187**, 342–351.

Carlson, E. C., Brendel, K., Hjelle, J. T., and Meezan, E. (1978). Ultrastructural and biochemical analyses of isolated basement membranes from kidney glomeruli and tubules and brain and retinal microvessels. *J. Ultrastruct. Res.* **62**, 26–53.

Carrel, A. (1912). On the permanent life of tissues outside the organism. *J. Exp. Med.* **15**, 516–528.

Chambard, M., Gabrion, J., and Mauchamp, J. (1981). Influence of collagen gel on the orientation of epithelial cell polarity: Follicle formation from isolated thyroid cells and from preformed monolayers. *J. Cell Biol.* **91**, 157–166.

Dehm, P., and Kefalides, N. A. (1978). The collagenous component of lens basement membrane. The isolation and characterization of an α chain size collagenous peptide and its relationship to newly synthesized lens components. *J. Biol. Chem.* **253**, 6680–6686.

Ehrmann, R. L., and Gey, G. O. (1956). The growth of cells on a transparent gel of reconstituted rat-tail collagen. *J. Natl. Cancer Inst.* **16**, 1375–1403.

Elsdale, T., and Bard, J. (1972). Collagen substrata for studies on cell behavior. *J. Cell Biol.* **54**, 626–637.

Emerman, J. T., Enami, J., Pitelka, D. R., and Nandi, S. (1977). Hormonal effects on intracellular and secreted casein in cultures of mouse mammary epithelial cells on floating collagen membranes. *Proc. Natl. Acad. Sci. U.S.A.* **74**, 4466–4470.

Emerman, J. T., Burwen, S. J., and Pitelka, D. R. (1979). Substrate properties influencing ultrastructural differentiation of mammary epithelial cells in culture. *Tissue Cell* **11**, 109–119.

End, P., and Engel, J. (1991). Multidomain proteins of the extracellular matrix and cellular growth. In "Receptors for Extracellular Matrix" (J. A. McDonald and R. P. Mecham, eds.), pp. 79–129. Academic Press, New York.

Engvall, E. (1995). Structure and function of basement membranes. *Int. J. Dev. Biol.* **39,** 781–787.

Farina, A. R., Tiberio, A., Tacconelli, A., Cappabianca, L., Gulino, A., and Mackay, A. R. (1996). Identification of plasminogen in Matrigel and its activation by reconstitution of this basement membrane extract. *BioTechniques* **21,** 904–909.

Freshney, R. I. (1994). "Culture of Animal Cells. A Manual of Basic Technique." Wiley-Liss, New York.

Grinnell, F. (1982). Cell–collagen interactions: Overview. *Methods Enzymol.* **82,** 499–503.

Hadley, M. A., Byers, S. W., Suárez-Quian, C. A., Kleinman, H. K., and Dym, M. (1985). Extracellular matrix regulates Sertoli cell differentiation, testicular cord formation, and germ cell development in vitro. *J. Cell Biol.* **101,** 1511–1522.

Harrison, R. G. (1907). Observations on the living developing nerve fiber. *Proc. Soc. Exp. Biol. Med.* **4,** 140–143.

Hay, E. D. (1993). Extracellular matrix alters epithelial differentiation. *Curr. Opin. Cell Biol.* **5,** 1029–1035.

Hodde, J. P., Badylak, S. F., Brightman, A. O., and Voytik-Harbin, S. L. (1996). Glycosaminoglycan content of small intestinal submucosa: A bioscaffold for tissue replacement. *Tissue Eng.* **2,** 209–217.

Ingber, D. E. (1992). Extracellular matrix as a solid-state regulator in angiogenesis: Identification of new targets for anti-cancer therapy. *Semin. Cancer Biol.* **3,** 57–63.

Kedeshian, P., Sternlicht, M. D., Nguyen, M., Shao, Z.-M., and Barsky, S. H. (1998). Humatrix, a novel myoepithelial matrical gel with unique biochemical and biological properties. *Cancer Lett.* **123,** 215–223.

Kefalides, N. A. (1971). Chemical properties of basement membranes. *Int. Rev. Exp. Pathol.* **10,** 1–39.

Kleinman, H. K., McGarvey, M. L., Liotta, L. A., Robey, P. G., Tryggvason, K., and Martin, G. R. (1982). Isolation and characterization of type IV procollagen, laminin, and heparan sulfate proteoglycan from the EHS sarcoma. *Biochemistry* **21,** 6188–6193.

Kleinman, H. K., McGarvey, M. L., Hassell, J. R., and Martin, G. R. (1983). Formation of a supramolecular complex is involved in the reconstitution of basement membrane components. *Biochemistry* **22,** 4969–4974.

Kleinman, H. K., McGarvey, M. L., Hassell, J. R., Star, V. L., Cannon, F. B., Laurie, G. W., and Martin, G. R. (1986). Basement membrane complexes with biological activity. *Biochemistry* **25,** 312–318.

Lee, E., Parry, G., and Bissell, M. J. (1984). Modulation of secreted proteins of mouse mammary epithelial cells by the collagenous substrata. *J. Cell Biol.* **98,** 146–155.

Lelievre, S., Weaver, V. M., and Bissell, M. J. (1996). Extracellular matrix signaling from the cellular membrane skeleton to the nuclear skeleton: A model of gene regulation. *Rec. Prog. Horm. Res.* **51,** 417–432.

Li, M. L., Aggeler, J., Farson, D. A., Hatier, C., Hassell, J., and Bissell, M. J. (1987). Influence of a reconstituted basement membrane and its components on casein gene expression and secretion in mouse mammary epithelial cells. *Proc. Natl. Acad. Sci. U.S.A.* **84,** 136–140.

Lin, C. Q., and Bissell, M. J. (1993). Multi-faceted regulation of cell differentiation by extracellular matrix. *FASEB J.* **7,** 737–743.

Liotta, L. A., Lee, C. W., and Morakis, D. J. (1980). New method for preparing large surfaces of intact human basement membrane for tumor invasion studies. *Cancer Lett.* **11,** 141–152.

McGuire, P. G., and Seeds, N. W. (1989). The interaction of plasminogen activator with a reconstituted basement membrane matrix and extracellular macromolecules produced by cultured epithelial cells. *J. Cell. Biochem.* **40,** 215–227.

Mackay, A. R., Gomez, D. E., Cottam, D. W., Rees, R. C., Nason, A., and Thorgeirsson, U. P. (1993). Identification of the 72-kDa (MMP-2) and 92-kDa (MMP-9) gelatinase/type IV collagenase in preparations of laminin and Matrigel. *BioTechniques* **15,** 1048–1051.

McPherson, T. B., and Badylak, S. F. (1998). Characterization of fibronectin derived from porcine small intestinal submucosa. *Tissue Eng.* **4,** 75–83.

Mareel, M. M. (1983). Invasion in vitro: Methods of analysis. *Cancer Metastasis Rev.* **2,** 201–218.

Meezan, E., Brendel, K., and Carlson, E. C. (1974). Isolation of a purified preparation of metabolically active retinal blood vessels. *Nature* **251,** 65–67.

Meezan, E., Hjelle, J. T., and Brendel, K. (1975). A simple, versatile, nondisruptive method for the isolation of morphologically and chemically pure basement membranes from several tissues. *Life Sci.* **17,** 1721–1732.

Michalopoulos, G., and Pitot, H. C. (1975). Primary culture of parenchymal liver cells on collagen membranes. *Exp. Cell Res.* **94,** 70–78.

Mignatti, P., Robbins, E., and Rifkin, D. B. (1986). Tumor invasion through the human amniotic membrane: Requirement for a proteinase cascade. *Cell* **47,** 487–498.

Miller, E., and Rhodes, K. (1982). Preparation and characterization of the different types of collagen. *Methods Enzymol.* **82,** 33–40.

Nicosia, R. F., and Ottinetti, A. (1990). Modulation of microvascular growth and morphogenesis by reconstituted basement membrane gel in three-dimensional cultures of rat aorta: A comparative study of angiogenesis in matrigel, collagen, fibrin, and plasma clot. *In Vitro Cell. Dev. Biol.* **26,** 119–128.

Olsen, B. R. (1997). Matrix molecules and their ligands. *In* "Principles of Tissue Engineering." R. G. Landes, Austin, Texas.

Orkin, R. W., Gehron, P., McGoodwin, E. B., Martin, G. R., Valentine, T., and Swarm, J. R. (1977). A murine tumor producing a matrix of basement membrane. *J. Exp. Med.* **145,** 204–220.

Paulsson, M. (1992). Basement membrane proteins: Structure, assembly, and cellular interactions. *Crit. Rev. Biochem. Mol. Biol.* **27,** 93–127.

Peel, S. A. F., Chen, H., Renlund, R., Badylak, S. F., and Kandel, R. A. (1998). Formation of a SIS-cartilage composite graft *in vitro* and its use in the repair of cartilage defects. *Tiss. Eng.* **4,** 143–155.

Schor, A. M., Schor, S. L., and Allen, T. D. (1983). Effects of culture conditions on the proliferation, morphology and migration of bovine aortic endothelial cells. *J. Cell Sci.* **62,** 267–285.

Schor, S. L., Schor, A. M., Winn, B., and Rushton, G. (1982). The use of three-dimensional collagen gels for the study of tumor cell invasion in vitro: Experimental parameters influencing cell migration into the gel matrix. *Int. J. Cancer* **29,** 57–62.

Siegal, G. P., Wang, M.-H., Rinehart, C. A. J., Kennedy, J. W., Goodly, L. J., Miller, Y., Kaufman, D. G., and Singh, R. K. (1993). Development of a novel human extracelluar matrix for quantitation of the invasiveness of human cells. *Cancer Lett.* **69,** 123–132.

Starkey, J. R. (1990). Cell matrix interactions during tumor invasion. *Cancer Metastasis Rev.* **9,** 113–123.

Stetler-Stevenson, W. G., Aznavoorian, S., and Liotta, L. A. (1993). Tumor cell interactions with the extracellular matrix during invasion and metastasis. *Annu. Rev. Cell Biol.* **9,** 541–573.

Sturgis, J. E., Robinson, J. P., and Voytik-Harbin, S. L. (1998). Three-dimensional (3-D) culture of human vascular cells in a complex extracellular matrix (ECM). *Mol. Biol. Cell* **9** (Suppl.), 168A.

Swarm, R. L. (1963). Transplantation of a murine chondrosarcoma in mice of different inbred strains. *J. Natl. Cancer Inst.* **31,** 953–975.

Thorgeirsson, U. P., Liotta, L. A., Kalebic, T., Margulies, I. M., Thomas, K., Rios-Candelore, M., and Russo, R. G. (1982). Effect of natural protease inhibitors and a chemoattractant on tumor cell invasion in vitro. *J. Natl. Cancer Inst.* **69,** 1049–1054.

Tryggvason, K., Hoyhtya, M., and Salo, T. (1987). Proteolytic degradation of extracellular matrix in tumor invasion. *Biochim. Biophys. Acta* **907,** 191–217.

Vernon, R. B., and Sage, E. H. (1995). Between molecules and morphology—extracellular matrix and creation of vascular form. *Am. J. Pathol.* **147,** 873–883.

Voyta, J. C., Via, D. P., Butterfield, C. E., and Zetter, B. R. (1984). Identification and isolation of endothelial cells based on their increased uptake of acetylated-low-density lipoprotein. *J. Cell Biol.* **99,** 2034–2040.

Voytik-Harbin, S. L., Brightman, A. O., Kraine, M. R., Waisner, B., and Badylak, S. F. (1997). Identification of FGF-2 and TGFβ as major extractable growth factors from small intestinal submucosa. *J. Cell. Biochem.* **67,** 478–491.

Voytik-Harbin, S. L., Brightman, A. O., Waisner, B. Z., Robinson, J. P., and Lamar, C. H. (1998). Small intestinal submucosa: A tissue-derived extracellular matrix that promotes tissue-specific growth and differentiation of cells *in vitro. Tissue Eng.* **4,** 157–174.

Vukicevic, S., Kleinman, H. K., Luyten, F. P., Roberts, A. B., Roche, N. S., and Reddi, A. H. (1992). Identification of multiple active growth factors in basement membrane matrigel suggests caution in interpretation of cellular activity related to extracellular matrix components. *Exp. Cell Res.* **202,** 1–8.

CHAPTER 27

Three-Dimensional Imaging of Extracellular Matrix and Extracellular Matrix–Cell Interactions

Sherry L. Voytik-Harbin,[*,†] Bartlomiej Rajwa,[‡] and J. Paul Robinson[*,†]

[*]Department of Basic Medical Sciences
School of Veterinary Medicine; and

[†]Department of Biomedical Engineering
Purdue University
West Lafayette, Indiana 47907

[‡]Department of Biophysics
Institute of Molecular Biology
Jagiellonian University
31-120 Krakow, Poland

I. Introduction
II. Three-Dimensional Imaging of Extracellular Matrix and Extracellular Matrix–Cell Interactions: Current Techniques and Their Limitations
    A. Light Microscopy
    B. Electron Microscopy
    C. Confocal Microscopy
    D. Multiphoton Microscopy
III. Three-Dimensional Microscopy of Living Systems: Extracellular Matrix and Extracellular Matrix–Cell Interactions
    A. Reflected Light Imaging
    B. Autofluorescence
IV. Summary
References

I. Introduction

In tissues, cells reside within a complex, three-dimensional (3-D) assembly of collagens, proteoglycans, glycosaminoglycans, and glycoproteins, otherwise known as the extracellular matrix (ECM). Reciprocal communication between cells and their ECM plays an important role in the modulation of critical physiological and pathological processes, including acquisition and maintenance of differentiated phenotypes during embryogenesis, the development of form (morphogenesis), vessel formation (angiogenesis), wound healing, and even tumor metastasis (Bissell et al., 1982). A major challenge to the biomedical community is to further understand the biophysical and biochemical aspects of ECM assembly and signaling as it relates to the structure and function of tissues and organs. One goal of our laboratory is to develop and identify approaches to visualize and quantitate the dynamic processes of the ECM and its interaction with cells within complex, 3-D, living biological systems. Herein we describe the integration of the principles and practices of microscopy with those of cell and extracellular matrix biology.

II. Three-Dimensional Imaging of Extracellular Matrix and Extracellular Matrix–Cell Interactions: Current Techniques and Their Limitations

Knowledge of the spatial distribution of biological components within cells and tissues often provides insight to their function and basic mechanisms of action. Unfortunately, many imaging techniques are either unable to provide insight into 3-D preparations or demand efforts that are often prohibitive to observations within living systems.

A. Light Microscopy

Light microscopy has been used routinely in conjunction with histochemical methods to visualize the components of cells, ECM, and tissues. Unfortunately, high quality images are often limited to thin, physically sectioned specimens or two-dimensional (2-D) culture systems in which cells grow along a translucent substrate (e.g., plastic). Thick slices or whole mount specimens cannot be readily studied because structures in the interior of the specimen are obscured by interferences from structures above and below the plane of focus. Although a third dimension can be reconstructed from hundreds of 2-D images generated from serial sections, this process is lengthy and tedious, and accurate image alignment is difficult. Likewise, physical sectioning requires extensive specimen processing including fixation, dehydration, and embedding, rendering this method inappropriate for viewing living systems. An alternative method for obtaining 3-D information involves recording a series of images within adjacent focal planes. Image degradation due to low-resolution "out-of-focus" light is then corrected by computer-based image processing known as deconvolution (for review, see Shaw, 1995). From a practical point of view, the major problems with deconvolution include the significant computing time and disk space needed for image processing as well as knowledge of an accurate point spread function. The latter requirement may be eliminated when so-called blind deconvolution is applied (Holmes *et al.*, 1995). Deconvolution computations can take from minutes to days, depending on the size of the image and the algorithm, computer, or number of iterations used.

B. Electron Microscopy

Electron microscopy (EM) uses a beam of electrons to form an image of a specimen. While offering much improved resolution, electron microscopy requires extensive specimen processing due to electron beam observation in vacuum. In fact, there is no real possibility of viewing biological specimens in a living, wet state, owing to the high vacuum operation of EM. Because of the very limited penetrating power of electrons, observations by transmission electron microscopy (TEM) require specimens that are cut into extremely thin sections before they can be viewed. As with light microscopy, EM visualization of samples in three dimensions can be achieved by collecting and aligning information gathered from a large number of serial sections. Alternatively, thick specimens may be imaged at two different tilt angles and viewed as a stereopair with a stereoviewer for purposes of 3-D visualization only. More recently, 3-D volume reconstruction of EM images has been achieved by electron tomography (Frank, 1992). This technique involves incrementally tilting the specimen through a range of angles, usually about 60°, and then backprojecting each of the tilt images.

In contrast, scanning electron microscopy (SEM) involves scanning the specimen with a very narrow beam of electrons and collecting the electrons scattered or emitted from the surface of the specimen. Since the amount of electron scattering depends on the angle of the surface relative to the beam, the SEM image has highlights and shadows and gives a 3-D appearance. Preparing samples for SEM usually requires fixation, dehydration, and drying whether by the critical point method or by freeze-drying. More recently, less well-known cryopreparation techniques have been used to preserve the native detail of biological structures, especially those with significant water content, with a high degree of fidelity and fewer artifacts (Voytik-Harbin *et al.*, 1998a).

Indeed, electron microscopy has been instrumental in elucidating the structural organization and binding interactions of individual protein and proteoglycan components of the ECM (for reviews, see Engel and Furthmayr, 1987; Engel, 1994). Despite the obvious advantages in resolution, these techniques do not provide true 3-D images and have been mostly limited to simple, nonliving systems often involving purified or partially purified ECM preparations.

C. Confocal Microscopy

Since the advent of confocal microscopy in the mid-1980s, 3-D spatial information can be collected from thick specimens by means of optical sectioning. The term confocal is derived from the optical platform in which the scanning point light source and the detector aperture share a common focus at the level of the specimen. This optical arrangement effectively eliminates much of the out-of-focus light from detection, thus improving the fidelity of focal sectioning in three dimensions (Sheppard, 1987; Wijnaendts van Resandt *et al.*, 1985). This technology offers the principal advantages of (1) thin optical "slices" through thick specimens, (2) rejection of out-of-focus light from other focal places, and (3) resolution in all three dimensions from multiple optical slices. This instrument faithfully images structures with dimensions as small as subcellular organelles up to whole tissue preparations (Brakenhoff *et al.*, 1979, 1988; Messerli and Perriard, 1995). However, a variety of technical constraints limit the maximal thickness of the objects that can be imaged. To obtain 3-D images that closely represent the geometry of the sample, the light path through the sample must be as short as possible, since imaging artifacts like astigmatism, spherical aberration, and intensity attenuation increase with path length (Aslund and Liljeborg, 1992; Hell *et al.*, 1993; Visser *et al.*, 1991). Coupled with digital reconstruction techniques, confocal microscopy can extract image information that, while present in the data, is not easily accessible by simply presenting the individual sections.

D. Multiphoton Microscopy

Three-dimensional optical sectioning is also possible with multiphoton microscopy. This technology, developed in 1989 (Denk *et al.*, 1990), is based on a well-known quantum mechanical concept presented for the first time by Maria Goeppert-Mayer in 1931. Specifically stated, multiphoton refers to the effect of two or three photons in a single quantum event (Goeppert-Mayer, 1931). This phenomenon allows two or more photons of long wavelength light to create the same excitation as one photon of shorter wavelength light provided the photons arrive simultaneously (Denk *et al.*, 1990). Several clear distinctions exist on comparison of confocal and multiphoton imaging technologies. These distinctions should be carefully considered when deciding which technology is most suited to a specific microscopic application. With multiphoton microscopy, the power density of the laser is only high enough to excite fluorescence in the focal volume. Therefore, unlike confocal microscopy, a pinhole aperture is not needed to exclude unwanted light. The elimination of the aperture results in an increase in the overall signal intensity detected with the multiphoton system. Because multiphoton excitation is restricted to a small focal volume, out-of-focus photodamage and photobleaching are almost eliminated. However, the longer wavelength source used in multiphoton excitation can have some potential drawbacks: system resolution will be less than a comparable laser scanning confocal system, the ultrashort pulsed source may also potentially excite intrinsic absorption in cells via two- and three-photon excitation, and sample heating may be a concern for excitation wavelengths near 1 μm (Wokosin *et al.*, 1996). A potential advantage of multiphoton microscopy is the increased tissue depth and the reduced damage at wavelengths in the ultraviolet (UV) range (Potter, 1996).

III. Three-Dimensional Microscopy of Living Systems: Extracellular Matrix and Extracellular Matrix–Cell Interactions

A. Reflected Light Imaging

1. Introduction

Reflected or back-scattered light is an intrinsic optical property of many materials that can be exploited to provide qualitative and quantitative microstructural information. When used in this mode, confocal microscopy, can provide 3-D structural details of unfixed and unstained biological specimens (Boyde and Jones, 1995). For example, Semler and co-workers applied confocal reflection microscopy (CRM) to visualize and quantitate the microtopography of porous biomaterials prepared from synthetic polymers (Semler et al., 1997). Likewise, we and others have applied this technique for surface and volume visualization of intact ECM biomaterials as well as 3-D reconstituted matrices consisting of individual (e.g., collagen) or mixtures of ECM components (Gunzer et al., 1997; Friedl et al., 1997; Brightman et al., 2000). Reflected light from collagen fibers has been collected simultaneously with fluorescence from cells stained with vital fluorochromes to monitor the dynamic process of cell migration through a 3-D collagen matrix (Friedl et al., 1997). More recently, we have provided the first account of CRM in a time-lapse mode for studying collagen fiber formation (fibrillogenesis) and ECM assembly in vitro (Voytik-Harbin et al., 1998b; Robinson et al., 1999; Brightman et al., 2000).

With time-lapse CRM both kinetic and 3-D structural information can be collected simultaneously as ECM components polymerize from a soluble to a gel phase. Taken together, CRM offers a useful technique for investigating biological processes in living systems involving the ECM and ECM–cell interactions that occur in multiple dimensions. Examples of such complex events include collagen fibrillogenesis, ECM assembly, morphogenesis, and cell migration. A detailed description of the application of CRM for visualization and analysis of collagen fibrillogenesis and ECM assembly is provided herein.

2. Methods

A soluble form of purified type I collagen, Vitrogen, was obtained from Collagen Aesthetics (Palo Alto, CA). Solubilized mixtures of ECM components were prepared from small intestinal submucosa, an intact interstitial ECM, as described previously (Voytik-Harbin et al., 1998a).

Unstained 3-D matrices of type I collagen or mixtures of ECM components were polymerized in Lab-Tek chambered coverglass (Nalge Nunc Int, Rochester, NY) and imaged using a Bio-Rad MRC1024 confocal microscope via a 60×, 1.4 NA (numerical aperature) oil immersion lens. Optical settings were established and optimized on reconstituted matrices after polymerization was completed. An aliquot of soluble collagen or ECM preparations then was placed onto the heated (37°C) stage of the microscope and fibrillogenesis and fibril assembly imaged. Samples were illuminated with 488 nm laser light, and the reflected light was detected with a photomultiplier tube (PMT) using a blue reflection filter. For the Bio-Rad MRC1024 confocal microscope, instrument setup involved a beam splitter placed in position D1 and a dichroic filter that reflects 488 nm light into PMT2 in position D2 (Fig. 1). Optical filters can be added in positions W1 and W3 for the simultaneous collection of fluorescence

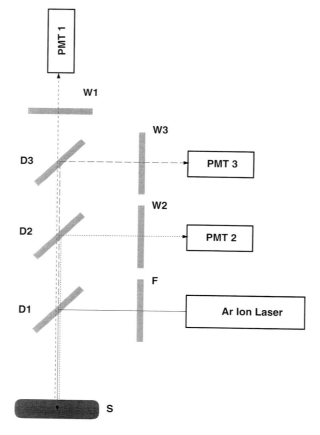

Fig. 1 Confocal microscope demonstrating filter (D, W, and F) and photomultiplier tube (PMT) positions.

information in PMT1 and PMT3, respectively. A z-step of 0.2 μm was used to optically section the samples. Because the resolution of the z-plane is less than that of the x–y plane, the sampling along the z-axis may be different from that of the x–y plane. 2-D (e.g., x–y) or 3-D (e.g., x–y–z) sections demonstrating fibrillogenesis and assembly events were recorded at step intervals ranging from 1 to 5 sec for up 1 hr. Minimum time intervals required for image collection are dependent on the section size, resolution, and scanning mode (point versus line). For standardization of scanning depth, the scan head was adjusted to a distance in the range of 20–50 μm from the upper surface of the coverglass.

Nonuniform background caused by interference and reflection from the optical pathway was removed from the images using a standard rank leveling procedure on each x–y section. Rank leveling consists of applying a multiple number of erosions followed by a Gaussian filter to each section to create a background approximation. This background approximation was then subtracted from the original section to enhance the signal to noise ratio. Three-dimensional images of reconstituted ECM biomaterials were either compiled into a single view projection using Laser Sharp image processing software (Bio-Rad, Hemel Hempstead, England) or compiled into a 3-D projection using Voxel-View reconstruction software (Vital Images, Minneapolis, MN). Rank leveling can also be performed on 3-D images prior to reconstruction into 3-D views to improve image quality. To perform rank leveling in two dimensions we used public domain Unix image processing toolsets: "imgstar" by Simon Winder and "pbmplus" by Jef Poskanzer. A "tcsh" script was used to automate the image processing tasks. Time-lapse image files were assembled into SGI movie format (sgimv) or MPEG movie format using "dmcovert," a standard converter for digital media, a part of the Irix 6.2 operating system on a Silicon Graphics Indigo2 workstation (SGI, Mountain View, CA). Because MPEG movie formatting involves lossy compression, this format was used only for visualization purposes.

3. Results

Confocal reflection images from unfixed, unstained 3-D matrices prepared from purified collagen or mixtures of ECM components revealed the morphology of individual fibrils as well as their density and orientation within the 3-D matrix (Fig. 2). Results obtained with CRM confirmed the dependence of matrix structure on composition as previously documented using SEM (Voytik-Harbin *et al.*, 1998a). Initial quantitative analyses of collagen matrices with CRM indicate fiber diameters ranging from 200 to 600 nm. These results are similar to those reported by Friedl using the same technique (Friedl *et al.*, 1997). For matrices prepared using air-drying and standard critical point drying techniques and visualized using SEM, average fiber diameters were significantly diminished and architectural discrepancies noted (Friedl *et al.*, 1997; Voytik-Harbin *et al.*, 1998a). Only with cryostage SEM, in which the specimen is viewed in a quick-frozen hydrated state, were fiber diameters similar to those observed with CRM. For

Fig. 2 Confocal reflection image of reconstituted 3-D matrix prepared from type I collagen (1.5 mg/ml). Scale bar = 5 μm.

reconstituted or intact ECM biomaterials, the depth resolution was inversely related to specimen opacity. For the intact ECM, intestinal submucosa, surface topography and depths of 10–20 μm could be effectively imaged with CRM without difficulty.

To date, spectrometry and dark-field microscopy techniques have been used for investigations of collagen fibrillogenesis and ECM assembly. Although these routine techniques provide either structural or kinetic information only, CRM allowed the collection of both types of data simultaneously. Time-lapse imaging of fibril polymerization and assembly demonstrated the dependence of kinetic parameters as well as architecture on matrix composition. For example, the addition of heparin to purified type I collagen or the complex ECM mixture resulted in an increase in the assembly (fibrillogenesis) rate, an increase in fibril thickness, and a decrease in fibril length. On the other hand, the addition of the proteoglycan decorin increased the lag phase and reduced the rate of collagen fibrillogenesis.

As with fluorescence-based imaging, CRM is a very sensitive technique. It benefits from the fact that although some very small reflecting particles or fibrils cannot be resolved (according to the Rayleigh criterion), they still can be detected and visualized. Based on the Rayleigh criterion, the calculated resolution for our CRM system was 212 nm. Interestingly, the x–y resolution of a microscope operating in reflected light mode is not improved by the confocal iris, but the contrast transfer characteristics for fine detail are much better than in the case of a nonconfocal microscope (Oldenbourg *et al.*, 1993). In these studies, optimized fiber contrast scanning distances ranged from 20 to 50 μm from the coverslip. Depth intensity of the fiber contrast was diminished at deeper focus levels, indicating substantial scattering and refraction of the laser intensity. Although offering resolution that is significantly less than that of SEM (2–8 nm), CRM readily provided 3-D spatial information and did not require extensive processing techniques (e.g., physical sectioning and staining). High throughput computer algorithms for quantitation of various fibril and matrix parameters, including diameter, length, degree of curvature, and orientation within the matrix, are currently under development.

B. Autofluorescence

1. Introduction

In the application of fluorescence and histochemical staining methods for microscopic investigation of ECM–cell interactions within *in vivo* and *in vitro* model systems, background autofluorescence has been observed and often deemed problematic. The autofluorescence has been attributed to naturally occurring fluorophores such as the prominent ECM components collagen and elastin. This autofluorescence represents the superposition of multiple intrinsic fluorophores within the ECM and is also dependent on the light absorption and scattering properties of the ECM. However, it should be noted that the physical environment, including pH, solvation, and oxidation state, affects the fluorescent properties of these molecules. Since the 1980s, characterization of tissue fluorescence in terms of the native fluorophores has been proposed as a method of differentiating between normal and diseased tissues (Schomacker *et al.*, 1992; Brennan, 1989; Deckelbaum *et al.*, 1987) as well as of studying the effect of aging (Kollias *et al.*, 1998; Leffell *et al.*, 1988). For these purposes, measurements have been conducted on solubilized or intact tissue specimens using standard spectrofluorometric or fiber optic systems. Here we describe the collection of autofluorescence from intact ECM biomaterials as well as 3-D tissue culture systems for the study of ECM architecture and ECM–cell interactions. Autofluorescence is an ideal optical property for visualization because it essentially requires no sample preparation and processing. When used in conjunction with vital dyes for cells, autofluorescence is an effective tool for investigating ECM–cell interactions in 3-D, living biological systems.

2. Methods

The intact ECM biomaterial, small intestinal submucosa (SIS), was obtained from Cook Biotech (West Lafayette, IN). This same biomaterial was used as a

3-D ECM substrate for the culture of human coronary artery smooth muscle cells *in vitro* as previously described (Voytik-Harbin *et al.*, 1998a; Sturgis *et al.*, 1998).

Confocal imaging of autofluorescence was performed on a Bio-Rad MRC1024 confocal microscope. The microscope was adapted with a dichroic filter that reflects 488 nm and passes all wavelengths greater than 520 nm in position D1 (UBHS, Bio-Rad) and with a dichroic filter that directed fluorescence to PMT2 in position D2. An argon ion laser was used to provide 488 nm excitation. To image ECM autofluorescence and fluorescence-labeled cells simultaneously, we equipped the Bio-Rad microscope with a dichroic mirror reflecting UV and 488 nm light in position D1 and a second dichroic in position D2 that split blue Hoechst 33342 fluorescence and yellow-green autofluorescence into PMT2 and PMT1, respectively.

For multiphoton autofluorescence imaging, an all-solid-state 1047-nm Nd:YLF laser (Microlase, Strathclyde, UK) and modified MRC-600 scan head (Bio-Rad, Hemel Hempstead, England) were configured as described previously (Wokosin *et al.*, 1996; Wokosin and White, 1997). The emission was detected by a standard S-20 PMT (Thorn 9828B, Electron Tubes, Rockaway, NJ) mounted external to the scan head directly beneath the objective of the microscope (direct detection).

3. Results

Autofluorescence imaging revealed the overall 3-D organization of collagen fibrils and bundles that form the structural backbone of intestinal submucosa (Figs. 3 and 4). Low intensity ECM autofluorescence could be collected with a confocal microscope over a broad range of UV and visible excitation wavelengths. For purposes of imaging 3-D culture systems consisting of ECM and cells, the vital dye Hoechst 33342 was used (Fig. 5). Hoechst 33342-labeled cells could be imaged with little or no detection of ECM autofluorescence with UV excitation. Likewise, excitation with visible 488 nm light was effective for imaging ECM autofluorescence, providing simultaneous collection of architectural details of the ECM.

Although high quality images could be obtained with both confocal and multi-photon excitation, the imaging penetration depth obtained with multiphoton microscopy and direct detection was greater than that obtained with confocal microscopy, as expected. This observation is consistent with those of Centonze and co-workers who reported that there was at least a twofold improvement in the maximum imaging penetration depth obtained with multiphoton excitation relative to confocal microscopy on a variety of biological samples (Centonze and White, 1998). The use of direct detection rather than signal collection followed by descanning likely contributed to improving the signal-to-noise ratio of the image, which in turn further increased the depth at which usable images could be obtained. The resolution of autofluorescence images compared to those based on fluorescence or reflected light was diminished. This reduction in resolution can be attributed to the use of longer wavelengths as well as to the low signal-

Fig. 3 3-D reconstructed confocal image demonstrating autofluorescence of a tissue-derived ECM biomaterial, small intestinal submucosa.

to-noise ratio characteristic of ECM autofluorescence. However, the relatively low signal provided by ECM autofluorescence provided an ideal background when imaging fluorescence-labeled cells on or within the intestinal submucosa substrate.

IV. Summary

In summary, noninvasive and nondestructive imaging modalities such as reflection and autofluorescence can readily be used in conjunction with the 3-D optical sectioning capabilities of confocal and multiphoton microscopy to investigate biological processes within living systems. The elimination of specimen fixation and extensive processing reduces the possibility of structural artifacts and facilitates repeat observations within a single sample. Therefore, information representing up to four dimensions (x, y, z, and time) can be readily collected and reconstructed for purposes of visualization and/or quantitative analysis. An advantage of using the techniques described in this chapter is the possibility of performing quantitative measurement of cell size, surface area, volume, depth (in matrix), orientation, receptor density, as well as fluorescence-based indicators of phenotype and function. At present, we are effectively utilizing these tech-

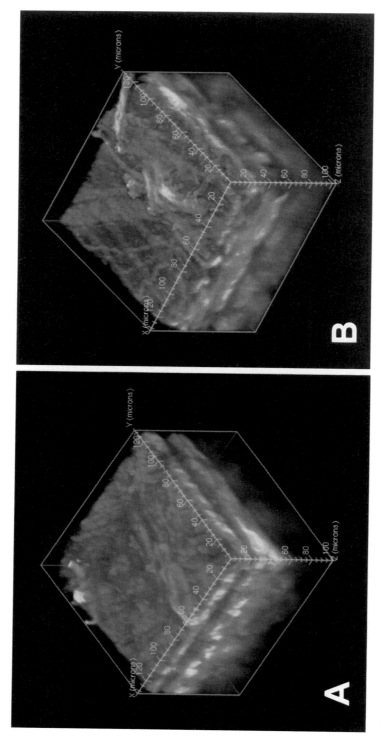

Fig. 4 3-D reconstructed multiphoton images demonstrating autofluorescence of a tissue-derived ECM biomaterial, small intestinal submucosa, sectioned optically from luminal (A) or abluminal (B) surfaces.

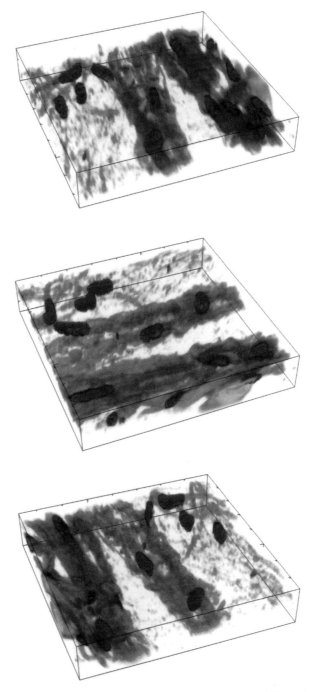

Fig. 5 3-D reconstructed confocal images of cells growing on and within a tissue-derived ECM substrate, small intestinal submucosa. Components of the ECM substrate were imaged based on their autofluorescence properties, and cell nuclei were stained with Hoechst 33342. (See color plates.)

niques to study collagen fibrillogenesis and ECM assembly, structural aspects of ECM-based biomaterials, as well as cell interactions within 3-D matrices (e.g., migration). New insights provided by these techniques regarding ECM and ECM–cell signaling will further the understanding of tissue structure and function and contribute to the development of new and improved strategies for tissue repair, replacement, and maintenance.

References

Aslund, N., and Liljeborg, A. (1992). A method to extract homogeneous regions in 3-D confocal microscopy enabling compensation for depth dependent light attenuation. *Micron Microsc. Acta* **23,** 463–479.

Bissell, M. J., Hall, H. G., and Parry, G. (1982). How does the extracellular matrix direct gene expression? *J. Theor. Biol.* **99,** 31–68.

Boyde, A., and Jones, S. J. (1995). Mapping and measuring surfaces using reflection confocal microscopy. *In* "Handbook of Biological Confocal Microscopy" (J. B. Pawley, ed.), pp. 255–266. Plenum, New York.

Brakenhoff, G. J., Blom, P., and Barends, P. J. (1979). Confocal scanning light microscopy with high aperture immersion lenses. *J. Microsc.* **117,** 219.

Brakenhoff, G. J., van der Voort, H., van Spronsen, E., and Nanninga, N. (1988). 3-dimensional imaging of biological structures by high resolution confocal scanning laser microscopy. *Scanning Microsc.* **2,** 33–40.

Brennan, M. (1989). Changes in solubility, non-enzymatic glycation, and fluorescence of collagen in tail tendons from diabetic rats. *J. Biol. Chem.* **264,** 20947–20952.

Brightman, A. O., Rajwa, B. P., Sturgis, J. E., McCallister, M. E., Robinson, J. P., and Voytik-Harbin, S. L. (2000). Time-lapse confocal reflection microscopy of collagen fibrillogenesis and ECM assembly *in vitro*. *Biopolymers,* in press.

Centonze, V. E., and White, J. G. (1998). Multiphoton excitation provides optical sections from deeper within scattering specimens than confocal imaging. *Biophys. J.* **75,** 2015–2024.

Deckelbaum, L. I., Lam, J. K., Cabin, H. S., Clubb, K. S., and Long, M. B. (1987). Discrimination of normal and atherosclerotic aorta by laser-induced fluorescence. *Lasers Surg. Med.* **7,** 330–335.

Denk, W., Strickler, J., and Webb, W. W. (1990). Two-photon laser scanning fluorescence microscopy. *Science* **248,** 73–76.

Engel, J., (1994). Electron microscopy of extracellular matrix components. *Methods Enzymol.* **245,** 469–488.

Engel, J., and Furthmayr, H. (1987). Electron microscopy and other physical methods for the characterization of extracellular matrix components: Laminin, fibronectin, collagen IV, collagen VI, and proteoglycans. *Methods Enzymol.* **145,** 3–78.

Frank, J. (1992). "Electron Tomography." Plenum, New York..

Friedl, P., Maaser, K., Klein, C. E., Niggemann, B., Krohne, G., and Zanker, K. S. (1997). Migration of highly aggressive MV3 melanoma cells in 3-dimensional collagen lattices results in local matrix reorganization and shedding of $\alpha 2$ and $\beta 1$ integrins and CD44. *Cancer Res.* **57,** 2061–2070.

Goeppert-Mayer, M. (1931). Ueber elementarakte mit quantenspruengen. *Ann. Phys.* **9,** 273–294.

Gunzer, M., Kampgen, E., Brocker, E., Zanker, K. S., and Friedl, P. (1997). Migration of dendritic cells in 3D-collagen lattices. Visualisation of dynamic interactions with the substratum and the distribution of surface structures via a novel confocal reflection imaging technique. *In* "Dendritic Cells in Fundamental and Clinical Immunology" (P. Ricciardi-Castagnoli, ed.), pp. 97–103. Plenum, New York.

Hell, S., Reiner, G., Cremer, C., and Stelzer, E. H. K. (1993). Aberrations in confocal fluorescence microscopy induced by mismatches in refractive index. *J. Microsc.* **169,** 391-405.

Holmes, T. J., Bhattacharyya, S., Cooper, J. A., Hanzel, D., Krishnamurthi, V., Lin, W.-C., Roysam, B., Szarowski, D. H., and Turner, J. N. (1995). Light microscopic images reconstructed by maximum likelihood deconvolution. *In* "Handbook of Biological Confocal Microcopy" (J. B. Pawley, ed.), pp. 389–402. Plenum, New York.

Kollias, N., Gillies, R., Moran, M., Kochevar, I. E., and Anderson, R. R. (1998). Endogenous skin fluorescence includes bands that may serve as quantitative markers of aging and photoaging. *J. Invest. Dermatol.* **111,** 776–780.

Leffell, D. J., Stetz, M. L., Milstone, L. M., and Deckelbaum, L. I. (1988). In vivo fluorescence of human skin: A potential marker of photoaging. *Arch. Dermatol.* **124,** 1514–1518.

Messerli, J. M., and Perriard, J.-C. (1995). Three-dimensional analysis and visualization of myofibrillogenesis in adult cardiomyocytes by confocal microscopy. *Microsc. Res. Technol.* **30,** 521–530.

Oldenbourg, R., Terada, H., Tiberio, R., and Inoue, S. (1993). Image sharpness and contrast transfer in coherent confocal microscopy. *J. Microsc.* **172,** 31–39.

Potter, S. M. (1996). Vital imaging: Two photons are better than one. *Curr. Biol.* **6,** 1595–1598.

Robinson, J. P., Brightman, A. O., Kelley, S., Rajwa, B., Sturgis, J., and Voytik-Harbin, S. L. (1999). Imaging technique for kinetic evaluation of collagen matrix assembly. *FASEB J.* **13,** A344 (Abstract).

Schomacker, K. T., Frisoli, J. K., Compton, C. C., Flotte, T. J., Richter, J. M., Nishioka, N. S., and Deutsch, T. F. (1992). Ultraviolet laser-induced fluorescence of colonic tissue: Basic biology and diagnostic potential. *Lasers Surg. Med.* **12,** 63–78.

Semler, E. J., Tjia, J. S., and Moghe, P. V. (1997). Analysis of surface microtopography of biodegradable polymer matrices using confocal reflection microscopy. *Biotechnol. Prog.* **13,** 630–634.

Shaw, P. J. (1995). Comparison of wide-field/deconvolution and confocal microscopy for 3D imaging. *In* "Handbook of Biological Confocal Microscopy" (J. B. Pawley, ed.), pp. 373–387. Plenum, New York.

Sheppard, C. J. R. (1987). Scanning optical microscopy. *In* "Advances in Optical and Electron Microscopy" (E. Barer and V. E. Cosslett, eds.), pp. 1–98. Academic Press, London.

Sturgis, J. E., Robinson, J. P., and Voytik-Harbin, S. L. (1998). Three-dimensional (3D) culture of human vascular cells in a complex extracellular matrix (ECM). *Mol. Biol. Cell* **9**(Suppl.), 168a [Abstract].

Visser, T. D., Groen, F. C. A., and Brakenhoff, G. J. (1991). Absorption and scattering correction in fluorescence confocal. *J. Microsc.* **163,** 189–200.

Voytik-Harbin, S. L., Brightman, A. O., Waisner, B. Z., Robinson, J. P., and Lamar, C. H. (1998a). Small intestinal submucosa: A tissue-derived extracellular matrix that promotes tissue-specific growth and differentiation of cells *in vitro*. *Tissue Eng.* **4,** 157–174.

Voytik-Harbin, S. L., Rajwa, B. P., Sturgis, J. E., McCallister, M. E., Brightman, A. O., and Robinson, J. P. (1998b). Time-lapse confocal reflection (TLR) imaging for the study of extracellular matrix assembly. *Mol. Biol. Cell* **9**(Suppl.), 61a(Abstract).

Wijnaendts van Resandt, R. W., Marsman, H. J. B., Kaplan, R., Davoust, J., Stelzer, E. K. H., and Stricker, R. (1985). Optical fluorescence microscopy in 3 dimensions: Microtomoscopy. *J. Microsc.* **138,** 29–34.

Wokosin, D. L., and White, J. G. (1997). Optimization of the design of a multiple-photon excitation laser scanning fluorescence imaging system. Three Dimensional Microscopy: Imaging Acquisition and Processing IV. *Proc. SPIE* **2984,** 25–29.

Wokosin, D. L., Centonze, V. E., White, J. G., Armstrong, D., Robertson, G., and Ferguson, A. I. (1996). All-solid-state ultrafast lasers facilitate multiphoton excitation fluorescence imaging. *IEEE J. Select. Top. Quant. Elec.* **2,** 1051–1065.

CHAPTER 28

Cytometric Analysis of Cell Contact and Adhesion

Michael S. Kinch

Department of Basic Medical Sciences
Purdue University
West Lafayette, Indiana 47907

I. Introduction and Application
II. General Strategies to Measure Cell–Cell Adhesions
    A. Aggregation Assays
    B. Monolayer Cell–Cell Adhesion Assays
    C. Reversed Centrifugation for Quantitation of Cell Adhesion
III. General Strategies to Measure Cell–Ligand Adhesions
    A. Monolayer Cell–Ligand Adhesion Assays
    B. Interference Reflection Microscopy
    C. Flow Measurements
IV. Specificity of Cell Adhesion
    A. Overview
    B. Controls
V. Optimization of Experimental Conditions
    A. Buffer Conditions
    B. Temperature
    C. Timing
    D. Cell Suspension and Resting
References

I. Introduction and Application

The physical placement of a cell within an organism is a complex process involving multiple, specific interactions among cells and their surrounding microenvironment (Farquhar and Palade, 1963; Staehlin and Hull, 1978). These adhesions occur with extraordinary specificity and facilitate the selective sorting of cells during tissue formation and remodeling. In addition to specificity, intermolecular interactions must be quite strong to overcome the considerable electrostatic repulsion of cells in close contact. For example, anastomosis of membranes within tight junctions requires that cellular adhesions overcome the electrostatic repulsion of apposing membranes in direct contact (Simons and Fuller, 1985). These adhesions must also be highly dynamic to facilitate the repeated cycles of disruption and reformation of cell contacts that occur often during development and tissue remodeling.

A number of adhesion molecule families have been identified that facilitate intercellular adhesions. They include integrins, cadherins, selectins, and the immunoglobulin superfamily. Each of these molecules has a unique adhesive specificity, although some overlapping specificities have been identified (e.g., different integrins that bind RGD peptide sequences; Ruoslahti and Pierschbacher, 1987). Although the extracellular domains of adhesion molecules define substrate specificity, their intracellular domains often are involved in linkages that regulate cell adhesion and signaling. For example, interactions of integrins or cadherins with components of the actin cytoskeleton often serve to stabilize adhesion or initiate intracellular signaling by associated kinases (Burridge *et al.*, 1988; Takeichi, 1994).

A variety of techniques have been developed to assess the molecular interactions that facilitate cell contact and adhesion. Herein, we describe some of the most commonly employed qualitative and quantitative techniques to measure cell adhesions. We will detail methods to measure cell–cell and cell–extracellular matrix (ECM) adhesions. We will also distinguish techniques that might be particularly useful for the study of adherent cells (e.g., fibroblasts, epithelia) from those used to study nonadherent cells (e.g., lymphoid cells).

II. General Strategies to Measure Cell–Cell Adhesions

A. Aggregation Assays

An early and eloquent demonstration of adhesive specificity was provided by Moscona and Hausman (1977). After cells were isolated from either embryonic retina or liver, the different populations were labeled with a specific marker (e.g., [^{3}H]thymidine) and mixed in suspension. The results revealed that cells derived from the same tissue selectively segregated into colonies. These assays provided some of the earliest evidence for the exquisite specificity of cellular adhesions (Roth and Weston, 1967). Moreover, this aggregation assay remains today one of the most popular methods to measure cell–cell adhesion.

1. Basic Aggregation Assays

Aggregation assays can measure the specificity and avidity of cell–cell adhesions. Often, aggregation assays are employed to test the ability of a transfected protein to promote cell–cell adhesion. In such instances, a positive outcome can be defined using microscopic observation of cell clusters in transfected cells, but not in properly controlled samples (as discussed). These assays can also be analyzed using flow cytometry or a Coulter counter, where aggregates are revealed by a shift in the forward angle light scatter that reflects the increased size of the aggregates (Lackie, 1991). An example of the successful use of aggregation assays is the aggregation of fibroblasts transfected with E-cadherin (Takeichi, 1994). In this situation, E-cadherin transfectants were found to aggregate with one another, but not with controls, thus providing evidence not only that E-cadherin was an adhesion molecule, but that this adhesion was homophilic in nature.

The setup and analysis of aggregation assays can become much more complicated when measuring heterophilic adhesions of different cell populations. If the two populations can be distinguished readily (on the basis of size or other morphological properties), then microscopic analysis is possible. Otherwise, microscopic analysis is precluded. As an alternative, accurate assessment can be accomplished using specific dyes and markers as described later.

For aggregation assays, the individual cell populations must be in single-cell suspension. Although tituration of nonadherent cells (e.g., leukocytes) is often necessary to minimize background clustering, more substantial measures may be necessary to suspend adherent cells (e.g., fibroblasts, epithelia, neutrophils). This is generally accomplished by incubating cell monolayers with chelating agents (e.g., 4 mM EDTA, 37°C, 20 min) to disrupt binding to the underlying ECM (as described later). Some cell types cannot be detached by treatment with EDTA alone and require protease treatment (e.g., trypsin/EDTA) for suspension. However, precautions may be necessary to avoid proteolytic digestion of the adhesion molecule, which could alter its function (as described later).

After establishing single-cell suspensions, the cells are mixed together in a minimal volume (20–50 μl) of assay buffer [generally phosphate-buffered saline (PBS) supplemented with calcium chloride and magnesium]. Adhesion rates in suspension are related the square of the cell population density, and thus it is important to increase cell density, usually using light centrifugation (50 g, 10 sec), to establish a loose pellet. After a determined period of incubation to promote cell–cell adhesions (normally seconds to minutes), the pellet is then gently loosened by gentle shaking. The aggregates are then retrieved using a wide-bore pipette inserted gently into the test tube to minimize shearing. The sample is then spread onto a glass slide for microscopic evaluation. It is our experience that commercial hemacytometers are often useful for such assays because cells can be deposited within a channel between the slide and coverglass. The proper use of hemacytometers for cell counting is described by Mishell and Shigii (1980).

a. Quantitation of Aggregation Assays

Light microscopy can be used to assess conjugate formation when studying homophilic aggregation (in a single population) or if heterophilic binding involves morphologically distinguishable populations. In these cases, cell–cell conjugates can be qualitatively assessed by comparing the size and frequency of cell clusters as described later.

For quantitative analysis, values must be assigned to the sample. Homophilic conjugates are sometimes evaluated by determining the average size or diameter of conjugates. Often, conjugate size is determined by dividing the number of total cells within a microscopic field of view by the number of aggregate clusters. Alternatively, the aggregates can be evaluated by estimating the number of cells associated with clusters of defined size classes (e.g., clusters with one to three cells versus five to eight). Although either technique can yield reliable results, evaluation of these types of assays can be highly subjective.

When evaluating conjugates between different populations of cells, additional criteria often must be addressed. First, the investigator must be able to reliably distinguish the two cell populations on the basis of differences in size or morphology. For example, when the adhesion of thymocytes and thymic epithelial cells is measured, the much larger size and unique morphology of thymic epithelial (TE) cells allows for accurate assessment of thymocyte binding (Vollger *et al.*, 1987). If morphological differences cannot distinguish the populations, specific identifiers may be employed as described later.

When studying the heterophilic binding of two different populations, it is often necessary to vary the ratio of two cell populations relative to one another. Returning to the example of TE cell adhesion to thymocytes, the investigators defined positive adhesion as a TE cell bound to at least three thymocytes. Optimal adhesion (specific/nonspecific binding) was obtained when the ratio of thymocytes to TE cells was 50:1. The samples were then quantified by scoring the adhesion of a fixed number of TE cells (e.g., 100) (Singer *et al.*, 1990).

b. Background

Reliable controls must be established, as most cells have some degree of nonspecific adhesions that can increase experimental background and thereby preclude low-avidity adhesive interactions. This background can be further complicated by experimental methods used to isolate and concentrate the sample, such as trypsin digestion. It is therefore critical to minimize the "nonspecific" adhesions (as described later) and establish specific, reproducible criteria to define negative and positive samples so as to minimize subjective analysis. As controls, it is often necessary to establish a negative control for adhesion. This could be mock-transfected cells (when studying an expressed protein) or experimental conditions that ablate adhesion. For example, if the adhesive molecules responsible for the adhesion are known (or suspected), function-blocking reagents (antibodies or peptides) are used to determine the contribution of specific adhesive events. An example of this is the use of RGD peptides to block integrin-mediated cell adhesions (Ruoslahti and Pierschbacher, 1987).

2. Fluorescent and Flow Cytometric Analyses of Cell–Cell Aggregation Assays

If two cell populations to be studied are phenotypically indistinguishable, modifications to the basic aggregation assay may be necessary. An early modification was to label one cell population with a radioactive marker before mixing the two populations (Moscona and Hausman, 1977). The resulting aggregates were then evaluated for the presence of radiolabeled cells. A modification of this strategy involves the use of fluorescent markers (Fishelson and Berke, 1979), which facilitates a more rapid and efficient analyses of cell–cell adhesion. For example, prior labeling of one population with a fluorescent dye (CMFDA, 5-chloromethylfluorescein diacetate) can allow for fluorescence microscopy to evaluate conjugates that are otherwise indistinguishable using light microscopy. We prefer CMFDA as it freely diffuses into cells and is activated intracellularly by cleavage of the acetate by cytosolic esterases. Once activated, CMFDA is retained at physiological temperatures for many hours and days and thus is compatible with long-term assays. The labeling of the second cell population with a second vital dye (e.g., blue fluorescent aminocoumarin, hydroethidine) can also be employed to visualize both cell populations simultaneously.

The use of fluorescent dyes also can facilitate the enumeration of cell conjugates using flow cytometry (Luce *et al.*, 1985; Storkus *et al.*, 1986). For this modification, the fluorescently labeled conjugates are prepared as described earlier and lightly suspended cells subjected to a flow cytometer. This technique was first devised to study the adhesion of lymphoid cells with their targets, but it can be modified to quantitate nonlymphoid adhesions as well. An advantage is that flow cytometric analysis is generally less subjective and allows for the rapid assessment of larger numbers of conjugates than microscopic analysis (10,000 versus 200 cells per sample, respectively). As negative controls, individual cell populations are first analyzed to assess the fluorescence of either cell type. After mixing the two populations and allowing for adhesion, the samples are then presented as a standard two-color fluorescence plot, with double-positive staining defined as conjugates (Fig. 1). The size and composition (ratio of the different cell types within a conjugate) can often be estimated by the relative fluorescence intensity. One important technical consideration, however, is to limit the size of conjugates (often by increasing the intensity of tituration) to prevent the clogging of lines by larger aggregates.

The use of flow cytometry provides the added advantage that additional markers (e.g., antibodies, dyes) can distinguish cell subpopulations (e.g., CD4 versus CD8 T cells) or physiological consequences (e.g., calcium fluxes). Finally, the use of flow cytometry imparts less vigorous shearing of conjugates than comparable microscopic techniques. This property can be particularly important for evaluating lower avidity interactions.

B. Monolayer Cell–Cell Adhesion Assays

Aggregation assays can accurately measure cell adhesion in suspended cells. However, many "adherent" cell types, including fibroblasts, epithelia, and endo-

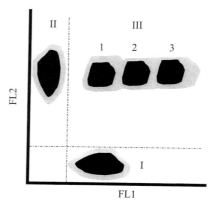

Fig. 1 Flow cytometric analysis of cell aggregates. Shown is an example of data obtained in a flow cytometric analysis of cell–cell conjugates. Two cell populations, each labeled with a different vital fluorescent dye (1 or 2), are mixed in suspension and allowed to form conjugates. The sample is then analyzed by flow cytometry with conjugates observed as events that display both colors (Region III). Often the binding of multiple cells can be distinguished by increasing color (e.g., Region III demonstrates one, two, or three cells from population 1 bound to a single cell from population 2).

thelia, are normally anchored to their surrounding microenvironment. This anchorage often provides critical signals that regulate normal cell behavior. Thus, the use of suspended cells may introduce undesirable properties that may bring into question the physiological relevance of the assay. Furthermore, some procedures necessary to suspend cells (e.g., trypsin/EDTA) can negatively impact the function of specific adhesion molecules. Finally, the architecture of the cell interior (especially the cytoskeleton) and membrane is radically altered on suspension. These facts can raise questions as to the applicability of aggregation assays for modeling cell–cell adhesion in adherent cells. Therefore, it is often more appropriate to model cell–cell binding using monolayer (coculture) systems.

As an overview, monolayer assays measure the adhesion of suspended cells plated onto a confluent monolayer of unlabeled cells (Lackie, 1991) (Fig. 2). Often, the monolayer cells have been transfected with a candidate adhesion molecule while untransfected (or parental) cells serve as a negative control. The suspended cells are labeled with either a radioactive (e.g., [^{3}H]thymidine) or fluorescent (e.g., CMFDA) marker before incubation with the monolayer. Before incubation, the overall activity (radioactive or luminescent) of a fixed number of cells is assessed so that the results of the adhesion assay can be translated into cell numbers. [We routinely measure the activity in 1×10^6 radiolabeled cells as a standard for each assay (Kinch *et al.*, 1993, 1994).] After incubation for a predetermined time in a minimal volume (to ensure that the two cell populations come into close contact), the monolayers are washed extensively to remove unbound cells, and bound cells are quantitated. If radioactively labeled cells are used, the samples can be either trypsinized or subjected to hypotonic

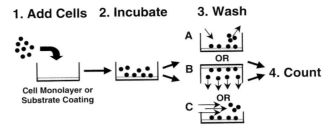

Fig. 2 Cell adhesion to monolayers. Different applications of the basic cell-substrate adhesion assay are shown. A population of suspended is cells is added to a test well coated with either a cell monolayer or purified substrate. After cells establish an interaction, different methods of removing unbound cells can yield different observations: (A) simple washing (addition and aspiration of medium) will remove unbound cells to reveal specific adhesions; (B) reversed centrifugation or (C) flow measurements can quantitate the strength of adhesion.

lysis by adding distilled water. If a fluorescent probe is used, the fluorescence of intact monolayers can be determined using a calibrated fluorimeter.

There are a number of technical considerations that must be addressed when performing these types of assays. First, the monolayers are usually formed in tissue-culture-treated multiwell plates. The size (6–96 wells per plate) of each monolayer needed for a particular assay may depend on the number of samples, sensitivity, or instrumentation (e.g., 96-well plate counter, fluorimeter). Similarly, the ratio of suspended to adherent cells must be tested to establish optimal conditions. It is usually essential to utilize confluent cell monolayers because exposed portions of tissue-culture-treated plastic can promote cell binding that increases the nonspecific background. It is also important to develop a washing procedure that minimizes background (nonspecific) adhesion while optimizing specific binding (as defined later). This will vary depending on the cell types and molecules being studied, but at least four to five repetitions of vigorous agitation in binding medium are recommended. Finally, it is our experience that the assays should be performed in triplicate to minimize experimental variability (Kinch *et al.*, 1993).

C. Reversed Centrifugation for Quantitation of Cell Adhesion

Reversed centrifugation is a modification of monolayer adhesion assays that allows for a quantitative assessment of the strength of cell–cell (or cell–substrate) adhesions (McClay *et al.*, 1981). The primary differences from the standard assay center on the washing steps to remove unbound cells. With this approach, detachment is achieved by inverting the sample and centrifuging at a controlled speed. The strength of cell adhesion is plotted as a function of the relative centrifugal force (g) necessary to detach bound cells. [The relative centrifugal force (RCF in g's) can be easily calculated if the rotor speed and radius are known.] These changes improve the reproducibility between samples and allow for quantitation of binding strength, especially low-avidity interactions that are otherwise precluded by the shear forces encountered in the standard monolayer assays. Indeed, these procedures were initially developed to distinguish the weak interactions of initial cell adhesion from the subsequent strengthening of cell–cell contacts as the adhesion sites mature (McClay *et al.*, 1981). Since the initial development of the assay, a number of modifications have been introduced to increase the ease and efficiency of sample preparation (Subiza *et al.*, 1991; Lotz *et al.*, 1989).

In brief, cell adhesion is allowed to take place, as described earlier, by incubating labeled cells with a confluent cell monolayer (McClay *et al.*, 1981). The sample is then gently immersed in a vessel containing binding medium (PBS), inverted 180°, and gently transferred to a centrifuge carrier (while remaining immersed) (Subiza *et al.*, 1991). After centrifuging at the desired rotor speed for 1–5 min to remove unbound cells, the sample is recovered and quick-frozen in an ethanol–dry ice bath. The bottom of each well is then clipped off, and the remaining bound cells are quantified as described earlier (using a scintillation counter or fluorimeter). For a quantitative assessment of adhesion strength, the procedure is repeated using known RCFs, ranging from 1 to 1000 g.

III. General Strategies to Measure Cell–Ligand Adhesions

A. Monolayer Cell–Ligand Adhesion Assays

The interaction of a cell with the surrounding extracellular matrix is understood to be a critical regulator of cell growth, survival, differentiation, and invasiveness. A variety of techniques have been developed to model cell binding to purified or semipurified basement membrane components (e.g., fibronectin, collagens, proteoglycans). Most are modifications of the monolayer adhesion assays described earlier:

1. Basic Assay

The standard cell monolayer assay used to study cell–ligand adhesion is a variation of the cell–cell adhesion assay. The major difference is that the cell monolayer is replaced by a purified (or partially purified) protein extract. For example, a tissue culture dish is coated with a solution of purified protein (often a 50 μg/ml protein solution in water or PBS) for at least 1 hr at 37°C (or overnight at 4°C) (Burridge *et al.*, 1992). After removing the suspension, the plates are washed and then blocked in 1 mg/ml bovine serum albumin (BSA) for 1 hr at room temperature before incubation with labeled cells. The cells, labeled with a radioactive or fluorescent marker, are then incubated with the monolayer in a buffer usually containing divalent cations and BSA. The cations are required if the adhesions to be studied are (or might be) mediated by integrins, cadherins, or other adhesion molecules that require the presence of divalent cations. BSA serves to minimize background binding to the plastic. As described later, serum and serum-derived products should be avoided when performing these types of assays. After extensive washing of the monolayers to remove unbound (or poorly bound) cells, the adhesion of remaining cells is enumerated by measuring the remaining fluorescence or radioactivity.

B. Interference Reflection Microscopy

"Adherent" cells (e.g., fibroblasts, epithelia) do not sit evenly along the surface of a solid support. The cell membrane rarely approaches closer than 50 nm to the support, except within specific sites known as focal adhesions where the membrane–matrix separation narrows to 10–15 nm (Burridge *et al.*, 1988). Immunostaining for specific focal adhesion proteins (e.g., vinculin, paxillin) can be used to assess focal adhesion formation but is not so useful for estimating the relative distance (and thus avidity) of cell–matrix adhesions. A more accurate assessment of the distances of cell–matrix interaction is achieved using interference reflection microscopy (IRM) (Zand and Albrecht-Buehler, 1989). IRM measures the gap of refractive index between a cell plated onto a glass substratum (usually a glass coverslip). In brief, light reflected from the basal cell membrane is absorbed at sites of close cell–ECM adhesion (within 10–20 nm) and appears as dark patches along the basal cell surface. Normal fluorescent microscopes can be adapted easily to perform IRM by replacing one of dichroic beam splitters with a neutral beam splitter. An iris diaphragm close to the lamp is kept mostly closed to maximize image contrast. Digitized images of IRM can then be acquired with a digital camera and be assessed using standard image analysis software packages (e.g., Adobe Photoshop, Optimus), using the ratio of light and dark areas of the membrane to provide an assessment of the distances of cell–matrix adhesions.

C. Flow Measurements

An increasingly popular method to measure cell–ligand adhesion involves the use of flow forces (Lackie, 1991). Whereas most techniques involve prolonged incubation times in which the cell can spread onto the substrate (and thus adhesion is complicated by changes in cell shape and cytoskeletal involvement), flow technology can measure cell–ligand interactions independent of these changes. In brief, cell adhesion to a ligand-coated surface is measured at a known flow rate. Assuming a laminar flow rate, the wall shear stress (which is a function of the fluid viscosity, velocity, and chamber size) exerts a force on the cell that is proportional to the square of the cell radius. Thus, the flow rate necessary to detach cells can be calculated as an estimate of the adhesion strength. Two modifications of flow measurements are common. In the first, cells are added and allowed to attach to the substrate in the presence of constant flow. Such systems are particularly useful for modeling blood cells within blood vessels (Lackie, 1991; Ballermann and Ott, 1995). Alternatively, cells are allowed to adhere to a surface (in the absence of flow), and the flow is then used to determine the force necessary to dislodge the cells. The data from these types of experiments are generally collected using videography and presented as a measure of cell attachment under different forces. The major disadvantages of flow systems at present are the inability to study large numbers of samples and the limited availability of specialized equipment and expertise.

IV. Specificity of Cell Adhesion

A. Overview

Determining the specificity of an adhesive event is critical for proper evaluation of cell adhesion assays. Background adhesions often preclude proper interpretations of experimental data, particularly when studying lower avidity adhesive interactions. One popular misconception is that background adhesion is nonspecific. On the contrary, background adhesion is often highly specific, but unwanted. In the next section we suggest methods that might be used to minimize background binding. Here, we will suggest methods to determine adhesive specificity.

B. Controls

The most critical parameter to confirm adhesive specificity is the identification of a proper negative control. This choice will vary relative to the experimental setup. If the goal is to test if a candidate gene product promotes cell adhesion, then the candidate gene can be expressed (or overexpressed) in a cell line that does not normally express it. "Specific" adhesion can then be defined as the adhesion levels of the transfectants less background binding of the negative control. This assumes that the transfected molecule does not impart differences in cell size or shape that alter background binding, although this assumption can sometimes be invalid. Similarly, when measuring cell–ligand binding, uncoated supports could be employed. Often, however, supports coated with an irrelevant ligand (e.g., BSA) are more suitable because unblocked tissue culture dishes often can promote high levels of background binding (as discussed later).

Another common control for adhesive specificity is to block the normal functioning of the molecule with specific reagents (function-blocking antibodies or peptides). Antibodies often can be successfully employed to disrupt specific adhesive events, but a number of potential problems must be considered. First, it must be known that the reagents block adhesive function. A variety of antibodies specific for adhesion molecules do not alter (and may even increase) adhesive function. This is particularly true for integrins, as conformational changes induced by some antibodies can increase ligand avidity. Another potential problem arises when studying cells with Fc receptors. Often "inhibitory" antibodies can promote adhesion by binding to Fc receptors, and this complex can promote, rather than discourage, adhesion through the antigen-binding domain. When such artifacts arise, one can employ methods to block Fc receptor binding (e.g., serum extracts, Fc receptor antibodies). Alternatively, the use of Fab fragments may be necessary. Fab fragments also prevent the activation of signaling molecules that can occur when using intact bivalent antibodies. Cell adhesiveness is often extremely sensitive to changes in intracellular signaling, and cross-linking of signaling molecules (that themselves have no intrinsic adhesive functions) with bivalent antibodies is sometimes sufficient to alter cell adhesiveness indirectly. This is further compounded by the fact that cross-linked adhesion molecules (e.g., cadherin, integrin) can also transduce intracellular signals that can alter cell adhesiveness (Guan *et al.*, 1991; Kinch *et al.*, 1997). Such observations support the use of Fab fragments as specific inhibitors of cell adhesion.

The experimental conditions can also dramatically alter adhesion specificity. Whenever possible, serum-free conditions should be utilized. Otherwise, it is likely that serum-derived vitronectin could coat all surfaces in which it comes into contact and promote the binding of cells with vitronectin-specific receptors (e.g., $\alpha_v\beta_3$). This can be a substantial problem when working with cell lines since these often have been selected for vitronectin binding (as serum is routinely used to establish and propagate cell cultures).

As mentioned earlier, the use of serum can provide an exogenous source of matrix materials that can confuse the interpretation of adhesion assays. Similarly, it is not always appreciated that most cells undergo constant remodeling of the surfaces in which they come into contact. For example, Fath *et al.* (1989) showed that the initial adhesion of cells onto fibronectin is rapidly (within minutes) remodeled to change to vitronectin binding. Although the fibroblasts initially bound to the plated fibronectin, this material was proteolytically digested, and subsequent adhesions were found to occur to endogenously produced vitronectin. It can therefore be impossible to attribute the resulting cell adhesion to the initial cell–ligand interaction. This ambiguity can compromise the interpretations of adhesion assays. This potential pitfall arising from the production of endogenous ligands can be avoided by treating cells with 25 μg/ml cycloheximide for 2 hr prior to the adhesion assay (Burridge *et al.*, 1992). If long periods of incubation (>4 hr) are necessary for the adhesion assay, protein inhibition by cycloheximide might be contraindicated.

V. Optimization of Experimental Conditions

The amount or specificity of cellular adhesions can be extremely sensitive to the assay conditions. Here, we will address some of the experimental details that may be altered to optimize the specificity of cell adhesion.

A. Buffer Conditions

The buffer employed for a specific adhesion assay can have a dramatic impact. The ion concentration of the buffer in particular is a critical regulator of cell adhesion. Adhesions mediated by cadherins require divalent calcium ions, whereas integrins can utilize Ca^{2+}, Mg^{2+}, or Mn^{2+}. These requirements can be exploited to increase the specific binding by either including these ions or depriving the buffer of these ions to minimize background binding by these molecules. The inclusion of a blocking protein in the buffer can also increase the sensitivity of the assays by blocking nonspecific protein binding that might otherwise increase background adhesion. Generally, 1 mg/ml BSA is employed, although other blocking proteins (e.g., gelatin) might also be preferred. As described earlier, the use of serum should be avoided whenever possible to minimize the likelihood that serum components (e.g., vitronectin) will contribute to background binding.

B. Temperature

Similar to the manipulation of buffer strength, the assay temperature can often be manipulated to maximize specific adhesion. As an example, optimal adhesion by the T lymphocyte adhesion molecule LFA-1 is optimal at 37°C, whereas CD2 binding to its counter-receptor (LFA-3) occurs equally well at 4° or 37°C (Springer, 1990). Thus, if an adhesive interaction to be studied is temperature independent, then the sensitivity of the assay may be optimal at lower temperatures that preclude temperature-dependent background adhesions.

C. Timing

The timing necessary for the establishment of a particular cell adhesion can vary widely. For example, adhesions mediated by integrins can occur within seconds, whereas the binding of CD4 to its counter-receptor [major histocompatibility complex (MHC) class II] is first detected after 30–60 min of incubation at 37°C (Kinch *et al.,* 1993). In addition, some adhesive interactions can be quite transient (i.e., a high off-rate), such as the binding of cytotoxic T lymphocytes to target cells, whereas the adhesion of helper T lymphocytes to antigen presenting cells can be stable for many days (Kupfer and Singer, 1989). It is therefore often necessary to vary the time of incubation over a wide range to determine optimal timing. In cases of very rapid adhesion (occurring within seconds or minutes), it is often helpful to gently centrifuge cells together (rather than relying on gravity) to synchronize the binding.

D. Cell Suspension and Resting

When suspension of adherent cells is required, the choice of procedures to suspend the cells can have a dramatic impact on the results. Ideally, cells can be detached by using ion chelators (e.g., 4 mM EDTA). Detachment can often take many minutes and may require vigorous pipetting and agitation. The chelating agents are then washed away by centrifuging the cells and decanting the supernatants. If chelation is insufficient (as is the case with many epithelial cells), then light treatment with, for example, 0.05% trypsin may be necessary. Trypsin mostly digests away the matrix material that promotes the attachment of a cell to tissue culture plastic, but it also acts on various adhesion and signaling molecules on the cells. This proteolysis of cell surface proteins can sometimes have a dramatic effect on cell adhesion. For example, it is well known that E-cadherin is cleaved by trypsin (Takeichi, 1977). To minimize potential artifacts arising from the suspension step, it is generally useful to wash the cells in serum-free medium containing 1 mg/ml turkey egg white trypsin inhibitor to inactivate the trypsin. The cells can then be suspended for 1–2 hr at 37°C to facilitate replacement of damaged membrane proteins before proceeding with adhesion assays. This incubation can also be used for cycloheximide treatment as described earlier.

References

Ballermann, B. J., and Ott, M. J. (1995). Adhesion and differentiation of endothelial cells by exposure to chronic shear stress: A vascular graft model. *Blood Purification* **13**, 125–134.

Burridge, K., Fath, K., Kelly, T., Nuckolls, G., and Turner, C. (1988). Focal adhesions: Transmembrane junctions between the extracellular matrix and the cytoskeleton. *Annu. Rev. Cell Biol.* **4**, 487–525.

Burridge, K., Turner, C. E., and Romer, L. H. (1992). Tyrosine phosphorylation of paxillin and pp125FAK accompanies cell adhesion to extracellular matrix: A role in cytoskeletal assembly. *J. Cell Biol.* **119**, 893–903.

Farquhar, M. G., and Palade, G. E. (1963). Junctional complexes in various epithelia. *J. Cell Biol.* **17**, 375–412.

Fath, K. R., Edgell, C. J., and Burridge, K. (1989). The distribution of distinct integrins in focal contacts is determined by the substratum composition. *J. Cell Sci.* **92**, 67–75.

Fishelson, Z., and Berke, G. (1979). *J. Immnol.* **120**, 1121–1126.

Guan, J. L., Trevithick, J. E., and Hynes, R. O. (1991). Fibronectin/integrin interaction induces tyrosine phosphorylation of a 120 kDa protein. *Cell Regul.* **2**, 951–964.

Kinch, M. S., Strominger, J. L., and Doyle, C. (1993). Cell adhesion mediated by CD4 and MHC class II proteins requires active cellular processes. *J. Immunol.* **151**, 4552–4561.

Kinch, M. S., Sanfridson, A., and Doyle, C. (1994). The protein tyrosine kinase p56lck regulates cell adhesion mediated by CD4 and major histocompatibility complex class II proteins. *J. Exp. Med.* **180**, 1729–1739.

Kinch, M. S., Petch, L., Zhong, C., and Burridge, K. (1997). E-cadherin engagement stimulates tyrosine phosphorylation. *Cell Adhesion Commun.* **4**, 425–437.

Kupfer, A., and Singer, S. J. (1989). Cell biology of cytotoxic and helper T cell functions: Immunofluorescence microscopic studies of single cells and cell couples. *Annu. Rev. Immunol.* **7**, 309–337.

Lackie, J. (1991). Adhesion from flow. *In* "Measuring Cell Adhesion" (A. S. G. Curtis and J. M. Lackie, eds.), pp. 41–66. Wiley, Chichester.

Lotz, M. M., Burdsal, C. A., Erickson, H. P., and McClay, D. R. (1989). Cell adhesion to fibronectin and tenascin: Quantitative measurements of initial binding and subsequent adhesion strengthening. *J. Cell Biol.* **109**, 1795–1805.

Luce, G. G., Sharrow, S. O., Shaw, S., and Gallop, P. M. (1985). Enumeration of cytotoxic cell–target cell conjugates by flow cytometry using internal fluorescent stains. *BioTechniques* **3**, 270–272.

McClay, D. R., Wessel, G. M., and Marchase, R. B. (1981). Intermolecular recognition: Quantitation of initial binding events. *Proc. Natl. Acad. Sci. U.S.A.* **78,** 4975.

Mishell, B. B., and Shigii, S. M. (1980). "Selected Methods in Cellular Immunology." W. H. Freeman, San Francisco.

Moscona, A. A., and Hausman, R. E. (1977). Biological and biochemical studies on embryonic cell–cell recognition. *In* "Cell and Tissue Interactions" (J. W. Lash and M. M. Burger, eds.), Society of General Physiologists Series, Vol. 32, pp. 173–185. Raven, New York.

Roth, S., and Weston, J. (1967). The measurement of intercellular adhesion. *Proc. Natl. Acad. Sci. U.S.A.* **58,** 974–980.

Ruoslahti, E., and Pierschbacher, M. D. (1987). New perspectives in cell adhesion: RGD and integrins. *Science* **238,** 491–497.

Simons, K., and Fuller, S. D. (1985). Cell surface polarity in epithelia. *Annu. Rev. Cell Biol.* **1,** 243–288.

Singer, K. H., Denning, S. M., Whichard, L. P., and Haynes, B. F. (1990). Thymocyte LFA-1 and thymic epithelial cell ICAM-1 molecules mediate binding of activated human thymocytes to thymic epithelial cells. *J. Immunol.* **144,** 2931–2939.

Springer, T. A. (1990). Adhesion receptors of the immune system. *Nature* **346,** 425–434.

Staehlin, L. A., and Hull, B. E. (1978). Junctions between living cells. *Sci. Am.* **238,** 141–152.

Storkus, W. J., Balber, A. E., and Dawson, J. R. (1986). Quantitation and sorting of vitally stained natural killer–target cell conjugates by dual beam flow cytometry. *Cytometry* **7,** 163–170.

Subiza, J. L., Gil, J., de Morales, R., Rodriguez, R., and De la Concha, E. G. (1991). Some improvements in the reversed centrifugation method for the quantitation of cell-to-cell adhesion. *J. Immunol. Methods* **140,** 127–129.

Takeichi, M. (1977). Functional correlation between cell adhesive properties and some cell surface proteins. *J. Cell Biol.* **75,** 464–474.

Takeichi, M. (1994). The cadherin cell adhesion receptor family: Roles in multicellular organization and neurogenesis. *Prog. Clin. Biol. Res.* **390,** 145–153.

Vollger, L. W., Tuck, D. T., Springer, T. A., Haynes, B. F., and Singer, K. H. (1987). Thymocyte binding to human thymic epithelial cells is inhibited by monoclonal antibodies to CD-2 and LFA-3 antigens. *J. Immunol.* **138,** 358–363.

Zand, M. S., and Albrecht-Buehler, G. (1989). Long-term observation of cultured cells by interference-reflection microscopy: Near-infrared illumination and Y-contrast image processing. *Cell Motil. Cytoskeleton* **13,** 94–103.

CHAPTER 29

Invadopodia: Unique Methods for Measurement of Extracellular Matrix Degradation *in Vitro*

Emma T. Bowden, Peter J. Coopman,[1] and Susette C. Mueller

Lombardi Cancer Center
Georgetown University Medical Center
Washington, DC 20007

I. Introduction
II. Invadopodia Activity, a Measurement for Localized Membrane Degradation
III. Fluorescent Activated Cell Sorting–Phagocytosis, a Measurement for Internalization of Proteolyzed Extracellular Matrix
IV. Protocols
  A. Preparation of FITC–Gelatin
  B. Invadopodia Holes Assay: Preparation of FITC–Gelatin Coated Coverslips
  C. Invadopodia Holes Assay: Image Capture and Analysis
  D. FACS–Phagocytosis Assay
References

I. Introduction

Disruption of the extracellular matrix (ECM) surrounding a tumor is a critical component during progression from benign tumor growth to invasive and metastatic tumor cell behavior. Simple assays to measure invasive processes are crucial to furthering an understanding of tumor invasion and metastasis. There are a number of *in vitro* assays to measure the invasion of tumor cells (Le and Bruyneel, 1996; Hendrix *et al.*, 1989; Waller *et al.*, 1986), and there are several well-described

[1] Present address: Dynamique Moleculaire des Interactions Membranaires, Montpellier University, Montpellier, France.

assays to measure total ECM degradation by cells (Shaw *et al.*, 1996; Zucker *et al.*, 1985; Bhatnagar and Decker, 1981). However, directed cellular invasion results from a coordination of cell migration, cell adhesion, and a localization of proteolytic activity to distinct regions of the cell surface (Montgomery *et al.*, 1994; Ruoslahti, 1992). Therefore, we suggest that local invasion has several component activities, including ECM proteolysis and removal of partially degraded matrix from the cell surface.

Our assays are specifically designed to measure ECM degradation mediated by the plasma membrane (Coopman *et al.*, 1996, 1998; Mueller and Chen, 1991; Mueller *et al.*, 1992; Chen *et al.*, 1984). The first assay measures site specific degradation that occurs in association with plasma membrane protrusions, known as invadopodia. This activity was originally described in chicken embryo fibroblasts transformed by Rous sarcoma virus (Chen *et al.*, 1984). The second measures phagocytosis of the matrix that is released by proteolysis and is mediated by both plasma membrane- and diffuse non-plasma membrane-mediated proteolysis of the ECM (Montcourrier *et al.*, 1994; Coopman *et al.*, 1996). Together these assays allow us to study several aspects of plasma membrane involvement in local invasion.

The morphology of invadopodia and phagosomes is shown in Fig. 1a, which illustrates an invasive cell growing on the surface of a gelatin bead. The cell is

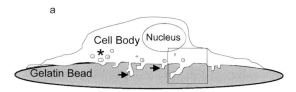

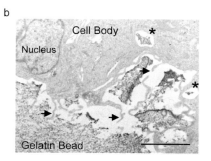

Fig. 1 Invasive breast cancer cells phagocytose partially proteolyzed gelatin matrix and extend invasive membrane protrusions, or invadopodia, into a gelatin bead. (a) Illustration depicting gelatin-containing phagosomes (star) and invadopodia (arrows) of an invasive breast cancer cell grown on a gelatin bead. The area in the box corresponds to an area similar to that shown in the electron micrograph (b). (b) Electron micrograph of an ultrathin vertical section through an MDA-MB-231 cell cultured on a gelatin cross-linked bead shows the ultrastructure of invadopodia (arrows) and the presence of gelatin-containing phagosomes (stars). Scale bar = 1 μm.

actively degrading the bead at its ventral surface and extends membrane protrusions, or invadopodia, into the bead as part of this activity (Fig. 1a, arrows). Gelatin is also actively taken up into phagosomes (Fig. 1a, star). The electron micrograph in Fig. 1b shows the ventral cell surface of an invasive breast cancer cell degrading a gelatin bead. Notice the presence of what appears to be matrix-containing phagosomes (Fig. 1b, star). Also notice the long, thin membrane extensions or invadopodia, which appear to have partially degraded the surrounding matrix (Fig. 1b, arrows). Our assays measure these activities and allow us to examine the mechanisms used to control local invasion. We, and others, have used these assays to identify several regulatory components critical to the invasive phenotype (Coopman *et al.*, 1996; Bowden *et al.*, 1999; Nakahara *et al.*, 1996, 1997, 1998; Chen, 1996; Chen *et al.*, 1985; Kelly *et al.*, 1994; Monsky *et al.*, 1993, 1994). Also, for breast cancer cells we have demonstrated a direct correlation between the invasive phenotype *in vivo*, the invasive phenotype as measured by several well-described assays, and our novel *in vitro* invasion assays (Coopman *et al.*, 1998).

II. Invadopodia Activity, a Measurement for Localized Membrane Degradation

Immunofluorescence, Nomarski illumination, and electron microscopy allow detection of invadopodia on the basis of visualization of membrane or invadopodia associated antigens. Using these methods it was observed that when invasive cells such as Rous sarcoma virus-transformed chicken embryo fibroblasts were cultured on ECM, invadopodia extended into the matrix (Monsky *et al.*, 1994; Mueller and Chen, 1991). Similarly, as described earlier, when invasive breast cancer cell lines were grown on the surface of a gelatin bead, they extended long protrusions, or invadopodia, into the surface of the bead from their ventral cell surface. This can be visualized in conventional electron microscopy (EM) sections as shown in Fig. 1b.

We have developed a quantitative assay to measure invadopodia activity. First, fluorescein isothiocyanate (FITC) is coupled to an ECM substrate, for example, gelatin. Then a thin coat of fluorescently conjugated matrix is coated and cross-linked on a glass coverslip. When cells are cultured on this labeled matrix, they extend invadopodia that degrade the matrix and leave dark areas in a bright field of fluorescence (Fig. 2a). Areas of invadopodia activity can be identified using conventional immunofluorescence microscopy. The area of matrix degradation in each field of view can be quantified after digital image capture using an image analysis program. Thus, there are two aspects to the assay: first, the visual, qualitative determination of the ability of a cell line to make invadopodia and, second, the quantitative measurement of this activity. We have demonstrated that cells will exhibit invadopodia activity when plated on FITC-labeled gelatin or on gelatin coated with FITC-labeled laminin or fibronectin (Monsky *et al.*,

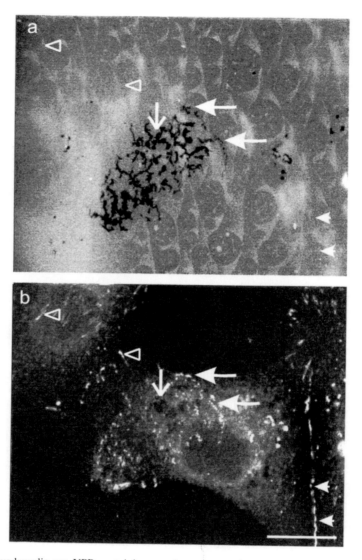

Fig. 2 Invadopodia are YPP containing membrane protrusions associated with sites of matrix proteolysis. (a) MDA-MB-231 cells were cultured on an FITC-labeled gelatin film. Degradation of the FITC–gelatin matrix leaves dark areas in the matrix (open and solid arrows). (b) Immunostaining identifies the localization of YPPs at focal adhesions (open arrowheads), cell–cell junctions (closed arrowheads) and punctate sites corresponding to active invadopodia (solid arrows). Compare the localization of YPP staining and "holes" in the matrix (a and b, solid arrows). Notice that there are sites of degradation in the FITC–gelatin matrix with no corresponding YPP staining (a and b, open arrows). Also note that there is no degradation of the FITC–gelatin matrix corresponding to the localization of either cell–cell adhesions (a and b, closed arrowheads) or focal adhesions (open arrowheads, a and b). Scale bar = 10 μm.

1994; Kelly *et al.*, 1994). FITC can easily be conjugated to a range of ECM substrates; however, because we find gelatin easier to handle, it remains our substrate of choice for this assay.

During the course of tissue culture on FITC–gelatin substrata, the cells move, form, and reform invadopodia, leaving a cumulative record of membrane-associated degradation (Fig. 2a). Thus, not all sites of ECM degradation under a cell will be associated with an invadopodium. Extremely high levels of tyrosine phosphorylated proteins (YPPs), up to 30 times enriched with respect to other cellular fractions, are present at invadopodia (Mueller *et al.*, 1992). Therefore, fluorescent immunolocalization of phosphotyrosine, in concert with visualization of FITC–gelatin matrix degradation, signal the presence of invadopodia and can identify sites of active matrix degradation. Thus, immunostaining can provide additional information about invadopodia activity. For example, Fig. 2b shows immunolocalization of YPPs in the cell lying over the area of matrix degradation in Fig. 2a. Note the presence of streaklike focal adhesions (open arrowheads), cell–cell junctions (closed arrowheads), and punctate staining (solid arrows). The punctate staining (Fig. 2b, closed arrows) corresponds to areas of matrix degradation (Fig. 2a, solid arrows) and provides a marker for sites of invadopodia localization. Notice also that there are areas of matrix degradation that have no punctate YPP staining associated with them (Fig. 2a and b, open arrow). These are sites from which an active invadopodia has retracted, highlighting the transient nature of these structures, in that they constantly extend and retract.

Similarly, cortactin, an actin-binding protein, colocalizes to invadopodia at sites of matrix degradation (Bowden *et al.*, 1999). We have also demonstrated that following microinjection, antibodies directed against cortactin block degradation of the ECM at invadopodia, suggesting a critical role for this molecule in invadopodia activity (Bowden *et al.*, 1999). Thus, combined detection of cortactin, phosphotyrosine, or actin and holes can be used to identify invadopodia.

Not all breast carcinoma cells exhibit this membrane-associated matrix degrading capability. We initially used this assay to assess breast cancer cell lines for their ability to locally degrade the matrix. Visual inspection provided a rapid method of detecting gross differences in degradation before undertaking time-consuming quantitative measurements. Using both qualitative assessment and extensive quantitation of invadopodia activity, we demonstrated that invasive but not noninvasive breast cancer cells lines exhibit invadopodia activity, and that invasive potential generally correlates with invadopodia activity (Coopman *et al.*, 1998; E. T. Bowden, P. J. Coopman, and S. C. Mueller, unpublished observations, 1999).

We have encountered invasive cell lines that do not behave as we expect. The most marked example is presented by the MDA-MB-435 cell line, an extremely aggressive cell line in most *in vitro* invasion assays and, in fact, a model cell system for metastatic behavior *in vivo*. In our hands, this cell line does not produce "holes" when plated onto an FITC–gelatin matrix, suggesting these cells do not have invadopodia. In other cell lines, it has been demonstrated that invadopodia activity can be induced by the addition of specific factors to the growth medium, for example, the addition of heregulin to SKBR3 cells (Staebler *et al.,* 1994; Xu *et al.,* 1997; E. T. Bowden, P. J. Coopman, and S. C. Mueller, unpublished observations, 1999). We hypothesize that the MDA-MB-435 cell line, although extremely invasive and metastatic *in vivo,* is lacking an unidentified factor, which is essential to invadopodia activity under our *in vitro* conditions.

When measuring invadopodia activity, the cumulative nature of the assay should be taken into account. We have characterized the behavior of cells in this assay over time and with varied cell densities, noting that there is little difference in the rate of matrix degradation (on a per cell basis) when cell density is varied, as long as cells are plated at subconfluence (Fig. 3). Invadopodia activity can be observed in as little as 4 hr after plating, since cells begin to degrade the matrix as soon as they attach. However, degradation continues at a steady rate between 10 and 48 hr (Fig. 3). Experimental considerations, including, for example, length of time required for cDNAs to be expressed or whether stimulation or inhibition of invadopodia are being investigated, are then the critical elements in determining the appropriate length of time to assay cells on FITC–gelatin films.

Interestingly, we have observed that even the most invasive breast cancer cell lines do not degrade the matrix in serum-free conditions (Bowden *et al.,* 1999; Imamura *et al.,* 1991; Itoh *et al.,* 1999; E. T. Bowden, P. J. Coopman, and S. C. Mueller, unpublished data, 1999). For example, we have plated cells in serum-free medium to allow cells to attach, before microinjecting them with either antibody or eukaryotic expression vectors. The cells were allowed to recover, and

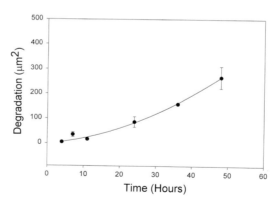

Fig. 3 Time course of FITC–gelatin matrix degradation. The area of degradation per cell was measured after plating MDA-MB-231 cells on FITC–gelatin for 4, 7, 10, 24, 36, and 48 hr. Data from three cell densities (20,000, 50,000, and 70,000 cells per well) were compared over the time course. There were no significant differences in degradation per cell between these cell densities. Data were pooled in this graph, and a third order regression curve was plotted. The slope of the curve indicates that there is a lag phase after cells are plated on matrix before they reach maximal rates of matrix degradation between 10 and 20 hr after plating.

then matrix degradation was allowed to begin by changing the growth medium to include serum (Bowden et al., 1999). This technique prevented an accumulation of matrix degradation that would otherwise have occurred during the cell spreading and microinjection phases of the experiment.

III. Fluorescent Activated Cell Sorting—Phagocytosis, a Measurement for Internalization of Proteolyzed Extracellular Matrix

Phagocytosis is the process whereby microorganisms or tissue debris are recognized by cell surface receptors, engulfed, and then internalized. Polymorphonuclear granulocytes, monocytes, and macrophages are predominantly responsible for this activity, and it is usually associated with infection, inflammation, and wound repair. However, the phagocytic capacity of tumor cells has been demonstrated in several systems, including human breast cancer cell lines (Montcourrier et al., 1994; Coopman et al., 1996), human epitheloid cervix carcinoma cells (Van Peteghem et al., 1980), and rat glioma cells (Bjerknes et al., 1987).

On the basis of this behavior, we have developed an assay to measure the ability of cells to phagocytose partially proteolyzed FITC–gelatin matrix, the FACS–phagocytosis assay. When cells are plated onto the surface of a layer of FITC–gelatin matrix, they begin to degrade and phagocytose the matrix almost immediately [we have demonstrated phagocytosis within 30 min of plating (Coopman et al., 1998)]. At the end of the assay, cells are released from the matrix by trypsinization. The amount of phagocytosed matrix per cell is then measured by fluorescent activated cell sorting (FACS). Data are standardized by calibration using fluorescent reference standards, expressed as mean MESF units (absolute calibrated fluorescence intensity), as the increase (-fold) over background (relative phagocytosis of FITC-labeled matrix as compared to unlabeled matrix), or as a percentage of stimulation or inhibition of phagocytosis as compared to controls in the absence of treatments. We have demonstrated that MDA-MB-231 cells will phagocytose FITC-labeled collagen type I, Matrigel, or cross-linked gelatin (Coopman et al., 1998). Because gelatin is easier to handle and less expensive than the other substrates listed, we use it for both this and the invadopodia assay.

The sensitivity of breast cancer cells to cytochalasin D suggests that phagocytosis measured by the FACS–phagocytosis assay is mediated by the actin cytoskeleton (Coopman et al., 1998). Phagocytosed gelatin is routed to actively acidified vesicles (Fig. 4) (Coopman et al., 1996). This indicates that the intracellular degradation of internalized matrix takes place in the lysosome. That extracellular matrix degradation is a prerequisite for phagocytosis can be demonstrated by its inhibition with various serine and metalloproteinase inhibitors (Coopman et al., 1998).

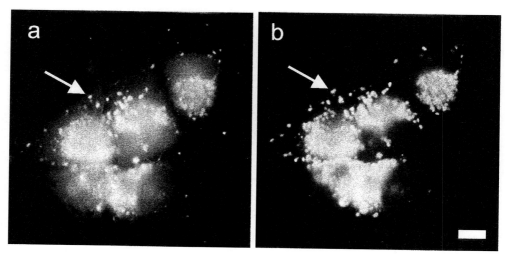

Fig. 4 FITC–gelatin is internalized and localizes to acidic vesicles. MDA-MB-231 cells were incubated on FITC–gelatin films for 3 hr in serum-free culture medium in the presence of 0.1 μm LysoTracker red (Molecular Probes, Eugene, OR). Cells were released from the matrix with trypsin, replated onto glass coverslips, and allowed to settle and spread before fixing (without permeabilization) and mounting for examination by immunofluorescence microscopy. Most FITC–gelatin (a, arrow) colocalized with acidic vesicles (b, arrow). Reprinted with permission from Coopman *et al.* (1998). Scale bar = 10 μm.

It should be noted that cell lines that do not exhibit measurable plasma membrane associated ECM degradation (invadopodia or hole producing activity) will phagocytose FITC–gelatin matrix, apparently due to diffuse degradation by secreted enzymes. The matrix itself is cross-linked, and, if the material is boiled in Laemmli sample buffer under reducing conditions, no protein is detected by SDS–PAGE (E. T. Bowden, P. J. Coopman, and S. C. Mueller, unpublished observations, 1999). Therefore, we conclude that degradation of the matrix must occur for phagocytosis to proceed. However, although both the invadopodia and the FACS–phagocytosis assays indirectly measure ECM degradation, only the invadopodia assay specifically measures membrane-associated degradation. On the other hand, holes formed in the FITC–gelatin matrix, and produced by invadopodia activity, may contribute to, but are not an absolute requirement for phagocytosis of matrix. Consistent with this is our observation that FACS–phagocytosis occurs in serum-free medium, whereas invadopodia activity and the formation of holes is completely suppressed in serum-free medium (Bowden *et al.*, 1999). In general, however, breast cancer cell lines that have high levels of phagocytosis also demonstrate invadopodia activity and form holes in FITC-labeled gelatin matrix (Fig. 5a) (Coopman *et al.*, 1998). These activities also correlate with their behavior in chemoinvasion and chemotaxis assays (Fig. 5b,c) (Coopman *et al.*, 1998).

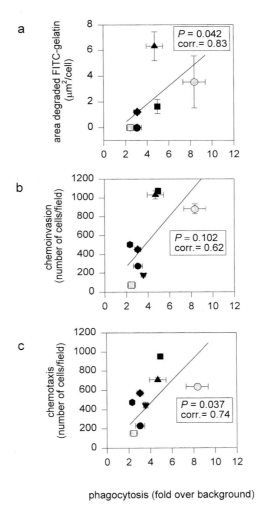

Fig. 5 The phagocytic capacity of breast cancer cell lines correlates with their invasive and migratory capacity. Phagocytosis is expressed as the increase (-fold) over background (mean fluorescence intensity of 10,000 cells incubated on FITC–gelatin/mean fluorescence intensity of 10,000 cells incubated on unlabeled gelatin). Phagocytic capacity is compared to (a) localized matrix degradation in the invadopodia holes assay, (b) chemoinvasion, and (c) chemotaxis. First order regression curves and Spearman rank correlation between the different assays are shown and are considered to be significant when $p \leq 0.05$. The following cell lines were tested: MDA-MB-231 (black squares), MDA-MB-436 (black diamonds), MDA-MB468 (black inverted triangles), BT549 (black triangles), MCF-7 (black circles), MCF-7-ADR (black hexagons), SKBR3 (shaded squares), and Hs578T (shaded circles). Reprinted with permission from Coopman *et al.* (1998).

Gelatin phagocytosis is both cell density and time dependent (Fig. 6). For example, FITC–gelatin uptake in our model cell line, MDA-MB-231, is inhibited when the cells are plated above 100–200,000 cells per well (12-well plate) (Fig. 6A). Also, for this cell line, the optimum time for phagocytosis assays is 6–24 hr (Fig. 6B). After 24 hr, increased incubation of cells with FITC–gelatin

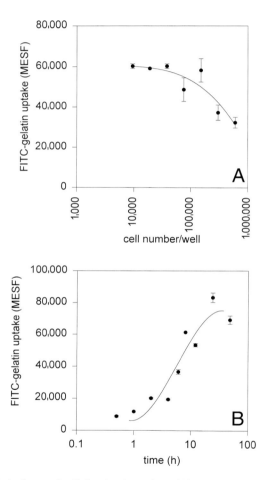

Fig. 6 Phagocytosis is time and cell density dependent. (A) MDA-MB-231 cells were incubated on FITC–gelatin in 24-well plates at varied cell concentrations (9400–600,000 cells per well), for 18 hr. The data is plotted as a third order regression curve and illustrates that as cell number exceeds 37,500 cells per well, phagocytosis decreases. (B) MDA-MB-231 cells were incubated on FITC–gelatin in 24-well plates at 150,000 cells per well for 0.5 to 48 hr. The data are plotted as a third order regression curve and illustrates that uptake is a rapid process, with phagocytosis detected after as little as 30 min. The rate of phagocytosis increases exponentially to reach a steady state between 24 and 48 hr. Reprinted with permission from Coopman *et al.* (1998).

does not give a proportional increase in FITC–gelatin uptake by cells. The data obtained for MDA-MB-231 can be used as a guideline to design assays for other cell lines. Also, we always include the invasive breast carcinoma cell line MDA-MB-231 as a positive control within each experiment we perform. This provides us with an internal standard for each experiment and allows direct comparison between experiments performed on different days. When using this assay to compare the phagocytic capacity of carcinoma cell lines, we tend to use longer time points, that is, 24 hr, to maximize the detection of differences. When examining stimulation or inhibition of phagocytic behavior in a particular cell line, we routinely use a shorter assay time, that is, 6 hr, because the assay is so reproducible.

In summary, the advantage of the FACS–phagocytosis assay is that it is rapid, reliable, and versatile and can be used with a variety of cell lines and matrices (Table I). It utilizes naturally occurring and endogenous type substrates as compared to the more common latex, polystyrene, or dextran bead based assays. It can be used as a relatively rapid throughput, low cost method to compare cell lines and to test the efficacy of treatments designed to stimulate or inhibit invasion. Even carcinoma cell lines that are not very invasive in other assays, such as MCF-7 breast cancer cells, show detectable activity in the FACS–phagocytosis assay (Coopman et al., 1998). Another advantage of this assay when screening the effects of exogenous growth factors, antibodies, etc., is that the FACS–phagocytosis assay is best carried out in serum-free medium. This is in contrast to the invadopodia assay. However, it is not an alternative to the invadopodia holes assay, but rather complements it.

IV. Protocols

A. Preparation of FITC–Gelatin

Add 6 mg of FITC (Sigma, St Louis, MO) to 200 ml buffer A (50 mM $Na_2B_4O_7$, pH 9.3). Dissolve 0.5 g 300 bloom gelatin (Sigma) in 20 ml buffer B (50 mM $Na_2B_4O_7$/40 mM NaCl, pH 9.3). Dialyze the gelatin against the FITC–buffer A solution in complete darkness for at least 1.5 hr at 37°C. Exchange the FITC–buffer A solution for prewarmed phosphate-buffered saline (PBS) at 37°C. The FITC–gelatin requires extensive washing against PBS, normally 2–3 days with two to three buffer changes per day. Before storing aliquots in the dark at 4°C, add sucrose to 2% (w/v).

B. Invadopodia Holes Assay: Preparation of FITC–Gelatin Coated Coverslips

Sterilize 18-mm glass coverslips by soaking in 70% ethanol for 15 min at room temperature. Air-dry coverslips and then coat each sterile coverslip with prewarmed FITC–gelatin, using enough to just cover the surface. Tilt the cover-

Table I
Summary

	FACs-phagocytosis	Invadopodia "holes" assay
Serum conditions	Serum independent	Serum requiring
Time required to perform assay	• 3–4 days FITC–gelatin preparation • 2–3 hr plate preparation • 6–24 hr cell incubation • 2–3 hr cell harvest • at 10^6 cells/ml, about 10 min/sample for flow cytometry • 1–2 hr data analysis	• 3–4 days FITC–gelatin preparation • 2–3 hr coverslip preparation • 6–24 hr cell incubation • 1 hr process coverslips • 2 hr microscopy and analysis/coverslip • 1 hr data processing/coverslip
Error	About 5%	About 5%
Major expense and equipment considerations	Requires access to flow cytometer	• Both qualitative and quantitative assays require access to epifluorescent or confocal microscope • Quantitative assay requires access to epifluorescent or confocal microscope with digital imaging capacity and image analysis software, e.g., Optimas
Technical difficulty	Straightforward, except operation of flow cytometer	• Qualitative assay is straightforward, results are obtained with a simple visual inspection of coverslips by epifluorescent microscopy • Quantitative assay is more complex, results require extensive data extraction, quantitation, and processing
Cell behavior contributing to measurement	• Cytoskeletal rearrangement • Nonmembrane, and membrane-associated matrix degradation • Phagocytosis (particle binding and internalization)	• Cytoskeletal rearrangement • Membrane-associated matrix degradation
Best used as	A straightforward, highly reproducible assay for invasive cell behavior	A specific assay to study the contribution of invadopodia activity to invasive cell behavior

slip and aspirate the excess gelatin. Invert each coverslip onto a 200 μl drop of 0.5% ice-cold glutaraldehyde in PBS and incubate at 4°C for 15 min. Transfer one coverslip to each well of a standard 12-well plate (2.3 cm²/well), gelatin coated side up. Gently wash coverslips three times with PBS at room temperature. Incubate with sodium borohydride (5 mg/ml in PBS), for 3 min at room temperature. Hydrogen bubbles are produced during this incubation time, be sure to keep the coverslip submerged. Wash coverslips with PBS, three times at room temperature. Resterilize the coverslips in 70% ethanol for 15–20 min; they should be handled in the tissue culture hood from now on. Wash coverslips in PBS, three times at room temperature, and finally quench in serum free medium for 1 hr at 37°C. Coverslips are now ready for seeding cells. The optimal density of cells must be determined for the cell type and experimental conditions under examination. We routinely use 70,000 cells per coverslip (approximately 70% confluence) for the invasive breast cancer cell line MDA-MB-231. After various culture times ranging from 4 to 72 hr, cells are fixed with 3.7% formaldehyde/0.1% Triton X-100 for 10 min at room temperature. Nuclei are stained with 4,6-diamidino-2-phenylindole (DAPI) at a final concentration of 80 ng/ml. Immunostaining for invadopodia associated antigens (YPPs or cortactin) is performed as previously described (Mueller *et al.*, 1992; Bowden *et al.*, 1999). Coverslips are mounted using the Prolong Antifade kit (Molecular Probes, Eugene, OR).

C. Invadopodia Holes Assay: Image Capture and Analysis

Images of nuclei (DAPI-stained) and degraded matrix (FITC–gelatin) are captured using Image Pro Plus software (Media Cybernetics, Silver Spring, MD) and a Toshiba integrating color camera supplied by I-Cube (Crofton, MD) attached to an Olympus Vanox-S microscope (Tokyo, Japan) equipped with the appropriate filter sets and a 40×, 0.65 numerical aperture lens. Image analysis was performed using the Optimas 5.2 image analysis program from Optimas (Bothwell, WA). FITC–gelatin film degradation experiments are carried out in triplicate, with at least 25 fields (100 cells, minimum) analyzed per coverslip. We have developed a semiautomated macro to aid in the extraction of data from these experiments. A copy of this macro can be obtained by request from the corresponding author (S.C.M.). For each field (0.01 mm^2) the area of degradation (μm^2) and the number of cells are summed and the areas of degradation per cell calculated.

In some cases, an all or nothing answer as to the ability of cells to exhibit invadopodia activity is expected. In these cases, there is an alternative, image analysis-independent means to assess matrix degradation. This method is based on a visual determination of the percentage of cells that are forming holes in the matrix (Bowden *et al.*, 1999; Nakahara *et al.*, 1998). This method is not sensitive to small changes in invasiveness on a per cell basis, but it gives an indication of the percentage of the population of cells exhibiting invadopodia activity.

D. FACS–Phagocytosis Assay

The base of each well of a chilled 24-well plate (Costar, Cambridge, MA) (1.8 cm^2/well) is coated with a thin film of warmed FITC–gelatin/sucrose or unlabeled gelatin/sucrose (2%/2%, w/w) (approximately 100 μl/well). After incubation at 4°C for 15 min, 1 ml of ice-cold 0.5% glutaraldehyde in PBS is added to each well for 15 min at 4°C to cross-link the gelatin. Films are washed three times in PBS and then sterilized by the addition of 1 ml of 70% ethanol for 15 min. After this, all manipulations are performed under sterile conditions. Films are washed three times with PBS before incubation for 1 hr at 37°C with serum free medium to quench free aldehyde groups. Approximately 150,000 cells are seeded per well in 0.5 ml of serum-free medium and incubated at 37°C. Cells are allowed to attach and spread for up to 72 hr. However, 6- and 24-hr incubation times have been used most routinely. After the cell layer is washed with PBS, cells are released from the matrix by incubation with approximately 200 μl trypsin/EDTA. Trypsin activity is halted by the addition of 1 ml of serum containing medium. Cells are harvested by centrifugation, washed in PBS, and fixed in 3.7% formaldehyde in PBS. The amount of phagocytosed matrix (mean value of the fluorescence intensity of 10,000 cells) is determined by fluorescent activated cell sorting (FACStar Plus, Becton Dickinson, San Jose, CA) and expressed as the increase (-fold) over background (cells incubated on unlabeled matrix).

References

Bhatnagar, R., and Decker, K. (1981). A collagenase assay using [^{3}H-methyl]collagen. *J. Biochem. Biophys. Methods* **5,** 147–152.

Bjerknes, R., Bjerkvig, R., and Laerum, O. D. (1987). Phagocytic capacity of normal and malignant rat glial cells in culture. *J. Natl. Cancer Inst.* **78,** 279–288.

Bowden, E. T., Barth, M., Thomas, D., Glazer, R. I., and Mueller, S. C. (1999). A novel association of cortactin, paxillin, and PKCμ with invasive and motile membranes. *Oncogene* **18,** 4440–4449.

Chen, W.-T. (1996). Proteases associated with invadopodia, and their role in degradation of extracellular matrix. *Enzyme Protein* **49,** 59–71.

Chen, W.-T., Olden, K., Bernard, B. A., and Chu, F.-F. (1984). Expression of transformation-associated protease(s) that degrade fibronectin at cell contact sites. *J. Cell Biol.* **98,** 1546–1555.

Chen, W.-T., Chen, J. M., Parsons, S. J., and Parsons, J. T. (1985). Local degradation of fibronectin at sites of expression of the transforming gene product pp60src. *Nature* **316,** 156–158.

Coopman, P. J., Thomas, D. M., Gehlsen, K. R., and Mueller, S. C. (1996). Integrin $\alpha 3\beta 1$ participates in the phagocytosis of extracellular matrix molecules by human breast cancer cells. *Mol. Biol. Cell* **7,** 1789–1804.

Coopman, P. J., Do, M. T. H., Thompson, E. W., and Mueller, S. C. (1998). Phagocytosis of cross-linked gelatin matrix by human breast carcinoma cells correlates with their invasive capacity. *Clin. Cancer Res.* **4,** 507–515.

Hendrix, M. J., Seftor, E. A., Seftor, R. E., Misiorowski, R. L., Saba, P. Z., Sundareshan, P., and Welch, D. R. (1989). Comparison of tumor cell invasion assays: Human amnion versus reconstituted basement membrane barriers. *Invasion Metastasis* **9,** 278–297.

Imamura, F., Horai, T., Mukai, M., Shinkai, K., and Akedo, H. (1991). Serum requirement for in vitro invasion by tumor cells. *Jpn. J. Cancer Res.* **82,** 493–496.

Itoh, K., Yoshioka, K., Akedo, H., Uehata, M., Ishizaki, T., and Narumiya, S. (1999). An essential part for Rho-associated kinase in the transcellular invasion of tumor cells. *Nat. Med.* **5,** 221–225.

Kelly, T., Mueller, S. C., Yeh, Y., and Chen, W.-T. (1994). Invadopodia promote proteolysis of a wide variety of extracellular matrix proteins. *J. Cell Physiol.* **158,** 299–308.

Le, M. N., and Bruyneel, E. (1996). Comparison of in vitro invasiveness of human breast carcinoma in early or late stage with their malignancy in vivo. *Anticancer Res.* **16,** 2767–2772.

Monsky, W. L., Kelly, T., Lin, C.-Y., Yeh, Y., Stetler-Stevenson, W. G., Mueller, S. C., and Chen, W.-T. (1993). Binding and localization of M_r 72,000 matrix metalloproteinase at cell surface invadopodia. *Cancer Res.* **53,** 3159–3164.

Monsky, W. L., Lin, C.-Y., Aoyama, A., Kelly, T., Mueller, S. C., Akiyama, S. K., and Chen, W.-T. (1994). A potential marker protease of invasiveness, seprase, is localized on invadopodia of human malignant melanoma cells. *Cancer Res.* **54,** 5702–5710.

Montcourrier, P., Mangeat, P. H., Valembois, C., Salazar, G., Sahuquet, A., Duperray, C., and Rochefort, H. (1994). Characterization of very acidic phagosomes in breast cancer cells and their association with invasion. *J. Cell Sci.* **107,** 2381–2391.

Montgomery, A. M. P., Reisfeld, R. A., and Cheresh, D. A. (1994). Integrin $\alpha_v\beta_3$ rescues melanoma cells from apoptosis in three-dimensional dermal collagen. *Proc. Natl. Acad. Sci. U.S.A.* **91,** 8856–8860.

Mueller, S. C., and Chen, W.-T. (1991). Cellular invasion into matrix beads: Localization of β 1 integrins and fibronectin to the invadopodia. *J. Cell Sci.* **99,** 213–225.

Mueller, S. C., Yeh, Y., and Chen, W.-T. (1992). Tyrosine phosphorylation of membrane proteins mediates cellular invasion by transformed cells. *J. Cell Biol.* **119,** 1309–1325.

Nakahara, H., Nomizu, M., Akiyama, S. K., Yamada, Y., Yeh, Y., and Chen, W.-T. (1996). A mechanism for regulation of melanoma invasion. Ligation of $\alpha 6\beta 1$ integrin by laminin G peptides. *J. Biol. Chem.* **271,** 27221–27224.

Nakahara, H., Howard, L., Thompson, E. W., Sato, H., Seiki, M., Yeh, Y. Y., and Chen, W. T. (1997). Transmembrane/cytoplasmic domain-mediated membrane type 1-matrix metalloprotease docking to invadopodia is required for cell invasion. *Proc. Natl. Acad. Sci. U.S.A.* **94,** 7959–7964.

Nakahara, H., Mueller, S. C., Nomizu, M., Yamada, Y., Yeh, Y. Y., and Chen, W. T. (1998). Activation of $\beta 1$ integrin signaling stimulates tyrosine phosphorylation of p190^{RhoGAP} and membrane-protrusive activities at invadopodia. *J. Biol. Chem.* **273,** 9–12.

Ruoslahti, E. (1992). The Walter Herbert Lecture. Control of cell motility and tumour invasion by extracellular matrix interactions. *Br. J. Cancer* **66,** 239–242.

Shaw, L. M., Chao, C., Wewer, U. M., and Mercurio, A. M. (1996). Function of the integrin $\alpha 6\beta 1$ in metastatic breast carcinoma cells assessed by expression of a dominant negative receptor. *Cancer Res.* **56,** 959–963.

Staebler, A., Sommers, C., Mueller, S. C., Byers, S., Thompson, E. W., and Lupu, R. (1994). Modulation of breast cancer progression and differentiation by the gp30/heregulin [correction of neregulin]. *Breast Cancer Res. Treat.* **31,** 175–182.

Van Peteghem, M. C., Mareel, M. M., and De Bruyne, G. K. (1980). Phagocytic capacity of invasive malignant cells in three-dimensional culture. *Virch. Arch. B. Cell Pathol. Incl. Mol. Pathol.* **34,** 193–204.

Waller, C. A., Braun, M., and Schirrmacher, V. (1986). Quantitative analysis of cancer invasion in vitro: Comparison of two new assays and of tumour sublines with different metastatic capacity. *Clin. Exp. Metastasis* **4,** 73–89.

Xu, F. J., Stack, S., Boyer, C., O'Briant, K., Whitaker, R., Mills, G. B., Yu, Y. H., and Bast, R. C. J. (1997). Heregulin and agonistic anti-p185(c-erbB2) antibodies inhibit proliferation but increase invasiveness of breast cancer cells that overexpress p185(c-erbB2): Increased invasiveness may contribute to poor prognosis. *Clin. Cancer Res.* **3,** 1629–1634.

Zucker, S., Lysik, R. M., Ramamurthy, N. S., Golub, L. M., Wieman, J. M., and Wilkie, D. P. (1985). Diversity of melanoma plasma membrane proteinases: Inhibition of collagenolytic and cytolytic activities by minocycline. *J. Natl. Cancer Inst.* **75,** 517–525.

INDEX

A

Aborting, of drop charging: purity protected by, 47–48
Absorption signals, flow cytometric, 112
Acetone, fixation of cell suspensions by, 405
N-Acetylcysteine, maintenance of $\Delta\Psi$, 479
Acridine orange, as lysosomal probe, 10
Activation antigens
  inclusion of Bcl-2, 435
  subset-specific differences, 445
  uptake, 445, 447
ADC, see Analog-to-digital converters
Aggregation assays, measurement of cell–cell adhesion, 600–603
Alcohols, cell fixation, 256, 277
Algorithms
  decision, 83
  LSCM software, 60–61
Allophycocyanin, as fluorescent label, 195
Amgel, preparation from amniotic membranes, 570
L-Aminopeptidase, kinetic activity measurement, 77–78
Amniotic membranes
  Amgel preparation from, 570
  intact, preparation, 565, 567
Amplification
  in situ, 84
  linear and logarithmic, 38–40
  logarithmic
    and fluorescence compensation, 124–127
    fluorescence signal, 534
Analog-to-digital conversion
  circuits with comparators, 155
  in confocal microscopy, 92
  flow cytometric data, 109
  in signal processing, 40
Analog-to-digital converters
  flash and hybrid, 157–158
  resolution, 158–159
  single slope, 156
  successive approximation, 156–157

Annexin V
  in differentiation of apoptosis and necrosis, 535–536
  staining for phosphatidylserine exposure, 493
Antibodies
  anti-PCNA, 422
  antitubulin, viability staining with, 491–492
  binding kinetics, 220–224
  bound per cell, 332–333
  calibrated measurements applied to, 330–331
  in detection of cyclins, 347
  directly conjugated, 232
  histone H3 plus, as marker of mitotic cells, 344–345
  MIB-1, staining pattern, 410–414
  one and two, staining of cells with, 239–240
  primary and secondary, in immunophenotyping, 228–229
  resolving discrete and nondiscrete populations, 245–247
  selection, 259
  specific binding, verification of, 247–248
  specificity, 287–292
  structure, 219–220
  three and four, staining of cells with, 241–242
  titering, 244–248
Antibody–antigen reactions
  kinetics
    in permeabilized cells, 282–287
    in solution, 281–282
  technical problems, 271–272
Antifade reagents, in confocal microscopy, 99
Antigen receptor, transgenic models, 395–396
Antigens, *see also* Activation antigens
  cell surface: simultaneous staining with
    APO2.7, 263
    Ki-67, 407–408
    PCNA, cytokeratin, DNA, and tubulin, 264–266
    T cell receptor z, 263–264
  expression, drug effects, 227
  immunolocalization, 292

intracellular, immunofluorescence, 12
quantitation per cell, 333–334
target, properties, 261–262
APO2.7
and cell surface antigens, simultaneous staining, 263
expression, staining cells for, 490–491
Apoptosis
application of LSCM cell position feature, 74
commercial detection kits, 538–539
duration, 530
flow cytometric quantification, 11
frequency, estimation, 531–532
LSCM analysis, 541–543
mitochondrial role during, 470–472
vs necrosis and vs necrotic stage of apoptosis, 535–536
plant cells, physiological changes during, 520–523
in plants, 506–509
flow cytometric analysis: problems, 509–512
thymus cell subpopulations, 143
triggering mechanisms, 468
Apoptotic bodies
engulfed by live cells, 538
misclassification as single apoptotic cells, 533–534
Apoptotic cells
DNA strand breaks, 352
identification
based on enzyme activity, 532
cell morphology for, 539, 541
by flow cytometry, 528
morphological changes, 506
selective loss during sample preparation, 537
single, misclassification of apoptotic bodies as, 533–534
subpopulations, ethidium bromide bound to, 143
and viable cells, distinguished, 492–493
Apoptotic index, correlation with incidence of cell death, 529–531
Arc epi-illumination, for fluorescence image analysis, 52
Arc light, mercury, 6
Area feature, LSCM list mode, 67–71
Argon ion laser
in combination with other lasers, 22
wavelengths, 21
Autofluorescence
during immunophenotyping, 438
plant cells: problems with, 511
reduction, 410
in studying ECM–cell interactions, 591–593
Avalanche photodiode, 121–122
Avidin, deglycosylated form, 192
Avidin/biotin, protein labeling, 191–193

B

Background approximation, in CRM, 589
Background binding, nonspecific antibodies, 283–284
Basement membranes
ECM, 563
intact, preparation from tissues, 565
Bcl-2 protein, apoptotic effects, 471
Beads
mixtures, in flow cytometry, 128
quantum simply cellular, 333
Rainbow
cross-calibration, 324–325
emission spectra, 319, 323–324
Beam
dimensions, 24–26
focusing lenses, 23–24
illuminating, particle centering in, 26–33
Binding kinetics, antibody, 220–224
Biotin/avidin, protein labeling, 191–193
Bivariate analysis, PCNA expression, 424–425
Bivariate distributions, mitotic cells, 348
Blank standard, defining, 317
Blocking
Fc receptors with IgG prior to staining, 229–232
with IgG, 239–240
Blood
collection, transport, and storage, 208–209
peripheral, erythrocyte lysis protocols, 214
Blood cells
early counters, 5–6
fixation and preservation, 210
preparation, 207–214
staining, 212
viability, 211–212
Blurring, reduction in confocal microscopy, 94
B lymphocytes, cultures, 390–392
Boyden chamber assay, cell migration, 554–557
BrdUrd, see Bromodeoxyuridine
Breast cancer
cell lines, under serum-free conditions, 618–619
JC-1 probe role, 478–479
Brefeldin A, effect on protein expression, 227

Index

Bromodeoxyuridine
　anti-BrdUrd fluorescence signal, 364
　anti-BrdUrd primary antibodies, 358
　cells labeled by, 356–357, 368–369
　in combination with CFSE labeling, 383–384
　content, vs DNA content: data displayed as, 365–366
　and general laboratory reagents, 359–360
　as radiation enhancer, 11
Buffers
　in cell adhesion assay, 610
　for cell division analysis with CFSE, 378
　optimal, for protein labeling, 187–189
Bulk culture, lymphocyte subsets, 448–449

C

Calcium, measurements in protoplasts, 520, 522
Calcium imaging, by confocal microscopy, 102
Calcium ions, Ca^{2+} waves, study with JC-1 probe, 478
Calibrated measurements, applications of, 330–332
Calibration
　common cytometry measurements, 312–330
　light scatter, 327–328
　material for, 311
Cameras, CCD
　in flow cytometric detection, 122–123
　in LSCM, 52, 57
Carboxyfluorescein diacetate succinimidyl ester, see CFDA-SE
Carboxyfluorescein succinimidyl ester, see CFSE
Caspases
　activation, identification with fluorogenic substrates, 495, 497
　cell death responses to, 488
　inhibition, followed by CD95 induction of Jurkat cells, 490
CCD, see Charge-coupled devices
CD3, and CD28, simultaneous binding, 435
CD15, conjugated to FITC, 80–81
CD69, identification of activated lymphocytes, 449, 451
CD95, induction of Jurkat cells, 500–501
　caspase inhibition followed by, 490
CD11b, conjugated to phycoerythrin, 80–81
ced-3 gene, required for programmed cell death, 487–488
Cell adhesion, specificity of, 608–609
　optimization of experimental conditions for, 609–611

Cell adhesion studies
　aggregation assays, 600–603
　by confocal microscopy, 102
　monolayer cell–cell adhesion assays, 603–605
　reversed centrifugation for, 605–606
Cell biology
　confocal microscopic applications, 100
　importance of extracellular matrix to, 561–562
Cell culture
　cell division asynchrony in, 384
　3-D ECM substrates for, 563–565
　establishment of 3-D systems, 574–575
Cell cycle
　histone H3 phosphorylation during, 343–344
　kinetics, estimation by BrdUrd incorporation, 357–372
Cell death
　associated changes in scattering properties, 512, 514
　flow cytometric analysis, 528
　incidence, correlation with apoptotic index, 529–531
　programmed, ced-3 required for, 487–488
　responses to specific caspases, 488
Cell division
　asynchrony, in cultures, 384
　CFSE distribution during, 376–377
Cell–ligand adhesion
　flow measurements, 607–608
　interference reflection microscopy, 607
　monolayer assays, 606–607
Cell migration
　Boyden chamber assays, 554–557
　multistep process, 549
　time-lapse microscopic measurement, 550–554
Cell morphology, for identification of apoptotic cells, 539, 541–542
Cell position, LSCM list mode feature, 71–76
Cell preparation
　B and T lymphocytes, 378
　blood cells, 207–214
　　universal precautions, 207–208
　cell suspensions, 394–395
　for identification of leukocytes, 217–250
　and immunochemical staining, 422–423
　by solid phase LSCM, 80–82
　and staining
　　for flow cytometry, 137–138
　　procedures, 237–244

Cell proliferation
  analysis with PCNA, 427–428
  effect on migration assays, 552
Cells
  activation, time application, 180–181
  birth rate and death rate, 530–531
  blood, see Blood cells
  constituents, LSC values, 59–61
  control, see Control cells
  cryopreservation, 406–407
  cultured
    labeling and fixation in vitro and in vivo, 360–361
    staining procedure, 361–362
  dead
    considered in data interpretation, 224
    tracking fate of, 386
  fixed, structure, 279–281
  flow rate past laser beam, 30–32
  interaction with light, in flow cytometry, 110–118
  living
    confocal microscopic applications, 101–102
    engulfing apoptotic bodies, 538
  mitotic, see Mitotic cells
  number per division, estimation, 388–390
  permeabilized
    antibody–antigen kinetics, 282–287
    structure, 274–281
  positive: subtracting, 231
  processing for antibody titration, 245
  structure, 274
Cell sorting, see also Presorting
  added to early flow cytometers, 8
  on CFSE peaks, 384, 386
  flow cytometric, 44–48
  fluorescent activated, 619–626
Cell suspensions
  acetone fixation, 405
  adherent cells, 610–611
  paraformaldehyde/methanol fixation, 406
  preparation, 394–395
  single-cell, for aggregation assay, 601
Centering, particles in illuminating beam, 26–33
Centrifugation
  antibodies, for aggregate removal, 221–223
  repeated, leading to cell loss, 537
  reversed, quantitation of cell adhesion, 605–606
Ceolomocytes, JC-1 staining technique applied to, 480

CFDA-SE, tracking of human lymphocytes, 453, 455
CFSE
  application to in vitro lymphocyte cultures, 390–392
  detection of intracellular components, 383
  distribution during cell division, 376–377
  labeling of
    cells, 378–379
    cell surface markers, 381–382
  monitoring of
    lymphocyte responses in vivo, 392–395
    position of undivided group, 386–387
  peaks, cell sorting on, 384, 386
  protein labeling, 191
  staining intensity, appropriate levels, 379–380
Channels per decade, and data accuracy, 305–309
Chaotic flow, at point of analysis, 173–174
Charge-coupled devices
  in flow cytometric detection, 122–123
  in LSCM, 52, 57
Chemokinesis assays, 555
Chemotactic assays, for cell migration, 552
  Boyden chamber, 555
Chromatin
  condensation
    in apoptosis, 507–509
    measurement in apoptotic plant cells, 514–516
  condensed, mitotic cells, 351
Circuits
  hybrid, pulse characterization with, 150–154
  peak and hold, 37
Circular Airy diffraction pattern, 91
Clogging, flow channel, 5–6, 28
Coaxial mixing device, 174–175
Coefficient of variation
  DNA content, 278–279, 293
  fluorescence intensity measurement, 313–314
  LSCM, 63
  measurement of resolution, 309–310
  for repeated samples, 287
Collagen
  component of ECM, 562–563
  fibrillogenesis, 590
  moldable gel, 564
  preparation of purified collagen gel, 568–569
Collection
  blood, prior to cell preparation, 208–209
  data, for Boyden chamber assays, 556–557

Index

fluorescent light, 32
light, optics for, 114–115
light signals from particles, 33–37
Colocalization studies, by confocal microscopy, 102–104
Color compensation, fluorochromes, 318–319
Colorimetry, application to Boyden chamber assay, 557
Commercial detection kits, for apoptosis, 538–539
Comparator, element in trigger circuit, 151–152
Compensation
  color, fluorochromes, 318–319
  fluorescence, and logarithmic amplification, 124–127
  in immunophenotyping of activated lymphocytes, 443–444
  phycoerythrin emission requiring, 235–236
Computers
  curve fitting programs, 388–389
  digital, data storage and display with, 161–167
  LSC, drawing of contours around cell event, 58–61
Confocal microscopy
  calcium imaging, 102
  cell adhesion studies, 102
  cell biology applications, 100
  colocalization studies, 102–104
  development, 90–92
  3-D imaging, 585–586
  fluorescence recovery after photobleaching, 104–105
  fluorescent probes for, 96–99
  image formation, 92–96
  living cells, 101–102
Confocal reflection microscopy, ECM and ECM–cell interactions, 587–591
Conformation, antigen, 262
Conjugation chemistry, protein labeling, 186–187
Contours
  around cell event, LSC computer-drawn, 58–61
  two-parameter histograms, 164–166
Control cells, 262
  mock-transfected, for adhesion studies, 602
  positive/negative, 289–290
Controls, to confirm cell adhesive specificity, 608–609
Coulter Counter, 5

Coupling
  fluorescent probes to proteins, 186–187
  principle, and use of heterobifunctional reagents, 195–201
Coverslips, FITC–gelatin coated, 623–625
CRM, see Confocal reflection microscopy
Cross-linkers
  cell fixation accompanied by, 254, 256
  fixatives: properties, 277–279
  heterobifunctional, 197, 200–201
Cryopreservation
  blood cells, 210
  single cell suspensions, 406–407
Current
  converting optical signals to, 118–123
  converting to voltage, 123–124
Curve-fitting analysis, estimating number of cells per division, 388–389
Cyclin A
  mitotic cells negative for, 349
  plus H3 phosphorylation, immunocytochemical detection, 345–346
Cyclin B1, maximally expressed during mitosis, 349
Cyclin D1, MAbs reacting with, 288–289
Cyclin E, immunoreactivity pattern, 286–287
Cyclins, immunocytochemical detection, 347
Cytochrome c, release without simultaneous drop in $\Delta\Psi$, 471
Cytograms, bivariate halogenated thymidine analog-DNA, 366
Cytokeratin, and surface antigen MC5: simultaneous staining, 264–266
Cytometers, bench-top, 22
Cytometry
  fluorescence lifetime image, 133
  image, analysis of PCNA, 425–426
  quantitative fluorescence, 311–312

D

Dark conditions, PECY5 reagents kept in, 236
Data collection, for Boyden chamber assays, 556–557
Data display
  analog, 154–155
  as histograms or by list mode, 176–178
  logarithmic, and dynamic range, 302–304
Data file
  continuum to information, 41–43
  from light signals to, 37–41

Data storage
  flow cytometry standard format, 161
  formats for, 175–176
Decay, heterogeneous fluorescence, 144–145
Degradation, membrane, invadopodia activity as measurement, 615–619
ΔΨ, see Mitochondrial membrane potential
Denaturation, DNA, 357
Density, cell, effect on migration assays, 550
Density plot, two-parameter histograms, 164–166
Detection threshold, 314–317
Detergents
  cell membrane dissolved with, 272
  nonionic, 276–277
  for permeabilization, 256, 258–259
4,6-Diamidino-2-phenylindole, 10
Dichlorotriazinylaminofluorescein, protein labeling, 191
Dichroic mirror
  long-pass and short-pass, 36–37
  in wavelength selection, 117–118
Differential nonlinearity, problem in ADCs, 157
Differentiation map analysis, data processing for, 390
Digital signal processing, pulse characterization by, 159–161
Digital-to-analog converter, 156–157
Digitonin, 490–491
Distance, particles separated by, 30–31
DNA
  altered stainability, 351
  cancer cells, propidium iodide-stained, 72
  and dsRNA, simultaneous measurement, 143
  ethidium bromide-bound, 135–136
  fluorescence staining, 9–10
  fragmentation
    in estimation of apoptosis frequency, 531–532
    studied by LSC, 543
  histograms, vs BrdUrd histograms, 371
  ladder
    detection with gel electrophoresis, 494–495, 499
    indicator of nDNA fragmentation, 516–517
  maximum pixel values in LSCM, 64
  nuclear, fragmentation, 506–509
    measuring techniques, 516–520
  staining, 266, 383–384
  strand breaks
    apoptotic cells, 352
    flow cytometric analysis, 493–494
DNA content
  fractional, 533–534
  in multiparametric analyses, 292–293
  vs BrdUrd content: data displayed as, 365–366
DNA repair, analysis with PCNA, 426
Dot plot
  flow cytometric: simple display, 161–164
  two-dimensional, 43
Drops, formation and delay time, 45–47
Drugs, effects on protein expression, 227
Dynamic range
  fluorescent particles, 318
  logarithmic amplifiers, 306
  and logarithmic display of data, 302–304

E

ECM, see Extracellular matrix
Electrochemical gradient, components, 468–469
Electron microscopy, ECM analysis, 585
Electrostatic cell sorters, 8
Electrostatic charge, applied to drops containing cells of interest, 45–46
Elliptical spot, laser beam illumination, 24
Ellipticine, DNA-binding, 136
Endothelial cells, expressing specific markers, 575–576
Engelbreth–Holm–Swarm matrix, preparation, 569–570
Enzyme activity, identification of apoptosis based on, 532
Enzyme kinetics, time application, 180
Enzyme-linked immunosorbent assay, 290
Epi-illumination, arc, for fluorescence image analysis, 52
Epitopes
  cross-reacting, 287–289
  effect on antibodies for immunophenotyping, 224–225
  expression, fixation effect on, 227
  masking, 291
Erythrocytes
  lysing, 237–238
  lysing solution: formulation, 214
  separation from leukocytes, 210–211
Estrogen receptors, application of LSCM cell position feature, 74

Ethidium bromide
  bound to apoptotic subpopulations, 143
  DNA-bound, 135–136
Ethidium monoazide, stock solution, 250
Excitation
  multiphoton, 586
  sources, in confocal microscopy, 94–95
Excitation beam
  early flow cytometers, 6–8
  for LSCM, 59–60
Excitation energy
  in flow cytometry compared to LSCM, 83
  LSC, 52
Extinction signals, flow cytometric, 112
Extracellular matrix
  classification, 562–563
  degradation, mediated by plasma membrane, 614
  and ECM–cell interactions
    autofluorescence, 591–593
    CRM, 587–591
    3-D imaging, 584–587
  importance to cell biology, 561–562
  intact, 565–568
  intestinal submucosa as 3-D substrate, 572–576
  proteolyzed, internalization of, 619–623
  reconstituted, 568–572
  3-D substrates for cell culture, 563–565
Extranuclear total value, LSCM list mode feature, 65–67

F

F(ab) fragments, use as second antibodies, 228–229
FACS, see Fluorescent activated cell sorting
Fc receptor binding domain, antibody, 219
Fc receptors, blocking with IgG prior to staining, 229–232
Fibrillogenesis, collagen, 590
Ficoll-Hypaque density gradient separation, 213
Filters, in wavelength selection, 115–118
FITC, see Fluorescein isothiocyanate
Fixation
  blood cells, 210
  cells
    after staining, 243
    and permeabilization, 254–259
  cell suspensions, by acetone, 405
  cultured cells *in vitro* and *in vivo*, 360–361
  paraformaldehyde, 404–405

properties of cross-linking fixatives, 277–279
in simultaneous staining procedure, 265
tested for different cell types, 408–409
Flow cytometers
  addition of cell sorting, 8
  analysis of plant cells: problems, 509–512
  BrdUrd antibodies, 11
  collection of light signals from particles, 37
  electronics, 123–124
  extinction and absorption signals, 112
  forward scatter signals, 112–113
  hemopoietic cell classification, 11–12
  laser light sources, 7
  light collection optics, 114–115
  measurement sensitivity, 127–129
  multiparameter, 6
  optics and illumination power, 110–112
  particle illumination, 20–26
  performance characteristics, 302–312
  time stamping of events, 171, 181
  wavelength selection, 115–118
Flow cytometry
  accommodating anti-BrdUrd signal, 364
  analysis of
    cell–cell adhesion assays, 603
    cell death, 528
    cellular fluorescence, 346
    PCNA expression, 423–425
  assessment of lymphocyte activation/proliferation, 447–453
  cell preparation and staining for, 137–138
  centering particle in illuminating beam, 26–33
  common measurements, standardization and calibration, 312–330
  and data acquisition, 370–371
  data display formats, 176–178
  from data to information, 41–43
  detection, 118–123
  distinguishing apoptotic cells, 492–493, 542
  fluorescence compensation and logarithmic amplification, 124–127
  fluorescence lifetime, 133–134
    excited-state, 132–133
    instrumentation, 138–141
  heterogeneous fluorescence decays, 144–145
  high discrimination preparative sorting, 83
  historical overview, 19–20
  homodyne vs heterodyne signal detection, 146
  identification of caspase activation sequence, 488–490

karyotyping, 12
from light signals to data file, 37–41
and LSCM: differences, 54
measuring particle concentration, 329–330
quantitative fluorescence, 334
removal of irrelevant cells, 80
sample mixing and delivery: time role, 171–175
signal, background, and sensitivity, 113–114
signal processing tasks, 108–110
sorting, 44–48
standardization, 300–302
  terminology, 301–302
time-resolved fluorescence measurements, 146–147, 170
zero resolution imaging in, 52
Flow cytometry standard
  data storage format, 161
  LSCM values stored as, 61–62
  support, 179
Flow rate, measurements for cell–ligand adhesion, 607–608
Fluidics, flow cytometric, 26–33
Fluorescein isothiocyanate
  bleaching by, 99
  CD15-conjugated, 80–81
  coupled to gelatin, 615–625
  misclassification of apoptotic cells due to, 539
  optimal labeling of proteins, 190–191
  photon flux, 95
  self-quenching, 134
Fluorescence
  average, as useful metric, 177
  background
    assessment for activation antigens, 439, 441, 443
    and flow cytometric sensitivity, 113–114
    principal components, 409
  compensation, and logarithmic amplification, 124–127
  decays, heterogeneous, 144–145
  from different types of particles, comparisons, 325–326
  dim, in washed specimen, 234
  energy transfer: problem in plant cell analysis, 512
  intensity, as indicator of substrate concentration, 180
  recovery, after photobleaching, 104–105
  standardization and calibration, 312–326
Fluorescence image analysis, arc epi-illumination for, 52

Fluorescence lifetime
  excited-state, 132–133
  flow cytometry, 133–134
    instrumentation, 138–141
  measurements, 134–136
    instrumentation, 140–141
    phase-resolved, 136–137
Fluorescence polarization, 325–326
Fluorescent activated cell sorting, phagocytosis assay, 619–623, 625–626
Fluorescent dyes
  in analysis of cell–cell adhesion assays, 603
  DNA staining, 9–10
  for flow cytometry, 7
  organic, protein labeling with, 186–194
  PKH26, 455
  propidium iodide, 66
Fluorescent particle standards
  commercially available, 322
  making of, 320
  spectrally matched and unmatched, 322–324
  tertiary, 324–325
Fluorescent probes
  for confocal microscopy, 96–99
  coupling to proteins, conjugation chemistry, 186–187
  MitoTracker, 475–476
  study of lymphocyte activation/proliferation, 453
Fluorochromes
  brightest, for immunophenotyping, 226–227
  color compensation, 318–319
  cycles of excitation and fluorescence, 32
  in epitope detection, 224–225
  photobleaching, in confocal microscopy, 97, 99
  photodetection, 37
  quantitation, issues regarding, 332–334
  selection, 259, 261
  in simultaneous multicolor applications, 201
  tandem, 234–237
Fluorophores, conjugation, 196
Formaldehyde
  as fixative: properties, 277–279
  plus methanol, fixation by, 280
Forward light scatter, 34, 36
  increased in necrotic plant cells, 514
  measurements, 11
  signals, in flow cytometry, 112–113
F/P ratio, in protein labeling, 189–190
Functional assay, verification of antibody specificity, 292

Index

G

Gating
  lymphocyte isolation by, 75
  procedures for lymphocytes, 439
  on scatter characteristics, 43
  scattergram, 55
Gelatin, coupled to FITC, 615–625
Gel electrophoresis, DNA ladder detection with, 494–495, 499
Granulocytes, peripheral blood, apoptosis in, 480

H

HeLa cells, Ki-67 expression, 411–414
Heterobifunctional reagents, in protein labeling, 195–201
Histograms
  bivariate DNA vs BrdUrd, 364–366, 371
  describing intensity distributions, 41–43
  DNA frequency, 534
  single- and two-parameter, 164–166
  three-dimensional plots and, 166
  univariate, 245, 247
Histone H3
  phosphorylation
    during cell cycle, 343–344
    monoclonal antibody, 350–353
    plus antibody, as marker of mitotic cells, 344–345
    plus cyclin A or B1, immunocytochemical detection, 345–346
Holes
  invadopodia assay, 623–625
  produced by invadopodia, 619–620
Humatrix, preparation from myoepithelial tumors, 570–571
Hydrolysis, and protein concentration, in protein labeling, 189–190
N-Hydroxysuccinimide ester
  plus maleimide, 197–198, 200
  protein labeling, 192

I

Illumination
  arc epi-illumination, 52
  flow cytometric, 110, 112
  in microscopy, 90
  particle, by flow cytometer, 20–26
  power, flow cytometric, 114

Image analysis, invadopodia holes assay, 625
Image formation, in confocal microscopy, 92–96
Imaging
  3-D, ECM and ECM–cell interactions, 584–587
  multiphoton autofluorescence, 592
  reflected light, 587–591
Immunization, with cognate antigen, 396
Immunochemical detection, PCNA, 421–423
Immunofluorescence
  intracellular, fixative solution affecting, 280–281
  quantification of intracellular antigens, 12
Immunofluorescence microscopy, mitotic cells, 347–349
Immunoglobulin
  IgG
    blocking of Fc receptors prior to staining, 229–232
    structure, 273
    removal by washing, 233
Immunolocalization, phosphotyrosine, 617
Immunophenotyping
  malignant cells, 224–225
  with multiple antibodies, 218
  problems during, 438
  with single color, 228
Immunoprecipitation, probing antibody specificity, 290–292
Indo-1, measurement of intracellular Ca, 520, 522
Initialization, instrument, for lifetime measurements, 141
Injection, cells into center of sheath stream, 28–30
In situ hybridization, fluorescence, 67–68, 70, 78
Instrumentation
  Boyden chamber, 554–555
  fluorescence lifetime flow cytometry, 138–141
  historical overview, 4–9
  performance in detecting low levels of fluorescence, 316–317
Integrators, timing and performance, 153–154
Intensity
  CFSE staining, appropriate levels, 379–380
  distributions, histograms describing, 41–43

fluorescence
    as indicator of substrate concentration, 180
    standardization, 319
    laser beam illumination, 23–24
Interference, optical, 116
Interference filters, short- and long-pass, 117–118
Interference reflection microscopy, 607
Interstitial connective tissues
    ECM, 563
    reconstituted ECM, preparation, 571–572
Interval analysis, estimating number of cells per division, 388
Intestinal submucosa
    as 3-D ECM substrate, 572–576
    intact, preparation, 567–568
    reconstituted interstitial ECM prepared from, 571–572
Invadopodia
    activity, as measurement for ECM degradation, 615–619
    holes assay, 623–625
    holes produced by, 619–620
Isometric plot, two-parameter histograms, 164–166
Isotype control, in Ki-67 protein assay, 410

J

JC-1 probe
    applications, 478–479
    detection of variations in $\Delta\Psi$, 469
    disadvantages of, 474
    green and orange fluorescence, 476–477
    staining with, 472, 477–478

K

Karyotyping, flow cytometric, 12
Kinetics
    L-aminopeptidase, 77–78
    antibody–antigen reaction
        in permeabilized cells, 282–287
        in solution, 281–282
    antibody binding, 220–224
    cell cycle, estimation by BrdUrd incorporation, 357–372
    enzyme, time application, 180
    parameter estimation, 366–369
Ki-67 protein
    and cell surface antigen, simultaneous staining, 407–408
    expression in HeLa cells, 410–414
    expression regulation, 401
    MAbs against, 402
    marker in neoplasms, 400

L

Labeling
    cells *in vitro* and *in vivo*
        critical aspects, 369
        methods, 360–361
        with CFSE, 378–379
Labeling index, 355–356
Laser light sources
    confocal microscopy, 90–91, 94–95
    introduction of, 7
    ion, 20–21
Laser scanning cytometer
    analysis of apoptosis, 541–543
    microscope plus PC, 55–57
    scan mirror and stepped stage, 57
Laser scanning cytometry
    area feature, 67–71
    cell position feature, 71–76
    extranuclear total value feature, 65–67
    and flow cytometry: differences, 54
    future directions, 82–84
    laser excitation energy, 52
    maximum pixel value feature, 64–65
    multiple cell exclusion feature, 79–80
    probe spot counting feature, 78
    solid phase, for cell preparation, 80–82
    time feature, 77–78
    total value feature, 62–63
Lenses
    laser beam-focusing, 23–24
    positioning for collection of light signals, 33–36
Leukocytes
    erythrocyte separation from, 210–211
    identification, cell preparation for, 217–250
Light
    absorbance, phycobiliproteins, 194–195
    collection optics, 114–115
    effect on migration assays, 551
    fluorescent, collection, 32
    interaction with cells, in flow cytometry, 110–118
    oscillation between front and rear mirrors, 20–21
Light microscopy, deconvolution, 584–585
Light pulse, processing, 31–32

Index

Light scatter
    changes in apoptotic plant cells, 512, 514
    elimination by dissolving particles, 321
    flow cytometric measurement, 493
    forward, see Forward light scatter
    gating on, 43
    standardization and calibration, 327–328
    studied with multiangle scatter sensor, 7
Light signals
    continuum to flow cytometric data file, 37–41
    from particles, collection, 33–37
Linearity
    and data accuracy, 304–309
    fluorescent particles, 318
Line scanners, confocal microscopy, 95–96
Logarithmic amplifier, 38–40, 42, 124–127
    conventional flow cytometry, 139
    and data accuracy, 304–309
Logarithmic scale, display of data, 302–304
LSC, see Laser scanning cytometer
LSCM, see Laser scanning cytometry
Lymphocytes
    activated/proliferating
        detection, 437–439
        flow cytometry
            advantages, 447–449
            shortcomings, 449–453
    activation, multiparametric analysis, 434
    application of LSCM cell position feature, 74–75
    gating procedures, 439
    intravenous transfer, 393–394
    in vivo responses, monitoring, 392–395
    percentage of positive cells, 445
    subsets, bulk culture, 448–449
Lymphoid tissue, enzymatic cell isolation from, 407
Lysing reagent, prepared daily, 249–250

M

MAb, see Monoclonal antibody
Macromolecular assembly, time application, 180
Magnification, electronic, in confocal microscopy, 92–94
Major histocompatibility complex, class II, expression, 451–452
Marker molecules
    cell surface staining: CFSE labeling method, 381–382
    empirical, PCNA as, 419–420

histone H3 plus antibody, 344–345
Ki-67 protein, 400
for LSCM total value, 63
mitotic cells, 350–353
specific, endothelial cells expressing, 575–576
Matrigel invasion assays, cell migration, 555–556
Maximum pixel value, LSCM list mode feature, 64–65, 68
MC5, in simultaneous staining, 264–266
Membrane invasion assays, cell migration, 556
Mercury arc light, 6
MESF, see Molecules of equivalent soluble fluorochrome
Methanol
    fixation of cell suspensions, 406
    plus formaldehyde, fixation by, 280
Microscopes
    developmental history, 89–90
    historical overview, 4–5
Migration
    antigen, 261
    cell, see Cell migration
    phosphatidylserine, in apoptotic protoplasts, 523
Mitochondria
    role during apoptosis, 470–472
    swelling, 471–472
Mitochondrial membrane potential
    in apoptotic protoplasts, 523
    collapse, 472–473
    in early apoptotic phases, 468
    fluorescent probes, 471
    measured with
        dye, 12
        JC-1 staining technique, 479–480
        TPP electrode, 469
Mitotic cells
    fluorescence analysis, 346
    immunofluorescence microscopy, 347–349
    markers, comparison with anti-H3-P monoclonal antibody, 350–353
    preparation and staining, 345–346
MitoTracker dye, detection of DY, 475–476
Mixing, sample, for flow cytometry, 171–175
Molecular biology, PCNA, 420–421
Molecules of equivalent soluble fluorochrome, 313, 315–317, 320–322, 324–325
Monoclonal antibody
    anti-H3-P, comparison with other mitotic cell markers, 350–353
    binding desired epitope, 229, 231

GNS1, 285–286
identification of lymphocyte subsets, 438
Ki-67, cell cycle distribution, 399–400
Monocytic cells, PECY5 specific binding to, 236–237
Monolayer assays
cell–cell adhesion, 603–605
cell–ligand adhesion, 606–607
Morphology, invadopodia and phagosomes, 614–615
MTT assay, applied to Boyden chamber assay, 557
Multiparametric analysis
DNA content in, 292–293
H3 phosphorylation in U937 cells, 348–349
lymphocyte activation, 434
using FITC conjugated MIB-1 antibody, 414
Multiphoton autofluorescence imaging, 592
Multiphoton microscopy, elimination of aperture in, 586–587
Multiple cell exclusion, LSCM list mode feature, 79–80
Murine hepatitis virus, 103

N

Necrosis, vs apoptosis, 535–536
Neoplasms, Ki-67 protein as prognostic marker, 400
Noise, background, and higher wavelengths, 93
Noise distribution, for low background condition, 314–315
Nonfluorescent standard, defining, 317
Nonlinearity, differential, problem in ADCs, 157
Nonyl acridine orange, detection of mitochondrial mass, 469, 473
Nuclear fragments, misclassification as apoptotic cells, 533–534
Nucleolar organizing region, 12
Nucleolus, protoplast, swelling, 507–508
Nucleoplasm, staining intensity, 410
Nucleus, bare, isolation, 409

O

Optical interference, 116
Optical signals, converting to current, 118–123
Optimization
conditions for protein labeling, 187–190
experimental conditions for cell adhesion, 609–611
Osmotic lysis, separation of erythrocytes from leukocytes, 211
Oxidative stress, in apoptotic protoplasts, 522–523
Oxygen free radicals, apoptosis triggered by, 468

P

p21, complex with PCNA, 421
PAB 416, affinity of, 284–287
Paraformaldehyde
fixation, 404–405
of cell suspensions, 406
preparation in PBS, 402, 404
Ultrapure formaldehyde, 243, 249
Particles
calibration in MESF units, 320–322
centering in illuminating beam, 26–33
comparing fluorescence from, 325–326
fluorochrome-stained, standards, 320
illumination, by flow cytometer, 20–26
light signals from, collection, 33–37
measurement, standardization, 301
sizing and concentration, 328–330
for standardizing fluorescence intensity, 319
weakly fluorescent, detection, 314–317
PBS, see Phosphate-buffered saline
PCNA, see Proliferating cell nuclear antigen
Peak detection
apoptotic, 516
device timing and performance, 153
in flow cytometry, 127
Peaks
CFSE, cell sorting on, 384, 386
individual positions, 387–388
Pepsin digestion, as variable step, 363, 370
Percentage of labeled mitotic figures, 355–356
Permeability transition pores, mitochondrial, 470
Permeabilization reagents
and cell fixation, 254–259
for intracellular staining, 242–243
properties, 276–277
pH, optimal, for protein labeling, 187–189
Phagocytosis, FACS assay, 619–623, 625–626
Phase-sensitive detector, two channels, 142
Phase suppression, 140
Phenotype, cell-specific, in 3-D ECM system, 575

Index

Phosphate-buffered saline
   paraformaldehyde preparation in, 402, 404
   with sodium azide, 249
Phosphatidylserine, migration, early indicator of apoptosis, 523
Phosphorylation, histone H3, during cell cycle, 343–344
Photobleaching, fluorochrome, in confocal microscopy, 97, 99
Photodetector, for side scatter detection, 36–37
Photodiode
   avalanche, 121–122
   lens focusing light onto, 33–36
Photodiodes, characteristics, 119
Photomultiplier tubes
   characteristics, 119–121
   in CRM, 588–589
   gain, 126
Photons
   reaching cell, 110, 112
   single, counting, 122
Phycobiliproteins, labeling of proteins, 194–201
Phycoerythrin
   activation/proliferation markers labeled with, 443
   CD11b-conjugated, 80–81
   complex with antibody, 136
   R and B forms, as fluorescent labels, 195
   tandem conjugate with Cy5, 109, 234–237
Phycoerythrinated goat anti-mouse Ig, 240
Pinhole diameter, achieving, 91
Pixel values, in LSCM, 58–61
Plant cells
   analysis with flow cytometer, 509–511
   autofluorescence, 511
   changes in scattering properties, 512, 514
   chromatin condensation, 514–516
   fluorescence energy transfer, 512
   nDNA fragmentation, 516–520
   physiological changes during apoptosis, 520–523
   probe loading problem, 511–512
Plants, apoptosis in, 506–509
Plasma membrane
   ECM degradation mediated by, 614
   integrity, changes during apoptosis, 520, 536
Presorting, improving resolution by, 380–381
Probe loading, problem in analysis of plant cells, 511–512

Probe spots
   contours, in LSCM, 61–62
   counting feature in LSCM, 78
Proliferating cell nuclear antigen
   analysis of
     cell proliferation, 427–428
     DNA repair, 426
   complex with
     p21, 421
     replication factor C, 420
   as empirical marker, 419–420
   flow cytometric analysis, 423–425
   image cytometric analysis, 425–426
   immunochemical detection, 421–423
   immunofluorescence, 280
   in simultaneous staining, 264–266
Propidium iodide
   characterization of nuclear changes in protoplasts, 507–508
   counterstaining with, 66
   fluorescence intensity decrease with cell age, 515
   stock solution, 250
Protein expression, brefeldin A effect, 227
Protein labeling
   with organic fluorescent dyes
     choice of reactive group, 193–194
     conjugation chemistry, 186–187
     optimizing conditions, 187–190
     specific dyes used, 190–193
   with phycobiliproteins
     heterobifunctional reagents, 195–201
     light absorbance, 194–195
Proteins
   intracellular, immunocytochemical detection, 346–347
   loss, from formaldehyde fixation, 279–280
Protoplasts
   isolation from whole tissue, 509–510
   nuclear changes in, 507–508
Pseudotime, flow cytometric, 179
Pulse characterization
   with analog hybrid circuits, 150–154
   by digital signal processing, 159–161
Pulse shaping, integration by, 154
Pulse width
   measurement circuits, timing and performance, 153–154
   measurement in flow cytometry, 127
Purity, protected by aborting of drop charging, 47–48

Q

Quality, antibody, K_a as descriptor, 223–224
Quality control, identifying time-dependent artifacts, 178–179
Quantification, coexistence with simplification, 161–167
Quantitation, aggregation assays, 601–602
Quantitation limit, 310–311, 315–316
Quantitative fluorescence cytometry, 311–312
Quantitative indirect immunofluorescence assay, 332–333
Quantum simply cellular beads, 333
Quenching, in protein labeling, 190

R

Reactive group, choice: for protein labeling, 193–194
Reference point, in time-lapse studies of cell migration, 553
Reimaging, green fluorescence, 293
Relocation functionality, LSCM, 71, 74
Resolution
    ADCs, 158–159
    coefficient of variation as measure, 309–310
    fluorescence, 313–317
    improvement by presorting, 380–381
    of time axis, 177
Rhodamine, protein labeling, 193
Rhodamine 123, study of lymphocyte activation/proliferation, 453

S

Sample protocol, Ficoll-Hypaque density gradient separation, 213
Samples, *see also* Specimens
    mixing and delivery: time role, 171–175
    preparation, 361–364
        critical aspects, 369–370
        selective loss of apoptotic cells during, 537
    time-point, single measurements, 367–369
Scaffolds, 3-D, in ECM systems, 564
Scanning electron microscope, cryostage, 589
Sensitivity
    applied to fluorescence measurements, 310–311
    and background fluorescence, 113–114
    fluorescence, 313–317
    LSC, evaluation, 63
    measurement, flow cytometers, 127–129
    proteins, to fixation/permeabilization, 291
    protoplasts, to changes in osmotica, 510
Sensors
    multiangle scatter, 7
    signal, simultaneously digitized, 57–58
Sheath stream
    center, 28–29
    velocity, 29–31
Signal processing
    amplification in, 38–39
    analog-to-digital conversion in, 40
    digital, pulse characterization by, 159–161
    in flow cytometry, 108–110
Signals
    absorption, flow cytometric, 112
    cross-talk, sources, 293–294
    emitted by cell passing through laser beam, 31–32
    extinction and absorption, 112
    fluorescence, 113
    fluorescence emission, 140
    homodyning and heterodyning, 146
    light
        continuum to flow cytometric data file, 37–41
        from particles: collection, 33–37
    optical, converting to current, 118–123
    sensor, simultaneously digitized, 57–58
Signal-to-noise ratio
    confocal microscopy, 94
    titration experiments, 283–284
Sizing, particle, 328–329
Slides, rerun and restained, 54
Sodium azide, phosphate-buffered saline with, 249
Software
    use of nominal channels per decade, 307
    WinCyte, for LSC, 58–62
Solubility, antigen, 261
Specimens, *see also* Samples
    handling and processing, 262
    washing, controversy, 232–234
Spectral response, photomultiplier tube, 120–121
Staining
    for APO2.7 expression, 490–491
    blood cells, 212
    cell surface marker: CFSE labeling method, 381–382
    cultured cells, 361–362
    double indirect, 294
    dual, 12

Index

for flow cytometry, 137–138
immunochemical, cell preparation and, 422–423
indirect, blocking for, 232
intracellular, 242–243
    CFSE method, 383
with JC-1, 472, 477–478
mitotic cells, 345–346
with one to four antibodies, 239–242
simultaneous
    APO2.7 and cell surface antigens, 263
    Ki-67 protein and cell surface antigen, 407–408
    T cell receptor z and cell surface antigens, 263–264
solid tumors and tissues, 362–364
for viability status with antitubulin antibody, 491–492
Standardization
    common cytometry measurements, 312–330
    flow cytometric, 300–302
    light scatter, 327–328
    quantitative fluorescence cytometry, 311–312
Storage, blood, prior to cell preparation, 208–209
Streptavidin, protein labeling, 192
Substitution, maximum, in protein labeling, 190
Synchrony experiments, BrdUrd-labeling in, 358
Syringe-driven mixing, 174–175

T

Tandem conjugate dyes, 201
T cell receptor z, and cell surface antigens: simultaneous staining, 263–264
Temperature
    for cell adhesion assay, 610
        effect on blood prior to cell preparation, 209
Tetraphenylphosphonium electrode, 469
Texas Red, protein labeling, 193
Three-dimensional plot, and histograms, 166
Threshold contours, LSC computer-drawn, 58–59
Thymic epithelial cells, cell adhesion studies, 602
Thymidine analogs, advantages of MAbs over, 356
Thymus cells, phase-resolved measurements, 141–143

Time, *see also* Pseudotime
artifacts dependent on: quality control, 178–179
list mode feature of LSCM, 77–78
measurements based on: applications, 179–181
particles separated by, 30–31
resolved fluorescence measurements, 146–147
role in sample mixing and delivery for flow cytometry, 171–175
Time-lapse microscopy
    data analysis, 553–554
    measurement of cell migration, 550–554
Time-point measurements, single and multiple, 367–369
Time-window device
    earliest time resolution by, 175
    for large number of measurements, 172–174
Time-zero device, for very early time point measurements, 174
Timing
    in cell adhesion assay, 610
    integrators and pulse width measurement circuits, 153–154
    peak detector, 153
Tissue dispersal, methods, 238
Titration experiments, in antibody–antigen reactions, 283–286
T lymphocytes
    cultures, 392
    MHC class II expression on, 451–452
Topography, in analysis of cell migration, 553
Total value, LSCM list mode feature, 62–63
TPP, *see* Tetraphenylphosphonium electrode
Transgenic models, antigen receptor, 395–396
Transmission electron microscopy, ultrastructural analysis with, 495
Transport, blood, prior to cell preparation, 208–209
Trigger circuit, 151–153
T system, device for mixing, 172–174
Tubulin, and surface antigen MC5: simultaneous staining, 264–266
Tumor cells, early DNA measurements, 10
Tumors
    myoepithelial, Humatrix preparation from, 570–571
    solid
        discrimination of apoptotic and necrotic cells, 536

sample labeling and fixation, 360–361
staining procedure, 362–364
TUNEL
   detection of DNA breaks by flow cytometry, 518–519
   in situ, 519–520
   study of apoptosis in plants, 506
Tyrosine phosphorylated proteins, 617

U

Ultraviolet absorption, early measurements, 4
Universal precautions, preparation of cells from blood, 207–208

V

Vascular smooth muscle cells
   propagation, 573–574
   spindle-shaped morphology, 576
Velocity
   fluid stream, 45, 47
   sheath stream, 29–31
Viability
   blood cells, 211–212
   cell, measuring, 244
Viability staining, 265
   with antitubulin antibody, 491–492
   with trypan blue, 495
Voltage, converting current to, 123–124
Volume, particle, measurement, 329

W

Washing, specimens, controversy, 232–234
Wavelengths
   higher, and background noise, 93
   laser light, in flow cytometry, 21
   selection, in flow cytometry, 115–118
Western blot, probing antibody specificity, 290–292
Wound healing assay, for cell migration, 551–552

VOLUMES IN SERIES

Founding Series Editor
DAVID M. PRESCOTT

Volume 1 (1964)
Methods in Cell Physiology
Edited by David M. Prescott

Volume 2 (1966)
Methods in Cell Physiology
Edited by David M. Prescott

Volume 3 (1968)
Methods in Cell Physiology
Edited by David M. Prescott

Volume 4 (1970)
Methods in Cell Physiology
Edited by David M. Prescott

Volume 5 (1972)
Methods in Cell Physiology
Edited by David M. Prescott

Volume 6 (1973)
Methods in Cell Physiology
Edited by David M. Prescott

Volume 7 (1973)
Methods in Cell Biology
Edited by David M. Prescott

Volume 8 (1974)
Methods in Cell Biology
Edited by David M. Prescott

Volume 9 (1975)
Methods in Cell Biology
Edited by David M. Prescott

Volume 10 (1975)
Methods in Cell Biology
Edited by David M. Prescott

Volume 11 (1975)
Yeast Cells
Edited by David M. Prescott

Volume 12 (1975)
Yeast Cells
Edited by David M. Prescott

Volume 13 (1976)
Methods in Cell Biology
Edited by David M. Prescott

Volume 14 (1976)
Methods in Cell Biology
Edited by David M. Prescott

Volume 15 (1977)
Methods in Cell Biology
Edited by David M. Prescott

Volume 16 (1977)
Chromatin and Chromosomal Protein Research I
Edited by Gary Stein, Janet Stein, and Lewis J. Kleinsmith

Volume 17 (1978)
Chromatin and Chromosomal Protein Research II
Edited by Gary Stein, Janet Stein, and Lewis J. Kleinsmith

Volume 18 (1978)
Chromatin and Chromosomal Protein Research III
Edited by Gary Stein, Janet Stein, and Lewis J. Kleinsmith

Volume 19 (1978)
Chromatin and Chromosomal Protein Research IV
Edited by Gary Stein, Janet Stein, and Lewis J. Kleinsmith

Volume 20 (1978)
Methods in Cell Biology
Edited by David M. Prescott

Advisory Board Chairman
KEITH R. PORTER

Volume 21A (1980)
Normal Human Tissue and Cell Culture, Part A: Respiratory, Cardiovascular, and Integumentary Systems
Edited by Curtis C. Harris, Benjamin F. Trump, and Gary D. Stoner

Volume 21B (1980)
Normal Human Tissue and Cell Culture, Part B: Endocrine, Urogenital, and Gastrointestinal Systems
Edited by Curtis C. Harris, Benjamin F. Trump, and Gary D. Stoner

Volume 22 (1981)
Three-Dimensional Ultrastructure in Biology
Edited by James N. Turner

Volume 23 (1981)
Basic Mechanisms of Cellular Secretion
Edited by Arthur R. Hand and Constance Oliver

Volume 24 (1982)
The Cytoskeleton, Part A: Cytoskeletal Proteins, Isolation and Characterization
Edited by Leslie Wilson

Volume 25 (1982)
The Cytoskeleton, Part B: Biological Systems and *in Vitro* Models
Edited by Leslie Wilson

Volume 26 (1982)
Prenatal Diagnosis: Cell Biological Approaches
Edited by Samuel A. Latt and Gretchen J. Darlington

Series Editor
LESLIE WILSON

Volume 27 (1986)
Echinoderm Gametes and Embryos
Edited by Thomas E. Schroeder

Volume 28 (1987)
***Dictyostelium discoideum*: Molecular Approaches to Cell Biology**
Edited by James A. Spudich

Volume 29 (1989)
Fluorescence Microscopy of Living Cells in Culture, Part A: Fluorescent Analogs, Labeling Cells, and Basic Microscopy
Edited by Yu-Li Wang and D. Lansing Taylor

Volume 30 (1989)
Fluorescence Microscopy of Living Cells in Culture, Part B: Quantitative Fluorescence Microscopy—Imaging and Spectroscopy
Edited by D. Lansing Taylor and Yu-Li Wang

Volume 31 (1989)
Vesicular Transport, Part A
Edited by Alan M. Tartakoff

Volume 32 (1989)
Vesicular Transport, Part B
Edited by Alan M. Tartakoff

Volume 33 (1990)
Flow Cytometry
Edited by Zbigniew Darzynkiewicz and Harry A. Crissman

Volume 34 (1991)
Vectorial Transport of Proteins into and across Membranes
Edited by Alan M. Tartakoff

Selected from Volumes 31, 32, and 34 (1991)
Laboratory Methods for Vesicular and Vectorial Transport
Edited by Alan M. Tartakoff

Volume 35 (1991)
Functional Organization of the Nucleus: A Laboratory Guide
Edited by Barbara A. Hamkalo and Sarah C. R. Elgin

Volume 36 (1991)
***Xenopus laevis:* Practical Uses in Cell and Molecular Biology**
Edited by Brian K. Kay and H. Benjamin Peng

Series Editors
LESLIE WILSON AND PAUL MATSUDAIRA

Volume 37 (1993)
Antibodies in Cell Biology
Edited by David J. Asai

Volume 38 (1993)
Cell Biological Applications of Confocal Microscopy
Edited by Brian Matsumoto

Volume 39 (1993)
Motility Assays for Motor Proteins
Edited by Jonathan M. Scholey

Volume 40 (1994)
A Practical Guide to the Study of Calcium in Living Cells
Edited by Richard Nuccitelli

Volume 41 (1994)
Flow Cytometry, Second Edition, Part A
Edited by Zbigniew Darzynkiewicz, J. Paul Robinson, and Harry A. Crissman

Volume 42 (1994)
Flow Cytometry, Second Edition, Part B
Edited by Zbigniew Darzynkiewicz, J. Paul Robinson, and Harry A. Crissman

Volume 43 (1994)
Protein Expression in Animal Cells
Edited by Michael G. Roth

Volume 44 (1994)
***Drosophila melanogaster:* Practical Uses in Cell and Molecular Biology**
Edited by Lawrence S. B. Goldstein and Eric A. Fyrberg

Volume 45 (1994)
Microbes as Tools for Cell Biology
Edited by David G. Russell

Volume 46 (1995)
Cell Death
Edited by Lawrence M. Schwartz and Barbara A. Osborne

Volume 47 (1995)
Cilia and Flagella
Edited by William Dentler and George Witman

Volume 48 (1995)
***Caenorhabditis elegans:* Modern Biological Analysis of an Organism**
Edited by Henry F. Epstein and Diane C. Shakes

Volume 49 (1995)
Methods in Plant Cell Biology, Part A
Edited by David W. Galbraith, Hans J. Bohnert, and Don P. Bourque

Volume 50 (1995)
Methods in Plant Cell Biology, Part B
Edited by David W. Galbraith, Don P. Bourque, and Hans J. Bohnert

Volume 51 (1996)
Methods in Avian Embryology
Edited by Marianne Bronner-Fraser

Volume 52 (1997)
Methods in Muscle Biology
Edited by Charles P. Emerson, Jr. and H. Lee Sweeney

Volume 53 (1997)
Nuclear Structure and Function
Edited by Miguel Berrios

Volume 54 (1997)
Cumulative Index

Volume 55 (1997)
Laser Tweezers in Cell Biology
Edited by Michael P. Sheez

Volume 56 (1998)
Video Microscopy
Edited by Greenfield Sluder and David E. Wolf

Volume 57 (1998)
Animal Cell Culture Methods
Edited by Jennie P. Mather and David Barnes

Volume 58 (1998)
Green Fluorescent Protein
Edited by Kevin F. Sullivan and Steve A. Kay

Volume 59 (1998)
The Zebrafish: Biology
Edited by H. William Detrich III, Monte Westerfield, and Leonard I. Zon

Volume 60 (1998)
The Zebrafish: Genetics and Genomics
Edited by H. William Detrich III, Monte Westerfield, and Leonard I. Zon

Volume 61 (1998)
Mitosis and Meiosis
Edited by Conly L. Rieder

Volume 62 (1999)
Tetrahymena Thermophila
Edited by David J. Asai and James D. Forney

Volume 63 (2000)
Cytometry, Third Edition, Part A
Edited by Zbigniew Darzynkiewicz, J. Paul Robinson, and Harry Crissman

Volume 64 (2000)
Cytometry, Third Edition, Part B
Edited by Zbigniew Darzynkiewicz, J. Paul Robinson, and Harry Crissman

ISBN 0-12-544166-5

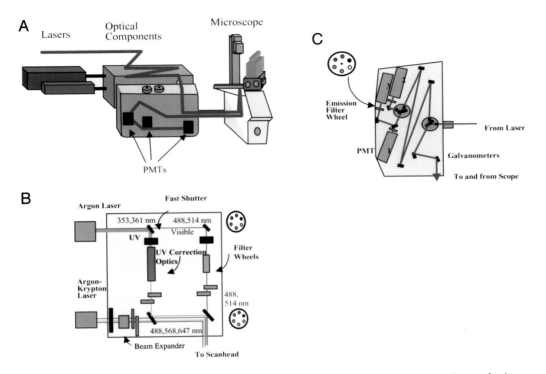

Chapter 4, Fig. 3 The light paths of a confocal microscope with multiple lasers and using an inverted microscope. This shows three components: (A) the light sources, (B) the optical components that manipulate the signal, and (C) the microscope system. Shown are several laser lines that can be used together or independently to excite an object. Resultant signals are collected in the detector region where several PMTs reside. A computer system controls the system and creates the images.

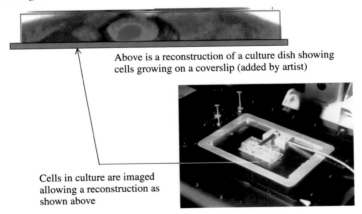

Chapter 4, Fig. 6 Endothelial cells growing on a coverslip–tissue culture chamber (shown in the photograph) were imaged using a confocal microscope mounted on an inverted scope. Approximately 60 sections were collected at 0.1 μm z-axis steps. The images were then analyzed using VoxelView (a three-dimensional software package, Vital Images, Inc., Plymouth, MN), and the cell image was reconstructed from an electronic slice through the cells. The cells appear to be sitting on the coverslip (which was added to the figure to demonstrate its position).

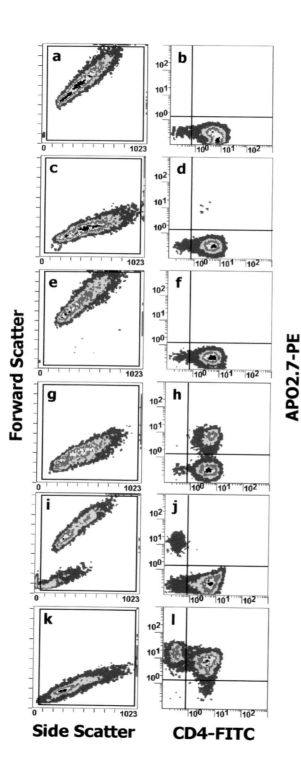

Chapter 12, Fig. 2 The response of CEM cells to induction of apoptosis by hypoxia or etoposide treatments. Apoptosis was induced in the cells by either hypoxia by pelleting the cells in tightly capped tubes or treating continuously in 0.5 μM etoposide followed by a 5-hr incubation. Cells were stained with CD4-FITC and APO2.7-phycoerythrin (PE) following treatment or no treatment in 100 $\mu g/ml$ digitonin in PBS. Histograms representing light scatter and dually stained distributions from cells without digitonin treatment are presented for noninduced control cells (a, b), 5-hr hypoxic cells (e, f), or 5-hr etoposide treated cells (i, j). Histograms representing light scatter and dually stained distributions from cells with digitonin treatment are presented for noninduced control cells (c, d), 5-hr hypoxic cells (g, h), or 5-hr etoposide treated cells (k, l).

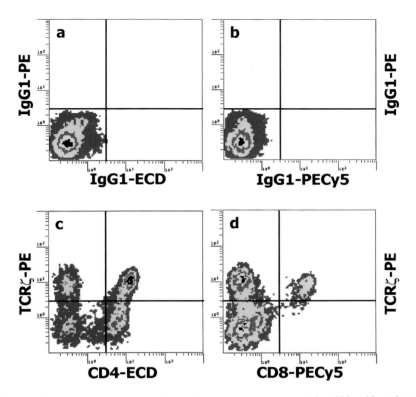

Chapter 12, Fig. 3 Simultaneous surface and internal staining of human peripheral blood lymphocytes (PBL) with anti–T cell receptor ζ (TCRζ), anti-CD4, and anti-CD8 monoclonal antibodies. Whole blood was collected from a normal healthy donor in EDTA anticoagulant, and PBL were harvested using Ficoll-Paque. Cells were stained with CD4-ECD and CD8-PECy5, followed by treatment and staining in a solution containing 500 μg/ml digitonin in PBS and TCRζ-PE. Histograms represent IgG1-PE/IgG1-ECD isotype controls (a), IgG1-PE/IgG1-PECy5 isotype controls (b), TCRζ-PE and CD4-ECD (c), and TCRz-PE and CD8-PECy5 (d).

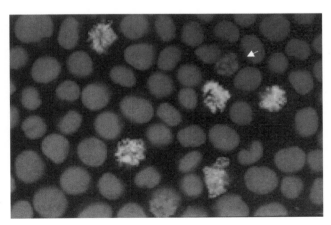

Chapter 15, Fig. 1 Fluorescence photomicrograph of U937 cells stained with anti-H3-P MAb and counterstained with 7-aminoactinomycin D. Cells in metaphase show strong reactivity toward anti-H3-P MAb. Note that the condensed chromatin of an apoptotic cell (spontaneous apoptosis) did not react with the antibody (arrow). Photographed with a Nikon Microphot-FXA flourescence microscope using a Plan Fluor 40× objective lens.

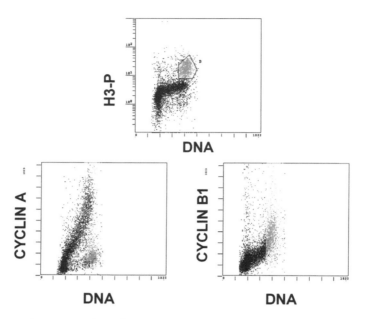

Chapter 15, Fig. 3 Multivariate analysis of U937 cells exposed for 6 hr to vinblastine and stained for DNA content, anti-H3-P reactivity (top), and for either cyclin A (bottom left) or B1 (bottom right) expression, detected immunocytochemically. U937 cells were fixed and stained for DNA content with DAPI (blue fluorescence emission), for H3-P with secondary antibody conjugated with PE (orange fluorescence), and either for cyclin A or B1 with the respective antibody directly conjugated with FITC (green). The cells reactive with anti-H3-P MAB are selected (gated) in the top plot and using a "paint-a-gate" analysis program marked, on these scattergrams, with green color. Thus, they appear green in the bottom scattergrams, which allows one to correlate expression of the respective cyclins with H3-P immunofluorescence on a cell-by-cell basis.

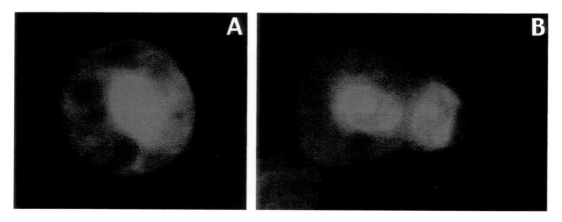

Chapter 23, Fig. 1 Apoptotic protoplasts when stained with ethidium bromide (B) show nuclear marginalization onto the nuclear membrane with chromatin loss, resulting in the enlarged nucleolus becoming prominent. Control protoplasts show dense nuclear staining (A). Micrographs are at 1000× magnification. [From O'Brien *et al.* (1998). Early stages of the apoptotic pathway in plant cells are reversible. *Plant Journal* **13**(6), 803–814, with permission from Blackwell Science Ltd.]

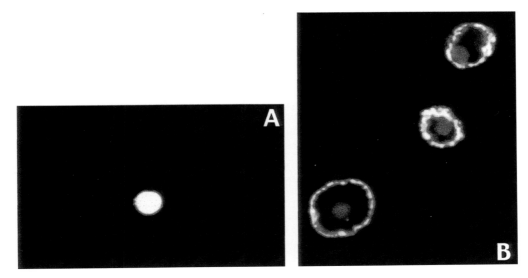

Chapter 23, Fig. 2 Confocal microscopy scanning of nuclei stained with PI shows a dense even nuclear structure in noncondensed nuclei (A), in contrast to a loss of DNA and marginalization of chromatin on the nuclear membrane in apoptotic nuclei (B). All micrographs are at 1000× magnification.

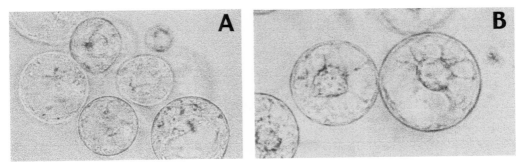

Chapter 23, Fig. 3 Micrographs of healthy protoplasts (A) in contrast to apoptotic protoplasts (B) as seen under light microscopy. The normal protoplasts are cytoplasmically dense, in contrast to the vacuolated apoptotic protoplasts in which the nucleus has become more distinct as the cytoplasm becomes more vacuolated. Micrographs are at 1000× magnification. [From O'Brien *et al.* (1998). Early stages of the apoptotic pathway in plant cells are reversible. *Plant Journal* **13**(6), 803–814, with permission from Blackwell Science Ltd.]

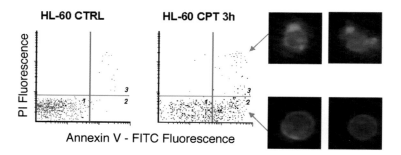

Chapter 24, Fig. 1 Discrimination between nonapoptotic, early apoptotic, and late apoptotic ("necrotic stage" of apoptosis) or necrotic cells following staining with annexin V-FITC conjugate and PI, and fluorescence measurement by laser scanning cytometry (LSC). The cells in quadrant 1 are live; they exclude PI and do not bind annexin V-FITC. They predominate in HL-60 untreated (control) cultures. The cells in quadrant 2 are early apoptotic; they bind annexin V-FITC but still exclude PI. The cells in quadrant 3 are late apoptotic (or necrotic), and they bind both PI and annexin V-FITC. The quadrant 2 and 3 cells are more frequent in the camptothecin-treated sample (CPT). The possibility of observing the morphology of the cells selected based on their fluorescence, as offered by LSC, helps to confirm the identity of apoptotic cells.

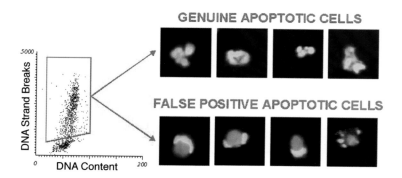

Chapter 24, Fig. 2 Discrimination between genuine and false positive apoptotic cells by LSC. Peripheral blood mononuclear cells were obtained from a leukemia patient during chemotherapy (Gorczyca et al., 1993a). DNA strand breaks were labeled with fluoresceinated dUTP using exogenous terminal deoxynucleotidyltransferase as described (Gorczyca et al., 1992; 1993a). Genuine apoptotic cells are characterized by morphology typical of apoptosis. However, the nonapoptotic cells resembling monocytes, which contain cytoplasmic inclusions stainable with PI and having DNA strand breaks, also are present in this sample. These cells (false positive) most likely represent monocytes that phagocytized apoptotic bodies. By flow cytometry, which does not allow morphological identification, such cells are mistakenly classified as apoptotic.

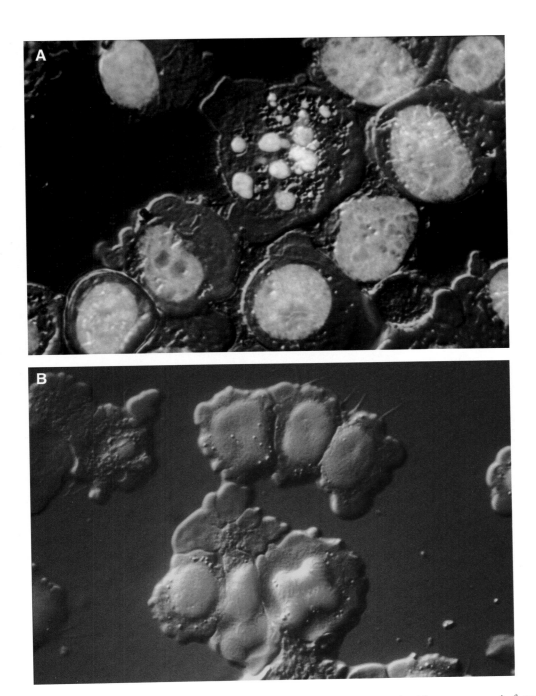

Chapter 24, Fig. 3 Morphology of apoptotic cells from U937 cultures treated with tumor necrosis factor α (TNF-α), cytocentrifuged, fixed in formaldehyde, and stained with 4,6-diamidino-2-phenylindole (DAPI) (A) or 7-AAD (B), as revealed by their examination by fluorescence microscopy combined with interference contrast (Nomarski) illumination. Photographs were taken with a Nikon Microphot-FXA, with a 40× objective.

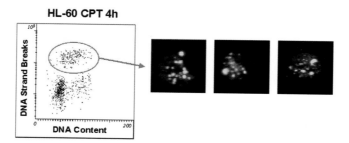

Chapter 24, Fig. 4 Morphological identification of apoptotic cells detected based on the prescence of DNA strand breaks, after relocation, by LSC. HL-60 cells treated with 0.15 μM CPT 4 hr were subjected to DNA strand break labeling with FITC-conjugated digoxigenin antibody (Gorczyca et al., 1992) and counterstained with PI. Note nuclear fragmentation and yellow fluorescence (due to colocalization and spectral overlap of FITC and PI) of the nuclear fragments.

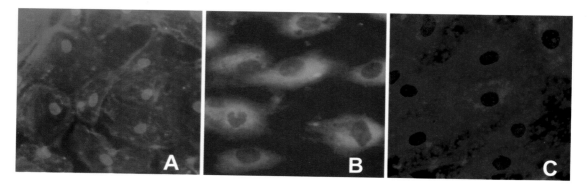

Chapter 26, Fig. 3 Endothelial cells cultured on intestinal submucosa formed a monolayer of polygonal-shaped cells along the surface of the substrate and continued to express endothelial cell-specific phenotype and function. (A) For morphological evaluation cells were dually labeled with Oregon Green-conjugated phalloidin (cytoplasmic stain) and Hoechst 33342 (nuclear stain). (B) Cells continued to express endothelial cell function as indicated by their ability to uptake DiI-Ac-LDL. (C) Cells stained positively with rhodamine-conjugated *Ulex europaeus* I lectin indicating the presence of an endothelial cell-specific surface marker. Magnification 40×.

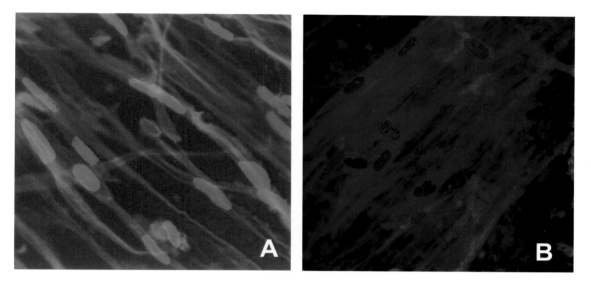

Chapter 26, Fig. 4 Smooth muscle cells cultured on intestinal submucosa demonstrate a spindle-shaped morphology and continue to express smooth muscle-specific phenotype. (A) For morphological evaluation cells were dually labeled with Oregon Green-conjugated phalloidin (cytoplasmic stain) and Hoechst 33342 (nuclear stain). (B) Cells continued to express the contractile-regulating protein calponin (red fluorescence). Magnification 40×.

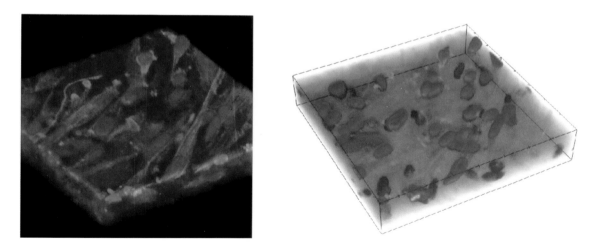

Chapter 26, Fig. 5 3-D reconstructed confocal images of human coronary artery smooth muscle cells grown on intestinal submucosa. Cells were dually labeled with Oregon Green-conjugated phalloidin and Hoechst 33342. The left image shows the cytoplasmic and nuclear aspects of the cells within the intestinal submucosa matrix, whereas the right image highlights the nucleus of each cell.

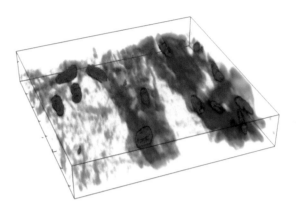

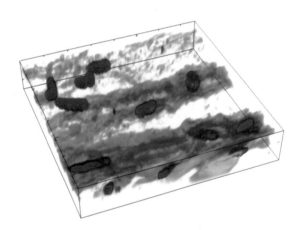

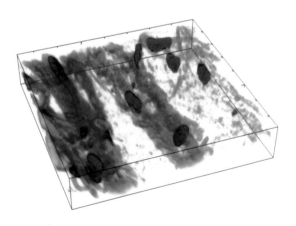

Chapter 27, Fig. 5 3-D reconstructed confocal images of cells growing on and within a tissue-derived ECM substrate, small intestinal submucosa. Components of the ECM substrate were imaged based on their autofluorescence properties, and cell nuclei were stained with Hoechst 33342.